British Plant Communities 5

British Plant Communities

VOLUME 5

MARITIME COMMUNITIES AND VEGETATION OF OPEN HABITATS

J. S. Rodwell (editor)
C. D. Pigott, D. A. Ratcliffe
A. J. C. Malloch, H. J. B. Birks
M. C. F. Proctor, D. W. Shimwell
J. P. Huntley, E. Radford
M. J. Wigginton, P. Wilkins

for the
U.K. Joint Nature Conservation Committee

CAMBRIDGE
UNIVERSITY PRESS

PUBLISHED BY THE PRESS SYNDICATE OF THE UNIVERSITY OF CAMBRIDGE
The Pitt Building, Trumpington Street, Cambridge, United Kingdom

CAMBRIDGE UNIVERSITY PRESS
The Edinburgh Building, Cambridge CB2 2RU, UK   http://www.cup.cam.ac.uk
40 West 20th Street, New York, NY 10011–4211, USA   http://www.cup.org
10 Stamford Road, Oakleigh, Melbourne 3166, Australia
Ruiz de Alarcón 13, 28014 Madrid, Spain

First published 2000

Printed in the United Kingdom at the University Press, Cambridge

Typeset in 9/11 pt Times, in QuarkXPress™ [SE]

A catalogue record for this book is available from the British Library

ISBN 0 521 39167 9 hardback
ISBN 0 521 64476 3 paperback

CONTENTS

Shingle, strandline and sand-dune communities

INTRODUCTION TO SHINGLE, STRANDLINE AND
SAND-DUNE COMMUNITIES

Vegetation of open habitats

Contents

FIGURES

PREFACE AND ACKNOWLEDGEMENTS

The appearance of this fifth volume of *British Plant Communities* brings to a close the publication of the National Vegetation Classification and, as Coordinator of the project, it is my privilege to put on record the gratitude of the whole research team and my own personal thanks to all who have been instrumental in the completion of the work.

For this volume, we were extremely fortunate in having access from the outset to the data which Dr Paul Adam had energetically assembled for his postgraduate research at Cambridge University into British salt-marsh vegetation. Extensive in its coverage and already developed into a classification scheme with highly informative vegetation descriptions, this work obviated the need for any further detailed survey on our part and more than laid a foundation for our own scheme. Such additional data as we did collect to fill any gaps was also supplemented by local surveys by Dr Pat Doody and Margaret Hill of the then NCC, Dr Malcolm Carter and Dr Judith Roper-Lindsay. In integrating this into the NVC framework and reviewing the progress of our synthesis Paul Adam continued to give of his time and expertise without demur.

For sea-cliffs we were equally blessed in inheriting large quantities of data from Andrew Malloch whose geographical and floristic coverage of this difficult and neglected habitat was adventurously wide and whose knowledge of the plant communities and their environmental relationships was second to none. For this section of the volume, Andrew generously provided the bulk of the material and commented on the final version of the community descriptions.

With sand-dunes, by contrast, we were more or less starting from scratch and the four research assistants, Jacqui Huntley (née Paice), Elaine Radford (née Grindey), Paul Wilkins, Martin Wigginton and myself, collected as broad a spread of data as we could from around the coasts of England and Wales. For Scotland, we were especially grateful here, as with other vegetation types covered by the project, for unhindered access

to data collected by Eric Birse and James Robertson, then of the Macaulay Institute in Aberdeen. More locally, but extremely helpful too, were data from Welsh dune slacks being assembled by Dr Peter Jones at Cardiff University. As Andrew Malloch developed the preliminary descriptions of the vegetation types, we were greatly assisted by comments from Drs Pat Doody and Geoff Radley, Imogen Crawford and Dr Tom Dargie and others involved in the country agencies' application of the NVC in their extensive survey and mapping of dune systems.

Also included in this volume are weed communities and other vegetation types of more open habitats like screes, rock outcrops, spoil, walls and pavement cracks, together with communities of periodically-inundated river banks, shoals, lake margins and trackways. For weeds of arable land and gardens, we were very grateful for access through David Shimwell to the data collected by Dr Alan Silverside, now of Paisley University, for his doctoral thesis at Durham. David Shimwell and Elaine Radford were responsible for the preliminary characterisation of most of the vegetation types in this section of the work and Elizabeth Cooper, then at Lancaster University, gave sterling help with the later stages of data analysis.

As with the other volumes of *British Plant Communities* then, there are many and diverse debts to acknowledge here. From the start, the NVC has been very much a collaborative effort and, in addition to the particular thanks paid above and in other volumes, I would like here to mark some of the more substantial contributions to the overall success of the project through the years.

First among them, I want to record my debts to Donald Pigott and Derek Ratcliffe, the two prime movers of the project and an inspiration throughout. The long progress of the work has seen the retirement of both from their final professional appointments but, in their continuing busy lives, they have gone on supporting the project and my own part in it with their concern and encouragement. The significance of their contributions to ecology and conservation are much wider, of course, than the NVC but their commitment to this particular task and their belief in the value of the work have been immensely sustaining and its results will stand especially as a testament to their role in the whole venture.

Continuously close at hand in Lancaster, always willing to help and support, even in the midst of his own heavy academic burdens, has been Andrew Malloch. Quite apart from his supervision of the project in the north-west of England and substantial contributions to this volume, Andrew has played a very significant role in developing the software used for much of the data analysis in the project, has provided welcome assistance, always patient and thoughtful, with the processing of numerous data sets and, throughout, has retained his original conviction of the worthwhile character of our task. Though also retired now, he has gone on gently pressing his concern for its success.

Among others in the team, Michael Proctor has been especially valuable in his contribution to the work on mires and tall-herb fens, but his supervision of the project field work in the south-west of Britain, his comments, always perceptive and enriching, on many sections of the text as they have progressed, and the humour with which he carries his depth of knowledge have been extremely helpful and entertaining.

John Birks provided, from the outset, an inspiration and model for the kind of industrious and painstaking science that we hope we have pursued throughout the project. With his formidable knowledge of the British flora and particular skill with bryophytes, he helped ensure a seriousness in our recording from the start, set a pace for data collection in his supervision of the south-east region of England and contributed especially to the characterisation of woodland and upland communities.

David Shimwell, like others among the team, had been a forerunner in developing a phytosociological approach to British vegetation and, in his oversight of survey in our Midlands region and particular contributions on the classification of heaths, swamps, weed and inundation communities, he played a key role, enlivened by his wry humour, in bringing the whole work to completion.

Under the supervision of this team, the burden of survey for the project fell on the four research assistants and me. The Coordinating Panel are immensely grateful to Jacqui Huntley, Elaine Radford, Paul Wilkins and Martin Wigginton for their energetic commitment to covering the ground in their own regions, their sustained accuracy in collecting samples, the data processing and preliminary characterisation of vegetation types which they carried out with their supervisors and with continuing good humour. From our earliest meeting in the field, when we gathered in 1975 at Preston Montford Field Centre to agree and test our sampling methodology and survey strategy, there was a lively team spirit which sustained our work to the end.

Particular individuals outside this group have played a variety of essential roles in the work. Katherine Hearn, now of the National Trust, and Ian Rotherham, now at Sheffield Hallam University, supplemented our survey effort in southern Scotland and the Yorkshire Dales. Then, we could never have acquitted ourselves adequately in the accounts for a number of vegetation types without access to substantial quantities of data given so generously by Dr Bryan Wheeler (mires), Dr Martin Page (mesotrophic grasslands), Dr Terry Wells (calcicolous grasslands) and Dr Paul Adam (salt-marsh vegetation) and for many plant communities in Scotland by Eric Birse and James Robertson. In data processing, we were particularly indebted to Dr Hilary Birks for her analysis of vast amounts of upland data and to Professor Brian Huntley for his ingenuity in developing software for processing our samples.

In projects of this kind, large, complex and generating substantial amounts of data and material, technical and secretarial assistance are crucial and from start to finish the research team has been admirably served by a series of outstanding colleagues. Philip Harper, Frances Rake, Beryl Fletcher, Sylvia Peglar, Mary Pettit, Margaret Pigott, Steve Ridgill and Joel Miller assisted with the laborious tasks of data coding and analysis and, in the early years, Jennie Ford and Claire Ashworth acted as secretaries to the team.

The bulk of the secretarial work for the project, though, fell on Carol Barlow who typed the great majority of the text, data tables and indexes for *British Plant Communities*, and this in days before the miracles of the word processor, helped prepare much of the manuscript for publication and serviced the operation of the entire task from soon after our start right through the middle years. She did so with unfailing efficiency and attention to detail and I am enormously grateful for the cheerfulness with which she accomplished this job.

Through the final stages of the work, when completion of the task was complicated by other ever-increasing responsibilities on my part, I have had outstanding support and assistance from Michelle Needham whose competence and skills have been vital to bringing the whole enterprise to its conclusion. Juggling this particular secretarial task – completing the typing of text, tables and indexes and helping prepare the last two volumes of *British Plant Communities* for the Press – with the demanding burden of all her other work at Lancaster has never exhausted her energy and ingenuity, nor her spirit.

The Nature Conservancy Council and later the Joint Nature Conservation Committee funded the NVC from beginning to end and, in these organisations, we had the benefit of a series of committed and enthusiastic staff involved, along with Derek Ratcliffe, in the advocacy and management of the project. Philip Oswald was of great assistance in the process of negotiating the detail of the publication process at the start and, as Chief Scientist of the NCC after Derek Ratcliffe, Professor Peter Bridgewater provided enthusiastic continuity of support. Also of critical importance in the middle years was Dr Tim Bines, whose own involvement with vegetation survey gave him a particular concern to ensure the success of the work. After him, Lynne Farrell and Margaret Palmer served us well as nominated officers and, in the last years, Dr John Hopkins and Debbie Jackson. I am grateful to all of these for their encouragement and patience and particularly to John for his elliptical wit and a kind of friendship which never compromised his professionalism.

At Cambridge University Press, too, we have benefited much from the ministrations of the Science team, first Dr Martin Walters, then Dr Maria Murphy and particularly Dr Alan Crowden – whose encouraging darts via fax and e-mail I shall greatly miss. Especially pleasing, also, is to thank Jane Bulleid, the sub-editor for all the volumes of *British Plant Communities*, whose enormous care in dealing with a vast quantity of complex manuscript and proofs has been greatly reassuring to me.

At Lancaster, the final stages of the project have taken place within the context of the Unit of Vegetation Science, much of whose work has been concerned with applications of the NVC among the now extremely wide and diverse user community. The various members of the Unit team, necessarily coming and going with the vagaries of funding and their own developing commitments, have provided an environment of great intellectual enrichment and entertaining companionship during this work. Among the training team, Kate Steele and Julia Milton have both contributed greatly to dissemination of NVC skills among a variety of environmental organisations and countless individuals; in NVC-related research, Sue Edwards, Sean Cooch, Kath Milnes and a series of masters students have broadened our understanding of vegetation types and their ecology; Deirdre Winstanley and, especially in later years, Julian Dring have put their energy and ingenuity at the service of NVC database development and computerised applications. Most of all, Elizabeth Cooper, in her exemplary NVC surveys and mapping, her determined commitment to applications in landscape characterisation and her energy, way beyond the call of duty, in helping others learn what the NVC is about, has been an inspiration. At Lancaster University but outside the Unit, I am personally grateful to Professors Terry Mansfield and Bill Davies, Robin Grove-White and Claire Waterton. When belief in the value of the work has

wavered at all, companionship from such as these has been a boon.

More widely among the community of NVC users in Britain, the project has been sustained over the years by the continuing interest, goading, impatience and disbelief of an enormous diversity of people. In the country agencies, I want to record my particular gratitude to Dr Keith Kirby, Mike Alexander, Dr Terry Rowell, Jane MacKintosh, Richard Tidswell, Dr Des Thompson, Dr Chris Sydes, David Horsfield, Alan Brown, Dr Wanda Fojt, Derek Wells, Dr Tim Blackstock, Dr David Stevens and Paul Corbett; also Dr Jonathan Mitchley, Jack Lavin, Geoffrey Wilmore, Dr Tony Whitbread, Reverend Gordon Graham and Dr Margaret Atherden. For those many excluded from this list who have made minor contributions that have accumulated in the various volumes to an impressive weight of help, our apologies, for it is not gratitude that is in short supply here.

In 1991, to mark the appearance of the first volume of *British Plant Communities* and the establishment of the Unit of Vegetation Science, we organised at Lancaster a conference on 'The Future of Phytosociology'. With the financial assistance of the British Ecological Society, involved from the start of the NVC in encouraging the funding of the project, and – especially pleasing – the Tansley Fund of the New Phytologist Trust, we were able to celebrate before an international audience of speakers and participants, the arrival on the European scene, albeit late, of what we hoped was a serious concern to join in the wider phytosociological community. The welcome our appearance received then and the continuing collaboration and friendship of colleagues across Europe and beyond has brought a rare joy and stimulus to the last years of the work and many valuable comments on the developing classification incorporated into the Conspectus included in this Volume. Among these, it is a privilege to single out Dr Joop Schaminée, Professor Sandro Pignatti, Professor Victor Westhoff, Professor Laco Mucina, Professor Hartmut Dierschke, Dr Milan Chytry, Professor Ayzik Solomeshch, Dr Nikolai Ermakov, Dr Petrit Hoda, Dr Milan Valachovic and Dr Mara Pakalne. My professor at Leeds University, the late (and extraordinary) Irene Manton, who did so much to inspire my early devotion to botany and who was kind enough to support my application to coordinate the National Vegetation Classification, always encouraged me to see the wider world as the proper intellectual framework for research and these new-found fellow-workers have more than borne out that conviction.

In conclusion, I want to go deep and straight in recording the extent of my gratitude to the most personal helpmates in my own contribution to this project. My parents always encouraged my enthusiasm for plants and, without their inheritance of determination and practicality, I would not have stayed this course. My two sons, Dominic and Peter, have grown up in the project's shadow, helping keep me sane with their devastating insights into my seriousness and their sidelong love. Primarily, though, it is my wife Rosemary who has borne the chief burden of my commitment to see it all through, shared intimately in the frustrations, exhaustion and excitements of the work and given all that she is to sustain my enthusiasm to the end.

John Rodwell
Lancaster

PREAMBLE

GENERAL INTRODUCTION

The background to the work

It is a tribute to the insight of our early ecologists that we can still return with profit to *Types of British Vegetation* which Tansley (1911) edited for the British Vegetation Committee as the first coordinated attempt to recognise and describe different kinds of plant community in this country. The contributors there wrote practically all they knew and a good deal that they guessed, as Tansley himself put it, but they were, on their own admission, far from comprehensive in their coverage. It was to provide this greater breadth, and much more detailed description of the structure and development of plant communities, that Tansley (1939) drew together the wealth of subsequent work in *The British Islands and their Vegetation*, and there must be few ecologists of the generations following who have not been inspired and challenged by the vision of this magisterial book.

Yet, partly because of its greater scope and the uneven understanding of different kinds of vegetation at the time, this is a less systematic work than *Types* in some respects: its narrative thread of explication is authoritative and engaging, but it lacks the light-handed framework of classification which made the earlier volume so very attractive, and within which the plant communities might be related one to another, and to the environmental variables which influence their composition and distribution. Indeed, for the most part, there is a rather self-conscious avoidance of the kind of rigorous taxonomy of vegetation types that had been developing for some time elsewhere in Europe, particularly under the leadership of Braun-Blanquet (1928) and Tüxen (1937). The difference in the scientific temperament of British ecologists that this reflected, their interest in how vegetation works, rather than in exactly what distinguishes plant communities from one another, though refreshing in itself, has been a lasting hindrance to the emergence in this country of any consensus as to how vegetation ought to be described, and whether it ought to be classified at all.

In fact, an impressive demonstration of the value of the traditional phytosociological approach to the

description of plant communities in the British Isles was published in German after an international excursion to Ireland in 1949 (Braun-Blanquet & Tüxen 1952), but more immediately productive was a critical test of the techniques among a range of Scottish mountain vegetation by Poore (1955*a*, *b*, *c*). From this, it seemed that the really valuable element in the phytosociological method might be not so much the hierarchical definition of plant associations, as the meticulous sampling of homogeneous stands of vegetation on which this was based, and the possibility of using this to provide a multidimensional framework for the presentation and study of ecological problems. Poore & McVean's (1957) subsequent exercise in the description and mapping of communities defined using this more flexible approach then proved just a prelude to the survey of huge tracts of mountain vegetation by McVean & Ratcliffe (1962), work sponsored and published by the Nature Conservancy (as it then was) as *Plant Communities of the Scottish Highlands*. Here, for the first time, was the application of a systematised sampling technique across the vegetation cover of an extensive and varied landscape in mainland Britain, with assemblages defined in a standard fashion from full floristic data, and interpreted in relation to a complex of climatic, edaphic and biotic factors. The opportunity was taken, too, to relate the classification to other European traditions of vegetation description, particularly that developed in Scandinavia (Nordhagen 1943, Dahl 1956).

McVean & Ratcliffe's study was to prove a continual stimulus to the academic investigation of our mountain vegetation and of abiding value to the development of conservation policy, but their methods were not extended to other parts of the country in any ambitious sponsored surveys in the years immediately following. Despite renewed attempts to commend traditional phytosociology, too (Moore 1962), the attraction of this whole approach was overwhelmed for many by the heated debates that preoccupied British plant ecologists in the 1960s, on the issues of objectivity in the sampling and sorting of data, and the respective values of classification or ordination as analytical techniques. Others, though, found it perfectly possible to integrate multivariate analysis into phytosociological survey, and demonstrated the advantage of computers for the display and interpretation of ecological data, rather than the simple testing of methodologies (Ivimey-Cook & Proctor 1966). New generations of research students also began to draw inspiration from the Scottish and Irish initiatives by applying phytosociology to the solving of particular descriptive and interpretative problems, such as variation among British calcicolous grasslands (Shimwell 1968a), heaths (Bridgewater 1970), rich fens (Wheeler 1975) and salt-marshes (Adam 1976), the vegetation of Skye (Birks 1969), Cornish cliffs (Malloch 1970) and Upper Teesdale (Bradshaw & Jones 1976). Meanwhile, too, workers at the Macaulay Institute in Aberdeen had been extending the survey of Scottish vegetation to the lowlands and the Southern Uplands (Birse & Robertson 1976, Birse, 1980, 1984).

With an accumulating volume of such data and the appearance of uncoordinated phytosociological perspectives on different kinds of British vegetation, the need for an overall framework of classification became ever more pressing. For some, it was also an increasingly urgent concern that it still proved impossible to integrate a wide variety of ecological research on plants within a generally accepted understanding of their vegetational context in this country. Dr Derek Ratcliffe, as Scientific Assessor of the Nature Conservancy's Reserves Review from the end of 1966, had encountered the problem of the lack of any comprehensive classification of British vegetation types on which to base a systematic selection of habitats for conservation. This same limitation was recognised by Professor Sir Harry Godwin, Professor Donald Pigott and Dr John Phillipson who, as members of the Nature Conservancy, had been asked to read and comment on the Reserves Review. The published version, *A Nature Conservation Review* (Ratcliffe 1977), was able to base the description of only the lowland and upland grasslands and heaths on a phytosociological treatment. In 1971, Dr Ratcliffe, then Deputy Director (Scientific) of the Nature Conservancy, in proposals for development of its research programme, drew attention to 'the need for a national and systematic phytosociological treatment of British vegetation, using standard methods in the field and in analysis/classification of the data'. The intention of setting up a group to examine the issue lapsed through the splitting of the Conservancy which was announced by the Government in 1972. Meanwhile, after discussions with Dr Ratcliffe, Professor Donald Pigott of the University of Lancaster proposed to the Nature Conservancy a programme of research to provide a systematic and comprehensive classification of British plant communities. The new

Nature Conservancy Council included it as a priority item within its proposed commissioned research programme. At its meeting on 24 March 1974, the Council of the British Ecological Society welcomed the proposal. Professor Pigott and Dr Andrew Malloch submitted specific plans for the project and a contract was awarded to Lancaster University, with sub-contractual arrangements with the Universities of Cambridge, Exeter and Manchester, with whom it was intended to share the early stages of the work. A coordinating panel was set up, jointly chaired by Professor Pigott and Dr Ratcliffe, and with research supervisors from the academic staff of the four universities, Drs John Birks, Michael Proctor and David Shimwell joining Dr Malloch. Later, Dr Tim Bines replaced Dr Ratcliffe as nominated officer for the NCC, then Lynne Farrell, Margaret Palmer and Dr John Hopkins.

With the appointment of Dr John Rodwell as full-time coordinator of the project, based at Lancaster, the National Vegetation Classification began its work officially in August 1975. Shortly afterwards, four full-time research assistants took up their posts, one based at each of the universities: Mr Martin Wigginton, Miss Jacqueline Paice (later Huntley), Mr Paul Wilkins and Dr Elaine Grindey (later Radford). These remained with the project until the close of the first stage of the work in 1980, sharing with the coordinator the tasks of data collection and analysis in different regions of the country, and beginning to prepare preliminary accounts of the major vegetation types. Drs Michael Lock and Hilary Birks and Miss Katherine Hearn were also able to join the research team for short periods of time. After the departure of the research assistants, the supervisors supplied Dr Rodwell with material for writing the final accounts of the plant communities and their integration within an overall framework. With the completion of this charge in 1989, the handover of the manuscript for publication by the Cambridge University Press began.

The scope and methods of data collection

The contract brief required the production of a classification with standardised descriptions of named and systematically arranged vegetation types and, from the beginning, this was conceived as something much more than an annotated list of interesting and unusual plant communities. It was to be comprehensive in its coverage, taking in the whole of Great Britain apart from Northern Ireland, and including vegetation from all natural, semi-natural and major artificial habitats. Around the maritime fringe, interest was to extend up to the start of the truly marine zone, and from there to the tops of our remotest mountains, covering virtually all terrestrial plant communities and those of brackish and fresh waters, except where non-vascular plants were the dominants. Only short-term leys were specifically excluded and, though care was to be taken to sample more pristine and long-established kinds of vegetation, no undue attention was to be given to assemblages of rare plants or to especially rich and varied sites. Thus widespread and dull communities from improved pastures, plantations, run-down mires and neglected heaths were to be extensively sampled, together with the vegetation of paths, verges and recreational swards, walls, man-made waterways and industrial and urban wasteland.

For some vegetation types, we hoped that we might be able to make use, from early on, of existing studies, where these had produced data compatible in style and quality with the requirements of the project. The contract envisaged the abstraction and collation of such material from both published and unpublished sources, and discussions with other workers involved in vegetation survey, so that we could ascertain the precise extent and character of existing coverage and plan our own sampling accordingly. Systematic searches of the literature and research reports revealed many data that we could use in some way and, with scarcely a single exception, the originators of such material allowed us unhindered access to it. Apart from the very few classic phytosociological accounts, the most important sources proved to be postgraduate theses, some of which had already amassed very comprehensive sets of samples of certain kinds of vegetation or from particular areas, and these we were generously permitted to incorporate directly.

Then, from the NCC and some other government agencies, or from individuals who had been engaged in earlier contracts for them, there were some generally smaller bodies of data, occasionally from reports of extensive surveys, more usually from investigations of localised areas. Published papers on particular localities, vegetation types or individual species also provided small numbers of samples. In addition to these sources, the project was able to benefit from and influence ongoing studies by institutions and individuals, and itself to stimulate new work with a similar kind of approach among university researchers, NCC surveyors, local flora recorders and a few suitably qualified amateurs. An initial assessment and annual monitoring of floristic and geographical coverage were designed to ensure that the accumulating data were fairly evenly spread, fully representative of the range of British vegetation, and of a consistently high quality. Full details of the sources of the material, and our acknowledgements of help, are given in the preface and introduction to each volume.

Our own approach to data collection was simple and pragmatic, and a brief period of training at the outset ensured standardisation among the team of five staff who were to carry out the bulk of the sampling for the project in the field seasons of the first four years, 1976–9. The thrust of the approach was phytosociological in its emphasis on the systematic recording of floristic information from stands of vegetation, though these were chosen solely on the basis of their relative homogeneity in composition and structure. Such selection took a little practice, but it was not nearly so difficult as some critics of this approach imply, even in complex vegetation, and not at all mysterious. Thus, crucial guidelines were to avoid obvious vegetation boundaries or unrepresentative floristic or physiognomic features. No prior judgements were necessary about the identity of the vegetation type, nor were stands ever selected because of the presence of species thought characteristic for one reason or another, nor by virtue of any observed uniformity of the environmental context.

From within such homogeneous stands of vegetation, the data were recorded in quadrats, generally square unless the peculiar shape of stands dictated otherwise. A relatively small number of possible sample sizes was used, determined not by any calculation of minimal areas, but by the experienced assessment of their appropriateness to the range of structural scale found among our plant communities. Thus plots of 2×2 m were used for most short, herbaceous vegetation and dwarf-shrub heaths, 4×4 m for taller or more open herb communities, sub-shrub heaths and low woodland field layers, 10×10 m for species-poor or very tall herbaceous vegetation or woodland field layers and dense scrub, and 50×50 m for sparse scrub, and woodland canopy and understorey. Linear vegetation, like that in streams and ditches, on walls or from hedgerow field layers, was sampled in 10 m strips, with 30 m strips for hedgerow shrubs and trees. Quadrats of 1×1 m were rejected as being generally inadequate for representative sampling, although some bodies of existing data were used where this, or other sizes different from our own, had been employed. Stands smaller than the relevant sample size were recorded in their entirety, and mosaics were treated as a single vegetation type where they were repeatedly encountered in the same form, or where their scale made it quite impossible to sample their elements separately.

Samples from all different kinds of vegetation were recorded on identical sheets (Figure 1). Priority was always given to the accurate scoring of all vascular plants, bryophytes and macrolichens (*sensu* Dahl 1968), a task which often required assiduous searching in dense and complex vegetation, and the determination of difficult plants in the laboratory or with the help of referees. Critical taxa were treated in as much detail as possible though, with the urgency of sampling, certain groups, like the brambles, hawkweeds, eyebrights and dandelions, often defeated us, and some awkward bryophytes and crusts of lichen squamules had to be referred to just a genus. It is more than likely, too, that some very diminutive mosses and especially hepatics escaped notice in the field and, with much sampling taking place in summer, winter annuals and vernal perennials might have been missed on occasion. In general, nomenclature for vascular plants follows *Flora Europaea* (Tutin *et al.* 1964 *et seq.*) with Corley & Hill (1981) providing the authority for bryophytes and Dahl (1968) for lichens. Any exceptions to this, and details of any difficulties with sampling or identifying particular plants, are given in the introductions to each of the major vegetation types.

A quantitative measure of the abundance of every taxon was recorded using the Domin scale (*sensu* Dahl & Hadač 1941), cover being assessed by eye as a vertical projection on to the ground of all the live, above-ground parts of the plants in the quadrat. On this scale:

Cover of 91–100% is recorded as Domin		10
76–90%		9
51–75%		8
34–50%		7
26–33%		6
11–25%		5
4–10%		4
<4%	with many individuals	3
	with several individuals	2
	with few individuals	1

In heaths, and more especially in woodlands, where the vegetation was obviously layered, the species in the

Figure 1. Standard NVC sample card.

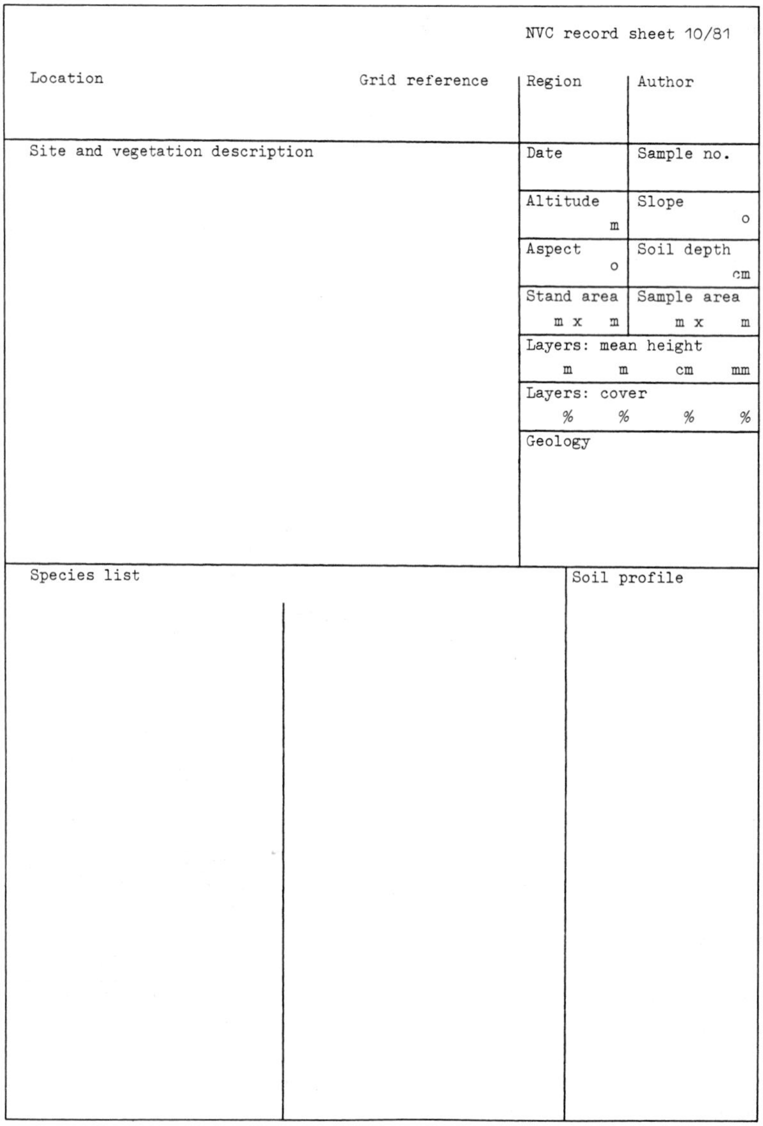
NVC record sheet 10/81
Location
Grid reference
Region
Author
Site and vegetation description
Date
Sample no.
Altitude m
Slope o
Aspect o
Soil depth cm
Stand area m x m
Sample area m x m
Layers: mean height m m cm mm
Layers: cover % % % %
Geology
Species list
Soil profile

different elements were listed separately as part of the same sample, and any different generations of seedling or saplings distinguished. A record was made of the total cover and height of the layers, together with the cover of any bare soil, litter, bare rock or open water. Where existing data had been collected using percentage cover or the Braun-Blanquet scale (Braun-Blanquet 1928), it was possible to convert the abundance values to the Domin scale, but we had to reject all samples where DAFOR scoring had been used, because of the inherent confusion within this scale of abundance and frequency.

Each sample was numbered and its location noted using a site name and full grid reference. Altitude was estimated in metres from the Ordnance Survey 1:50000 series maps, slope estimated by eye or measured using a hand level to the nearest degree, and aspect measured to the nearest degree using a compass. For terrestrial samples, soil depth was measured in centimetres using a probe, and in many cases a soil pit was dug sufficient to allocate the profile to a major soil group (*sensu* Avery 1980). From such profiles, a superficial soil sample was removed for pH determination as soon as possible thereafter using an electric meter on a 1:5 soil:water paste. With aquatic vegetation, water depth was measured in centimetres wherever possible, and some indication of the character of the bottom noted. Details of bedrock and superficial geology were obtained from Geological Survey maps and by field observation.

This basic information was supplemented by notes, with sketches and diagrams where appropriate, on any aspects of the vegetation and the habitat thought likely to help with interpretation of the data. In many cases, for example, the quantitative records for the species were filled out by details of the growth form and patterns of dominance among the plants and an indication of how they related structurally one to another in finely organised layers, mosaics or phenological sequences within the vegetation. Then, there was often valuable information about the environment to be gained by simple observation of the gross landscape or microrelief, the drainage pattern, signs of erosion or deposition and patterning among rock outcrops, talus slopes or stony soils. Often, too, there were indications of biotic effects including treatments of the vegetation by man, with evidence of grazing or browsing, trampling, dunging, mowing, timber extraction or amenity use. Sometimes, it was possible to detect obvious signs of ongoing change in the vegetation, natural cycles of senescence and regeneration among the plants, or successional shifts consequent upon invasion or particular environmental impacts. In many cases, also, the spatial relationships between the stand and neighbouring vegetation types were highly informative and, where a number of samples were taken from an especially varied or complex site, it often proved useful to draw a map indicating how the various elements in the pattern were interrelated.

The approach to data analysis

At the close of the programme of data collection, we had assembled, through the efforts of the survey team and by the generosity of others, a total of about 35000 samples of the same basic type, originating from more than 80% of the 10 × 10 km grid squares of the British mainland and many islands (Figure 2). Thereafter began a coordinated phase of data processing, with each of the four universities taking responsibility for producing preliminary analyses from data sets crudely separated into major vegetation types – mires, calcicolous grasslands, sand-dunes and so on – and liaising with the others where there was a shared interest. We were briefed in the contract to produce accounts of discrete plant communities which could be named and mapped, so our attention was naturally concentrated on techniques of multivariate classification, with the help of computers to sort the very numerous and often complex samples on the basis of their similarity. We were concerned to employ reputable methods of analysis, but the considerable experience of the team in this kind of work led us to resolve at the outset to concentrate on the ecological

Figure 2. Distribution of samples available for analysis.

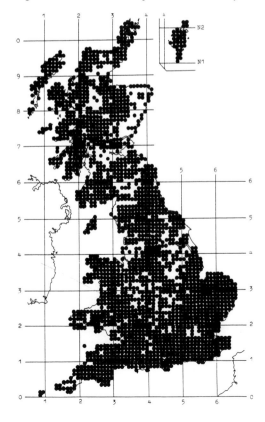

integrity of the results, rather than on the minutiae of mathematical technique. In fact, each centre was free to some extent to make its own contribution to the development of computer programs for the task, Exeter concentrating on Association and Information Analysis (Ivimey-Cook et al. 1975), Cambridge and Manchester on cluster analysis (Huntley et al. 1981), Lancaster on Indicator Species Analysis, later Twinspan (Hill et al. 1975, Hill 1979), a technique which came to form the core of the VESPAN package, designed, using the experience of the project, to be particularly appropriate for this kind of vegetation survey (Malloch 1988).

Throughout this phase of the work, however, we had some important guiding principles. First, this was to be a new classification, and not an attempt to employ computational analysis to fit groups of samples to some existing scheme, whether phytosociological or otherwise. Second, we were to produce a classification of vegetation types, not of habitats, so only the quantitative floristic records were used to test for similarity between the samples, and not any of the environmental information: this would be reserved, rather, to provide one valuable correlative check on the ecological meaning of the sample groups. Third, no samples were to be rejected at the outset because they appeared nondescript or troublesome, nor removed during the course of analysis or data presentation where they seemed to confuse an otherwise crisply-defined result. Fourth, though, there was to be no slavish adherence to the products of a single analyses using arbitrary cut-off points when convenient numbers of end-groups had been produced. In fact, the whole scheme was to be the outcome of many rounds of sorting, with data being pooled and reanalysed repeatedly until optimum stability and sense were achieved within each of the major vegetation types. An important part of the coordination at this stage was to ensure roughly comparable scales of definition among the emerging classifications and to mesh together the work of the separate centres so as to avoid any omissions in the processing or wasteful overlaps.

With the departure from the team of the four research assistants in 1980, the academic supervisors were left to continue the preparation of the preliminary accounts of the vegetation types for the coordinator to bring to completion and integrate into a coherent whole. Throughout the periods of field work and data analysis, we had all been conscious of the charge in the contract that the whole project must gain wide support among ecologists with different attitudes to the descriptive analysis of vegetation. Great efforts were therefore made to establish a regular exchange of information and ideas through the production of progress reports, which gained a wide circulation in Britain and overseas, via contacts with NCC staff and those of other research agencies, and the giving of papers at scientific meetings. This meant that, as we

approached the presentation of the results of the project, we were well informed about the needs of prospective users, and in a good position to offer that balance of concise terminology and broadly-based description that the NCC considered would commend the work, not only to their own personnel, but to others engaged in the assessment and management of vegetation, to plant and animal ecologists in universities and colleges, and to those concerned with land use and planning.

The style of presentation

The presentation of our results thus gives priority to the definition of the vegetation types, rather than to the construction of a hierarchical classification. We have striven to characterise the basic units of the scheme on roughly the same scale as a Braun-Blanquet association, but these have been ordered finally not by any rigid adherence to the higher phytosociological categories of alliance, order and class, but in sections akin to the formations long familiar to British ecologists. In some respects, this is a more untidy arrangement, and even those who find the general approach congenial may be surprised to discover what they have always considered to be, say, a heath, grouped here among the mires, or to search in vain for what they are used to calling 'marsh'. The five volumes of the work gather the major vegetation types into what seem like sensible combinations and provide introductions to the range of communities included: aquatic vegetation, swamps and tall-herb fens; grasslands and montane vegetation; heaths and mires; woodlands and scrub; salt-marsh, sand-dune and sea-cliff communities and weed vegetation. The order of appearance of the volumes, however, reflects more the exigencies of publishing than any ecological viewpoint.

The bulk of the material in the volumes comprises the descriptions of the vegetation types. After much consideration, we decided to call the basic units of the scheme by the rather non-committal term 'community', using 'sub-community' for the first-order sub-groups which could often be distinguished within these, and 'variant' in those very exceptional cases where we have defined a further tier of variation below this. We have also refrained from erecting any novel scheme of complicated nomenclature for the vegetation types, invoking existing names where there is an undisputed phytosociological synonym already in widespread use, but generally using the Latin names of one, two or occasionally three of the most frequent species. Among the mesotrophic swards, for example, we have distinguished a *Centaurea nigra-Cynosurus cristatus* grassland, which is fairly obviously identical to what Braun-Blanquet & Tüxen (1952) called *Centaureo-Cynosuretum cristati*, and within which, from our data, we have characterised three sub-communities. For the convenience of shorthand description and mapping, every vegetation type has been given a code letter and number, so that *Centaurea-Cynosurus* grassland for example is MG5, MG referring to its place among the mesotrophic grasslands. The *Galium verum* sub-community of this vegetation type, the second to be distinguished within the description, is thus MG5b.

Vegetation being as variable as it is, it is sometimes expedient to allocate a sample to a community even though the name species are themselves absent. What defines a community as unique are rarely just the plants used to name it, but the particular combination of frequency and abundance values for all the species found in the samples. It is this information which is presented in summary form in the floristic tables for each of the communities in the scheme. Figure 3, for example, shows such a table for MG5 *Centaurea-Cynosurus* grassland. Like all the tables in the volumes, it includes such vascular plants, bryophytes and lichens as occur with a frequency of 5% or more in any one of the sub-communities (or, for vegetation types with no sub-communities, in the community as a whole). Early tests showed that records of species below this level of frequency could be largely considered as noise, but cutting off at any higher level meant that valuable floristic information was lost. The vascular species are not separated from the cryptogams on the table though, for woodlands and scrub, the vegetation is sufficiently complex for it to be sensible to tabulate the species in a way which reflects the layered structure.

Every table has the frequency and abundance values arranged in columns for the species. Here, 'frequency' refers to how often a plant is found on moving from one sample of the vegetation to the next, irrespective of how much of that species is present in each sample. This is summarised in the tables as classes denoted by the Roman numerals I to V: 1–20% frequency (that is, up to one sample in five)=I, 21–40%=II, 41–60%=III, 61–80%=IV and 81–100%=V. We have followed the usual phytosociological convention of referring to species of frequency classes IV and V in a particular community as its constants, and in the text usually refer to those of class III as common or frequent species, of class II as occasional and of class I as scarce. The term 'abundance' on the other hand, is used to describe how much of a plant is present in a sample, irrespective of how frequent or rare it is among the samples, and it is summarised on the tables as bracketed numbers for the Domin ranges, and denoted in the text using terms such as dominant, abundant, plentiful and sparse. Where there are sub-communities, as in this case, the data for these are listed first, with a final column summarising the records for the community as a whole.

The species are arranged in blocks according to their pattern of occurrence among the different sub-communities and within these blocks are generally ordered by decreasing frequency. The first group, *Festuca rubra* to *Trifolium pratense* in this case, is made up of the commu-

nity constants, that is those species which have an overall frequency IV or V. Generally speaking, such plants tend to maintain their high frequency in each of the sub-communities, though there may be some measure of variation in their representation from one to the next: here, for example, *Plantago lanceolata* is somewhat less common in the last sub-community than the first two, with *Holcus lanatus* and a number of others showing the reverse pattern. More often, there are considerable differences in the abundance of these most frequent species: many of the constants can have very high covers, while others are more consistently sparse, and plants which are not constant can sometimes be numbered among the dominants.

The last group of species on a table, *Ranunculus acris* to *Festuca arundinacea* here, lists the general associates of the community, sometimes referred to as companions. These are plants which occur in the community as a whole with frequencies of III or less, though sometimes they rise to constancy on one or other of the sub-communities, as with *R. acris* in this vegetation. Certain of the companions are consistently common overall like *Rumex acetosa*, some are more occasional throughout as with *Rhinanthus minor*, some are always scarce, for example *Calliergon cuspidatum*. Others, though, are more unevenly represented, like *R. acris*, *Heracleum sphondylium* or *Poa trivialis*, though they do not show any marked affiliation to any partiucular sub-community. Again, there can be marked variation in the abundance of these associates: *Rumex acetosa*, for example, though quite frequent, is usually of low cover, while *Arrhenatherum elatius* and some of the bryophytes, though more occasional, can be patchily abundant; *Alchemilla xanthochlora* is both uncommon among the samples and sparse within them.

The intervening blocks comprise those species which are distinctly more frequent within one or more of the sub-communities than the others, plants which are referred to as preferential, or differential where their affiliation is more exclusive. For example, the group *Lolium perenne* to *Juncus inflexus* is particularly characteristics of the first sub-community of *Centaurea-Cynosurus* grassland, although some species, like *Leucanthemum vulgare* and, even more so, *Lathyrus pratensis*, are more strongly preferential than others, such as *Lolium*, which continues to be frequent in the second sub-community. Even uncommon plants can be good preferentials, as with *Festuca pratensis* here: it is not often found in *Centaurea-Cynosurus* grassland but, when it does occur, it is generally in this first sub-type.

The species group *Galium verum* to *Festuca ovina* helps to distinguish the second sub-community from the first, though again there is some variation in the strength of association between these preferentials and the vegetation type, with *Achillea millefolium* being less markedly

Floristic table MG5

	a	b	c	MG5
Festuca rubra	V (1–8)	V (2–8)	V (2–7)	V (1–8)
Cynosurus cristatus	V (1–8)	V (1–7)	V (1–7)	V (1–8)
Lotus corniculatus	V (1–7)	V (1–5)	V (2–4)	V (1–7)
Plantago lanceolata	V (1–7)	V (1–5)	IV (1–4)	V (1–7)
Holcus lanatus	IV (1–6)	IV (1–6)	V (1–5)	IV (1–6)
Dactylis glomerata	IV (1–7)	IV (1–6)	V (1–6)	IV (1–7)
Trifolium repens	IV (1–9)	IV (1–6)	V (1–4)	IV (1–9)
Centaurea nigra	IV (1–5)	IV (1–4)	V (2–4)	IV (1–5)
Agrostis capillaris	IV (1–7)	IV (1–7)	V (3–8)	IV (1–8)
Anthoxanthum odoratum	IV (1–7)	IV (1–8)	V (1–4)	IV (1–8)
Trifolium pratense	IV (1–5)	IV (1–4)	IV (1–3)	IV (1–5)
Lolium perenne	IV (1–8)	III (1–7)	I (2–3)	III (1–8)
Bellis perennis	III (1–7)	II (1–7)	I (4)	II (1–7)
Lathyrus pratensis	III (1–5)	I (1–3)	I (1)	II (1–5)
Leucanthemum vulgare	III (1–3)	I (1–3)	II (1–3)	II (1–3)
Festuca pratensis	II (1–5)	I (2–5)	I (1)	I (1–5)
Knautia arvensis	I (4)			I (4)
Juncus inflexus	I (3–5)			I (3–5)
Galium verum	I (1–6)	V (1–6)		II (1–6)
Trisetum flavescens	II (1–4)	IV (1–6)	II (1–3)	III (1–6)
Achillea millefolium	III (1–6)	V (1–4)	III (1–4)	III (1–6)
Carex flacca	I (1–4)	II (1–4)	I (1)	I (1–4)
Sanguisorba minor	I (4)	II (3–5)		I (3–5)
Koeleria macrantha	I (1)	II (1–6)		I (1–6)
Agrostis stolonifera	I (1–7)	II (1–6)	I (6)	I (1–7)
Festuca ovina		II (1–6)		I (1–6)
Prunella vulgaris	III (1–4)	III (1–4)	IV (1–3)	III (1–4)
Leontodon autumnalis	II (1–5)	II (1–3)	IV (1–4)	III (1–5)
Luzula campestris	II (1–4)	II (1–6)	IV (1–4)	III (1–6)
Danthonia decumbens	I (2–5)	I (1–3)	V (2–5)	I (1–5)
Potentilla erecta	I (1–4)	I (3)	V (1–4)	I (1–4)
Succisa pratensis	I (1–4)	I (1–5)	V (1–4)	I (1–5)
Pimpinella saxifraga	I (1–4)	I (1–4)	III (1–4)	I (1–4)
Stachys betonica	I (1–5)	I (1–4)	III (1–4)	I (1–5)
Carex caryophyllea	I (1–4)	I (1–3)	II (1–2)	I (1–4)
Conopodium majus	I (1–4)	I (1–5)	II (2–3)	I (1–5)
Ranunculus acris	IV (1–4)	II (1–4)	IV (2–4)	III (1–4)
Rumex acetosa	III (1–4)	III (1–3)	III (1–3)	III (1–4)
Hypochoeris radicata	III (1–5)	II (2–4)	III (1–4)	III (1–5)
Ranunculus bulbosus	III (1–7)	II (1–5)	III (1–2)	III (1–7)
Taraxacum officinale agg.	III (1–4)	III (1–4)	III (1–3)	III (1–4)
Brachythecium rutabulum	II (1–6)	III (1–4)	II (2)	III (1–6)
Cerastium fontanum	III (1–3)	II (1–3)	II (1–3)	II (1–3)
Leontodon hispidus	II (1–6)	III (2–4)	III (1–5)	II (1–6)
Rhinanthus minor	II (1–5)	II (1–4)	II (1–3)	II (1–5)
Briza media	II (1–6)	III (1–4)	II (2–3)	II (1–6)
Heracleum spondylium	II (1–5)	II (1–3)	III (1–3)	II (1–5)
Trifolium dubium	II (1–8)	II (1–5)	I (2)	II (1–8)
Primula veris	II (1–4)	II (2–4)	I (2)	II (1–4)
Arrhenatherum elatius	II (1–6)	II (1–7)	I (3–4)	II (1–7)
Cirsium arvense	II (1–3)	II (1–4)	I (1)	II (1–4)
Eurhynchium praelongum	II (1–5)	II (1–4)	I (1–2)	II (1–5)
Rhytidiadelphus squarrosus	II (1–7)	II (1–5)	III (1–4)	II (1–7)
Poa pratensis	II (1–6)	II (2–5)		II (1–6)
Poa trivialis	II (1–8)	I (1–3)	I (1–2)	II (1–8)
Veronica chamaedrys	II (1–4)	I (1–4)	I (1)	II (1–4)
Alopecurus pratensis	I (1–6)	I (1–4)	I (1)	I (1–6)
Cardamine pratensis	I (1–3)	I (1)	I (3)	I (1–3)
Vicia cracca	I (1–4)	I (1–3)	I (1–2)	I (1–4)
Bromus hordeaceus hordeaceus	I (1–6)	I (2–3)	I (3)	I (1–6)
Phleum pratense pratense	I (1–6)	I (1–5)	I (1)	I (1–6)
Juncus effusus	I (2–3)	I (3)	I (1–2)	I (1–3)
Phleum pratense bertolonii	I (1–3)	I (1–3)	I (1)	I (1–3)
Calliergon cuspidatum	I (1–5)	I (2–4)	II (3)	I (1–5)
Ranunculus repens	II (1–7)	I (2)	II (1–4)	I (1–7)
Pseudoscleropodium purum	I (1–5)	I (3–4)	II (2)	I (1–5)
Ophioglossum vulgatum	I (1–5)	I (1)		I (1–5)
Silaum silaus	I (1–5)	I (1–3)		I (1–5)
Agrimonia eupatoria	I (1–5)	I (1–3)		I (1–5)
Avenula pubescens	I (1–3)	I (2–5)		I (1–5)
Plantago media	I (1–4)	I (1–4)		I (1–5)
Alchemilla glabra	I (2)	I (3)		I (2–3)
Alchemilla filicaulis vestita	I (1–3)	I (3)		I (1–3)
Alchemilla xanthochlora	I (1–3)	I (2)		I (1–3)
Carex panicea	I (1–4)	I (2–4)		I (1–4)
Colchicum autumnale	I (3–4)	I (1–3)		I (1–4)
Crepis capillaris	I (1–5)	I (3)		I (1–5)
Festuca arundinacea	I (1–5)	I (3–5)		I (1–5)

Figure 3. Floristic table for NVC community MG5 *Centaurea nigra-Cynosurus cristatus* grassland.

diagnostic than *Trisetum flavescens* and, particularly, *G. verum*. There are also important negative features, too, because, although some plants typical of the first and third sub-communities, such as *Lolium* and *Prunella vulgaris*, remain quite common here, the disappearance of others, like *Lathyrus pratensis*, *Danthonia decumbens*, *Potentilla erecta* and *Succisa pratensis* is strongly diagnostic. Similarly, with the third sub-community, there is that same mixture of positive and negative characteristics, and there is, among all the groups of preferentials, that same variation in abundance as is found among the constants and companions. Thus, some plants which can be very marked preferentials are always of rather low cover, as with *Prunella*, whereas others, like *Agrostis stolonifera*, though diagnostic at low frequency, can be locally plentiful.

For the naming of the sub-communities, we have generally used the most strongly preferential species, not necessarily those most frequent in the vegetation type. Sometimes, sub-communities are characterised by no floristic features over and above those of the community as a whole, in which case there will be no block of preferentials on the table. Usually, such vegetation types have been called Typical, although we have tried to avoid this epithet where the sub-community has a very restricted or eccentric distribution.

The tables organise and summarise the floristic variation which we encountered in the vegetation sampled: the text of the community accounts attempts to expound and interpret it in a standardised descriptive format. For each community, there is first a synonymy section which lists those names applied to that particular kind of vegetation where it has figured in some form or another in previous surveys, together with the name of the author and the date of ascription. The list is arranged chronologically, and it includes references to important unpublished studies and to accounts of Irish and Continental associations where these are obviously very similar. It is important to realise that very many synonyms are inexact, our communities corresponding to just part of a previously described vegetation type, in which case the initials *p.p.* (for *pro parte*) follow the name, or being subsumed within an older, more broadly-defined unit. Despite this complexity, however, we hope that this section, together with that on the affinities of the vegetation (see below), will help readers translate our scheme into terms with which they may have been long familiar. A special attempt has been made to indicate correspondence with popular existing schemes and to make sense of venerable but ill-defined terms like 'herb-rich meadow', 'oakwood' or 'general salt-marsh'.

There then follow a list of the constant species of the community, and a list of the rare vascular plants, bryophytes and lichens which have been encountered in the particular vegetation type, or which are reliably known to occur in it. In this context, 'rare' means, for vascular plants, an A rating in the *Atlas of the British Flora* (Perring & Walters 1962), where scarcity is measured by occurrence in vice-counties, or inclusion on lists compiled by the NCC of plants found in less than one hundred 10×10 km squares. For bryophytes, recorded presence in under 20 vice-counties has been used as a criterion (Corley & Hill 1981), with a necessarily more subjective estimate for lichens.

The first substantial section of text in each community description is an account of the physiognomy, which attempts to communicate the feel of the vegetation in a way which a tabulation of data can never do. Thus, the patterns of frequency and abundance of the different species which characterise the community are here filled out by details of the appearance and structure, variation in dominance and the growth form of the prominent elements of the vegetation, the physiognomic contribution of subordinate plants, and how all these components relate to one another. There is information, too, on important phenological changes that can affect the vegetation through the seasons and an indication of the structural and floristic implications of the progress of the life cycle of the dominants, any patterns of regeneration within the community or obvious signs of competitive interaction between plants. Much of this material is based on observations made during sampling, but it has often been possible to incorporate insights from previous studies, sometimes as brief interpretative notes, in other cases as extended treatments of, say, the biology of particular species such as *Phragmites australis* or *Ammophila arenaria*, the phenology of winter annuals or the demography of turf perennials. We trust that this will help demonstrate the value of this kind of descriptive classification as a framework for integrating all manner of autecological studies (Pigott 1984).

Some indication of the range of floristic and structural variation within each community is given in the discussion of general physiognomy, but where distinct sub-communities have been recognised these are each given a descriptive section of their own. The sub-community name is followed by any synonyms from previous studies, and by a text which concentrates on pointing up the particular features of composition and organisation which distinguish it from the other sub-communities.

Passing reference is often made in these portions of the community accounts to the ways in which the nature of the vegetation reflects the influence on environmental factors upon it, but extended treatment of this is reserved for a section devoted to the habitat. An opening paragraph here attempts to summarise the typical conditions which favour the development and maintenance of the vegetation types, and the major factors which control floristic and structural variation within it. This is followed by as much detail as we have at the present time about the impact of particular climatic, edaphic and biotic variables on the community, or as we suppose to be important to its essential character and distribution. With climate, for example, reference is very frequently made to the influence on the vegetation of the amount and disposition of rainfall through the year, the variation in temperature season by season, differences in cloud cover and sunshine, and how these factors interact in the maintenance of regimes of humidity, drought or frosts. Then, there can be notes of effects attributable to the extent and duration of snow-lie or to the direction and strength of winds, especially where these are icy or salt-laden. In each of these cases, we have tried to draw upon reputable sources of data for interpretation, and to be fully sensitive to the complex operation of topographic climates, where features like aspect and altitude can be of great importance, and of regional patterns, where concepts like continental, oceanic, montane and maritime climates can be of enormous help in understanding vegetation patterns.

Commonly, too, there are interactions between climate and geology that are best perceived in terms of variations in soils. Here again, we have tried to give full weight to the impact of the character of the landscape and its rocks and superficials, their lithology and the ways in which they weather and erode in the processes of pedogenesis. As far as possible, we have employed standardised terminology in the description of soils, trying at least to distinguish the major profile types with which each community is associated, and to draw attention to the influence of its floristics and structure of processes like leaching and podzolisation, gleying and waterlogging, parching, freeze-thaw and solifluction, and inundation by fresh- or salt-waters.

With very many of the communities we have distinguished, it is combinations of climatic and edaphic factors that determine the general character and possible range of the vegetation, but we have often also been able to discern biotic influences, such as the effects of wild herbivores or agents of dispersal, and there are very few instances where the impact of man cannot be seen in the present composition and distribution of the plant communities. Thus, there is frequent reference to the role which treatments such as grazing, mowing and burning have on the floristics and physiognomy of the vegetation, to the influence of manuring and other kinds of eutrophication, of draining and re-seeding for agriculture, of the cropping and planting of trees, of trampling or other disturbance, and of various kinds of recreation.

The amount and quality of the environmental information on which we have been able to draw for interpreting such effects has been very variable. Our own sampling provided just a spare outline of the physical and edaphic conditions at each location, data which we have summarised where appropriate at the foot of the floristic tables; existing sources of samples sometimes offered next to nothing, in other cases very full soil analysis or precise specifications of treatments. In general, we have used what we had, at the risk of great unevenness of understanding, but have tried to bring some shape to the accounts by dealing with the environmental variables in what seems to be their order of importance, irrespective of the amount of detail available, and by pointing up what can already be identified as environmental threats. We have also benefited by being able to draw on the substantial literature on the physiology and reproductive biology of individual species, on the taxonomy and demography of plants, on vegetation history and on farming and forestry techniques. Sometimes, this information provides little more than a provisional substantiation of what must remain for the moment an interpretive hunch. In other cases, it has enabled us to incorporate what amount to small essays on, for example, the past and present role of *Tilia cordata* in our woodlands with variation in climate, the diverse effects of dunging by rabbit, sheep and cattle on calcicolous swards, or the impact of burning on *Calluna-Arctostaphylos* heath on different soils in a boreal climate. Debts of this kind are always acknowledged in the text and, for our part, we hope that the accounts indicate the benefits of being able to locate experimental and historical studies on vegetation within the context of an understanding of plant communities (Pigott 1982).

Mention is often made in the discussion of the habitat of the ways in which stands of communities can show signs of variation in relation to spatial environmental differences, or the beginnings of a response to temporal changes in conditions. Fuller discussion of zonations to other vegetation types follows, with a detailed indication of how shifts in soil, microclimate or treatment affect the composition and structure of each community, and descriptions of the commonest patterns and particularly distinctive ecotones, mosaics and site types in which it and any sub-communities are found. It has also often been possible to give some fuller and more ordered account of the ways in which vegetation types can change through time, with invasion of newly available ground, the progression of communities to maturity, and their regeneration and replacement. Some attempt has been made to identify climax vegetation types and major lines of succession, but we have always been wary of the temptation to extrapolate from spatial patterns to temporal sequences. Once more, we have tried to incorporate the results of existing observational and experimental studies, including some of the classic accounts of patterns and processes among British vegetation, and to point up the great advantages of a reliable scheme of classification as a basis for the monitoring and management of plant communities (Pigott 1977).

Throughout the accounts, we have referred to particular sites and regions wherever we could, many of these visited and sampled by the team, some the location of previous surveys, the results of which we have now been able to redescribe in the terms of the classification we have erected. In this way, we hope that we have begun to make real a scheme which might otherwise remain abstract. We have also tried in the habitat section to provide some indications of how the overall ranges of the vegetation types are determined by environmental conditions. A separate paragraph on distribution summarises what we know of the ranges of the communities and sub-communities, then maps show the location, on the 10×10 km national grid, of the samples that are available to us for each. Much ground, of course, has been thinly covered, and sometimes a dense clustering of samples can reflect intensive sampling rather than locally high frequency of a vegetation type. However, we believe that all the maps we have included are accurate in their general indication of distributions, and we hope that this exercise might encourage the production of a comprehensive atlas of British plant communities.

The last section of each community description considers the floristic affinities of the vegetation types in the scheme, and expands on any particular problems of synonymy with previously described assemblages. Here, too, reference is often given to the equivalent or most closely-related association in Continental phytosociological classifications and an attempt made to locate each community in an existing alliance. Where the fuller account of British vegetation that we have been able to provide necessitates a revision of the perspective on European plant communities as a whole, some suggestions are made as to how this might be achieved.

Meanwhile, each reader will bring his or her own needs and commitment to this scheme and perhaps be dismayed by its sheer size and apparent complexity. For those requiring some guidance as to the scope of each volume and the shape of that part of the classification with which it deals, the introductions to the major vegetation types will provide an outline of the variation and how it has been treated. The contents page will then give directions to the particular communities of interest. For readers less sure of the identity of the vegetation types with which they are dealing, a key is provided to each major group of communities which should enable a set of similar samples organised into a constancy table to be taken through a series of questions to a reasonably secure diagnosis. The keys, though, are not infallible short cuts to identification and must be used in conjunction with the floristic tables and community descriptions. An alternative entry to the scheme is provided by the species index which lists the occurrences of all taxa in the communities in which we have recorded them. There is also an index of synonyms which should help readers find the equivalents in our classification of vegetation types already familiar to them.

Finally, we hope that whatever the needs, commitments or even prejudices of those who open these volumes, there will be something here to inform and challenge everyone with an interest in vegetation. We never thought of this work as providing the last word on the classification of British plant communities: indeed, with the limited resources at our disposal, we knew it could offer little more than a first approximation. However, we do feel able to commend the scheme as essentially reliable. We hope that the broad outlines will find wide acceptance and stand the test of time, and that our approach will contribute to setting new standards of vegetation description. At the same time, we have tried to be honest about admitting deficiencies of coverage and recognising much unexplained floristic variation, attempting to make the accounts sufficiently open-textured that new data might be readily incorporated and ecological puzzles clearly seen and pursued. For the classification is meant to be not a static edifice, but a working tool for the description, assessment and study of vegetation. We hope that we have acquitted ourselves of the responsibilities of the contract brief and the expectations of all those who have encouraged us in the task, such that the work might be thought worthy of standing in the tradition of British ecology. Most of all, we trust that our efforts do justice to the vegetation which, for its own sake, deserves understanding and care.

SALT-MARSH COMMUNITIES

INTRODUCTION TO SALT-MARSH COMMUNITIES

The sampling and analysis of salt-marsh vegetation

The herbaceous vascular vegetation on the intertidal silts and sands of salt-marshes is one of the most frequently used illustrations of ecological pattern but there are considerable difficulties in producing an adequate national classification of the plant communities of this distinctive habitat.

First, much salt-marsh vegetation is species-poor. There is little problem in sampling and sorting monospecific stands but, in many cases, a small number of species occur with varying abundance in a wide variety of combinations on salt-marshes. Early accounts of this vegetation (e.g. Tansley 1911, 1939) relied heavily on dominance in an attempt to make sense of such variation, but, as Dalby (1970) noted, this may obscure patterning among less conspicuous species that it is sensible to try and interpret. Furthermore, there has been a tendency in Britain to lump more complex vegetation, less susceptible to analysis, into a 'general salt-marsh' community. This term has sometimes been applied in its original, broad sense (Tansley 1911) to vegetation 'not dominated by any single species, except locally' and varying 'from place to place according to local conditions and to the accidents of colonisation by different species'; on other occasions (e.g. Chapman 1934), it has been used to denote a more clearly-defined community.

Second, on many salt-marshes there is a site-related element in the floristic variation among the communities which reflects particular local histories of marsh use or unique combinations of environmental conditions. Detailed studies of limited areas of salt-marsh (e.g. Yapp & Johns 1917, Chapman 1934, Dalby 1970, Packham & Liddle 1970, Gray & Bunce 1972) can be particularly valuable in elucidating such local patterns of variation but the use of a single suite of salt-marshes as a reference point for interpreting floristic variation throughout the country can be misleading. The especially attractive and varied salt-marshes of the north Norfolk coast have been frequently employed in such a way and this has bequeathed to us a perspective in which the salt-marshes of the north and west tend to be underrated. On the other hand, to treat all local variation on an equal level would produce a very cumbersome national classification.

A third point is that 'salt-marsh' is as much a habitat as a group of plant communities and, although the vegetation itself plays some part in salt-marsh development, the physiographic boundaries of the habitat do not exactly coincide with a well-circumscribed range of communities. The salt-marsh flora has two major components: a halophyte element more or less confined to this particular kind of saline environment and an element comprising species which are widespread in inland, non-saline habitats. The latter species are commonly referred to as glycophytes, although it is possible that they include some distinct ecotypes which differ markedly from their inland counterparts in their physiological tolerances. Communities consisting predominantly or entirely of halophytes can sensibly be termed salt-marsh vegetation types but, towards the upper marsh limit and, in some areas (like the grazed marshes of the north and west) more extensively, communities consisting mainly or exclusively of glycophytes also occur in the salt-marsh habitat. These may extend well into the zone of tidal influence but they are often far from the common conception of salt-marsh vegetation. Some are perhaps best seen as highly modified forms of more typical salt-marsh communities produced by specialised treatments. Others probably reflect coincidences of environmental conditions which, though not especially coastal, occur only on salt-marshes. Deciding whether a particular vegetation type is more closely related to a mainstream salt-marsh community or a predominantly inland community is sometimes very difficult.

Finally, algae are often a conspicuous feature of salt-marsh vegetation and a decision has to be taken about whether or not to record them with the vascular flora and employ them in the analysis of data. Although there are some difficulties of identification with these taxa, especially among the microscopic species, a number of schemes have been proposed for the classification of algal communities on salt-marshes (e.g. Cotton 1912, Carter 1932, 1933*a*, *b*, Chapman 1974, Polderman 1979, Polderman & Polderman-Hall 1980). These suggest that the concordance of algal assemblages with vascular plant communities may not be precise. In addition, algal communities appear to be subject to greater seasonal changes and, at least where the smaller species are concerned, to be organised on a finer scale. For the most part, therefore, vascular communities appear to be superimposed upon a distinct, more changeable and finer pattern of algal vegetation.

In an attempt to take account of such difficulties, Adam (1976, 1981) collected almost 3000 new samples of vegetation from British salt-marshes and his classification forms the basis of the scheme presented here. Adam's geographical coverage was extensive but some stretches of coastline were sparsely sampled (the Hampshire coast and south-west England) and others unvisited by him (the Thames estuary, the Humber and eastern Scotland). Where possible, this under-representation has been rectified by our own sampling pro-

Figure 4. Distribution of samples available from salt-marshes.

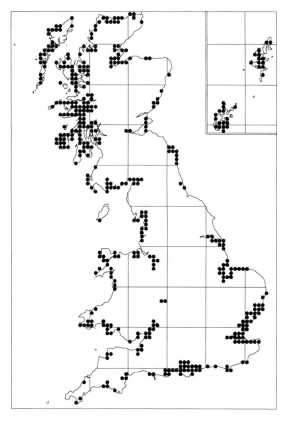

gramme and by the generous donation of external data, most notably from Birse & Robertson (1976), Hilliam (1977) and Birse (1980), which considerably extended coverage, particularly in eastern Scotland, Orkney and Shetland (Figure 4).

Adam did not include the *Zostera* vegetation of flats in his survey but some very limited sampling by the survey team and a good deal of qualitative information forms the basis of an outline description included here. The work of Lee (1975, 1977) has also enabled a fuller account to be given of those communities represented in inland saline habitats. It has been possible, too, to integrate Adam's data with samples of swamps, mires and mesotrophic grasslands widely distributed inland and so produce a coherent account of some of the communities of the upper salt-marsh. However, the vegetation of brackish pools and ditches and the grassy sea-banks and walls characteristic of many reclaimed sites (Beeftink 1975, Gray 1977, Adam & Akeroyd 1978) remains under-sampled.

As with other sections of the National Vegetation Classification, floristic data alone were used to characterise the vegetation types, any available environmental or site information being employed afterwards to help provide an ecological interpretation to the various sample groups distinguished.

The description of salt-marsh communities

Adam warned about the difficulty of generalising from a national scheme to a particular local situation and the same caution should be applied to this expanded and modified classification. Two of the communities distinguished are especially problematic in this respect: the *Puccinellia maritima* salt-marsh (SM13, *Puccinellietum maritimae* (Warming 1906) Christiansen 1927) and the *Festuca rubra* salt-marsh (SM16, *Juncetum gerardi* Warming 1906). These both encompass a very wide range of floristic variation, the internal differences between the sub-communities being almost as great as those features which distinguish these vegetation types from other salt-marsh communities. Although the sub-communities characterised should be useful in discussing national variation, they may well be of less value in local small-scale studies and, in certain cases, it might be appropriate to devise *ad hoc* classifications within these major types for particular sites.

Although Adam's approach was phytosociological, it was an important feature of his work that he classified the samples without prior reference to existing schemes devised for salt-marshes in other parts of Europe (e.g. Beeftink 1962, 1965, 1966, Géhu 1975). Nevertheless, there is a striking similarity between many of his final groups and the salt-marsh associations of Continental classifications and much British vegetation of this kind can be seen as extending the known distribution of previously-described communities. It is less easy to relate the vegetation types characterised here to those in earlier descriptive accounts of British salt-marshes where floristic definition was sometimes vague and units often rather heterogeneous.

A total of 28 communities of salt-marsh vegetation has been characterised from the available data (Figure 5). These can be conveniently reviewed under four main heads: eel-grass and tassel-weed communities of tidal flats, pools and ditches (3 communities), communities of the lower salt-marsh (13), communities of the middle salt-marsh (9) and communities of the upper salt-marsh (3). Brief mention is also made below of vegetation types that are treated in other volumes but which sometimes figure prominently on salt-marshes.

Eel-grass and tassel-weed communities of tidal flats, pools and ditches

Our three native species of *Zostera* (*Z. marina, Z. angustifolia* and *Z. noltii*) are prominent, usually with very few other vascular species but often with abundant algae, in vegetation that occurs on the eu-littoral and sub-littoral zones of sand and silt flats. Without extensive floristic

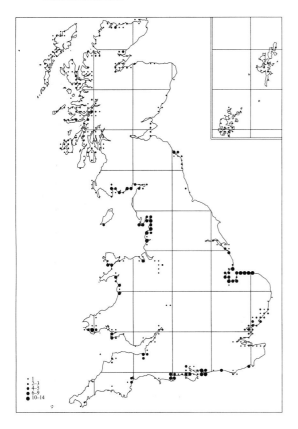

Figure 5. Distribution of vegetation types characterised from salt-marshes.

data, separate communities have not been defined here but these vegetation types are the British representatives of the eel-grass communities of the Mediterranean, west European and Baltic coasts placed in the alliance Zosterion Christiansen 1934 of the class Zosteretea marinae.

Ruppia maritima and the much rarer *Eleocharis parvula* can each occur prominently in communities of brackish pools, pans, and creeks and, in certain parts of their ranges in Britain, on the open surface of salt-marshes. Comparable vegetation elsewhere in Europe has been characterised as a *Ruppietum maritimae* Hocquette 1927 (SM2) and an *Eleocharetum parvulae* (Preuss 1911/12) Gillner 1960 (SM3) and grouped in the alliance Ruppion maritimae Br.-Bl. 1931 of the class Ruppietea maritimae J. Tüxen 1960.

Lower salt-marsh communities

Although the distinction between low, mid- and upper marsh is not a simple one, it is convenient to group together thirteen communities in which either *Spartina* spp., annual *Salicornia* spp., *Suaeda maritima* and/or *Puccinellia maritima* generally form a prominent component of the vegetation with, more unevenly, *Aster tripolium* and *Halimione portulacoides*.

Three communities dominated by *Spartina* spp. occur in Britain. By far the commonest is the *Spartina anglica* salt-marsh (SM6, *Spartinetum townsendii* (Tansley 1939) Corillion 1953), dominated by *S. townsendii sensu lato* (generally the fertile amphidiploid *S. anglica* but also occasionally with its male sterile F_1 precursor *S. × townsendii*). Although *S. anglica* can be found as a scattered associate in almost every salt-marsh community, the spread over the last 100 years of dense stands of this species is one of the most spectacular recent changes in the vegetation of the maritime zone of Britain. Communities dominated by its presumed parents, the native *S. maritima* (SM4, *Spartinetum maritimae* (Emb. & Regn. 1926) Corillion 1953) and the naturalised alien *S. alterniflora* (SM5, *Spartinetum alterniflorae* Corillion 1953), appear to be declining and are now much restricted in their distribution. Cord-grass vegetation of these types through western Europe and on the east coast of North America has been placed in the alliance Spartinion Conrad 1933 of the class Spartinetea maritimae R.Tx. 1961.

Annual *Salicornia* spp., *Suaeda maritima* and *Puccinellia maritima* occur together in various combinations as colonising vegetation towards the lower limit of salt-marshes and in open and disturbed areas at higher levels. Three communities have been distinguished according to the balance of the various components: the annual *Salicornia* salt-marsh (SM8, *Salicornietum europaeae* Warming 1906), *Suaeda maritima* salt-marsh (SM9, *Suaedetum maritimae* (Conrad 1935) Pignatti 1953) and the transitional *Puccinellia-Salicornia-Suaeda* salt-marsh (SM10). Predominantly low-marsh vegetation of these kinds in which annual chenopods are prominent is grouped in the alliance Thero-Salicornion strictae Br.-Bl. 1933 *emend.* R.Tx. 1950 of the class Thero-Salicornietea Pignatti 1953 *emend.* R.Tx. in R.Tx. & Oberdorfer 1958.

The perennial relative of *Salicornia*, now termed *Arthrocnemum perenne*, occurs occasionally in Britain in a variety of salt-marsh communities but locally forms dense stands which are best treated as a distinct vegetation type similar to the *Salicornietum radicantis* Br.-Bl. 1931, traditionally separated off from the annual chenopod communities into the Salicornion fruticosae Br.-Bl. 1931 alliance of the Salicornietea fruticosae.

The five remaining low-marsh communities are all richer and more varied vegetation types than these, composed largely of perennial halophytes and, among these, the grass *Puccinellia maritimae* is of prime importance with, less frequently and more unevenly throughout, *Aster tripolium*, *Halimione portulacoides*, *Glaux maritima*, *Plantago maritima*, *Limonium* cf. *vulgare*, *Triglochin maritima*, *Armeria maritima* and *Spergularia media*.

The general relationship of these communities to phytosociological units defined from mainland Europe is fairly clear. They fall within the class Juncetea maritimae R.Tx. & Oberdorfer 1958 which also takes in much perennial mid-marsh and sea-cliff vegetation extending from the Arctic to the Mediterranean. West European salt-marsh communities are assigned to the order Glauco-Puccinellietalia Beeftink & Westhoff 1962 but, in Britain, the floristic distinction between the two major alliances, the Puccinellion maritimae Christiansen 1927 of the low-marsh and the Armerion maritimae Br.-Bl. & de Leuuw 1936 of the mid-marsh, is not as clear as on the Continent. In this country, *Armeria maritima* and, to a lesser extent, *Glaux maritima*, both considered good diagnostic species for the Armerion elsewhere in Europe, extend on to the low marsh and, indeed, are important components of some of the Puccinellion communities. Within Britain, a better general distinction between low- and mid-marsh vegetation types is the separation between the dominant role of *Puccinellia maritima* on the one hand and *Festuca rubra* and *Juncus gerardii* on the other, although, in particular situations, this too may be an unclear criterion.

In this scheme, the bulk of this remaining low-marsh vegetation is included in a single large and varied community, the *Puccinellia maritima* salt-marsh (SM13, *Puccinellietum maritimae* (Warming 1906) Christiansen 1927). This is the most widespread of all British salt-marsh vegetation types and it spans swards which grade, in one direction, to the Thero-Salicornion through an increased representation of annual chenopods and, in

another, to the Armerion communities with a switch in dominance to *F. rubra*, *J. gerardii* and *Agrostis stolonifera*. As well as some rather species-poor *Puccinellia*-dominated swards, it also includes a variety of richer vegetation types, some previously considered within the ambit of a 'general salt-marsh' community and others representing local variation in which individual species attain prominence.

A second major community, especially on ungrazed sites to the south and east, is the *Halimione portulacoides* salt-marsh (SM14, *Halimionetum portulacoidis* (Kuhnholtz-Lordat 1927) Des Abbayes & Corillion 1949). This shares many species with the *Puccinellietum* and grades floristically to it, but it is generally distinct in the partial or total dominance of *H. portulacoides*. Also predominantly on ungrazed south-eastern sites, though somewhat more restricted in its distribution, is the *Aster tripolium* var. *discoideus* salt-marsh (SM11, *Asteretum tripolii* Tansley 1939). Like the *Halimionetum*, this community is often prominent on creek-sides, though it is also frequent low down on salt-marshes and shows some floristic overlap with Thero-Salicornion vegetation. Variation within *A. tripolium* is complex but a provisional community has been erected to contain stands dominated by the rayed form (SM12, cf. Sociatie van *Aster tripolium* Beeftink 1962). This is of local distribution and it shows some affinities with vegetation of brackish waters but further sampling is needed to establish its exact status and relationships. With a similar range and also showing close floristic relationships to the *Halimionetum* is vegetation with a striking local dominance of *Inula crithmoides* (SM26), a plant more geographically confined on salt-marshes than its occurrences on sea cliffs in Britain.

Finally, Puccinellion species form an understorey to one of the British salt-marsh communities in which *Juncus maritimus* is a physiognomic dominant. The classification of these vegetation types is problematic (Adam 1977): *J. maritimus* is dominant in certain mid-marsh communities as well as in sub-communities of the *Halimionetum* and the upper-marsh *Atriplici-Elymetum pycnanthi* (see below). However, the *Juncus maritimus-Triglochin maritima* salt-marsh (SM15) is a distinct type floristically, is the most widespread of all British *J. maritimus* communities and satisfactorily incorporates those stands in which *J. maritimus* reaches its lowest limit around our coasts. Vegetation of this type has sometimes been separated off into a separate alliance, the Halo-Scirpion (Dahl & Hadač 1971) den Held & Westhoff 1969 *nom. nov.*

Middle salt-marsh communities

Eight communities are distinguished from the middle salt-marsh zone (Figure 6). Three have a generally high frequency of *Festuca rubra*, *Juncus gerardii* and *Agrostis stolonifera* with *Glaux maritima* and *Plantago maritima* and, more unevenly, *Armeria maritima* and *Triglochin maritima*. *Cochlearia officinalis*, *Plantago coronopus*, *Carex extensa* and *C. distans* occur patchily throughout

Figure 6. Generalised salt-marsh zonations in the south-east and west of Britain. The figure shows the relative extent of the major communities with an indication of the clarity of distinctions between low, mid and upper marsh.

SM6	*Spartinetum townsendii*
SM8	*Salicornietum europaeae*
SM11	*Asteretum tripolii*
SM13	*Puccinellietum maritimae*
SM14	*Halimionetum portulacoidis*
SM15	*Juncus maritimus-Triglochin maritima* salt-marsh
SM16	*Juncetum gerardi*
SM17	*Artemisietum maritimae*
SM18	*Juncus maritimus* salt-marsh
SM24	*Atriplici-Elymetum pycnanthi*
SM28	*Elymetum repentis*

South-east

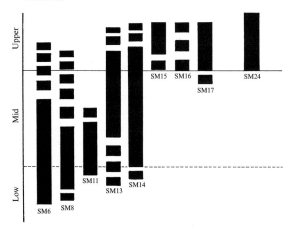

West

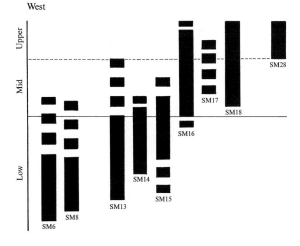

and a variety of glycophytes, notably *Trifolium repens*, *Potentilla anserina*, *Holcus lanatus* and *Leontodon autumnalis*, attain prominence in some communities. Thero-Salicornion species such as *Puccinellia maritima*, *Halimione portulacoides* and *Limonium* cf. *vulgare* are infrequent. These communities correspond approximately to the Armerion alliance within the Glauco-Puccinellietalia.

The *Festuca rubra* salt-marsh (SM16, *Juncetum gerardi* Warming 1906) is, like the *Puccinellietum*, a large and varied vegetation type with a wide distribution, especially on the grazed marshes of the north and west of Britain. It includes swards which, on the one hand, grade to the *Puccinellietum* and, on the other, show diverse affinities with brackish and freshwater inundation communities of the Elymo-Rumicion crispi, Cynosurion pastures and Caricion davallianae mires. These reflect its considerable vertical range on salt-marshes, its widespread use for grazing and turf-cutting and the disturbance and freshwater flushing which it often experiences at higher levels.

Juncetum gerardi species form an understorey to a second community, the *Juncus maritimus* salt-marsh (SM18) which includes the bulk of those mid-marsh stands in which *J. maritimus* is dominant, often with abundant *Oenanthe lachenalii* and Elymo-Rumicion species such as *Elymus repens*, *Rumex crispus* and *Atriplex prostrata*. An *Artemisia maritima* salt-marsh (SM17, *Artemisietum maritimae* Hocquette 1927) has not traditionally been characterised in British accounts but it is a distinct vegetation type which shows affinities with both the Armerion communities (especially the *F. rubra*-dominated form of the *Juncetum gerardi*) and the Puccinellion (particularly the *Halimionetum*).

The six remaining mid-marsh communities are rather specialised vegetation types of either widespread but local occurrence or restricted geographical distribution. Two are dominated by colonial members of the Cyperaceae and are especially characteristic of damp depressions and brackish sites, especially to the north and west. The *Blysmus rufus* salt-marsh (SM19, *Blysmetum rufi* (G. E. & G. Du Rietz 1925) Gillner 1960) and the *Eleocharis uniglumis* salt-marsh (SM20, *Eleocharitetum uniglumis* Nordhagen 1923) are sometimes accommodated within a distinct alliance, the Eleocharion uniglumis, and they represent a phytogeographical affinity with predominantly north European salt-marsh communities.

Two further communities are unique to Britain, and within the country confined to north Norfolk and Sussex, and are characterised by the occurrence of Mediterranean plants in highly distinctive salt-marsh/sand-dune/shingle transitions. The *Suaeda vera-Limonium binervosum* salt-marsh and the *Halimione portulacoides-Frankenia laevis* salt-marsh (SM22, perhaps equivalent to the *Limonio vulgaris-Frankenietum laevis* Géhu & Géhu-Franck 1975) can be seen as the northernmost outpost of the vegetation of the Frankenio-Armerenion, proposed as a sub-alliance of the Armerion.

Also, within the Glauco-Puccinellietalia is the *Spergularia marina-Puccinellia distans* salt-marsh (SM23, *Puccinellietum distantis* Feekes (1934) 1945), a community which is especially characteristic of the hypersaline conditions developing in drying pans and depressions on salt-marshes, in inland saline sites and, increasingly now, along the edges of inland roads which have received heavy applications of rock-salt in frosty weather. This kind of vegetation is usually placed in the alliance Puccinellio-Spergularion salinae Beeftink 1965. Finally, among the mid-marsh communities, it is sensible to include ephemeral vegetation with *Sagina maritima, S. nodosa* and various local annuals which finds a place in turf-cuttings and other breaks in the salt-marsh swards of the *Puccinellietum* and *Juncetum gerardi* as an early stage in recolonisation. We have not characterised any separate communities here but referred the assemblages (SM27) more generally to the Saginion maritimae Westhoff, van Leeuwen & Adriani 1962, an alliance placed in its own class.

Upper salt-marsh communities

Of the considerable variety of vegetation types which occur on salt-marshes towards the upper limit of tidal influence, three are described in this volume. They are characterised by the general prominence of *Elymus pycnanthus* and/or *E. repens*, patchy representation of Puccinellion and Armerion species and the scattered occurrence throughout of nitrophilous weeds and plants of fresh-water inundation communities. Predominantly perennial vegetation of this kind, characteristic of European drift-lines, has been variously placed in the alliance Elymion pycnanthi of the Elymetea pycnanthi or the Elymo-Rumicion crispi Nordhagen 1940 of the Molinio-Arrhenatheretea.

Two of the communities are grass-dominated. The *Elymus pycnanthus* salt-marsh (SM24, *Atriplici-Elymetum pycnanthi* Beeftink & Westhoff 1962) is the vegetation type which commonly terminates the salt-marsh zonation in the south and east. To the north and west, it is replaced by the *Elymus repens* salt-marsh (SM28, *Elymetum repentis maritimum* Nordhagen 1940) which has a less conspicuous representation of salt-marsh species and which shows more obvious floristic affinities with the halophyte forms of Elymo-Rumicion vegetation. On drift-lines on the salt-marsh/sand-dune transi-

tion at scattered localities in the south-east, a third vegetation type, the *Suaeda vera* community (SM25, *Elymo pycnanthi-Suaedetum verae* (Arènes 1933) Géhu 1975), is characteristic.

Other vegetation types on salt-marshes

A variety of vegetation types described fully in other sections of *British Plant Communities* occurs on salt-marshes where there is a combination of little tidal influence and low soil salinity with either some influence of fresh-water or types of treatment and/or disturbance characteristic of other habitats (Figure 9).

Three mesotrophic grasslands occur commonly on salt-marshes (see Rodwell 1992). Both the *Festuca rubra-Agrostis stolonifera-Potentilla anserina* inundation community (MG11) and the *Festuca arundinacea* coarse grassland (MG12, *Potentillo-Festucetum arundinaceae* Nordhagen 1940) have distinct halophyte sub-communities and are found at scattered localities, mainly on the west coast, the former sometimes extensively on the grazed open marsh, the latter more patchily on ungrazed sites and on ditch-banks where there is some brackish influence. The *Agrostis stolonifera-Alopecurus geniculatus* inundation grassland (MG13) is also widespread as small stands in areas where there is some brackish influence and poaching by stock.

Swamp vegetation may occur in estuaries and in salt-marsh ditches and pools where there is slow-moving or standing brackish water and is also occasionally encountered on the open surface of salt-marshes and around saline springs inland (Rodwell 1994a). Two such communities are largely confined to such situations: the *Scirpus maritimus* swamp (S21, *Scirpetum maritimi* (Br.-Bl. 1931) R.Tx. 1937), which is widespread and sometimes extensive and the *Scirpus lacustris* ssp. *tabernaemontani* swamp (S20, *Scirpetum tabernaemontani* Passarge 1964) which is more local. The *Phragmites australis* swamp (S4, *Phragmitetum australis* Gams (1927) Schmale 1939), the *Typha latifolia* swamp (S12, *Typhetum latifoliae* Soó 1927) and the *Phalaris arundinacea* tall-herb fen (S28, *Phalaridetum arundinaceae* Libbert 1931) are much more widely distributed in freshwater habitats and salt-marsh stands are often only marginally halophyte in character.

Finally here, the *Iris pseudacorus-Filipendula ulmaria* tall-herb fen (M28, *Filipendulo-Iridetum pseudacori* Adam 1976) is a very conspicuous feature of the upper-marsh and some raised beaches on the west coast of Scotland where stands may be extensive and rich around freshwater flushes.

KEY TO SALT-MARSH COMMUNITIES

With something as complex and variable as vegetation, no key can pretend to offer an infallible short cut to diagnosis. The following should thus be seen as simply as a crude guide to identifying the types of vegetation found on salt-marshes and must always be used in conjunction with the data tables and community descriptions. It relies on floristic (and, to a lesser extent, physiognomic) features of the vegetation and demands a knowledge of the British vascular flora. It does not make primary use of any habitat features, though these may provide a valuable confirmation of a diagnosis.

Because the major distinctions between the vegetation types in the classification are based on inter-stand frequency, the key works best when sufficient samples of similar composition are available to construct a constancy table. It is the frequency values in this (and, in some cases, the ranges of abundance) which are then subject to interrogation with the key.

Samples should always be taken from homogeneous stands and be 2×2 m or 4×4 m according to the scale of the vegetation or, where stands are irregular, of identical size but different shape.

---

1   Open or closed vegetation of, or overwhelmingly dominated by, a single species   2

Vegetation with two or more co-dominants or, if with a single dominant, then some other species with cover values of Domin 4–7   3

2   Open or closed vegetation of, or overwhelming dominated by:

Zostera marina, Z. angustifolia or *Z. noltii* on sub- or eu-littoral flats, often with no other vascular plants but commonly with some fucoids and green algae

>   **SM1** *Zostera* communities
>   Zosterion   Christiansen 1934

Ruppia maritima, sometimes with *Potamogeton pectinatus, Zannichellia palustris* and/or *Ranunculus baudotii* as submerged vegetation in brackish pools, in dried-up pans or, more rarely, on open flats

>   **SM2** *Ruppia maritima* salt-marsh
>   Ruppietum maritimae   Hocquette 1927

Suaeda maritima in usually somewhat open vegetation and often in small stands

>   **SM9** *Suaeda maritima* salt-marsh
>   Suaedetum maritimae   (Conrad 1935) Pignatti 1953

Aster tripolium var. *discoideus*

>   **SM11** *Aster tripolium* var. *discoideus* salt-marsh
>   Asteretum tripolii   Tansley 1939

Rayed *Aster tripolium*

>   **SM12** Rayed *Aster tripolium* stands

Variation within *Aster tripolium* is complex and the phytosociological relationships of the forms are unclear.

Puccinellia maritima in low, open or closed vegetation or occasionally in dense, tall swards but with no extensive understorey of turf fucoids

>   **SM13** *Puccinellia maritima* salt-marsh
>   Puccinellietum maritimae (Warming 1906) Christiansen 1927
>   *Puccinellia maritima* dominated sub-community

In hot dry summers on the upper marsh, when the shoots of *Glaux maritima* may become shrivelled, some stands of the *Puccinellietum maritimae, Glaux maritima* sub-community may key out here.

Eleocharis parvula in a very diminutive sward, some-times obscured by algae or freshly-deposited silt

> **SM3** *Eleocharis parvula* salt-marsh
> *Eleocharetum parvulae* (Preuss 1911/12) Gillner 1960

Spartina maritima in isolated clumps or as extensive stands

> **SM4** *Spartina maritima* salt-marsh
> *Spartinetum maritimae* (Emb. & Regn. 1926) Corillion 1963

Spartina alterniflora in a dense cover with a little *S. anglica*, *Puccinellia maritima* and *Aster tripolium*

> **SM5** *Spartina alterniflora* salt-marsh
> *Spartinetum alterniflorae* Corillion 1953

Spartina anglica, sometimes with *S. × townsendii*, often in very extensive stands

> **SM6** *Spartina anglica* salt-marsh
> *Spartinetum townsendii* (Tansley 1939) Corillion 1953

Arthrocnemum perenne in dense pure stands or as open mosaic with *Halimione portulacoides*, *Puccinellia maritima* and *Suaeda maritima*

> **SM7** *Arthrocnemum perenne* stands

Annual *Salicornia* spp. in usually somewhat open vegetation

> **SM8** Annual *Salicornia* spp. salt-marsh
> *Salicornietum europaeae* Warming 1906

Puccinellia maritima or *Plantago maritima* with an extensive understorey of diminutive turf fucoids

> **SM13** *Puccinellietum maritimae*
> *Puccinellia maritima*-turf fucoid sub-community

Glaux maritima in often small and fragmentary stands

> **SM13** *Puccinellietum maritimae*
> *Glaux maritima* sub-community

Halimione portulacoides as an even-topped bushy canopy or discrete hemispherical bushes in species-poor vegetation without *Juncus maritimus*

> **SM14** *Halimione portulacoides* salt-marsh
> *Halimionetum portulacoidis* (Kuhnholtz-Lordat 1927) Des Abbayes & Corillion 1949
> *Halimione portulacoides*-dominated sub-community

Halimione portulacoides with some *Juncus maritimus* as scattered shoots or small dense patches

> **SM14** *Halimionetum portulacoidis*
> *Juncus maritimus* sub-community

Small but discrete patches of these two sub-communities of the *Halimionetum* may occur in mosaics with the *Puccinellietum maritimae* and these should be distinguished from the intimate mixtures of *H. portulacoides* and *P. maritima* that characterise the *Puccinellia maritima* sub-community of the *Halimionetum*.

Juncus maritimus as small dense patches within a ground of *Halimione portulacoides*

> **SM14** *Halimionetum portulacoidis*
> *Juncus maritimus* sub-community

Juncus maritimus as tall dense patches with little or no *Halimione portulacoides*

> **SM15** *Juncus maritimus-Triglochin maritima* salt-marsh

Juncus maritimus may also be locally dominant in the *Juncus maritimus* salt-marsh but the consistent presence there of *Festuca rubra*, *Agrostis stolonifera* and *Juncus gerardii* as an often thick understorey usually serves to separate this vegetation from the two above. *J. maritimus* may also be locally abundant in the *Atriplici-Elymetum pycnanthi* but there *Elymus pycnanthus* is consistently dominant.

Juncus gerardii as generally small and often roughly circular patches of sometimes tall vegetation

> **SM16** *Festuca rubra* salt-marsh
> *Juncetum gerardi* Warming 1906
> *Juncus gerardii*-dominated sub-community

Juncus gerardii may also be locally abundant in other sub-communities of the *Juncetum gerardi*.

Festuca rubra as a thick springy mattress of tall and dense vegetation

> **SM16** *Juncetum gerardi*
> Sub-community with tall *Festuca rubra* dominant

Festuca rubra may also be locally abundant in the shorter swards of other sub-communities of the *Juncetum gerardi*.

Blysmus rufus in often small stands of sometimes open vegetation

> **SM19** *Blysmus rufus* salt-marsh
> *Blysmetum rufi* (G. E. & G. Du Rietz 1925) Gillner 1960

Eleocharis uniglumis in often small stands of sometimes open vegetation

> **SM20** *Eleocharis uniglumis* salt-marsh
> *Eleocharitetum uniglumis*   Nordhagen 1923

Suaeda vera as an open bushy canopy with one or more of *Limonium binervosum, L. bellidifolium* or *Frankenia laevis* beneath

> **SM21** *Suaeda vera-Limonium binervosum* salt-marsh

Suaeda vera as a more or less closed canopy in strand-line vegetation without the above species

> **SM25** *Suaeda vera* salt-marsh
> *Elymo pycnanthi-Suaedetum verae* (Arènes 1933) Géhu 1975

Spergularia marina or *Puccinellia distans* in often small stands of usually somewhat open vegetation

> **SM23** *Spergularia marina-Puccinellia distans* salt-marsh
> *Puccinellietum distantis*   Feekes (1934) 1945

Elymus pycnanthus as stiff clumps, usually without any *Suaeda vera* or *Inula crithmoides*

> **SM24** *Elymus pycnanthus* salt-marsh
> *Atriplici-Elymetum pycnanthi*   Beeftink & Westhoff 1962

Elymus repens in a closed grassy sward

> **SM28** *Elymus repens* salt-marsh
> *Elymetum repentis maritimum*   Nordhagen 1940

Inula crithmoides, usually with some *Halimione portulacoides*

> **SM26** *Inula crithmoides* stands

Sagina maritima or *Plantago coronopus* in often open or fragmentary vegetation in breaks within swards of other communities, especially the *Juncetum gerardi*

> **SM27** Ephemeral *Sagina maritima* vegetation
> Saginion maritimae   Westhoff, van Leeuwen & Adriani 1962

Potentilla anserina as small stands colonising breaks within swards of other mid- and upper-marsh vegetation

> **SM16** *Juncetum gerardi* phase of sward regeneration in turf-cuttings

---

Vegetation dominated by swamp species such as *Scirpus maritimus, S. lacustris* ssp. *tabernaemontani, Phragmites australis, Typha latifolia* and *Phalaris arundinacea* may be encountered on salt-marshes but these communities are included in Rodwell (1994*a*).

---

3   Low swards, sometimes rather open, dominated by various mixtures of annual *Salicornia* spp., *Suaeda maritima* and *Puccinellia maritima*   4

Annual *Salicornia* spp. and *Suaeda maritima* not dominant or co-dominant   5

4   *Suaeda maritima* and annual *Salicornia* spp. co-dominant with less than 10% *Puccinellia maritima*

> **SM9** *Suaedetum maritimae*

Annual *Salicornia* spp., *Suaeda maritima* and *Puccinellia maritima* co-dominant in various proportions, often with a little *Aster tripolium*

> **SM10** Transitional low-marsh vegetation

Vegetation of this kind frequently occurs as mosaics between the *Salicornietum europaeae* and the *Puccinellietum maritimae, Spartinetum maritimae* and, especially in the south-east, the *Asteretum tripolii* and *Halimionetum portulacoidis.*

---

5   *Aster tripolium* var. *discoideus* or rayed *Aster tripolium* dominant   6

Aster tripolium absent or present in small amounts   7

6   *Aster tripolium* var. *discoideus* dominant

> **SM11** *Asteretum tripolii*

Rayed *Aster tripolium* dominant

> **SM12** Rayed *Aster tripolium* stands

Variation within *Aster tripolium* is complex and the phytosociological relationships of the different forms are unclear.

---

7   *Arthrocnemum perenne* co-dominant with *Halimione portulacoides* and some *Puccinellia maritima*

> **SM7** *Arthrocnemum perenne* stands

Arthrocnemum perenne absent or present in small amounts   8

8   *Puccinellia maritima* a major constituent of the vegetation   9

Puccinellia maritima absent or present in small amounts   16

9   Any of *Festuca rubra, Agrostis stolonifera* and *Juncus gerardii* present in more than a trace and often co-dominant with *Puccinellia maritima*

SM16 *Juncetum gerardi*
Puccinellia maritima sub-community

Above species usually comprising less than 10% of the
sward   10

10   *Limonium binervosum* and/or *Frankenia laevis*
present with *Halimione portulacoides*   11

Neither *Limonium binervosum* nor *Frankenia laevis*
present   12

11   *Suaeda vera* present as a conspicuous component

SM21 *Suaeda vera-Limonium binervosum* salt-
marsh

Suaeda vera absent

SM22 *Halimione portulacoides-Frankenia laevis*
salt-marsh
Limonio vulgaris-Frankenietum laevis   Géhu &
Géhu-Franck 1975

12   *Puccinellia maritima* dominant or co-dominant
with *Plantago maritima* and/or *Armeria maritima* with a
conspicuous understorey of diminutive turf fucoids

SM13 *Puccinellietum maritimae*
Puccinellia maritima-turf fucoid sub-community

Turf fucoids absent or with low cover   13

13   *Spartina maritima* present

SM13 *Puccinellietum maritimae*
Puccinellia maritima-Spartina maritima sub-community

Spartina maritima absent   14

14   *Halimione portulacoides* co-dominant with *Puccinellia maritima* in intimate mixtures in which shoots of
the latter emerge through an open network of shoots of
the former; *Festuca rubra* rare and never abundant

SM14 *Halimionetum portulacoidis*
Puccinellia maritima sub-community

Prostrate *Halimione portulacoides* is also sometimes abundant in the *Limonium vulgare-Armeria
maritima* sub-community of the *Puccinellietum
maritimae* but other dicotyledons are usually co-
dominant there and *P. maritima* itself rarely comprises more than 10% of the swards. Intimate
mixtures of *Halimione portulacoides* and *Puccinellia maritima* such as are included here should

be distinguished from mosaics of discrete patches
of the *Halimionetum portulacoidis* and the *Puccinellietum maritimae*.

Halimione portulacoides infrequent and never co-domi-
nant   15

15   *Puccinellia maritima* and *Glaux maritima* co-
dominant in species-poor vegetation usually in small
stands

SM13 *Puccinellietum maritimae*
Glaux maritima sub-community

Puccinellia maritima dominant in open vegetation with
Spergularia marina and/or *Puccinellia distans*

SM23 *Puccinellietum distantis*

16   Varied swards dominated by mixtures of dicotyledons including *Armeria maritima*, *Triglochin maritima*
and *Plantago maritima* with usually less than 10% *Puccinellia maritima* and without *Frankenia laevis*, *Limonium
binervosum*, *L. bellidifolium* and *Suaeda vera*   17

Vegetation not dominated by mixtures of the listed dicotyledons or, if so, then some of *Frankenia laevis*, *Limonium
binervosum*, *L. bellidifolium* and *Suaeda vera* also present
                                                                                        18

17   *Limonium vulgare* (or, locally, *L. humile*), *Halimione portulacoides* and annual *Salicornia* spp. present and
sometimes abundant

SM13 *Puccinellietum maritimae*
Limonium vulgare-Armeria maritima sub-community

Glaux maritima and rayed *Aster tripolium* constant and
sometimes abundant with no *Limonium vulgare* and little
Halimione portulacoides

SM13 *Puccinellietum maritimae*
Plantago maritima-Armeria maritima sub-community

18   *Suaeda vera* and *Limonium binervosum* present
and/or *Frankenia laevis*   19

Not as above   21

19   *Frankenia laevis* present   20

Frankenia laevis absent

SM21 *Suaeda vera-Limonium binervosum* salt-
marsh
Typical sub-community

20   *Suaeda vera* present

SM21 *Suaeda vera-Limonium binervosum* salt-marsh
Frankenia laevis sub-community

Suaeda vera absent

SM22 *Limonio vulgaris-Frankenietum laevis*

21   *Artemisia maritima* prominent in usually small stands of somewhat variable vegetation ranging from rank grassy swards with much *Festuca rubra* to open bushy canopy of *A. maritima* over low *Halimione portulacoides*

SM17 *Artemisietum maritimae*

Artemisia maritima absent or inconspicuous   22

22   Grassy swards in which *Festuca rubra*, *Agrostis stolonifera* and *Juncus gerardii* are generally important components in the absence of *Juncus maritimus*   23

Juncus maritimus an important component of the vegetation   25

23   *Trifolium repens*, *Leontodon autumnalis* and *Potentilla anserina* present and often abundant in various combinations, sometimes with *Carex distans* and/or *C. flacca*   24

Short swards of very variable composition but usually dominated by *Festuca rubra* and *Agrostis stolonifera* with some *Juncus gerardii*, *Glaux maritima*, *Triglochin maritima*, *Armeria maritima* and *Plantago maritima* and with the above species absent or at less than 10% cover

SM16 *Juncetum gerardi*
Festuca rubra-Glaux maritima sub-community

On heavily-grazed marshes, especially in north-west England, swards lacking *Trifolium repens*, *Leontodon autumnalis* and *Potentilla anserina* may also have a very low cover of either *Festuca rubra* or *Agrostis stolonifera* or *Juncus gerardii*. These are best considered as derivatives of the *Festuca-Glaux* sub-community of the *Juncetum gerardi*.

---

24   *Carex flacca* constant and sometimes abundant

SM16 *Juncetum gerardi*
Carex flacca sub-community

Carex flacca infrequent

SM16 *Juncetum gerardi*
Leontodon autumnalis sub-community

On heavily-grazed marshes, especially in north-west England, swards lacking *Carex flacca* but also poor in *Leontodon autumnalis* and *Potentilla anserina* may be encountered. *Trifolium repens* remains a conspicuous component and such swards are best considered as derivatives of the *Leontodon autumnalis* sub-community of the *Juncetum gerardi*.

---

25   *Oenanthe lachenalii* constant and often abundant
26

Oenanthe lachenalii rare and never abundant but *Plantago maritima* and rayed *Aster tripolium* often conspicuous

SM18 *Juncus maritimus* salt-marsh
Plantago maritima sub-community

26   *Festuca arundinacea* constant and often co-dominant with *Juncus maritimus*

SM18 *Juncus maritimus* salt-marsh
Festuca arundinacea sub-community

Festuca arundinacea infrequent and never abundant

SM18 *Juncus maritimus* salt-marsh
Oenanthe lachenalii sub-community

A variety of other vegetation types encountered on salt-marshes may fail to key out here. These are most likely to be certain kinds of driftline vegetation, of mires and of mesotrophic grasslands. The mesotrophic grasslands are likely to be the most troublesome to distinguish as they often grade into forms of the *Juncetum gerardi* which have been much altered by agricultural treatment or into the communities of brackish pools with an increase in soil water salinity.

COMMUNITY DESCRIPTIONS

SM1
Zostera communities
Zosterion   Christiansen 1934

In Britain, three species of eel-grass, *Zostera marina, Z. angustifolia* and *Z. noltii*, form distinctive stands in the sub-littoral and eu-littoral zones of sand and mud flats. Very few samples of this vegetation were taken and the following account relies heavily on published and unpublished material relating in particular to The Solent (C. R. & J. M. Tubbs), the Thames estuary and Essex (Wyer & Waters 1975; Charman 1975, 1977*b*, 1979), north Norfolk (Ranwell & Downing 1959, Charman & Macey 1978), Lindisfarne (D. O'Connor), the Moray Firth (Rae 1979), and the west coast of Scotland (A. Currie). There are two difficulties in making use of existing information. First, *Z. angustifolia* is not consistently distinguished from narrow-leaved forms of *Z. marina*: this partly reflects the long-standing discussion on the taxonomic status of plants variously described as *Z. marina* var. *angustifolia, Z. hornemanniana* or *Z. angustifolia*. Second, eu-littoral stands have often been described simply as '*Zostera*' irrespective of whether they comprise *Z. angustifolia, Z. noltii* or both these species. This has been particularly true of accounts of the grazing of *Zostera* spp. by wildfowl and a separate note on this important aspect of the conservation value of the vegetation has therefore been appended.

Zostera marina stands
Zosteretum marinae   Harmsen 1936

Zostera marina forms stands with a cover of trailing leaves up to 1 m long. Algae, especially *Enteromorpha* spp., are usually the sole associates. *Z. marina* is essentially a sub-littoral species, extending from 1–4 m below to just above low water of spring tides, although it also occurs in lagoons. The lower salinity limit for the species is about 35 g l^{-1} (chloridity 24 g l^{-1}) but the exact limits of its distribution may be controlled by light requirement below and susceptibility to dessication above. Around The Solent, plants are exposed for only 1½ hours even at low water of spring tides.

Z. marina shows considerable morphological variation with a decrease in leaf size and density upshore. Narrow-leaved plants from the lower eu-littoral have been described as *Z. marina* var. *angustifolia* or confused with *Z. angustifolia*. There also appears to be some variation in phenology in relation to the position of the plants on the shore. *Z. marina* shows considerable leaf loss in autumn and early winter but this may be much more apparent in eu-littoral plants than in those which are permanently submerged where a dense cover is maintained throughout the winter. Regrowth occurs in all plants in spring and early summer. Flowering seems to be most frequent in eu-littoral plants and in those sheltered from wave action with larger sub-littoral plants reproducing vegetatively.

In Britain, *Z. marina* always grows on a firm substrate, usually sand or sandy mud, though sometimes with an admixture of fine gravel.

Where their ranges overlap, as in The Solent, *Z. marina* passes upshore to *Z. noltii*; elsewhere *Z. marina* stands may be separated by a considerable expanse of bare substrate from salt-marsh vegetation proper. In The Solent, *Z. marina* may have a potential competitor in the sub-littoral brown alga *Sargassum muticum*, a native of Japan which has colonised some sites once occupied by *Z. marina*.

Z. marina was much reduced in the early 1930s by a wasting disease which seems to have been a combination of attack by a protozoan and an ascomycete fungus. Butcher (1934, 1941) catalogued the most substantial decrease on the East Anglian and north Kent coasts and around The Solent. In recent years, the species has certainly reappeared in abundance in The Solent but seems to have remained rare elsewhere in the south-east. Butcher (1934) did not examine changes on the Scottish coast but *Z. marina* is now abundant down the western coast of the mainland and the Outer Hebrides and also in the Moray Firth. The map shows the distribution of the species in Perring & Walters (1962) with modifications.

Zosteretum marinae has been widely reported from throughout Europe though its exact status following the 1930s disease and subsequent erosion of substrates is uncertain. In The Netherlands, Beeftink (1962) records the association as rare; in France it appears to have recovered somewhat (Géhu 1975).

Zostera angustifolia stands

Zostera angustifolia forms stands with a cover of trailing leaves up to about 25 cm long. It may occur pure, though it is often mixed with the smaller *Z. noltii* and with a variety of algae among which species of *Ulva*, *Chaetomorpha* and *Enteromorpha* are often abundant. The table lists some samples of mixed *Zostera* vegetation from the Exe estuary, Devon. On the extensive estuarine flats of the Cromarty Firth, it occurs with *Ruppia maritima* and annual *Salicornia* spp.

Z. angustifolia can behave as a short-lived perennial. Around the Moray Firth, Rae (1979) noted that few plants lasted longer than two years and, throughout its British range, the species seems to suffer heavy leaf loss in autumn and early winter by a combination of natural shedding, storm damage and wildfowl grazing. Regrowth in spring can be largely by seedling germination (Ranwell & Downing 1959, Wyer & Waters 1975, Rae 1979) though good regeneration from existing rhizomes has also been reported.

Z. angustifolia is a plant of the lower and middle eu-littoral zone, extending to well above low water of neap tides and sometimes to high water of neap tides. Its optimal salinity is about 25–34 g l^{-1} (chloridity 16–20 g l^{-1}; Proctor 1980) and, as with *Z. marina*, its exact limits seem to be controlled by light requirement below and susceptibility to desiccation above. In The Solent, it is exposed for a maximum of about 6½ hours on the spring tides. It certainly grows best in sites which are never deeply submerged at high tide nor ever fully dry at low tide and is particularly characteristic of shallow depressions on tidal flats, often with some standing water at low tide. In such situations, it may form distinctive mosaics with *Z. noltii* which prefers the drier tops of low marsh ridges (Tutin, 1942, Wyer & Waters 1975, Rae 1979). It also occurs in the wet bottoms of deep marsh creeks (Chapman 1959).

Z. angustifolia is most characteristic of muds and muddy sands. These may be quite firm and contain some fine gravel but the species is typically associated with very sloppy mud on which even duck boards are an unsuccessful aid to sampling.

Z. angustifolia may pass upshore to stands of *Z. noltii* through mosaics of the two species; elsewhere it may give way to salt-marsh vegetation proper with an expanse of bare substrate between or through *Salicornietum europaeae*. In the Exe estuary, *Z. angustifolia* is replaced upshore by *Spartinetum townsendii* (Proctor 1980).

The disease of the 1930s seems to have left *Z. angustifolia* largely untouched and, at present, the species is widespread along the south and east coasts of England and the east coast of Scotland (Perring & Walters 1962). It is all but absent from the west coast of Scotland. There are very extensive stands in the Cromarty Firth (Figure 7) and also along the Essex and north Kent coasts.

In Europe, the equivalent community *Zosteretum marinae stenophyllae* Harmsen 1936 has been recorded from The Netherlands (Beeftink 1962) and France (Géhu 1975).

Zostera noltii stands
Zosteretum noltii   Harmsen 1936

Zostera noltii forms stands with a cover of delicate trailing narrow leaves up to about 20 cm long. It may occur pure or with *Z. angustifolia* (see table) and occasional plants of lower salt-marsh species such as annual *Salicornia* spp. or *Spartina anglica*. *Ruppia maritima* occurs with *Z. noltii* on the estuarine flats of the Cromarty Firth (Rae 1979).

Like *Z. angustifolia*, *Z. noltii* experiences considerable leaf loss in autumn and early winter through natural shedding, storm damage and wildfowl grazing but plants towards the lower limit may remain winter-green (Wyer & Waters 1975, Rae 1979). Unlike *Z. angustifolia*, expansion in spring seems to occur more consistently by the regrowth of existing rhizomes (Wyer & Waters 1975, Rae 1979) as well as by the germination of seed, production of which may be prolific, especially at higher levels.

In general, *Z. noltii* is a species of the middle and upper eu-littoral zone and its lower salinity limit is about 15 g l^{-1} (chloridity 9 g l^{-1}; Mathiesen & Nielsen 1956). It occurs on mud/sand mixtures of a variety of consistencies from very soft to quite firm. It is most characteristic of situations where the substrate dries out somewhat on exposure and on flats with a gentle bar/hollow topography it forms distinctive mosaics with *Z. angustifolia*. It can also occur in shallow standing water.

Stands of *Z. noltii* pass downshore to *Z. angustifolia* and above may grade to communities of the lower salt-marsh, notably the *Salicornietum europaeae*. *Spartina anglica* is known to have invaded stands of *Z. noltii* at various sites (Chapman 1959, Goodman *et al.* 1959, Bird & Ranwell 1964, Hubbard & Stebbings 1968).

The British distribution of *Z. noltii* is similar to that of *Z. angustifolia* (Perring & Walters 1962) and there are particularly extensive stands in the Cromarty Firth (Rae 1979: Figure 7) and along the Essex and north Kent coasts (Wyer & Waters 1975).

In Europe the *Zosteretum noltii* is widespread in similar situations to those in Britain (e.g. Beeftink 1962, Géhu 1975).

Figure 7. Distribution of mud-flat and salt-marsh vegetation in the Cromarty Firth, Scotland.

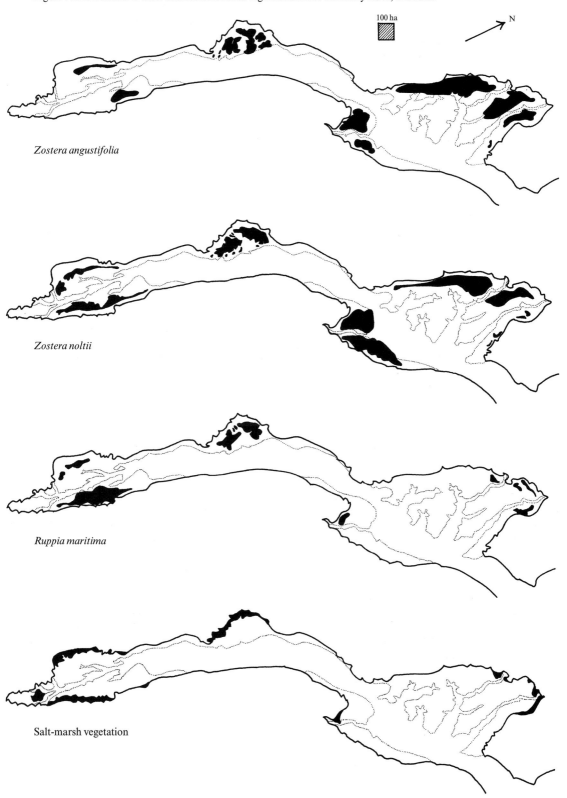

Zostera angustifolia

Zostera noltii

Ruppia maritima

Salt-marsh vegetation

Zostera and wildfowl grazing

Zostera spp. provide an important source of food for certain wildfowl, notably in Britain for overwintering brent goose (*Branta bernicla*) and wigeon (*Anas penelope*) and, to a lesser extent, of mute swan (*Cygnus olor*) and whooper swan (*Cygnus cygnus*).

The early wildfowling literature and some recent studies (e.g. Charman 1977*a*) consider *Z. marina* to have been the species most frequently eaten by brent in the past but it seems likely that, at the present time at least, *Z. angustifolia* and *Z. noltii* account for the bulk of the *Zostera* consumed. There is some suggestion (e.g. Ranwell & Downing 1959; Charman 1977*a*, 1979) that of these *Z. noltii* is the preferred species for brent. This may reflect its generally longer periods of exposure on flats but *Z. noltii* appears to reach its standing crop maximum later in the year than *Z. angustifolia*, around September/October (Wyer & Waters 1975, Rae 1979) just when brent are beginning to gather in their winter haunts. *Z. angustifolia* may be preferentially grazed by wigeon: its standing crop peak, in July/August, coincides with the gathering of that species.

A number of studies (Ranwell & Downing 1959; Charman 1975, 1977*a*, *b*, 1979; Charman & Macey 1978) have demonstrated a distinctive sequential exploitation of flat and salt-marsh food sources by brent. The accumulating birds begin feeding on *Zostera* in September/October and only when their numbers reach a peak and the *Zostera* is largely consumed do they move on, first to *Enteromorpha*, then to salt-marsh vegetation and sometimes to arable and pasture. This timing coincides to some extent with the maximum availability of nutritious food, though Charman (1979) has suggested that, among the various foods, only *Zostera* can provide an adequate daily energy requirement for brent.

Zostera stands therefore provide what seems to be an indispensible resource for some wintering wildfowl and vast numbers of birds exploit the larger beds. The 820 ha of *Zostera* along the coasts of south Suffolk, Essex and north Kent (Wyer & Waters 1975) receive about 30000 dark-bellied brent (*Branta bernicla bernicla*) (Ogilvie 1978), the expanding stands of The Solent foreshore and harbours about 23000 (figure for 1979/80), north Norfolk about 5000 and The Wash about 6000 (Ogilvie 1978), in total about half of the world population of this race. Smaller numbers of light-bellied brent (*Branta bernicla*

Floristic table SM1

Zostera angustifolia	V (2–8)
Zostera noltii	V (4–10)
Fucus spiralis	IV (1–4)
Enteromorpha cf. *E. marginata*	IV (1–7)
Ulva lactuca	III (1–4)
Chaetomorpha linum	II (1–6)
Polysiphonia cf. *P. insidiosa*	I (2)
Ceramium rubrum	I (1–2)
Polyneura gmelinii	I (1)
Fucus vesiculosus	I (1–2)
Cladophora sp.	I (4)
Chondria dasyphylla	I (1)
Polysiphonia cf. *P. nigrescens*	I (1)
Enteromorpha intestinalis	I (1–2)
Ectocarpus sp.	I (1)
Chaetomorpha cf. *C. tortuosa*	I (2)
Porphyra umbilicalis	I (2)
Spartina anglica	I (4)
Number of samples	15

Shells of the cockle (*Cerastoderma edule*) and common periwinkle (*Littorina littorea*) and casts of the lugworm (*Arenicola marina*) occasional to very abundant in the samples; spire shell (*Hydrobia ulvae*), mussel (*Mytilus edulis*) and shore crab (*Carcinus maenas*) recorded less frequently.

hrota), between 200 and 1100, winter at Lindisfarne NNR. What is probably the largest total area of *Z. noltii* and *Z. angustifolia* in Britain, the 1200 ha in the Cromarty Firth, is outside the winter range of the brent goose but the estuary is visited by enormous numbers of wigeon.

Although wildfowl sometimes uproot *Zostera* while feeding they seem mostly to eat the leaves and flowering shoots. Beds appear able to recover even from very heavy grazing and the resource to renew itself adequately from year to year by vegetative expansion and/or seed germination.

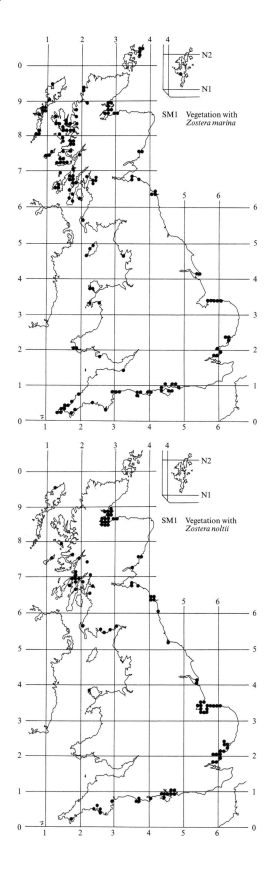

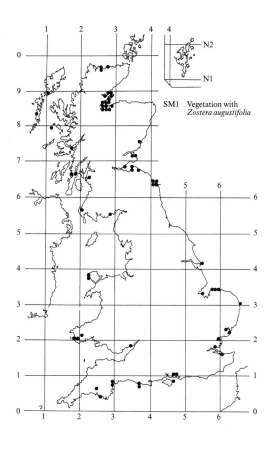

SM2
Ruppia maritima salt-marsh community
Ruppietum maritimae   Hocquette 1927

Ruppia maritima is a monocotyledonous perennial which can occur as the dominant in a submerged aquatic community with *Zannichellia palustris*, *Potamogeton pectinatus*, *Ranunculus baudotii* and, within its rather restricted range, *Ruppia spiralis*. The community occurs locally in permanently-filled pans and creeks on coastal salt-marshes, at some inland saline sites (Lee 1977) and also in brackish counter-dykes behind sea walls (Rose & Géhu 1964, Jermyn 1974). In the tidal portion of the outlet stream of the Loch of Wester in Caithness there is a zonation from *R. maritima*-dominated vegetation through *Potamogeton pectinatus* to *Hippuris vulgaris* at the tidal limit.

R. maritima can also occur as a plant of estuarine flats and it is particularly abundant in this habitat in the Cromarty Firth (A. Currie, P. Steele, pers. comm.: Figure 7) where it forms a belt of varying width between the salt-marsh proper, sometimes overlapping with *Salicornietum europaeae*, and stands of *Zostera noltii*. Here *R. maritima* seems to behave as an annual (P. Steele, pers. comm.), disappearing very rapidly from September onwards. It is known to be a food source for wigeon (*Anas penelope*) but frost sensitivity may also play a part in its behaviour.

The *Ruppietum maritimae* has been described from The Netherlands (Beeftink 1962) and from France (Géhu 1975) and in the latter it occurs on coastal flats.

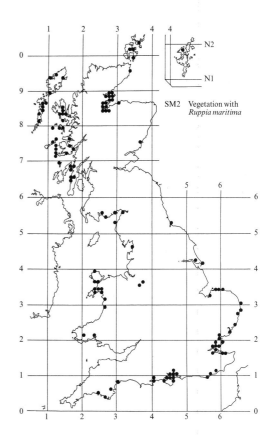

SM2   Vegetation with *Ruppia maritima*

SM3

Eleocharis parvula salt-marsh community
Eleocharitetum parvulae   (Preuss 1911/12) Gillner 1960

Eleocharis parvula occurs as a short open sward which is 'physiognomically the least conspicuous of all sea-shore communities' (Tyler 1969*b*). The diminutive shoots, only 1–2 cm tall, are frequently matted with (mainly green) algae and obscured by freshly-deposited silt (cf. Praeger 1934). At Beaulieu in Hampshire, stands occur at the limit of tidal influence with some input of fresh-water from land drainage at low tide (Géhu 1973*a*) but in Ireland the species may extend further downshore (Praeger 1934; C. D. Pigott, pers. comm.).

E. parvula is a very rare species in Britain with records for Beaulieu, Poole Harbour in Dorset, Bigbury Bay in Devon and Tremadoc Bay in Gwynedd. It has a similarly disjunct distribution throughout much of Europe (Beeftink 1972) but the *Eleocharitetum parvulae* has been recorded from the Biscay coast of France and Spain, from northern Portugal and the Mediterranean and Black Seas. The association is widespread in the Baltic where it frequently contains *Ruppia maritima* and *Zannichellia palustris* (Gillner 1960, Tyler 1969*a*). This led Gillner (1960) to place the association alongside the *Ruppietum maritimae* in the Ruppion maritimae, a view which is now generally accepted.

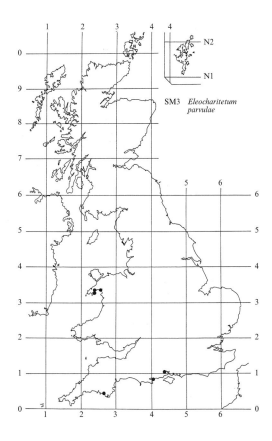

SM3 *Eleocharitetum parvulae*

SM4
Spartina maritima salt-marsh community
Spartinetum maritimae   (Emb. & Regn. 1926)
Corillion 1953

Spartina maritima is a native species which seems to have reached a peak of abundance in the late nineteenth and early twentieth centuries when it grew in every harbour between Beaulieu in Hampshire to Chichester in Sussex and plentifully along the coasts of Kent, Essex and south Suffolk and around The Wash (Marchant & Goodman 1969*a*). It declined rapidly thereafter and now survives often as isolated clumps around The Solent and on the north Isle of Wight coast (Marchant & Goodman 1969*a*) though extensive stands remain in parts of Essex (Jermyn 1974, Boorman & Ranwell 1977). The population at Scolt Head Island in Norfolk (Deighton & Clapham 1925, Chapman 1934) is now extinct.

   S. maritima grows as clumps of stiff shoots and at Wittering in Sussex it has some *Spartina anglica* and a little *Arthrocnemum perenne*, *Puccinellia maritima*, *Suaeda maritima* and *Salicornia* agg. (Géhu & Delzenne 1975). It is a pioneer community throughout its European range which runs south from The Netherlands to Portugal (Beeftink & Géhu 1973). The cause of its demise in Britain is not fully understood. It may partly be due to competition with *S. anglica*: Some former *S. maritima* sites are now occupied by *S. anglica* and the former seems to survive best where the latter is least aggressive, on drier sites above mean high water of spring tides (Marchant & Goodman 1969*a*). However, *S. maritima* is at the northern limit of its range in Britain and small climatic fluctuations may have played a part in its reduction (Marchant 1967). Certainly, little viable seed is produced at the present time (Marchant & Goodman 1969*a*).

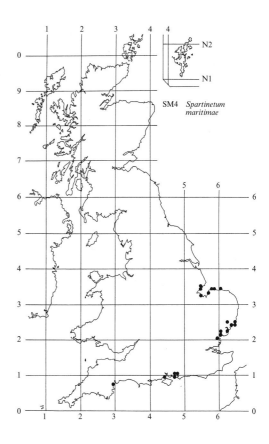

SM4 *Spartinetum maritimae*

SM5

Spartina alterniflora salt-marsh community
Spartinetum alterniflorae   Corillion 1955

Spartina alterniflora is a naturalised alien in Europe first recorded in Britain in 1829 from the river Itchen, Hampshire (Marchant & Goodman 1969*b*). By the turn of the century, it had spread to occupy extensive areas of Southampton Water and occurred as far east as Chichester Harbour in Sussex (Rankin in Tansley 1911, Marchant & Goodman 1969*b*). Since then it has declined, at least partly in response to land reclamation (Marchant 1967) though perhaps also as a result of invasion by *Spartina anglica*, the fertile amphidiploid which arose from hybridisation between *S. alterniflora* and *S. maritima*.

S. *alterniflora* now survives only at Marchwood, Hampshire, and as transplanted clumps in the *Spartina* Garden in Poole Harbour, Dorset (Marchant & Goodman 1969*b*). The natural stand comprises a dense cover of *S. alterniflora* shoots with some *Spartina anglica*, *Puccinellia maritima* and *Aster tripolium* (Beeftink & Géhu 1973, Géhu & Delzenne 1975). The association occurs in scattered localities down the Atlantic coast of western Europe and is characteristic of situations with a lower and more variable salinity than other *Spartina*-dominated communities (Beeftink & Géhu 1973). The Marchwood stand has been placed in the sub-association *asteretosum tripolii* which has also been recorded from Brittany and Spain (Beeftink & Géhu 1973, Géhu & Delzenne 1975, Kortekaas *et al.* 1976).

SM6

Spartina anglica salt-marsh community
Spartinetum townsendii (Tansley 1939) Corillion 1953

Constant species

Spartina townsendii sensu lato comprises the male sterile F_1 hybrid *S.* × *townsendii* and the much commoner fertile amphidiploid from the same cross, *S. anglica*. The latter alone was recorded in our samples and is the sole constant of the community.

Rare species

Arthrocnemum perenne

Physiognomy

S. anglica always dominates as scattered tussocks, coalescing clumps or a continuous sward up to 1 m in height. The community is species-poor though the associates are somewhat varied. *Puccinellia maritima* and annual Salicornias occur frequently and may account for up to 50% cover and beneath them there is often an algal mat. Free-living fucoids such as *Fucus vesiculosus* ecad *caespitosus* and ecad *volubilis* may be locally abundant. *S.* × *townsendii* may be recognised within *S. anglica* swards by its denser tussocks of shorter shoots. It sometimes forms extensive swards as at Hythe and Poole (Marchant 1967) and may occur as a landward fringe to *S. anglica* (Hubbard 1965). In such cases the associates of *S.* × *townsendii* are the same as in the community as a whole.

Sub-communities

Beeftink & Géhu (1973) and Kortekaas *et al.* (1976) have characterised a variety of sub-communities within the European *Spartinetum townsendii*. Some corresponding distinction could be made within the British stands but their general species poverty, their capacity for rapid floristic change and the unchallenged dominance of *S. anglica* throughout argue for retaining a single community at national level. Sub-division may be more appropriate in detailed studies of particular marshes: a few sites, for example, have a distinctive phase with *Atriplex hastata* and *Suaeda maritima* conspicuous.

Habitat

Although scattered plants of *S. anglica* can be found in almost every salt-marsh community, the distinctive situations of the community are towards the seaward fringes of marshes, on creek sides, colonising old pans in the upper-marsh zone and, more rarely, in brackish seepage areas behind sea walls. The *Atriplex hastata-Suaeda maritima* phase is characteristic of tidal drift.

A wide variety of substrates is colonised, from extremely soft mud to shingle. There is some evidence of more rapid vegetative spread through finer material (Chater & Jones 1957). The pH is generally above 7.0 and loss-on-ignition varies from 0.2% to 36.3% (Adam 1976). The sediment is generally strongly reduced although there may be a narrow zone of oxidation around the roots. *S. anglica* is extremely tolerant of tidal submergence. In general, its lower limit seems to be around mean high water of neap tides, which implies about 6 hours' submersion/day at spring tides (Goodman *et al.* 1969, Dalby 1970, Morley 1973, Proctor 1980) but in exceptional situations, as in the shelter of Poole Harbour with its narrow tidal range, the community may extend down to mean low water of neap tides, with as much as 23½ hours' submersion/day at neap tides (Hubbard 1969). The lower limit of colonisation is therefore probably controlled by factors other than submersion tolerance and it appears most likely that exposure to the physical effects of wave or tidal action is responsible. Goodman *et al.* (1959) and Goodman (1960) have shown how *S. anglica* may become dwarfed by such a process where the maximum fetch is short. Such stunted plants resume normal growth under culture (Goodman *et al.* 1969) and should be distinguished from the dwarf brown mutants described from certain sites (e.g. Chater 1965). It is also possible that the nature of the substrate may influence the seaward limit of the species but data are lacking. The landward limit of extension may be controlled by the juxtaposition of existing communities up the shore. *S. anglica* can tolerate up to about 2.5% chloridity

(Ranwell *et al.* 1964, Proctor 1980) and this may give it more competitive advantage in the hypersaline conditions when dry weather follows high tides on the upper marsh.

The community occurs on both grazed and ungrazed marshes. Although less heavily exploited than some marsh communities, it appears to be eaten by rabbits, sheep and cattle and *S.* × *townsendii* may actually be encouraged or maintained by grazing (Hubbard 1965) and its growth favoured against invasion by *S. anglica* where the two species occur contiguously. Hubbard & Ranwell (1966) showed that cut and ensilaged *S. anglica* had a similar intake, digestibility and crude protein content to medium quality hay when fed *ad libitum* to Halfbred × Suffolk wethers, though they recognised that marsh physiography would usually militate against its widespread exploitation as a crop. Despite its dramatic spread, *S. anglica* rarely seems to pose a navigational hazard though its presence in coastal resorts or nature reserves may be undesirable. Some control has been achieved by helicopter spraying with dalapon (Ranwell 1967) though assiduous digging is usually necessary to ensure eradication.

Zonation and succession

S. × *townsendii* was first noticed around Southampton Water in the 1870s and is presumed to be a natural hybrid between the native *Spartina maritima* and the introduced *S. alterniflora* (Marchant 1967). Its slow natural spread was quickly overtaken by the much more vigorous fertile amphidiploid *S. anglica* which, in the 20 years after its first appearance in Southampton Water and The Solent probably around 1890, had colonised every estuary and salt-marsh between Chichester and Poole (see, for example, Goodman *et al.* 1959, 1969, Hubbard 1965, Ranwell 1967, Hubbard & Stebbings 1967). Widespread planting for reclamation after 1910 extended the distribution of *S. anglica* in Britain and natural spread from these centres has filled many gaps.

S. × *townsendii* appears to have been transmitted to a number of scattered localities by inclusion of sterile material in transplant consignments but this species probably accounted for less than 20 ha out of a total of over 12 000 ha of marsh dominated by *S. townsendii s. l.* in Great Britain at the last detailed survey (Hubbard & Stebbings 1967).

Once established, the plants perennate and reproduce naturally by rhizome fragmentation, especially where the tidal run is fast, as at Bridgwater Bay (Ranwell 1964a) and, in the case of *S. anglica*, by seed, the set of which is regular though variable in quantity (Goodman *et al.* 1969). Spread from such fragments or seedlings can be rapid in *S. anglica* with clonal patches expanding and coalescing into clumps and then closing to a sward; in other cases discrete patches may persist for long periods.

In many places *S. anglica* has become established on previously bare substrates and initiated the development of new marshes. Accretion of material has been found to vary between 0.5 and 10 cm/year (Ranwell 1964a, Bird & Ranwell 1964); at the higher rates something like 500 cm/ha/year of material is deposited. Accretion rate may depend on local climate, the tidal pattern and perhaps the seasonal microflora (Ranwell 1964a) and the subsequent marsh drainage pattern may be influenced by the slope, tidal range and substrate type (Braybrooks & J. M. Lambert, unpublished). Accretion eventually raises the marsh surface to a level at which other species can theoretically compete with the *S. anglica* but, although species from *Puccinellietum maritimae* and *Juncetum gerardi* communities occur occasionally within swards, they are never particularly abundant and competition with the tall and vigorous *S. anglica* may prevent overtopping and the replacement of the community. Litter accumulation or frost action (Hubbard & Stebbings 1967, Ranwell 1972) and grazing (Ranwell 1961, Goodman *et al.* 1969) may initiate the opening up of the *S. anglica* sward and allow the spread of *Puccinellia maritima*.

On the higher parts of ungrazed marshes there is evidence of invasion by a variety of plants. At Bridgwater Bay, a short-period sub-seral alternation of *S. anglica* and *Atriplex hastata* has been observed on accumulated drift with sand (Ranwell 1961, 1964b). *A. hastata* has also invaded the upper part of a *S. anglica* marsh at Lytham on the Ribble estuary. At Keysworth in Poole Harbour, *Elymus pycnanthus* has invaded (Hubbard & Stebbings 1968). There and at Bridgwater, *Scirpus maritimus* and *Phragmites australis* have also appeared at higher levels and replaced about 50% of the *S. anglica* sward in 12 years. Ranwell (1972) has suggested that such a process is favoured by the development of less saline conditions consequent upon land-drainage seepage but this has not been widely investigated. *Halimione portulacoides* can establish itself along creek margins within *S. anglica* marshes (Goodman *et al.* 1959).

S. anglica has also become established on existing marshes. Invasion of *Zostera noltii* swards has been recorded on the south and east coasts (Goodman *et al.* 1959; Bird & Ranwell 1964; Hubbard & Stebbings 1968) and Chapman (1959) mapped such a process in North Cockle Bight at Scolt Head between 1932 and 1959. At that site, *Z. angustifolia* on very soft mud was not invaded; neither is there any evidence that *S. anglica* has anywhere replaced *Z. marina* which occurs at lower levels than *Z. noltii*. At Keysworth in Poole Harbour, a *Ruppia maritima-Potamogeton pectinatus* community has been replaced by *S. anglica* (Hubbard & Stebbings 1968). *S. anglica* can flourish at the same level as the *Salicornietum europaeae* and, as a result, pioneer vegetation of this kind is now of local occurrence throughout south-east England.

The extent to which *S. anglica* invades other existing marsh communities is uncertain. Pans and creeks in vegetation higher up the marsh may be grown over and scattered plants are widespread throughout marshes, but the wholesale replacement of other communities is not well documented. Chater & Jones (1957) provide some evidence for a slow advance into *Puccinellietum maritimae* and *Juncetum gerardi* in the Dovey estuary but this is not apparent at many sites. Similarly there is little evidence as to how much invasion into *S. anglica* swards takes place from contiguous communities. On grazed marshes in the Dovey, creek levees have become colonised by *Festuca rubra* which has eventually ousted *S. anglica*; because of the frequency of creeks there, the total area of marsh affected is considerable (Chater 1973). Heavily grazed and poached upper levels of *S. anglica* have elsewhere been invaded by *Puccinellia maritima*. It is not known how far the development of a *S. anglica* community to seaward of an existing marsh complex affects the overall nature of the marsh but it might be expected that alterations in drainage would be of prime importance.

In general, *S. anglica* has consolidated its early initial spread but the patchy degeneration of sward which became known as 'die-back' was noticed as early as 1928 and in some sites has made a considerable impression. It is still mainly restricted to Channel coast marshes and its exact cause remains unknown. Pathogens and pollution have been ruled out (Goodman *et al.* 1959) and it seems possible that the process is caused by a toxic reduced inorganic ion (perhaps sulphide) produced in anaerobic waterlogged root environments (Goodman & Williams 1961). Alternatively, the switch from accretion to ablation under *S. anglica* may be responsible for its demise.

Distribution

Spartinetum townsendii is widespread around the English and Welsh coasts and is still expanding vigorously at a number of sites along the Scottish shore of the Solway. *S. × townsendii* in itself present in abundance only below Hythe in Southampton Water (Hubbard & Stebbings 1967) though there is F$_1$ material scattered through *S. anglica* swards from Poole to Wittering and on the Isle of Wight. This natural limit is probably set by the slow vegetative spread of the species. There are also small quantities, probably transmitted with *S. anglica* for transplant, in Norfolk, Somerset, Merioneth and

Dublin. The limit of *S. anglica* and of the association as a whole, may be related to temperature: in the northern hemisphere, really successful plantings occur south of the 13–18 °C July isotherms (Goodman *et al.* 1969). The far northern stations of the species in Argyll and Harris (Hubbard & Stebbings 1967) have not been checked but it is known that growth in these localities is very slow and seedling establishment poor because of winter storms, cold and bird damage (Shaw, *pers. comm.* in Goodman *et al.* 1969). The European distribution of the community is discussed by Beeftink (1972), Géhu (1972) and Beeftink & Géhu (1973).

Affinities

The association is easily defined floristically by the dominance of *S. anglica* and in phytosociological schemes the community has been placed in a separate class, the Spartinetea, with other communities based on *S. maritima* and *S. alterniflora*.

Floristic table SM6

Spartina anglica	V (5–10)
Algal mat	III (2–9)
Puccinellia maritima	III (1–7)
Salicornia agg.	III (1–7)
Suaeda maritima	II (1–5)
Aster tripolium var. *discoideus*	I (1–7)
Aster tripolium (rayed)	I (2–6)
Aster tripolium	I (1–5)
Atriplex prostrata	I (2–7)
Limonium cf. *L. vulgare*	I (1–3)
Plantago maritima	I (3–5)
Fucus vesiculosus ecad *caespitosus*	I (2–5)
Fucus vesiculosus ecad *volubilis*	I (2–6)
Catenella repens	I (2)
Cochlearia anglica	I (2–5)
Spergularia media	I (1–5)
Halimione portulacoides	I (2–6)
Number of samples	136
Mean number of species/sample	3 (1–10)
Mean vegetation height (cm)	34 (8–10)
Mean total cover (%)	84 (25–100)

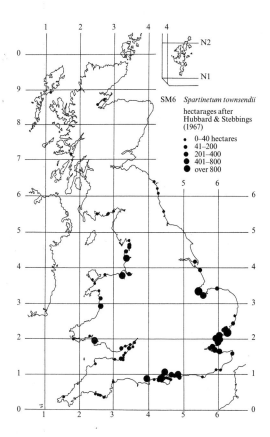

SM6   *Spartinetum townsendii*

hectarages after
Hubbard & Stebbings
(1967)

- 0–40 hectares
- 41–200
- 201–400
- 401–800
- over 800

SM7
Arthrocnemum perenne stands

Arthrocnemum perenne is a perennial halophyte with a restricted distribution in Britain: it occurs around the coast of south-east England from The Wash to Poole Harbour with isolated records from North Wales and Teesmouth (Perring & Walters 1962). It is encountered as an occasional in a variety of communities from both low and high marsh and only very locally is it an important constituent of salt-marsh vegetation.

At a number of sites in north Norfolk, *A. perenne* forms an open mosaic with *Halimione portulacoides*, *Puccinellia maritima* and *Suaeda maritima* at the lower limit of *Halimionetum* on sand or firm silt with abundant gravel and shell fragments. A similar community occurs very locally on firm clays with shell fragments elsewhere in south-east England.

Dense pure stands of *A. perenne* are found on drift litter over shell banks at a few sites, particularly around Chichester Harbour, Hampshire. Scattered bushes of *A. perenne* are associated with local erosion within low-marsh communities, especially where these occur on gravel-rich substrates (Beeftink 1965, 1977*a*; Beeftink & Géhu 1973).

Although Beeftink (1965, 1977*a*) recognises the occurrence of *Salicornietum radicantis* Br.-Bl. 1931 in Britain we have insufficient data to characterise a community. The *A. perenne-Halimione portulacoides* mosaics could perhaps be regarded as an extreme form of *Halimionetum*.

Floristic table SM7

Arthrocnemum perenne	V (2–9)
Halimione portulacoides	V (2–9)
Puccinellia maritima	V (2–6)
Suaeda maritima	IV (2–4)
Salicornia agg.	III (2–6)
Algal mat	III (4–8)
Limonium cf. *L. vulgare*	III (2–5)
Aster tripolium var. *discoideus*	III (2–7)
Aster tripolium	I (1–3)
Bostrychia scorpioides	I (4–5)
Spergularia media	I (1)
Number of samples	12
Mean number of species/sample	6 (3–9)
Mean vegetation height (cm)	19 (8–30)
Mean total cover (%)	76 (40–100)

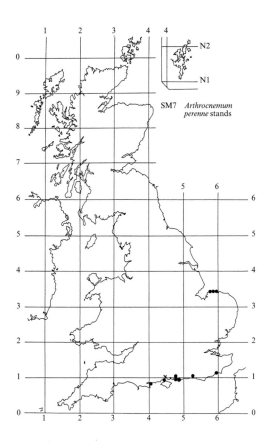

SM7   *Arthrocnemum*
       perenne stands

SM8

Annual *Salicornia* salt-marsh community
Salicornietum europaeae   Warming 1906

Constant species

Several distinct taxa can be recognised among the British annual Salicornias but diagnosis below the level of the three groups *S. europaea*, *S. procumbens* and *S. pusilla* is difficult for non-specialists. Here, all annual taxa encountered are described under *Salicornia* agg. and this is the sole constant of the community.

Rare species

Arthrocnemum perenne.

Physiognomy

The community comprises ephemeral stands of annual Salicornias sometimes with no other species. The vegetation is invariably somewhat open and though *Salicornia* agg. is always conspicuous, the density of plants is variable: around The Wash, for example, *Salicornia* agg. cover is high whereas at sites with sandier substrates the density can be very low. There is often an algal mat over the substrate surface but vascular companions are usually very few. Scattered plants of *Puccinellia maritima*, *Suaeda maritima* and *Spartina anglica* occur frequently with occasional records for a variety of other lower marsh species. At a few sites (Blakeney Point, for example), *Fucus vesiculosus* ecad *caespitosus* is abundant.

Sub-communities

With careful identification of distinct taxa, it may be possible to define a range of communities within this broad general unit as a number of Continental authors have done (see, for example, Géhu & Delzenne 1975). Ball & Tutin (1959) recommend collecting a dozen specimens from populations in September/October when the characteristic colours have developed. Where particular taxa have been ascribed distinct ecological preferences *in litt.* these have been noted below.

Habitat

Annual Salicornias germinate in May from seeds widely dispersed over whole marsh surfaces. The lower limit of establishment appears to be set by the time necessary for the seedlings to become firmly anchored: Wiehe (1935) showed that, in the Dovey estuary, two to three days' exposure between tidal flooding was necessary for sufficient root growth to take place. The speedier radical growth of *Salicornia dolichostachya* over *S. europaea sensu stricto* may give the former an establishment advantage in such situations: in the Dee estuary, *S. dolichostachya* is certainly the commoner species in the open habitats of the lower marsh (Ball & Brown 1970; see also Ball & Tutin 1959).

Salicornia agg. is tolerant of frequent tidal submersion, enduring around 600 flooding tides/year at its lower limits where it forms the familiar pioneer stands. The community is also characteristic of other bare marsh habitats such as creek sides, borrow pits and other disturbed areas in the upper marsh. Here seedlings grow rapidly and by August the plants are bushy, green and up to 15–20 cm high. Although certain taxa, *S. europaea s. s.* for example, appear less susceptible than others to competition from perennial grasses (Ball & Brown 1970), growth in the upper marsh is generally slow and the restrictions may be due to the lack of competition for sediment nutrients, especially nitrogen, with established perennials (Pigott 1969, Stewart *et al*. 1972). Addition of nutrients to *Salicornia* plants within the other high marsh communities stimulates growth to levels characteristic of the lower marsh stands, though a lag in response is suggestive of a determinate growth pattern genetically adapted to an environment with a cyclical but delayed suitability: such sites are not flooded again after *Salicornia* germination until the autumn equinox (Jefferies *et al*. 1979).

Within suitable sites, the community can flourish on a variety of substrates from hard clay to shelly sand, occasionally even on shingle but only rarely are very soft sediments colonised. Where *Spartina anglica* has become well established, the low-marsh *Salicornia* stands tend to be restricted to thin gravels or shingles over hard clay (Perraton 1953). Where wind-blown sand is abundant, plants tend to adopt a decumbent habit and can survive virtual burial for most of the summer (Tüxen 1974). In brackish areas behind sea walls, which we have not examined closely but where annual Salicornias may flourish, *S. ramosissima* and *S. prostrata* appear to be the common taxa (Ball & Tutin 1959).

Annual *Salicornia* stands occur on both grazed and ungrazed marshes. All species are highly susceptible to oil and refinery effluent spills being killed by a single inundation (Baker 1979). On a few marshes in south-east England (principally around The Wash) Salicornias are harvested as 'samphire' for human consumption.

Zonation and succession

Salicornia stands may form a distinct zone in the lower marsh, sometimes hundreds of metres deep. At some sites, particularly those on sandy substrates, patches of *Salicornia* may be separated from the main marsh front by several hundred metres of bare flat. The community can also occur in a mosaic with the *Puccinellietum maritimae* or with the *Spartinetum townsendii*. The expansion of the latter has much reduced the area of lower marsh available for pioneer *Salicornia* establishment especially in south-east England.

Although the annual *Salicornietum* is generally the lowest marsh community proper it may rarely initiate a succession because summer accretion can be offset by ablation after the stands have disappeared in the winter. Chapman (1957) has, however, produced a cartographic record of changes in the distribution of *Salicornia* marsh at Scolt Head, Norfolk with ageing of open and closed marshes.

Distribution

The community is widely distributed around the British coastline. On the sandy marshes of the west coast occurrences are local, though extensive open stands occur in some estuaries. The very local distribution in western Scotland is largely a reflection of the lack of suitable habitats: many loch-head marshes are fronted by cobble beaches rather than sand flats and these carry a dense cover of free-living fucoids.

Affinities

Equivalent communities have a widespead distribution in Europe. Although various divisions have been made with the *Salicornietum europaeae*, the general composition and habitat relationships of the vegetation types accord with the British community (see, for example, Beeftink 1962, 1965, 1972, 1977*a*).

Floristic table SM8

Salicornia agg.	V (4–9)
Algal mat	III (3–8)
Puccinellia maritima	III (1–7)
Suaeda maritima	II (1–5)
Spartina anglica	II (1–5)
Halimione portulacoides	I (1–3)
Aster tripolium var. *discoideus*	I (1–4)
Aster tripolium (rayed)	I (1–3)
Number of samples	81
Mean number of species/sample	3 (1–7)
Mean vegetation height (cm)	7 (2–20)
Mean total cover (%)	53 (5–95)

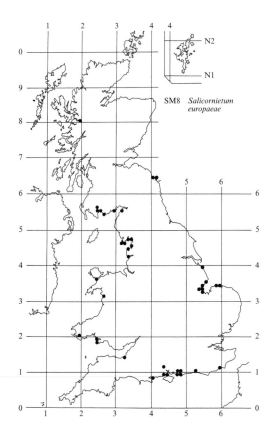

SM8   *Salicornietum europaeae*

SM9
Suaeda maritima salt-marsh community
Suaedetum maritimae   (Conrad 1935) Pignatti 1953

Synonymy
Suaeda maritima nodum   Adam 1976; *Salicornietum auct. p.p.*

Constant species
Suaeda maritima is a variable taxon within which a number of distinct forms have been recognised. These are sometimes treated as varieties (e.g. Chapman 1947, Clapham *et al*. 1962) or as species. Ball (1964) has a single sub-species *S. maritima maritima* to include all British material. This is the only constant taxon of the community.

Physiognomy
This is a species-poor community, generally open, though always dominated by *Suaeda maritima* the density of which is normally high. There is sometimes a little annual *Salicornia*, *Puccinellia maritima*, *Spartina anglica*, *Halimione portulacoides* and *Aster tripolium* var. *discoideus*. An algal mat is quite common and Chapman (1947) lists seven different algal assemblages associated with abundant *Suaeda maritima*.

Sub-communities
Variation within the community is continuous, though the predominance in particular stands of different forms of *Suaeda maritima* has been used by some (e.g. Géhu 1975) to assign such stands to different communities. However, diagnosis is often difficult and there seem to be few consistent ecological differences between the taxa.

Habitat
Suaeda maritima is an annual and it is tolerant of a wide range of soil types subject to various submersion regimes: Chapman (1947) reported it dominant on Norfolk marshes with between 290 and 430 submergences/year. Like the annual Salicornias, its growth appears heavily dependent upon sediment nutrients, especially nitrogen (Pigott 1969, Stewart *et al*. 1972), and it is particularly characteristic of open situations free of competition from established perennials. On the lower marsh it is especially distinctive of rather gravelly mud where it forms mosaics with stands of annual Salicornias. Fragmentary stands are found around the base of the shell banks which occur at low levels in a few sites. Pure stands of *S. maritima* are a distinctive feature of disturbed situations such as the piles of sediment dumped on marshes during the construction of sea walls and drainage channels. Creek sides can also carry the community. Two further distinctive habitats are the accumulations of drift litter that occur at the foot of sea walls where dense stands can exploit the release of nutrients upon decomposition of the litter (see Beeftink 1966) and brackish areas behind sea walls where prostrate forms of *S. maritima* are common.

Zonation and succession
The habitat diversity of the community makes it difficult to generalise about the successional status of the community. In situations subject to repeated disturbance it can recur every year but increased stabilisation leads to replacement by the community appropriate for the particular level of the marsh.

Distribution
The community is widespread but many stands are fragmentary. It is most frequent in south-east England and very local in west Scotland.

Affinities
Although sometimes considered as part of a *Salicornietum*, stands dominated by *Suaeda maritima* are sufficiently distinctive to be worthy of considering as a separate community. Certain authors (e.g. Beeftink 1962, 1965, 1977a; Westhoff & den Held 1969) consider the nitrophilous character of the vegetation warrants placing the community with the ephemeral driftline associations of the Cakiletea but the floristic affinities to that class are few. The low-marsh occurrences are seen by others (e.g. Géhu 1975, Géhu & Delzenne 1975) as indicating a similarity to the communities of the Thero-Salicornieta which then becomes the class for all ephemeral chenopod-dominated vegetation types of the low marsh.

Floristic table SM9

Suaeda maritima	V (6–10)
Salicornia agg.	IV (2–7)
Puccinellia maritima	II (2–4)
Algal mat	II (5–8)
Spartina anglica	II (2–3)
Halimione portulacoides	II (1–4)
Aster tripolium var. *discoideus*	II (1–3)
Aster tripolium (rayed)	I (2)
Number of samples	18
Mean number of species/sample	3 (2–8)
Mean vegetation height (cm)	27 (8–50)
Mean total cover (%)	69 (30–100)

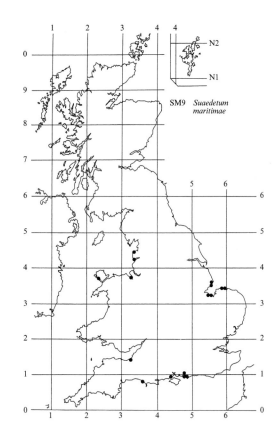

SM9   *Suaedetum maritimae*

SM10

Transitional low-marsh vegetation with *Puccinellia maritima*, annual *Salicornia* species and *Suaeda maritima*

Adam (1976) recognised a number of vegetation types of the low marsh in which *Puccinellia maritima* was co-dominant with annual *Salicornia* species and/or *Suaeda maritima* during the growing season but which during the winter took on the appearance of very open *Puccinellia maritima* swards. Such transitional vegetation can be regarded as one extreme of variation within the *Puccinellietum maritimae* but, particularly in detailed studies of individual marshes, separate recognition might be appropriate. The description below refers to stands in which all three taxa are present.

Synonymy
Puccinellietum maritimae (Warming 1906), W. Christiansen 1927 *auct. p.p.*; *Puccinellia-Salicornia-Suaeda* nodum Adam 1976; *Suaedetum maritimae auct. p.p.*

Constant species
Puccinellia maritima, annual *Salicornia* spp., *Suaeda maritima*.

Rare species
Arthrocnemum perenne.

Physiognomy
Stands of the community are invariably species-poor and always dominated by complementary proportions of the three constants which during the growing season form a fairly low sward of rather variable total cover. Rayed *Aster tripolium* and *A. tripolium* var. *discoideus* are quite frequent though never abundant. There is sometimes an algal mat which can cover up to 50% of the substrate surface.

Habitat
At its lower limit the number of tides flooding the community is probably similar to that experienced by the lower part of the *Puccinellietum maritimae*. Soils vary from firm clays to coarse sands with a pH range of 7.0–8.0 and high levels of free calcium carbonate.

On sandy substrates, the community may occur as a pioneer. It is then rarely extensive, forming patches in a mosaic with the *Salicornietum europaeae*, the *Spartinetum townsendii* or the *Puccinellietum maritimae*. Where the lower marsh consists of a hummocky *Puccinellia maritima* community, a situation confined to sandy marshes which are normally heavily grazed, this transitional community may be found on the hummock tops (cf. Oliver 1907, Hill 1909, Tansley 1911).

On muddier marshes in south-east England, the community behaves in the contrary fashion, occurring in slight depressions within the *Puccinellietum maritimae*, *Spartinetum townsendii*, *Asteretum tripolii* and *Halimionetum portulacoidis*.

The community is also widespread on the sides of large creeks where it occupies a distinct zone above the *Salicornietum europaeae*. The majority of such occurrences are on ungrazed or cattle-grazed marshes; on sheep-grazed marshes, the community is confined to inaccessible creek sides.

Zonation and succession
In the low marsh, the community will be replaced by others as accretion progresses: in the south-east most probably by the *Puccinellietum maritimae* or the *Halimionetum portulacoidis*, in the west by the former or, more rarely, by the *Juncetum gerardi*. Creekside occurrences are part of what is probably a static zonation rather than a successional sequence.

Distribution
Apart from along the western Scottish coast, where occurrences are relatively rare, the community is widespread, although stands are often small.

Floristic table SM10

Puccinellia maritima	V (2–9)
Salicornia agg.	V (2–8)
Suaeda maritima	V (2–8)
Aster tripolium var. *discoideus*	III (2–5)
Aster tripolium (rayed)	III (1–4)
Algal mat	II (4–7)
Spartina anglica	I (1–4)
Halimione portulacoides	I (1–2)
Triglochin maritima	I (3–4)
Spergularia media	I (2–3)
Limonium cf. *L. vulgare*	I (2)
Armeria maritima	I (1–4)
Number of samples	50
Mean number of species/sample	5 (3–8)
Mean vegetation height (cm)	15 (4–40)
Mean total cover (%)	88 (30–100)

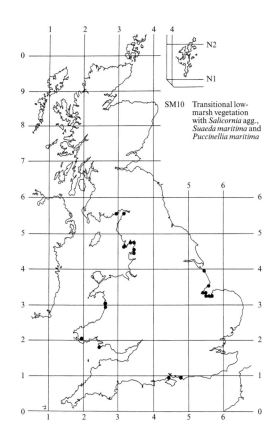

SM10 Transitional low-marsh vegetation with *Salicornia* agg., *Suaeda maritima* and *Puccinellia maritima*

SM11

Aster tripolium var. discoideus salt-marsh community
Asteretum tripolii   Tansley 1939

Synonymy

Asteretum and Creek *Asteretum* Chapman 1934; *Aster tripolium* var. *discoideus* nodum Adam 1976; descriptions of Great *Aster* marsh, Scolt Head, Norfolk.

Constant species

Aster tripolium var. *discoideus, Puccinellia maritima, Salicornia* agg.

Rare species

Arthrocnemum perenne.

Physiognomy

The association is dominated by the rayless *Aster tripolium* var. *discoideus* which is especially distinctive in the late summer–early autumn flowering season when its stems may attain a height of about 1 m; at other times the vegetation is 10–20 cm tall. The stands are generally rather species-poor though there is usually some *Salicornia* agg., *Puccinellia maritima* and *Suaeda maritima*. The substrate surface is frequently dissected by small drainage runnels threading between the *A. tripolium* rootstocks and locally may be carpeted by free-living fucoids, mainly *Fucus vesiculosus* ecad *caespitosus* and *Pelvetia canaliculata* ecad *libera*.

Although there is a floristic gradation between low-level stands with abundant *Salicornia* agg. and those at higher levels with abundant *Puccinellia maritima*, no sensible subdivisions can be made within the association. The floristic disinctions catalogued by Chapman (1934) between a low-marsh *Asteretum* and a Creek *Asteretum* are not borne out in the samples.

Habitat

The association occurs as an extensive zone in the low marsh or on creek sides at varying levels in the marsh. At its lower limits, the association seems able to tolerate upwards of 500 submergences/year (Chapman 1960*a*) with a maximum development around 350 submergences/year (Clapham *et al.* 1942). The sediments are predominantly firm clays or silts low in organic matter but with a high proportion of fine shell fragments and a pH between 7.0 and 8.0. Most of the sites are ungrazed or only lightly cattle-grazed.

Zonation and succession

In the low marsh the association forms a distinct zone above the *Salicornietum europaeae* or the *Spartinetum townsendii* or, occasionally, at the most seaward limit. Upwards it passes into the *Puccinellietum maritimae* or the *Halimionetum portulacoidis*. Landward boundaries are diffuse and patches of the association are frequently found in the lower part of the zone above.

Although *A. tripolium* var. *discoideus* can grow at lower levels on the shore than many salt-marsh species, it is not a successful primary coloniser. Gray (1971) has suggested that it has spread in recent years, possibly following *Spartina anglica* invasion of bare substrates, but the evidence for this is inconclusive. At Scolt Head, Norfolk, Chapman (1959) has shown the association developing from the *Salicornietum europaeae* in about 25 years.

Distribution

The association is predominantly south-eastern in its distribution, being frequent in The Wash, north Norfolk and Essex. Old records have *A. tripolium*-dominated communities on Canvey Island (Carter 1932) and in the Humber (Good & Waugh 1934). It is local on the south coast and in the Bristol Channel (but see Thompson 1922, 1930) and its general absence from the west may reflect climatic limitations, the scarcity of muddy marshes or the higher incidence of grazing there.

Affinities

The position of the association in the salt-marsh zonation places it between the annual communities of the Salicornietea and the perennial communities of the Asteretea but the perennial nature of *A. tripolium* var. *discoideus* itself suggests that it is best seen alongside the *Puccinellietum maritimae* and the *Juncetum gerardi* of the latter class.

Floristic table SM11 & SM12

	11	12a	12b
Aster tripolium var. *discoideus*	V (4–10)	III (4–7)	
Puccinellia maritima	V (1–9)	V (3–7)	
Salicornia agg.	V (3–9)	III (2–4)	
Suaeda maritima	III (2–8)	III (2–4)	
Algal mat	II (2–8)	I (6)	
Halimione portulacoides	II (1–5)		
Arthrocnemum perenne	I (1–5)		
Fucus vesiculosus ecad *caespitosus*	I (5–8)		
Pelvetia canaliculata	I (4–9)		
Bostrychia scorpioides	I (3–7)		
Aster tripolium (rayed)		V (5–8)	V (7–10)
Spartina anglica	II (1–6)	IV (2–3)	
Plantago maritima	I (4)	III (2–6)	
Spergularia media	I (3–4)	III (4–6)	
Triglochin maritima		II (5)	
Puccinellia distans			V (3–4)
Spergularia marina			III (1–3)
Atriplex prostrata		I (5)	III (2–3)
Scirpus lacustris tabernaemontani			II (2–7)
Juncus bufonius			II (2–3)
Number of samples	53	7	9
Mean number of species/sample	5 (3–8)	6 (4–9)	4 (2–6)
Mean vegetation height (cm)	28 (5–150)	43 (15–80)	68 (60–100)
Mean total cover (%)	80 (45–100)	81 (50–90)	99 (90–100)

11   *Aster tripolium* var. *discoideus* salt-marsh
12a Coastal stands of rayed *Aster tripolium*
12b Inland stands of rayed *Aster tripolium*

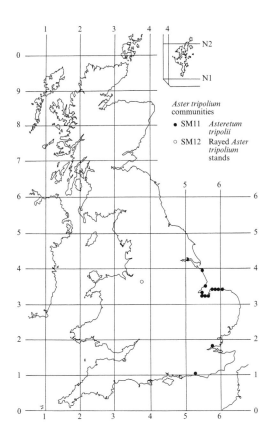

Aster tripolium
communities
● SM11 Asteretum tripolii
○ SM12 Rayed Aster tripolium stands

SM12
Rayed *Aster tripolium* on salt-marshes

Stands dominated by rayed *Aster tripolium* have been encountered in situations with some freshwater influence such as brackish ditches behind sea walls where *Spartina anglica* and *Puccinellia maritima* are frequent associates. Beeftink (1962, 1965) has described similar vegetation from sites with local freshwater flushing as a distinct *Aster tripolium* sociatie. Rayed *A. tripolium* is also abundant on periodically-flooded saline muds in inland salt-marshes with *Spergularia marina* and *Puccinellia distans* (see the *Puccinellietum distantis astereto-sum* of Lee 1977; also Edees 1972).

Although some floras note a certain habitat distinction between rayed forms of *A. tripolium* and the var. *discoideus* (e.g. Petch & Swann 1968, Jermyn 1974,

Gibbons 1975), the situation is far from simple. The var. *discoideus* can also occur in brackish habitats and Jermy & Crabbe (1978) have recorded vegetation rich in dwarf rayed *A. tripolium* and *Suaeda maritima* from Salen Marsh, Mull, where there is little freshwater influence (cf. the intertidal *Suaedeto maritimae-Asteretum tripolii* Hocquette & Géhu 1965 in Ghestem 1972). Furthermore, even within the rayed form there appears to be a complex of genetically determined variation adapted to different environmental conditions (Gray 1971, 1974; Gray *et al*. 1979). Further sampling is needed to establish the ecological implications of this variation and to check the validity of erecting separate communities for the different forms.

SM13
Puccinellia maritima salt-marsh community
Puccinellietum maritimae (Warming 1906) Christiansen 1927

Synonymy

The *Puccinellietum maritimae* includes a considerable range of closely-related vegetation types. The limits of the association as a whole are similar to those adopted in recent Continental accounts (e.g. Beeftink 1965, Westhoff & den Held 1969) and, as such, would include all or part of a variety of salt-marsh types in earlier British descriptions. The synonmy of the communities is complex and, as many synonyms would be partial, a complete list is not attempted here. Where a particular sub-community has a clear counterpart *in litt.* a note is given below.

Constant species

Puccinellia maritima.

Rare species

The association does not provide the sole, or even major, context for any national rarity but the following occur occasionally: *Arthrocnemum perenne, Limonium bellidifolium, L. binervosum, Salicornia pusilla, Spartina maritima* and *Suaeda vera.*

Physiognomy

Mostly, the association occurs as a closed species-poor grassland but the complete range includes very open pioneer vegetation and herb-dominated stands in which *Puccinellia maritima* is of minor importance or even absent. The sward varies from a tight low turf 1–2 cm high to a rank mattress up to 50 cm tall. Although grazing is important in controlling the physiognomy (see below), a considerable range of genetically determined morphotypes of *P. maritima* is present in Britain (Gray & Scott 1977*a*; 1980). Although species from low- and high-marsh communities occur as associates, the *Puccinellietum maritimae* as a whole is differentiated from the low-marsh communities by the reduced frequency and cover/abundance of annual *Salicornia* species and *Spartina anglica* and from the high marsh by low levels of *Festuca rubra, Agrostis stolonifera* and *Juncus gerardii.*

The most common associates throughout are *Triglochin maritima, Plantago maritima* and *Armeria maritima* and there is frequently an algal mat, often floristically varied and comprising a number of distinct species assemblages (Carter 1932, 1933*a, b*; Chapman 1934, 1937; Polderman 1979).

Sub-communities

Although the association is of widespread occurrence and individual stands are often great in extent and highly distinctive, the general species-poverty of the vegetation and the site-specific nature of much variation makes it difficult to attain an entirely satisfactory national sub-division. The sub-communities described below should be regarded as provisional and independent local schemes may sometimes be preferable. Alternative national classifications would also be possible: the scheme of Beeftink (1962, 1965) could, for example, be applied here with the addition of units to accommodate communities of northern and western Britain.

Sub-community with *Puccinellia maritima* dominant: *Puccinellietum (Glycerietum) maritimae* Tansley 1911. This is the most extensive and widespread perennial community of the lower salt-marsh in the British Isles. *P. maritima* is constant and dominant throughout forming a fairly closed sward in which other species are generally poorly represented. Adam (1976) recognised two noda, making a distinction between samples which are less and more species-rich but such a division is somewhat arbitrary and it is probably preferable to recognise a single rather diverse unit. The associate species vary with the level in the marsh and the geographical locality of the sub-community. At the lowest levels, annual *Salicornia* spp. and *Suaeda maritima* are most frequent: at higher levels, *Triglochin maritima, Plantago maritima* and *Aster tripolium* are found. *Limonium vulgare* is more frequent in the south and east and *Armeria maritima* in the west. *Glaux maritima* is virtually absent from this vegetation in south-east England. Locally, very dense stands of tall

Puccinellia maritima occur at relatively high levels in the marsh, most notably around The Wash; these stands are often monospecific but may have *Atriplex hastata*.

***Glaux maritima* sub-community:** *Glaucetum maritimae* Dahl & Hadač 1941; *Glaux maritima* isozion Dahlbeck 1945; *Glaux maritima* sociatie Beeftink 1962. Although *G. maritima* occurs at varying levels throughout the association, it is here constant and co-dominant with *Puccinellia maritima* in a low generally species-poor sward. The frequency and cover/abundance of *Triglochin maritima*, *Plantago maritima* and *Armeria maritima* stand between their low levels in the *Puccinellia*-dominated sub-community and the high values they attain in the *Limonium-Armeria* sub-community. *L. vulgare* itself and *Halimione portulacoides* are rare. Stands of this sub-community are often small and fragmentary but at some sites cover large areas. During hot dry summers, the shoots of *Glaux maritima* may become shrivelled by late July in upper-marsh sites. Further, although *G. maritima* is a perennial, its aerial parts die back completely in the winter when stands may appear virtually devoid of vegetation.

***Limonium vulgare-Armeria maritima* sub-community:** General Salt Marsh Tansley 1911 & Chapman 1934 (but not Chapman 1960); *Puccinellietum* (*Glycerietum*) *maritimae* Tansley 1911 *p.p.*; *Statice & Armeria* societies Marsh 1915; *Plagntag(in)etum* Chapman 1934 *p.p.*; *Limonietum & Armerietum* Tansley 1939 (but not *Armerietum* Yapp & Johns 1917); *Plantagini-Limonietum* Westhoff & Segal 1961; forb salt marsh Dalby 1970. This is one of the most distinctive communities of British salt-marshes with a varied sward dominated by herbaceous dicotyledons which present a colourful spectacle when flowering. *Limonium vulgare*, *Armeria maritima*, *Triglochin maritima*, *Plantago maritima*, *Halimione portulacoides* and annual *Salicornia* spp. (including locally the uncommon *S. pusilla*) are all constant and can be abundant. *Puccinellia maritima*, though also constant, rarely comprises more than 10% of the sward. There is often an algal mat and frequent scattered plants of *Spergularia media* and *Suaeda maritima*. *Limonium humile*, though of only occasional occurrence in this sub-community in south-east England, sometimes replaces *Limonium vulgare*, as in Milford Haven (Dalby 1970) and, to a lesser extent, in south-west Scotland. At the highest levels to which the sub-community extends, *Festuca rubra* and *Juncus gerardii* may occur.

The vitality of the common species is variable. Both *Limonium vulgare* and *Armeria maritima* flower profusely but *Aster tripolium* and, at some sites, *Plantago maritima* flower infrequently. *Halimione portulacoides* occurs not in shrubby form but as scattered prostrate shoots with small fleshy leaves which tend to be shed in late summer. Annuals, such as the *Salicornia* spp. and

Suaeda maritima, persist as small plants which often turn red in early summer, possibly reflecting their inability to compete with established perennials for nutrients, particularly nitrogen.

***Plantago maritima-Armeria maritima* sub-community:** *Armerietum* Yapp & Johns 1917 *p.p.*; *Plantag(in)etum* Chapman 1934 *p.p.*; *Plantago maritima* isozion Dahlbeck 1945. This resembles the *Limonium-Armeria* sub-community in being dominated by herbaceous dicotyledons. *Plantago maritima*, *Armeria maritima* and *Triglochin maritima* are again constant and abundant and *Puccinellia maritima* is, as there, a relatively inconspicuous contributor to the sward. Here, however, *Halimione portulacoides* is rare and *Limonium vulgare* absent while *Glaux maritima* and rayed *Aster tripolium* are constant. Although always virtually closed, the vegetation exists in two physiognomic forms based on variation in *Plantago maritima* which sometimes has long leaves and an upright habit (Chapman 1934), sometimes short leaves appressed to the soil surface.

***Puccinellia maritima*-turf fucoid sub-community.** Turf fucoids occur at low levels in various types of *P. maritima* salt-marsh, but here they comprise an extensive dense understorey of diminutive plants beneath a sward often dominated by *Plantago maritima* but also with constant and abundant *Puccinellia maritima*, *Glaux maritima* and *Armeria maritima*. *Fucus vesiculosus* ecad *muscoides* is the principal fucoid and Cotton (1912) and Polderman & Polderman-Hall (1980) have both described the understorey as a discrete algal assemblage.

***Puccinellia maritima-Spartina maritima* sub-community:** *Puccinellietum maritimae typicum*, phase with *Spartina maritima* Beeftink 1962. Here, *P. maritima* dominates with variable amounts of *Spartina maritima*, annual *Salicornia* spp., *Limonium vulgare*, *Suaeda maritima* and *Aster tripolium* var. *discoideus*. Stands of this sub-community are generally a few tens of square metres in extent.

Habitat

The *Puccinellietum maritimae* is the most widespread and extensive perennial community of the lower salt-marsh in Britain. It occurs both as a discontinuous pioneer zone and as a continuous sward in the zonation above the pioneer vegetation. It is also common on slumped creek-sides, in old pans and on disturbed sites in the upper marsh. Fragmentary stands of the association are found infrequently on very exposed maritime cliffs, for example on the Butt of Lewis, Outer Hebrides. (see also Praeger 1911).

The association occurs on a wide range of substrates including various clays and silts, highly calcareous sands and soils of high organic content; more rarely, it is found on gravel and shingle. Its importance as a colonising community is very much increased on sandier substrates: it is the most frequent pioneer on the sandy marshes of western England and Wales and commoner, for example, on the north-west as against the south-east shore of The Wash (Anon. 1976). The pH is usually basic with most soils in the range 6.0–8.5 (Adam 1976, Bridges 1977, Gray & Scott 1977a). Sediments in pioneer and lower marsh zones are generally higher in calcium content and lower in organic matter than those higher up the marsh (Gray & Bunce 1972, Adam 1976). Soils are often intermittently waterlogged and poorly aerated and share a moderate to high submergence rate and salinity. Data on submergence are limited but suggest that the lower limit of the *Puccinellietum* may experience more than 350 submergences/year; Gray & Scott (1977b) recorded a mean rate for their Morecambe Bay samples with *Puccinellia maritima* of 220 submergences/year while on Scolt Head, Norfolk, Chapman's General Salt Marsh extended from 150 to 225 submergences/year (Chapman 1960b). Proctor (1980) measured salinities of 12–30 g l^{-1} for *Puccinellia maritima* in the Exe estuary, Devon, but levels well in excess of those of sea-water may develop in the higher marsh because of evaporation in the absence of submersion.

Grazing is of undoubted importance in the maintenance of the association though its effect is complex and there is evidence that the response of species varies between sites. Many marshes are heavily used for pasturing stock, most frequently sheep but also cattle and horses; wildfowl, rabbit, hare and vole grazing may also be intensive. Grazing affects the species composition of the sward. It may be important in maintaining the dominance of perennial grasses as against herbaceous dicotyledons (Gray 1972) or in controlling the balance between *Puccinellia maritima* and *Festuca rubra* (Gray & Scott 1977b): *P. maritima* responds to grazing by the production of small, prostrate, short-leaved and rapidly tillering forms (Gray & Scott 1977a, 1980). With intensive grazing *Limonium* spp. and *Halimione portulacoides* may be reduced in abundance (Boorman 1967, Ranwell 1968, Rojanavipart & Kay 1977). On silt and clay marshes heavy cattle-trampling can lead to widespread poaching.

Moderate grazing helps maintain a sward which can support considerable populations of wintering wildfowl. There is evidence of a preference for *Puccinellia maritima* as against rank swards, such as those of the *Juncetum gerardi*, in wigeon (Cadwalladr *et al.* 1972, Cadwalladr & Morley 1974) and the brent goose (Charman & Macey 1978). For brent, the *Puccinellietum* provides a valuable food source after *Zostera* and *Enteromorpha* and there is heavy use in January–March (Ranwell & Downing 1959; Charman 1975, 1977b, 1979; Charman & Macey 1978).

Some of the species in the association are resistant to oil spillage by virtue of their underground storage organs, e.g. *Plantago maritima*, *Armeria maritima* and *Triglochin maritima*, but *Puccinellia maritima* itself declines rapidly with repeated oiling (Baker 1979).

The particular environmental relationships of the sub-communities are as follows. The *Puccinellia-Glaux* sub-community occurs in a number of different habitats all of which are open to rapid disturbance: old turf-cuttings, former pans, creeks, old cart tracks. It is also found in situations where the boundary between salt-marsh and dune becomes blurred: where sand is blown on to the upper marsh, where dune lows are subject to tidal flooding (see Lambert & Davis 1940) and where salt-marsh/dune interfaces are subject to disturbance by trampling or car-parking. There are small stands on gravel and shingle on the upper marsh at some sites and, at others, large stands in the open areas behind sea walls.

The *Limonium-Armeria* sub-community is found at relatively high levels in the salt-marsh zonation. Frequently it does not form a continuous belt but occurs as a series of small discrete stands separated from each other by creek levees. The soil in these inter-creek basins is normally a heavy clay with a considerable quantity of organic matter (loss on ignition >30%) in the upper few centimetres of the profile. The development of creek levees restricts the drainage in the basins (the concave stage of marsh development after Beeftink 1966): water may be retained there after submergence (Perraton 1953) and the soils are often strongly gleyed. On many salt-marshes in south-east England such stands have the maximum pan density within the sites (Pethwick 1974). The pan edges are often marked by a narrow fringe of more vigorous vegetation in which *Triglochin maritima* is particularly prominent.

At some sites, where salt-marsh abuts onto dunes, there is an unbroken zone of this sub-community. Here the sediments are sands or alternating bands of sand and clay and pans and creeks are relatively few; such creeks as do occur lack pronounced levees. At the higher parts of such stands *Festuca rubra* and *Juncus gerardii* occur. This habitat seems to be that described for the *Plantagini-Limonietum* Westhoff & Segal 1961.

The *Plantago-Armeria* sub-community is also found at comparatively high levels. The form with tall *Plantago maritima* occurs above the *Limonium-Armeria* sub-community and the form with short *P. maritima* in shallow depressions throughout the upper *Puccinellietum* and sometimes in the higher *Juncetum gerardi* of grazed salt-marshes. Extensive stands are found in some re-vegetated turf-cuttings in Morecambe Bay.

The *Puccinellia*-turf fucoid sub-community is rare or absent from sandy salt-marshes and is especially characteristic of loch-head sites in west Scotland where shallow soils (20–30 cm deep) develop over rock or shingle. Such soils tend to have a high organic content, to be reddish in colour and to contain coarse gravel throughout the profile. Although high salinities can be attained during drought (Gillham 1957*b*), *Festuca rubra* may occur even at the lowest levels attained by this sub-community, perhaps reflecting the influence of high regional rainfall. Small stands of the sub-community are also widespread among coastal rocks in the lower splash zone.

The *Puccinellia-Spartina maritima* sub-community is very local but it has been recorded from mid-marsh depressions and upper-marsh borrow pits with soft mud. Beeftink (1962) considered the vegetation characteristic of mud-flat/salt-marsh transitions.

Zonation and succession

Where the *Puccinellietum* is a pioneer community, as on sandier substrates, it appears to establish itself mainly by the rooting of vegetative fragments of *P. maritima* uprooted from existing swards by grazing stock and carried by tides (Ranwell 1961, Brereton 1971, Adam 1976, Gray & Scott 1977*a*). *P. maritima* can set abundant seed (Gray & Scott 1977*a*) but it has no special dispersal mechanism and, though caryopses can be washed away, seedling establishment in the pioneer zone seems uncommon.

Once established, the scattered plants produce numerous radiating stolons and accrete sediment into a series of hummocks (see Plate 3 in Ranwell 1972). Hummock size varies from shore to shore: some hummocks never exceed 50 cm in height but others are taller and attain a diameter of several metres. The hummock tops may carry the transitional *Puccinellia-Salicornia-Suaeda* community. Yapp & Johns (1917) postulated that the intervening hollows developed into pans but Pethwick (1974) showed that such a model could not account for the majority of upper-marsh pans.

At some sites a narrow zone of very scattered hummocks gives way quickly to a continuous sward of the *Puccinellia*-dominated sub-community. In other cases the hummocky topography persists much higher upshore (see the striking photographs in Yapp & Johns 1917) and eventually passes to fairly smooth swards of some *Puccinellietum* vegetation or, in the mid- and upper marsh, to *Juncetum gerardi* which is the usual high level vegetation of the grazed marshes of the west coast (Figure 8). On ungrazed west coast sites, the *Limonium-Armeria* sub-community may occur in the upper marsh.

In the south-east, the *Puccinellietum* is rarely a pioneer community. Its position in the zonation varies, the *Puccinellia*-dominated sub-community appearing either below or above the *Halimionetum portulacoidis*. In this region, the *Puccinellietum* can be found right up to the tidal limit, either as the *Puccinellia*-dominated sub-community, as around The Wash, or as the very characteristic high marsh *Limonium-Armeria* sub-community.

In the upper reaches of estuaries, where the soil salinity in the lower marsh is kept constantly low by freshwater dilution, an inversion of the normal zonation may be found with the *Puccinellietum* in upper marsh depressions where evaporation produces high salinities (Gillham 1957*a*, Adam 1976). Disturbance of upper marsh sites frequently results in the association appearing as a secondary pioneer, especially in the form of the *Puccinellia-Glaux* sub-community (cf. Beeftink 1962).

In the loch-head sites where the *Puccinellia*-turf fucoid sub-community is characteristic, it is frequently the lowest vegetation but it seems only rarely to be actively expanding. To seaward, there is usually a very low cliff or the vegetation cover is discontinuous with discrete patches on isolated rock or gravel plinths.

Distribution

The association is the most widespread community on British salt-marshes and probably no site lacks at least a fragmentary stand. The *Puccinellia*-dominated sub-community is the most widespread of the types, being frequent on all coasts except those of west Scotland and

Figure 8. Zonation on an eroding salt-marsh. The intact marsh carries various kinds of SM16 *Juncetum gerardi*, a ground of the *Armeria* variant of the *Festuca-Glaux* sub-community with patches of the *Juncus gerardii* sub-community. Running down below, on material slumped from the sides of the simple 'herring-bone' creeks, is a narrow zone of the SM13 *Puccinellietum maritimae*. The sequence terminates above in fragmentary SM24 *Atriplici-Elymetum pycnanthi*.

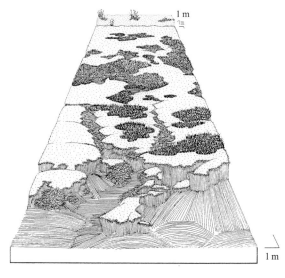

the northern Isles where it is largely replaced by the *Puccinellia*-turf fucoid sub-community. The *Plantago-Armeria* sub-community is also widespread, though local, and the *Puccinellia-Glaux* sub-community is commoner on the west coast. The *Limonium-Armeria* subcommunity is widespread in the south-east but much less frequent on the west coast where it is confined to lightly grazed and ungrazed sites. *Spartina maritima* is declining throughout northern Europe and the *Puccinellia-S. maritima* sub-community is restricted to Essex (and perhaps north Kent?).

Affinities
The vegetation types within the *Puccinellietum maritimae* can be seen as a floristic transition between the open annual communities of the lower marsh dominated by *Salicornia* spp. and *Suaeda maritima* and the *Juncetum gerardi* swards of the mid- and high marsh. The association grades floristically to the former through the more species-poor forms of the *Puccinellia*-dominated subcommunity and the transitional *Puccinellia-Salicornia-Suaeda* vegetation.

Floristic table SM13

	a	b	c	d	e	f	13
Puccinellia maritima	V (4–10)	V (2–10)	V (1–8)	III (2–7)	V (4–9)	V (9–10)	V (1–10)
Glaux maritima	II (1–7)	V (5–9)	I (2–5)	V (2–5)	V (3–6)		II (1–9)
Armeria maritima	II (1–7)	III (2–7)	IV (3–8)	V (3–8)	IV (2–8)	I (1–5)	II (1–8)
Plantago maritima	II (1–7)	III (2–5)	IV (3–9)	V (5–9)	V (4–7)		II (1–9)
Salicornia agg.	III (2–6)	III (2–6)	IV (2–8)	II (2–3)	II (2–4)	V (4–10)	II (2–10)
Algal mat	II (3–10)	II (5–8)	IV (3–9)	II (4–8)	IV (4–7)		II (3–10)
Triglochin maritima	II (1–8)	III (2–4)	V (2–8)	IV (2–7)	II (1–3)	II (1–6)	II (1–8)
Aster tripolium (rayed)	III (1–8)	II (2–6)	II (2–7)	IV (2–7)	II (2–3)		II (1–8)
Suaeda maritima	III (1–6)	I (1–3)	III (1–3)	I (2–3)		IV (1–10)	II (1–10)
Halimione portulacoides	II (1–5)	I (2)	IV (1–6)	I (2–4)		III (4–10)	II (1–10)
Limonium cf. *L. vulgare*	II (1–8)	I (3)	V (2–8)			V (1–10)	II (1–10)
Turf fucoids*	I (2–3)		I (3)		V (4–9)		I (2–9)
Spartina maritima						V (1–10)	I (1–10)
Aster tripolium var. *discoideus*	I (1–5)					IV (1–8)	I (1–8)
Spergularia media	II (1–5)	II (2–3)	III (1–4)	III (1–3)	II (2–4)	I (3–5)	II (1–5)
Spartina anglica	I (1–7)	I (3)	I (2)	I (1–2)		I (5)	I (1–7)
Festuca rubra	I (2–5)	II (2–5)	II (2–6)	III (2–6)	II (2–4)		I (2–6)
Agrostis stolonifera	I (3–5)	I (2–3)	I (3–4)	II (3–6)	I (3–4)		I (2–6)
Limonium humile	I (2–5)	I (1–3)	I (1–8)	I (1–3)		I (4)	I (1–8)
Cochlearia officinalis	I (1–3)		I (1–4)		I (3)		I (1–4)
Juncus gerardii	I (2–3)	I (2)	I (2–6)	II (2–6)	I (3–4)		I (2–6)
Bostrychia scorpioides	I (4–7)		I (3–8)	I (6)	I (3)	I (8)	I (3–8)
Arthrocnemum perenne	I (1–3)		I (1–4)			II (1–6)	I (1–6)
Aster tripolium	I (1–6)		II (1–5)	I (3–4)			I (1–6)
Cochlearia anglica	I (1–6)		I (1–4)	I (2–3)			I (1–6)
Atriplex prostrata	I (1–7)						I (1–7)

	319	23	89	26	17	20	494
Number of samples							
Mean number of species/sample	6 (1–13)	6 (3–12)	9 (3–14)	8 (5–11)	7 (5–12)	7 (4–10)	7 (1–14)
Mean vegetation height (cm)	11 (2–50)	6 (2–15)	11 (3–25)	6 (2–15)	5 (2–15)	no data	6 (2–50)
Mean total cover (%)	88 (20–100)	85 (45–100)	91 (50–100)	96 (90–100)	91 (75–100)	no data	90 (20–100)

*   Includes *Ascophyllum nodosum* ecad *mackaii*, *Fucus vesiculosus* ecad *caespitosus* and ecad *volubilis* and *Pelvetia canaliculata*.

a   Sub-community with *Puccinellia maritima* dominant

b   *Glaux maritima* sub-community

c   *Limonium vulgare–Armeria maritima* sub-community

d   *Plantago maritima–Armeria maritima* sub-community

e   *Puccinellia maritima*-turf fucoid sub-community

f   *Puccinellia maritima–Spartina maritima* sub-community

13   *Puccinellietum maritimae* (total)

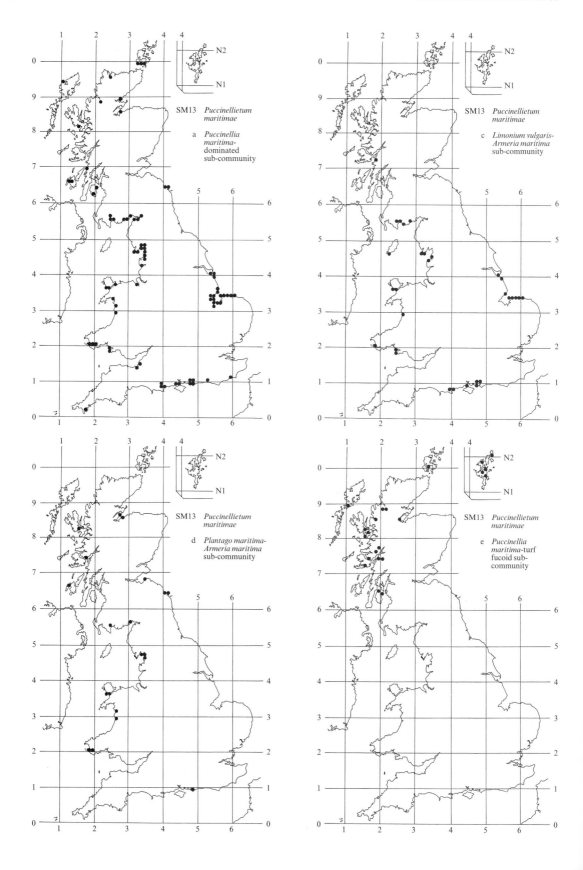

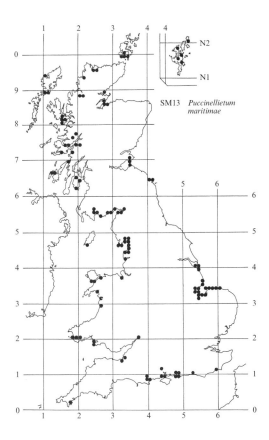

SM13   *Puccinellietum
maritimae*

SM14

Halimione portulacoides salt-marsh community
Halimionetum portulacoidis (Kuhnholtz-Lordat 1927) Des Abbayes et Corillion 1949

Synonymy
Obionetum and *Halimionetum auct. angl.*

Constant species
Halimione portulacoides, Puccinellia maritima.

Rare species
Arthrocnemum perenne, Frankenia laevis, Inula crithmoides, Limonium bellidifolium, Suaeda vera.

Physiognomy
This is a closed, species-poor association in which *Halimione portulacoides* is constant and physiognomically conspicuous as a bushy canopy up to 50 cm high or as a virtually prostrate carpet. *Puccinellia maritima* is also constant and there is frequently a little *Suaeda maritima* and sparse records for a variety of species from both low and upper marsh. Epiphytic algae are often abundant on the lower stems of the *H. portulacoides* and at some sites small patches of fucoids are present beneath canopy gaps.

Sub-communities
Géhu & Delzenne (1975) allocated samples of *Halimione portulacoides* vegetation with the red alga *Bostrychia scorpioides* to the separate association *Bostrychio-Halimionetum portulacoidis* (Corillion 1953) R.Tx. 1963. Although *B. scorpioides* was recorded occasionally here, there is no substantial floristic reason for distinguishing samples containing this species as even a sub-community within the *Halimionetum*. The three following sub-communities are, however, quite distinct.

Sub-community with *Halimione portulacoides* dominant.
In this most species-poor sub-community, *H. portulacoides* always has high cover values (usually >90%) and there is usually a sparse undercover of *Puccinellia maritima* and some *Suaeda maritima*. The *H. portulacoides* may be present as a low or tall even-topped canopy or as discrete hemispherical bushes 1–2 m diameter and up to

50 cm high. Though this last form is developed on sandy substrates, there is no floristic basis for distinguishing a Sandy *Obionetum* (*sensu* Chapman 1934) within this sub-community.

Juncus maritimus **sub-community:** *Halimione-Juncus maritimus* nodum Adam 1976. Here *H. portulacoides* is somewhat reduced in cover and varying amounts of *J. maritimus* are present as scattered shoots emerging through the shrubby canopy or as small dense patches, presumably of clonal origin. *Puccinellia maritima* remains constant with low cover but here *Plantago maritima, Limonium vulgare* and *Triglochin maritima* are also frequent. *Elymus pycnanthus* is an uncommon but distinctive associate.

Puccinellia maritima **sub-community:** *Puccinellio-Halimionetum portulacoidis* Sea Meadow Chapman 1934; *Puccinellietum maritimae typicum*, terminal phase with *H. portulacoides* Beeftink 1962. *H. portulacoides* and *P. maritima* are co-dominant in an intimate mixture with the shoots of the former making a diffuse open network (cf. mosaics with discrete patches of *Halimionetum* and *Puccinellietum*). *Suaeda maritima, Triglochin maritima, Plantago maritima* and *Limonium vulgare* are frequent and sometimes abundant and, at high levels on the marsh, *Festuca rubra* may be common.

Habitat
The association occurs on a variety of substrates including clays, sands, shingle and occasionally soils of high organic content (Chapman 1950, O'Reilly & Pantin 1957, Adam 1976). Most commonly, it is developed on silty clay of low organic content, with some free calcium carbonate and a pH in the range 7.0–8.0. It appears tolerant of a range of submersion regimes: at Scolt Head, Norfolk, the *Halimionetum* extends from about 100 to 400 submergences/year (Chapman 1950, 1960a; cf. O'Reilly & Pantin 1957). Proctor (1980) has shown that, in the Exe salt-marshes, Devon, *H. portulacoides*

tolerates chloride levels at 10–24 g l^{-1} (salinity 16–36 g l^{-1}). Within these rather wide limits, the association occurs in two distinct situations, as an extensive belt of variable position in the general zonation or as narrow ribbons on creek levees (the 'Great *Obione* Fringe' of Chapman 1934) and low ridges on the marsh surface (Proctor 1980). The occurrences may reflect a need in *H. portulacoides* for a well-drained aerobic soil environment, at least for seed germination (Chapman 1950). Creek levees offer such conditions and, even in intervening basins where soils may be strongly reduced a few centimetres below the surface, the shallow adventitious roots of *H. portulacoides* may avoid the more severe effects of waterlogging (see Figure 28 in Chapman 1960*b*). Alternatively, levee occurrences may reflect a preference for a good supply of soil nutrients, particularly nitrogen and phosphate.

The *Juncus maritimus* and *Puccinellia maritima* sub-communities occur throughout the habitat range of the association but the bushy form of the *H. portulacoides*-dominated sub-community is confined to sandy substrates where salt-marsh abuts dunes or, less frequently, on the lower marsh.

Halimionetum is generally absent from sheep-grazed marshes (e.g. Yapp & Johns 1917) except for those creek-sides which are inaccessible to the stock. It is, however, found on a number of cattle-grazed marshes, notably around The Wash, and it will tolerate a certain amount of rabbit grazing (Chapman 1950). Brent geese do not graze extensively on *H. portulacoides* when feeding on saltings (Charman & Macey 1978).

Zonation and succession

Where *Halimionetum* occurs within the marsh zonation, its position is variable. It can be either above or below the *Puccinellietum maritimae* and boundaries between the two associations can be marked by mosaics (see Corillion 1953). At some sites, *Halimionetum* may run right from the upper limit of the pioneer zone to the sea wall. Where it does extend far down the marsh there is sometimes an open mosaic of *H. portulacoides* and *Arthrocnemum perenne* at its lower limit.

The association can occur on creek levees whether or not there is a nearby inter-creek zone of *Halimionetum*. Where it occurs in both situations on the same marsh, the creek *Halimionetum* may be above or below the inter-creek zone. Usually the creek *Halimionetum* cuts across the boundaries of a number of marsh communities.

On the high marsh, *Halimionetum* in both situations

may give way to a zone of *Atriplici-Elymetum pycnanthi*, sometimes with an intervening but patchy zone of *Artemisietum maritimae*. This zonation may indicate a successional sequence consequent upon sediment accretion.

The origin and successional status of the *Puccinellia maritima* sub-community is obscure. Its distinctive physiognomy may arise by invasion of the *H. portulacoides*-dominated sub-community by *P. maritima* when the canopy opens with ageing of the bushes or as a result of grazing or by invasion of *Puccinellietum* by *H. portulacoides*. Alternatively the co-dominants may simultaneously invade some other salt-marsh community. Only long-term observation can elucidate the process(es) involved here.

Other changes can occur within *Halimionetum* as a result of frost or human disturbance (Beeftink 1977*a*, *b*; Beeftink *et al*. 1978). Killing of *H. portulacoides* on creek levees by frost can result in the temporary replacement of the association by *Artemisietum maritimae* for 4–5 years. Disturbance in inter-creek basins produces a phase characterised by *Suaeda maritima* and *Aster tripolium*.

Distribution

The *Halimionetum* is most widespread and extensive in south-east England: it is estimated that the association covers 30% of the salt-marshes of The Wash (Anon. 1976). It reaches its northern limit in south Scotland and this may be related to the incidence of severe frosts rather than to any effect of low mean summer temperatures (Ranwell 1972, Beeftink 1977*a*, *b*; cf. Chapman 1950). Sensitivity to grazing restricts its occurrences on the west coast. There is evidence of a recent expansion of the community within Europe (Beeftink 1959, 1977*a*).

Affinities

Some authorities (e.g. Beeftink 1962) expand the *Halimionetum* to take in the *Artemisia maritima*-dominated vegetation of the high marsh and there may also be a case for considering some *Arthrocnemum perenne* stands as part of the association. Whatever its precise limits, the *Halimionetum* is a distinctive community of widespread occurrence on European coasts. It is usually placed alongside the *Puccinellietum maritimae* in the Asteretea but Géhu (1975) has erected an alliance Halimionion within the Arthrocnemetea to emphasise its affinities with the dwarf chenopod communities best developed around the Mediterranean.

Floristic table SM14

	a	b	c	14
Halimione portulacoides	V (7–10)	V (6–9)	V (5–9)	V (5–10)
Puccinellia maritima	IV (2–5)	IV (3–6)	V (2–8)	V (2–8)
Juncus maritimus		V (4–8)	I (2)	I (2–8)
Plantago maritima	I (1–5)	IV (2–5)	III (1–7)	II (1–7)
Limonium cf. *L. vulgare*	I (1–4)	IV (2–5)	III (1–7)	II (1–7)
Suaeda maritima	III (2–4)	I (2–3)	III (1–6)	III (1–6)
Triglochin maritima	I (1–4)	III (2–6)	III (1–6)	II (1–6)
Algal mat	I (5–8)	II (4–6)	I (3–8)	I (3–8)
Aster tripolium (rayed)	I (2–3)	II (2–3)	II (2–5)	I (2–5)
Salicornia agg.	I (1–5)	I (2–3)	III (2–5)	II (1–5)
Aster tripolium	I (1–3)	I (2–3)	II (1–4)	I (1–4)
Armeria maritima	I (5)	I (3)	I (2–4)	I (2–5)
Artemisia maritima	I (2)	I (2–5)	I (1–5)	I (1–5)
Arthrocnemum perenne	I (1–5)	I (1–2)	I (1–4)	I (1–5)
Spartina anglica	I (1–3)	I (2)	I (1–3)	I (1–3)
Spergularia media	I (1–5)	I (2)	I (1–4)	I (1–5)
Inula crithmoides	I (2–5)			I (2–5)
Elymus pycnanthus		II (1–6)		I (1–6)
Aster tripolium var. *discoideus*	I (1–3)		II (1–6)	I (1–6)
Number of samples	91	19	64	174
Mean number of species/sample	4 (1–10)	7 (4–11)	6 (4–10)	6 (1–10)
Mean vegetation height (cm)	24 (4–40)	46 (35–70)	25 (8–45)	27 (4–70)
Mean total cover (%)	97 (50–100)	95 (80–100)	97 (80–100)	97 (50–100)

a   Sub-community with *Halimione portulacoides* dominant

b   *Juncus maritimus* sub-community

c   *Puccinellia maritima* sub-community

14   *Halimionetum portulacoidis* (total)

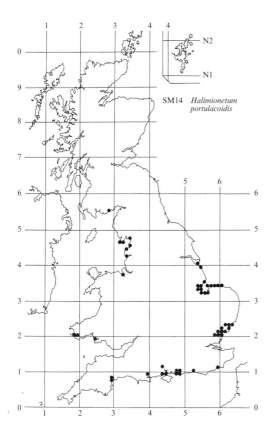

SM14 *Halimionetum portulacoidis*

SM15

Juncus maritimus–Triglochin maritima salt-marsh community

Synonymy
Juncetum maritimi auct. angl. p.p.; *Triglochin-Juncus maritimus* nodum   Adam 1976.

Constant species
Juncus maritimus, Plantago maritima, Triglochin maritima.

Physiognomy
Tall tussocks of *Juncus maritimus* are always overwhelmingly dominant in this association and the associates are rather variable. However, *Triglochin maritima* and *Plantago maritima* are constant in usually small amounts in the understorey and various Puccinellion species, such as *Puccinellia maritima*, rayed *Aster tripolium*, *Armeria maritima* and *Glaux maritima*, occur frequently throughout. The association differs from the *Juncus maritimus* salt-marsh in the relative infrequency of *Agrostis stolonifera*, *Festuca rubra* and *Juncus gerardii*. Commonly the bases of the *J. maritimus* shoots support a variety of epiphytic algae, notably *Bostrychia scorpioides* and *Catenella repens*, and there may be an extensive algal mat, locally rich in dwarf free-living fucoids, on the substrate surface. Stands are often based upon discrete and sometimes large clones of *J. maritimus* but may also form a distinct zone within the marsh.

Sub-communities
Adam (1977) suggested that there are three centres of variation within the association around which sub-communities might be erected: stands which are very species-poor, sometimes pure *J. maritimus* in vigorous, tall and dense patches; stands in which *Halimione portulacoides* and *Limonium* cf. *vulgare* are conspicuous and fairly rich stands lacking these two species.

Habitat
J. maritimus is tolerant of a wide range of salinities and soil moisture conditions (Ranwell *et al.* 1964, Gillham 1957*b*) and the association occurs at all levels on salt-marshes and on a variety of substrates. Soil pH is generally around 7.0 but loss-on-ignition varies from 3% to more than 40%.

The most species-poor stands are found on the low marsh, usually on soft anaerobic mud (Gillham 1957*a*, Chater 1973, Adam 1976, 1977, Proctor 1980), though sometimes, as in Scottish sites, on shallow peaty soils over shingle (Gillham 1957*b*, Adam *et al.* 1977). The lowest stand for which accurate data are available experiences 220 submergences/year but many stands seem to occur at lower levels. Richer stands lacking *H. portulacoides* and *Limonium* cf. *vulgare* also occur at low levels, on western salt-marshes frequently along the foot of small erosion cliffs where there is perhaps water-seepage. Stands with these two species are found as a narrow zone in the upper parts of salt-marshes in south-east England (the *Juncetum maritimi* Chapman 1934).

The association occurs on both grazed and ungrazed salt-marshes but, even where there is grazing, stands tend to be avoided by stock.

Zonation and succession
There is a marked difference in the relative position of the association on salt-marshes in south-east England and those elsewhere but lack of submersion data makes it difficult to assess these variations in terms of absolute relationships to tidal levels. On the west and Channel coasts, the association generally occurs at relatively low levels in association with the *Spartinetum townsendii* or more usually within or at the upper limit of the *Puccinellietum maritimae*. In the south-east, a narrow belt of the association occurs normally between the *Puccinellietum maritimae*, *Limonium-Armeria* sub-community, and the *Atriplici-Elymetum pycnanthi* or the tall *Festuca rubra* sub-community of the *Juncetum gerardi*. The association grades smoothly into the *Puccinellietum* which effectively constitutes the understorey of the *Juncus-Triglochin* vegetation. On those few ungrazed western marshes where the *Puccinellietum maritimae*,

Limonium-Armeria sub-community occurs, the association occupies the position typical of south-east salt-marshes.

In at least one site, *Juncus maritimus* has been seen as a coloniser with *Spartina anglica*.

Distribution
The association is the most widespread community dominated by *J. maritimus* in Great Britain. It is common on the west coast and is the major *J. maritimus* community in south-east England. One of the most extensive stands in the country is at Cefni Marsh, Anglesey where the association forms mosaics with *Scirpetum maritimi* over much of the marsh (Packham & Liddle 1970).

Affinities
The association can be seen as the northern extremity of a continuum of vegetation types in which *J. maritimus*, *Triglochin maritima*, *Limonium vulgare* and *Aster tripolium* are important components and which reaches down to the Mediterranean in the *Junco maritimi-Triglochinetum maritimi* Br.-Bl. 1931 (Braun-Blanquet & de Ramm 1957, Adam 1977). Such a range of vegetation types could be accommodated within the Puccinellion of the Asteretea which would also allow some weight to be given to the interesting low-level occurrences of *J. maritimus* vegetation.

Floristic table SM15

Juncus maritimus	V (5–10)
Triglochin maritima	IV (2–6)
Plantago maritima	IV (2–8)
Aster tripolium (rayed)	III (2–4)
Puccinellia maritima	III (2–7)
Armeria maritima	III (2–5)
Glaux maritima	III (2–5)
Algal mat	III (2–8)
Limonium cf. *L. vulgare*	II (1–6)
Cochlearia anglica	II (1–4)
Juncus gerardii	II (3–7)
Halimione portulacoides	II (1–6)
Festuca rubra	II (2–8)
Agrostis stolonifera	II (2–6)
Aster tripolium	I (1–5)
Cochlearia officinalis	I (2–3)
Salicornia agg.	I (2–5)
Suaeda maritima	I (2–3)
Spartina anglica	I (1–5)
Spergularia media	I (2–4)
Turf fucoids	I (3–6)
Phragmites australis	I (2–6)
Carex extensa	I (1–3)
Atriplex prostrata	I (1–3)
Limonium humile	I (1–3)
Oenanthe lachenalii	I (3–4)
Artemisia maritima	I (1–4)
Number of samples	63
Mean number of species/sample	10 (2–14)
Mean vegetation height (cm)	57 (25–100)
Mean total cover (%)	76 (70–100)

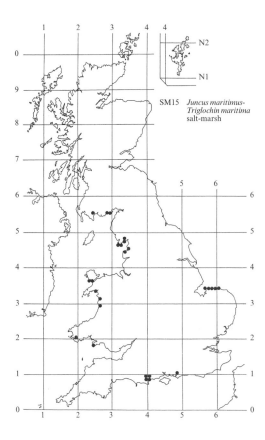

SM15   *Juncus maritimus-*
       Triglochin maritima
       salt-marsh

SM16
Festuca rubra salt-marsh community
Juncetum gerardi   Warming 1906

Synonymy
Festucetum (rubrae) auct. angl.

Constant species
Festuca rubra, Plantago maritima, Glaux maritima.

Physiognomy
The closed grasslands of the *Juncetum gerardi* are normally dominated by mixtures of *Festuca rubra* and *Agrostis stolonifera* with a variety of herbaceous associates among which *Plantago maritima*, *Glaux maritima*, *Armeria maritima* and *Triglochin maritima* are generally the most frequent and abundant. *Juncus gerardii* itself is present in varying amounts: it is usually constant through all but the most anomalous of the *Juncetum* swards and in some cases is dominant or co-dominant. In certain sub-communities, there are frequent records for low-marsh species and an algal mat is often conspicuous over the substrate surface. In other sub-communities, a group of mesotrophic grassland and flush species are well-represented. The *Juncetum gerardi* is the community within which bryophytes reach their lowest limit on salt-marshes.

Sub-communities
As in the other major British salt-marsh association, the *Puccinellietum maritimae*, variation is virtually continuous, largely based on quantitative differences among relatively few species and frequently including a site-specific element reflecting local histories of marsh use. The following sub-communities should therefore be seen as foci of national variation with somewhat hazy boundaries.

***Puccinellia maritima* sub-community:** *Juncus gerardii-Puccinellia maritima* nodum   Adam 1976; *Puccinellietum maritimae agrostidetosum* Beeftink 1962. This sub-community comprises generally short swards which are floristically transitional between the *Juncetum gerardi* and the *Puccinellietum maritimae. J. gerardi, Puccinellia maritima, Festuca rubra, Plantago maritima, Glaux maritima* and *Triglochin maritima* are constant and varying proportions of these species co-dominate. *Agrostis stolonifera, Armeria maritima* and rayed *Aster tripolium* are less frequent but each may be abundant in particular stands.

Sub-community with *Juncus gerardii* dominant: *Juncetum gerardi, J. gerardii* variant   Beeftink 1962; (not *Juncetum gerardi juncetosum* Tyler 1969). *Juncus gerardii* always dominates in the tall swards of this sub-community, the stands of which are rarely extensive, 2–3 m diameter at most, and probably vegetatively expanding clones. Even on heavily-grazed marshes, *J. gerardii* remains largely untouched by stock and clumps remain tall and conspicuous though such clumps are often surrounded by a short cropped turf in which *J. gerardii* is still abundant: this perhaps indicates the existence of genotypes of *J. gerardii* of differing palatability. *Plantago maritima, Glaux maritima, Triglochin maritima* and rayed *Aster tripolium* are also constant though rarely of great abundance. *Festuca rubra* and *Agrostis stolonifera* are reduced in frequency compared with the association as a whole.

***Festuca rubra-Glaux maritima* sub-community:** *Festucetum rubrae*   Yapp & Johns 1917; *Juncus gerardii-Glaux maritima-Agrostis stolonifera* Association   Nordhagen 1923; *Festuca-Glaux, Festuca-Agrostis* and *Festuca-Armeria* noda   Adam 1976; *Juncetum gerardi*, variant with *Festuca rubra* f. *littoralis*   Beeftink 1962 *p.p.*; *Juncetum gerardi festucetosum* Tyler 1969 *p.p. Festuca rubra* and *Agrostis stolonifera* are usually co-dominant in the low swards of this sub-community. *Plantago maritima, Glaux maritima, Triglochin maritima* and *Armeria maritima* are also constant and may each be abundant. *Juncus gerardii* is somewhat variable in amount and even when abundant may be difficult to detect in close-cropped turf. This is the lowest vegetation in which bryophytes are typically encountered on salt-marshes: *Rhytidiadelphus squarrosus, Hypnum cupressiforme* and *Eurhynchium praelongum* are the most frequent species. Algae are uncommon.

Within the sub-community stands may be encountered in which either *A. stolonifera* or *F. rubra* are sparsely represented. In other cases, these two species are overwhelmingly co-dominant in short swards in which *J. gerardii* is very poorly represented (the *Agrostis stolonifera* variant). *J. gerardii* is also sparse in some stands where *F. rubra* and *Armeria maritima* are co-dominant in the absence of *A. stolonifera* (the *Armeria maritima* variant). There is good evidence to see these very distinct communities as extreme forms of *Juncetum gerardi* derived as a result of particular marsh management regimes (see below).

Leontodon autumnalis sub-community: *Juncus gerardii-Trifolium repens-Leontodon autumnalis* Association Nordhagen 1923; *Juncetum gerardi leontodetosum* Raabe 1950; *Juncetum gerardi leontodetosum* and *odontitosum* Gillner 1960; *Carex distans-Plantago maritima* Association Ivimey-Cook & Proctor 1966 *p.p.*; *Juncetum gerardi festucetosum* Tyler 1969 *p.p.* This sub-community has much the same physiognomy as the last and here too *Festuca rubra*, *Agrostis stolonifera* and *Juncus gerardii* can all be well-represented in the short, smooth swards. *Plantago maritima* and *Glaux maritima* remain constant but here there are also frequent records for a variety of species characteristic of non-maritime vegetation. Among these, *Trifolium repens* is constant but *Potentilla anserina*, *Leontodon autumnalis* and *Carex flacca* can each be frequent and abundant. On cattle-grazed marshes, where the vegetation is normally not so shortly cropped as under sheep-grazing, a number of species flower and *L. autumnalis* may be particularly conspicuous. *Carex distans* may also be abundant in this sub-community but this species has different habitat preferences across its British range. It is uncommon on salt-marshes in the south-east but frequent in western England and in Wales; in Scotland, it again becomes rare on salt-marshes though it remains quite common among low coastal rocks (see Jermy & Tutin 1968).

Within the belts occupied by this sub-community there is sometimes a zonation of *T. repens*, *L. autumnalis*, *C. distans* and *P. anserina* in order of lowest occurrence but this is not universal and, indeed, all these species can occur occasionally in the lower marsh *Festuca-Glaux* sub-community. At the highest levels occupied by the *L. autumnalis* sub-community *Lolium perenne*, *Cynosurus cristatus*, *Bromus hordeaceus* ssp. *hordeaceus*, *Elymus repens* and *Poa pratensis* are sometimes found. It is possible that these species seed in from adjacent sea-banks where grassland mixtures have been sown. Another occasional species in the upper-marsh sites is *Trifolium fragiferum* which becomes restricted to coastal communities at the northern limits of its British range. It is rarely extensive, tending to occur in discrete patches often associated with freshwater seepage onto the upper marsh.

As with the *Festuca-Glaux* sub-community there are rather extreme forms of salt-marsh swards which are perhaps best seen in relation to a more central type of *Juncetum gerardi* vegetation. A *Trifolium repens* variant is very similar to the *L. autumnalis* sub-community except for its lower levels of *J. gerardii*.

Low turf very similar in floristics to this sub-community is of common occurrence on some sea cliffs. Here *F. rubra*, *Plantago maritima*, *Armeria maritima* and more rarely *Glaux maritima* are generally co-dominant but *J. gerardii*, *C. distans*, *T. repens* and *L. autumnalis* may all be conspicuous.

Carex flacca sub-community. *Juncetum gerardi festuco-caricetosum nigrae* Tyler 1969; ? *Danthonia decumbens-Agrostis canina* community Tyler 1969; *Agrostis tenuis-Festuca ovina* community Tyler 1969 *p.p.* The floristics and physiognomy of this sub-community are generally similar to the last except that here *C. flacca* is much more frequent and sometimes co-dominant with the grasses and herbaceous halophytes. Bryophytes may also be more conspicuous: *Campylium polygamum*, *Amblystegium serpens*, *Grimmia maritima*, *Cratoneuron filicinum*, *Amblystegium riparium*, *Calliergon cuspidatum*, *Rhytidiadelphus squarrosus*, *Hypnum cupressiforme* and *Eurhynchium praelongum* all occur occasionally and each may be abundant in particular samples.

Sometimes the turf of this sub-community is broken by flushed gravelly patches and here *Blysmus rufus*, *Eleocharis uniglumis*, *E. palustris* and *E. quinqueflora* may be locally abundant.

Sub-community with tall Festuca rubra dominant: *Festucetum littoralis* Corillion 1953; Tall *Festuca rubra* nodum Adam 1976; *Juncetum gerardi*, variant with *Festuca rubra* f. *littoralis* Beeftink 1962 *p.p.*; includes *Festuca rubra-Agrostis stolonifera-Hordeum secalinum* associes Ranwell 1961; (not *Festucetum rubrae* Yapp & Johns 1917). The very distinctive springy mattresses of this sub-community are perhaps best seen as a physiognomic variant of the *Juncetum gerardi*. *F. rubra* is consistently dominant. It grows tall and dense and, after tidal inundation, presents a bedraggled appearance. Although all of the species frequent in the association as a whole occur here, most are reduced in frequency and rarely make a major contribution to the sward. *Plantago maritima* and *Agrostis stolonifera* are the most common associates. Some stands are distinctive in the presence of conspicuous amounts of *Halimione portulacoides*; others may have *Elymus pycnanthus* and, in Somerset and the upper Severn estuary, *Hordeum secalinum* occurs in this sub-community (Ranwell 1961, Owen 1972). Flowering appears to be rare in British stands (cf. Gravesen & Vestegaard 1969 in Denmark). In winter, when the vegetation may remain flattened for long periods, seedlings of *Atriplex* spp. and *Cochlearia* spp. may appear in profusion on top of the matted grass.

Habitat

The *Juncetum gerardi* covers extensive areas of salt-marsh especially in the north and west of Britain where it is the predominant community of the mid- and upper marsh. It occurs on a range of substrates from marsh levels experiencing several hundred submergences/year to the upper tidal limit. It is usually grazed and provides swards that are valuable for commercial turf-cutting.

Regimes of salt-marsh grazing are very variable. The stock involved, the stocking rates, the pattern of use through the year may all vary from marsh to marsh and through time and all these factors might be expected to influence the appearance of the vegetation. Much of the site-specific variation within the *Juncetum gerardi* is probably related to the unique grazing history of every site.

The general effect of grazing is to maintain a fine short sward, preventing the overwhelming dominance of (a) particular species (Dahlbeck 1945, Gillner 1960, Beeftink 1977*a*). It is probably important in controlling the proportions of *Puccinellia maritima*, *Agrostis stolonifera* and *Festuca rubra* in the sward and thus influences the position and the nature of the boundary between the communities of the *Puccinellietum maritimae* and the *Juncetum gerardi* and the extent and composition of the transitional vegetation classified here as the *Puccinellia* sub-community (Ranwell 1968, Gray & Scott 1977*b*). If grazing pressure is generally low or if grazing ceases, *F. rubra* is particularly responsive, growing tall and rank, excluding most potential competitors and eventually producing the sort of tussocky, species-poor grassland that is characteristic of the tall *F. rubra* sub-community. Such vegetation is unpalatable to wildfowl (Cadwalladr *et al.* 1971, Cadwalladr & Morley 1974, Charman & Macey 1978) and to re-introduced sheep.

The preparation and cutting of 'sea-washed' turf is important at a number of salt-marsh sites (e.g. Morecambe Bay; see Gray 1972). The grass-dominated swards of the *Festuca-Glaux* sub-community (the *Agrostis stolonifera* variant) are most favoured and are prepared over a number of years by mowing during the growing season, the application of fertiliser and sometimes of selective herbicides. This produces a virtually pure turf of fine-leaved *F. rubra* and *A. stolonifera*. Cutting is now highly mechanised and involves the removal of shallow (*c*. 3.5 cm deep) turves often over considerable areas. Recolonisation of cuttings produces a diverse and irregular succession (see below) and may involve the development of the transitional *Puccinellia* sub-community.

There is a broad correlation between variation in the sub-communities of the *Juncetum gerardi* and the incidence of tidal submersion. The transitional *Puccinellia* sub-community usually extends furthest down-marsh and it may be subject to more than 250 submergences/year, though it can also occur in very slight hollows in the upper marsh. Where it extends down into the *Puccinellietum* it is found on knolls and creek levees. The *Festuca-Glaux* sub-community is also found in such situations though the lower limit of continuous swards experiences between 150 and 200 submergences/year. The *Leontodon* sub-community occurs at higher levels which are subject to up to 100–120 submergences/year. Where vegetation virtually identical to this sub-community occurs on sea cliffs, it is found in situations which receive very considerable amounts of sea-spray and its soils show some of the highest values of Na/organic matter encountered in that habitat. The *Carex flacca* sub-community is best developed at the the storm-tide level where there are usually only one or two flooding tides per annum and perhaps at extremes up to 25 submergences/year. Despite the frequent seepage of freshwater into sites occupied by this sub-community, the soil salinity during droughts may reach quite high values (Gillham 1957*b*).

Substrates on which the *Juncetum gerardi* occurs include clays, silts, sands, shingle and soils of high organic content. The *Puccinellia* sub-community spans the entire range of substrate variation. Other sub-communities are more restricted: the tall *Festuca rubra* sub-community tends to occur on clays, silts and sands while the *Festuca-Glaux* and *Leontodon* sub-communities are generally confined to sandier material with some occurrences on more organic soils. Although the *Juncus gerardii* sub-community occurs on various substrates, its occurrences in south-east England frequently indicate the presence of shingle below the top soil horizon and, in some cases, this vegetation can develop directly on shingle banks. The *Carex flacca* sub-community is most frequently found on soils with high organic content, at least in the upper part of the profile. The pH of the substrates on which the *Juncetum* occurs varies between 5.0 and 7.0, with finer material without organic enrichment being more basic.

A combined effect of tidal inundation and substrate type is mediated through soil permeability. The degree of waterlogging probably has some effect on the proportions of *F. rubra*, *A. stolonifera* and *Puccinellia maritima* in the vegetation. *F. rubra* may suffer competitively against *P. maritima* under waterlogged and more saline conditions and against *A. stolonifera* in waterlogged and less saline situations (Gray & Scott 1977*a*). On cliffs, the factor which favours the development of the *Leontodon* sub-community of the *Juncetum gerardi* rather than some form of *Festuca-Armeria* sward is probably the retention of water in the heavy gleyed soils: both vegetation types receive similar amounts of salt-spray and both are grazed.

Among the grasses of the *Juncetum gerardi*, *A. stolonifera* seems more resistant to oil and refinery effluent spillage than either *F. rubra* or *P. maritima* and it may gain a competitive advantage in vegetation recovering from such pollution. *Armeria maritima*, *Plantago maritima* and *Triglochin maritima* are able to resist considerable amounts of spillage by virtue of their underground storage organs (Baker 1979).

Zonation and succession

In general, the *Juncetum gerardi* occupies a position above the *Puccinellietum maritimae* in the salt-marsh zonation but the extent of the *Juncetum* in the south-east differs strikingly from its contribution to salt-marshes elsewhere. In the south-east, the association is of very limited extent and occurs only at high levels in the marsh, most often forming a discontinuous zone in contact with the *Limonium-Armeria* sub-community of the *Puccinellietum*. In the north and west, the *Juncetum* is usually very extensive in both the mid- and upper-marsh. The exact reasons for this difference, and for the more seaward extension of particular species in the west, are unknown (Beeftink 1977*a*, *b*, Adam 1978) but the major factor controlling the relative positions of the two associations is the degee of submersion. In upper estuaries, where there is freshwater dilution, the positions of the *Juncetum* and *Puccinellietum* are reversed. The location and nature of the junction between the associations is also affected markedly by the extent and nature of the grazing.

Within the *Juncetum*, there is usually a zonation of the different sub-communities in relation to their tolerance of submersion. The detailed pattern varies from site to site and, though the *Festuca-Glaux* sub-community usually gives way to the *Leontodon* sub-community up-marsh, the relative depth of the zones is very variable (Figure 9). In some cases, there is a complex mosaic of the two communities over the mid-marsh. The *Leontodon* sub-community may, in turn, pass into the *Carex flacca* sub-community. Provided that the upper limit of the salt-marsh is not an artificial boundary, the topmost zone grades into non-maritime grassland or mire (e.g. Gillham 1957*b*). The tall *Festuca rubra* sub-community often forms part of zonations with the *Halimionetum*, sometimes occupying a position between this association and fragmentary stands of the *Artemisietum*.

The zonation of the sub-communities may represent a successional sequence in response to substrate accretion and the gradual raising of salt-marsh surfaces. Grazing too, can, be responsible for considerable temporal changes within the *Juncetum gerardi* (see above) and may shift the succession towards the development of other

associations. Very heavy grazing, particularly by cattle and horses on clay and silt substrates, can lead to poaching and the appearance of *Puccinellietum maritimae* or to communities characteristic of disturbed saline sites such as the *Puccinellietum distantis* and the *Agrostis stolonifera-Alopecurus geniculatus* community.

Turf-cutting opens up areas for colonisation by a variety of species. In the early stages a variety of annuals and short-lived perennials predominate: *Spergularia marina*, *Juncus bufonius*, *Plantago coronopus* and *Sagina maritima*, for example, often with *Pottia heimii*. Diverse assemblages of such species have sometimes been classified within the Saginetea. *Puccinellia maritima* is frequently an early colonist and a closed *Puccinellietum maritimae* may develop. *P. maritima* may persist within a *Festuca rubra* sward to produce a patchwork of the transitional vegetation of the *Puccinellia* sub-community of the *Juncetum gerardi*. It is this pattern of recolonisation which helps make grazing and turf-cutting compatible activities on the same salt-marsh.

Distribution

The *Juncetum gerardi* is widespread except in the south-east where it is local and where the *J. gerardii*-dominated sub-community is the most frequent representative of the association. The *Festuca-Glaux* and *Leontodon* sub-communities are virtually ubiquitous in western Britain but very sparsely distributed in the south-east. In north Norfolk, for example, their only extensive occurrence is at Brancaster which, interestingly, is the only marsh in the area still subject to regular grazing. Both sub-communities have been reported from brackish reclaimed pastures and they may be more widespread in this habitat. Where the *Leontodon* sub-community occurs on sea cliffs it is chiefly northern with some isolated occurrences in Wales and Cornwall where its distribution may be related to localised flushing rather than a generally high precipitation. It is commonest in west Scotland, the Outer Isles, Orkney and Shetland. The grass-dominated swards of the *Festuca-Glaux* and *Leontodon* sub-communities have been encountered chiefly in those areas where sheep-grazing and turf-cutting are most intensive. The *Carex flacca* sub-community is widespread in the west but most frequent in west Scotland.

Affinities

The *Juncetum gerardi* is one of the most important communities on British salt-marshes but its internal diversity and its affinities have been little discussed. The view of the *Juncetum* adopted here is a broad one, roughly comparable to that of Tyler (1969*b*). A similar range of vegetation types to that included here occurs widely in northern Europe and numerical studies (Adam 1977) have emphasised the close relationship between the British and European communities.

Figure 9. Complex of upper marsh communities at Bolton-le-Sands, Morecambe Bay.

The bulk of the marsh vegetation comprises various kinds of SM16 *Juncetum gerardi*. In the foreground, as a mosaic around the largely dried-up pans, are the *Festuca-Glaux* and *Juncus gerardii* sub-communities, with a small patch of the tall *Festuca rubra* sub-community to the left. Above, these give way to the *Leontodon* sub-community. Scattered through the *Juncetum gerardi* are dense clumps of the SM18 *Juncus maritimus* salt-marsh. On the slope below the road, the *Juncetum* gives way to a narrow zone of the MG11 *Festuca-Agrostis-Potentilla* grassland with small stands of OV25 *Urtica-Cirsium* vegetation on rotting horse faeces. On the flushed ground below are small stands of the SM20 *Eleocharitetum uniglumis* and some larger areas of S21 *Scirpetum maritimi* and S20 *Scirpetum tabernaemontani*; from one of these, a small stand of MG13 *Agrostis-Alopecurus* grassland runs down the marsh towards a large pan. The flooded pans to the right have thick festoons of the SM2 *Ruppietum maritimae*.

Salt-marsh communities

SM16 *Juncetum gerardi*
  Leontodon sub-community
  Trifolium variant

SM16 *Juncetum gerardi*
  Festuca-Glaux sub-community
  Armeria variant

SM16 *Juncetum gerardi*
  Juncus gerardii-dominated
  sub-community

SM16 *Juncetum gerardi*
  Sub-community with tall
  Festuca rubra dominant

SM18 *Juncus maritimus* salt-marsh
  Oenanthe sub-community

SM20 *Eleocharitetum uniglumis*

Swamps

S21   *Scirpetum maritimi*

S20   *Scirpetum tabernaemontani*

Mesotrophic grasslands

MG13 *Agrostis-Alopecurus*
  community

MG11 *Festuca-Agrostis-Potentilla*
  community
  (with *Urtica* patches)

MG6 *Lolio-Cynosuretum*

The *Festuca-Glaux* sub-community can be regarded as the core of the British *Juncetum gerardi*. Floristic transitions from this sub-community to the *Puccinellietum maritimae* are obvious and the major difficulty is deciding where exactly to draw the line between the two associations. In other directions, floristic affinities are more diverse and contentious. Certain authorities would see some of the samples included here within the *Leontodon* sub-community as part of the Elymo-Rumicion crispi, emphasising the transitional nature of the vegetation (see the *Agrostidetum stoloniferae* sub-association of var. *salina* and *Trifolium fragiferum* Westhoff 1947 and *Ononis spinosa-Carex distans* Association Runge 1966 in Westhoff & den Held 1969; Géhu 1973*b*). An alternative treatment of the high level stands of the *Leontodon* sub-community with pasture grasses would be to place such vegetation in a maritime sub-community of the *Lolio-Cynosuretum cristati* (e.g. Raabe 1953; see also Gillner 1960).

Carex-rich upper marsh grasslands similar to those included here within the *Carex flacca* sub-community have been described from Scandinavia (Nordhagen 1923, Du Rietz & Du Rietz 1925, Almquist 1929, Gillner 1960, Tyler 1969*b*), Germany (Tüxen 1937) and The Netherlands (Westhoff 1947). Some would place these again within the Elymo-Rumicion crispi, while others see them as variants of inland mire types occasionally encountered in maritime or paramaritime situations (cf. the *Isolepsis setacea* variant of the *Schoeno-Juncetum serratuletosum* in Wheeler 1980*b* and the Caricion davallianae dune-slack communities of Westhoff & den Held 1969).

The *J. gerardii*-dominated sub-community bears some resemblance to communities of the Eleocharion. It should, however, be distinguished from the *Juncetum gerardi juncetosum* (Tyler 1969*b*) which possesses a distinctive suite of bryophytes not represented here.

The tall *Festuca rubra* sub-community is a somewhat diverse assemblage united by the overwhelming dominance of *F. rubra*. It could be divided on a strict floristic basis between the *Juncetum gerardi*, the *Halimionetum* and the *Atriplici-Elymetum pycnanthi*. Alternatively, the entire sub-community could be separated entirely from the *Juncetum* as part of the *Festucetum littoralis* Corillion 1953 (e.g. Géhu 1975, Géhu & Delzenne 1975).

Floristic table SM16

	a	b	c	d	e	f	16
Festuca rubra	IV (2–7)	III (2–7)	V (2–10)	V (5–10)	V (2–9)	V (4–8)	V (2–10)
Juncus gerardii	V (3–7)	V (6–10)	V (2–7)	I (2–3)	V (2–7)	V (3–6)	V (2–10)
Glaux maritima	V (2–8)	IV (2–7)	V (2–8)	II (2–6)	V (2–7)	IV (2–6)	IV (2–8)
Plantago maritima	V (3–8)	IV (2–6)	V (2–8)	IV (1–5)	IV (2–7)	V (2–6)	IV (2–8)
Agrostis stolonifera	III (2–9)	II (3–7)	IV (2–8)	III (2–8)	V (2–8)	V (3–8)	IV (2–9)
Triglochin maritima	V (1–7)	IV (2–5)	IV (1–6)	II (1–5)	III (1–6)	III (1–5)	III (1–7)
Armeria maritima	III (2–6)	III (2–5)	IV (2–8)	I (2–3)	III (2–5)	III (2–6)	III (2–8)
Aster tripolium (rayed)	III (1–5)	IV (1–4)	III (1–5)	II (2–6)	I (1–4)	I (3)	II (1–6)
Puccinellia maritima	V (2–9)	I (3–6)	I (2–6)	II (2–5)			II (2–9)
Algal mat	II (3–8)	II (3–8)	I (4–8)	I (3)	I (4–5)		I (3–8)
Spergularia media	II (1–4)	I (2)	I (2–3)	II (1–4)	I (2–3)		I (1–4)
Aster tripolium	II (2–4)	I (3)	I (1–3)	I (1–4)			I (1–4)
Salicornia agg.	II (2–5)	I (2)	I (1–3)		I (2)		I (1–5)
Atriplex prostrata	I (3)	II (2–3)	I (1–3)	II (1–4)	I (2–3)	I (2)	I (1–4)
Cochlearia anglica	I (2–4)	II (2–3)	I (2–3)		I (2)		I (2–4)
Halimione portulacoides	I (2–4)	II (2–3)	I (1–2)	II (1–8)			I (1–8)
Limonium cf. *L. vulgare*	I (2–6)	II (2–4)	I (1–8)	II (1–5)	I (2)		I (1–8)
Trifolium repens		I (2–4)	I (2–5)	I (2–4)	IV (2–7)	V (1–7)	II (1–7)
Leontodon autumnalis		I (2)	I (2–5)	I (2–4)	III (1–6)	V (2–5)	II (1–6)
Carex flacca	I (3)				I (2–6)	IV (1–7)	I (1–7)
Carex distans		I (2)	I (1–3)	I (2–3)	III (1–7)	II (1–5)	I (1–7)
Potentilla anserina			I (2)	I (3–6)	II (2–8)	III (3–7)	I (2–8)
Holcus lanatus				I (3–6)	I (2–5)	I (2–6)	I (2–6)
Lotus corniculatus			I (4)	I (2–4)	I (2–5)	I (2–6)	I (2–6)
Cerastium fontanum			I (2)	I (2)	I (2–3)	II (2–3)	I (2–3)
Sagina procumbens			I (2–4)		I (2–5)	II (2–5)	I (2–5)
Eurhynchium praelongum			I (3–4)		I (2–7)	I (3–6)	I (2–7)
Rhytidiadelphus squarrosus			I (3)		I (3–5)	I (3–8)	I (3–8)
Hypnum cupressiforme			I (2)		I (2–3)	I (4–6)	I (2–6)
Anthoxanthum odoratum				I (3)	I (3)	I (2–5)	I (2–5)
Plantago lanceolata				I (2)	I (2–3)	I (1–3)	I (1–3)

Floristic table SM16 (*cont.*)

	a	b	c	d	e	f	16
Cynosurus cristatus					I (2–3)	I (5–6)	I (2–6)
Ranunculus acris					I (2)	I (1–4)	I (1–4)
Number of samples	34	49	150	85	149	46	513
Mean number of species/sample	9 (6–13)	8 (3–12)	9 (5–16)	7 (3–12)	11 (7–18)	16 (9–31)	10 (3–31)
Mean vegetation height (cm)	7 (2–25)	26 (2–40)	11 (2–75)	26 (10–60)	10 (2–50)	10 (2–60)	14 (2–75)
Mean total cover (%)	95 (70–100)	96 (50–100)	96 (50–100)	99 (80–100)	100 (90–100)	100 (90–100)	98 (50–100)

a   *Puccinellia maritima* sub-community
b   Sub-community with *Juncus gerardii* dominant
c   *Festuca rubra-Glaux maritima* sub-community
d   Sub-community with tall *Festuca rubra* dominant
e   *Leontodon autumnalis* sub-community
f   *Carex flacca* sub-community
16   *Juncetum gerardi* (total)

SM16 sub-communities

	c	ci	cii	e	ei
Festuca rubra	V (2–10)	V (5–10)	V (5–10)	V (2–9)	V (4–9)
Juncus gerardii	V (2–7)	I (2)	I (1)	V (2–7)	I (3–4)
Glaux maritima	V (2–8)	IV (2–7)	V (3–7)	V (2–7)	IV (2–7)
Plantago maritima	V (2–8)	IV (1–8)	V (2–7)	IV (2–7)	IV (2–5)
Agrostis stolonifera	IV (2–8)	V (3–8)	V (2–8)	V (2–8)	V (4–8)
Triglochin maritima	IV (1–6)	III (2–5)	IV (1–7)	III (1–6)	II (2–4)
Armeria maritima	IV (2–8)	IV (2–8)	V (2–8)	III (2–5)	III (2–5)
Aster tripolium (rayed)	III (1–5)	II (1–4)	III (2–7)	I (1–4)	I (2)
Puccinellia maritima	I (2–6)	I (2–6)	I (2–5)		
Algal mat	I (4–8)	I (3–6)	II (3–7)	I (4–5)	
Spergularia media	I (2–3)	II (1–3)	II (2–3)	I (2–3)	I (1)
Aster tripolium	I (1–3)	I (3)	I (2–5)		
Salicornia agg.	I (1–3)	I (2–3)	II (2–3)	I (2)	
Atriplex prostrata	I (1–3)	I (1–2)	I (1–2)	I (2–3)	I (2–3)
Cochlearia anglica	I (2–3)	I (2)	I (2–3)	I (2)	I (2)
Halimione portulacoides	I (1–2)	I (4)	I (1–2)		
Limonium cf. L. vulgare	I (1–8)	I (1–2)	I (1–2)	I (2)	
Suaeda maritima	I (1–3)	I (2)	II (1–4)	I (2)	I (1)
Trifolium repens	I (2–5)			IV (2–7)	V (2–8)
Leontodon autumnalis	I (2–5)	I (2–5)		III (1–6)	II (2–5)
Carex flacca				I (2–6)	I (5)
Carex distans	I (1–3)			III (1–7)	I (2–4)
Potentilla anserina	I (2)	I (5)		II (2–8)	I (2–6)
Holcus lanatus				I (2–5)	I (2–3)
Lotus corniculatus	I (4)			I (2–5)	I (3–4)
Cerastium fontanum	I (2)			I (2–3)	I (2–3)
Sagina procumbens	I (2–4)	I (4)		I (2–5)	I (2)
Eurhynchium praelongum	I (3–4)	I (4)		I (2–7)	I (3)
Rhytidiadelphus squarrosus	I (3)			I (3–5)	I (3–4)
Hypnum cupressiforme	I (2)			I (2–3)	
Anthoxanthum odoratum				I (3)	

SM16 sub-communities (*cont.*)

	c	ci	cii	e	ei
Plantago lanceolata				I (2–3)	
Cynosurus cristatus				I (2–3)	
Ranunculus acris				I (2)	
Number of samples	150	68	51	149	30
Mean number of species/sample	9 (5–16)	6 (2–12)	7 (4–12)	11 (7–18)	9 (5–13)
Mean vegetation height (cm)	11 (2–75)	5 (2–20)	5 (2–15)	10 (2–50)	7 (2–25)
Mean total cover (%)	96 (50–100)	98 (80–100)	96 (80–100)	100 (90–100)	99 (85–100)

c   *Festuca rubra-Glaux maritima* sub-community
ci   *Agrostis stolonifera* variant
cii   *Armeria maritima* variant
e   *Leontodon autumnalis* sub-community
ei   *Trifolium repens* variant

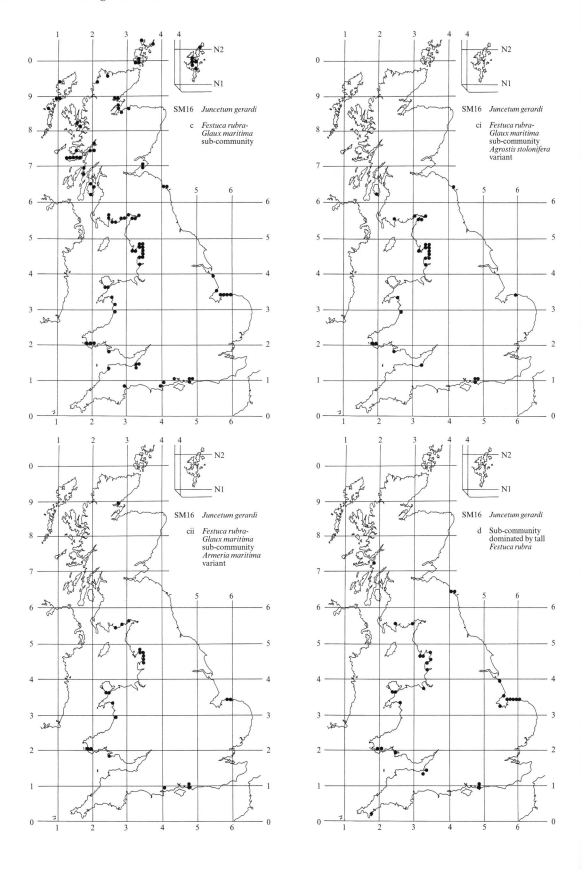

SM16   *Juncetum gerardi*

c   *Festuca rubra-Glaux maritima* sub-community

SM16   *Juncetum gerardi*

ci   *Festuca rubra-Glaux maritima* sub-community *Agrostis stolonifera* variant

SM16   *Juncetum gerardi*

cii   *Festuca rubra-Glaux maritima* sub-community *Armeria maritima* variant

SM16   *Juncetum gerardi*

d   Sub-community dominated by tall *Festuca rubra*

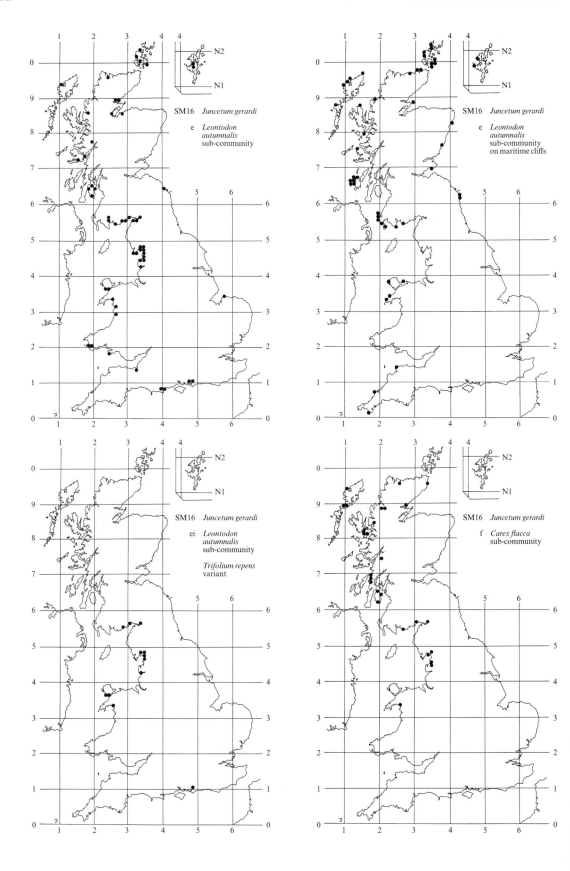

SM16 *Juncetum gerardi*

e *Leontodon autumnalis* sub-community

SM16 *Juncetum gerardi*

e *Leontodon autumnalis* sub-community on maritime cliffs

SM16 *Juncetum gerardi*

ei *Leontodon autumnalis* sub-community

Trifolium repens variant

SM16 *Juncetum gerardi*

f *Carex flacca* sub-community

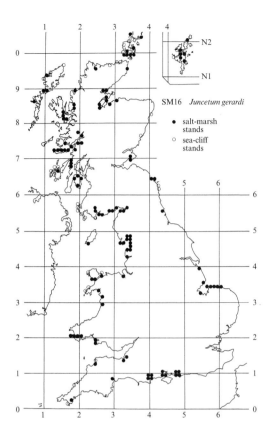

SM16   *Juncetum gerardi*

•   salt-marsh
     stands

○   sea-cliff
     stands

SM17

Artemisia maritima salt-marsh community
Artemisietum maritimae   Hocquette 1927

Synonymy
Festucetum littoralis artemisietosum Ghestem 1972; includes *Halimionetum portulacoidis*, terminal phase with *Artemisia maritima*   Beeftink 1962.

Constant species
Artemisia maritima, Festuca rubra, Halimione portulacoides, Plantago maritima.

Rare species
Limonium binervosum, L. humile, Suaeda vera.

Physiognomy
This is a species-poor community of somewhat variable physiognomy. Stands are generally small and fragmentary but they range from rank grassy patches dominated by *Festuca rubra* with prominent *Artemisia maritima* to open bushy vegetation with *A. maritima* over low *Halimione portulacoides*. There is usually a little *Plantago maritima*; *Limonium* cf. *vulgare* and *Puccinellia maritima* are also frequent.

Habitat
The *Artemisietum maritimae* is an upper-marsh community occurring on a variety of substrates, though often in association with tidal litter and inwashed shell fragments. Its most usual habitat is on creek levees where it forms small patches on the usually heavy clay soils which frequently show organic enrichment in the upper part of the profile. It is also found on ridges and mounds on the upper marsh and sometimes forms a fringe along the foot of sea walls or around stands of *Atriplici-Elymetum pycnanthi*. Where such zones occur at the marsh/dune interface the substrate is often a coarse sand.

Most occurrences are on ungrazed marshes but this may be a reflection of the predominantly south-eastern distribution of the association.

Zonation and succession
The association is normally found as a patchy zone between the *Halimionetum* and *Atriplici-Elymetum pycnanthi*. Junctions with the former may be marked by an intermediate zone with abundant *Festuca rubra*. This zonation may reflect a succession related to increase in marsh height with sediment accretion.

Distribution
The *Artemisietum* is widespread in East Anglia and along the south coast and it extends north into Scotland. West coast occurrences are scattered and restricted mainly to ungrazed marshes.

Affinities
Traditionally, an *Artemisia maritima* community has not been distinguished in British accounts of salt-marsh vegetation (e.g. Chapman 1934, Tansley 1939) and some Continental authorities regard the community as a sub-division of the *Halimionetum* or *Festucetum littoralis*. Nonetheless, though British stands are generally fragmentary and united mainly by the prominence of *Artemisia maritima*, the major associates represent a distinctive assemblage of species. Furthermore, Continental stands of this vegetation are frequently far more extensive and floristically distinct than those in Britain. A British *Artemisietum* could thus be sensibly seen as a somewhat impoverished extension of an association which occupies a noteworthy floristic transition between a number of Puccinellion and Armerion communities.

Floristic table SM17

Artemisia maritima	V (3–9)
Festuca rubra	V (2–9)
Halimione portulacoides	IV (1–8)
Plantago maritima	IV (1–5)
Limonium cf. *L. vulgare*	III (2–6)
Puccinellia maritima	III (2–7)
Armeria maritima	II (2–4)
Glaux maritima	II (1–4)
Triglochin maritima	II (2–5)
Aster tripolium (rayed)	II (2–4)
Aster tripolium	I (1–3)
Cochlearia anglica	I (1–3)
Juncus gerardii	I (3–5)
Agrostis stolonifera	I (2–5)
Atriplex prostrata	I (2–3)
Elymus pycnanthus	I (2–4)
Suaeda vera	I (1–3)
Spergularia media	I (1–4)
Plantago coronopus	I (2–5)
Cochlearia officinalis	I (2)
Elymus repens	I (2–3)
Juncus maritimus	I (3–5)
Suaeda maritima	I (2–4)
Salicornia agg.	I (2)
Parapholis strigosa	I (2–4)
Hordeum marinum	I (3–4)

Number of samples	42
Mean number of species/sample	7 (2–12)
Mean vegetation height (cm)	19 (7–35)
Mean total cover (%)	98 (50–100)

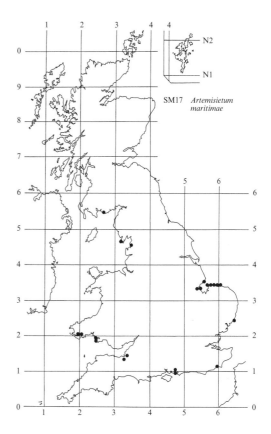

SM17 *Artemisietum maritimae*

SM18
Juncus maritimus salt-marsh community

Synonymy
Juncetum maritimi auct. angl. p.p., includes *Juncus maritimus-Oenanthe lachenalii* ass. R.Tx. 1937

Constant species
Agrostis stolonifera, Festuca rubra, Glaux maritima, Juncus gerardii, J. maritimus.

Physiognomy
The association is dominated by tall dense clumps of *Juncus maritimus* with an understorey of *Agrostis stolonifera, Festuca rubra, Glaux maritima* and *Juncus gerardii*. There is a rich subsidiary flora in which mesotrophic grassland species (notably *Leontodon autumnalis* and *Trifolium repens*) and weed species (for example, *Atriplex hastata, Elymus repens* and *Rumex crispus*) are conspicuous. Bryophytes may be locally abundant with patches of *Calliergon cuspidatum, Amblystegium riparium, A. serpens* and *Eurhynchium praelongum*. Stands of the association may be based on individual clones of *J. maritimus*, in some cases up to 15 m in diameter, or occur as an extensive zone.

Sub-communities

***Plantago maritima* sub-community:** *Juncetum maritimi* Yapp & Johns 1917; *Festuca rubra-Juncus maritimus* nodum Adam 1976. *Plantago maritima* and *Triglochin maritima* attain constancy in the often luxuriant vegetation of this sub-community and there are frequent records for *Leontodon autumnalis*, rayed *Aster tripolium* and *Armeria maritima*. It sometimes occupies extensive areas, notably at Ynys Hir in the Dovey estuary, Dyfed.

***Oenanthe lachenalii* sub-community:** *Juncus maritimus-Oenanthe lachenalii* ass. R.Tx. 1937; *Oenanthe lachenalii-Juncus maritimus* nodum Adam 1976. *Triglochin maritima, Leontodon autumnalis* and *Oenanthe lachenalii* are constant here, the last giving a particularly distinctive appearance to the vegetation in the flowering season, though it is rarely present in abundance. Individual stands, even those in close proximity, may have singular characteristics and different weed species, germinating in trapped drift litter, are especially variable. Some stands have abundant *Cirsium arvense*, others *Atriplex hastata, Sonchus arvensis* or *Urtica dioica*.

***Festuca arundinacea* sub-community:** *Festuca arundinacea-Juncus maritimus* nodum Adam 1976. *Oenanthe lachenalii*, remains constant in this sub-community but salt-marsh species like *Glaux maritima, Juncus gerardii, Plantago maritima* and *Triglochin maritima* are less frequent. However, the most obvious feature here is the constancy of *Festuca arundinacea*, the large tussocks of which may be co-dominant with the *Juncus maritimus*, and of *Leontodon autumnalis, Potentilla anserina* and *Trifolium repens* in the understorey. Other mesotrophic grassland species such as *Holcus lanatus, Lotus corniculatus, Ranunculus acris* and *Vicia cracca* are also frequent.

Habitat
The association is predominantly an upper-marsh community but the sub-communities differ in their tolerance of tidal submersion. The lowest recorded site for the *Festuca arundinacea* sub-community experienced 25 submergences/year while the *Oenanthe* sub-community seems to be able to tolerate at least 150 submergences/year. As the *Plantago* sub-community is normally found seaward of the *Oenanthe* sub-community, its tolerance is presumably even greater.

The association occurs on a variety of substrates but the pH is generally around 7.0 (cf. Bridges 1977 who recorded values down to 5.1). There is normally an appreciable accumulation of organic matter in the top 10–20 cm of the soil and superficial litter trapping may be considerable. This material provides a suitable substrate for colonisation by weed species.

Although the association is common on grazed marshes, *Juncus maritimus* is itself unpalatable and its dense tall growth confers protection on the associated species. Yapp & Johns (1917) and Tansley (1939) suggested that the luxuriance of vegetation within *J. maritimus* stands may also be due to the higher and more constant humidity levels attained there. Extensive spread of the association on grazed marshes reduces their agricultural value but eradication has been attempted on only a small local scale. Mowing, draining, the use of herbicides and physical removal of *J. maritimus* have all been attempted. Packham & Liddle (1970) have reported some success in control on Cefni Marsh, Anglesey, by cutting close to the ground in early summer.

Oenanthe lachenalii is remarkably resistant to oil and refinery effluent spillage (Baker 1979). Even after repeated oiling, plants respond simply by producing new shoots.

Zonation and succession

At a few sites, there is a zonation within single extensive stands of the association from the *Plantago* sub-community through the *Oenanthe* sub-community to the *Festuca* sub-community. More generally, isolated stands of each of the sub-communities occur within other communities, the *Oenanthe* and *Plantago* sub-communities usually within the *Juncetum gerardi*, though the *Plantago* sub-community may also extend down-marsh into the upper part of the *Puccinellietum maritimae*. Unlike these two sub-communities where stands are sharply defined the *Festuca* sub-community often has rather diffuse boundaries with its neighbouring communities on the upper marsh.

Juncus maritimus can be an aggressive invader. Packham & Liddle (1970) reported the transformation of an area of *Puccinellietum maritimae* within the space of 20 years. It has been conventional in British accounts to regard *Juncus maritimus* salt-marsh, if not as the true climax of succession, then at least as a very stable stage in upper marsh development which can be considered for most purposes as the climax. The association is clearly important on those marshes where it occurs but it is probably better to see it as part of a seral sequence parallel to that involving the *Juncetum gerardi*.

Distribution

The association is widespread on the west coast as far north as Arran but very local in south-east England, though it may occur there on derelict reclaimed land. In Norfolk, the association is replaced by the *Juncus maritimus-Triglochin maritima* salt-marsh.

Affinities

The classification of *Juncus maritimus* vegetation poses a number of problems (Adam 1977). *J. maritimus* occurs widely in British salt-marshes but those vegetation types in which it is dominant or co-dominant are nonetheless distinct. Two of these are best considered as sub-communities of other well-defined associations: the *Halimionetum portulacoidis* and the *Atriplici-Elymetum pycnanthi*. A further community forms the distinctive, partly low-marsh, *Juncus maritimus-Triglochin maritima* association.

The three remaining types are those grouped here as the *Juncus maritimus* salt-marsh. The *Juncetum maritimi* of British authors has not been taken up in Continental studies but, although this partly reflects a different approach to classification, it is probably also an indication of the less important role which *J. maritimus* plays outside Great Britain. An alternative approach to these three types of *J. maritimus* salt-marsh would be to consider them as variants of the *Juncetum gerardi* (see Braun-Blanquet & Tüxen 1952, Ivimey-Cook & Proctor 1966, Moore *et al.* 1970 and Moore & O'Reilly 1977 in Ireland). However, although the two associations share a considerable number of species, the *Juncus maritimus* salt-marsh is distinct in containing conspicuous weed species and also in its striking physiognomy. The representation of ruderals has led some workers to place *J. maritimus* vegetation in various taxa of the Elymo-Rumicion crispi (e.g. Westhoff & den Held 1969).

There is no single well-described phytosociological equivalent of the association diagnosed here. Tüxen (1937) reported a *Juncus maritimus-Oenanthe lachenalii* association from north Germany and similar communities have been encountered from other sites in that region (Libbert 1940, Voderberg 1955, Passarge 1964), from The Netherlands (Westhoff & den Held 1969) and from north Spain (Tüxen & Oberdorfer 1958, Bellot 1966). These have not been fully described or related to British *J. maritimus* vegetation but it may eventually be sensible to incorporate them into a single association.

The general floristic similarities to the *Juncetum gerardi* suggest that the *Juncus maritimus* salt-marsh is best placed within the Armerion maritimae of the Asteretea.

Floristic table SM18

	a	b	c	18
Juncus maritimus	V (7–9)	V (2–9)	V (5–8)	V (2–9)
Agrostis stolonifera	V (3–8)	V (3–8)	V (4–7)	V (3–8)
Festuca rubra	V (1–8)	V (3–8)	V (5–7)	V (1–8)
Glaux maritima	IV (2–7)	IV (2–6)	III (2–4)	IV (2–7)
Juncus gerardii	IV (3–5)	IV (2–6)	III (3–5)	IV (2–6)
Triglochin maritima	IV (2–5)	IV (2–5)	II (2–3)	III (2–5)
Plantago maritima	V (2–6)	III (2–4)	II (2–3)	III (2–6)
Oenanthe lachenalii	I (1–3)	V (2–5)	V (2–5)	III (1–5)
Leontodon autumnalis	III (2–4)	IV (2–5)	IV (1–4)	II (1–5)
Festuca arundinacea	I (2)	I (2–3)	V (3–7)	II (2–7)
Potentilla anserina	I (3–5)	II (2–7)	IV (2–7)	II (2–7)
Trifolium repens	II (2–6)	III (3–8)	IV (3–5)	III (2–8)
Aster tripolium (rayed)	III (2–4)	II (2–3)	I (2)	II (2–4)
Armeria maritima	III (2–5)	I (2–4)	I (2–3)	II (2–5)
Algal mat	II (4–8)	I (4–5)		I (4–8)
Atriplex prostrata	II (1–5)	III (2–5)	II (2–3)	II (1–5)
Elymus repens	I (4)	II (2–6)	III (3–6)	II (2–6)
Carex distans	I (2–3)	II (2–5)	III (1–4)	II (1–5)
Lotus corniculatus	I (4)	I (2–4)	III (3–5)	I (2–5)
Eurhynchium praelongum	I (3–5)	II (3–6)	III (3–7)	II (3–7)
Carex extensa	II (3–4)	I (1–4)	I (3)	I (1–4)
Poa pratensis	I (3–5)	II (2–5)	I (2–4)	I (2–5)
Rumex crispus	I (1–3)	II (2–3)	I (1–3)	I (1–3)
Samolus valerandi	I (3)	I (2–4)	II (2–4)	I (2–4)
Cirsium arvense		I (2–7)	II (2–3)	I (2–7)
Holcus lanatus		I (2–7)	II (2–7)	I (2–7)
Ranunculus acris		I (2–3)	II (2–4)	I (2–4)
Vicia cracca		I (2–4)	II (1–6)	I (1–6)
Carex otrubae		I (2–4)	I (2–4)	I (2–4)
Cochlearia officinalis	II (2–4)	II (2–3)	II (1–3)	II (1–4)
Amblystegium serpens	I (3–5)	I (3)	I (3–5)	I (3–5)
Galium palustre	I (3)	I (2–4)	I (3)	I (2–4)
Lychnis flos-cuculi	I (3)	I (2–3)	I (3)	I (2–3)
Sonchus arvensis	I (3)	I (2–3)	I (2–5)	I (2–5)
Number of samples	51	71	33	155
Mean number of species/sample	11 (5–20)	13 (7–25)	15 (8–32)	13 (5–32)
Mean vegetation height (cm)	51 (5–100)	49 (30–100)	53 (30–100)	50 (5–100)
Mean total cover (%)	98 (85–100)	99 (70–100)	99 (95–100)	98 (70–100)

a   *Plantago maritima* sub-community
b   *Oenanthe lachenalii* sub-community
c   *Festuca arundinacea* sub-community
18   *Juncus maritimus* salt-marsh (total)

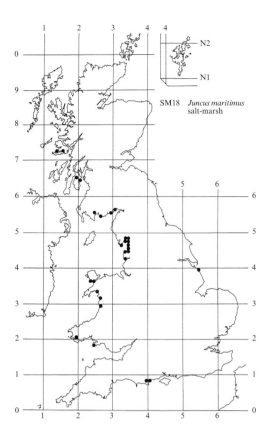

SM18 *Juncus maritimus*
salt-marsh

SM19

Blysmus rufus salt-marsh community
Blysmetum rufi (G. E. & G. Du Rietz 1925) Gillner 1960

Synonymy
Juncus gerardii-Carex extensa Association Birks 1973 *p.p.*

Constant species
Blysmus rufus, Agrostis stolonifera, Glaux maritima, Juncus gerardii, Triglochin maritima.

Rare species
Blysmus rufus.

Physiognomy
The *Blysmetum rufi* is a species-poor association, generally dominated by *Blysmus rufus* but often with abundant *Agrostis stolonifera, Glaux maritima* and *Juncus gerardii. Triglochin maritima, Festuca rubra, Plantago maritima* and *Carex extensa* are all frequent but rarely present in quantity. Some stands may have an extensive algal mat and certain bryophytes may be abundant: *Amblystegium riparium, A. serpens, Calliergon cuspidatum, Campylium stellatum, C. polygamum, Drepanocladus aduncus* and *Cratoneuron filicinum.* Cover may be somewhat open, especially on gravelly or rocky substrates, and stands are usually small (10–20 m^2), though at a number of Scottish sites the association covers hundreds of square metres.

Habitat
The association occurs on a variety of substrates but sites are often either poorly-drained or subject to flushing by brackish or fresh-water. The characteristic situation is in small depressions in the upper marsh. In some cases, the *Blysmetum* may develop in old upper-marsh pans, especially where these have a shingle base, and it is sometimes present along path edges (Gillner 1960, Tyler 1969b, Beeftink 1977a). In west Scotland, small stands are widespread within rocky flushes in the salt-marsh/mire transition on raised beaches and also among coastal rocks (Gillham 1957b, Birks 1973, Adam *et al.*

1977). The majority of occurrences are on grazed salt-marshes, although *B. rufus* itself does not appear to be much eaten.

Zonation and succession
Although *B. rufus* sometimes occurs as scattered shoots within the *Juncetum gerardi* (notably at Caerlaverock NNR, Dumfries & Galloway; Martin 1977), stands of the *Blysmetum* are usually rather sharply defined from the *Juncetum gerardi* which is the usual surrounding vegetation. Freshwater flushing over gravel sometimes allows the association to develop at fairly low levels in the marsh but it is unlikely that the *Blysmetum* plays any role in succession.

Distribution
B. rufus is one of the very few species constituting a northern element in the British salt-marsh flora (Ratcliffe 1977) and the association is locally distributed along the west coast from mid-Wales northwards, being commonest in west Scotland. Its generally small stands render it vulnerable to local extinction following habitat disturbance but, though there is some evidence for the loss of *B. rufus* in the southern part of its range (Perring & Walters 1962, Martin 1977, Ratcliffe 1977), there is nothing to suggest a large-scale contraction in distribution.

Affinities
The *Blysmetum rufi* is not discussed in the early descriptions of British salt-marshes which were mainly concerned with communities in south-east England. It has, however, been referred to in more recent accounts from northern and western Britain (Gillham 1957b, Greenwood 1972, Birks 1973, Adam *et al.* 1977) and the association is widely described on the Continent, where it is especially widespread in southern Scandinavia.

The *Blysmetum* shows clear affinities with the *Juncetum gerardi* and some accounts of the community regard it as part of that association (e.g. Birks 1973) or as a closely-related association within the Armerion maritimae (e.g. Beeftink 1965, 1977*a*). An alternative treatment is to place the association alongside the *Eleocharitetum uniglumis*, which is similar in its physiognomy and its habitat to the *Blysmetum*, and which is itself the centre of a separate alliance, the Eleocharion uniglumis (Siira 1970, Tyler *et al.* 1971).

There are some ecological similarities between the *Blysmetum* of the Scottish raised-beach flushes and certain of the Caricion davallianae communities *sensu* Wheeler (1980*b*).

Floristic table SM19

Blysmus rufus	V (3–9)
Agrostis stolonifera	V (2–7)
Glaux maritima	V (2–7)
Juncus gerardii	V (2–6)
Triglochin maritima	IV (1–5)
Festuca rubra	III (2–6)
Plantago maritima	III (2–5)
Carex extensa	III (2–5)
Aster tripolium (rayed)	II (2–3)
Armeria maritima	II (2–5)
Algal mat	II (3–7)
Trifolium repens	II (2–5)
Juncus articulatus	II (1–6)
Eleocharis uniglumis	II (4–9)
Alopecurus geniculatus	I (3–5)
Potentilla anserina	I (2–3)
Leontodon autumnalis	I (2–3)
Eleocharis quinqueflora	I (5)
Triglochin palustris	I (2–3)
Amblystegium riparium	I (3–4)
Amblystegium serpens	I (2–4)
Carex nigra	I (3–4)
Calliergon cuspidatum	I (3–7)
Cochlearia anglica	I (2)
Campylium polygamum	I (2–5)
Oenanthe lachenalii	I (3)
Puccinellia maritima	I (1–4)
Carex lepidocarpa	I (2–3)
Campylium stellatum	I (3–5)
Number of samples	23
Mean number of species/sample	10 (5–17)
Mean vegetation height (cm)	17 (6–25)
Mean total cover (%)	90 (50–100)

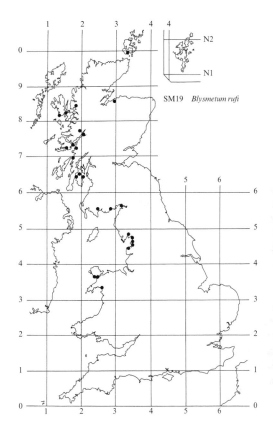

SM19 *Blysmetum rufi*

SM20
Eleocharis uniglumis salt-marsh community
Eleocharitetum uniglumis   Nordhagen 1923

Constant species

Eleocharis uniglumis, Agrostis stolonifera.

Physiognomy

Although *Eleocharis uniglumis* is a widespread minor constituent of a variety of damp transitional communities along the upper marsh fringes (Birks 1973, Adam 1976), it is here dominant in a species-poor association, similar in floristics and appearance to the *Blysmetum rufi*. There is often an extensive undercarpet of *Agrostis stolonifera* and *Juncus gerardii*, *Glaux maritima*, *Festuca rubra*, *Triglochin maritima*, *Potentilla anserina* and *Alopecurus geniculatus* all occur frequently and may be abundant in particular stands. As in the *Blysmetum*, cover is variable and algae and bryophytes may form a patchy carpet over the substrate surface.

Habitat

In Britain, the *Eleocharitetum* occurs most frequently in depressions in the upper marsh. Some of the most extensive stands occur in brackish marshes by the River Gilpin, Cumbria. Rarely, it occurs in what is a widespread habitat in Scandinavia, as a fringe of emergent vegetation, as around the brackish and atidal Loch an Amadain in Skye.

Zonation and succession

The *Eleocharitetum* occurs patchily within other upper-marsh associations such as the *Juncetum gerardi* and it does not play a major role in salt-marsh succession.

Distribution

The association is a rare community on British salt-marshes occurring locally along the west coast from the Dovey estuary northwards.

Affinities

British vegetation dominated by *E. uniglumis* is clearly closely related to that described from Scandinavia (Gillner 1960, Tyler 1969*b*, Siira 1970) although the emergent stands in Britain lack the aquatic species characteristic of the *Eleocharetum* of, for example, the Baltic. Siira (1970) and Tyler *et al.* (1971) assign the association to the alliance Eleocharion uniglumis, and a sensible treatment of both the *Eleocharitetum* and the *Blysmetum rufi* would be to regard them as constituting, in this alliance, the brackish end of variation within the Asteretea.

Floristic table SM20

Eleocharis uniglumis	V (5–10)
Agrostis stolonifera	V (3–7)
Glaux maritima	III (2–5)
Juncus gerardii	III (3–6)
Festuca rubra	II (2–7)
Triglochin maritima	II (2–5)
Alopecurus geniculatus	II (2–6)
Potentilla anserina	II (2–7)
Aster tripolium (rayed)	I (2–3)
Plantago maritima	I (4)
Armeria maritima	I (2–3)
Algal mat	I (8)
Carex extensa	I (3)
Trifolium repens	I (3–6)
Leontodon autumnalis	I (2–3)
Juncus articulatus	I (3–6)
Eleocharis quinqueflora	I (2–3)
Triglochin palustris	I (2)
Amblystegium riparium	I (3)
Carex nigra	I (2–4)
Hydrocotyle vulgaris	I (3–8)
Atriplex prostrata	I (2–3)
Samolus valerandi	I (3–4)
Carex distans	I (3–4)
Galium palustre	I (3–4)
Oenanthe lachenalii	I (1–4)
Scirpus maritimus	I (2–3)
Number of samples	17
Mean number of species/sample	8 (4–22)
Mean vegetation height (cm)	22 (10–45)
Mean total cover (%)	90 (60–100)

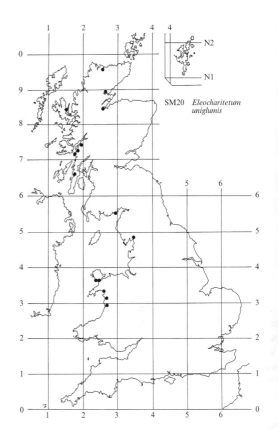

SM20 *Eleocharitetum uniglumis*

SM21
Suaeda vera–Limonium binervosum salt-marsh community

Synonymy
Suaedeto-Limonietum Chapman 1934 *p.p.*; *Halimioneto-Limonietum* Chapman 1934 *p.p.*; *? Suaedetum fruticosae* Tansley 1939 *p.p.*; *Suaedeto-Limonietum binervosi* Adam 1976; *Halimiono-Frankenietum laevis* Adam 1976 *emend.*; Norfolk *Frankenia laevis* stands Brightmore 1979.

Constant species
Armeria maritima, Halimione portulacoides, Limonium binervosum, Puccinellia maritima, Suaeda vera.

Rare species
Frankenia laevis, Limonium bellidifolium, L. binervosum, Suaeda vera.

Physiognomy
The rather open vegetation of this community is generally dominated by scattered bushes of *Suaeda vera* and *Halimione portulacoides* up to 40 cm high with a patchy cover of herbaceous halophytes between. Among the constants, *Puccinellia maritima* and *Limonium binervosum* are usually most abundant with smaller amounts of *Armeria maritima*. *Suaeda maritima* is fairly frequent throughout.

Sub-communities

Typical sub-community: *Suaedeto-Limonietum binervosi* Adam 1976. *Festuca rubra, Plantago maritima* and *Artemisia maritima* are frequent in this sub-community, the first sometimes in abundance. There are occasional records for a variety of species characteristic of disturbed places on the upper marsh and of strandlines.

Frankenia laevis sub-community: *Halimiono-Frankenietum laevis* Adam 1976 *emend. Frankenia laevis* and *Limonium bellidifolium* are constant in this sub-community which is more species-poor than the above.

Habitat
The community is characteristic of salt-marsh/dune interfaces, spit laterals, eroded dunes and some sand-dune lows where there is a base of shingle covered with varying amounts of blown sand and inwashed silt (Chapman 1934, 1960*b*, Tansley 1939).

The sub-communities differ in their tolerance of tidal inundation. The Typical sub-community is most frequently encountered at or above the tidal limit where there is inundation only during severe storms. The *Frankenia* sub-community extends further down-marsh and at its lower limit there may be a thick layer of heavy clay over the shingle base. During the summer, high soil salinities may be experienced with a salt crust forming on the soil surface (see also Brightmore 1979).

Grazing, especially by rabbits, is, or has been, of considerable importance in the maintenance of this community. Heavy grazing of low *Suaeda vera* and *Halimione portulacoides* helps to maintain an open cover but reduction of grazing or resumed grazing of taller bushes of these species (Brightmore 1979) may allow the development of a closed or more erect canopy which can shade out *Frankenia laevis* from the *Frankenia* sub-community.

Zonation and succession
In general the community occupies a stable position at the uppermost end of the salt-marsh zonation but reduction of grazing may lead eventually to the development of the *Elymo-Suaedetum verae*, especially at lower levels.

Distribution
The community is endemic to Great Britain and is restricted to the north Norfolk coast. *Frankenia laevis, Limonium bellidifolium, L. binervosum* and *Suaeda vera* are all members of the Mediterranean element in the British flora (Matthews 1955) but, though of restricted

occurrence, they are not, apart from *L. bellidifolium*, confined to north Norfolk and their distributions overlap elsewhere. Neither is the distinctive salt-marsh/sand-dune interface habitat restricted to that area. Yet there is no evidence to suggest that this particular species assemblage has ever had a more widespread distribution (Adam 1978). Both *Frankenia laevis* and *Suaeda vera* are conspicuous members of other communities in a similar habitat.

Affinities

Together with the *Limonio vulgaris-Frankenietum laevis*, this community represents vegetation which has floristic affinities with the Puccinellion communities yet which stands alongside the Armerion communities in its high position on the salt-marsh. Géhu & Géhu-Franck (1975) erected a new taxon, the Frankenio-Armerion, for similar vegetation described from France and suggested that this might be regarded as a sub-alliance within the Armerion.

Floristic table SM21

	a	b	21
Suaeda vera	V (1–4)	V (1–7)	V (1–7)
Puccinellia maritima	V (3–7)	V (2–6)	V (2–7)
Armeria maritima	V (2–7)	IV (2–5)	IV (1–7)
Halimione portulacoides	V (1–6)	IV (2–7)	IV (1–7)
Limonium binervosum	V (2–7)	IV (1–5)	IV (1–7)
Suaeda maritima	II (3)	III (2–6)	II (2–6)
Artemisia maritima	III (1–4)		I (1–4)
Festuca rubra	III (5–6)	I (3)	II (3–6)
Plantago maritima	III (2–4)	I (2–3)	II (2–4)
Elymus pycnanthus	II (1–2)		I (1–2)
Limonium cf. *L. vulgare*	II (2–5)		I (2–5)
Sagina maritima	II (2–3)	I (2)	I (2–3)
Spergularia marina	II (2–3)	I (2)	I (2–3)
Glaux maritima	I (3)		I (3)
Spergularia media	I (2)	I (1)	I (1–2)
Frankenia laevis		V (2–5)	III (2–5)
Limonium bellidifolium	II (1–4)	IV (1–6)	III (1–6)
Cochlearia anglica	I (2–3)	II (2)	I (2–3)
Cochlearia danica	I (2–3)	II (3)	I (2–3)
Salicornia agg.	I (3)	II (2–3)	I (2–3)
Cochlearia officinalis		I (1–2)	I (1–2)
Number of samples	11	14	25
Mean number of species/sample	10 (5–13)	8 (5–12)	8 (5–13)
Mean vegetation height (cm)	15 (4–40)	18 (3–40)	17 (3–40)
Mean total cover (%)	63 (20–90)	53 (20–80)	57 (20–90)

a   Typical sub-community
b   *Frankenia laevis* sub-community
21   *Suaeda vera-Limonium binervosum* salt-marsh (total)

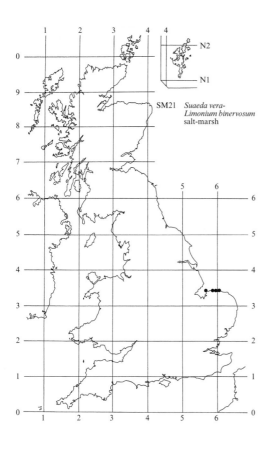

SM21 *Suaeda vera-*
Limonium binervosum
salt-marsh

SM22

Halimione portulacoides-Frankenia laevis salt-marsh community
Limonio vulgaris-Frankenietum laevis   Géhu & Géhu-Franck 1975

Synonymy
Halimiono-Frankenietum laevis   Adam 1976 *p.p*; Sussex *Frankenia laevis* stands   Brightmore 1979.

Constant species
Armeria maritima, Frankenia laevis, Halimione portulacoides.

Rare species
Frankenia laevis, Inula crithmoides, Arthrocnemum perenne.

Physiognomy
This community has a short uneven open sward dominated by *Halimione portulacoides* and *Frankenia laevis* with scattered plants of *Armeria maritima* and *Puccinellia maritima*. *Festuca rubra, Limonium vulgare*, annual *Salicornia* spp. and *Spergularia media* are frequent at low cover values and at some sites *Inula crithmoides* is conspicuous.

Habitat
The community generally occurs on mixtures of silt, sand and shingle at salt-marsh/sand-dune interfaces. Similar vegetation but lacking *H. portulacoides* has been recorded from Chalk undercliffs and rubble (Brightmore 1979).

Distribution
The community is confined to the south coast of Sussex where the best stands are developed at East Head, Chichester Harbour. *Frankenia laevis* has recently been reported from Anglesey (Roberts 1975) where it has become well established (after original planting?) among *Festuca rubra, Armeria maritima* and *Puccinellia maritima* to produce vegetation rather similar to this community.

Affinities
The community is distinguished from the *Suaeda vera-Limonium binervosum* salt-marsh by the absence here of *S. vera* and the replacement of *Limonium bellidifolium* and *L. binervosum* by *L. vulgare*. Géhu & Géhu-Franck (1975) and Géhu & Delzenne (1975) regard the Sussex stands as representing a species-poor parallel to the *Frankenio-Limonietum lychnidifolii* of north-west France and have provisionally assigned them (and the Anglesey vegetation) to the *Limonio vulgaris-Frankenietum laevis*. This association and the *Suaeda vera-Limonium binervosum* community could be placed together in the Frankenio-Armerion.

Floristic table SM22

Frankenia laevis	V (3–8)
Halimione portulacoides	IV (3–9)
Puccinellia maritima	IV (2–5)
Armeria maritima	III (2–4)
Spergularia media	III (1–2)
Festuca rubra	II (3–5)
Salicornia agg.	II (1–3)
Elymus farctus	II (3–4)
Limonium vulgare	II (1–5)
Parapholis strigosa	II (1–4)
Suaeda maritima	I (1–2)
Arthrocnemum perenne	I (1–2)
Plantago maritima	I (1–2)
Plantago coronopus	I (1–5)
Number of samples	18
Mean number of species/sample	8 (6–13)
Mean vegetation height (cm)*	3 (2–5)
Mean total cover (%)	86 (80–100)

*   Data of four samples only.

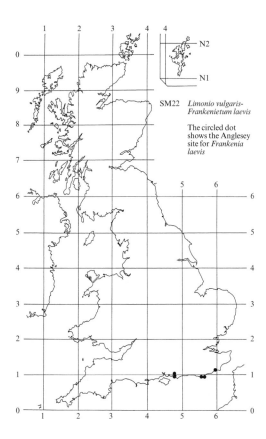

SM22 *Limonio vulgaris-*
Frankenietum laevis

The circled dot
shows the Anglesey
site for *Frankenia*
laevis

SM23

Spergularia marina-Puccinellia distans salt-marsh community
Puccinellietum distantis   Feekes (1934) 1945

Synonymy
Sperguletum marinae   Tyler 1969.

Constant species
Spergularia marina, Puccinellia distans, P. maritima.

Physiognomy
The *Puccinellietum distantis* is a generally open association of scattered but often abundant individuals of *Spergularia marina, Puccinellia distans* and *P. maritima* with rather variable amounts of *Agrostis stolonifera* and sparse records for a variety of salt-marsh species (especially in coastal sites) and ruderal glycophytes. An algal mat is sometimes conspicuous in coastal stands but bryophytes are always rare.

Sub-communities
Sampling of coastal stands of the association has been insufficient to detect the existence of well-defined sub-communities but individual stands bear some resemblance to the *Puccinellietum distantis polygonetosum* R.Tx. 1956 *emend.* Beeftink 1962 (with *Polygonum aviculare*) and the *Puccinellietum distantis pholiuretosum* (with *Parapholis strigosa*) described from The Netherlands (Beeftink 1962, 1965, 1977*a*). Lee (1977) encountered inland stands similar to the 'initial' (with *Atriplex hastata*) and 'degraded' (without *A. hastata* and *Spergularia marina*) sub-associations recognised on Polish coastal marshes (Piotrowski 1974) and to the *Puccinellietum distantis juncetosum* Westhoff 1947 (with *Juncus ambiguus* Guss.) described from the Netherlands (Beeftink 1962). Lee's (1977) *asteretosum* is probably best considered in relation to the *Aster tripolium* communities of salt-marshes and brackish habitats. Further sampling is necessary to establish the validity of these sub-divisions in Britain.

Habitat
The association is characteristic of disturbed situations with soils of variable but generally high salinity. On coastal marshes, it is found in dried-up pans in the upper marsh, in old turf-cuttings, along paths and (particularly in The Wash) in cattle-poached areas. It also occurs on and behind sea walls.

Inland, *Puccinellietum distantis* has been described (Lee 1977) from both natural brine springs and marshes, where it is best developed on the most saline, cattle-poached soils, and from the artificial habitats associated with the salt and alkali industries. Disturbance helps maintain and extend the association, especially through the establishment of the prolifically-seeding annual *Spergularia marina*. Lee found this species best able to tolerate the most saline conditions, though it appeared to suffer from competition with *Puccinellia distans* on soils of lower salinity. Differential response of these two species formed the basis of small-scale mosaics over uneven spoil and soil surfaces.

In these inland sites, the so-called 'initial' sub-association (after Piotrowski 1974) persisted provided soil salinity remained high. The 'degraded' sub-association, dominated by *P. distans*, was characteristic of drier soils and the *juncetosum* of wetter soils, both of lower salinity.

Zonation and succession
Mosaics of the various sub-communities appear to develop in relation to differences in soil salinity levels and the height of the water-table. In coastal sites, the association is usually rather sharply marked off from the surrounding vegetation, often *Juncetum gerardi* or *Puccinellietum maritimae*, though in some cases there may be a more gradual transition to a *Puccinellietum maritimae* with large amounts of *Puccinellia distans* (e.g. the upper reaches of the tidal Nene; Adam & Akeroyd 1978).

The association is maintained by continued disturbance. If this ceases, then a closed sward of the surrounding vegetation is likely to extend into the *Puccinellietum distantis*.

Distribution

Fragmentary stands occur on coastal marshes through-out the country. Inland, the association is confined to areas with percolation of saline waters or accumulation of salt and alkali waste. The most extensive sites are in Cheshire, though some of these have been lost by recla-mation (Lee 1975, 1977).

Affinities

The British stands are typical of the *Puccinellietum dis-tantis* widely described from Europe. The association is most closely related to the *Puccinellietum maritimae* but its distinctive ephemeral nature has led some authorities to place it in a separate alliance, the Puccinellion distan-tis, within the Asteretea.

Floristic table SM23

	a	b
Spergularia marina	V (4–8)	V (2–9)
Puccinellia maritima	V (1–7)	IV (4–5)
Puccinellia distans	II (1–6)	IV (2–8)
Agrostis stolonifera	II (4–6)	IV (2–7)
Salicornia agg.	III (2–7)	
Suaeda maritima	III (2–5)	
Glaux maritima	II (2–5)	I (1)
Algal mat	II (5–8)	
Parapholis strigosa	II (4–8)	
Halimione portulacoides	II (1–3)	
Atriplex prostrata	II (1–5)	I (2–3)
Triglochin maritima	II (2–3)	I (1)
Plantago maritima	II (2–4)	
Elymus pycnanthus	II (1–4)	
Aster tripolium (rayed)	I (3–4)	I (1–9)
Alopecurus geniculatus	I (2–5)	I (2–3)
Juncus bufonius	I (3–4)	I (2–9)
Polygonum aviculare	I (3–4)	
Festuca rubra	I (3–4)	
Spergularia media	I (3–8)	
Number of samples	13	180
Mean number of species/sample	7 (2–11)	6 (3–8)*
Mean vegetation height (cm)	6 (3–15)	no data
Mean total cover (%)	70 (50–90)	71 (30–100)*

*   Means of 25 samples only.

a   Coastal stands

b   Inland stands (Lee 1977)

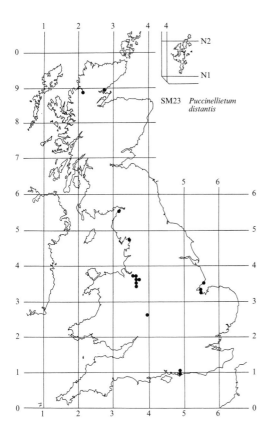

SM23   *Puccinellietum*
        distantis

SM24

Elymus pycnanthus salt-marsh community
Atriplici-Elymetum pycnanthi   Beeftink & Westhoff 1962

Synonymy
Agropyretum pungentis   Perraton 1953; includes *Agropyron pungens-Juncus maritimus* nodum   Adam 1976.

Constant species
Elymus pycnanthus.

Physiognomy
The association is invariably dominated by the stiff clumps of *Elymus pycnanthus* and this may be the sole species. Usually, however, there are a few associates though these are somewhat varied and individual stands may be rendered distinctive by the abundance of (a) particular species. Sometimes there is a patchy or extensive understorey of *Halimione portulacoides*, *Artemisia maritima* and/or *Puccinellia maritima*. In other cases, *Atriplex prostrata* and/or *Festuca rubra* may be conspicuous. *Juncus maritimus* is sometimes abundant though stands with this species are not worthy of distinction as a subcommunity (cf. Adam 1976, 1977). Other stands have a prominent umbelliferous element with *Conium maculatum*, *Foeniculum vulgare* and *Smyrnium olusatrum* and, more locally, *Petroselinum segetum* and *Sison amomum*.

Habitat
The *Atriplici-Elymetum* is an upper-marsh community occurring on a variety of substrates including organically-enriched clay, sand (where *Festuca rubra* is often abundant) and shingle. Substrates are generally well-drained and there is often considerable free calcium carbonate derived from inwashed shell fragments. The pH is generally above 7.0. The association also occurs on older, partly decayed drift litter, where *Atriplex prostrata* flourishes, but in many stands there is little litter except at the seaward edge.

The association may be confined to a narrow strip around the tidal limit or form extensive stands in the upper marsh; occasionally it forms mosaics with other communities. It may extend down the marsh on creek levees and reach above the tidal limit, sometimes covering unmown sea walls where umbellifer-rich stands are characteristic.

Most stands occur on ungrazed or cattle-grazed marshes. Where sheep are admitted to sites with established stands, these are avoided by stock (Cadwalladr & Morley 1973) but the rarity of the association on marshes with a long tradition of sheep-grazing suggests that establishment may not be possible under such a management regime.

Zonation and succession
The association often terminates the zonation at the upper limit of British salt-marshes. A common pattern, seen on many Essex salt-marshes and around the Exe in Devon (Proctor 1980), runs from *Spartinetum townsendii* or *Asteretum tripolii* through *Halimionetum portulacoidis* to the *Atriplici-Elymetum*. The largest stands of the association appear to have developed from the *Halimionetum*.

On creek levees, the association may develop from the *Artemisietum maritimae* or, more locally, the *Spartinetum townsendii*. On high-level drift, there is sometimes a succession from the *Atriplex* strand-lines to the association. In sites inundated by only very exceptional storms, the association may be invaded by shrubs and trees but succession to woodland is likely to be prevented by the occasional subjection to saline waters.

Distribution
The association is most abundant in south-east England and stands on the west coast are local and small. *Elymus pycnanthus* reaches its northern limit in Britain at the Solway.

Affinities
Westhoff & den Held (1969) emphasise the nitrophilous character of the *Atriplici-Elymetum* by assigning it to the Angelicion litoralis in the Artemisietea but the similarities here are weaker than those between the association and other clearly maritime communities of the upper marsh and strand-line. A better solution is to place the association with the *Elymo pycnanthi-Suaedetum verae* in the Elymion pycnanthi of the Elymetea (Géhu & Géhu 1969).

Floristic table SM24

Elymus pycnanthus	V (2–10)
Halimione portulacoides	III (1–8)
Festuca rubra	III (1–9)
Atriplex prostrata	II (1–5)
Glaux maritima	II (1–5)
Puccinellia maritima	I (2–7)
Juncus maritimus	I (2–8)
Agrostis stolonifera	I (3–8)
Artemisia maritima	I (2–7)
Suaeda vera	I (1–7)
Juncus gerardii	I (2–5)
Limonium cf. *L. vulgare*	I (1–5)
Plantago maritima	I (1–5)
Armeria maritima	I (1–4)
Beta maritima	I (1–3)
Atriplex littoralis	I (1–3)
Sonchus arvensis	I (2–3)
Parapholis strigosa	I (2–4)
Potentilla anserina	I (2–3)
Aster tripolium var. *discoideus*	I (1–3)
Aster tripolium (rayed)	I (1–3)
Galium aparine	I (2–3)
Phragmites australis	I (2–6)
Ammophila arenaria	I (1–6)
Hypnum cupressiforme var. *lacunosum*	I (4–6)

Number of samples	110
Mean number of species/sample	6 (1–16)
Mean vegetation height (cm)	61 (30–100)
Mean total cover (%)	98 (75–100)

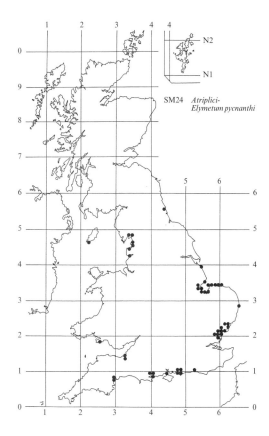

SM24 *Atriplici-Elymetum pycnanthi*

SM25
Suaeda vera drift-line community
Elymo pycnanthi-Suaedetum verae   (Arènes 1933)
Géhu 1975

Synonymy
Suaedetum fruticosae Tansley 1939 *p.p.*; *Agropyro-Suaedetum fruticosae* Adam 1976; *Halimiono-Suaedetum fruticosae* Adam 1976.

Constant species
Halimione portulacoides, Suaeda vera.

Rare species
Arthrocnemum perenne, Suaeda vera.

Physiognomy
The *Elymo pycnanthi-Suaedetum verae* is of variable appearance. The two association constants are sometimes co-dominant as a more or less closed shrubby cover; in other cases they occur as scattered bushes in a grassy ground. The *S. vera* shoots provide a niche for a varied flora of epiphytic lichens (Ellis 1960).

Sub-communities

***Elymus pycnanthus* sub-community:** *Elymo pycnanthi-Suaedetum verae typicum* Géhu & Delzenne 1975; *Agropyro-Suaedetum fruticosae* Adam 1976. Here the cover of *H. portulacoides* is low and the vegetation is dominated by complementary proportions of *S. vera* bushes and stiff clumps of *Elymus pycnanthus* with usually a little *Festuca rubra*. The upper edge of stands may be marked in spring by a narrow band of winter annuals such as *Cochlearia danica*, *Myosotis ramosissima*, *Sagina maritima* and *Valerianella locusta* germinating on drift debris.

***Halimione portulacoides* sub-community:** *Elymo pycnanthi-Suaedetum verae halimionetosum* Géhu & Delzenne 1975; *Halimiono-Suaedetum fruticosae* Adam 1976. *S. vera* and *H. portulacoides* are co-dominant as a shrubby canopy of variable height over a ground of scattered *Puccinellia maritima* and *Limonium* cf. *vulgare*, often with a little *Aster tripolium*, annual *Salicornia* spp., *Spergularia media* and *Suaeda maritima*. The sub-com-

munity provides an occasional high-level context for *Bostrychia scorpioides* and *Pelvetia canaliculata*.

Habitat
The association is most characteristic of drift-lines at salt-marsh/shingle interfaces with a tendency for the *Halimione* sub-community to be associated with superficial smears of sticky yellow-brown clay. The *Elymus* sub-community often runs down-marsh on ridges of drier silt and shorter-growing stands of the *Halimione* sub-community can tolerate up to about 120 submergences/year.

Zonation and succession
The association marks a particular type of transition from the upper marsh to other maritime communities and the driftline stands are probably stable in time.

The low-level stands of the *Halimione* sub-community overlap the habitat of the *Frankenia laevis* sub-community of the *Suaeda vera-Limonium binervosum* salt-marsh and at some sites in north Norfolk there is a mosaic of the two communities. The balance between them could be controlled by rabbit-grazing. Chapman (1960*b*) suggested that *Limonium bellidifolium* declined on Hut Marsh, Scolt Head Island, Norfolk because of the increased vigour of *H. portulacoides* following the reduction there of rabbit-grazing.

Distribution
The association occurs in north Norfolk and Essex.

Affinities
Vegetation similar to the British stands of the *Elymo-Suaedetum* has been described from western France (Corillion 1953, Vanden Berghen 1965*a*, Géhu & Géhu 1969, Géhu 1972, 1975). Géhu (1975) and Géhu & Delzenne (1975) have emphasised the Mediterranean affinities of *Suaeda vera* by assigning the association to the Halimionion in the Arthrocnemetea fruticosae. An alternative view would be to stress the drift-line character of the vegetation and place the association with the *Atriplici-Elymetum pycnanthi* in the Elymion pycnanthi.

Floristic table SM25

	a	b	25
Suaeda vera	V (1–8)	V (3–8)	V (1–8)
Halimione portulacoides	V (2–4)	V (6–9)	V (2–9)
Elymus pycnanthus	V (4–10)		IV (4–10)
Festuca rubra	III (2–5)		II (2–5)
Artemisia maritima	II (1–2)		I (1–2)
Cochlearia anglica	II (1–3)	I (1)	I (1–3)
Cochlearia danica	II (1–3)		I (1–3)
Glaux maritima	II (2–3)		I (2–3)
Plantago maritima	II (1–5)		I (1–5)
Atriplex littoralis	I (2)		I (2)
Puccinellia maritima	I (2)	V (1–7)	III (1–7)
Limonium cf. *L. vulgare*	I (1–4)	V (2–3)	II (1–4)
Suaeda maritima	I (2–3)	III (2–3)	II (2–3)
Aster tripolium	I (2)	III (1–2)	I (1–2)
Bostrychia scorpioides		III (2–5)	I (2–5)
Salicornia agg.		III (3)	I (3)
Spergularia media		III (3–4)	I (3–4)
Arthrocnemum perenne		II (1–2)	I (1–2)
Cochlearia officinalis		II (2–3)	I (2–3)
Pelvetia canaliculata		II (4)	I (4)
Triglochin maritima		II (2–3)	I (2–3)
Algal mat		I (5)	I (5)
Number of samples	13	7	20
Mean number of species/sample	6 (4–8)	8 (5–11)	7 (4–11)
Mean vegetation height (cm)	70 (50–100)	43 (20–73)	61 (20–100)
Mean total cover (%)	95 (70–100)	84 (50–100)	91 (50–100)

a   *Elymus pycnanthus* sub-community
b   *Halimione portulacoides* sub-community
25   *Elymo pycnanthi-Suaedetum verae* salt-marsh (total)

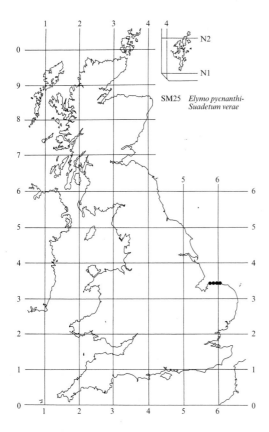

N2

N1

SM25   *Elymo pycnanthi-*
       Suadetum verae

SM26
Inula crithmoides on salt-marshes

Inula crithmoides is a maritime perennial largely confined to southern England and Wales: it is recorded from Essex round to Anglesey with an isolated station in south-west Scotland (Perring & Walters 1962). Although it occurs in maritime cliff communities throughout its range, occurrences in salt-marsh vegetation are restricted to south-east England from Essex to Hampshire.

Here it is an occasional in various associations but it is sometimes encountered in abundance, usually with *Halimione portulacoides* as a co-dominant. In the few available samples there is a distinction between stands where *Puccinellia maritima*, annual *Salicornia* spp. and *Limonium* cf. *vulgare* are constant in generally small amounts and those which have abundant *Elymus pycnanthus*. The former occur on low-marsh sites with coarse sand; the latter on moderately organic soils with much drift litter on the upper marsh.

Ranwell (1972; Ranwell & Boorman 1977) has correlated the distribution of *I. crithmoides* on salt-marshes with the occurrence of lime-rich freshwater influence and the presence of Chalk bedrock near the surface. Though this may be true, it is difficult to see the ecological significance of the observation: most salt-marshes are alkaline to some degree and *I. crithmoides* may be climatically restricted to salt-marshes which are coincidentally particularly base-rich.

Within its limited range on salt-marshes, *I. crithmoides* tends to be more confined to high-marsh occurrences with *Elymus pycnanthus* in Essex (see Rose 1964, Rose & Géhu 1964).

The most obvious affinities of the *I. crithmoides* salt-marsh vegetation are with the *Halimionetum portulacoidis*. Vanden Berghen (1965*a*) has suggested that, along the Biscay coast of France, *I. crithmoides* and *Arthrocnemum perenne* characterise a southern variant of the *Halimionetum*. Stands with abundant *Elymus pycnanthus* will perhaps find a place within the *Atriplici-Elymetum pycnanthi* but could also be seen as the northern limit of the Mediterranean association *Elymo pycnanthi-Inuletum crithmoidis* Br.-Bl. 1952 (Molinier & Tallon 1974).

The rather striking difference in distribution between salt-marsh and maritime cliff communities with *I. crithmoides* points to the possibility of there being distinct ecotypes of the species in Great Britain.

Floristic table SM26

	a	b	26
Inula crithmoides	V (4–7)	IV (5–9)	V (4–9)
Halimione portulacoides	V (5–9)	IV (3–8)	V (3–9)
Puccinellia maritima	V (2–4)		III (2–4)
Salicornia agg.	V (3–5)		III (3–5)
Limonium cf. *L. vulgare*	V (2–5)		III (2–5)
Plantago maritima	III (3)		II (3)
Armeria maritima	III (3–7)		II (3–7)
Suaeda maritima	III (2–4)	I (3)	III (2–4)
Arthrocnemum perenne	II (3–5)		II (3–5)
Algal mat	II (5)		II (5)
Spergularia media	II (2)		II (2)
Spartina anglica	I (2)		I (2)
Festuca rubra	I (2)		I (2)
Aster tripolium (rayed)	I (4)		I (4)
Elymus pycnanthus		IV (5–9)	III (5–9)
Bostrychia scorpioides		II (4)	I (4)
Number of samples	6	4	10
Mean number of species/sample	8 (3–10)	4 (3–5)	6 (3–10)
Mean vegetation height (cm)	26 (8–50)	50 (35–60)	36 (8–60)
Mean total cover (%)	93 (80–100)	100	96 (80–100)

a Stands with *Puccinellia maritima*, *Salicornia* agg. and *Limonium* cf. *L. vulgare*

b Stands with *Elymus pycnanthus*

26 *Inula crithmoides* stands

SM27

Ephemeral salt-marsh vegetation with *Sagina maritima*

Saginion maritimae Westhoff, van Leeuwen & Adriani 1962

Small stands of ephemeral vegetation with an often open cover of annuals and short-lived perennials occur patchily on British salt-marshes. Recurrent assemblages are rare and there seems to be a large element of chance in the floristic composition, early arrivals frequently pre-empting the niche. Such vegetation may include *Sagina maritima*, *S. nodosa* and *Plantago coronopus*, more rarely *Bupleurum tenuissimum* (to the south-east) and *Centaurium littorale* (to the north) and provide a salt-marsh context for ephemerals such as *Cochlearia danica* and *Desmazeria marina* which also occur in other maritime habitats.

Breaks in the turf of mid- and upper-marsh communities provide the most usual habitat for such species and they are especially characteristic of old turf-cuttings where they form part of the sequence of recolonising vegetation giving way to mixtures of *Festuca rubra*, *Agrostis stolonifera*, *Puccinellia maritima* and *Potentilla anserina*, which come to approximate to the *Puccinellietum maritimae*, the *Juncetum gerardi* or the *Festuca-Agrostis-Potentilla* mesotrophic grassland. Such ephemerals also occur in disturbed situations around reclamation banks (e.g. Gray 1977, 1979, Adam & Akeroyd 1978).

This kind of vegetation is the nearest equivalent in Britain to similarly diverse assemblages on Continental salt-marshes which have been assigned to the Saginion alliance in the Saginetea maritimae Westhoff, van Leeuwen & Adriani 1962 (e.g. Beeftink 1962, 1965, 1975, 1977*a*; Tüxen & Westhoff 1963; Westhodd & den Held 1969). There, too, the vegetation is characteristic of upper-marsh situations, being especially associated with salt-marsh/sand-dune transitions where there is a contact between more and less maritime sediments of low soil moisture content but fluctuating salinity.

SM28

Elymus repens salt-marsh community
Elymetum repentis maritimum   Nordhagen 1940

Synonymy
Elymetum repentis maritimum, Elymus repens, Potentilla anserina-Elymus repens-Vicia and *Elymus repens-Potentilla anserina* soziations   ? Nordhagen 1940.

Constant species
Agrostis stolonifera, Atriplex prostrata, Elymus repens, Festuca rubra.

Rare species
Allium scorodoprasum, Hordeum marinum.

Physiognomy
The *Elymetum repentis* has a closed grassy sward up to about 1 m tall, generally dominated by *Elymus repens* with usually smaller amounts of *Festuca rubra* and *Agrostis stolonifera* and, beneath, scattered plants of *Atriplex prostrata* and an open ground cover of *Potentilla anserina. Oenanthe lachenalii, Sonchus arvensis, Rumex crispus* and *Cirsium arvense* are occasional and often give a scruffy appearance to the vegetation and tussocks of *Juncus gerardii* or *Festuca arundinacea* may be locally prominent. The community is generally richer and more varied than the *Atriplici-Elymetum* with a wide range of occasionals of low frequency, some characteristic of other disturbed upper-marsh vegetation of strand-lines and reclamation banks, others more typical of rank inland grasslands. *Allium scorodoprasum* has been recorded in vegetation of this kind on the north Solway coast and *Hordeum marinum* from Somerset. Bryophytes occur occasionally with *Eurhynchium praelongum, Amblystegium riparium, Funaria hygrometrica, Pottia heimii* and *Bryum* spp.

Habitat
The community is characteristic of similar situations to those occupied by the *Atriplici-Elymetum*: upper-marsh areas where there is often a combination of disturbance, drift-litter deposition and some freshwater influence. It is, however, less consistently confined to well-drained sites, occasionally growing on heavy waterlogged clays. At Cefni salt-marsh in Anglesey, it occupies the areas marked as 'drift' on the map of Packham & Liddle (1970). The community also occurs on the recently-excavated material thrown on to the banks of drainage channels while, on some brackish marshes, such as those at the tidal limit in estuaries (as in the Lune in Lancashire), it may form extensive stands.

Zonation and succession
Like the *Atriplici-Elymetum*, this community is often part of the vegetation which terminates the salt-marsh vegetation at its upper limit and in such situations it may occur in clear zonations or confused mosaics with such communities as the *Juncetum gerardi*, the *Juncus maritimus* salt-marsh, the *Potentillo-Festucetum arundinaceae*, the *Festuca rubra-Agrostis stolonifera-Potentilla anserina* grassland and various of the vegetation types in which Cyperaceae or tall swamp helophytes predominate in brackish pools and ditches.

Distribution
The community can be seen as the north-western equivalent of the *Atriplici-Elymetum*, being especially frequent around the Irish Sea coast. It is probably more widespread in eastern Scotland than the map suggests.

Affinities
Although there are clear floristic similarities between this community and *Elymus repens* vegetation of fore-dunes and shingle strand-lines, salt-marsh *Elymetum repentis* is sufficiently distinct to be considered as a separate vegetation type. *Elymus repens* growing on salt-marshes is morphologically distinct and may represent a separate ecotype.

As defined here, the community is synonymous with the vegetation described by Nordhagen (1940) which is frequent in Scandinavia and northern Germany (see also Störmer 1938, Tüxen 1950, Gillner 1960, Tyler 1969b). Authors differ as to whether the community is best placed in a narrowly-defined Elymo-Rumicion crispi (Nordhagen 1940), in that alliance as expanded by Tüxen (1950) or alongside the *Atriplici-Elymetum* in the Elymion pungentis (Géhu & Géhu 1969).

Floristic table SM28

Elymus repens	V (4–10)	*Plantago maritima*	I (2)
Festuca rubra	V (3–8)	*Arrhenatherum elatius*	I (2–7)
Agrostis stolonifera	IV (3–8)	*Stellaria media*	I (2–6)
Atriplex prostrata	IV (2–6)	*Cirsium vulgare*	I (1–2)
Potentilla anserina	III (2–8)	*Scirpus maritimus*	I (4)
Oenanthe lachenalii	II (1–4)	*Puccinellia maritima*	I (3)
Sonchus arvensis	II (2–6)	*Beta vulgaris* ssp. *maritima*	I (2–5)
Rumex crispus	II (1–5)	*Holcus lanatus*	I (2–4)
Festuca arundinacea	II (1–9)	*Taraxacum* sp.	I (2–3)
Cirsium arvense	II (1–4)	*Trifolium repens*	I (2–4)
Juncus gerardii	II (2–6)	*Plantago lanceolata*	I (1–2)
Vicia cracca	I (2–5)	Algal mat	I (4–6)
Matricaria maritima	I (1–4)	*Anthriscus sylvestris*	I (1–4)
Carex otrubae	I (1–6)	*Aster tripolium*	I (2–3)
Cochlearia officinalis	I (2–4)	*Torilis japonica*	I (2)
Glaux maritima	I (2–4)	*Odontites verna*	I (2–3)
Atriplex littoralis	I (2–5)	*Alopecurus geniculatus*	I (2–6)
Galium aparine	I (1–4)	*Eleocharis uniglumis*	I (4)
Aster tripolium (rayed)	I (2–4)	*Rumex conglomeratus*	I (2–3)
Oenanthe crocata	I (1–5)	*Dactylis glomerata*	I (3)
Triglochin maritima	I (2)	*Medicago lupulina*	I (2–3)
Cochlearia anglica	I (2–4)	*Silene vulgaris* ssp. *maritima*	I (1–4)
Eurhynchium praelongum	I (2–5)	*Centaurea nigra*	I (2–3)
Lotus corniculatus	I (3–4)	*Heracleum sphondylium*	I (1–2)
Poa pratensis	I (3–4)	*Apium graveolens*	I (2)
Melilotus altissima	I (2–5)	*Calystegia sepium*	I (4–6)
Deschampsia cespitosa	I (3–5)	Number of samples	62
Leontodon autumnalis	I (1–2)	Number of species/sample	9 (2–27)
Urtica dioica	I (3–5)	Vegetation height (cm)	66 (30–120)
Lolium perenne	I (2–3)	Total cover (%)	99 (70–100)

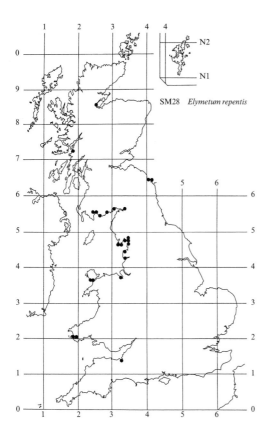

SM28   *Elymetum repentis*

SHINGLE, STRANDLINE AND SAND-DUNE COMMUNITIES

INTRODUCTION TO SHINGLE, STRANDLINE AND SAND-DUNE COMMUNITIES

The sampling and analysis of the vegetation

Sand dunes occur widely around the British coast and, from the beginnings of plant ecology in this country, attracted sufficient attention for Tansley (1939) to be able to outline some of the major vegetation types encountered there and to sketch out a successional line that remains a commonly invoked example of seral development.

Necessarily, perhaps, the understanding of pattern and process in this distinctive landscape was dominated in these early years by studies of a few particular sites like Braunton Burrows on the Somerset coast, the Southport dunes and north Norfolk (e.g. Oliver 1913; Watson 1918; Salisbury 1922, 1925; Pearsall 1934). With the vegetation of coastal shingle which figured from the start in these enquiries, it was again striking sites like Blakeney and Chesil Beach which received attention (Oliver 1911, 1913; Watson 1922; Richards 1929). Further painstaking investigation of some of these and a very few other sites, like Newborough Warren on Anglesey, has yielded an informative understanding of the ecology of important aspects of the dune habitat (e.g. Ranwell 1959, 1960*a*, *b*; Willis *et al*. 1959*a*, *b*; Willis & Yemm 1961) but without any wider descriptive framework of dunes throughout Britain in which to contextualise the picture of vegetation types emerging. More broadly, Scottish sand dunes figured little in our view of these communities until Gimingham's (1964*a*) overview.

The emphasis of some early work was as much on the successional processes in dunes and their relationship to physiography on the broader scale as on the systematic characterisation of vegetation types. Later, it was the physiological implications of nutrient shortage and the vagaries of water supply for dune plants that tended to preoccupy investigators; either that or the challenge of understanding the ecology of such a consummate performer as *Ammophila arenaria* (e.g. Huiskes 1977*a*, *b*, 1979).

In so far as vegetation types were defined on British dune systems, the importance of a few grass species and the prevailing tradition among ecologists here to characterise assemblages by dominance or physiognomy meant that early accounts recognised relatively few broad vegetation types often related to the major stages in succession like 'yellow' and 'grey' dunes. Certain kinds of dune habitat like slacks and especially machair also came to acquire a significance not supported by detailed floristic descriptions.

In our survey for this project, we therefore began virtually from scratch in the assembly of systematic data, reliant initially on just the efforts of the NVC team. As with all other vegetation types, within the limits of time and personnel, we aimed for a representative cover of British dune systems, not concentrating on renowned or especially rich or diverse sites and also including such inland sands as remained in this country. We were especially grateful in Scotland to make use of the data which were eventually published in Birse & Robertson (1976) and Birse (1980) and, for dune-slack vegetation, to have access to a wealth of samples from Wales together with some community descriptions from Dr Peter Jones, then at Cardiff University. As our own work progressed, a broader perspective was opening up on dune vegetation through the surveys which the then NCC initiated, applying early accounts of these communities provided from the NVC to assess the character and distribution of the whole resource. These surveys, summarised in Dargie (1993, 1995) and Radley (1994), have gone on to accumulate many more data and numerous site descriptions and maps. We were not able to draw on all of these but they, and the continuing dune survey in Scotland, greatly fill out our understanding of these vegetation types. They also add further interesting detail to the definition of variation, particularly in north-west and eastern Scotland, though this appears to be at the level of sub-communities. Overall, the definition of the communities as characterised here seems to stand.

Our coverage of the vegetation of shingle features around the British coast was less adequate and, though we were kindly able to see developing surveys of

Dungeness by Dr Brian Ferry and of very many shingle beaches around the coasts of England and Wales by Pippa Sneddon working with Dr Roland Randall (Sneddon & Randall 1992*a*, *b*, 1993*a*, *b*), we did not incorporate their data or community characterisations here. Probably, at least one further community could be added to our account from these surveys.

As with other vegetation types, we used only samples located on the basis of the floristic and structural homogeneity of the vegetation, recording all vascular plants, bryophytes and macrolichens and their cover/abundance using the Domin scale. Sometimes, as with strandline and dune annual vegetation, where stands were linear or small and irregular, some ingenuity was needed in deciding sample size and shape but such problems were never insuperable. Information on the structure of the vegetation – the dominance and vigour of the various important dune grasses, for example, the various contributions of perennial and ephemeral associates, the extent of a bryophyte or lichen carpet – was recorded to fill out our understanding of the character of the communities and notes were made on the wider vegetation context of each sample, remarking on any obvious zonations or mosaics, and any signs of seral changes in train.

The basic environmental information on altitude, slope and aspect was supplemented by observations on the apparent extent of erosion or accretion, the stability of the sand or shingle surface, the maturity of the soil profile and the wetness of the surface. Any signs of grazing by stock or rabbits were noted, impacts of cropping for hay or mowing on golf courses, trampling or disturbance by those using foreshore or dunes for recreation or other activities.

A total of 2304 samples was available for analysis (Figure 10) and, as usual, these were processed using only floristic information to characterise the vegetation types.

In all, 19 communities have been characterised from these data (Figure 11) and these can sensibly be summarised under four main headings: strandline and shingle vegetation (3 communities), foredune and mobile dune communities (6), fixed dune grasslands (4), dune-slack communities (5) and dune scrub (1). Also included below is a note on vegetation types treated elsewhere in this scheme but which can occur with some prominence on sand-dunes.

Strandline and shingle communities

Two assemblages, dominated by ephemeral, nitrophilous herbs, make a brief, often fragmentary, appearance during the growing season on beach-top sands and fine shingle where organic detritus has been dumped along the strandline. Around our warmer southern coasts, it is *Honkenya peploides*, *Cakile maritima*, *Salsola kali* and various *Atriplex* spp. that are most prominent in the

Honkenya-Cakile community (SD2). Towards northern Britain, this tends to be replaced by the *Matricaria maritima-Galium aparine* community (SD3) where *Atriplex* spp., particularly *A. glabriuscula*, remain important but where *M. maritima*, *G. aparine* and *Stellaria media* are additional constants. These vegetation types are pioneer communities, though especially high tides and storm surges usually overwhelm them, setting back any tendency to succession and leaving them to re-establish periodically as conditions again become suitable. They are the main British representatives of the Salsolo-Honkenyion peploidis alliance, though further sampling looks likely to characterise other *Atriplex* assemblages, which would traditionally be placed in the Atriplicion littoralis, another Atlantic coast alliance of the Cakiletea maritimae, the class of nitrophilous strandline vegetation.

It is also sensible to include here the only assemblage characterised in this scheme from coastal shingle. The *Rumex crispus-Glaucium flavum* community (SD1) occurs on coarser sediments than the above – sharply-draining pebbles and gravel beyond the reach of all but exceptional tides but still vulnerable to such occasional inundation and redistribution as sets back any

Figure 10. Distribution of samples available from sand-dunes.

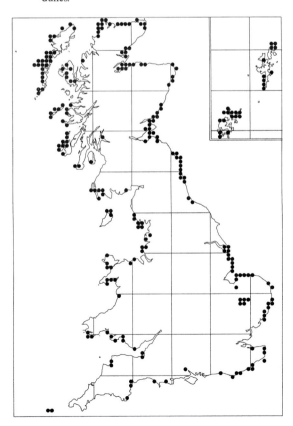

progressive colonisation. *R. crispus, G. flavum, Beta vulgaris* ssp. *maritima* and the rare *Crambe maritima* are the most characteristic species of this vegetation, the thermophilous character of *Glaucium* and *Crambe* making this an essentially southern British community. Around the south-east coast, *Lathyrus japonicus* is an additional striking associate in this vegetation, the major British representative of the Honkenyo-Crambion, an alliance generally placed in a distinct class of enriched shingle and rocky cliff vegetation, the Honkenyo-Elymetea.

Foredune and mobile dune communities

Five vegetation types have been characterised from more mobile coastal sand, and those very few places where such sediments occur inland. Here, it is plants such as *Elymus farctus* ssp. *boreali-atlanticus, Leymus arenarius* and pre-eminently, of course, *Ammophila arenaria* which characterise the early and middle phases of dune building and stabilisation above the limit of frequent tides. *Carex arenaria* can also be included among this group, though its role is mainly in areas of secondary erosion like blow-outs and it is also sensible to consider here

Figure 11. Distribution of vegetation types characterised from sand-dunes.

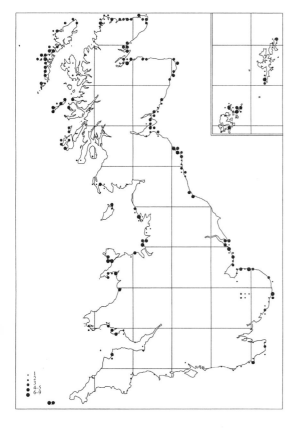

vegetation dominated by ephemerals, including *Phleum arenarium* that occurs where more stable dune swards become locally disturbed and opened up.

The *Elymus farctus* foredune community (SD4) is the vegetation type that generally begins this sequence, developing where *E. farctus* gets a hold through colonisation by seed or rhizome fragments on beach-top accumulations of sand, the young plants rooting and branching below and encouraging accretion around the aerial shoots. Strandline plants like *Honkenya, Cakile* and *Atriplex* spp. persist here as occasionals but periodic immersion by sea-water in exceptional tides helps keep *Ammophila* in check and sets back progressive succession. Traditionally, such vegetation as this has been placed in an Elymo-Honkenyion peploidis alliance as part of the class Ammophiletea.

Four further communities of mobile and semi-fixed dunes are the main British representatives of the larger Ammophiletea alliance, the Ammophilion. Where *Ammophila* itself does get a hold, on embryo dunes as they rise above the limit of inundation, it quickly encourages such substantial accretion that *Elymus* is overwhelmed as the *Ammophila arenaria* community (SD6) develops. This is the most widespread and extensive colonising community on mobile coastal sands all around the British coast, playing an integral role in the development of young dune ridges and prevailing until reduced accretion itself curtails the vigour of the marram. Diversity in this relatively species-poor vegetation reflects the waning influence of tidal inundation at the seaward fringe and the increasing stability of the sand surface as deposition slows. The plants like *Festuca rubra, Poa pratensis* (probably mostly *P. subcaerulea*), *Senecio jacobaea, S. vulgaris, Sonchus arvensis, S. asper, Cirsium vulgare, C. arvense* and *Rumex crispus* become characteristic with *Eryngium maritimum, Calystegia soldanella, Euphorbia paralias* and *E. portlandica* in more southerly stands providing a link with dune vegetation on warmer coasts of Europe.

Locally, *Leymus arenarius*, an Oceanic Northern Grass in its natural distribution but one planted outside its range to encourage accretion, complicates the early stages of this dune succession. Though it does not extend so close to the tidal limit as *Elymus*, it is more tolerant of salt than *Ammophila*, so can interpose itself among the embryo and mobile dunes as the *Leymus arenarius* community (SD5), succumbing in vigour as accretion progresses.

As accretion slows but where edaphic changes beneath the stabilising sand surface are not yet appreciable, the *Ammophila-Festuca* community (SD7) is characteristic. Here, there is a waning in the vigour of the marram and an increased opportunity for a more extensive and varied cover of associates. *Festuca rubra* and *Poa pratensis* become more consistently abundant with *Hypochoeris radicata, Taraxacum officinale* agg., *Leontodon taraxa-coides* and *Lotus corniculatus* among the more frequent companions. Especially distinctive in this vegetation around our warmer southern coasts is *Ononis repens* but many stands also show an increasing contribution from bryophytes on a sand surface that is now upbuilding only very slowly if at all: among these *Hypnum cupressiforme*, *Brachythecium albicans* and *Tortula ruralis* ssp. *ruraliformis* are particularly striking.

Where disturbance among such dunes is more substantial provoking renewed erosion in stabilising dunes, the *Carex arenaria* community (SD10) may become prominent, the far-creeping rhizomes of the sedge extending into the freshly-disturbed sand, and the shoots thickening up as the surface stabilises. *F. rubra* is a common associate on coastal dunes, with *Ammophila* eventually reasserting its hold, but this kind of vegetation is also found inland, in those very few places now, like Breckland and Lincolnshire, where mobile sand is still an element of the landscape, and where *F. ovina* and *Koeleria macrantha* are characteristic associates. In a very few localities, this community can provide a locus for the nationally rare *Corynephorus canescens*, a grass which elsewhere in more Continental parts of Europe represents an important early coloniser of mobile acid sands.

Smaller ephemerals can make a sporadic appearance among the grasses and herbs of the semi-fixed *Ammophila-Festuca* vegetation but, where drought or disturbance by rabbits creates bigger gaps in which the sand yet remains essentially stable, the *Phleum arenarium-Arenaria serpyllifolia* community (SD19 *Tortulo-Phleetum arenariae* (Massart 1908) Br.-Bl. & de Leeuw 1938) can develop. Here *P. arenarium* and a variety of annuals such as *A. serpyllifolia*, *Cerastium diffusum* ssp. *diffusum*, *Aira praecox* and *Viola tricolor* ssp. *curtisii* occur more consistently with *Tortula ruralis* ssp. *ruraliformis* as the most distinctive bryophyte. Of brief prominence, the diminutive preferentials of this community are often shrivelled by early summer, dispersing to take advantage of any new openings in the cover. This distinctive vegetation is the main British representative of the Koelerion arenariae alliance, ephemeral assemblages of bare but stable calcareous sands.

Fixed dune communities

Where calcareous sands have become more or less completely stabilised on coastal dune systems and sand plains all around the British coast, the *Festuca rubra-Galium verum* community (SD8) is characteristic. *Ammophila* can be quite frequent here but it is typically of sparse cover, its shoots moribund among a grassy sward dominated by *F. rubra* and *Poa pratensis* with *G. verum*, *Plantago lanceolata*, *Lotus corniculatus*, *Trifolium repens*, *Achillea millefolium*, *Thalictrum minus*, *Cerastium fontanum*, *Bellis perennis* and *Ranunculus acris*. The vegetation here takes much of its character from the somewhat less droughty and impoverished conditions that come with long stability of the sand surface but climatic contrasts across the country and grazing by stock also play an important part in determining the nature of the soil and the composition of the sward. Particularly distinctive is the part played by this community in the machair landscapes of north-west Scotland and the Outer Isles where it makes a major contribution to the township pastures. Closed swards of this type have been located by Dutch phytosociologists in a Plantagini-Festucion alliance in the Koelerio-Corynephoretea, a class of pioneer communities and grasslands on dry infertile soils.

Wherever it occurs, grazing by stock or rabbits is essential to maintain the *Festuca-Galium* community. If such predation does not occur, stabilised dunes in Britain often carry the *Ammophila arenaria-Arrhenatherum elatius* community (SD9). Here, *F. rubra*, *Ammophila* and *Poa pratensis* all remain constant but *A. elatius* is a distinctive and often very abundant constituent of swards that are generally tussocky and rank. Frequent associates here include *Heracleum sphondylium*, *Dactylis glomerata*, *Veronica chamaedrys*, *Achillea millefolium* and *Plantago lanceolata* and, though this grassland could be placed in the Ammophilion, these species emphasise the floristic links with the Arrhenatherion. However, a striking enrichment from thermophilous plants like *Geranium sanguineum* and *Rosa pimpinellifolia* on warmer south-facing dune faces can give this vegetation the look of a Geranion sanguinei assemblage.

Where grazing of stabilised dunes has been neglected for even relatively short periods, shrubs and trees can begin to seed in from nearby broadleaf woodland or coniferous plantations and various scrub and woodland communities, described elsewhere in the NVC, can develop. Included here, however, is a distinctive vegetation type dominated by sea buckthorn, a plant native with us but sometimes planted and often now naturalised on coastal dunes. The *Hippophae rhamnoides* scrub (SD18) is a local community able to develop on sand that is still somewhat mobile, the buckthorn invading initially by seed but then spreading vigorously by suckering from its horizontal root mat and quickly forming a murderously spiny canopy. Early stages can see the survival of species such as *Ammophila*, *F. rubra*, *Senecio jacobaea* and *Cirsium arvense* but more established stands often have a sparser cover of *Urtica dioica*, *Arrhenatherum elatius*, *Galium aparine* and *Solanum dulcamara* that reflects the patchy shade but also the more eutrophic conditions at least partly originating from nitrogen-fixation by the buckthorn. The occasional presence of *Sambucus nigra* among these canopies is a reminder that elsewhere on coasts of north-west Europe, this scrub is typically overwhelmed by elder and then by *Acer pseudo-platanus*. It has been placed in phytosociological schemes with the scrubs of the Salicion repentis areanariae in the Rhamno-Prunetea.

Where stabilised sands on dunes and sand-plains open to grazing become somewhat acid superficially, through long-continued leaching, the *Festuca-Galium* grassland can have a modest contingent of calcifuges but it is in the *Carex arenaria-Festuca ovina-Agrostis capillaris* community that this distinctive aspect of dune grasslands are best seen. Here, *F. rubra* is frequently accompanied by *F. ovina*, *A. capillaris*, *Anthoxanthum odoratum*, *Galium saxatile*, *Luzula campestris*, *Dicranum scoparium*, *Hylocomium splendens* and *Pleurozium schreberi* in generally closed swards where *Ammophila*, though common, is sparse and debilitated. Strongly leached or initially acid sands provide the characteristic substrate here but this vegetation is also strongly dependent on grazing, often by stock as well as or rather than rabbits. It represents the nearest approach among dune communities to the Nardo-Galion alliance but, on balance, it seems best to place it in the Corynephorion canescentis, colonising vegetation and grasslands of acid sands.

Long-continued predation by rabbits is probably a key factor in the development on fixed acid sands of another Corynephorion assemblage, the *Carex arenaria-Cornicularia aculeata* community (SD11). Here, it is an extensive and diverse carpet of lichens – species like *Cladonia arbuscula*, *C. foliacea*, *C. impexa*, *C. pyxidata*, *C. uncialis*, *C. gracilis* and *C. furcata* as well as *Cornicularia aculeata* – that give the swards their distinctive stamp and lend older dunes where they are prominent the epithet 'grey'. Close cropping of the herbage and impoverishment of the system are the critical determinative features here in maintaining this striking vegetation that is best placed in the Corynephorion alliance and which locally provides a context for *Corynephorus canescens*.

Dune-slack communities

Though dependent on the physiography of dune systems for their origin and maintainance, the vegetation types found in dune slacks (or dune valleys as they are known elsewhere in Europe) are floristically of quite a different character. Some of their distinctive species reflect the sandy nature of the substrate and provide a measure of continuity with the swards of more fixed dune sands and plains, but their overall floristic affinities are with various kinds of mires and other vegetation types of periodically-flooded habitats inland.

Five communities of dune-slack vegetation have been characterised and their differences in composition and physiognomy can be related to variations in the frequency and extent of ground-water fluctuations, the time since colonisation of the bare sand began and the intensity of grazing.

The most open and immature dune-slack vegetation distinguished from the data is seen in the *Sagina nodosa-Bryum pseudotriquetrum* community (SD13), a rare assemblage of young and perpetually rejuvenated slacks which are inundated to shallow depth in winter but rather dry in summer. The periodic wetting provides ideal conditions for a variety of ephemeral plants like *S. nodosa*, *Centaurium erythraea*, *Poa annua* and *Blackstonia perfoliata* among a patchy cover of perennials such as *Salix repens*, *Juncus articulatus*, *Agrostis stolonifera*, *Leontodon hispidus* and *Hydrocotyle vulgaris*, along with a contingent of bryophytes which benefit from the lack of shade – *B. pseudotriquetrum*, *Aneura pinguis*, *Moerckia hibernica*, *Pellia endiviifolia* and *Petalophyllum ralfsii*. Although older stands show transitions to drier slack vegetation, the affinities of this community are clearly with the alliance Nanocyperion and, more particularly with the association *Centaurio-Saginetum moniliformis* Diemont, Sissingh & Westhoff 1940.

Young and moderately old slacks flooded to some depth in winter and kept moist in summer by base-rich ground waters share frequent records with the above for *S. repens*, *J. articulatus*, *H. vulgaris* and *A. stolonifera*. Here, though, there are also more calcicolous plants like *Equisetum variegatum*, *Epipactis palustris* and *Carex flacca* and a striking carpet of bryophytes in which *Campylium stellatum* and *Calliergon cuspidatum* are the commonest members with more light-demanding thallose liverworts prominent in younger stands. This *Salix repens-Campylium stellatum* community (SD14) can be seen as the most obvious coastal representative of the Caricion davallianae Klika 1934, the alliance of small-sedge rich fens of calcareous flushes. It is a scarce community, often species-rich and providing a locus for various national rarities including *Liparis loeselii*.

Older wet slacks, even where the ground waters are not so base-rich, have the *Salix repens-Calliergon cuspidatum* community (SD15), a vegetation type which has much in common with the *Salix-Campylium* community but where the deepening shade beneath the often dense willow canopy favours the more tolerant *C. cuspidatum* and where the frequency in some sub-communities of plants like *Carex nigra*, *Galium palustre* and *Lotus uliginosus* show transitions to poor-fen vegetation.

A continuation of this trend can be seen in the *Potentilla anserina-Carex nigra* dune-slack community (SD17) but, as well as being characteristic of situations kept moist by more base-poor ground waters in southern dune systems, this vegetation type is especially frequent in the wetter climate of northern Britain on older sand plains where heavier rainfall enhances leaching but also keeps the swards free of droughting in summer. Often, too, this kind of vegetation is grazed by stock. *S. repens*, then, is less frequent and abundant and it is plants like *Carex nigra*, *Agrostis stolonifera*, *Potentilla anserina*,

Cardamine pratensis and *Holcus lanatus* which give the community its distinctive character. Grading floristically into damp grasslands of fixed dune sands, this vegetation is probably best accommodated in the Elymo-Rumicion crispi, an alliance of diverse vegetation types from inundated habitats.

Drier slack vegetation, where *S. repens* is often dominant but with frequent *Festuca rubra*, *Holcus lanatus*, *Poa pratensis*, *Lotus corniculatus*, *Carex flacca*, and occasional *Euphrasia officinalis* agg. and *Hieracium pilosella* is all included in a *Salix repens-Holcus lanatus* community (SD16). This is a widespread kind of vegetation in ageing, often now ungrazed, slacks, a product of succession where dune systems are drying out and quite often having young *Betula pubescens* and *Salix* spp. which presage the development of woodland. It can be placed in the Salicion repentis arenariae.

Other vegetation types on sand-dunes

A variety of other vegetation types described elsewhere in *British Plant Communities* can be seen on sand-dune systems around our coasts. Most closely related to those included here is the *Atriplex-Beta* community (MC6), dealt with in this volume among the vegetation of sea-cliffs because of its frequent association with sea-bird colonies but very similar assemblages to which are also found on strandline detritus around the southern coasts of Britain.

Among more stable dunes and on sand plains, where the long-felt influence of low-input agricultural use tends to be more important than the physiographic processes of erosion and accretion, there are often transitions from *Festuca-Galium* swards to various kinds of mesotrophic grasslands as a result of pasturing of stock or cropping for hay. On drier dunes, the *Centaurea-Cynosurus* grassland (MG5 *Centaureo-Cynosuretum* Br.-Bl. & R.Tx. 1952) can figure prominently in drier areas treated as meadows or mown as golf-course rough. Very locally, now, in some areas of machair on the Outer Isles, such traditional meadows occur among low-input

arable land, cropped in rotation for potatoes or oats, and carrying the *Chrysanthemum segetum-Spergula arvensis* community (OV4, *Spergulo-Chrysanthemetum* (Br.-Bl. & de Leeuw 1936) R.Tx. 1937) in the tumble down to fallow.

Much more widely, the impact of human activity is now seen in the occurrence of vegetation types related to gross disturbance or neglect – weedy assemblages like the *Epilobium angustifolium* community (OV27) and a range of sub-scrubs, scrubs and woodlands.

Where dunes are kept moist by ground water other mesotrophic grasslands can figure with the fixed grasslands of stabilised sands – like the *Festuca-Agrostis-Potentilla* inundation grassland (MG11) for example, and, more locally, the *Holcus-Juncus* rush pasture (MG10 *Holco-Juncetum* Page 1980) and the *Caltha-Cynosurus* grassland (MG8). Particularly in northern Britain, where heavier rainfall helps replenish the water supply on extensive dune plains with areas of open water, mosaics of such communities are very striking.

It is there, too, that aquatic and swamp vegetation is found in close proximity to dune slacks and grasslands with the *Eleocharis palustris* swamp (S19 *Eleocharitetum palustris* Schennikow 1919) and *Phragmites australis* swamp (S4 *Phragmitetum australis*) being particularly frequent. Peat accumulation where waters are stagnant sees the occurrence of *Scirpus-Erica* wet heath (M15) and *Erica-Sphagnum* wet heath (M16 *Ericetum tetralicis*), these sometimes grading to or forming complex mosaics with dune-slack vegetation. Then there is the sort of situation where highly acid sands now sustain the *Calluna-Carex arenaria* heath (H11), a community very similar in floristics to the *Carex-Cornicularia* and *Carex-Festuca-Agrostis* swards but with *Calluna vulgaris* and sometimes also either *Erica cinerea* or *Empetrum nigrum* ssp. *nigrum* dominant. There is little doubt that grazing, by stock and often also, or instead, rabbits mediates much of the variation among these vegetation types on ancient dune surfaces leached of surface lime or in those few localities where acid sands prevail.

KEY TO SHINGLE, STRANDLINE AND SAND-DUNE COMMUNITIES

With something as complex and variable as vegetation, no key can pretend to offer an infallible short cut to diagnosis. The following should thus be seen as simply a crude guide to identifying the types of vegetation found on shingle, strandline and sand-dunes and must always be used in conjunction with the data tables and community descriptions. It relies on floristic (and, to a lesser extent, physiognomic) features of the vegetation and demands a knowledge of the British vascular flora and some bryophytes and lichens. It does not make primary use of any habitat features, though these may provide a valuable confirmation of a diagnosis.

Because the major distinctions between the vegetation types in the classification are based on inter-stand frequency, the key works best when sufficient samples of similar composition are available to construct a constancy table. It is the frequency values in this (and, in some cases, the ranges of abundance) which are then subject to interrogation with the key.

Samples should always be taken from homogeneous stands and be 2×2 m or 4×4 m according to the scale of the vegetation or, where stands are irregular, of identical size but different shape.

---

1   Generally species-poor and usually open, but sometimes locally luxuriant vegetation of strandlines or shingle with some of *Honkenya peploides*, *Cakile maritima*, *Crambe maritima*, *Glaucium flavum*, *Atriplex* spp., *Matricaria maritima* and *Beta vulgaris* ssp. *maritima* frequent   2

Generally species-poor and often open vegetation of more mobile sand on foredunes and dunes with one or more of *Elymus farctus*, *Ammophila arenaria*, *Leymus arenarius* and *Carex arenaria* constant and dominant   5

More species-rich and generally closed vegetation of less mobile and fixed sand on dunes and sand plains with *Ammophila* sometimes frequent but typically moribund, *Festuca rubra* common and at least some of *Galium verum*, *Plantago lanceolata*, *Trifolium repens*, *Lotus corniculatus*, *Poa pratensis*, *Festuca ovina* and *Agrostis capillaris* frequent   15

Open or closed grassy or bushy vegetation in dune slacks, often with *Salix repens*, and at least some of *Hydrocotyle vulgaris*, *Agrostis stolonifera*, *Mentha aquatica*, *Potentilla anserina*, *Epipactis palustris*, *Carex flacca*, *C. serotina*, *C. nigra*, *C. panicea*, *Juncus articulatus*, *Campylium stellatum*, *Calliergon cuspidatum*, *Bryum pseudotriquetrum* and *Aneura pinguis*   26

Scrub with an open or closed canopy of *Hippophae rhamnoides*   43

Open and sometimes fragmentary vegetation among semi-fixed and fixed dunes with constant *Phleum arenarium* and *Arenaria serpyllifolia* and frequent records for some of *Cerastium diffusum* ssp. *diffusum*, *Aira praecox*, *Viola tricolor* ssp. *curtisii*, *Sedum acre* and *Tortula ruralis* ssp. *ruraliformis*

> **SD19** *Phleum arenarium-Arenaria serpyllifolia* dune annual community
> *Tortulo-Phleetum arenariae* (Massart 1908) Br.-Bl. & de Leeuw 1936

2   *Rumex crispus* and *Glaucium flavum* constant with *Beta vulgaris* ssp. *maritima* and *Crambe maritima* frequent

> **SD1** *Rumex crispus-Glaucium flavum* shingle community   3

Rumex crispus can be frequent but other listed species scarce   4

3   *Lathyrus japonicus* constant and *Arrhenatherum elatius* frequent

SD1 *Rumex crispus-Glaucium flavum* shingle community
Lathyrus japonicus sub-community

L. japonicus and *A. elatius* scarce

SD1 *Rumex crispus-Glaucium flavum* shingle community
Typical sub-community

4   *Honkenya peploides* and *Cakile maritima* constant, often with *Atriplex* spp. and *Elymus farctus*, in ephemeral and often patchy strandline vegetation

SD2 *Honkenya peploides-Cakile maritima* strandline community

Atriplex spp. can be prominent and *C. maritima* and *H. peploides* occasional in open, patchy strandline vegetation but *Matricaria maritima* and *Galium aparine* are constant

SD3 *Matricaria maritima-Galium aparine* strandline community

5   *Elymus farctus* constant and dominant with *Leymus arenarius, Ammophila arenaria* and *Carex arenaria* at most occasional and subordinate in foredune vegetation

SD4 *Elymus farctus* ssp. *boreali-atlanticus* foredune community

E. farctus can be frequent and abundant but then *L. arenarius, A. arenaria* or *C. arenaria* also constant and dominant   6

6   *Leymus arenarius* constant and dominant with *Ammophila arenaria* sparse and only locally of moderate abundance

SD5 *Leymus arenarius* mobile dune community 7

A. arenaria or *Carex arenaria* constant and abundant   8

7   *Elymus farctus* constant in small quantities

SD5 *Leymus arenarius* mobile dune community
Elymus farctus sub-community

Festuca rubra frequent and locally abundant among the *Leymus arenarius*

SD5 *Leymus arenarius* mobile dune community
Festuca rubra sub-community

E. farctus and *F. rubra* occasional at most

SD5 *Leymus arenarius* mobile dune community
Species-poor sub-community

8   *Carex arenaria* the most abundant plant in open or closed swards on mobile sand

SD10 *Carex arenaria* dune community   9

C. arenaria can be frequent but the main constant and dominant is *Ammophila arenaria*

SD6 *Ammophila arenaria* mobile dune community   10

9   *Festuca rubra* frequent with no *F. ovina*

SD10 *Carex arenaria* dune community
Festuca rubra sub-community

F. ovina constant with no *F. rubra* but occasional *Senecio jacobaea* and *Holcus lanatus*

SD10 *Carex arenaria* dune community
Festuca ovina sub-community

10   *Festuca rubra* constant and sometimes abundant with occasional *Senecio jacobaea* and *Cirsium arvense*   11

F. rubra very scarce   12

11   *Poa pratensis* constant with occasional *Heracleum sphondylium* and *Sonchus arvensis*

SD6 *Ammophila arenaria* mobile dune community
Poa pratensis sub-community

P. pratensis and these other species very scarce at most

SD6 *Ammophila arenaria* mobile dune community
Festuca rubra sub-community

12   *Elymus farctus* constant with *Honkenya peploides* occasional   13

Above species very scarce at most   14

13   *Leymus arenarius* constant

SD6 *Ammophila arenaria* mobile dune community
Elymus farctus-Leymus arenarius sub-community

L. arenarius very scarce

SD6 *Ammophila arenaria* mobile dune community
Elymus farctus sub-community

14   *L. arenarius* constant

    SD6 *Ammophila arenaria* mobile dune community
    Leymus arenarius sub-community

Carex arenaria constant

    SD6 *Ammophila arenaria* mobile dune community
    Carex arenaria sub-community

L. arenarius and *C. arenaria* both very scarce

    SD6 *Ammophila arenaria* mobile dune community
    Ammophila arenaria sub-community

15   *Ammophila arenaria* generally dominant and vigorous with *Festuca rubra* constant, though usually subordinate in cover; *Poa pratensis* and *Hypochoeris radicata* frequent but *Lotus corniculatus*, *Trifolium repens* and *Galium verum* not common throughout.

    SD7 *Ammophila arenaria-Festuca rubra* semi-fixed dune community   16

A. arenaria constant and vigorous but accompanied by and often co-dominant with *Arrhenatherum elatius* as well as frequent *F. rubra* and *P. pratensis*

    SD9 *Ammophila arenaria-Arrhenatherum elatius* dune grassland   19

A. arenaria very frequent but usually moribund and exceeded in cover by *F. rubra*; *P. pratensis* also constant and generally accompanied by *Galium verum*, *Lotus corniculatus* and *Trifolium repens*

    SD8 *Festuca rubra-Galium verum* fixed dune grassland   20

A. arenaria can be frequent but lacking vigour and subordinate to *Carex arenaria* in swards with a grassy or lichen-rich cover   23

16   *Elymus pycnanthus* constant and locally abundant with *Taraxacum officinale* agg., *Senecio jacobaea* and *Carex arenaria* at most occasional

    SD7 *Ammophila arenaria-Festuca rubra* semi-fixed dune grassland
    Elymus pycnanthus sub-community

E. pycnanthus very scarce but other listed species can be frequent   17

17   *Ononis repens* constant

    SD7 *Ammophila arenaria-Festuca rubra* semi-fixed dune grassland
    Ononis repens sub-community

O. repens very scarce   18

18   *Hieracium pilosella*, *Galium verum* and *Lotus corniculatus* frequent with occasional annuals such as *Myosotis ramosissima*, *Aira praecox*, *Valerianella locusta* and, locally, *Acaena novae-zelandiae* and sometimes extensive moss carpet with *Hypnum cupressiforme* especially frequent

    SD7 *Ammophila arenaria-Festuca rubra* semi-fixed dune grassland
    Hypnum cupressiforme sub-community

Above species at most occasional in usually rather species-poor swards

    SD7 *Ammophila arenaria-Festuca rubra* semi-fixed dune grassland
    Typical sub-community

19   *Geranium sanguineum* constant with frequent *Plantago lanceolata* and *Lotus corniculatus*

    SD9 *Ammophila arenaria-Arrhenatherum elatius* dune grassland
    Geranium sanguineum sub-community

Above species occasional at most but *Cirsium arvense* common

    SD9 *Ammophila arenaria-Arrhenatherum elatius* dune grassland
    Typical sub-community

20   *Euphrasia officinalis* agg. and *Holcus lanatus* frequent, *Trifolium pratense*, *Carex flacca* and *Calliergon cuspidatum* occasional   21

These species generally scarce   22

21   *Bellis perennis* and *Ranunculus acris* constant

    SD8 *Festuca rubra-Galium verum* fixed dune grassland
    Bellis perennis-Ranunculus acris sub-community

Above-named species occasional at most but *Prunella vulgaris*, *Campanula rotundifolia* and *Linum catharticum* constant

    SD8 *Festuca rubra-Galium verum* fixed dune grassland
    Prunella vulgaris sub-community

22   *Cerastium fontanum*, *Luzula campestris*, *Hieracium pilosella*, *Veronica chamaedrys* and *Agrostis capillaris* frequent

    SD8 *Festuca rubra-Galium verum* fixed dune grassland
    Luzula campestris sub-community

Above species occasional at most but *Tortula ruralis* ssp. *ruraliformis* and *Homalothecium lutescens* constant with occasional *Sedum acre*

> **SD8** *Festuca rubra-Galium verum* fixed dune grassland
> *Tortula ruralis* ssp. *ruraliformis* sub-community

Not as above

> **SD8** *Festuca rubra-Galium verum* fixed dune grassland
> Typical sub-community

23   Lichens forming an extensive and diverse carpet among the *C. arenaria* sward with frequent records for *Cornicularia aculeata*, *Cladonia arbuscula*, *C. foliacea* and *C. impexa*

> **SD11** *Carex arenaria-Cornicularia aculeata* dune community   24

Lichens at most occasional and patchy among a more grassy *C. arenaria* sward

> **SD12** *Carex arenaria-Festuca ovina-Agrostis capillaris* dune grassland   25

24   *Ammophila arenaria* constant in small amounts with frequent *Cladonia furcata* and occasional *Festuca rubra*, *Aira praecox*, *Sedum acre*, *Hypochoeris radicata* and (locally) *Corynephorus canescens*

> **SD11** *Carex arenaria-Cornicularia aculeata* dune community
> *Ammophila arenaria* sub-community

Festuca ovina constant with frequent *C. pyxidata*, *C. uncialis* and *C. gracilis* and occasional *Calluna vulgaris* at low cover

> **SD11** *Carex arenaria-Cornicularia aculeata* dune community
> *Festuca ovina* sub-community

25   *Anthoxanthum odoratum* and *Luzula campestris* constant with occasional to frequent *Hypochoeris radicata*, *Koeleria macrantha*, *Hieracium pilosella* and *Thymus praecox*

> **SD12** *Carex arenaria-Festuca ovina-Agrostis capillaris* dune grassland
> *Anthoxanthum odoratum* sub-community

Above species occasional at most but *Holcus lanatus*, *Campanula rotundifolia* and *Viola riviniana* frequent

> **SD12** *Carex arenaria-Festuca ovina-Agrostis capillaris* dune grassland
> *Holcus lanatus* sub-community

26   *Salix repens* cover very short and often patchy or sparse with constant *Sagina nodosa*, *Juncus articulatus*, *Leontodon hispidus*, *Bryum pseudotriquetrum* and *Arneura pinguis*

> **SD13** *Sagina nodosa-Bryum pseudotriquetrum* dune-slack community   27

Salix repens can be frequent and often abundant but, if so, then not with the other listed associates   28

27   *Poa annua*, *Hydrocotyle vulgaris*, *Blackstonia perfoliata* and *Moerckia hibernica* frequent, *Campylium stellatum* and *Petalophyllum ralfsii* occasional

> **SD13** *Sagina nodosa-Bryum pseudotriquetrum* dune-slack community
> *Poa annua-Moerckia hibernica* sub-community

Agrostis stolonifera, *Holcus lanatus*, *Poa pratensis*, *Festuca rubra*, *Lotus corniculatus* and *Anthyllis vulneraria* constant

> **SD13** *Sagina nodosa-Bryum pseudotriquetrum* dune-slack community
> *Holcus lanatus-Festuca rubra* sub-community

28   *Salix repens* constantly associated with *Campylium stellatum*, *Equisetum variegatum*, *Carex flacca* and usually *Juncus articulatus* and *Epipactis palustris*

> **SD14** *Salix repens-Campylium stellatum* dune-slack community   29

Salix repens frequent but not with above associates constant   32

29   *Lotus corniculatus*, *Poa pratensis*, *Holcus lanatus* and *Prunella vulgaris* all very frequent with *Carex arenaria* less common   30

Carex arenaria constant with above listed associates more occasional   31

30   *Leontodon hispidus*, *Pellia endiviifolia*, *Aneura pinguis* and *Bryum pseudotriquetrum* constant

> **SD14** *Salix repens-Campylium stellatum* dune-slack community
> *Bryum pseudotriquetrum-Aneura pinguis* sub-community

Festuca rubra, *Pulicaria dysenterica*, *Trifolium pratense* and *T. repens* all very frequent with above associates not consistently common

> **SD14** *Salix repens-Campylium stellatum* dune-slack community
> *Festuca rubra* sub-community

31 *Carex serotina* and *Drepanocladus sendtneri* frequent, *Epipactis palustris* occasional

> **SD14** *Salix repens-Campylium stellatum* dune-slack community
> *Carex serotina-Drepanocladus sendtneri* sub-community

E. palustris constant with *Leontodon autumnalis, Ranunculus flammula, Rubus caesius* and *Galium palustre* frequent

> **SD14** *Salix repens-Campylium stellatum* dune-slack community
> *Rubus caesius-Galium palustre* sub-community

32 *Salix repens* only occasional but *Potentilla anserina, Carex nigra* and *Agrostis stolonifera* constant associates

> **SD17** *Potentilla anserina-Carex nigra* dune-slack community   33

P. anserina and *C. nigra* can occur but *Salix repens* is a constant and often abundant plant   36

33 *Festuca rubra* and *Ranunculus repens* very frequent, *Bellis perennis* occasional   34

Above species not consistently present   35

34 *Trifolium repens, Carex flacca, Poa pratensis* and *Prunella vulgaris* all frequent

> **SD17** *Potentilla anserina-Carex nigra* dune-slack community
> *Carex flacca* sub-community

Above species not consistently present

> **SD17** *Potentilla anserina-Carex nigra* dune-slack community
> *Festuca rubra-Ranunculus repens* sub-community

35 *Hydrocotyle vulgaris, Ranunculus flammula, Eleocharis palustris* and *Galium palustre* frequent

> **SD17** *Potentilla anserina-Carex nigra* dune-slack community
> *Hydrocotyle vulgaris-Ranunculus flammula* sub-community

Caltha palustris, Cynosurus cristatus, Rhinanthus minor and *Lychnis flos-cuculi* frequent

> **SD17** *Potentilla anserina-Carex nigra* dune-slack community
> *Caltha palustris* sub-community

36 *Salix repens* constantly associated with *Hydrocotyle vulgaris, Mentha aquatica* and *Calliergon cuspidatum, Lotus corniculatus* scarce

> **SD15** *Salix repens-Calliergon cuspidatum* dune-slack community   37

S. repens constantly associated with *Lotus corniculatus, Carex flacca, Holcus lanatus* and *Festuca rubra* in drier vegetation

> **SD16** *Salix repens-Holcus lanatus* dune-slack community   40

37 *Agrostis stolonifera, Equisetum variegatum* and *Carex arenaria* frequent   38

Above species occasional at most   39

38 *Rubus caesius, Galium palustre* and *Carex nigra* frequent

> **SD15** *Salix repens-Calliergon cuspidatum* dune-slack community
> *Equisetum variegatum* sub-community

Carex flacca, Epipactis palustris and *Pulicaria dysenterica* frequent

> **SD15** *Salix repens-Calliergon cuspidatum* dune-slack community
> *Carex flacca-Pulicaria dysenterica* sub-community

39 *Rubus caesius, Galium palustre* and *Carex nigra* frequent

> **SD15** *Salix repens-Calliergon cuspidatum* dune-slack community
> *Carex nigra* sub-community

Holcus lanatus, Angelica sylvestris, Phragmites australis, Molinia caerulea and *Succisa pratensis* frequent

> **SD15** *Salix repens-Calliergon cuspidatum* dune-slack community
> *Holcus lanatus-Angelica sylvestris* sub-community

40 *Leontodon hispidus, Equisetum variegatum, Pyrola rotundifolia* and *Trifolium pratense* frequent

> **SD16** *Salix repens-Holcus lanatus* dune-slack community
> *Prunella vulgaris-Equisetum variegatum* sub-community

Above species not consistently present   41

41   *Ononis repens, Carex arenaria, Hypochoeris radic-ata* and *Salix caprea* frequent

> **SD16** *Salix repens-Holcus lanatus* dune-slack community
> *Ononis repens* sub-community

Above species not consistently present   42

42   *Agrostis stolonifera, Hydrocotyle vulgaris, Juncus articulatus* and *Leontodon taraxacoides* frequent

> **SD16** *Salix repens-Holcus lanatus* dune-slack community
> *Agrostis stolonifera* sub-community

Rubus caesius frequent but above species not consistently present

> **SD16** *Salix repens-Holcus lanatus* dune-slack community
> *Rubus caesius* sub-community

43   More open canopies of *Hippophae rhamnoides* with grassy field layer dominated by *Ammophila arenaria* and *Festuca rubra*

> **SD18** *Hippophae rhamnoides* dune scrub
> *Festuca rubra* sub-community

Denser canopies of *H. rhamnoides* with constant *Urtica dioica, Galium aparine, Arrhenatherum elatius* and *Solanum dulcamara*

> **SD18** *Hippophae rhamnoides* dune scrub
> *Urtica dioica-Arrhenatherum elatius* sub-community

COMMUNITY DESCRIPTIONS

SD1

Rumex crispus-Glaucium flavum shingle community

Synonymy
Shingle beach community Oliver 1911, Oliver & Salisbury 1913, Tansley 1939, all *p.p*; *Glaucium flavum* sites Scott 1963 *p.p.*; *Crithmo-Crambetum maritimae* (Géhu 1960) Géhu & Géhu 1969 *p.p.*; *Lathryo-Crambetum maritimae* Géhu & Géhu 1969; *Crambe maritima* sites Scott & Randall 1976 *p.p.*; *Rumici-Lathyretum maritimi* Géhu & Géhu-Franck 1979.

Constant species
Glaucium flavum, Rumex crispus.

Rare species
Crambe maritima.

Physiognomy
The *Rumex crispus-Glaucium flavum* community comprises more or less open assemblages of rather coarse hemicryptophytes, usually few in number and with none consistently dominant, but together giving a highly distinctive character to the stretches of bare shingle or gravel that form the typical habitat, especially when the plants are fully grown in flower or fruit. The commonest species overall is *Rumex crispus*, generally with the obviously tri-tubercular perianths and dense panicles of var. *littoreus* Hardy (which now includes var. *trigranulatus* Syme: Lousley & Kent 1981, Rich & Rich 1988), the tall inflorescences remaining upstanding, brown and brittle, at the close of the season. *Glaucium flavum* is also constant and sometimes very abundant. It behaves as a short-lived perennial, each plant having one to many biennial leaf-rosettes, these persisting through their first winter, with the tall, branched flowering stems growing up in spring. Both these inflorescences and the foliage die after fruiting, but several new rosettes replace each old one before winter, such that colonies expand progressively from one season to the next (Scott 1963*b*).

The nationally rare *Crambe maritima* is also very characteristic of this vegetation at many sites throughout its range, particularly on rather more exposed and shifting shingle, though it can remain totally unseen outside the growing season. It is a rosette plant, but the foliage is deciduous, the much-branched vertical stem being generally buried among the pebbles in winter and perennating by means of terminal buds. In spring, these produce sometimes massive cabbage-leaves, deep purple-crimson at first, then dark glaucous green, with erect and bushy corymbose panicles. After flowering, which continues until August, the fruits ripen, with the whole inflorescences drying and eventually breaking off entirely, being bowled along the beaches in winter storms and often getting caught around the persistent dock and poppy stalks (Scott & Randall 1976).

Scattered among these plants are a number of occasionals, sometimes themselves locally abundant. *Silene vulgaris* ssp. *maritima* is one of the most frequent of these, its creeping stems buried in the shingle and putting up often quite extensive patches or carpets of low bushy shoots. Then, there is commonly some *Beta vulgaris* ssp. *maritima*, this becoming especially prominent on accumulations of organic detritus, like decaying wrack, cast up on the beaches. In such places, it often behaves as an annual and can be accompanied by *Atriplex prostrata*. More occasional overall, though abundant in some sites, are *Senecio jacobaea*, the probably native *S. viscosus*, *Cirsium arvense*, *C. vulgare*, *Sonchus arvensis*, *S. asper* and *Lactuca serriola*, some of these species overwintering just one season as rosettes, others being more long-lived. Some stands, often closer to the sea and with a little sand among the shingle, have *Honkenya peploides*, but *Cakile maritima* and *Salsola kali*, its companions on sandy strand-lines, are not generally found here. *Crithmum maritimum*, too, is only very occasional, in contrast to the shingle vegetation with *R. crispus* and *Glaucium* described from northern France (Géhu & Géhu 1969, Géhu & Géhu-Franck 1979). *Matricaria maritima* can be quite frequent, and there is sometimes a little *Geranium robertianum*, *Euphorbia paralias* and small patches of *Sedum acre*, *Cerastium fontanum* and *Potentilla anserina*.

Certain grasses can be found in the community, though these are usually only occasional and never of high cover, particularly among more exposed and mobile shingle. Scattered tussocks of *Festuca rubra*, *Elymus pycnanthus*, *E. farctus* ssp. *boreali-atlanticus* and *Ammophila arenaria* occur at low frequency throughout, but it is only where there is considerable shelter and stability that species such as *Arrhenatherum elatius* or *Holcus lanatus* make any obvious contribution to the vegetation and, even then, their cover is limited. It is in these situations, though, that the *Rumex-Glacium* community provides the major locus for the remaining British colonies of the attractive rarity *Lathyrus japonicus*.

Sub-communities

Typical sub-community: *Crithmo-Crambetum maritimae* (Géhu 1960)   Géhu & Géhu 1969 *p.p.* This is generally the more open and diverse kind of *Rumex-Glaucium* vegetation with *R. crispus* and *G. flavum* both constant, *B. vulgaris* ssp. *maritima* and *S. vulgaris* ssp. *maritima* common, and *Crambe* and *Honkenya* preferentially frequent. Then, along occasional *Senecio viscosus* and *Cirsium arvense*, there are quite often some scattered plants of *S. jacobaea*, *C. vulgare*, *Sonchus* spp., with *Lactuca serriola* and *Picris echioides* occurring more rarely. *Atriplex prostrata*, *Solanum dulcamara* and *Matricaria maritima* are also preferential at low frequency and it is in this sub-community that very occasional *Crithmum* and *Plantago coronopus* are recorded. Apart from sparse tussocks of *Festuca rubra*, *Ammophila arenaria* and *E. pycnanthus*, grasses tend to be poorly represented.

Lathyrus japonicus sub-community: *Lathyro-Crambetum* Géhu & Géhu 1969; *Rumici-Lathyretum maritimi* Géhu & Géhu-Franck 1979. *R. crispus* remains very common here, but *G. flavum* is somewhat reduced in frequency and *Crambe* is only occasional. *B. vulgaris* ssp. *maritima* and *S. vulgaris* ssp. *maritima* occur quite often, with *Senecio viscosus* and *Cirsium arvense* occasional, but the variety among these weedy hemicryptophytes, particularly the more short-lived ones, is less than in the Typical sub-community. The most striking feature, however, is the frequent occurrence of *Lathyrus japonicus* whose new shoots begin to appear above ground, among the dried remains of the previous year's stems, from April onwards, growing out procumbently to form quite extensive patches, adjacent plants sometimes touching to form a discontinuous carpet (Brightmore & White 1963). *Arrhenatherum elatius* is also preferentially common and quite abundant at some sites, with scattered tussocks of *Holcus lanatus* and *Festuca rubra* occurring occasionally, but grasses like *E. pycnanthus*, *E.*

farctus ssp. *boreali-atlanticus* and *Ammophila* are very scarce, except where there is local accumulation of blown sand among the herbage. Scattered plants of *Plantago lanceolata* and *Hypochoeris radicata* can sometimes be found along with *Cerastium fontanum*, *Sedum acre* and *Leontodon taraxacoides*.

Habitat

The *Rumex-Glaucium* community is the characteristic pioneer vegetation of maritime shingle around the coast of the warmer south of Britain. Within this climatic zone, it is widely distributed but distinctly local, being more or less confined to stretches of sharply-draining pebbles and gravel accumulating just beyond the reach of all but exceptional tides, though not so stable and sheltered as to support progressive colonisation by plants. Fragmentary stands can be found just on or behind the seaward crests of many shingle beaches, and indeed some way inland, where suitable open habitats occur, but the community is best developed on the more extensive spits and apposition features that are found around the coasts of East Anglia and southern England.

Of the most striking plants of this kind of vegetation, *R. crispus* var. *littoreus* occurs widely around the coasts of northern Europe, but both *G. flavum* and *Crambe* are of more restricted distribution. *G. flavum* is the more extensive to the south, being a Continental Southern species (Matthews 1955) found right through the Mediterranean (Scott 1963b), while *Crambe* is an Oceanic West European plant, limited to the more humid north-west Atlantic coast (Scott & Randall 1976). To the north, however, their limits are more or less co-terminous, stopping in Britain at around the Forth–Clyde line (Perring & Walters 1962) and, on the Continent, in southern Scandinavia. The availability of suitable habitats beyond these areas might play some part in restricting the range of this community, but bare shingle is not entirely absent further north, and *R. crispus* var. *littoreus* continues to make a contribution to vegetation on it. Just as important is likely to be the vulnerability of the more temperate species to the cooler climate. With us, for example, *G. flavum* and *Crambe* do not penetrate into those latitudes where the July maxima fall below 17.5 °C (Conolly & Dahl 1970) and, for *Crambe*, Eklund (1931) has actually recorded poorer fruiting at the Baltic limit of the range. Another frequent member of the community, *Beta vulgaris* ssp. *maritima*, though it is not so strictly confined to shingle as *G. flavum* or *Crambe*, is an Oceanic Southern plant with an almost identical overall distribution to theirs in mainland Britain. This vegetation also provides an occasional locus for a number of other rather thermophilous plants with similar European ranges: *Crithmum maritimum*, *Euphorbia paralias* and *Elymus pycnanthus*.

Around these warmer coasts of southern Britain, the

Rumex-Glaucium community is probably confined to more exposed shingle habitats as much by the vulnerability of the establishing plants to competition away from open and disturbed ground, as by virtue of any obligate dependence on strong maritime influence. In fact, were the seeds of *G. flavum* and *Crambe* to be more readily dispersed inland, assemblages of this kind would probably colonise open river shingle and ballast away from the coast, where non-maritime *R. crispus* and some of the other associates of this vegetation are often prominent invaders. *G. flavum*, with its lighter wind-borne seeds, is very occasionally found in such situations (Scott 1963*b*), but *Crambe* has bigger sea-dispersed fruits and, though the plant is cultivated inland as the vegetable seakale, it does not seem to escape into the wild away from the coast in this country (Eklund 1931, Scott & Randall 1976). Apart from extremely local and fragmentary stands on inland pebbles and spoil, then, the community occurs very close to the sea and there many of its most distinctive species, like *R. crispus* var. *littoreus*, *G. flavum*, *Crambe*, *B. vulgaris* ssp. *maritima* and *Silene vulgaris* ssp. *maritima* have the advantage over inland colonists of bare shingle by virtue of their salt-tolerance (Scott 1963*b*, Cavers & Harper 1964, Malloch 1972, Scott & Randall 1976). For, though this vegetation usually develops out of reach of all but the highest spring tides and storm surges, it is very occasionally inundated by seawater and receives a lot of salt-spray from onshore winds throughout the year.

Winds and tides are also of great importance in influencing the character of the vegetation by their control of the disposition, size and stability of the beach material. Shingle is very much the preferred substrate of the *Rumex-Glaucium* community but the size of the material varies quite considerably from fine gravel less than 1 cm in diameter up to coarse pebbles 5 or 10 cm across and, though many beaches colonised by this vegetation show longshore sorting, some substrates are more mixed. In general, gravelly material supports richer and denser stands of the community, particularly where there is some sand or comminuted organic detritus mixed in: indeed, it is possible that the presence of this much finer fraction is more or less essential for many shingle plants to develop the extensive absorptive roots they need to thrive (Oliver 1912, Tansley 1939, Scott 1963*a*). Characteristically, *Rumex-Glaucium* vegetation avoids those stretches of shingle subject to flooding or waterlogging, most of the moisture for plant growth on the raised and sharply-draining parts of beaches probably coming from rain (Scott 1963*a*), so the occurrence of some more retentive material is perhaps crucial for the establishment and maintenance of the community.

Local variety in the substrate controls some of the diversity in cover and composition that can be seen here within and between stands. Where there is a little more

sand blown in from nearby strands, for example, or where the community has developed on shingle exposed by the erosion of dunes developed on top, associates like *Festuca rubra*, *Ammophila arenaria*, *Elymus farctus*, *E. pycnanthus*, *Honkenya* and *Sedum acre* tend to become more prominent. By contrast, where decaying wrack or other driftline detritus is thrown up by the waves, more nitrophilous plants, such as *Atriplex prostrata*, *B. vulgaris* ssp. *maritima* and *Sonchus* spp., can increase in abundance, and such inputs of nutrients are probably of some significance for the community as a whole, much-rotted material providing a good seed-bed for germination and periodic additions perhaps prolonging the life of particular stands (Géhu 1960, Scott 1963*a*, Scott & Randall 1976). In general, however, the organic content of the substrate is slight and there is never any development of an integrated soil profile, even in more stable stands. In most cases, the shingle is calcareous, with the pH of the finer detritus being usually above 7, but this may reflect the fortuitous fact that most of the beaches that are physically suitable for the establishment of this vegetation are fed by exposures of lime-rich bedrocks or superficials.

The accumulation of such material in more or less stable beaches is a complex function of current and tidal movements and coastal form (e.g. Steers 1953) but the crucial factor for the development of the *Rumex-Glaucium* vegetation appears to be that the shingle should be out of reach of destructive waves during the growing season, and that only between autumn and spring is the zone of movement brought up into the area occupied by the community (Scott 1963*a*). Then, there may be some seaward erosion with shingle slipping down the steepened beach face, or some throwing up of material on to the vegetated zone. Well-established stands, with their predominance of deep-rooted hemicryptophytes, seem well able to tolerate some modest shifting of the substrate and *Crambe* in particular is especially resistant to burial and able to put up shoots from beneath 50 cm or more of shingle. Both root and stems have densely-packed starch reserves which appear normally not to be drawn upon at the start of the season, and these perhaps provide an emergency source of food for recovery in such circumstances (Scott & Randall 1976).

Although the *Rumex-Glaucium* vegetation never really develops on those beaches which are only periodically stable between spring and autumn, short-lived and fragmentary stands can be found coming and going on many stretches of shingle that persist largely intact for a few years at a time. The bigger and more striking stands, however, are characteristic of extensive and permanent features like the beaches and spits of the East Anglian coast, as on Scolt Head and at Kessingland, Walberswick, Sizewell, Thorpeness, Orford Ness and Shingle Street, on Chesil Beach in Dorset and around the seaward fringe of the enormous apposition beach at Dungeness (Oliver 1913, 1915, Scott 1963*a*, Ratcliffe 1977). Throughout the range, the Typical sub-community is the commoner form, indeed the only type of *Rumex-Glaucium* vegetation found on more exposed shingle and on the less substantial beaches that occur away from the south and east of England. Its more open and diverse character reflect the less stable conditions, with opportunity for colonisation by a variety of widely-distributed plants more or less tolerant of some exposure to salt-spray. Even where it grades to more persistent shingle vegetation, it can form a zone but a few metres deep.

The *Lathyrus* sub-community is much more restricted in its distribution, being almost entirely confined to the south-east coast, with a few outlying stands to the south-west. Here, it persists on more sheltered stretches of shingle, away from periodic tidal erosion and onshore winds, often set further back on wider beaches. Such conditions are very congenial for *L. japonicus*, a circumpolar seashore plant (Hultén 1950) which favours stable and well-drained pebbles or gravel, often with some sand (Brightmore & White 1963), but also allow the invasion of species of ranker inland swards such as *Arrhenatherum* and *H. lanatus*. *L. japonicus* may itself aid the development of such assemblages, or the local appearance of dune plants, by the accumulation of decaying organic matter and wind-blown sand among its patches (Brightmore & White 1963). Where consolidation is advanced, species such as *G. flavum* and *Crambe* can persist for some considerable time, but do not establish anew at all readily (Scott 1963*b*), and *L. japonicus* itself eventually disappears.

Apart from its susceptibility to tidal erosion of the substrate, the *Rumex-Glaucium* community is vulnerable to human disturbance of shingle beaches, where sediments are shifted for sea-defence works or for coastal development. Trampling may also damage stands which are left largely intact, with *Crambe* and *L. japonicus* in particular appearing to suffer from this effect: their decline has been especially marked where tourism has prospered (Scott & Randall 1976, Randall 1977).

Zonation and succession
The *Rumex-Glaucium* community is sometimes the only vegetation to be seen on shingle beaches, forming an open zone isolated from the hinterland of the shore, where dune or inland communities sometimes occur, but where there is often now a sharp boundary with agricultural land or settlements, frequently behind sea-defences. With a shift to less stable conditions, however, or where lines of drift are deposited on the beach, the *Rumex-Glaucium* community can pass to annual strandline assemblages, and on wider shores there can be gradations to vegetation of more consolidated beach

substrates, either grasslands on shingle or dune communities. Where shingle accumulation is progressive, such zonations may represent a succession, but usually the *Rumex-Glaucium* community persists as a perpetually renewed pioneer vegetation.

Throughout the range of the community, the associated strandline vegetation is commonly of the *Atriplex-Beta* type, an assemblage typically found as strips on lines of decaying wrack and other detritus deposited at the tidal limits. *G. flavum* and *Crambe* are sometimes recorded among such vegetation on shingle, and *R. crispus* var. *littoreus* remains quite common, but it is *Beta vulgaris* ssp. *maritima* and various *Atriplex* spp., together with *Matricaria maritima*, that provide much of its distinctive character. Such assemblages may come and go along the strandlines in successive seasons, or replace the *Rumex-Glaucium* community where stretches of beach do not remain stable from one year to the next.

Sandier strandlines can see transitions from the shingle vegetation to the *Honkenya-Cakile* community. Here, again, *Atriplex* spp. are very common, and often locally abundant, and these, together with occasional *R. crispus* var. *littoreus*, *B. vulgaris* ssp. *maritima* and *Silene vulgaris* ssp. *maritima*, can provide some floristic continuity with the *Rumex-Glaucium* community, but it is patches of *Honkenya peploides*, with *Cakile maritima* and *Salsola kali*, that usually mark out the broken strips of this vegetation that develop at extreme high water mark. Local accumulations of sand around the patches are prone to invasion by *Elymus farctus*, *Leymus arenarius* and *Ammophila arenaria*, and this can initiate a temporary or progressive succession to fore-dunes. A very few sites, notably the beaches of north Norfolk, show a patchy but complete zonation from Typical *Rumex-Glaucium* vegetation, through the *Honkenya-Cakile* community and *Elymus* fore-dunes, to *Ammophila* vegetation of varying degrees of maturity (Oliver 1913, 1915, Ratcliffe 1977).

In other places around the coast of east and southern England, where there is an increase in the stability and shelter of the shingle habitat, without any marked accumulation of wind-blown sand, the Typical form of the *Rumex-Glaucium* vegetation can pass landwards to the *Lathyrus* sub-community, and this in turn gives way to some distinctive grasslands in which dominance passes to *Arrhenatherum*, *Festuca rubra* and *Silene vulgaris* ssp. *maritima*, with an almost total occlusion of shingle hemicryptophytes. Orford Ness and Dungeness show the best development of this *Arrhenatherum-Silene* community with, in the latter site, zonations to non-maritime *Arrhenatheretum* and uniquely extensive mosaics on compacted sandy shingle with *Festuca-Agrostis-Rumex* calcifugous grassland and scrub dominated by *Cytisus scoparius*.

Another species which, very locally around the south-east coast, assumes prominence on sheltered shingle, especially on the leeward site of spits, is *Suaeda vera*. Its water-borne seed germinates very readily on accumulations of drift, the plants soon anchoring firmly with deep tap roots and then growing from shoots held horizontally under the pebbles, spreading outwards or extending in a single direction as shingle is shifted over the bushes. On Chesil Beach, for example, a patchy zone of open *Elymo-Suaedetum*, dominated by *S. vera*, replaces the *Rumex-Glaucium* community along the landward side of the bar, with its complex of fans and flats looking out over The Fleet (Oliver 1912, Tansley 1939). And, at Blakeney Point in Norfolk, this kind of vegetation runs around the drift-line of the sheltered embayments between the shingle laterals, replacing the *Rumex-Glaucium* community in the narrow strip that forms a transition on sandy pebbles from the dunes of the spit to the salt-marsh behind (Oliver 1913, Oliver & Salisbury 1913*a*, *b*).

Distribution

The *Rumex-Glaucium* community occurs from north Norfolk around the coasts of eastern and southern England and then, more fragmentarily, up the west coast as far as the Firth of Forth. The more extensive stands of the Typical form and all occurrences of the *Lathyrus* sub-community are found in south-east England from Chesil round to Scolt Head.

Affinities

In early accounts of British maritime vegetation (Oliver 1911, Tansley 1939), this assemblage was generally included with annual strandline plants in a broadly-defined shingle community. In fact, although it is usually very open in structure and varied in its composition, often grading to such more ephemeral mixtures, it is a well-characterised kind of vegetation worth separate recognition as the major community of pioneer shingle perennials in southern Britain. Géhu & Géhu-Franck (1979; see also Géhu & Géhu 1969) acknowledged this in their diagnosis from south-east England of a *Rumici-Lathyretum*, although this took in just the *Lathyrus* sub-community of our vegetation, and not the more widely-distributed Typical form. Apart from the absence of *Crithmum maritimum* from most of the British stands of this kind of shingle vegetation, the *Rumex-Glaucium* community is very similar to some forms of the *Crithmo-Crambetum* described from Breton shingle (Géhu 1960, Géhu & Géhu 1969, Géhu & Géhu-Franck 1979) and forms part of a sequence which runs on into the Baltic and northern Europe. Géhu & Géhu (1969) erected a new Class, the Elymetea pycnanthi, to hold this suite of associations, along with various strandline assemblages. Alternatively, their relevant alliance, the Honkenyo-Crambion could be located within Tüxen's (1966) proposed class, the Honkenyo-Leymeetea.

Floristic table SD1

	a	b	1
Rumex crispus	IV (1–4)	IV (1–4)	IV (1–4)
Glaucium flavum	IV (1–4)	III (1–5)	IV (1–5)
Crambe maritima	III (1–6)	II (1–3)	III (1–6)
Honkenya peploides	II (1–7)	I (2–4)	II (1–7)
Sonchus arvensis	II (1–4)	I (1–5)	II (1–5)
Senecio jacobaea	II (1–4)	I (1–4)	I (1–4)
Sonchus asper	II (1–4)	I (1–4)	I (1–4)
Ammophila arenaria	II (1–5)	I (1–3)	I (1–5)
Atriplex prostrata	II (1–6)		I (1–6)
Solanum dulcamara	II (1–4)		I (1–4)
Matricaria maritima	II (1–4)		I (1–4)
Cirsium vulgare	I (1–4)		I (1–4)
Crithmum maritimum	I (2–4)		I (2–4)
Plantago coronopus	I (2–4)		I (2–4)
Sagina apetala erecta	I (1–2)		I (1–2)
Lactuca serriola	I (1–4)		I (1–4)
Picris echioides	I (1–2)		I (1–2)
Lathyrus japonicus	I (1–6)	IV (1–8)	III (1–8)
Arrhenatherum elatius	I (1–2)	III (2–7)	II (1–7)
Plantago lanceolata	I (3)	II (1–5)	I (1–5)
Holcus lanatus		II (1–6)	I (1–6)
Hypochoeris radicata		I (1–4)	I (1–4)
Beta vulgaris maritima	III (1–4)	III (1–5)	III (1–5)
Silene vulgaris maritima	II (1–6)	II (1–6)	II (1–6)
Festuca rubra	II (1–5)	II (1–6)	II (1–6)
Senecio viscosus	II (1–4)	II (1–4)	II (1–4)
Cirsium arvense	II (1–4)	II (1–4)	II (1–4)
Elymus pycnanthus	I (1–4)	I (1)	I (1–4)
Euphorbia paralias	I (1–2)	I (1)	I (1–2)
Geranium robertianum	I (4)	I (1–4)	I (1–4)
Sedum acre	I (1–4)	I (4)	I (1–4)

Floristic table SD1 (*cont.*)

	a	b	1
Lolium perenne	I (1)	I (1)	I (1)
Cerastium fontanum	I (1–4)	I (1–5)	I (1–5)
Potentilla anserina	I (3)	I (2)	I (2–3)
Leontodon taraxacoides	I (2)	I (2)	I (2)
Elymus farctus	I (3)	I (1)	I (1–3)
Number of samples	62	36	98
Number of species/sample	8 (2–13)	5 (1–12)	7 (1–13)

a   Typical sub-community

b   *Lathyrus japonicus* sub-community

1   *Rumex crispus-Glaucium flavum* shingle community (total)

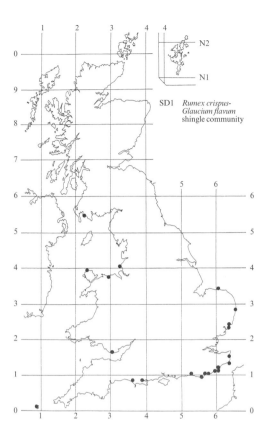

SD1 *Rumex crispus-*
Glaucium flavum
shingle community

SD2
Honkenya peploides-Cakile maritima strandline community

Synonymy
Strand plants association Tansley 1911; Foreshore communities Tansley 1939; *Salsola kali-Atriplex glabriuscula* Association Tx. 1950; *Atriplici-Polygonetum raii* Tx. 1950; *Cakile maritima*-sociatie Boerboom 1960; Sociation à *Salsola kali* Géhu & Géhu 1969; Sociatie van *Honkenya peploides* Westhoff & den Held 1969; *Honkenietum peploidis* Géhu & Géhu 1969.

Constant species
Cakile maritima, Honkenya peploides.

Rare species
Polygonum oxyspermum ssp. *raii.*

Physiognomy
The *Honkenya peploides-Cakile maritima* community occurs as patchy strips of strandline vegetation in which *Honkenya peploides, Cakile maritima* and various *Atriplex* spp. are the most consistent elements. *Honkenya* is a perennial and low clumps of its succulent, creeping shoots can be a conspicuous feature here all the year round, being firmly anchored in the sandy gravel substrate, able to accumulate a little wind-blown sand and tolerant of the very occasional, brief inundations by seawater that come with extreme high tides and winter storms. Other characteristic species are more ephemeral, but they too can become abundant in summer, especially where organic detritus has been deposited along the tidal limit or incorporated beneath a shallow covering of beach material, and they may give a distinctive appearance to particular stretches of the community. Among these annuals, *Cakile* is very frequent and sometimes plentiful and there is usually one or other, or a mixture, of the common *Atriplex* spp. of strandline habitats, quite often in locally dense populations which have sometimes been treated as the basis of separate vegetation types (e.g. Birse 1980). These plants can be difficult to identify (Taschereau 1985), but *A. prostrata* seems to be the most widespread species in the community, apart

from in northern Scotland where it tends to be replaced by *A. glabriuscula*, with *A. laciniata* found more locally in this vegetation right around the coast. *A. patula* and also *A. littoralis*, which is more usually a salt-marsh plant, are more occasional, and in a few localities around the heads of sheltered inlets in north-west Scotland and Shetland, the rare *A. praecox* can be seen in intimate association with the *Honkenya-Cakile* community, colonising downshore towards the tidal algal zones (Taschereau 1985).

Common associates of *Atriplex* spp. in other kinds of strandline vegetation, such as *Beta vulgaris* ssp. *maritima* and *Matricaria maritima*, are usually no more than occasional in this community, though they can be moderately abundant and sometimes overwinter establish as perennials. More distinctive and fairly common here except in the far north is *Salsola kali*, another annual chenopod, whose prickly decumbent shoots can form quite far-spreading open patches. This vegetation also provides an important locus along our southern and western coasts for *Polygonum oxyspermum* ssp. *raii*, a rare and decreasing species with us (Lousley & Kent 1981) that has sometimes been regarded as characteristic of a discrete assemblage of strandline annuals (Géhu & Géhu 1969, Birse 1980). Then, there can be occasional plants of other ephemerals like *Senecio vulgaris, Cerastium diffusum* ssp. *diffusum, Stellaria media* and *Poa annua*.

Scattered individuals of bigger perennial herbs such as *Rumex crispus* var. *littoreus, Silene vulgaris* ssp. *maritima, Eryngium maritimum, Sonchus arvensis* and *Cirsium arvense* are sometimes seen in this vegetation but, among the longer established plants, a number of grasses are more characteristic. *Elymus farctus* ssp. *boreali-atlanticus* is especially common and often quite abundant, particularly where the *Honkenya-Cakile* strandline merges imperceptibly with lines of incipient foredunes. *Leymus arenarius* and *Ammophila arenaria* are also quite frequent, and more occasionally there can be tussocks of *Festuca rubra, Elymus repens* and *Agrostis stolonifera*.

Habitat

The *Honkenya-Cakile* community is the characteristic pioneer vegetation of sand and fine shingle strandlines on flat or gently-sloping beach tops all around the British coast. Periodic additions of organic detritus along the tidal limit encourage the development of the vegetation, particularly the more nitrophilous ephemerals which are able to exploit the warmer and more settled conditions in summer. Local accretion of sand may favour the invasion of dune-building grasses, but very often the exposure to strong salt-laden winds and occasional tidal inundations keep the vegetation in a perpetually immature state.

Some of the plants found in this community, such as *Eryngium maritimum*, *Beta vulgaris* ssp. *maritima* and *Polygonum oxyspermum* ssp. *raii*, have an Oceanic Southern distribution through Europe (Matthews 1955) and lend a distinctive character to stands in the south and west, while others, like *Atriplex glabriuscula*, become more important towards the cooler north. Most of the commoner species in this vegetation, however, extend right around the British seaboard, or nearly so, and the occurrence of the community along particular stretches of coast is more a reflection of the suitability of the beaches than of any direct influence of regional climate on plant growth. Sandy and fine shingle substrates are much preferred, and these must be beyond the reach of all but the most extreme high tides and storm surges to support the more than a fleeting and fragmentary development of the vegetation. Typically, then, the *Honkenya-Cakile* community is found along the flatter tops of beaches in what is often a very narrow zone between the tidal limit and the stable or accreting hinterland of the shore. On long strands, such as border some stretches of soft, low coastline, interrupted strips of this vegetation can extend for considerable distances, but frequently the stands are much more patchy, even here, and around the tidal margins of little bays and in the heads of sandy creeks the assemblages can be very fragmentary, though the shelter in such situations may favour locally luxuriant growth. Especially well developed stands are sometimes seen along the strandlines on the lee side of barrier islands or spits (Chapman 1976).

Wind and water are both important in the dissemination of propagules into such strandline habitats, but continuing exposure to salt-spray and occasional inundation by the sea exert a strong limitation on the kinds of plants that can gain a hold on the raw sand and gravel soils. This accounts for the strongly halophytic, or at least salt-tolerant, nature of this vegetation, with important species here having provision in their leaf tissues (*Honkenya*, *Cakile*) or special hairs (*Atriplex* spp.) for the retention of moisture while their roots subsist in markedly saline ground water (Salisbury 1952, Chapman 1976). Also very important for the develop-

ment of the community is the ability of many species, especially the summer annuals, to capitalise upon the periodic additions of nutrients, particularly nitrogen, that come with the driftline detritus. Often, the *Honkenya-Cakile* vegetation clearly marks out the highest and most stable of a series of lines of tidal debris cast up on the beach top, and even where such material cannot be seen, excavation frequently reveals that the plants are rooted in a layer of rotting wrack from an old driftline that has been buried beneath a few centimetres of sand (Gimingham *et al.* 1948, Gimingham 1951).

The dead remains of the annuals, as well as the perennial clumps of *Honkenya*, are able to accumulate a little wind-blown sand (Tansley 1939) and patches of this may present the appearance of tiny low dunes along the strandline. These offer congenial sites for invasion by *Elymus farctus* which, like the plants of the *Honkenya-Cakile* community, can stand occasional, brief submergence by the sea, but which more readily encourages the accretion of sand and so may initiate the development of foredunes, within which *Ammophila* can gain a hold. Frequently, though, this process is offset by tidal and wind erosion, the firmly-anchored *Honkenya* surviving to form the basis of a newly-developing stand of the community around deposited drift. Trampling by humans, scuffing their way along the strandline, may in the end be more damaging, particularly to the survival of rare plants like *P. oxyspermum* ssp. *raii* (Lousley & Kent 1981).

Zonation and succession

Along some stretches of shore, the *Honkenya-Cakile* community is the only kind of maritime vegetation to be seen above the tidal limit, but elsewhere it occurs in zonations with other sand and shingle communities, the patterns being strongly dependent on the deposition and erosion of beach material of different grades, and the frequency of addition of organic detritus. Most existing stands can be regarded as perpetually renewed pioneer vegetation, although where opportunity arises, the progressive accumulation of sand over and around the community can initiate a dune succession.

On beaches that are naturally narrow, or where the shore hinterland has become occupied by settlements or converted to agricultural or recreational use, often behind some sort of sea defence, the *Honkenya-Cakile* community can occur isolated as a narrow open zone of vegetation just above the tidal limit. In such situations, stands may be very fragmentary and, along any one stretch of beach, not very long-lived, with the additional threat of trampling where beach use is heavy. Even here, however, there can be considerable floristic differences between and along stands, particularly in the varying development of the summer annuals, with species such as *P. oxyspermum* ssp. *raii* and *A. glabriuscula* giving some measure of regional contrast. And, where there is a shift from sand to coarser shingle in moving along beaches, a common feature where longshore drift operates on extensive lengths of coast, the community can give way to other kinds of strandline vegetation. In southern Britain, its counterpart on drift-enriched pebbles is the *Atriplex-Beta* community, where *Atriplex* spp., particularly in this case *A. prostrata*, remain common and abundant, but where *Cakile* and *Honkenya* are much reduced in frequency, *B. vulgaris* ssp. *maritima* and *Matricaria maritima* greatly increased. To the north, along sheltered shingly strandlines in Scotland, there is an analogous switch to the *Matricaria-Galium* community, with *A. glabriuscula* this time providing some floristic similarity, but where *M. maritima*, *Galium aparine* and *Stellaria media* become a more or less constant feature.

Where a sandy beach top runs along the front of shingle deposits that are beyond the reach of all but very extreme tides, the *Honkenya-Cakile* community can be replaced by open *Rumex-Glaucium* vegetation. There, *Honkenya* can persist with some vigour on sandy patches, its stolons sometimes burrowing down into the underlying pebbles (Tansley 1939) and *Atriplex* spp. may thrive on local accumulations of organic detritus, but it is coarse perennial hemicryptophytes like *R. crispus* var. *littoreus*, *Glaucium* and *Crambe* that provide most of the character of the shingle assemblage.

In other places, a shingle spine or low hinterland behind a beach top has provided a suitable base for the development of dunes by the accretion of sand blown in from flats exposed at low tide. In such situations, the *Honkenya-Cakile* vegetation can form a narrow, interrupted front to sequences of dune communities, the sandy patches around the plants giving way to a broken line of *Elymus farctus* foredunes, these in turn passing to *Ammophila yellow* dunes, with *Leymus arenarius* playing an important part in these younger stages at sites down the east coast. *Honkenya*, *Cakile*, *Salsola* and *Atriplex* spp. can persist for some time in these kinds of vegetation as accretion progresses, but they are quickly overwhelmed as bigger mobile yellow dunes develop. In many of our dune systems, however, it is clear that both the *Honkenya-Cakile* stands and the *Elymus* foredunes are repeatedly renewed as bouts of fierce wind erosion

and occasional tidal surges destroy the beach-top vegetation and set back any advance of the sere.

Distribution
The *Honkenya-Cakile* community occurs around all parts of the British coast where suitable substrates exist.

Affinities
As characterised in this scheme, the *Honkenya-Cakile* community corresponds to the rather broadly-defined assemblages of sandy strandlines described in early accounts of British coastal vegetation (Tansley 1911, 1939). It thus subsumes most of the fine variation, often reflected in the local frequency and abundance of individual species, particularly the annual plants, that has been used in Continental schemes to distinguish a range of separate vegetation types. In mainland Britain, only Birse (1980, 1984) has pursued this approach although from Ireland Braun-Blanquet & Tüxen (1952), Ivimey-Cook & Proctor (1966) and Beckers *et al.* (1976) have described various assemblages (summarised in White & Doyle 1982) and Géhu & Géhu (1969), in their account of French communities of this kind, made passing reference to the occurrence of some of them around the coasts of the British Isles.

Essentially, the core of the *Honkenya-Cakile* community is equivalent to what European ecologists have traditionally defined as a perennial *Honkenietum* (Géhu & Géhu 1969), together with populations of annuals variously grouped into sociations or simple associations of *Cakile* (Boerboom 1960, Westhoff & de Held 1969), *Salsola* (Géhu & Géhu 1969), different *Atriplex* spp. (Tüxen 1950, Westhoff & Beeftink 1950, Géhu & Géhu 1969), *Polygonum oxyspermum* ssp. *raii*, or mixtures of these (Nordhagen 1940, Tüxen 1950, Braun-Blanquet & Tüxen 1952, Ivimey-Cook & Proctor 1966, Géhu & Géhu 1969, Birse 1980, 1984). A case has been made (Géhu & Géhu 1969) for locating strandlines with prominent *Honkenya* among the perennial foredune vegetation of the Elymo-Honkenion, an Ammophiletea alliance, but the *Honkenya-Cakile* community as defined here clearly belongs among the annual Salsolo-Honkenion assemblages, within the Cakiletea foreshore vegetation.

Floristic table SD2

Honkenya peploides	V (1–8)
Cakile maritima	IV (1–8)
Elymus farctus	III (2–8)
Atriplex prostrata	III (1–8)
Atriplex glabriuscula	II (1–6)
Leymus arenarius	II (1–4)
Atriplex laciniata	II (1–5)
Ammophila arenaria	II (1–5)
Salsola kali	II (1–5)
Matricaria maritima	II (1–5)
Rumex crispus	I (1–4)
Eryngium maritium	I (1–4)
Festuca rubra	I (1–4)
Elymus repens	I (1–5)
Potentilla anserina	I (1–6)
Sonchus arvensis	I (1–5)
Agrostis stolonifera	I (1–6)
Atriplex patula	I (1–3)
Silene vulgaris maritima	I (1–5)
Beta vulgaris maritima	I (3–4)
Hypochoeris radicata	I (1)
Senecio vulgaris	I (1–4)
Achillea millefolium	I (1–4)
Arrhenatherum elatius	I (1–2)
Atriplex littoralis	I (2)
Cerastium diffusum diffusum	I (3)
Cirsium arvense	I (1–3)
Poa annua	I (1–4)
Polygonum oxyspermum	I (1–2)
Sedum acre	I (3–5)
Stellaria media	I (1–3)
Number of samples	39
Number of species/sample	5 (2–10)

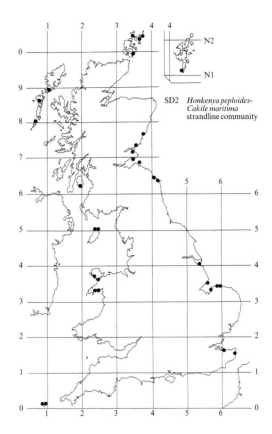

SD2 *Honkenya peploides-Cakile maritima* strandline community

SD3
Matricaria maritima-Galium aparine strandline community

Synonymy
Mertensia maritima localities Scott 1963c; *Atriplex glabriuscula-Rumex crispus* Association Birks 1973.

Constant species
Galium aparine, Matricaria maritima.

Rare species
Mertensia maritima, Polygonum oxyspermum spp. *raii.*

Physiognomy
The *Matricaria maritima-Galium aparine* community consists of generally open and often patchy strandline vegetation in which *Matricaria maritima*, various *Atriplex* spp. and the annual weeds *Galium aparine* and *Stellaria media* are the most frequent and prominent elements. Among the oraches, *A. glabriuscula* is especially common here, but *A. prostrata* and *A. patula* also occur occasionally, and each of these can be found in some abundance. The more local *A. laciniata* is sometimes recorded, too, though usually as sparse scattered individuals, and the rare *A. praecox* can be seen in close association with this vegetation at some of its few localities around sea lochs in western Scotland (Taschereau 1985).

Beta vulgaris ssp. *maritima*, a very characteristic plant of shingle strandlines in southern Britain, is hardly ever found here, and *Honkenya peploides* and *Cakile maritima*, which are a constant feature of sandy foreshore vegetation around our coasts, tend to be only infrequent. *Rumex crispus* var. *littoreus*, however, remains fairly common and it can be conspicuous with its tall flowering shoots, and there may be occasional patches of *Sonchus asper, S. arvensis, Cochlearia officinalis, Chamomilla suaveolens* and *Silene vulgaris* ssp. *maritima*. In some stands, small amounts of *Elymus repens, Festuca rubra, Agrostis stolonifera, Glaux maritima* and *Armeria maritima* bring the vegetation close to salt-marsh strandline assemblages of northern and western Britain.

Ligusticum scoticum, generally speaking a plant of sea-cliff crevices, very occasionally finds a place here and, along the west coast of Scotland, the community can provide a locus for the rare annual *Polygonum oxyspermum* ssp. *raii*. More striking, though, is the occurrence in this vegetation of *Mertensia maritima*, a scarce and declining species, but one which may nevertheless still appear in great abundance in its remaining localities (Randall 1988, Farrell 1989). It is a perennial, dying back above ground each winter to dormant buds, but producing through the spring and summer sometimes very numerous leafy and flowering shoots, fleshy, glaucous and procumbent, spreading to form patches up to 1 m or more across (Scott 1963c).

Habitat
The *Matricaria-Galium* community is the characteristic vegetation of sandy shingle strandlines with drift detritus around more sheltered shores in the cooler, wetter north of Britain.

Like the *Atriplex-Beta* community, this assemblage appears to favour periodically-disturbed beach deposits of a mixed nature, rather than pure shingle, being found most often along strandlines where pebbles occur with coarse sand, occasionally with some silt, or with shell fragments, which later can bring a lime-rich element to material that is otherwise frequently completely siliceous. Generally, though, the lithology of the substrate is of little consequence to the vegetation, except in so far as the development of suitable beaches depends on the occurrence of more readily-weathering rocks to feed them.

Of greater importance to the sustenance of the community is the periodic addition of organic detritus, especially masses of seaweed torn off in storms, but also fragments of driftwood and agricultural debris, cast up on winter high tides and then rotting on the surface or becoming incorporated among the pebbles and sand, coating them and the plant roots with a slimy decaying mass (Scott 1963c, Birks 1973). Typically, the *Matricaria-Galium* community develops as a patchy strip of plants colonising such a driftline, or the topmost of a series of driftlines, or their decaying remains hidden now beneath shifted beach material. Among such generally barren and sharply-draining shingle and sand as accumulate here, the supply of nutrients released from such organic matter, and the moisture that the material helps retain, are of major significance to the development and composition of the vegetation. In particular, among the more persistent plants like *Matricaria maritima* and *R. crispus* var. *littoreus*, they favour the occurrence of nitrophilous ephemerals such as *Atriplex* spp., *Galium aparine* and *Stellaria media*, which are able to capitalise quickly on the enrichment in the summer months.

Species like *Matricaria*, *Rumex* and, more occasionally, *Sonchus* and *Cochlearia* spp., provide strong floristic continuity with the *Atriplex-Beta* strandline, but the ranges of the two vegetation types are largely mutually exclusive, the *Matricaria-Galium* community replacing the other assemblage north of a line running roughly from the Mull of Galloway across to the Forth. Around these northern Scottish coasts, mean annual maximum temperatures fall below 25 °C (Conolly & Dahl 1970) and the climate is generally more humid, particularly in the west, where annual precipitation often exceeds 1200 mm (*Climatological Atlas* 1952) with over 160 wet days yr^{-1} (Ratcliffe 1968). *Beta vulgaris* ssp. *maritima*, which gives much of the distinctive character to strandline vegetation around the warmer, drier coasts of southern Britain, only just extends into this zone, but increasingly northwards the Oceanic Northern *Atriplex glabriuscula*, with the more local *A. laciniata*, replaces *A. prostrata* in the *Matricaria-Galium* community, and conditions become suitable for the Arctic–Subarctic *Mertensia*. There seems little doubt that the southern limit of this plant's range is controlled by climate, although quite precise factors may be involved, such as a dependence on a certain degree or duration of cold for seed germination, or a vulnerability to summer drought of particular severity (Scott 1963*c*).

The moister climate which our northern strandlines experience is also probably important in the frequency and luxuriance in their vegetation of non-maritime plants like *Galium aparine* and *Stellaria media*, species which are typically excluded from the *Atriplex-Beta* community. Often here, too, their vigour is encouraged by relatively sheltered conditions, for the richest and most luxuriant stands of the *Matricaria-Galium* vegetation, sometimes forming a strip 2–3 m wide, are found around the heads of sea lochs and big inlets such as characterise the west coast of Scotland and certain of the Isles, situations where there is some relief from fierce salt-laden winds and more violent tidal inundations. Even in more exposed places, where wind and spray may attenuate the plant cover to just a sparse scatter of more salt-tolerant plants, it is unlikely that the highest winter

tides flood stands every year and, since much of the vegetation consists of summer annuals, the effects of this may be minimal anyway.

Among the perennials, the rare *Mertensia* is, in fact, quite well adapted to disturbance of the beach material. It can tolerate inundation by sea-water, its seedlings and new shoots growing from established individuals can push up through a considerable thickness of sand, shingle and drift and, to a lesser extent, plants can withstand excavation of material from around them (Scott 1963*c*). Its robust habit is probably related to a peculiar pattern of root growth, in which the tap root splits each season, the strands becoming secondarily thickened and braided into a massive cable-like structure, often binding pebbles within it, with abundant laterals spreading sideways through the shingle (Skutch 1930). Despite this, however, more catastrophic shifts of beach material in exceptional tides and storms along more exposed shores can obliterate whole colonies of *Mertensia*, and the plant is rather striking in the way it comes and goes at particular stations within its range (Scott 1963*c*, Farrell 1989). Underlying such local changes, the distribution of *Mertensia* also appears to be contracting northwards, both in Britain and elsewhere in Europe, perhaps in response to a continuing movement of temperature zones towards the Arctic pole (Scott 1963*c*, Randall 1988).

Zonation and succession

The *Matricaria-Galium* community often occurs as an isolated strip of vegetation running along the strandline, although it is sometimes contiguous downshore with marine algal swards and, with a shift to finer beach material, can grade to upper salt-marsh vegetation. It is essentially a repeatedly-renewed pioneer assemblage and, even where it occurs on drift thrown high on to beaches, forming a front to dunes or various non-maritime communities, it cannot be seen as a seral precursor to them.

Along very sheltered shores, the lower edge of *Matricaria-Galium* stands can be regularly lapped by high tides and there the community may give way directly below to a zone of fucoids, usually *Fucus vesiculosus* or *F. spiralis*, with *Mertensia* in some of its localities, as on the Treshnish Isles off Mull, extending a little way into the littoral (Jermy & Crabbe 1978). In other places along the north-west coast of Scotland and on Shetland, the rare *Atriplex praecox* is abundant in a zone interposed between the strandline and the marine algae (Taschereau 1985). Elsewhere, with a shift to finer substrates or over the silt- or sand-smeared shingle that often forms a base for salt-marsh development around the hands of Scottish sea-lochs, the *Matricaria-Galium* community can peter out among upper salt-marsh vegetation. The most similar assemblage among these is the *Elymetum repentis*, a community of moist, drift-strewn soils around high tide mark, in which *Atriplex* spp., *R. crispus* and *Matricaria* continue to find a place but where the grasses occasionally seen in the *Matricria-Galium* vegetation, such as *Elymus repens*, *Festuca rubra* and *Agrostis stolonifera*, form the more extensive basis of a rank, weedy sward.

A further, very distinctive zonation can be seen around some Scottish sea-lochs where the *Matricaria-Galium* community has developed on drift abutting freshwater seepage zones along the beach top. Here, it can form a low weedy front to the *Filipendulo-Iridetum*, plants like the *Atriplex* spp., *Matricaria* and *Galium aparine* running in as a sparse understorey, together with *Agrostis stolonifera* and *Poa trivialis*, to *Iris pseudacorus*, *Filipendula ulmaria* and *Oenanthe crocata* in the *Urtica-Galium* or *Atriplex* sub-communities of the tall-herb fen.

Distribution

This kind of strandline vegetation is confined to Scotland and is more common along sheltered shores in the west.

Affinities

Apart from the community which Birks (1973) defined from Skye beaches and the rather particular stands with *Mertensia* which Scott (1963c) recorded in Arran, this distinctive vegetation has received only rather brief passing reference in the British literature (e.g. Jermy & Crabbe 1978). It is clearly a phytogeographic counterpart around our northern shores to the *Atriplex-Beta* community and similar assemblages have been recorded from Scandinavia (Nordhagen 1940, Dahl & Hadač 1941) and Iceland (Hadač 1970) but the exact affinities of these are disputed. In broad terms, this kind of vegetation is probably best placed among the Cakiletea strandlines, perhaps with the Salsolo-Honkenion communities among which Nordhagen (1940) characterised an *Atriplicetum laciniatae*, or with the Atriplicion littoralis strandlines of more silty substrates (Westhoff & Beeftink 1950, Géhu & Géhu 1969, Beckers *et al.* 1976), but it also has floristic links with the more brackish inundation communities of the Elymo-Rumicion crispi.

Floristic table SD3

Matricaria maritima	IV (1–8)
Galium aparine	IV (1–5)
Stellaria media	III (1–6)
Rumex crispus	III (1–7)
Atriplex glabriuscula	II (2–8)
Elymus repens	II (2–5)
Cakile maritima	II (1–6)
Atriplex patula	II (2–7)
Agrostis stolonifera	II (2–4)
Festuca rubra	II (2–4)
Atriplex prostrata	I (2–6)
Glaux maritima	I (1–4)
Armeria maritima	I (2–4)
Leymus arenarius	I (1–8)
Sonchus asper	I (1–2)
Cochlearia officinalis	I (1–3)
Holcus lanatus	I (1–3)
Honkenya peploides	I (1–3)
Juncus bufonius	I (1–3)
Mertensia maritima	I (1–5)
Poa annua	I (1–3)
Juncus gerardii	I (3)
Ligusticum scoticum	I (1–5)
Chamomilla suaveolens	I (1)
Plantago maritima	I (1–3)
Sonchus arvensis	I (2–5)
Scirpus maritimus	I (6–7)
Triglochin maritima	I (2)
Number of species	23
Number of species/sample	7 (2–15)

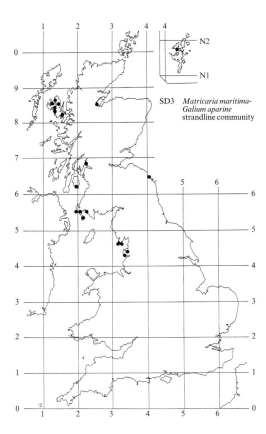

SD3   *Matricaria maritima-*
       Galium aparine
       strandline community

SD4
Elymus farctus ssp. *boreali-atlanticus* foredune community

Synonymy
Agropyretum juncei Moss 1906, Tansley 1911, 1939; *Agropyretum boreo-atlanticum* (Warming 1909) Br.-Bl. & De Leeuw 1936; *Elymo-Agropyretum junceiforme* Tx. 1955; *Agropyron junceiforme* stands Gimingham 1964*a*; *Elymo-Agropyretum boreo-atlanticum* Tx. (1937) 1967; Sociation à *Agropyron junceiforme* Géhu & Géhu 1969.

Constant species
Elymus farctus ssp. *boreali-atlanticus*.

Rare species
Euphorbia paralias.

Physiognomy
The *Elymus farctus* community comprises generally open, though often locally dense, vegetation in stretches of wind-blown sand, in which the dominant is the perennial grass long familiar as *Agropyron junceiforme*, but now known as *Elymus farctus* ssp. *boreali-atlanticus*. It is a rhizomatous plant, growing in the early stages after colonisation as small rosettes of shoots, often appressed to the surface, but then spreading outwards by means of its long and wiry underground stems and putting up vertical sympodial branches which by repeated tillering can keep pace with rapid though quite modest accumulation of sand (Nicholson 1952). Young colonies of the grass often have little more than small, low domes of sand around them but, where accretion progresses, distinct dunes, sometimes 1 m or more high, can develop. The grass shoots grow 20–60 cm tall and may be closely massed where the plants are especially vigorous, the glaucous foliage contrasting sharply with the bright yellow of the mobile sand.

In some stands, particularly around our northern coasts, *Leymus arenarius* invades together with or subsequent to *E. farctus*, its robust shoots often attaining more than a metre in height, but its cover here is always subordinate and, where it begins to dominate, the vegetation should be considered part of the *Leymus* foredune community. Other species are no more than occasional overall, though some can be locally frequent and abundant. *Honkenya peploides*, for example, is quite common and, provided sand accumulation is not too rapid, it will persist for some time among developing dunes and readily regenerate its patches of low shoots where beach-top stands are inundated by exceptional tides or the sand eroded in gales. Then, there can be scattered plants of a number of annuals characteristic of strandline vegetation, such as *Cakile maritima*, *Salsola kali*, *Atriplex prostrata*, *A. glabriuscula* and, more locally, *A. laciniata*, with very occasional *Matricaria maritima* and coarse weedy species like *Senecio jacobaea*, *S. squalidus*, *Cirsium arvense* and *Sonchus arvensis*. Around our more southerly coasts, there are sparse records in the community for *Chamomilla recutita* and the Oceanic Southern *Eryngium maritimum* and nationally rare *Euphorbia paralias*.

Where more substantial foredunes raise the level of the beach, *Ammophila arenaria* can invade the *Elymus farctus* community, and there is a continuous floristic transition between such vegetation and the *Ammophila* dune community where marram dominates.

Habitat
The *Elymus farctus* community is a pioneer vegetation type of wind-blown sand on foreshores around most of the British coast, developing along and above the strandline or among distinct foredunes.

E. farctus is well adapted to survival in raw sand soils close to the tidal limit, but its initial establishment in more exposed situations can be a somewhat precarious affair. It can colonise by both seed and rhizome fragments washed or blown on to the small patches of sand that accumulate around living strandline plants and their dead remains, or where slight irregularities along the beach top encourage accretion. Where there is less shelter from very high tides and wind erosion, however, and particularly through the more disturbed autumn and winter months, its persistence may depend on repeated invasion from already established stands nearby, as on foredunes behind (Harris & Davy 1986*a*). Burial of propagules under shifting beach material can also be a problem, their depth when growth starts being critical for successful shoot emergence. Both seedlings and single-noded pieces of rhizome, for example, have been shown to survive burial beneath 13 cm or so of sand, but to succumb under 18 cm (Harris & Davy 1986*b*). Multi-node fragments survive better and from greater depths, probably because of their larger reserves of accumulated nutrients and because the dormancy of subordinate buds makes more resources available for the dominant shoot, while retaining some flexibility of response should this apex be lost or overwhelmed. Even then, though, there is considerable variation in the regenerative ability of such propagules, perhaps because of fluctuations in nitrogen and carbohydrate reserves in the parent plants from which they break (Harris & Davy 1986*b*), the former nutrient being of especial importance in a habitat where exogenous supplies are unpredictable and patchily distributed (Lee *et al.* 1983). Tiller apices of *E. farctus* also have a vernalisation requirement and do not normally initiate flowers until their second year, so the poor survival of plants in the pioneer strandline stands inevitably results in low seed production. Rabbit-grazing can be a further important factor limiting inflorescence production, both in the more vulnerable colonising zone and among the foredune stands that can keep it supplied with seed (White 1961, Harris & Davy 1986*a*).

If these hazards are overcome, the young plants are able to consolidate their hold by putting out rhizome branches and roots into the underlying sand and even into the shingle that often forms a base to beaches (Tansley 1939, Nicholson 1952). *E. farctus* is also tolerant of periodic, brief immersion in sea-water (Gimingham 1964*a*, Chapman 1976) so, provided plants are strongly anchored, they are not damaged by any occasional extreme high tides that subsequently wash over the strandline, something which gives this species an important advantage over *Ammophila* in the colonising zone. Quite quickly, too, the plants themselves are able to help offset any loss of beach material through sea and wind erosion by encouraging accretion of sand among the virtually prostrate early shoots (Nicholson 1952). Where there is a net gain in material, *E. farctus* can keep pace to some extent by horizontal and vertical growth of the rhizome and shoot system, actually stimulating the formation of, first, low mounds of sand, then, if the process continues without any drastic erosive setback, small dunes. In such circumstances, growth of the plant can be very vigorous: at Blakeney, for example, Oliver (1929) reported that a single seedling gave rise to a dune more than 1 m high and 6 m across after just a few years,

and especially where adjacent colonies coalesce shoots can become densely crowded.

Some of the characteristic associates of the *Elymus farctus* community are survivors of the strandline assemblages among which the grass gains a hold and eventually comes to dominate, the perennial *Honkenya* able to grow through more shallow coverings of sand, annuals like *Cakile*, *Salsola* and *Atriplex* spp. continuing to invade afresh in spring and summer where the substrate remains sufficiently stable, and particularly where patches of drift detritus thrown high up on to the beach top give a firmer footing and a flush of nutrients. Where the sand is accreting quickly, though, these species can soon be reduced to a very sparse element in the vegetation, and few other dicotyledonous herbs invade along with the *E. farctus* to maintain any richness in the community. In the warmer south of the country, however, where mean annual maxima around our coasts are in excess of 25 °C (Conolly & Dahl 1970), the occasional appearance of more thermophilous dune perennials like *Eryngium maritimum* and *Euphorbia paralias*, which can keep a hold in quite mobile sand, brings some stands close to the more diverse foreshore assemblages found along the French Atlantic coast (Géhu & Géhu 1969).

More characteristic of this kind of vegetation around the cooler seaboard of north-west Europe and particularly well seen with us on beaches in northern England and Scotland, is the presence of *Leymus arenarius*. The natural southern boundary of the range of this Oceanic Northern grass is uncertain (Bond 1952) but probably related to temperature and, in congenial climatic conditions, it can invade the strandline along with *E. farctus*. It has been variously suggested, however, that it does not penetrate so close to the tidal limit as *E. farctus* (Turner 1977), that its taller shoots are more susceptible to wind damage (Bond 1952) and that it favours sands richer in organic matter (Géhu & Géhu 1969), each or all of which might play some part in restricting its role here as opposed to the *Leymus* foreshore community where it is dominant.

The contribution of *Ammophila* to the *Elymus farctus* vegetation, on the other hand, is often clearly limited by its greater susceptibility to tidal flooding, this grass appearing as an occasional here only where foredunes have raised the general level of the beach above the limit of inundation. Then, it may further encourage accretion beyond the limit of tolerance of *Elymus* itself, the original pioneer grass losing vigour and persisting with reduced cover.

Zonation and succession

The *Elymus farctus* community typically occupies a distinct zone on the foreshore, occurring above the strandline vegetation and fronting such other dune assemblages as are present. Along more exposed stretches of beach, perhaps in most of our dune systems now, it functions as a repeatedly renewed pioneer assemblage, but theoretically it can initiate a dune succession.

Very often, the *Elymus farctus* community occurs in close association with the *Honkenya-Cakile* vegetation and, quite commonly, the two form an ill-defined linear mosaic of vegetation strung out along and just above strandlines subject to varying periods of erosion and accretion. The two assemblages intergrade continuously, being distinguished according to whether dominance lies with *Honkenya* and the strandline annuals on the one hand or with *E. farctus* on the other, this reflecting the balance between continuing disturbance by sea and wind along and just above the tidal limit and progressive accumulation of sand. In other places, where accretion has been able to continue unchecked, a better-defined and wider zone of *Elymus farctus* vegetation can be seen above the strandline, distributed over gently undulating stretches of sand or on distinct dunes, these sometimes very few and irregularly distributed, in other places numerous, rising in height towards the beach top and developed in lines parallel to the shore. Particularly along the coasts of north-east England and in Scotland, this simple pattern can be complicated by transitions between *Elymus* vegetation and foreshore *Leymus* stands.

Along some coasts, such may be the limit of dune vegetation, the sequence of communities being abruptly terminated inland by a switch to agricultural enclosures or golf-course rough and greens on reclaimed and improved soils or to settlements and industrial developments. More extensive sequences occur quite widely, however, and here the *Elymus* community typically occurs on foredunes which front larger yellow dunes with sometimes, behind these, immobile dunes and stretches of undulating sandy ground. Usually, in such situations, the *Elymus* vegetation passes to *Ammophila* dune stands, *E. farctus* and, to a lesser extent, plants such as *Honkenya*, *Cakile* and *Atriplex* spp. retaining some representation in the transitional zone. Patterns of this kind can be seen all around the coasts of England and Wales, with *Leymus* vegetation and intermediate stands also figuring in the zonations along the coasts of Northumberland and at some sites in Lincolnshire. In its turn, the *Ammophila* community may give way to *Ammophila-Festuca* vegetation with *Festuca-Galium* assemblages becoming important in northern Britain.

The studies of Harris & Davy (1968a, b) along the north Norfolk coast have demonstrated very clearly that strandline stands of *Elymus farctus* are maintained in more disturbed situations only by continual replenishment from such vegetation on foredunes behind and suggested that the classic seral progression might not occur very readily.

Obviously, this has happened in past times, when it is likely that *Elymus* foredunes have been succeeded by *Ammophila* dunes and these in their turn by more stable tracts of vegetated sand. Along the outer fringes of our dune systems at the present time, however, the *Elymus* community may persist widely as a pioneer vegetation that is continually set back by more disturbed periods of wind and wave erosion.

Distribution
The community is found around most of the British coastline on suitably sandy beaches.

Affinities
The distinctive place of *E. farctus* among British foreshore vegetation was acknowledge in early studies (Moss 1906, Tansley 1911, 1939) by the characterisation of assemblages defined by its dominance, and an essentially similar approach has been adopted in more recent accounts, both descriptive (Gimingham 1964a) and phytosociological (Birse 1980, 1984). Here, too, the separation of the *Elymus farctus* community from other strandline and dune vegetation is by the cover contribution of the plant, but this is a reasonable diagnostic feature in view of its important ecological role on the foreshore.

Phytosociologically, our *E. farctus* vegetation is part of what was early defined as an *Agropyretum boreo-atlanticum* (Braun-Blanquet & de Leeuw 1936). More recent schemes have reserved this name (Westhoff & den Held 1969, White & Doyle 1982) or *Elymo-Agropyretum junceiformis* (Géhu & Géhu 1969, Birse 1980, 1984) to describe *E. farctus* stands from more northerly parts of Europe in which *Leymus arenarius* is a common feature, as with much of our vegetation of this kind. Along the warmer Atlantic coast of France, this community is seen as being replaced by a *Euphorbio-Agropyretum junceiformis* in which such species as *Eryngium maritimum*, *Euphorbia paralias* and *Calystegia soldanella* become constant (Géhu & Géhu 1969). These plants are generally no more than occasional among British stands, but locally high frequencies may bring the composition close to the *Euphorbio-Agropyretum* and Braun-Blanquet & Tüxen (1952) and Schouten & Nooren (1977) allocated some Irish dune vegetation to this association. Géhu & Géhu (1969) also characterised an *E. farctus* sociation comprising impoverished stands derived from either of the richer communities, but in this scheme these are simply subsumed. Traditionally, assemblages dominated by *E. farctus* have been grouped among the dune vegetation of the Ammophiletalia in an Elymo-Honkenion (Braun-Blanquet & Tüxen 1952), although Géhu & Géhu (1969) proposed redefining this alliance as what they termed an Agropyrion boreo-atlanticum, so as to avoid confusion with strandline *Honkenya* vegetation of the Salsolo-Honkenion.

Floristic table SD4

Elymus farctus	V (5–9)
Leymus arenarius	II (1–7)
Honkenya peploides	II (1–4)
Ammophila arenaria	II (2–5)
Cakile maritima	II (1–5)
Atriplex prostrata	I (1–3)
Atriplex glabriuscula	I (1–2)
Atriplex laciniata	I (1–4)
Cirsium arvense	I (2–5)
Eryngium maritimum	I (1–4)
Hypochoeris radicata	I (2–4)
Sonchus arvensis	I (1–5)
Elymus pycnanthus	I (3–7)
Festuca rubra	I (3–6)
Rumex crispus	I (2–4)
Senecio squalidus	I (2–5)
Matricaria maritima	I (1–4)
Taraxacum officinale	I (1–2)
Agrostis stolonifera	I (1–5)
Cirsium vulgare	I (1–4)
Chamomilla recutita	I (4–5)
Salsola kali	I (3–5)
Senecio jacobaea	I (1–2)
Senecio vulgaris	I (1–4)
Sonchus asper	I (1–2)
Cerastium diffusum diffusum	I (2–3)
Euphorbia paralias	I (1)
Polygonum oxyspermum	I (2)
Artemisia maritima	I (3)
Number of samples	51
Number of species/sample	4 (1–11)

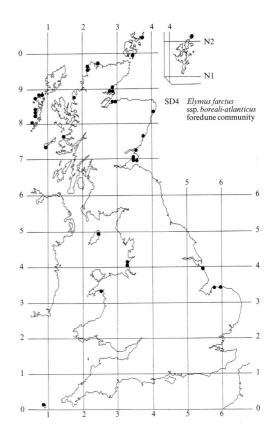

SD4 *Elymus farctus* ssp. *boreali-atlanticus* foredune community

SD5
Leymus arenarius mobile dune community

Synonymy
Ammophiletum arenariae Tansley 1911, 1939 *p.p.*;
Elymo-Ammophiletum arenariae Br.-Bl. & De Leeuw
1936 *p.p.*; *Leymus arenarius* consocies Bond 1952;
Ammophila arenaria stands Gimingham 1964*a p.p.*;
Potentillo-Elymetum arenariae (Raunkiaer 1965) Tx.
1966 *sensu* Birse 1980.

Constant species
Leymus arenarius.

Physiognomy
The *Leymus arenarius* community consists of open to
locally dense stands of dune vegetation dominated by
the tall and tussocky perennial grass *Leymus arenarius*.
It is a rhizomatous plant, able to colonise and fix mobile
sand and keep pace with substantial accumulation by
upward and outward extension of its stout buried stems,
such that the robust glaucous shoots, often well over 1 m
high, can be found emerging from dunes that have grown
to several metres tall (Trail 1904, Bond 1952).

No other species is constant throughout, but *Elymus
farctus* is locally common, invading foreshore sand
ahead of or together with the *Leymus* and sometimes
remaining as a subsidiary and shorter element to the
cover here, though only thickening up in vegetation
which is best regarded as transitional to *Elymus farctus*
foredunes. *Ammophila arenaria* is likewise never a prominent feature here, but it is a characteristic colonist of
wind-blown sand along with or subsequent to *Leymus*
and can be found as an occasional in the community,
going on to exceed the lyme grass in frequency and abundance among the various kinds of *Ammophila* dunes.
Then, in some stands, *Festuca rubra* is quite common
with *Elymus repens* sometimes marking out places where
there has been incorporation of organic detritus.

Other associates are sparse, and are mostly strandline
plants persisting among the developing dunes. Patches
of *Honkenya peploides* occur occasionally, for example,
and there may be scattered individuals of *Cakile marit-*
ima, *Atriplex prostrata*, *A. glabriuscula*, *Sonchus arvensis*, *S. asper*, *Cirsium arvense*, *Galium aparine* and *Rumex
crispus*, but sand accretion rapidly overwhelms the more
vulnerable of these. Where somewhat more stable
patches of ground develop, species such as *Cerastium
fontanum*, *Hypochoeris radicata*, *Poa pratensis*, *Taraxacum officinale* and some bryophytes can figure in small
amounts.

Sub-communities

Species-poor sub-community. In many stands of this sub-
community, there is nothing but *Leymus*, sometimes
sparse in newly-colonising or moribund vegetation, in
other cases very vigorous and dense. Even where other
species are found, there is no real consistency in their
occurrence, although big patches of *Honkenya* can be a
distinctive feature where this vegetation is developing
close to the strandline, with tussocks of *Ammophila*
occasionally colonising stands further up the beach.

Elymus farctus sub-community. Although *Leymus*
remains a clear dominant here, small amounts of *E.
farctus* are a constant feature, with very occasional *Honkenya*, *Atriplex prostrata*, *Sonchus arvensis* and *Ammophila*.

Festuca rubra sub-community: *Potentillo-Elymetum arenariae* (Raunkiaer 1965) Tx. 1966 *sensu* Birse 1980. *E.
farctus* can be occasional in this sub-community too, but
much more characteristic is the high frequency of
Festuca rubra beneath the *Leymus*, thickening up in
places where the canopy of the dominant is a little less
dense. Some other grasses also play a minor role in the
ground cover, with *Elymus repens* quite common, *Poa
pratensis* and *Holcus lanatus* more sparse, and *Ammophila* again occurring as scattered tussocks in some
stands. Among the dicotyledons, *Sonchus arvensis* is
especially frequent, with *Atriplex glabriuscula* often
commoner than *A. prostrata*, and occasional plants of

Cirsium arvense, Cakile, Sonchus asper and, on more consolidated sand, *Cerastium fontanum, Hypochoeris radicata* and *Taraxacum officinale*. It is among this kind of *Leymus* vegetation, too, that bryophytes can make a sparse contribution, with records for *Brachythecium rutabulum, Ceratodon purpureus, Tortula ruralis* ssp. *ruraliformis* and *Bryum algovicum* ssp. *rutheanum*.

Habitat

The *Leymus* community is a locally important early colonising vegetation of wind-blown sand above the strandline and on young dunes around the more northerly coasts of Britain.

Leymus is an Oceanic Northern plant (Matthews 1955), widely distributed around the seaboard of north-west Europe with occasional apparently natural stations inland, though not in Britain. The true southern limits of its range are uncertain, but it has been reported only as a casual or an introduction from Portugal and around the Mediterranean (Tutin *et al*. 1980), and is probably restricted at lower latitudes by the warmer and drier climate. In Britain, too, it is noticeably more common and abundant around our cooler and more humid northern shores, being rare on or altogether absent from some major dune systems in the south and west of the country, though successfully planted at certain localities in Cornwall and Sussex (Bond 1952, Perring & Walters 1962). The community becomes locally prominent in moving up the Irish Sea coast, but it is on our somewhat more sheltered eastern shores that its increasing contribution towards the north is more obvious, particularly from Northumberland up into Scotland, Orkney and Shetland. Even at some places in these parts of the country, though, there is evidence or suspicion of planting (Bond 1952).

In Britain, *Leymus* is almost exclusively a colonist of sandy substrates, occurring only very occasionally on coarser beach material (Tansley 1939), a preference which probably restricts its invasion along some shores within its overall range, as around much of the north-west Scottish mainland. It can colonise from seed (Graham 1938) or from rhizome fragments, even small ones (Bond 1952), readily getting a hold in patches of sand that have accumulated around strandline plants or their remains, in slight hollows along the beach top and on already established foredunes. Here, its habitat overlaps considerably with that of *E. farctus*, and the *Elymus* sub-community contains those stands where this smaller grass has not been overwhelmed by sand accretion. In general, however, *Leymus* does not seem to extend as close to the tidal limit as does *E. farctus* (Turner 1977), perhaps because of the greater vulnerability of its taller shoots to physical damage by wind (Bond 1952).

It has also been suggested that the *Leymus* community prefers sands rich in organic matter (Géhu & Géhu 1969). Certainly, the *Festuca* sub-community is espe-cially associated with sites where seaweed and other debris has been deposited on or incorporated into the surface of the beach (Birse 1980), the varied associated flora of ephemerals and weedy herbs reflecting the nutrient-rich conditions. The short extension inland of *Leymus* vegetation along the sandy banks of some streams that debouch on to the foreshore has also been adduced as evidence of a nitrophilous tendency in the grass itself (Bond 1952), as well as for its tolerance of more brackish conditions (Géhu & Géhu 1969).

Sand accumulation among *Leymus* tussocks is not always a progressive phenomenon and bouts of wind erosion, exceptional high tides or storm surges can set back the process. Where accretion is more rapid, however, and where there is no replenishment of drift detritus, any strandline survivors are quickly overwhelmed, the Species-poor sub-community tending to develop. There is some evidence that the buds of *Leymus* themselves are swamped by a covering of only 6–8 cm of sand (Ranwell 1959), but internode differentiation during autumn and winter and rhizome elongation in spring and summer seem to be paced according to the rate of deposition, such that growth can keep up with considerable increase in height of the developing dunes, with new buds being differentiated at the base of the current year's shoots ready for extension upwards and outwards in the following season (Bond 1952). Provided the water supply remains adequate and the dune system stays stable, the *Leymus* community appears to retain its vigour on sand hills up to 5 m high (Trail 1904, Bond 1952).

Zonation and succession

In many dune systems, particularly in southern and western Britain, *Leymus* has at most a minor role among younger dunes carrying the *Elymus farctus* and *Ammophila* communities but, along those coasts where *Leymus* vegetation is better developed, it usually dominates in a more or less well-defined zone between these two other assemblages, or tends to replace the *Ammophila* community as the most important builder of mobile dunes. In some areas, the *Leymus* community has increased its extent considerably within living memory but, even where it forms part of more extensive sequences, backed by other vegetation types on shifting or fixed dunes, such zonations do not necessarily represent straightforward successional developments.

Through much of its range, *Leymus* vegetation occurs in close association with the *Elymus farctus* foreshore community, sometimes replacing it in a clear zonation with a shift a little way up the beach away from the strandline with its more severe bouts of wind erosion and tidal disturbance, in other cases occurring intermixed with it in a complex linear mosaic over small foredunes running along the shore. Floristically, the two assemblages intergrade continuously through the *Elymus* sub-community of the *Leymus* vegetation, but changes in dominance from the one grass to the other effectively distinguish them, the *E. farctus* community also generally retaining a more consistent scattering of strandline survivors like *Honkenya*, *Cakile* and *Atriplex* spp.

Where *Ammophila* also invades mobile sand in which *Leymus*, and sometimes *E. farctus* too, have got a hold above the limit of even the extreme tides, the zonations are more complex. In such situations, the *Leymus* community may give way behind to a distinct zone of *Ammophila* vegetation on bigger shifting dunes, but there can be transitional stretches where *Leymus* remains as an important subsidiary element among the marram. This *Leymus* sub-community of the *Ammophila* vegetation has been recorded in these sequences at scattered localities up the east coast of Britain and occasionally on western shores. In other places, particularly over foredunes along the Lincolnshire and Humberside coasts and more locally in Northumberland, *E. farctus* also plays a prominent role right through a compressed zonation, with the *Leymus-Elymus* sub-community of *Ammophila* vegetation occurring behind the *Elymus* type of *Leymus* dune.

In shifting further north up the eastern coast of Britain, the *Leymus* community tends to become more important among the vegetation of the younger mobile dunes, giving way behind to the somewhat more stable kinds of *Ammophila* assemblage, of the *Festuca* and *Poa* sub-communities, on sand that is a little more fixed. *Leymus* can persist as an occasional there and the *Festuca* sub-community of *Leymus* vegetation, with its scattered plants of *Elymus repens*, *Sonchus arvensis* and *Hypochoeris radicata* sometimes forms a transitional zone. But, eventually, such lyme grass as has not been overwhelmed by sand accretion loses its vigour and dies out, the plant hardly ever extending into the *Ammophila-Poa* or *Festuca-Galium* communities that cover much of the fixed sand of dunes and stretches of machair around our northern coasts.

Distribution
The *Leymus* community is scarce around our southern shores from Suffolk to North Wales, but it occurs at scattered localities around the Irish Sea and becomes increasingly common up the east coast, especially from Northumberland into Scotland.

Affinities
It has not been customary, in either descriptive or phytosociological schemes, to recognise a distinct *Leymus* dune community, stands with abundant, even dominant, lyme grass being subsumed into the kind of *Ammophiletum* of mobile dunes familiar from Tansley (1911, 1939) or its equivalent in Continental classifications, the *Elymo-Ammophiletum* (Braun-Blanquet & de Leeuw 1936, Birse 1980, 1984), where *Leymus* is considered a good indicator of the cool, oceanic conditions of the north-west European seaboard (Géhu & Géhu 1969). Some authors, however, have acknowledged the existence of pure *Leymus* stands in Britain (Bond 1952, Gimingham 1964*a*) and it seems sensible to retain a separate unit for that vegetation in which the plant is overwhelmingly abundant, parallel to the *Ammophila* community, intergrading with it, though extending closer to the tidal limit and having a distinctly northern distribution around our coasts. Such vegetation type could readily take in transitions to *Elymus farctus* foredunes and also the swards described from drift-rich Scottish shores by Birse (1980) as part of the *Potentillo-Elymetum arenariae* (Raunkiaer 1965) Tx. 1966, an association with clear affinities with the inudation communities of the Elymo repentis-Rumicion. As defined here, the *Leymus* community is best placed with our *Ammophila* vegetation of mobile sands in the Ammophilion borealis, the alliance of assemblages from younger coastal dunes. An alternative view would stress the occurrence of *Leymus* on often drift-enriched foreshores and locate it with other strandline communities in a Honkenyo-Crambion (Géhu & Géhu 1969), which White & Doyle (1982) thought sensibly placed in the Honkenyo-Elymetea arenariae, a class created by Tüxen (1966) to accommodate maritime inundation vegetation, with *Leymus arenarius* as a characteristic species. This would hardly be borne out by the behaviour of the plant in Britain.

Floristic table SD5

	a	b	c	5
Leymus arenarius	V (7–9)	V (5–8)	V (2–9)	V (2–9)
Elymus farctus		V (3–7)	II (1–5)	III (1–7)
Honkenya peploides	I (5–6)	I (3)		I (3–6)
Festuca rubra			V (1–8)	II (1–8)
Sonchus arvensis	I (1–4)	I (1–2)	III (1–5)	II (1–5)
Elymus repens	I (2–5)		II (3–5)	I (2–5)
Atriplex glabriuscula	I (4)		II (1–5)	I (1–5)
Cakile maritima			I (1–4)	I (1–4)
Cirsium arvense			I (1–3)	I (1–3)
Hypochoeris radicata			I (1–5)	I (1–5)
Poa pratensis			I (2–5)	I (2–5)
Bryum algovicum rutheanum			I (2–5)	I (2–5)
Taraxacum officinale			I (1–3)	I (1–3)
Epilobium angustifolium			I (2–4)	I (2–4)
Holcus lanatus			I (1–2)	I (1–2)
Sonchus asper			I (1–3)	I (1–3)
Brachythecium rutabulum			I (3–5)	I (3–5)
Ceratodon purpureus			I (1–5)	I (1–5)
Tortula ruralis ruraliformis			I (4–5)	I (4–5)
Atriplex prostrata	I (6)	I (1–2)	I (1–2)	I (1–6)
Ammophila arenaria	I (7)	I (5)	I (5)	I (5–7)
Cerastium fontanum	I (1)		I (1–2)	I (1–2)
Galium aparine	I (1)		I (1)	I (1)
Elymus pycnanthus		I (2)	I (2)	I (2)
Rumex crispus		I (1)	I (1)	I (1)
Number of samples	7	6	15	28
Number of species/sample	2 (2–5)	3 (2–5)	8 (4–17)	5 (2–17)

a Species-poor sub-community
b *Elymus farctus* sub-community
c *Festuca rubra* sub-community
5 *Leymus arenarius* mobile dune community (total)

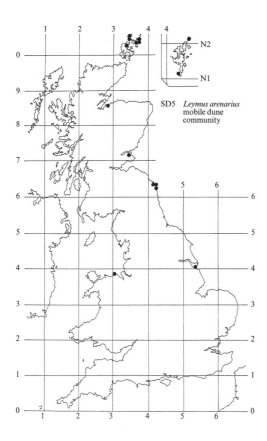

SD5 *Leymus arenarius*
     mobile dune
     community

SD6
Ammophila arenaria mobile dune community

Synonymy
Ammophiletum arenariae Moss 1906, Tansley 1911, 1939, all *p.p.*; *Elymo-Ammophiletum arenariae* Br.-Bl. & De Leeuw 1936; *Ammophila arenaria* stands Gimingham 1964*a p.p.*

Constant species
Ammophila arenaria.

Rare species
Allium ampeloprasum ssp. *babingtonii, Corynephorus canescens, Euphorbia paralias, E. portlandica, Scrophularia scorodonia.*

Physiognomy
The *Ammophila arenaria* community includes virtually all the vegetation of more mobile coastal sands in which the robust perennial grass *Ammophila arenaria* plays a dominant role in the dune-building process. In young stands, or where newly eroded surfaces among established dunes are being recolonised, the cover of the plant is often low, with just a few sparse shoots emerging from freshly-blown sand that may be otherwise quite bare. Growth can become very vigorous, however, in actively-accreting material where *Ammophila* can spread its horizontal rhizomes far and keep pace with burial of up to 1 m a year, the expanding clones putting up densely-branching tillers that become aggregated into the characteristic tussocks (Gemmell *et al.* 1953, Greig-Smith 1961, Ranwell 1972). The shoots enhance accretion even more, continuing to proliferate among the upbuilding sand, until they can be packed at 150–200 m^{-2}. Adjacent tussocks, with their spreading foliage reaching 1 m or more in height, can thus enlarge to form an extensive cover of the grass over stretches of still shifting sand that is being built into often substantial dunes. As the substrate becomes more fixed, however, the vigour of the *Ammophila* in older stands of the community declines, with a reduction in shoot density to about half that in mobile sand, although bouts of erosion can stimulate

fresh bursts of growth and a renewal of this vegetation in blow-outs (Tansley 1939, Ranwell 1960*a*, Huiskes 1979).

Almost all British *Ammophila* is *A. arenaria* ssp. *arenaria*, although the North American *A. breviligulata* (differing only in its truncate ligule and sometimes regarded as *A. arenaria* ssp. *breviligulata*: Maire 1953) was planted on some our dune systems following the 1953 coastal floods (Hubbard 1968) and persists as an element in the community at Newborough in Gwynedd (Huiskes 1979). The hybrid marram × *Ammocalamagrostis baltica*, a sterile intergeneric cross with *Calamagrostis epigejos* and a bulkier plant even than *Ammophila*, can also be found among this vegetation on dunes in Norfolk and Northumberland. These populations, like that on the cliffs of Handa in Sutherland seem to be of natural origin, although this grass, too, is regarded as a valuable sand-binder (Huiskes 1979) and was introduced at some new sites along the East Anglian coast after 1953 (Hubbard 1968).

In many stands, particularly where the sand remains very mobile, marram is the only plant in this vegetation and it is the sole constant of the community. Some other grasses, though, have a subordinate but locally prominent place here, and they provide important links with different kinds of dune vegetation. In younger stands, for example, where *Ammophila* is invading foredunes, *Elymus farctus* can persist for some time before being overwhelmed by the accumulating sand. Then, especially along the north-east coast of Britain, *Leymus arenarius* is a common colonist ahead of or along with *Ammophila* and it can remain as dense tufts or clusters of shoots, often a little taller than the marram and a more striking grey-green, in stands of the community on mobile dunes of up to moderately large size. Much less structurally important, but noticeable here as an invader of more open areas among the *Ammophila* where sand has recently settled is *Carex arenaria*. On somewhat more stable dunes, on the other hand, *Festuca rubra* often becomes the most frequent associate in this vegetation, being commonly recorded as the var. or ssp. *arenaria* with its stiff tufts of rather glaucous leaves and extensively creeping rhizomes (Tutin *et al.* 1980). In the moister and cooler climate of northern Britain, this is frequently accompanied by what is grouped here as *Poa pratensis* agg., but which is probably often *P. subcaerulea*, a plant that is not so strongly tufted as *P. pratensis s.s.* and where both glumes have three veins (Tutin *et al.* 1980). Even these commoner grasses never form anything like a continuous turf beneath the *Ammophila* in this community, but much more occasional, and usually only of very patchy abundance even on less mobile sand, are *Elymus repens*, *E. pycnanthus*, *Holcus lanatus*, *Dactylis glomerata*, *Agrostis stolonifera* and *Poa annua*. A further interesting species which sometimes finds a place in this vegetation at a few localities along the East Anglian coast is *Corynephorus canescens*, a nationally rare perennial grass that can become locally abundant among the *Ammophila* provided accretion is not too rapid, at least when the plant is establishing (Marshall 1967).

As with the grasses, no other associates occur at all frequently throughout the community, though the commonest group comprises more short-lived herbs, often of a rather coarse and weedy character, which can take advantage of periodically stable conditions among the developing dunes. Closer to the strandline, for example, where this vegetation can develop on foredunes, plants like *Cakile maritima* and various *Atriplex* spp. may figure, together with surviving patches of the perennial *Honkenya peploides*. Then, *Sonchus arvensis*, *S. asper*, *Senecio vulgaris*, *S. jacobaea*, *S. squalidus*, *Cirsium vulgare*, *C. arvense* and *Rumex crispus* all occur with varying degrees of frequency through the community, sometimes becoming locally abundant over particular stretches of dune, only to disappear completely in the next season. More especially characteristic of the sandy maritime habitat, though confined to stands in the warmer, southerly part of the range of this vegetation, are *Eryngium maritimum*, *Calystegia soldanella* and the rarities *Euphorbia paralias* and *E. portlandica*, these often establishing as scattered individuals but, except for the last, able to persist for some years where the sand is not too mobile. The community can also provide one of the loci for the rare *Scrophularia scorodonia* and *Allium ampeloprasum* var. *babingtonii* on some dune systems in south-west England (Perring & Farrell 1977).

Other plants found here, though usually where the sand has become somewhat stabilized, include *Hypochoeris radicata*, *Taraxacum officinale*, *Ranunculus repens*, *Trifolium repens*, *Galium verum*, *Tussilago farfara*, *Cynoglossum officinale*, *Linaria vulgaris* and *Erodium cicutarium*. More local, though usually noticeable by virtue of their tall stature are *Heracleum sphondylium*, *Angelica sylvestris* and *Epilobium angustifolium* and *Rubus fruticosus* agg., this last sometimes forming small patches among the *Ammophila*. The Californian

Lupinus arboreus has also become naturalised in this community on various dune systems around our southern coasts in dense bushy growth 2 or 3 m high (Nicholson 1985).

Bryophytes are much more a characteristic feature of the *Ammophila-Festuca* vegetation that develops on less mobile sand than is usual here, but a few species begin to make an appearance at very low overall frequencies but occasionally with some abundance where the surface has become more stable: *Ceratodon purpureus*, *Tortula ruralis* ssp. *ruraliformis* and *Brachythecium albicans*.

Sub-communities

Elymus farctus sub-community. Small amounts of *E. farctus* are a constant feature among the *Ammophila* here in what is often very open vegetation, with occasional low patches of *Honkenya* but otherwise just sparse individuals of *Cakile maritima*, *Senecio jacobaea*, *S. vulgaris*, *Sonchus arvensis*, *S. asper*, *Cirsium vulgare*, *C. arvense*, *Eryngium maritimum* and *Calystegia soldanella*.

Elymus farctus-Leymus arenarius sub-community: *Elymo-Ammophiletum typicum* Br.-Bl. & De Leeuw 1936 *sensu* Birse 1980 *p.p.* Both *E. farctus* and *L. arenarius* accompany *Ammophila* in this sub-community, one or other or both sometimes sub-dominant, with occasional records for *Atriplex* spp. and very much the same associates as in the above.

Leymus arenarius sub-community: *Ammophila* with scattered *Elymus* Gimingham 1964a; *Elymo-Ammophiletum typicum* Br.-Bl. & De Leeuw 1936 *sensu* Birse 1980 *p.p.* *L. arenarius* is a constant with *Ammophila* here and sometimes quite abundant, but other plants are usually very few, just sparse individuals of the weedy community associates, often without strandline survivors.

Ammophila arenaria sub-community. *Ammophila* is often overwhelmingly dominant in this sub-community with scattered records for a wide variety of associates, but no consistency of enrichment from stand to stand. *Senecio jacobaea* is the commonest companion, but *Sonchus asper* and *Cirsium arvense* are also fairly frequent and each may be locally quite abundant. Less common, but rather striking on some of our more southerly dune systems, are *Calystegia soldanella*, *Eryngium maritimum*, *Euphorbia paralias* and *E. portlandica*. *Rumex crispus*, *Hypochoeris radicata* and *Festuca rubra* occur occasionally, too, and there can be very sparse bryophytes.

Festuca rubra sub-community. *Ammophila* is still very much the dominant here, accompanied in some stands by smaller amounts of *Leymus*, but the vegetation cover between the tussocks and the variety of the associated flora are considerably increased. Most obviously, *F. rubra* is a constant companion, usually growing as scattered plants, though sometimes occurring with moderate local abundance, with *Senecio jacobaea*, *Hypochoeris radicata* and *Cirsium arvense* becoming quite common. *Eryngium*, *C. soldanella* and the *Euphorbia* spp. continue to add a distinctive character to more southern stands and then, along with occasional records for community associates like *Sonchus* spp., *Senecio vulgaris* and *Rumex crispus*, there can be scattered individuals of *Taraxacum officinale* agg., *Ranunculus repens*, *Galium verum*, *Trifolium repens* and *Tussilago farfara*. Rather more striking on this kind of dune, though never very frequent, are *Cynoglossum officinale*, *Erodium cicutarium*, *Linaria vulgaris* and *Trifolium arvense*. Then, there can be very occasional taller herbs like *Angelica*, *Heracleum* and *Epilobium angustifolium* with sparse patches of *Rubus fruticosus* agg. and, at some sites, bushes of invading *Lupinus*. Other grasses apart from *F. rubra* can sometimes occur, such as *Holcus lanatus*, *Dactylis glomerata*, *Elymus repens* and *E. pycnanthus*, but these are usually no more than scattered tussocks, and *Poa pratensis* is very scarce. Bryophytes can form locally quite extensive patches.

Poa pratensis sub-community: *Elymo-Ammophiletum*, *Festuca rubra* Subassociation Br.-Bl. & De Leeuw 1936 *sensu* Birse 1980. *F. rubra* remains very frequent between the dominant *Ammophila*, but *P. pratensis* is also constant in this sub-community, these two grasses, occasionally with *Elymus repens* and a little *Holcus lanatus* or *Dactylis glomerata*, sometimes forming a moderately extensive, though still patchy, cover. *Senecio jacobaea*, *Cirsium arvense* and *Sonchus arvensis* are again quite common, with occasional *Heracleum sphondylium*, *Cynoglossum officinale*, *Angelica sylvestris*, *Epilobium angustifolium*, *Ranunculus repens*, *Trifolium repens* and *Galium verum*, but more thermophilous herbs of southern dunes are very scarce. As in the *Festuca* sub-community, bryophytes may be locally quite abundant.

Carex arenaria sub-community. Usually this vegetation comprises little more than abundant *Ammophila* with sparse to quite plentiful *C. arenaria*, the sedge often extending out in its distinctive runs of shoots over the bare sand among the marram.

Habitat

The *Ammophila* community is the most widespread and extensive colonising vegetation of mobile sands above the limit of tidal inundation all around the British coast, dominating on young dune ridges and in blow-outs until such time as reduced accretion limits the vigour of the

marram and allows an increase in the extent and diversity of the associated flora. Variation in the floristics and structure of the vegetation relates largely to the waning influence of tidal disturbance towards the lower limit of marram invasion and, in the opposite direction, to the increased stability of the sand surface.

Ammophila is the most important colonist of wind-blown sand above the tidal limit around the entire coast of Britain and the plant best adapted to the inhospitable environment characteristic of the mobile dunes that can develop there. Our own ssp. *arenaria* is widely distributed around the seaboard of north-west Europe, its northern limit at the 0 °C January isotherm (Salisbury 1952) occurring well beyond our shores in the Faeroes and in southern Norway, and only very locally in this country is its invasive and dune-building potential on suitable substrates out of reach of the tides rivalled. Planted × *Ammocalamagrostis* can be more vigorous, perhaps because, being sterile, it needs to spare no resources for seed production (Huiskes 1979), but *Leymus arenarius*, though a vigorous colonist along with *Ammophila* along our north-eastern shores, is eventually overwhelmed by more than a few metres of accumulated sand. The *Leymus* sub-community here is thus generally characteristic of younger mobile dunes, or the lower seaward faces of bigger sandhills carrying this kind of vegetation.

Leymus is also tolerant of occasional tidal inundation and can invade sandy patches closer to the strandline than is possible for *Ammophila*. Marram needs a substrate with less than 1% sea salt (Huiskes 1979), indeed experimentally it is almost always killed by salinities of only 1.5% (Benecke 1930), so it only becomes an important colonist where the beach level has already been raised out of reach of even extreme tides. However, where *Ammophila* invades foredunes, more salt-tolerant strandline survivors can persist for some time, *Honkenya*, *Cakile* and *Atriplex* spp., as well as *Elymus farctus*, the grass most tolerant of tidal submersion, providing a distinctive character to the *Elymus* and *Elymus-Leymus* sub-communities.

In such situations, *Ammophila* can colonise from both seed and rhizome fragments. Seed production in well-established plants can be prolific, Salisbury (1952) estimating an average of 20000 caryopses annually from vigorous tussocks, and dispersal occurs readily by wind. Viability is high, and germination succeeds best in the strongly-fluctuating temperature regime characteristic of mobile dunes (Kinzel 1926, Willis *et al.* 1959b, Stoutjesdijk 1961) with a marked flush in April and May after a winter chilling (Huiskes 1979). Even slight burial, however, beneath just a few centimetres of sand, greatly reduces germination and subsequent disturbance of the surface or desiccation kills many survivors (Huiskes 1979). Dampening of the ground greatly increases the chance of seedlings getting away (Huiskes 1977b), particularly on the more open surfaces in younger dunes, although even then, it is at least two years before inflorescences are produced.

Meanwhile, however, the young plant can get a firm hold in the substrate, a main shoot developing from the caryopsis or bud on the rhizome fragment, axillary buds giving rise to daughter tillers or new rhizomes, these growing upwards and outwards with repeated sympodial branching (Greig-Smith et al. 1947, Gemmell et al. 1953, Greig-Smith 1961). Once established, vegetative spread of the clones by such rhizome extension and proliferative shoot production is far more important within stands of the community than reproduction from seed (Huiskes 1979), though where conditions are suitable new seedlings can continue to find a place among the older tussocks, helping to infill the cover. However, only in foredune stands, or where the community develops around young slacks, does shoot production continue to result in a net increase in tillers per unit area, a process which continues until what is apparently the shoot-carrying capacity of the ground is reached. The tillers are, in fact, monocarpic although, even in mobile sand where inflorescences are mainly found, many tillers die from other causes and, in established stands, shoot production involves mainly a regenerative replacement of tillers that have died (Huiskes 1977a, 1979).

The real key to the success of Ammophila in this environment, however, is its impressive ability to keep pace with burial by the blown sand. The rhizomes and roots help bind the substrate against erosion and the densely-crowded tillers enhance accretion by reducing the speed and sand-carrying capacity of the wind (Willis et al. 1959b) but, when buried, the internodes on the shoots elongate, effectively converting them into vertical rhizomes which are able to maintain upward growth through 1 m or so of sand a year (Grieg-Smith et al. 1947, Ranwell 1972). By producing more daughter tillers and horizontal rhizomes, the plant can thus re-establish dominance on the upbuilding surface it has helped to create. Conversely, where erosion strips back the substrate or where whole sections of dune shift to expose deeper layers of sand, dormant buds on buried rhizomes can regenerate a cover of tillers on the freshly-exposed surface.

Accretion and ablation on the scale characteristic of mobile dunes are inimical to most potential competitors of Ammophila and help maintain the sands in a raw and uncongenial state. The material freshly blown from the beach is of varying composition, though often very calcareous where the proportion of shell fragments is high: pH is generally within the range 6–9 and, with any tendency to leaching outpaced by fresh deposition, fairly constant with depth (Willis et al., 1959b, Marshall 1967, Huiskes 1979). In fact, although Ammophila itself is reported as absent from soils with a pH below 4.5 or so (Lux 1964, 1966) and shows depressed shoot densities where this vegetation occurs on more acid dunes, as at Winterton in Norfolk (Boorman in Huiskes 1979), its growth seems to be little affected by differences above neutral, and the community associates rarely reflect variations in this factor.

More limiting to the character and vigour of the vegetation is the impoverished nutrient content of the mobile sands, whether these are alkaline or acidic (Willis et al. 1959a, Wilson 1960, Willis & Yemm 1961). Away from the foot of foredune stands, where more nitrophilous herbs can still capitalise on flushing from strandline debris in the Elymus and Elymus-Leymus sub-communities, organic carbon is in extremely short supply, with very low amounts, too, of total nitrogen and available N, P and K. The growth of Ammophila itself is markedly affected by such deficiences and much stimulated by experimental additions of mineral nutrients (Lux 1964, Willis, 1965). After a two-year period of additions at Braunton Burrows in Devon, for example, marram showed a three-fold increase in fresh weight of the shoot system with enhanced tillering, taller shoots and wider leaves, and a four-fold increase in the root systems, with dense mats of fine roots developing. Glasshouse sand culture revealed here that nitrogen shortage was the major limitation and also that adding purified sand to the established cultures continued to promote growth by providing an opportunity for new root formation by the buried nodes. Other authors, like Hassouna & Wareing (1964) have suggested that Ammophila may be able to avoid some of the effects of nitrogen-limitation by capitalising on the activity of non-symbiotic nitrogen-fixers in the rhizosphere, and it is known that Azotobacter occurs in mobile dune sand (Géhu 1960). Abdel-Wahab (1969, 1975) found that Bacillus spp. also played a role in nitrogen-fixation in mobile dunes, although Pegtel (1976) and Woldendorp (in Huiskes 1979) have expressed doubts about the contribution of this process, dependent as it is on organic carbon, itself in very short supply in this habitat.

The second important factor limiting the cover and composition of this vegetation on mobile sand is water, particularly as the dune surface and the rooting zone of the plants are raised by accretion to what is often a very considerable distance above the ground water-table, over 26 m at Braunton, for example (Willis et al. 1959a). Most of the water available to the community comes from rain and the level of the water-table is strongly dependent on the amount of precipitation in the preceding period. Where dunes develop on an elevated rocky base and where there is substantial drainage in from the surrounding area, the water regime can be very complex. If they accumulate on low ground with its own catchment, then the hydrology is that of a virtually isolated granular deposit, the water-table is domed and there is an essentially rectilinear relationship between its level and rainfall (Willis *et al.* 1959*b*, Willis 1985*a*). Close to the water-table, the sand is maintained near saturation capacity but capillary rise is poor in dune sand, reaching only 30–50 cm at Braunton, for example (Willis *et al.* 1959*b*; see also Olsson-Seffer 1909, Ranwell 1959). Thus, although *Ammophila* regularly puts down roots below 2 m depth (Salisbury 1952, Willis in Huiskes 1979), with other associates like *Euphorbia portlandica* and *Hypochoeris radicata* often reaching 50 cm below the surface (Willis 1985*a*), the free water-table can be of little significance for growth on even moderately high dunes.

Much of the available water is thus probably intercepted as it drains through the sand after episodes of rain which can temporarily increase the moisture content tenfold. But drainage is generally very sharp, especially through the superficial layers, even in finer sands, because the amount of retentive organic matter is so small, less than 1% among mobile dunes under this vegetation at Braunton, for example (Willis & Jefferies 1963). Field moisture capacity here may thus be less than one-quarter of that in fixed dune sand (Salisbury 1952) and much lower than that of most other kinds of soil (Willis 1985*a*). In summer in particular, when the water-table is lower (Ranwell 1959, Willis *et al.* 1959*b*) and the climate generally drier and warmer, drought can be severe, with sand more than 50 cm or so above the water-table having a water-content of less than 5%. *Ammophila*, however, is strongly drought-tolerant: it transpires at a comparatively low rate, protecting itself by inrolling of the leaves and wilts only when water is in extremely short supply, below 0.5% (Salisbury 1952, Willis & Jefferies 1963). In the exceptionally dry summer of 1976, for example, Huiskes (1979) detected only a slight decrease in the rate of leaf production in the field, a marked contrast with *Festuca rubra* where growth ceased completely. These plants showed the same contrast in their tolerance of high temperatures in greenhouse experiments, and these different responses probably contribute to the restricted role which *F. rubra* has even on somewhat more stable sands here in the *Festuca* sub-community. Probably, too, the occurrence of *Poa pratensis* in the *Ammophila* vegetation is related to the sand moisture, the *Poa* sub-community becoming common on less mobile dunes only around our northern coasts, where potential water deficits are smaller and where the summers especially are wetter and cooler than further south (*Climatological Atlas* 1952, Chandler & Gregory 1976).

Particular episodes of rainfall and, in periods of drier weather, the formation of dew, are likely to be also of considerable significance for the establishment of the various associates able to find a limited place on the somewhat more stable sands typical of the *Festuca* and *Poa* sub-communities. Wide diurnal fluctuations in soil temperatures, from over 40 °C at midday in summer to only 10 °C in the night, are a striking feature of the more open dune environment (Willis *et al.* 1959*b*, Stoutjesdijk 1961) and the moist air of onshore breezes may condense in appreciable amounts on the surfaces of sand-grains and herbage. Willis (1985*a*) drew attention to the dew that can be seen running down marram leaves at dawn after clear summer nights, and Salisbury (1952) recorded a nightly increase in water content of almost 1 ml per 10 ml soil in cloudless summer weather. More important perhaps than this advective humidity is the redistribution of moisture that may occur within the sand from deeper warmed levels to cooling superficial layers (Willis *et al.* 1959*b*), thus bringing water within the reach of more shallow-rooted plants. Even this moisture, though, may be largely lost as the sand surface warms up during the day (Huiskes 1979).

The *Ammophila* community maintains its ascendancy among mobile coastal sands until accretion slows enough for associates to compete more effectively for water and nutrients in the stabilising environment. The marram itself can retain spatial dominance for some time in the vegetation that succeeds the community and, where erosion strips back more consolidated sand surfaces, releasing material for renewed bouts of local accretion in blow-outs or precipitating more catastrophic shifts of entire tracts of dune, *Ammophila* vegetation can be found reinvigorated on the mobilising sand. Such stands are usually of the species-poor *Ammophila* sub-community, but the *Carex* sub-community is also characteristically associated with newly-deposited sand, particularly where the ground is a little moister, as around slacks and pools.

The impressive sand-binding ability of *Ammophila* has led to its being widely planted around the British coast, as elsewhere in Europe, to stabilise eroding stretches of dune, something which is especially important where dunes form an integral part of soft sea-defences (Kidson & Carr 1960, Lux 1964, 1966). Cuttings are used, often in conjunction with some kind of brushwood thatch or fencing to help arrest sand movement, and can quickly take a hold to form the basis of stands of this vegetation soon indistinguishable from those of natural origin. Remedial action of this kind may also be necessary where disturbance by visitors damages the vegetation cover, whether in young stands along the beach top, where trampling may open up the foredunes to destruction in winter gales, or further in the dune hinterland, where camping or picnicking are commonplace. Enclosure is often necessary to ensure the re-establishment of the community in such circumstances.

Zonation and succession

Where there is little opportunity for accretion of wind-blown sand above the tidal limit, fragmentary stands of the *Ammophila* community may be the only kind of vegetation maintaining itself on a narrow and ill-defined zone of low dunes exposed to the constant threat of erosion. In more extensive systems, though, this community can occupy wide tracts of mobile sand accumulating behind strandline and foredunes, and giving way inland to stretches of more stable material deposited in fixed dunes or on sand plains. The zonations of vegetation types across such landscapes have a generalised form in which the *Ammophila* community represents an immature phase between the pioneer foreshore assemblages and the more permanent products of succession on ground where accretion and erosion have come to a halt. However, the dynamics of dune development, and the accompanying vegetation changes, are often complex and the communities represented in the sequences also vary considerably according to the phytogeographic influence of regional climate on the dune flora. Later stages in the successions are increasingly affected, too, by biotic factors like grazing and land treatment, which is often intensive and destructive of natural patterns. Throughout, there may be additional variation in the vegetation types related to the hydrology of the dune system, slacks or pools interrupting the basic zonation.

The *Ammophila* sub-community itself sometimes comprises the pioneer zone of vegetation on sand deposited above the tidal limit, helping to build up foredunes set back a little way from some kind of assemblage on the strandline and usually clearly marked off from it floristically. Along many shores around Britain, however, the *Elymus farctus* community precedes *Ammophila* vegetation, establishing itself within the zone of periodic tidal inundation, colonising the strandline itself and then stimulating modest accretion of wind-blown sand. Where this raises the beach level sufficiently, it can be instrumental in allowing *Ammophila* to invade, such that the low foredunes are occupied by the transitional *Elymus* sub-community, with its strandline survivors. Particularly down the eastern coast as far as Norfolk, *Leymus arenarius* complicates this basic zonation, being able to colonise a little further down the beach than *Ammophila*, sometimes invading an *Elymus* zone or coming to dominate in virtually pure stands on young dunes, when the *Elymus-Leymus* or *Leymus* sub-communities form a transition to the *Ammophila* sub-community on the bigger dunes behind. The Lincolnshire and Humberside coasts have some complex zonations in which all three of these grasses play a part.

Except locally in Northumberland and parts of north-east Scotland, Orkney and Shetland, it is the *Ammophila* community rather than the *Leymus* vegetation which is the more important on younger mobile dunes and, even there, the latter begins to be excluded with accretion

above a certain speed and height. Around the whole coast, the species-poor *Ammophila* sub-community is the usual vegetation type where sand mobility is at its peak closer to the coastline or in the recolonisation of more catastrophic blow-outs in the dune hinterland, with the *Carex* sub-community figuring very locally on freshly-stabilised sand or where bared ground is a little moister. Further back in the dunes, where accretion has begun to slow down, the *Ammophila* sub-community typically gives way to the slightly richer and more varied *Festuca* sub-community or, increasingly towards the moister and cooler north of Britain, the *Poa* sub-community.

On dune systems where there is a more extended gradation from mobile to fixed sand, one or other of these latter kinds of *Ammophila* vegetation generally forms a transition to the *Ammophila-Festuca* community on dunes where accretion is appreciably slower and erosion increasingly rare (Figure 12). Here in addition to constant *P. pratensis* and *F. rubra*, which can increase their cover appreciably among the *Ammophila*, species such as *Senecio jacobaea*, *Hypochoeris radicata*, *Taraxacum officinale* agg., *Galium verum* and *Lotus corniculatus*, at most only moderately frequent in the *Ammophila* community, become very common. In addition, *Carex arenaria* and *Hieracium pilosella* are often found, while there are many occasional associates, including numerous plants hardly ever found on mobile dunes, such as *Plantago lanceolata*, *Leontodon taraxacoides*, *Luzula campestris* and *Cerastium fontanum*. A number of bryophytes, too, increase their frequency and cover, with *Brachythecium albicans*, *Tortula ruralis* ssp. *ruraliformis* and *Hypnum cupressiforme s.l.* becoming especially important, while on more open patches of the essentially stable surface, a variety of diminutive annuals come and go. Around our more southerly shores, *Ononis repens* is a further distinctive newcomer with the shift from what is usually the *Festuca* sub-community of the *Ammophila* vegetation to the *Ononis* or *Viola* sub-communities of *Ammophila-Festuca* vegetation. North of the Solway–Forth line, it is the *Poa* type of *Ammophila* dune which generally passes behind to the Typical or *Hypnum* sub-communities of the *Ammophila-Festuca* dune.

In its turn, the *Ammophila-Festuca* dune can grade on fixed sand where the surface pH is often a little below neutral, though less drought-prone, to the *Festuca-Galium* community, a vegetation type widely distributed on older dunes all around the British coast, though especially extensive on the sand-plains of the machair along the north-west Scottish seaboard. Bouts of erosion in both the *Ammophila-Festuca* community and, less commonly, in the *Festuca-Galium* vegetation can lead to a local regeneration of the *Ammophila* community with sharper floristic boundaries at the switch to the newly mobile sand and, where there are more substantial shifts in dune dynamics, whole areas of partly stabilised dune may be overtaken by freshly deposited sand on which the more immature vegetation gains a hold again. In this way, stands of the *Ammophila* community can be found developing next to calcifugous grassland, heath or other vegetation types of the dune hinterland before the sand finally comes to a halt.

Other variations on the basic zonation reflect edaphic and treatment differences. Where the sands are somewhat more fixed, and particularly where there is little grazing, the *Ammophila* community may give way inland, not to the *Ammophila-Festuca* vegetation, but to the *Ammophila-Arrhenatherum* community, the *Geranium sanguineum* sub-community of warm, south-facing dune slopes being particularly distinctive with its mixtures of *Arrhenatherum*, *G. sanguineum*, *Ononis repens*, *Dactylis glomerata*, *Galium verum* and *Veronica chamaedrys*, together with occasional low bushes of *Rosa pimpinellifolia*, among the *Ammophila*, *P. pratensis* and *F. rubra*. Especially good examples of this kind of pattern can be seen on the Northumbrian dune systems. More species-poor rank mesophytic swards of the *Ammophila-Arrhenatherum* type can also sometimes form an intermediate zone between the *Ammophila* community and dune-slack surrounds but these wetter areas

more often occur in depressions among more stable dune ridges where *Ammophila-Poa* or *Festuca-Galium* vegetation covers the surrounding slopes.

Distribution

The *Ammophila* community occurs widely on suitably mobile sands in dune systems all around the British coast. *Leymus* figures most prominently in the younger stands of the *Leymus* and *Elymus-Leymus* sub-communities on the east coast, with the *Poa* sub-community much more common north of the Solway–Forth line, but otherwise the presence of the different kinds of *Ammophila* vegetation is a reflection of the local stability of the dune sands.

Affinities

As defined here, the *Ammophila* community represents the younger vegetation among that generally subsumed in a broadly-defined *Ammophiletum* in British descriptive schemes (Moss 1906, Tansley 1911, 1939), a category which also takes in much of what we have separated off as the *Ammophila-Festuca* community, an assemblage of less mobile sands, and even on occasion some of our *Festuca-Galium* vegetation where marram can remain of high cover, though much reduced vitality. As Birse (1980, 1984) proposed, this kind of vegetation is best seen as equivalent to part of the *Elymo-Ammophiletum* which Braun-Blanquet & De Leeuw (1936) characterised as the major association of more mobile sands around the seaboard of north-west Europe, and which has since been described from Belgium (LeBrun *et al.* 1949), The Netherlands (Westhoff & den Held 1969), France (Géhu & Géhu 1969) and Ireland (Braun-Blanquet & Tüxen 1952, Ivimey-Cook & Proctor 1966, Beckers *et al.* 1976, Schouten & Nooren 1977 and Ni Lamhna 1982).

Figure 12. Simplified zonation of vegetation types on strandline, embryo, semi-fixed and fixed dunes in southern Britain.
The strandline has fragmentary SD2 *Honkenya-Cakile* vegetation with embryo dunes building around SD4 *Elymys farctus* community behind. Then, there are mobile dunes with SD6 *Ammophila* vegetation and, behind, semi-fixed dunes with the SD7 *Ammophila-Festuca* community. A large blow-out on one ridge has the SD10 *Carex arenaria* community. To the rear, the fixed dunes have SD8 *Festuca-Galium* grassland.

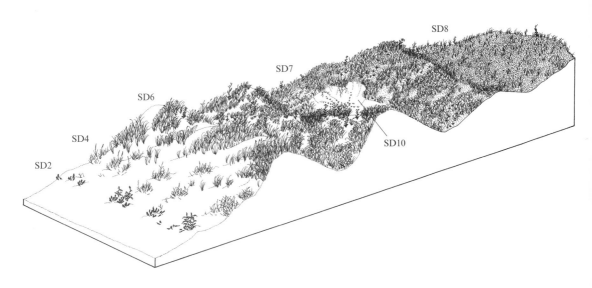

Floristic table SD6

	a	b	c	d	e	f	g	6
Ammophila arenaria	V (4–10)	V (2–8)	V (4–8)	V (3–10)	V (3–8)	V (2–9)	V (2–10)	
Elymus farctus	V (1–7)	V (2–5)			I (2–4)	I (2–4)	I (2–4)	II (1–7)
Honkenya peploides	II (3–4)	II (1–6)		I (3)	I (1–4)	I (2–4)		I (1–6)
Leymus arenarius		V (2–7)	V (1–7)		II (2–5)	II (1–5)		II (1–7)
Atriplex glabriuscula		I (3)						I (3)
Atriplex laciniata		I (1)						I (1)
Carex arenaria	I (1–3)				I (1–4)	I (1–4)	V (2–5)	I (1–5)
Festuca rubra		I (3)		I (1)	V (2–9)	V (1–7)	I (5)	III (1–9)
Senecio jacobaea	I (2–3)	I (1)		II (1–5)	II (1–3)	II (1–3)		II (1–5)
Cirsium arvense	I (1–4)	I (1–3)		I (1–3)	II (1–4)	II (1–6)		II (1–6)
Hypochoeris radicata	I (1–2)	I (1–3)	I (1)	I (2–4)	II (1–5)	I (1–5)	I (3)	I (1–5)
Trifolium arvense					I (3–5)			I (3–5)
Agrostis capillaris					I (2–5)			I (2–5)
Desmazeria marina					I (1–2)			I (1–2)
Urtica dioica					I (2–3)			I (2–3)
Poa pratensis					I (3)	V (1–7)		II (1–7)
Elymus repens			I (1)	I (3)	I (1–5)	II (2–3)		I (1–5)
Heracleum sphondylium		I (2)			I (1–4)	II (1–6)	I (3)	I (1–6)
Sonchus arvensis	I (3–4)	I (1–3)			I (1–4)	II (1–6)		I (1–6)
Sonchus asper	I (4)	I (1)	I (3)	I (2–3)	I (1–3)	I (1)		I (1–4)
Rumex crispus	I (2)	I (1)		I (2–3)	I (1–3)	I (1–5)		I (1–5)
Senecio vulgaris	I (1)	I (1–2)	I (1)		I (2–3)	I (3)		I (1–3)
Calystegia soldanella	I (3–5)		I (4)	I (3–4)	I (2–5)			I (2–5)
Cakile maritima	I (3)			I (3)	I (1–6)			I (1–6)
Eryngium maritimum	I (2–4)			I (3)	I (3–5)	I (1)		I (2–5)
Cirsium vulgare	I (4)			I (1)	I (1)			I (1–4)
Ceratodon purpureus	I (6)			I (1–3)	I (1–9)	I (2–5)		I (1–9)
Tortula ruralis ruraliformis		I (5)		I (1–3)	I (3–6)	I (1–5)		I (1–6)
Elymus pycnanthus		I (2–7)		I (2–4)	I (3–8)			I (2–8)
Brachythecium albicans				I (2)	I (1–6)	I (2–6)		I (1–6)

	a	b	c	d	e	f	g	6
Holcus lanatus		I (1)			I (1–6)	I (2–5)	I (2)	I (1–6)
Euphorbia paralias		I (1)		I (1–3)	I (2–7)			I (1–7)
Euphorbia portlandica		I (3)		I (1–4)	I (4)	I (3)		I (1–4)
Taraxacum officinale agg.				I (2)	I (1–7)	I (1–5)		I (1–7)
Senecio squalidus		I (3)			I (1–2)	I (1)		I (1–3)
Tussilago farfara	I (5)	I (4)			I (2–5)	I (1–4)		I (1–5)
Cynoglossum officinale		I (3)			I (2)	I (2–3)		I (2–3)
Ranunculus repens		I (3)			I (3–5)	I (2–3)		I (2–5)
Rubus fruticosus agg.		I (3)		I (2)	I (2–4)		I (4)	I (2–4)
Lupinus arboreus				I (2–3)	I (4–9)			I (2–9)
Epilobium angustifolium					I (2–8)	I (2–4)	I (4)	I (2–8)
Agrostis stolonifera					I (3)	I (1)		I (1–3)
Galium verum					I (2–4)	I (1–4)		I (1–4)
Linaria vulgaris					I (1–3)	I (3–5)		I (1–5)
Angelica sylvestris					I (2–3)	I (1–3)		I (1–3)
Dactylis glomerata					I (3–4)	I (3)		I (3–4)
Erodium cicutarium					I (1–3)	I (4)		I (1–4)
Poa annua					I (2–3)	I (2)		I (2–3)
Trifolium repens					I (2–5)	I (2–3)		I (2–5)
Number of samples	28	18	10	35	62	36	10	199
Number of species/sample	4 (2–10)	5 (3–10)	3 (2–7)	3 (1–9)	7 (4–14)	9 (5–15)	3 (2–7)	5 (1–15)

a   *Elymus farctus* sub-community

b   *Elymus farctus-Leymus arenarius* sub-community

c   *Leymus arenarius* sub-community

d   Typical sub-community

e   *Festuca rubra* sub-community

f   *Poa pratensis* sub-community

g   *Carex arenaria* sub-community

6   *Ammophila arenaria* mobile dune community (total)

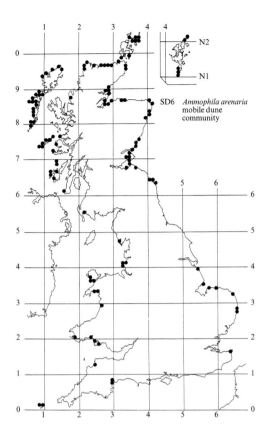

SD6   *Ammophila arenaria*
mobile dune
community

SD7
Ammophila arenaria-Festuca rubra semi-fixed dune community

Synonymy
Ammophiletum arenariae Moss 1906, Tansley 1911, 1939, all *p.p.*; *Elymo-Ammophiletum arenariae* Br.-Bl. & De Leeuw 1936.

Constant species
Ammophila arenaria, Festuca rubra, Hypochoeris radicata, Poa pratensis s.l.

Rare species
Acaena novae-zelandiae, Epipactis dunensis, Euphorbia paralias, Mibora minima, Oenothera stricta, Vulpia fasciculata.

Physiognomy
The *Ammophila arenaria-Festuca rubra* community is the major vegetation type of less mobile coastal sands where *Ammophila arenaria* is still usually the dominant, but where conditions are such as to allow the development of a fairly rich and often abundant associated flora over the stabilising dune surface. The marram tussocks may be big, with the shoots generally reaching 60–100 cm in height, but the tillers are not so densely packed or vigorous as in younger plants, and the accretion of bare sand among them much less rapid than in the earlier stages of colonisation. Where the tussocks are growing close together, the canopy of arching foliage can be quite extensive, but beneath this and particularly among more widely spaced marram, a variety of other herbs and also bryophytes can gain a hold, forming a patchy carpet of shorter vegetation that may be of locally high cover.

Along with the *Ammophila*, other sand-binding grasses are of only minor significant in this community. The hybrid × *Ammocalamagrostis baltica* occurs in some places, a natural associate in stands in Northumberland and on some Norfolk dunes, and of planted origin elsewhere in East Anglia (Hubbard 1968, Huiskes 1979), and *Leymus arenarius* remains of occasional importance here on dunes of moderate size and in disturbed situations, mostly along the east coast. *Elymus farctus*, however, a fairly common survivor in the *Ammophila* community where marram has invaded foredunes, is very scarce on this less mobile sand. Much more obvious throughout this kind of vegetation is *Festuca rubra*, often recorded as the var. or ssp. *arenaria* (Hubbard 1968, Tutin *et al.* 1980), here a constant companion of *Ammophila*, often sub-dominant to it and occasionally, in more open stretches of dune, the more abundant grass. *Poa pratensis s.l.* (probably *P. subcaerulea* in many cases: Hubbard 1968, Tutin *et al.* 1980) is also very common in the community, not usually of such high cover as *F. rubra*, though particularly noticeable in stands around our northern coasts. Other perennial grasses found much more occasionally are *Holcus lanatus, Arrhenatherum elatius* and *Dactylis glomerata*, the tussocks of which can be quite prominent, *Agrostis capillaris* and *Elymus repens*. At some sites, *E. pycnanthus* is locally abundant. The sand-sedge *Carex arenaria* also becomes very frequent here as a component of the sward among the marram, though it is never dominant.

Among the grassy matrix, a variety of perennial dicotyledons are characteristic of the *Ammophila-Festuca* community. The most frequent of these are *Hypochoeris radicata, Taraxacum officinale* agg. and *Senecio jacobaea* (sometimes behaving as a biennial) with *Hieracium pilosella, Lotus corniculatus* and *Galium verum* more unevenly distributed among the various sub-communities, but altogether commoner than in the *Ammophila* vegetation of more mobile dunes. In southern Britain and locally up the eastern Scottish coast, a further constant, *Ononis repens*, becomes very characteristic of the community, its patches of procumbent sticky shoots sometimes quite abundant.

More occasional associates of the *Ammophila-Festuca* vegetation, usually represented as scattered plants, are *Cerastium fontanum, Plantago lanceolata, Leontodon taraxacoides, L. hispidus, Luzula campestris, Achillea millefolium, Viola riviniana, Veronica chamaedrys, Trifolium repens* and *Anthyllis vulneraria*, with *Calystegia soldanella, Eryngium maritimum, Euphorbia paralias* and

E. portlandica a distinctive feature of some stands around our southern coasts. More patchily distributed and never really very common are *Viola canina, V. tricolor, Thymus praecox, Sedum acre, Campanula rotundifolia, Rhinanthus minor, Leontodon autumnalis* and *Ranunculus bulbosus*. Then, very locally, there can be clumps of *Rubus fruticosus* agg., *R. idaeus, R. caesius* or *Epilobium angustifolium*. Among more unusual perennials, the Australasian wool-alien *Acaena novae-zelandiae* (at one time thought to be *A. anserinifolia*) has become well established in this vegetation at certain of its coastal localities, notably in Northumberland, since its accidental introduction into Britain around the turn of the century (Gynn & Richards 1985). The rare and rather nondescript orchid *Epipactis dunensis* is also sometimes found in the community close to transitions to dune-slack surrounds.

Coarse weedy herbs tend to be rather less noticeable here than in the *Ammophila* vegetation but, along with *S. jacobaea*, there are occasional plants of *Cirsium arvense, C. vulgare, Rumex crispus, Senecio vulgaris, Sonchus arvensis, S. asper* and *Tragopogon pratensis*. Trailing masses of *Vicia sativa* ssp. *nigra* are also prominent in some stands. The South American evening primrose *Oenothera stricta* has also become well established in this vegetation on a number of dune systems around our southern coasts where it occasionally overwinters.

Compared with the *Ammophila* community, bryophytes are a more consistent feature here and they can be locally very extensive among the herbs, although the common species are rather few and none of these occurs frequently throughout. However, *Tortula ruralis* ssp. *ruraliformis* and *Hypnum cupressiforme s.l.* are characteristic of different kinds of *Ammophila-Festuca* vegetation, with *Brachythecium albicans, Homalothecium lutescens* and *Ceratodon purpureus* more occasional, *Brachythecium rutabulum, Eurhynchium praelongum, Pseudoscleropodium purum, Bryum capillare* and *B. argenteum* sparse. Lichens can also occur in small amounts, with records for *Cladonia fimbriata, C. impexa, C. furcata, C. rangiformis, C. pyxidata, Peltigera canina* and *P. rufescens*.

Sub-communities

Typical sub-community. *Ammophila*, or mixtures of *Ammophila* and *F. rubra*, dominate here, with very frequent and locally quite abundant *P. pratensis*, and in some stands patches of *Leymus arenarius* and occasional tussocks of *Dactylis*. Scattered plants of *H. radicata, T. officinale* agg. and *S. jacobaea* occur very commonly and there is often some *Carex arenaria*, but no other plants are frequent and the variety of associates is limited, so the vegetation is often rather species-poor and the cover of herbs among the grasses fairly open. Rather

strikingly, there is never any *Ononis repens* in this sub-community.

Occasionals include *Hieracium pilosella, Galium verum, Lotus corniculatus, Cerastium fontanum, Plantago lanceolata* and *Cirsium arvense* with only rather sparse records for the various annuals found in the community. Bryophytes, too, are few in number but *T. ruralis* ssp. *ruraliformis, B. albicans* and *O. purpureus* occur quite commonly, with more infrequent *H. cupressiforme* and *Homalothecium lutescens*.

Hypnum cupressiforme sub-community. *Ammophila*, or occasionally mixtures of *Ammophila* and *F. rubra*, are dominant in this sub-community, with *P. pratensis* still constant but generally subordinate. As in Typical *Ammophila-Festuca* vegetation, *H. radicata, T. officinale* agg., *S. jacobaea* and *C. arenaria* are all very common, but *Hieracium pilosella, Lotus corniculatus* and *Galium verum* all increase in frequency, and there is also occasional *Cerastium fontanum, Leontodon taraxacoides* and *Luzula campestris* in a cover of perennials that is usually more extensive.

Again, *Ononis repens* is not a characteristic plant but a number of preferentials give a distinct stamp to the vegetation. *Viola canina*, for example, is commoner here than elsewhere in the community and in Northumberland, prostrate patches of the alien *Acaena* can be very abundant. Then, along with annuals of the community like *Aira praecox* and *Cerastium semidecandrum, Arenaria serpyllifolia* and *Phleum arenarium*, others such as *Myosotis ramosissima, Valerianella locusta* and *Veronica arvensis* have been commonly recorded with more occasional *Vicia lathyroides* and *Erodium cicutarium*.

Equally striking in this sub-community is the frequency and often the abundance of *Hypnum cupressiforme s.l.* which, together with more patchy *Rhytidiadelphus squarrosus, R. triquetrus, Dicranum scoparium, Brachythecium albicans* and *Tortula ruralis* ssp. *ruraliformis*, can form an extensive carpet over the sand. Lichens, too, are often rather more conspicuous and varied here than elsewhere in the community, with fairly common *Cladonia fimbriata, C. impexa, Peltigera canina* and *P. rufescens*.

Ononis repens sub-community. *Ammophila* usually dominates in this vegetation, with *F. rubra* constant though typically subordinate and *P. pratensis* rather reduced in frequency and rarely of appreciable cover. The commonest herbs are those of the Typical sub-community, *H. radicata, T. officinale* agg., *S. jacobaea* and *C. arenaria*, though even these can be rather patchy. Very distinctive here, though, is the constant occurrence, sometimes in considerable abundance, of *Ononis repens*.

Among the occasional associates, more thermophilous dune plants, like *Calystegia soldanella, Eryngium maritimum, Euphorbia paralias* and *E. portlandica*, are also found more commonly here, and there can be occasional *Galium verum, Lotus corniculatus, Plantago lanceolata, Leontodon taraxacoides* and *Cirsium arvense*. On more open areas, annuals such as *Aira praecox, Vulpia fasciculata, Phleum arenarium, Arenaria serpyllifolia, Cerastium semidecandrum* and *C. diffusum* ssp. *diffusum* make an occasional appearance. Bryophytes are less frequent here though scattered patches of *Brachythecium albicans* and *Tortula ruralis* ssp. *ruraliformis* can sometimes be found.

***Elymus pycnanthus* sub-community.** *Ammophila* remains abundant in this sub-community together with usually smaller amounts of *F. rubra* and *P. pratensis* but it is consistently accompanied by *Elymus pycnanthus* which can itself attain high cover. In this grassy ground, frequent associated herbs are few with even species like *Taraxacum officinale, Senecio jacobaea* and *Carex arenaria* being only occasional. Bryophyte cover is low, with just *Brachythecium albicans* showing any frequency.

Habitat
The *Ammophila-Festuca* community is the characteristic vegetation type of less mobile sands on dune systems all around the British coast, developing where accretion has become appreciably slower and erosion more rare, but where edaphic changes beneath the stabilising surface are not yet very great. Some of the floristic and structural variation in the community reflects the degree of fixity of the sand, but climatic differences through the season of growth and across the range of this vegetation are also influential with temperature and rainfall in particular having important effects on the representation of associates here on substrates that are still generally drought-prone and impoverished.

The rate and disposition of sand movement within particular dune systems around our coasts are complex functions of sediment supply and the development of dune-building vegetation on whatever material is able to accumulate behind the beach. In the establishment of an *Ammophila* cover, which is the controlling element in the colonisation of wind-blown coastal sands with us, the grass grows best where accretion is rapid and substantial, itself encouraging accumulation and keeping pace with burial of up to 1 m a year (Ranwell 1972). Eventually, though, there comes a point when reduced deposition of sand is matched by a waning in the vigour of the marram and, in this *Ammophila-Festuca* community, where accretion is probably less than 5 cm annually, we can see the early effects of that process. Only in young mobile dunes, or where there is some local spur to rejuvenation, is shoot production in the grass proliferative and, as death of the tillers comes to outstrip their replacement, the density of live shoots falls: here, there

may be just half or so of the 150–200 tillers m^{-2} characteristic of the *Ammophila* community (Huiskes 1979). For some reason, too, flowering is not so free among the more fixed sand so, though the number of spikelets per inflorescence remains fairly constant, the output of ripe caryopses per unit area is reduced (Huiskes 1977*a*). Burial of the seed by wind-blown sand, which is highly inimical to germination (Huiskes 1972), is much less likely among the *Ammophila-Festuca* vegetation, but areas of bare ground in which seedlings might get a hold are that much sparser, and germination seems to be best adapted to the strongly-fluctuating soil temperatures typical of the open sandy surfaces of mobile dunes (Willis *et al.* 1959*b*, Stoutjesdijk 1960). In the field, seedlings of marram are rarely found away from the youngest dune ridges (Huiskes 1977*b*) and the tussocks of the *Ammophila-Festuca* community are probably almost always the thinning remnants of clones established early in the colonisation process (Huiskes 1979).

The reasons for the decline in *Ammophila* vitality are unclear. It has been suggested (Marshall 1965) that reduced accretion deprives the grass of the opportunity to put out new roots on freshly-buried internodes, but this seems an inadequate explanation (Hope-Simpson & Jefferies 1966, Huiskes 1977*a*). More likely are the impacts of increased competition for the very limited amounts of nutrients and water with other plants that are now able to establish on the less mobile surface, and the edaphic changes consequent upon the greater stability of the sand and the development of the vegetation upon it (Carey & Oliver 1918, Benecke 1930, Tansley 1939, Salisbury 1952, Willis *et al.* 1959*a, b*, Huiskes 1977*a*). In general terms, such changes involve an increase in the amount of organic matter, a leaching of carbonate and a reduction in pH within the surface layers of sand, but such developments can take several centuries and are much influenced by the initial character of the wind-blown sediment, particularly by the proportion of shell fragments, and the local climate (Salisbury 1952, Wilson 1960, Ranwell 1972, Willis 1985*b*). We do not know how far the changes have progressed with the shift from the *Ammophila* community of very mobile dunes to this next stage of *Ammophila-Festuca* vegetation, but perhaps not much. There is generally very little incorporated humus, even where there is an extensive moss carpet, and the superficial pH is usually between 7 and 8. The main response here may thus be to the increased physical stability of the sand surface.

The earlier stages of vegetation change are seen in the Typical and *Ononis* sub-communities characteristic of the lee faces of young dunes and tracts of semi-mobile sand. These can appear little different in general composition form the *Festuca* and *Poa* types of the *Ammophila* community on more shifting sand, but the increased cover of the two smaller grasses, together with the more consistent occurrence of herbs like *Hypochoeris radicata*, *Taraxacum* and *Senecio jacobaea*, is testimony to the reduced accretion among marram that is not quite so vigorous nor so influential on its environment and associates. Both *F. rubra* ssp. *arenaria* and *P. subcaerulea* have far-creeping rhizome systems that can spread extensively among the *Ammophila* tussocks, but their shoots are easily overwhelmed by the deep and more sustained burial characteristic of earlier stages of dune growth. For *Carex arenaria*, too, it is only now that conditions begin to become more generally congenial: this sedge can readily accommodate to some measure of accretion by producing tiered lateral rhizomes (Tidmarsh 1939), but its increasing prominence in these kinds of vegetation is best seen as an indication of recently-declining sand mobility (Noble 1982). For the first time, also, the substrate is sufficiently stable for *Ononis repens* to find a frequent place in stands of the *Ammophila-Festuca* community around our warmer southern shores, and for the occasional appearance of plants like *Lotus corniculatus*, *Galium verum*, *Plantago lanceolata*, *Leontodon taraxacoides*, *Luzula campestris*, *Thymus praecox* and *Sedum acre*, smaller hemicryptophytes and chamaephytes that can never gain a permanent hold on more shifting sand.

In the *Hypnum* sub-community, the further increase in frequency of these herbs, the sharp rise to constancy of *Hieracium pilosella*, a plant well adapted to dry, nutrient-poor but stable soils, and the much more important contribution to the vegetation cover from mosses, all reflect the continuing reduction in accretion and the further fixing of the sand surface on more sheltered situations in semi-mobile dunes and altogether older ridges. Here, there may be but a very gentle upbuilding of the surface: *Tortula ruralis* ssp. *ruraliformis* and *Brachythecium albicans*, for example, can grow with their shoots crammed around with sand right to their tops, but rapid burial overwhelms them.

Throughout the development of the *Ammophila-Festuca* community, however, the composition of the vegetation reflects interactions between this changing physical character of the sand surface and the climatic conditions under which the dunes are being built. In the first place, among both younger and older stands of the community, the major floristic contrasts betray some effect of regional differences in rainfall and temperature. The Typical and *Hypnum* sub-communities, for example, occur most extensively north of the Solway–Forth line, with more local occurrences down to Northumberland and East Anglia and, in the West, to Cumbria and North Wales. Around these coasts, the combination of a cooler climate, with mean annual maxima usually below 25 °C (Conolly & Dahl 1970), and generally higher rainfall, especially in the west,

where there are more than 160 wet days yr^{-1} (Ratcliffe 1968), results in a more humid atmosphere and reduced droughtiness in the dune sands. The greater abundance of *Poa subcaerulea* in these kinds of *Ammophila-Festuca* vegetation is probably a response to these conditions and continues a trend that has begun to appear in northern stands of the *Ammophila* community. The persistence of the Oceanic Northern *Leymus arenarius* in the Typical sub-community is also related to the cooler and moister climate around these coasts, although it is a scarce plant even there away from smaller and usually younger semi-mobile dunes or disturbed places.

By contrast, the *Ononis* sub-community is predominantly southern in its distribution, extending very little beyond the Solway–Forth line, particularly in the west. Here, *P. subcaerulea* has reduced frequency and cover, but more noticeably there is the common occurrence in this kind of *Ammophila-Festuca* vegetation of *Ononis repens*, a plant that is largely confined to warmer and drier parts of Britain, though not of course to coastal situations, except in the far north of its range, where suitably dry soils become scarce inland (Perring & Walters 1962). Other thermophilous associates of the *Ammophila-Festuca* community like the Oceanic Southern *Eryngium maritimum*, *Euphorbia paralias*, *Phleum arenarium* and the rare *Mibora minima* and *Vulpia fasciculata*, and the Oceanic West European *Euphorbia portlandica*, are also largely confined to the *Ononis* sub-community and, where numbers of these occur together, they create a very different impact from that seen in the Typical and *Hypnum* sub-communities.

The second kind of interaction between climate and soils which influences the *Ammophila-Festuca* community is evident among both northern and southern stands in the appearance of a range of diminutive ephemerals, more especially winter annuals. These usually have their best representation on our dunes in the later stages in the development of this kind of vegetation, providing some of the character of the *Hypnum* sub-community and representing a transition to the *Tortulo-Phleetum* which often occurs among it, though their contribution is by nature rather varied, local and sporadic.

Zonation and succession
In ageing dune systems, the *Ammophila-Festuca* community can occupy the first zone inland of the strandline on sand which has become more or less fixed but, where growth is still active, this vegetation is characteristically set back behind banks of foredunes and mobile dunes supporting different communities, only finding a place further towards the shore where there is a measure of local shelter. Inland, it can give way over ridges and sand-plains to dune grasslands and heath typical of

largely stabilised surfaces, the sequence of vegetation types sometimes clearly representing a succession. However, the patterns are often interrupted by zonations to slack communities, where the height of the water-table mediates the vegetation gradient, and are much influenced landwards by treatments and land improvement or reclamation.

In the most abbreviated beach-top sequences, where accretion and erosion are much reduced, there can be a sharp transition from the shore, sometimes with a strip of strandline vegetation, to more open stands of the *Ammophila-Festuca* community. Generally, though, even where the sand supply has become much reduced, the exposed face of the most seaward dunes carries the *Ammophila* community, the *Ammophila-Festuca* vegetation being limited to the lee slopes where the material is much less mobile. In southern Britain, it is the *Ononis* sub-community which is characteristic of these younger sheltered faces, with a zone of the *Festuca* sub-community of the *Ammophila* vegetation sometimes interposed in the transition to very open marram on the exposed, shifting sand. Around northern coasts, Typical *Ammophila-Festuca* vegetation replaces the *Ononis* sub-community, with the *Poa* type of *Ammophila* vegetation occurring where there is a more gradual shift to the immature marram. Then, especially down the east coast, and where the dunes remain low, these early sequences can be complicated by the persistence of *Leymus arenarius*, the *Festuca* sub-community of *Leymus* vegetation sometimes occurring on more mobile or disturbed surfaces among tracts of Typical *Ammophila-Festuca* grassland.

Where the zonations continue on to sand which is becoming more fixed, the *Ononis* and Typical sub-communities usually give way to the *Hypnum* type, the different vegetation types sometimes being found as distinct zones but more usually disposed in complex mosaics over surfaces of individual dunes or ridges where there is a diversity of aspects variously sheltered from the wind and sun. Locally severe erosion may result in a rejuvenation of the *Ammophila* community or the spread of *Carex arenaria* vegetation in blow-outs within such patterns. More local appearance of bare but still quite stable sand is often marked by patches of the *Tortulo-Phleetum*, particular on dune systems in the warmer and drier south of Britain.

In bigger dune systems, where there are expanses of fixed sand which have not been reclaimed, the more mature kinds of *Ammophila-Festuca* vegetation are frequently transitional inland to closed dune grassland, usually of the *Festuca-Galium* type. Here, *Ammophila* often remains very common, but it is generally much reduced in vigour, with *Festuca rubra* and *Poa pratensis* now typically providing the bulk of the grass cover. By this stage, the sands have become somewhat richer in organic matter and nutrients, and are more water-reten-

tive, so the swards fill up and take on a more mesophytic character. Plants that are generally only occasional in the *Ammophila-Festuca* community become very frequent with *Galium verum, Plantago lanceolata, Trifolium repens, Lotus corniculatus* and *Achillea millefolium* especially distinctive, *Ranunculus acris, Bellis perennis* and *Euphrasia officinalis* agg. sometimes important, and a variety of other grasses, sedges and dicotyledons appearing that hardly ever figure on the more mobile, droughty and impoverished sands. Mosses such as *Tortula ruralis* ssp. *ruraliformis* and *Homalothecium lutescens* remain common but some pleurocarpous species like *Rhytidiadelphus squarrosus, R. triquetrus, Calliergon cuspidatum* and *Pseudoscleropodium purum* also occur among the turf. Grazing also has a more obvious effect upon the appearance of the vegetation than is the case with the communities of more mobile dunes sometimes trimming down the herbage to a short sward, but open patches are scarce and, where scuffing or droughting of the surface creates a gap, the appearance of assemblages of annuals is usually best seen as a transition to recurring *Ammophila-Festuca* vegetation. As with the communities of more mobile sand, there are some regional differences among the various kinds of *Festuca-Galium* grasslands, some of the most striking and extensive zonations occurring over the sand-plains along the very wet north-west seaboard of Scotland. There, the Typical and *Hypnum* sub-communities of *Ammophila-Festuca* vegetation on the semi-mobile dunes give way on the stretches of machair to the *Ranunculus-Bellis* and *Prunella* sub-communities of the *Festuca-Galium* grassland.

More locally, rather different zonations can be seen. Where there is little or no grazing on less mobile sands, a rather rare kind of situation on British dunes, the *Ammophila-Festuca* community can give way to the *Ammophila-Arrhenatherum* vegetation. This is a generally rather rank sward which has a number of characteristic plants in common with other assemblages of surfaces that are becoming fixed, like *Festuca rubra, Poa pratensis, Ononis repens, Galium verum, Lotus corniculatus* and *Achillea millefolium*, but the transition differs from zonations to the *Festuca-Galium* grassland in the prominence of *Arrhenatherum elatius* and *Dactylis glomerata*. These, and plants such as *Heracleum sphondylium* and *Lathyrus pratensis* often bring the vegetation close to an inland Arrhenatherion sward, but in some places the *Ammophila-Festuca* community passes on sunny, south-facing dune faces to a more striking kind of *Ammophila-Arrhenatherum* in which *Geranium sanguineum* is a constant feature, with scattered bushes of *Rosa pimpinellifolia*.

A more common variation occurs where the sands have become more sharply surface-leached on the fixed dune surfaces, when the *Ammophila-Festuca* community can give way to the *Carex-Festuca-Agrostis* grass-land. Here, again, *Ammophila* is often reduced to a minor and not very vigorous component in the swards, but *F. rubra* and *P. pratensis* are now usually matched in frequency and often in cover by *F. ovina*, *A. capillaris* and *C. arenaria*. Plants like *Hypochoeris radicata*, *Lotus corniculatus* and *Galium verum* remain fairly common but, with the increased importance of *Galium saxatile*, *Luzula campestris*, *Rumex acetosella* and acidophilous bryophytes, the vegetation approaches a dry calcifuge grassland in its appearance, especially where it is grazed, as is often the case. Transitions of this kind may also involve the local occurrence of patches of the *Carex-Cladonia* community on areas of open but compacted sand and, where ericoids gain a hold, there can be stands of *Calluna-Carex* heath.

All these different kinds of zonations may be interrupted by the occurrence of slack vegetation where the ground water table comes sufficiently close to the surface between the dunes to keep the sand permanently moist or at least seasonally flooded. On more calcareous sands, the *Salix-Holcus* community often occurs around the slack edges, with drier kinds of *Potentilla-Carex* vegetation on substrates that are less base-rich but, in both cases, boundaries are usually much sharper where the *Ammophila-Festuca* community occurs over the surrounding dune slopes, than where the *Festuca-Galium* vegetation provides the context. *F. rubra* often maintains some representation in the slacks, with herbs like *Lotus corniculatus*, *Ranunculus acris*, *Trifolium repens* and *Euphrasia officinalis* variously recorded, but *Ammophila* disappears sharply at the slack surrounds and plants such as *Poa pratensis* and *Carex arenaria* sometimes penetrate little further.

Distribution
The *Ammophila-Festuca* community occurs all around the British coast where there are suitable semi-mobile

sands, with the *Ononis* sub-community found largely south of the Solway–Forth line, the Typical and *Hypnum* forms to the north, with an area of overlap along the coast of Northumberland.

Affinities
Although many accounts of the development of dune vegetation in Britain draw attention to the changes occurring with the reduction in sand mobility, there have never been any convincing attempts to systematically distinguish the various assemblages occurring between the foredunes and the fixed dunes in different parts of the country. In early schemes, then, the *Ammophila-Festuca* community would have been subsumed within a general *Ammophiletum* (Moss 1906, Tansley 1911, 1939), though it is sufficiently distinct in its floristics, structure and ecological relationships to warrant separation from the more immature *Ammophila* community, as well as from the *Festuca-Galium* vegetation of fixed dunes.

Phytosociologically, some of our southern stands in the *Ononis* sub-community approach the *Euphorbio-Ammophiletum* Tx. 1945, a more thermophilous association reported from semi-mobile dunes along the French Atlantic coast (Géhu & Géhu 1969) and from Eire (Braun-Blanquet & Tüxen 1952, Ivimey-Cook & Proctor 1966, Beckers *et al.* 1976, Schouten & Nooren 1977, Ni Lamhna 1982). With us, however, *O. repens* tends to be a better distinguishing species of this kind of dune vegetation around our warmer coasts than plants like *Calystegia soldanella*, *Euphorbia paralias* or *Eryngium maritimum*. Further north, the absence of these plants brings the Typical and *Hypnum* sub-communities close to the *Elymo-Ammophiletum* Br.-Bl. & De Leeuw 1936. Here again, though, there is no attempt in Continental schemes to distinguish earlier and later stages in the development of either of these communities.

Floristic table SD7

	a	b	c	d	7
Ammophila arenaria	V (1–9)	V (3–8)	V (1–10)	IV (1–9)	V (1–10)
Festuca rubra	V (2–9)	V (1–8)	V (1–10)	V (2–10)	V (1–10)
Poa pratensis	V (1–7)	V (1–8)	III (1–7)	V (1–7)	IV (1–8)
Hypochoeris radicata	IV (1–7)	IV (1–6)	III (1–4)	III (1–7)	IV (1–7)
Taraxacum officinale agg.	IV (1–5)	IV (1–3)	III (1–5)	II (1–2)	III (1–5)
Senecio jacobaea	IV (1–3)	IV (1–3)	III (1–5)	II (1–4)	III (1–5)
Carex arenaria	III (1–5)	IV (1–4)	II (1–5)	II (1–7)	III (1–7)
Ceratodon purpureus	II (1–6)	I (3)	I (1–4)	I (1–5)	I (1–6)
Leymus arenarius	II (1–5)		I (1–6)	I (2–5)	I (1–6)
Dactylis glomerata	II (1–5)		I (1–5)	I (1–4)	I (1–5)
Hypnum cupressiforme	I (1–8)	V (1–8)	I (1–4)	I (6)	II (1–8)
Hieracium pilosella	II (1–5)	IV (1–6)	I (1–3)	I (1–5)	II (1–6)
Galium verum	II (1–4)	III (1–6)	II (1–7)	I (3)	II (1–7)
Acaena novae-zelandiae	I (2–8)	III (1–6)	I (1–5)		I (1–8)
Myosotis ramosissima	I (1–2)	III (1–3)	I (1–2)	I (1)	I (1–5)
Valerianella locusta	I (1–2)	III (1–3)	I (1–2)	I (1–3)	I (1–4)
Viola canina	I (1–3)	III (1–3)	I (1–2)	I (3)	I (1–3)
Epilobium angustifolium	I (2–6)	II (1–3)	I (2–5)	I (2–4)	I (1–6)
Rhytidiadelphus squarrosus	I (2–3)	II (2–6)	I (1–3)		I (1–6)
Cladonia fimbriata	I (1–4)	II (1–3)		I (1–2)	I (1–5)
Dicranum scoparium	I (1–4)	II (1–5)		I (2)	I (1–6)
Veronica arvensis	I (1)	II (2–3)		I (1–2)	I (1–3)
Rhytidiadelphus triquetrus	I (1)	II (2–9)		I (3)	I (1–9)
Cladonia impexa		I (2–4)			I (2–4)
Vicia lathyroides		I (1–2)			I (1–2)
Erodium cicutarium		I (1–2)			I (1–2)
Peltigera rufescens		I (1–2)			I (1–2)
Ononis repens		I (2)	V (1–8)	I (2–6)	III (1–8)
Calystegia soldanella			I (3–4)		I (3–4)
Elymus pycnanthus	I (1–3)		I (2–6)	V (4–10)	II (1–10)

Floristic table SD7 (cont.)

	a	b	c	d	7
Lotus corniculatus	II (1–9)	III (1–7)	II (1–8)	II (1–6)	II (1–9)
Cerastium fontanum	II (1–4)	II (1–3)	I (1–3)	I (1–4)	II (1–4)
Brachythecium albicans	II (1–9)	II (2–4)	I (1–5)	III (1–7)	II (1–9)
Plantago lanceolata	II (1–5)	I (3)	II (1–6)	I (2–6)	II (1–6)
Cirsium arvense	II (1–4)	I (2)	II (1–4)	II (1–3)	II (1–4)
Tortula ruralis ruraliformis	II (1–8)	II (2–7)	I (1–9)	II (1–7)	II (1–9)
Viola tricolor	I (1–3)	I (1–2)	I (1–4)		I (1–5)
Thymus praecox	I (1)	I (1–7)	I (1–8)		I (1–8)
Homalothecium lutescens	I (1–8)	I (1–5)	I (2)		I (1–8)
Arenaria serpyllifolia	I (1–3)	I (3)	I (2–4)	I (1–3)	I (1–4)
Phleum arenarium	I (1–3)	I (1–4)	I (1–4)	I (1–3)	I (1–5)
Sedum acre	I (1–4)	I (1–2)	I (1–5)	I (5)	I (1–5)
Cerastium diffusum diffusum	I (1–4)	I (1–2)	I (1–3)	I (1–2)	I (1–4)
Leontodon taraxacoides	I (1–4)	II (1–2)	II (1–5)	I (3–4)	I (1–5)
Aira praecox	I (2–4)	II (2–3)	I (1–4)	I (1–3)	I (1–5)
Luzula campestris	I (1–3)	II (1–3)	I (2)	I (1)	I (1–4)
Cerastium semidecandrum	I (1–2)	II (1–3)	I (1–4)	II (2–4)	I (1–6)
Peltigera canina	I (1–7)	II (1–3)			I (1–7)
Brachythecium rutabulum	I (1–8)	I (1)	I (1–6)	I (1–4)	I (1–8)
Holcus lanatus	I (1–4)	I (2–5)	I (1–5)	I (6)	I (1–6)
Arrhenatherum elatius	I (1–3)	I (3)	I (1–6)	I (3–6)	I (1–6)
Achillea millefolium	I (2–3)	I (3)	I (1–3)	I (1)	I (1–3)
Viola riviniana	I (1–5)	I (1–2)	I (1–2)		I (1–5)
Leontodon hispidus	I (3–4)	I (1–3)	I (1–2)	I (2–3)	I (1–5)
Eurhynchium praelongum	I (1–5)	I (4)	I (1)	I (3)	I (1–5)
Agrostis capillaris	I (1–7)	I (1)	I (2)		I (1–7)
Veronica chamaedrys	I (1–2)	I (1)	I (3–4)		I (1–4)
Linum catharticum	I (3)	I (1–3)	I (1)	I (1–3)	I (1–5)
Trifolium repens	I (1–5)	I (1–3)	I (1)		I (1–5)
Anthyllis vulneraria	I (1–3)		I (2–7)		I (1–7)
Trifolium arvense	I (2–5)		I (1–2)	I (2–6)	I (1–6)
Crepis capillaris	I (1–3)		I (1–4)	I (2–3)	I (1–4)

Species					
Euphorbia paralias	I (1)		I (1–5)	I (2)	I (1–5)
Eryngium maritimum	I (1)		I (2–6)	I (2–3)	I (1–6)
Campanula rotundifolia	I (1)		I (1–2)	I (2)	I (1–3)
Rumex crispus	I (1)		I (1–2)	I (2–4)	I (1–4)
Rubus fruticosus agg.	I (2)		I (1–2)	I (3–4)	I (1–4)
Centaurium erythraea	I (2–4)	I (1)		I (1–3)	I (1–4)
Pseudoscleropodium purum	I (1–4)	I (2–5)		I (3–7)	I (1–7)
Cladonia furcata	I (1)	I (1–3)		I (5)	I (1–5)
Cladonia rangiformis	I (1)	I (1–4)		I (2–6)	I (1–6)
Rhinanthus minor	I (1–4)	I (3)	I (1–3)	I (1–3)	I (1–4)
Elymus repens	I (1)	I (3)	I (3–6)		I (1–6)
Leontodon autumnalis	I (1–4)	I (3)	I (2)	I (1–2)	I (1–4)
Senecio vulgaris	I (1–2)	I (3)	I (2)		I (1–3)
Sonchus arvensis	I (4)	I (1)	I (1–2)	I (1–4)	I (1–4)
Vicia sativa nigra	I (2–4)	I (3)	I (2–5)	I (3)	I (2–5)
Tragopogon pratensis	I (1)	I (3)	I (1–2)	I (1)	I (1–3)
Erophila verna		I (2–3)	I (2)	I (1)	I (1–3)
Ranunculus bulbosus		I (1)	I (1)		I (1)
Trifolium campestre	I (2–3)			I (1–3)	I (1–3)
Bryum capillare	I (4–6)			I (5–9)	I (4–9)
Lophocolea bidentata	I (1–3)				I (1–3)
Tussilago farfara	I (1–4)		I (2)	I (3–4)	I (1–4)
Centaurea nigra	I (1–2)		I (1)		I (1–2)
Elymus farctus	I (1–4)		I (1–5)	I (1–8)	I (1–8)
Plantago coronopus	I (1–2)		I (4)	I (2)	I (1–4)
Bryum argenteum	I (1–5)		I (1)		I (1–5)
Heracleum sphondylium	I (1)	I (1)	I (1–5)	I (1–2)	I (1–5)
Ranunculus repens	I (2)	I (1–2)			I (1–2)
Cladonia pyxidata	I (1)	I (1)			I (1–2)
Cirsium vulgare			I (1)	I (1–3)	I (1–3)
Vulpia fasciculata			I (1–7)		I (1–7)
Rubus caesius			I (1–9)	I (1)	I (1–9)
Euphorbia portlandica			I (1–3)		I (1–3)
Sonchus asper			I (1–3)	I (1)	I (1–3)

Floristic table SD7 (*cont.*)

	a	b	c	d	7
Number of samples	52	21	74	54	201
Number of species/sample	15 (7–28)	19 (14–30)	13 (7–25)	11 (4–26)	14 (4–30)
Herb height (cm)	27 (1–80)	17 (3–50)	32 (2–100)	no data	27 (1–100)
Herb cover (%)	66 (40–100)	89 (75–98)	79 (25–100)	no data	80 (25–100)
Bryophyte height (mm)	30 (10–50)	19 (15–20)	9 (4–30)	no data	13 (4–50)
Bryophte cover (%)	39 (5–85)	21 (10–40)	7 (1–80)	no data	17 (1–85)
Slope (°)	9 (0–40)	12 (0–40)	11 (0–70)	no data	11 (0–70)
Soil pH	7.8 (5.3–8.8)	7.8 (7.0–8.9)	7.7 (5.5–8.7)	no data	7.8 (5.3–8.9)

a   Typical sub-community
b   *Hypnum cupressiforme* sub-community
c   *Ononis repens* sub-community
d   *Elymus pycnanthus* sub-community
7   *Ammophila arenaria-Festuca rubra* semi-fixed dune community (total)

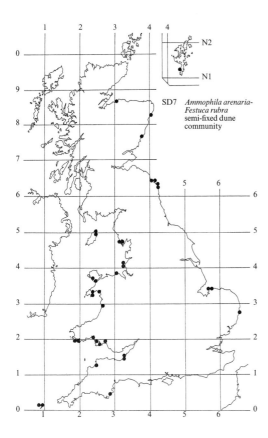

SD7   *Ammophila arenaria-*
       Festuca rubra
       semi-fixed dune
       community

SD8
Festuca rubra-Galium verum fixed dune grassland

Synonymy
Dune grassland Tansley 1911, 1939, Gimingham 1964*a*; Dune pasture Gimingham 1964*a p.p.*; Machair Gimingham 1964*a*, 1974, Ranwell 1974; *Euphrasio-Festucetum arenariae* Birse 1980; *Astragalo-Festucetum arenariae* Birse 1980 *p.p.*

Constant species
Festuca rubra, Galium verum, Lotus corniculatus, Plantago lanceolata, Poa pratensis, Trifolium repens.

Rare species
Acaena novae-zelandiae, Astragalus danicus, Dianthus deltoides, Epipactis atrorubens, Mibora minima, Oxytropis halleri, Primula scotica.

Physiognomy
The *Festuca rubra-Galium verum* community consists of dune vegetation in which *Festuca rubra* and a variety of other grasses, dicotyledons and mosses make up a generally closed sward, occasionally rank, but usually just a decimetre or two tall, and sometimes closely cropped to a short, tussocky turf. *Ammophila arenaria*, the usual dominant with us on more mobile sand, remains common overall, but it is no longer a constant feature of the vegetation: indeed, in some sub-communities here, it is only occasional. Moreover, even where it is still frequent, the cover is rarely extensive and often reduced to small tufts of shoots noticeably lacking in vigour. *F. rubra*, by contrast, is usually abundant, the typical dominant in the sward, and commonly recorded as var. or ssp. *arenaria* with its far-creeping rhizome systems and rather lax tussocks of stiff leaves (Hubbard 1968, Tutin *et al.* 1980). Other grasses are generally subordinate but some can be quite common. *Poa pratensis* agg., for example, often obviously *P. subcaerulea*, is a constant of the community, though only exceptionally of more than moderate cover, and in regions of moister climate *Holcus lanatus* becomes very frequent. *Dactylis glomerata*, too, is occasionally seen and there are sometimes

records for *Avenula pubescens*, though *Arrhenatherum elatius*, another bigger tussocky grass that is sometimes seen on dunes, is characteristically rare here. *Leymus arenarius* very occasionally maintains a place in the community but, along with *Elymus farctus* ssp. *borealiatlanticus*, *E. pycnanthus* and *E. repens*, it is not really a typical feature of this vegetation.

Among smaller grasses, *Koeleria macrantha* can be quite frequent, this community providing an important locus for this species towards the north of its range where suitably dry soils are scarce. *Festuca ovina* may also sometimes accompany *F. rubra*, though this grass, together with *Agrostis capillaris* and *Anthoxanthum odoratum*, has a restricted role here compared with more calcifuge dune swards. *Agrostis stolonifera* can occur in damper places and becomes increasingly important in transitions to slacks. In stands close to improved areas of dune pasture, *Lolium perenne* and *Cynosurus cristatus* may seed-in in small amounts. In contrast to this diversity of grasses, sedges are few in number, with only *Carex arenaria* occurring at all frequently, although *C. flacca* becomes quite common in some sub-communities and very occasionally there is a little *C. caryophyllea*. *Luzula campestris* is also a frequent plant in certain situations here.

Along with these graminoids, there is in the *Festuca-Galium* community a characteristic variety of dicotyledons, more numerous and usually more abundant than in the earlier stages of dune vegetation, with the commonest among them reinforcing the impression of a mesophytic sward. *Galium verum, Plantago lanceolata, Trifolium repens* and *Lotus corniculatus* now become constant, while *Hypochoeris radicata* persists only very occasionally and *Ononis repens* is scarce, even among stands in southern Britain. Also distinctive, though somewhat more unevenly represented in the different sub-communities, are *Cerastium fontanum, Bellis perennis, Ranunculus acris, Euphrasia officinalis* agg., *Senecio jacobaea* and *Hieracium pilosella*. Then, occasional to frequent throughout, there are *Achillea millefolium*,

Thymus praecox, *Viola riviniana*, *Heracleum sphondy-lium* and *Thalictrum minus*, this last often showing the shorter stature and low-branching panicle that have sometimes been used to segregate a ssp. *arenarium* (Clapham *et al.* 1962). Among other perennials recorded at low frequency in the community are *Taraxacum officinale* agg., *Ranunculus bulbosus*, *Plantago major*, *P. maritima*, *P. coronopus*, *Primula vulgaris*, *P. veris* and *Equisetum arvense*, with *Veronica chamaedrys*, *Linum catharticum*, *Campanula rotundifolia*, *Prunella vulgaris* and *Trifolium pratense* becoming common in different kinds of *Festuca-Galium* vegetation. The community also provides an important northern locus for the nationally rare *Astragalus danicus* and around the north-west coast of Scotland in particular, stands can show a local profusion of orchids, with *Coeloglossum viride*, *Gymnadenia conopsea*, *Listera ovata*, *Dactylorhiza fuchsii*, *D. majalis* ssp. *purpurella*, *D. incarnata* and the rare *Epipactis atrorubens* all having been recorded here. In this part of Britain, too, where montane plants can be found virtually at sea-level, *Dryas octopetala* and *Oxytropis halleri* are very occasionally seen among *Festuca-Galium* swards, together with *Primula scotica*.

Apart from coarser ephemerals like *Senecio jacobaea* and occasional *Cirsium vulgare* and *Sonchus oleraceus*, short-lived plants tend not to be a prominent feature of this vegetation but *Rhinanthus minor* and *Odontites verna* are sometimes seen growing semi-parasitically among the sward, and scattered places where the turf is broken can provide an opportunity for very occasional records of *Gentianella amarella*, *Viola tricolor*, *Erodium cicutarium*, *Cerastium diffusum* ssp. *diffusum*, *C. semidecandrum*, *C. arvense*, *Medicago lupulina*, *Trifolium campestre*, *Vicia lathyroides*, *Veronica arvensis*, *Myosotis arvensis*, *M. ramosissima*, *Aira praecox* and the rare *Mibora minima*. Typically, however, the richer assemblages of the winter annuals seen among *Ammophila-Festuca* vegetation are rarely found here.

Mosses are quite often a prominent feature of the sward with locally high cover, though the species involved vary somewhat in the different sub-communities. Plants such as *Tortula ruralis* ssp. *ruraliformis*, *Homalothecium lutescens* and *Brachythecium albicans* can remain quite common but frequently it is pleurocarps like *Rhytidiadelphus squarrosus*, *R. triquetrus*, *Pseudoscleropodium purum*, *Calliergon cuspidatum* and *Hylocomium splendens* that give a distinctive look to this element of the vegetation, with *Hypnum cupressiforme*, *Plagiomnium rostratum*, *P. undulatum*, *Eurhynchium praelongum*, *Thuidium tamariscinum*, *T. delicatulum* and *Entodon concinnus* occurring more occasionally. Lichens are far fewer in number and only rarely abundant, but *Peltigera canina* can be quite frequent, with *P. rufescens* less common.

Sub-communities

Typical sub-community. *F. rubra* is generally an obvious dominant in this kind of *Festuca-Galium* vegetation, with *P. pratensis* very common but almost always of low cover, and *Ammophila* only moderately frequent, though sometimes quite abundant and vigorous where there is still some small measure of sand movement. *Dactylis* occurs occasionally and its tussocks can be quite prominent, but other grasses are rather sparse. *C. arenaria* is frequently found, though hardly ever in any abundance, and there is only rarely any *Luzula campestris*.

Dicotyledonous associates also tend to be fewer here than in most other kinds of *Festuca-Galium* vegetation. The community constants *G. verum*, *P. lanceolata*, *T. repens* and *L. corniculatus* are all well represented, and there is commonly some *A. millefolium* and *S. jacobaea* but, apart from these, it is usually just occasional *C. fontanum*, *Bellis*, *R. acris*, *Heracleum* and *T. minus* that provide variety in the sward. Mosses, too, are generally not very prominent, with just occasional *R. squarrosus* and rather infrequent *B. albicans* and *T. ruralis* ssp. *ruraliformis*.

Luzula campestris **sub-community:** *Astragalo-Festucetum arenariae*, Typical subassociation Birse 1980. *F. rubra* is still usually the most abundant plant here, but the sward is considerably richer and more diverse than in the Typical sub-community and various associates can attain quite high cover. Among other grasses, tussocks of *Ammophila* are very frequent and locally abundant, and *P. pratensis* too can be moderately plentiful. Then, there is occasional *H. lanatus* and *K. macrantha* but, more distinctive, is the rather common occurrence of *Agrostis capillaris*, *Anthoxanthum* and *F. ovina* which, with very frequent *Luzula campestris*, can give a quite fine-grained character to much of the turf. *C. arenaria* is also often found.

Then, among shorter stretches of the sward, the typical dicotyledons of the community are frequently accompanied by chamaephytes or smaller rosette plants like *T. praecox*, *H. pilosella*, *Veronica chamaedrys* and *Hypochoeris radicata*, all of which tend to have their best representation in this kind of *Festuca-Galium* vegetation. *Astragalus danicus* is also occasionally found here along the east coast of Scotland. Less strikingly, there are sometimes records for *V. riviniana*, *T. officinale*, *S. jacobaea*, *L. catharticum* and *C. rotundifolia*, with *Rumex acetosella*, *Cerastium arvense*, *Myosotis arvensis* and *Veronica arvensis* seen in a few stands.

Moss cover can be quite high among these herbs, with *R. squarrosus* occurring commonly, *R. triquetrus*, *B. albicans*, *P. purum*, *H. lutescens* and *T. ruralis* ssp. *ruraliformis* more occasional, but all able to form extensive patches, particularly where the swards are close grazed. *Peltigera canina* also occurs quite frequently.

Tortula ruralis ssp. ruraliformis sub-community. In its vascular component, this vegetation is similar to the Typical sub-community in the quite impoverished and unvarying flora, although *Ammophila* is somewhat more common and abundant here, *P. trivialis* rather less so, and there are more frequent records for *Hieracium pilosella* and *Thymus*. *Sedum acre* and *Anthyllis* occur as preferential occasionals too and there can be a modest local abundance of annuals. Much more distinctive, however, is the constant occurrence and patchily high cover of *H. lutescens* and *T. ruralis* ssp. *ruraliformis*, with *R. squarrosus* also very common and locally abundant. Both *Peltigera canina* and *P. rufescens* are occasionally found.

Bellis perennis-Ranunculus acris sub-community: *Euphrasio-Festucetum arenariae*, Typical subassociation Birse 1980. *F. rubra* is almost always the most abundant plant in this kind of *Festuca-Galium* vegetation, and is that much more noticeable with *Ammophila* reduced here to a usually low-cover occasional. However, *P. pratensis* and *H. lanatus* are also very common and there is quite often some *A. stolonifera*, particularly where this sub-community extends on to moister ground. *C. arenaria* is also sometimes accompanied by *C. flacca* and *L. campestris*. Also rather striking is the vigorous contribution of dicotyledons to the sward which, though commonly cropped quite short, is generally closed. Thus, along with the community constants, there is frequently much *E. officinalis* agg., *B. perennis* and *R. acris* with *S. jacobaea*, *A. millefolium*, *C. fontanum* all common and, more occasional, *L. catharticum*, *Prunella*, *Trifolium pratense*, *Heracleum* and *Thalictrum*. Bryophyte cover is somewhat patchy, but *R. squarrosus* is very frequent and *H. lutescens*, *C. cuspidatum* and *T. ruralis* ssp. *ruraliformis* also fairly common.

Prunella vulgaris sub-community: *Euphrasio-Festucetum arenariae*, *Linum* subassociation Birse 1980. Some important features of this vegetation are similar to the *Bellis-Ranunculus* sub-community, such as the general prominence of *F. rubra*, the reduced contribution from *Ammophila*, and the frequency of *H. lanatus*, *C. flacca*, *E. officinalis* and *S. jacobaea*. Here, though, both *Bellis* and *R. acris* are only occasional and the most striking associates of the community constants are *L. catharticum*, *C. rotundifolia*, *T. pratense*, *G. amarella* and, most strongly preferential, *Prunella vulgaris*. More occasionally, there are records for *Centaurea nigra*, *Daucus carota*, *Leucanthemum vulgare* and *Ranunculus repens* with *Thalictrum*, *Thymus* and *C. fontanum* well represented among the community companions.

As for bryophytes, *R. squarrosus* remains the most frequent moss and it is often abundant, but *P. purum* is quite common, *H. splendens* occasional, and more obviously preferential are *R. triquetrus*, *C. cuspidatum*, *Plagiomnium undulatum* and the leafy liverwort *Lophocolea bidentata s.l.*

Habitat

The *Festuca-Galium* community is the characteristic grassland of more calcareous fixed sands on dunes and coastal plains all around Britain. Its floristics and structure are strongly influenced by the rather less droughty and impoverished conditions that come with long stability of the lime-rich sand surface, but climatic variation across the range of the community affects the processes of soil development, and grazing also often plays an important part in enhancing fertility and maintaining the physiognomy and variety of the sward. Especially striking and extensive stands of this vegetation contribute to the machair landscape of north-west Scotland and the Isles, though it is these same general factors, albeit in a particular combination, that give them their distinctive character.

The *Festuca-Galium* community cannot become permanently established on accumulations of wind-blown sand around our coasts until accretion has come to a virtual halt, and it is therefore typically found where distance or shelter put the ground beyond the reach of freshly-deposited material derived from beach sources, occurring on stable and usually gentle dune slopes and over stretches of low-relief sand plain. Localised areas of erosion and renewed deposition can develop within tracts of this kind of vegetation, but these generally support rejuvenated stages in dune colonisation, and it is a distinguishing feature of the *Festuca-Galium* community that invaders of mobile sand, so important on young dunes and secondarily exposed areas, now play a much less significant role. *Leymus arenarius*, for example, along those coasts where it assumes a prominent place in early invasion, and much more widely obvious, *Ammophila*, are past their peak of vigour here and no longer exert a dominating influence on either the physical environment or the character and disposition of the other elements of the vegetation. With marram, where shoot production ceases to be proliferative as accretion declines (Huiskes 1979) and flowering becomes less free (Huiskes 1977a), with but sparse regeneration from seed (Huiskes 1977b), the effects are seen among *Festuca-Galium* vegetation in a fall in tiller density and loss of the strong tussock habit in many stands, the frequent reduction of the clones to scattered, delibitated groups of shoots and the eventual loss of the plant altogether (Gimingham 1964a, Huiskes 1979).

The reasons for this decline in vitality are uncertain but probably relate to the edaphic changes that are set in train with increased stability of the sand surface and the greater competition from other plants that can develop upon it (Carey & Oliver 1918, Benecke 1930, Tansley 1939, Salisbury 1952, Willis *et al.* 1959a, b; Huiskes 1977a, 1979). Very few quantitative data are available but, compared with the more immature sands beneath the *Ammophila-Festuca* community, the major developments here are an accumulation of organic matter in the upper few centimetres, an increased capacity for retention of moisture, still derived mostly from rain, and some enhancement of the trophic state (Salisbury 1952, Willis & Yemm 1961, Willis *et al.* 1959a, Ranwell 1972, Chapman 1976, Willis 1985b). Even with the passage of centuries, however, it seems that the changes may be relatively modest. Thus, although the surface layers of sand under the *Festuca-Galium* community are usually noticeably darkened by the incorporation of decaying plant material and humus staining, the amount of organic matter can remain as little as 2–3% (Salisbury 1952, Knox 1974). Major nutrients, particularly nitrogen and phosphorus, also continue to limit plant growth for very considerable periods of time (Willis *et al.* 1959a, Willis & Yemm 1961), and some trace elements, like copper and cobalt, can be in short supply (Knox 1974). As with the more open *Ammophila-Festuca* vegetation, then, the addition of balanced fertiliser to this kind of sward results in a marked response in growth, the turf filling up, fresh weight and height of the herbage increasing, though here there is an accompanying decline in diversity, particularly among dicotyledons and mosses (Willis 1963, 1985b). Furthermore, although there is increased leaching of calcium carbonate from the upper layers with time, specially where the *Festuca-Galium* community extends into regions of higher rainfall, the sands are generally so rich in lime from the outset that the effects of this are negligible. Typically, then, this is a vegetation type of dunes and sand-plains where shell fragments make up a considerable proportion of the beach sediment that has fed them. Where the *Festuca-Galium* swards occur on machair, for example, the amount of calcium carbonate is commonly more than 50% of the sand, sometimes well over 75% (Gimingham *et al.* 1949, Vose *et al.* 1957, Ritchie 1974) and, even where there is more siliceous sand, reduction of the surface pH may take a very long time (Ranwell 1972, Willis 1985b). Usually, then, the pH here is not very different from that beneath the *Ammophila-Festuca* community, being mostly between 6.5 and 8.5.

These general edaphic conditions are reflected through the *Festuca-Galium* vegetation as a whole in a number of ways. With the decline in accretion and the waning of the dominance of *Ammophila*, other plants can capitalise on the expanses of stabilised sand surface but, while moisture and nutrients remain limiting, they are unable to thicken up into anything like a luxuriant sward. In particular, although the rhizomatous grasses *F. rubra* and *P. pratensis*, and the far-creeping sedge *C. arenaria*, increase their cover here compared with most *Ammophila-Festuca* vegetation, they are still held in check, and coarser tussock species like *H. lanatus* and *D. glomerata*, or *Agrostis stolonifera*, only make any prominent contribution where the ground is kept a little moister. There thus remains ample room among them for the establishment of the numerous herbs characteristic of the *Festuca-Galium* community, many of them smaller rosette plants or low-growing chamaephytes susceptible to crowding out, together with the mosses that can find patchy representation among the herbage. With the maintenance of high pH, however, more calcifuge species are very scarce in this vegetation and, only in the *Luzula* sub-community, with its preferential records for *Agrostis capillaris* and *Anthoxanthum*, along with *L. campestris*, does the sward come at all close to the *Carex-Festuca-Agrostis* vegetation characteristic of siliceous or strongly surface-leached sands. Even in the *Luzula* sub-community, with the pH usually remaining about 6, the flora continues to be mixed, and the suites of more acidophilous mosses and lichens that are so striking a feature on acidic fixed dunes still do not make an appearance.

For the most part, then, the *Festuca-Galium* community has the look of a calcicolous sward, though one in which there is some modest amelioration of a harsh edaphic environment. The commonest plants are thus species like *G. verum*, *T. repens*, *L. corniculatus*, *P. lanceolata*, *A. millefolium*, *C. fontanum* and *E. officinalis* agg. which have a broad tolerance of fairly dry, quite nutrient-poor, base-rich soils and provide strong floristic continuity with a variety of inland grasslands of a less improved character. The ground is sufficiently limey and sharply-draining for the vegetation to provide an occasional place for the small tussock grass *K. macrantha*, and the frequent occurrence in some sub-communities of *T. praecox* enhances the similarity of the sward to the kinds of Mesobromion grasslands found on limestones in the warmer and drier south of Britain. However, it is interesting that, apart from very occasional *Ranunculus bulbosus*, very few of the other widely distributed calcicoles typical of rendziniform soils are found in the *Festuca-Galium* community and, and even where this sort of dune vegetation occurs around our warmer southern coasts, more thermophilous Mesobromion plants are likewise very scarce. With *Ononis repens*, which is very diagnostic of more southerly *Ammophila-Festuca* vegetation, this may have something to do with the fact that the *Festuca-Galium* swards are often grazed, but this would scarcely eliminate many of the pasture calcicoles. Only in the *Tortula* sub-community, which extends on to some of the most base-rich soils, with a pH often above 8, does the sward take on a little more of the appearance of an open Mesobromion turf and, even then, with the patchy abundance of *T. ruralis* ssp. *ruraliformis* and *H. lutescens*, and scattered occurrence of *S. acre* and *A. vulneraria*, the resemblance is, not surprisingly, to the rather distinctive grasslands of some of the sandiest inland rendzinas, like those of Breckland.

Differences in regional climate have a marked effect on this general edaphic environment of the fixed sand surface, continuing and accentuating influences that have developed during earlier stages of dune colonisation and helping to distinguish the various sub-communities. Around our warmer and drier coasts, for example, south of the Solway–Forth line, mean annual maximum temperatures are usually above 25 °C (Conolly & Dahl 1970) and rainfall often as low as 1000 mm annually (*Climatological Atlas* 1952) with sometimes less than 140 wet days yr^{-1}, particularly to the south and east (Ratcliffe 1968). In these conditions, the fixed sands remain more drought-prone and the most widely-distributed sub-community in this part of the country, Typical *Festuca-Galium* vegetation, is often only a little less open and impoverished in its flora than the *Ammophila-Festuca* community of more mobile sands. More locally around these coasts, the *Luzula* and *Tortula* sub-communities bring some enrichment to the fixed dune swards, the first on the somewhat less base-rich surfaces, the second on those that are rather more so, but even here the herbage remains thin. It is among these kinds of *Festuca-Galium* vegetation, where open patches are more likely to develop in the sward in drier summers, particularly where there is heavy grazing by rabbits and locally by sheep, and where winter rains are insufficient to pose any threat of rotting to small rosettes, that winter annuals retain a somewhat better representation. Occasionally, then, species such as *Aira praecox*, *Erodium cicutarium*, *Cerastium diffusum* ssp. *diffusum*, *Viola tricolor*, *Vicia lathyroides* and *Trifolium campestre* bring additional diversity to the swards here.

Both the Typical and the *Luzula* sub-communities extend their range around the northern coasts of Britain, particularly to the east where, though mean annual maximum temperatures can fall below 24 °C (Conolly & Dahl 1970), with very cold winters (Chandler & Gregory 1976), the precipitation remains low (*Climatological Atlas* 1952, Ratcliffe 1968). The particular combination of climatic conditions favours the occurrence in the *Festuca-Galium* vegetation of this part of Britain of the Continental Northern *Astragalus danicus*, but apart from this there is often not much to distinguish these swards from more southerly stands. Along the west coast of northern Britain, however, the rainfall and temperature regimes are very different, with stretches of fixed dune often experiencing over 1200 mm precipitation annually (*Climatological Atlas* 1952) with more than 200 wet days yr^{-1} (Ratcliffe 1968), cool, cloud-ridden summers and relatively mild winters (Chandler & Gregory 1976, Page 1982). The influence of this is seen among the *Festuca-Galium* swards in this part of Britain, not so much in any striking phytogeographical response to the cool, oceanic conditions, but in the increased prominence of mesophytic plants benefiting from the more consistently moist character of the sand surface, even on ground that is well removed from the water-table. Thus, in the *Bellis-Ranunculus* and *Prunella* sub-communities, which make up much of the *Festuca-Galium* vegetation on the fixed sands of the western and northern Scottish coasts, through the Hebrides and on Orkney and Shetland, it is grasses like *H. lanatus* and, to a lesser extent, *A. stolonifera* and *Dactylis*, and dicotyledons such as *B. perennis*, *R. acris*, *P. vulgaris*, *E. officinalis* agg., *C. rotundifolia*, *T. pratense* and *R. repens* that give much of the distinctive character to the swards. Stands of these sub-communities can be all the more striking because they are often disposed over extensive stretches of the gently-undulating machair landscape, developed perhaps over many centuries where the profile of deposited sand has become closely adjusted to a low reception surface on hindshore rock platforms, raised beaches or terraces of drift (Ritchie & Mather 1974).

Throughout the range of the *Festuca-Galium* vegetation, grazing by rabbits, and often by stock, also has important effects on the composition and structure of the swards. For one thing, continual close cropping helps keep the herbage short, maintaining the diversity of smaller plants sensitive to shading by those able to make bulkier growth, and ultimately hindering any tendency to succession to ranker grasslands or scrub where soil conditions would favour this. Even on somewhat more fertile and moister sands, then, the community only locally takes on the look of Arrhenatherion or Rubion vegetation, with species like *Arrhenatherum elatius*, *Heracleum sphondylium*, *Daucus carota* and *Centaurea nigra* generally infrequent and nibbled back to short tufts or rosettes. Grazing animals also trample the sward and can disrupt the vegetation cover, making room for the spread of mosses or the fleeting appearance of annual plants in the community. More drastic disturbance, as by burrowing rabbits, can destroy stretches of the *Festuca-Galium* vegetation, precipitating renewed erosion of the sand, and perhaps a local rejuvenation of earlier stages in colonisation.

Sometimes, grazing works together with the edaphic and climatic conditions to maintain a generally harsh environment for the community. This is particularly the case where rabbit predation is heavy in *Festuca-Galium* vegetation around our drier southern coasts, when substantial removal of nutrients from the sward with the concentration of dung and urine latrines, can lead to a run-down of already droughty and impoverished soils, favouring a spread of plants like *Hieracium pilosella* and

Homalothecium lutescens, which contribute to the distinctive character of the *Tortula* sub-community. In other cases, however, grazing animals may play an important part in enhancing the general trophic state of fixed dune soils by the distribution of urine and faeces across the sward. Sheep and, in some regions, cattle have been frequently pastured on stretches of *Festuca-Galium* vegetation, especially round our northern coasts, where enrichment from their manuring has combined with the moister climatic conditions to encourage the development of the more mesophytic character of the *Bellis-Ranunculus* and *Prunella* sub-communities. It is this particular kind of pastoral dune economy that has helped make the machair *Festuca-Galium* stands so distinctive because, through the Isles in particular, they have long provided important grazing on the township commons (Fraser Darling & Morton Boyd 1969, Knox 1974, Ranwell 1974). Under the old souming system, as it was called, these traditionally carried only moderate numbers of cattle, though they sometimes wintered larger burdens of stock, thus benefiting whole local areas by relieving the pressure on improved hill pasture. With an increasing switch to sheep since the early 1800s and the supplementing of natural manuring by the use of chemical fertilisers, the style and intensity of machair grazing has been much altered, and different patterns of past treatment may contribute to the floristic variations seen in the *Bellis-Ranunculus* and *Prunella* sub-communities.

It is also likely that machair stands of *Festuca-Galium* vegetation have often been influenced by the arable cultivation that began sporadically with the Viking occupation and which, over recent centuries, has brought large areas into rotational use, mainly for oats and potatoes, with resting under grass (Knox 1974, Ranwell 1974; see also Figure 13). Traditionally, seaweed has been spread on such fields, adding valuable bulk and nutrients to the light, infertile sands and such manuring may well have occurred on ground at present occupied by the community: certainly, in some places, the soils show a much deeper than usual dark loamy layer with beach cobbles betraying their past enrichment with loads of wrack (Fraser-Darling & Morton Boyd 1969). Also, there may be a local addition to the flora of *Lolium perenne* and *Cynosurus cristatus* from the seeded leys.

Zonation and succession
Where wind-blown sand has become stabilised over the surface of raised beaches, low shelves of rock or terraces of superficials set back but a little way from the shore, the *Festuca-Galium* community can occupy the first vegetated zone behind the beach top or cliffs. In more extensive dune systems, however, it typically occurs on older ridges or plains inland from tracts of somewhat more mobile sand, on which it is generally replaced by

Ammophila-Festuca vegetation. There, with continuing modest accretion, *Ammophila* is more consistently prominent, retaining some vigour and holding its own against the smaller rhizomatous grasses, with more sporadic representation of the range of perennial herbs that only become really common as the surface is finally fixed. Often, though, the two communities intergrade, the boundaries between them depending on the varying proportions of marram and *F. rubra*, and the differing frequencies of associates such as *G. verum*, *L. corniculatus*, *P. lanceolata*, *C. fontanum*, *T. repens*, *H. radicata* and

Figure 13. Vegetation pattern in the machair landscape of the Outer Isles.
The dune system is fronted by narrow zones of SD2 *Honkenya-Cakile* and SD4 *Elymus farctus* vegetation and the SD7 *Ammophila-Festuca* community. Behind, the sand-plain has extensive areas of various kinds of SD8 *Festuca-Galium* grassland with scattered fields cultivated for oats and rye or potatoes or reverting as fallow to MG11 *Festuca-Agrostis-Potentilla* grassland. Behind large MG7 *Lolium* leys, are fields with mosaics of MG10 *Holco-Juncetum*, MG13 *Agrostis-Alopecurus* grassland and M28 *Filipendulo-Iridetum*, interspersed with purer stands of the last. (Redrawn from Dargie 1998, by permission of Scottish Natural Heritage).

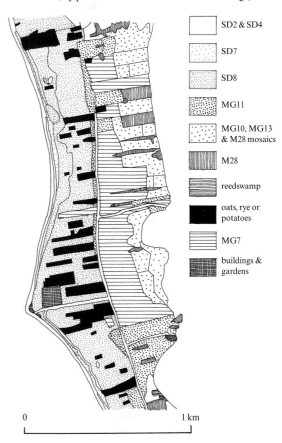

☐	SD2 & SD4
⋮	SD7
▦	SD8
▦	MG11
∴	MG10, MG13 & M28 mosaics
▥	M28
☰	reedswamp
■	oats, rye or potatoes
☰	MG7
▦	buildings & gardens

0   1 km

Taraxacum. Around our southern coasts, the usual transition is from the *Ononis* sub-community of *Ammophila-Festuca* vegetation to the Typical or, more locally, the *Luzula* or *Tortula* sub-communities of the *Festuca-Galium* vegetation. Towards the north-west of Britain where, on the machair, stretches of stabilised sand are especially extensive, these generally carry the *Bellis-Ranunculus* or *Prunella* sub-communities, passing on younger dunes to Typical *Ammophila-Festuca* vegetation.

Particularly in southern Britain, locally bare areas within the *Festuca-Galium* Community often have patches of the *Tortulo-Phleetum*. With more abrupt transitions to areas of highly mobile sand, as where severe erosion has been precipitated among stable dunes following surface disturbance of some kind, stretches of *Festuca-Galium* vegetation can be punctuated by stands of the *Ammophila* community, where rejuvenated marram may be accompanied by little else at first among the shifting substrate. Or, where sand eventually settles in and around such blow-outs, the *Carex arenaria* community may develop, sharply marked off from the surrounding *Festuca-Galium* swards or grading to them as tillers of *F. rubra* and *Ammophila* spread in among the sedge.

Over fixed dunes where there is some variation in base-richness of the sand surface, the *Festuca-Galium* community is often accompanied by other swards, the zonations between the vegetation types being especially gradual where, as is frequently the case, grazing by stock or rabbits helps keep all the herbage short and diverse. Sometimes, it is the varied intensity of leaching of what seem to be fairly uniform sands that has produced differences in surface pH, but very commonly such patterns are influenced by contrasts in the lime-content of the wind-blown sediments, something which is usually dependent on the proportion of shell fragments to siliceous material. With a shift on to more acid sands, where the pH can fall from near 8 down to 5 or less, the *Festuca-Galium* community is often replaced by the *Carex-Festuca-Agrostis* grassland. There, marram remains similarly moribund but *F. rubra* is generally accompanied and sometimes replaced by *F. ovina* and, along with frequent and often abundant *C. arenaria*, there is commonly some *Agrostis capillaris* and *Anthoxanthum*, grasses which make only an occasional contribution to the *Festuca-Galium* community. Herbs such as *G. verum*, *L. corniculatus*, *C. fontanum*, *T. repens* and *Campanula rotundifolia* can remain fairly frequent, but *Galium saxatile* now becomes very common and, among the mosses, *Dicranum scoparium*, *Hylocomium splendens* and *Pleurozium schreberi* accompany *Rhytidiadelphus squarrosus* and *Pseudoscleropodium purum*. Generally, then, the sward has the look of a Nardo-Galion community rather than some kind of mesotrophic grassland

although, again, transitions can be gradual, particularly where the *Luzula* sub-community of *Festuca-Galium* grassland passes to the *Anthoxanthum* sub-community of the *Carex-Festuca-Agrostis* vegetation, a zonation that is especially common in dune systems down the eastern Scottish and Northumberland coasts. In some places, such patterns are further complicated by the occurrence on compacted sand, or on sand–shingle mixtures, of the *Carex-Cornicularia* community, where the turf is much more open and where lichens such as *Cornicularia aculeata*, *Cladonia arbuscula*, *C. foliacea*, *C. impexa*, *C. furcata* and *C. fimbriata* are a very prominent feature.

Transitions from the *Festuca-Galium* community to more obviously calcicolous swards are much more local, but they can be seen where shell-sand has been deposited over exposures of limestone or calcareous drift with rendziniform soils. Along parts of the south Wales coast, for example, Carboniferous Limestone underlies some stretches of fixed dune, and in a few places the *Tortula* sub-community of *Festuca-Galium* vegetation passes to *Festuca-Hieracium-Thymus* grassland with the shift on to sandy rendzinas. *Ammophila*, *C. arenaria* and *F. rubra* largely disappear with this transition but among a grassy turf of *F. ovina* and *K. macrantha*, plants like *T. praecox*, *H. pilosella* and *S. acre* provide some continuity, together with *H. lutescens* and *P. purum* in what is often an extensive moss layer. Far to the north, where wind-blown sand has been deposited among exposures of Durness Limestone along the Sutherland coast, comparable zonations can be seen very locally between the *Bellis-Ranunculus* and *Prunella* types of *Festuca-Galium* grassland and the *Dryas-Carex* heath. Again, *Festuca-Galium* grassland and the *Dryas-Carex* heath. Again, *Ammophila* and *C. arenaria* drop out, while plants like *Dryas octopetala*, *Carex flacca*, *C. panicea*, *Plantago maritima* and *Antennaria dioica* become very common, but *L. corniculatus*, *B. perennis*, *G. verum*, *K. macrantha*, *T. praecox*, *H. pilosella* and *Homalothecium lutescens* continue to give character to many stands of the heath and some fine mosaics of the vegetation types have ill-defined boundaries.

The other important kind of edaphic variation that influences vegetation patterns on the fixed sands where the *Festuca-Galium* community occurs is related to the height of the ground water table. In depressions among undulating sand-plains or between ridges of immobile dunes, this can come close to the surface, keeping the ground very moist or waterlogged through much of the year, or even giving some flooding in winter. Then, the *Festuca-Galium* swards typically give way to some kind of slack vegetation, usually around the drier margins of more base-rich slacks, of the *Salix-Holcus* type. That community is quite varied, but some of the more moisture-tolerant plants of the *Festuca-Galium* vegetation can run on into slacks with some frequency, *F. rubra, H. lanatus, C. arenaria, L. corniculatus, T. repens, Euphrasia officinalis* agg. and *Prunella* commonly making a contribution to the sward. With the appearance of *Salix repens*, however, and such associates as *Epipactis palustris, Carex panicea* and *Hydrocotyle vulgaris*, together with varied suites of other herbs and bryophytes, there is often little difficulty in discerning boundaries between the vegetation types, especially where sudden transitions to wetter ground occur among dunes in drier parts of the country with Typical *Festuca-Galium* forming the usual slack surround. In the wetter north-west of Britain, and particularly over the gently rolling sand-plains of the machair, the zonations can be less well defined because more moisture-demanding herbs extend further into the *Festuca-Galium* community and the slack vegetation tends not to be so strictly confined to lower depressions. Here, then, stretches of the more mesophytic *Bellis-Ranunculus* and *Prunella* sub-communities often pass more gradually into *Salix-Holcus* vegetation, sometimes with an intervening zone of the *Festuca-Agrostis-Potentilla* grassland. Wetter slacks then see a transition to the *Salix-Calliergon* or *Salix-Campylium* community. Where transitions to wetter ground also involve a reduction in the base-richness of the substrate, as with a shift on to moist acid sands, the *Potentilla-Carex* slack replaces the *Salix-Holcus* and these other communities in such sequences. There, it is the presence of mixtures of *P. anserina, C. nigra, S. repens, Galium palustre, Ranunculus flammula* and *Cardamine pratensis* that distinguish the damp swards, giving the look of a poor fen.

With the increased stability of the surface among fixed dunes, and the more hospitable nutrient and moisture regimes beneath the *Festuca-Galium* vegetation, there are enhanced possibilities of seral progression to scrub or woodland. Very commonly, though, such succession is held in check by the grazing of stock or rabbits, so that the community is maintained as a plagioclimax. Where there is some relief from the predations of herbivores, *Festuca-Galium* swards can grow more rank, grasses such as *F. rubra, H. lanatus* and *Dactylis* taking on a more tussocky appearance, and herbs like *Heracleum, Centaurea nigra* and *Daucus carota* growing up tall from their basal rosettes, producing something like a *Centaureo-Cynosuretum*. More locally, but especially along the north-east coast of England, the *Ammophila-Arrhenatherum* community can occur among *Festuca-Galium* vegetation where there is little or no grazing. Here, *F. rubra* and *Ammophila* can both persist in some quantity, with *P. pratensis, H. lanatus, A. millefolium, G. verum* and *L. corniculatus* also often present, but *Arrhenatherum* is a very common and sometimes abundant feature, with frequent *Dactylis, Veronica chamaedrys* and *Heracleum* confirming the character of an Arrhenatherion sward. On warmer dune slopes, the additional

presence of *Geranium sanguineum* and patches of *Rosa pimpinellifolia* can mark out such transitions even more strikingly.

Continued freedom from grazing can allow the invasion of *Rubus fruticosus* agg. among the *Festuca-Galium* grassland producing patches of *Rubus-Holcus* underscrub, with rank growth of *F. rubra, P. pratensis, H. lanatus, Dactylis, Arrhenatherum* and umbellifers around the bramble, or bracken may spread in stands of *Pteridium-Rubus* vegetation. In other cases, the removal or demise of herbivores has allowed the direct invasion of trees, birch frequently figuring prominently in such successions, with conifers sometimes seeding in from nearby plantations. Where reduction of grazing has taken place on patchworks of *Festuca-Galium* grassland and *Carex-Festuca-Agrostis* swards on more acid sands, rank derivatives of the former often persist among some kind of *Calluna-Carex* heath, where mixtures of *Calluna vulgaris* with *Erica cinerea* or *Empetrum nigrum* ssp. *nigrum* are characteristically dominant among calcifuge herbs, bryophytes and lichens.

Landward patterns among stretches of *Festuca-Galium* vegetation are often further confused by various kinds of dune reclamation or improvement. Sometimes, but a small zone of natural fixed dune persists as a fringe to a golf course, on which the *Festuca-Galium* swards may survive only in a modified form in periodically mown rough, or to pasture where the community has been entirely replaced by sown *Lolio-Cynosuretum* or Lolio-Plantaginion leys. More strikingly, on the machair, *Festuca-Galium* grassland can be seen among extensive patchworks of rotational arable land, having escaped cultivation or being in various stages of reversion after short periods under the plough. Finally, on many dune systems, stands of the community give way abruptly to conifer plantations established on the stable sand.

Distribution

The *Festuca-Galium* grassland can be found on suitable stable dunes and sand plains all around the British coast. The Typical sub-community occurs throughout the range, and along our more southerly coasts it is the major type, with the *Luzula* and especially the *Tortula* sub-communities more locally represented. The *Bellis-Ranunculus* and *Prunella* swards, by contrast, are almost wholly confined to the north-west Scottish coast and the Isles.

Affinities

Although reference was made in early descriptions of British dune vegetation to grassland of this general type (Tansley 1911, 1939, Gimingham 1964a), there was no systematic attempt to characterise a distinct community or define the floristic variation within it. Likewise, accounts of machair vegetation, while stressing the peculiarity of the habitat, have often been rather vague, doing little to distinguish this kind of grassland from other swards represented there, or to compare it with other assemblages of fixed dunes from elsewhere in Britain (Gimingham 1964*a*, 1974, Ranwell 1974). Only with Birse's (1980, 1984) scheme do we have anything like a broadly-based definition, and even then this gives us just a Scottish perspective. Birse also splits his samples of this sort of sward into two associations, his *Euphrasio-Festucetum* corresponding with our *Bellis-Ranunculus* and *Prunella* sub-communities, while what is here retained as a *Luzula* sub-community is separated off into an *Astragalo-Festucetum*. This also includes dune grassland which, while retaining frequent records for *A. danicus*, is more calcifuge in character than our *Festuca-Galium* vegetation, being included in our scheme in the *Carex-Festuca-Agrostis* community.

Birse (1980, 1984) placed his grassland of this kind within the Koelerion albescentis of the Sedo-Scleranthetea (or Galio-Koelerion of the Koelerio-Corynephoretea as Westhoff & den Held (1969) have it), but affinities with the emphemeral-rich swards of sandy soils are really only clearly seen here in the *Luzula* and *Tortula* sub-communities. An alternative view would be to retain the *Festuca-Galium* grassland within the Ammophilion, although *Ammophila* and other pioneer dune plants are by this time patchy in their representation. The other obvious affinity of this vegetation, especially well seen in the *Bellis-Ranunculus* and *Prunella* sub-communities, is with the grazed swards of among the Arrhenatheretalia. For all the particular character of the machair environment, the combination of habitat factors operative there tends to move the composition of the fixed dune grasslands close to some of the richer mesotrophic pastures of unimproved soils in lowland Britain.

Floristic table SD8

	a	b	c	d	e	8
Festuca rubra	V (1–10)	V (2–9)	V (2–9)	V (3–9)	V (5–10)	V (1–10)
Galium verum	V (1–7)	V (1–8)	V (2–6)	IV (1–7)	V (2–7)	V (1–8)
Plantago lanceolata	V (1–7)	IV (1–7)	IV (2–4)	V (1–8)	V (1–5)	V (1–8)
Trifolium repens	IV (1–8)	IV (1–6)	IV (2–6)	V (1–8)	IV (2–7)	IV (1–8)
Lotus corniculatus	IV (1–9)	V (1–7)	III (2–5)	IV (1–7)	IV (2–6)	IV (1–9)
Poa pratensis	IV (1–7)	V (1–7)	II (2–5)	IV (1–8)	IV (2–7)	IV (1–8)
Cerastium fontanum	II (1–4)	IV (1–5)	II (1–3)	III (1–4)	III (1–3)	III (1–5)
Luzula campestris	I (1–4)	IV (1–5)	I (2–3)	II (1–5)	II (2–3)	II (1–5)
Hieracium pilosella	I (1–5)	III (1–7)	II (3–4)	I (1–3)	II (2–3)	I (1–7)
Veronica chamaedrys	I (1–5)	III (1–5)	I (3)	I (2)	I (2)	I (1–5)
Agrostis capillaris	I (1–7)	III (1–7)	I (3)	I (1–6)		I (1–7)
Anthoxanthum odoratum	I (2–4)	II (1–8)	I (5)	I (2–5)	I (4–7)	I (1–8)
Brachythecium albicans	I (1–6)	II (1–7)	I (1)	I (1–5)		I (1–7)
Hypochoeris radicata	I (1–4)	II (1–6)		I (2)		I (1–6)
Festuca ovina	I (1–5)	II (1–7)		I (1–5)		I (1–7)
Climacium dendroides	I (2–4)	II (1–5)		I (2–7)		I (1–7)
Astragalus danicus	I (3–4)	II (2–7)				I (2–7)
Rumex acetosella		I (1–4)				I (1–4)
Cerastium arvense		I (1–4)				I (1–4)
Myosotis arvensis		I (1–3)				I (1–3)
Veronica arvensis		I (1–4)				I (1–4)
Homalothecium lutescens	I (1–4)	I (2–6)	V (2–7)	III (1–9)	I (2–3)	II (1–7)
Tortula ruralis ruraliformis	I (1–9)	I (1–6)	V (2–9)	I (1–8)	I (3–4)	I (1–9)
Polygala vulgaris	I (1–4)	I (1–4)	II (2–4)	I (1–5)	I (2–3)	I (1–5)
Sedum acre	I (1–4)	I (1–4)	II (2–4)	I (1–2)		I (1–4)
Anthyllis vulneraria	I (4)	I (1–7)	II (2–6)	I (1–4)		I (1–7)
Bellis perennis	II (1–8)	I (1–7)	III (2–6)	IV (1–6)	II (1–4)	III (1–8)
Ranunculus acris	II (1–5)	I (1–3)	III (1–4)	IV (1–5)	II (1–3)	III (1–5)
Agrostis stolonifera	I (2–5)	I (1–5)	I (2–3)	II (1–7)	I (3)	I (1–7)
Vicia cracca				I (1–3)		I (1–3)

Floristic table SD8 (*cont.*)

	a	b	c	d	e	8
Senecio jacobaea	III (1–5)	II (1–4)	III (1–5)	III (1–6)	IV (1–5)	III (1–6)
Euphrasia officinalis agg.	I (1–5)	I (1–6)	II (1–5)	IV (1–7)	V (1–6)	II (1–7)
Holcus lanatus	I (1–6)	II (1–8)	I (2–3)	III (1–6)	V (3–7)	II (1–8)
Linum catharticum	I (2–4)	II (1–5)	I (1–3)	II (1–4)	IV (1–5)	II (1–5)
Campanula rotundifolia	I (1–5)	II (1–5)	I (2–3)	I (1–2)	IV (1–4)	I (1–5)
Prunella vulgaris	I (2)	I (1–3)	I (2–6)	II (1–5)	IV (1–5)	I (1–6)
Rhytidiadelphus triquetrus	I (1–9)	II (1–9)	I (3–5)	I (1–8)	III (2–7)	I (1–9)
Calliergon cuspidatum	I (1)	I (1–6)	I (2–3)	II (1–8)	III (2–7)	I (1–8)
Trifolium pratense	I (1–4)	I (2–6)		II (1–6)	III (2–5)	I (1–6)
Carex flacca	I (1–6)	I (1–6)		II (1–6)	III (3–5)	I (1–6)
Gentianella amarella	I (3)	I (1–4)		I (1–4)	III (1–4)	I (1–4)
Centaurea nigra	I (1–7)	I (2–6)	I (1–4)	I (2–5)	II (2–7)	I (1–7)
Ranunculus repens	I (1–4)	I (1–4)	I (2–3)	I (1–6)	II (2–6)	I (1–6)
Lophocolea bidentata	I (1–4)	I (1–5)	I (3–5)	I (2–5)	II (2–7)	I (1–7)
Daucus carota	I (1–5)	I (3)	I (2–5)	I (1–4)	II (2–8)	I (1–8)
Plagiomnium undulatum	I (1–4)	I (1–6)		I (1–5)	II (2–4)	I (1–6)
Salix repens					I (3–8)	I (3–8)
Leucanthemum vulgare					I (3–8)	I (3–8)
Ammophila arenaria	III (2–9)	IV (2–8)	V (2–8)	II (1–6)	III (3–8)	III (1–9)
Rhytidiadelphus squarrosus	II (1–9)	III (1–8)	IV (2–7)	IV (1–8)	IV (2–8)	III (1–9)
Carex arenaria	III (1–7)	III (1–7)	III (2–5)	II (1–9)	II (1–3)	III (1–9)
Achillea millefolium	III (1–5)	II (1–4)	II (1–3)	III (1–8)	III (1–3)	III (1–8)
Thalictrum minus	II (1–7)	I (1–6)	III (2–6)	II (1–7)	III (1–7)	III (1–7)
Thymus praecox	I (1–8)	III (1–8)	II (2–7)	I (1–8)	III (1–7)	II (1–8)
Pseudoscleropodium purum	I (1–3)	III (1–7)	I (3)	I (1–4)	III (2–6)	II (1–7)
Koeleria macrantha	I (1–7)	II (1–7)	I (3–6)	I (2–6)	II (1–6)	II (1–7)
Viola riviniana	I (1–6)	II (1–6)	I (1–5)	I (1–5)	II (1–5)	I (1–6)
Heracleum sphondylium	II (1–8)	I (1–5)	I (2–3)	II (1–8)	II (2–5)	I (1–8)
Peltigera canina	I (5)	II (1–5)	II (2–3)	I (1–4)	I (1–3)	I (1–5)
Dactylis glomerata	II (1–8)	I (1–5)		I (1–6)	II (2–4)	I (1–8)
Viola tricolor	I (4)	I (1–7)	II (1–4)	I (1–4)	II (2–3)	I (1–7)
Hylocomium splendens	I (2–6)	II (1–8)		I (1–2)	II (1–5)	I (1–8)

Species					
Taraxacum officinale agg.	II (1–5)	II (1–4)	I (1–5)	I (1–4)	I (1–5)
Ranunculus bulbosus	I (1–3)	I (1)	I (1–4)	I (1–4)	I (1–5)
Hypnum cupressiforme	I (1–7)	I (3)	I (1–5)	I (2)	I (1–9)
Elymus repens	I (1–6)	I (2)	I (2–4)	I (2)	I (1–6)
Succisa pratensis	I (4)	I (3)	I (2–4)	I (1–3)	I (1–4)
Elymus farctus	I (1–4)	I (1–2)	I (1–3)	I (3)	I (1–4)
Avenula pubescens	I (2–8)	I (2–4)	I (2–8)	I (2–4)	I (2–8)
Plagiomnium rostratum	I (2)	I (1–3)	I (2–4)	I (2–3)	I (1–4)
Primula vulgaris	I (1–7)	I (1)	I (2–4)	I (2–3)	I (1–7)
Plantago major	I (1–4)	I (2)	I (2)	I (2–3)	I (1–4)
Plantago coronopus	I (1–4)	I (2–4)	I (1–3)		I (1–4)
Cirsium vulgare	I (1–3)	I (1–2)	I (1)		I (1–3)
Erodium cicutarium	I (1–4)	I (3–4)	I (2–3)		I (1–5)
Rhinanthus minor	I (1–7)		I (1–6)	I (2–4)	I (1–7)
Lolium perenne	I (1–6)		I (1–5)	I (3–5)	I (1–6)
Primula veris	I (4)		I (1–2)	I (3–4)	I (1–6)
Eurhynchium praelongum	I (1–3)		I (1–3)	I (2–4)	I (1–4)
Equisetum arvense	I (1–2)		I (1–3)	I (1–3)	I (1–4)
Crepis capillaris	I (1–4)	I (2–3)	I (1–4)	I (3)	I (1–4)
Entodon concinnus	I (6)	I (3–8)	I (2–3)	I (2–8)	I (2–8)
Cirsium arvense	I (1–3)	I (2)	I (1–3)	I (1)	I (1–3)
Plantago maritima	I (2–4)		I (1–3)	I (2–3)	I (1–5)
Trisetum flavescens	I (1–5)		I (1–5)		I (1–5)
Leymus arenarius	I (1–6)		I (1)		I (1–6)
Potentilla anserina	I (2–3)		I (1–7)		I (1–7)
Poa trivialis	I (1–3)		I (1–5)		I (1–5)
Aira praecox	I (3–7)		I (1)		I (1–7)
Leontodon autumnalis	I (2–4)		I (1–6)		I (1–6)
Rumex acetosa	I (1–3)	I (1–3)	I (1–4)		I (1–4)
Peltigera rufescens	I (1)	I (2–3)	I (2–4)		I (1–4)
Cerastium semidecandrum	I (1–3)	I (2–3)	I (1–3)		I (1–3)
Angelica sylvestris	I (1–5)	I (3)		I (3)	I (1–5)
Odontites verna	I (2–4)		I (2–4)	I (2–5)	I (2–5)
Ditrichum flexicaule			I (2–4)	I (3–5)	I (2–5)
Thuidium tamariscinum	I (1–3)	I (2–4)	I (3–4)	I (3–4)	I (1–4)
Arrhenatherum elatius	I (1–4)		I (3)	I (3)	I (1–4)

Floristic table SD8 (*cont.*)

	a	b	c	d	e	8
Cerastium diffusum diffusum	I (1–3)				I (2)	I (1–3)
Medicago lupulina	I (1–5)				I (3)	I (1–5)
Cynosurus cristatus	I (1–4)				I (2–3)	I (1–4)
Elymus pycnanthus	I (1–4)	I (2)				I (1–4)
Carex caryophyllea	I (1)	I (1–4)				I (1–4)
Rosa pimpinellifolia	I (1–5)	I (2–8)				I (1–8)
Trifolium campestre	I (1–3)	I (1–4)				I (1–4)
Vicia lathyroides	I (1–3)	I (1–4)				I (1–4)
Ononis repens	I (1–6)		I (1–7)			I (1–7)
Pleurozium schreberi		I (2–5)		I (2)		I (2–5)
Cladonia rangiformis		I (1–5)			I (2)	I (1–5)
Myosotis ramosissima		I (1–4)			I (2)	I (1–4)
Sonchus oleraceus			I (1–2)	I (2)		I (1–2)
Thuidium delicatulum		I (1–2)			I (3–4)	I (1–4)
Listera ovata		I (2)			I (1–4)	I (1–4)
Coeloglossum viride				I (1–3)	I (2–3)	I (1–3)
Number of samples	111	117	36	129	25	418
Number of species/sample	15 (7–26)	23 (15–33)	18 (9–25)	20 (14–30)	24 (16–32)	20 (7–33)
Vegetation height (cm)	19 (1–80)	16 (2–70)	18 (2–50)	12 (2–84)	19 (3–63)	16 (1–84)
Vegetation cover (%)	85 (30–100)	82 (40–100)	No data	88 (60–100)	No data	84 (30–100)
Slope (°)	7 (0–40)	7 (0–45)	25 (0–60)	7 (0–40)	8 (0–50)	8 (0–60)
Soil pH	7.8 (4.6–9.3)	7.2 (4.7–9.0)	8.5 (8.3–9.2)	8.2 (7.5–8.9)	8.2 (7.7–8.6)	7.8 (4.6–9.3)

a   Typical sub-community
b   *Luzula campestris* sub-community
c   *Tortula ruralis* ssp. *ruraliformis* sub-community
d   *Bellis perennis*–*Ranunculus acris* sub-community
e   *Prunella vulgaris* sub-community
8   *Festuca rubra*–*Galium verum* fixed dune community (total)

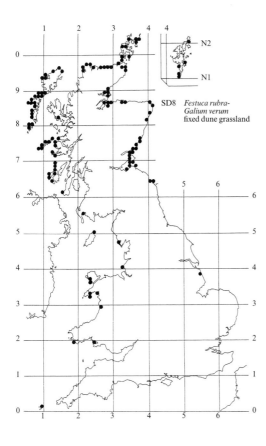

SD8 *Festuca rubra-
Galium verum*
fixed dune grassland

SD9

Ammophila arenaria-Arrhenatherum elatius dune grassland

Synonymy
Dune grassland   Gimingham 1964*a*.

Constant species
Achillea millefolium, Ammophila arenaria, Arrhenatherum elatius, Festuca rubra, Poa pratensis.

Rare species
Acaena novae-zelandiae, Astragalus danicus.

Physiognomy
The *Ammophila arenaria-Arrhenatherum elatius* grassland includes rank, tussocky swards in which both *Festuca rubra* and *Ammophila* remain very common, the former especially being often abundant, indeed the most frequent dominant overall, the latter rather more patchily represented, but plentiful and vigorous in some stands. Compared with other dune grasslands, however, a noticeable difference here is the common occurrence of *Arrhenatherum elatius*, often growing in some abundance and quite frequently co-dominant. Other grasses figure, too, though their contribution to the cover is generally small. *Poa pratensis* agg. (probably *P. subcaerulea* in many cases) is constant in small amounts, for example, and *Dactylis glomerata* is often found, though only exceptionally with any abundance. More occasionally, there can be some *Holcus lanatus, Trisetum flavescens, Elymus pycnanthus, E. farctus* and *E. repens*, but smaller species like *Koeleria macrantha, Agrostis capillaris* and *Anthoxanthum odoratum* tend to be very scarce. *Carex arenaria* is only moderately common, and *Luzula campestris* rather infrequent.

Quite a variety of dicotyledonous herbs occur in the community, the commonest able to grow up tall through the grasses or form a loose, bushy understorey in more open parts of the sward. Most frequent among these plants are *Heracleum sphondylium, Achillea millefolium, Veronica chamaedrys, Galium verum, Ononis repens* and *Plantago lanceolata*, with *Cirsium arvense, Senecio jacobaea* and *Geranium sanguineum* preferential to particular

sub-communities. Less often, there can be some *Lotus corniculatus, Trifolium repens, Cerastium fontanum, Taraxacum officinale* agg. *Campanula rotundifolia, Centaurea nigra, Cruciata laevipes, Primula veris* and *Thalictrum minus* with occasional *Hypochoeris radicata* and *Hieracium pilosella* where the herbage is less rank. Among rarer plants, *Astragalus danicus* and the introduced *Acaena novae-zelandiae* have been recorded here.

In the denser swards, bryophytes are usually few and of patchy cover, but *Pseudoscleropodium purum* and *Brachythecium rutabulum* occur occasionally and *Rhytidiadelphus triquetrus, R. squarrosus* and *Hypnum cupressiforme s.l.* more rarely though sometimes with local abundance. Mosses like *Tortula ruralis* ssp. *ruraliformis* and *Homalothecium lutescens* are rare.

Sub-communities

Typical sub-community. Mixtures of *Arrhenatherum, F. rubra* and *Ammophila* form the bulk of the cover here, with *P. pratensis* generally very subordinate, *Dactylis, H. lanatus* and *C. arenaria* only occasional and locally abundant and *E. pycnanthus* very sparse. Among bigger associates, *Cirsium arvense* and *Senecio jacobaea* often accompany *Heracleum* giving a somewhat weedy appearance to the swards, but other distinctive features are few. *Achillea* remains frequent and there is occasional *V. chamaedrys, G. verum, O. repens* and *P. lanceolata*, with *Myosotis arvensis* and *Crepis capillaris* recorded rarely. Bryophytes are generally very sparse, but *P. purum* and *R. triquetrus* can be locally abundant, with occasional *B. rutabulum*.

Geranium sanguineum sub-community. *F. rubra* is the usual dominant in this sub-community, with *Arrhenatherum* and *Ammophila* somewhat less frequent and only patchily abundant. Along with *P. pratensis* and occasional *H. latatus*, there is also very often some *Dactylis* and, less commonly, *Trisetum*, with *Koeleria* and *Carex flacca* occurring sparsely in more open areas. More striking, though, is the constancy of *Geranium sanguineum*, sometimes growing in a procumbent form, though occasionally abundant and very eye-catching in summer with its red-purple flowers. Then, together with frequent *A. millefolium*, *V. chamaedrys*, *G. verum*, *O. repens* and *Heracleum*, there is often some *L. corniculatus*, *T. repens*, *P. veris* and *L. pratensis*, with occasional *C. fontanum*, *T. officinale*, *C. rotundifolia* and *T. minus*. Also preferential at low frequency are *Thymus praecox*, *Pimpinella saxifraga*, *Ranunculus bulbosus* and *Sanguisorba minor*, while at some sites this vegetation provides a locus for the rare *Astragalus danicus*. Occasionally, low bushes of *Rosa pimpinellifolia* can be found scattered through the rank sward. Bryophytes are again often sparse, but patches of *T. ruralis* ssp. *ruraliformis* sometimes occur with the occasional *P. purum*.

Habitat

The *Ammophila-Arrhenatherum* grassland is typically confined to less heavily grazed stretches of more fixed, calcareous coastal sands, occurring rather locally on dune systems all around Britain, though much more commonly along the seaboard of north-east England.

Like the *Festuca-Galium* grassland and the more well established *Ammophila-Festuca* swards, this is a vegetation type of wind-blown sands that have become more or less stabilised on sheltered slopes of coastal dunes, older ridges or tracts of sand-plain set back some distance from the shore. With accretion reduced to near negligible levels, *Ammophila* is past the peak of its vegetative and reproductive vigour here (Gimingham 1964*a*, Huiskes 1977*a*, *b*, 1979), though it often retains a quite strongly tussocky form in the community, rather than being reduced to the scattered tufts of debilitated shoots in which state it usually lingers on in the pastures of fixed dunes. Free of its overwhelmingly dominant influence, however, the smaller rhizomatous grasses *F. rubra* and *P. pratensis*, which begin to colonise less mobile sands, can make a more substantial contribution while, on the stabilising surface, there is opportunity for the appearance of such other herbs as can tolerate the base-rich and probably still quite impoverished and droughty conditions. For this is a vegetation type of more calcareous sands, where shell fragments often comprise a high proportion of the dune sediments or where leaching has not yet had a marked impact, superficial pH generally remaining between 6 and 8. And, though edaphic changes are in train with the more extensive colonisation of the fixed sand surface, there is perhaps but little accumulation yet of organic matter or nutrients in the substrate, and a still poor retention of moisture (Willis *et al.* 1959*a*, Willis 1985*b*). Such features thus probably continue to have a limiting effect on the composition and luxuriance of the associated flora.

Equally important with this vegetation, however, is the absence or at least the low intensity of grazing, something which much restricts the distribution of the community on fixed dunes around our coasts, and has a variety of effects on the floristics and structure of those stands which can develop in the rather scarce localities where there is little or no predation by either stock or wild herbivores. In the first place, there is the prominence here, along with the usual perennial grasses of more stable coastal sands, of *Arrhenatherum*, a plant of wide distribution around most of the British coast, one tolerant of some quite harsh edaphic environments, but rare in grazed swards, including those on dunes, because of its great palatability. To a lesser extent, the increased frequency of *Dactylis* in this community, as compared with fixed dune pastures, probably reflects this same factor, and there are some dicotyledons here too, notably *Ononis repens* and *Geranium sanguineum*, which fare much better in ungrazed swards.

The second influence is felt through the generally rank growth which the grasses in particular are able to make in this vegetation with freedom from grazing, for this greatly increases competition for the available water and nutrients and also, very importantly, for light. The commonest associates are thus those plants which can maintain growth among the tussocky dominants, by putting up elongated leaves or tall shoots from rosettes, as with *Heracleum*, *A. millefolium*, *P. lanceolata*, *S. jacobaea* or *C. arvense*, or producing more straggling masses of shoots among the herbage, like *V. chamaedrys*, *G. verum*, *L. corniculatus* and *G. sanguineum*. By contrast, *C. arenaria* and small tussock grasses, more diminutive hemicryptophyte dicotyledons and chamaephytes, which can maintain themselves in some variety in close-cropped *Festuca-Galium* swards, find little place here except where the bulkier perennials are not so dense. There is also very little opportunity for small ephemerals or bryophytes to make any consistent contribution on the sand surface.

Particular treatment histories, with local neglect of grazing on dunes, or vagaries in wild herbivore predation, as after the myxomatosis epidemic in the 1950s, may play some part in the marked concentration of the *Ammophila-Arrhenatherum* community along the Northumberland coast and around the Humber. However, although this kind of vegetation can be found on other ungrazed dune systems elsewhere in Britain, there are many places where it has not developed, despite low numbers of stock or rabbits. It is possible, therefore, that some other factors also influence its development, perhaps a requirement for sands that are not quite so droughty or impoverished as usual, or a dependence on certain climatic conditions. Interestingly, *Arrhenatherum* becomes prominent in cliff grasslands down the north-east coast of England where, with prevailing offshore winds, salt-spray deposition is that much reduced compared with similar situations on western cliffs, and, further south, around parts of East Anglia, *Arrhenatherum-Silene* swards are a distinctive feature at the top of some shingle beaches. A measure of protection from such maritime influence may thus be essential for the vigorous development of *Ammophila-Arrhenatherum* grassland even on ungrazed tracts of stabilised dunes.

Climate also certainly affects this vegetation on a smaller scale in the association of the *Geranium* sub-community with warmer dune slopes. This kind of *Ammophila-Arrhenatherum* grassland is more local than the Typical form and is usually found on south-facing slopes, often quite steep, where insolation is at its maximum, benefiting the thermophilous *G. sanguineum*.

Zonation and succession

The *Ammophila-Arrhenatherum* grassland is usually found as a local replacement for other swards on more stabilised calcareous dunes, occurring in zonations and mosaics which are presumably influenced, at least in part, by variations in grazing intensity. In many places, the community could probably progress very readily to scrub with the invasion of shrubs and trees, or to bracken, and stands are sometimes found among these kinds of vegetation.

In the generalised sequence of communities on British dunes, the *Ammophila-Arrhenatherum* grassland generally occurs where more mature *Ammophila-Festuca* swards or *Festuca-Galium* vegetation would otherwise be found and, on more varied systems, it often grades to these. Compared with the former community, the most obvious differences can be the abundance of *Arrhenatherum* and, particularly with the shift to younger dune ridges, Typical *Ammophila-Arrhenatherum* grassland may pass quite gradually to the *Ononis* sub-community of *Ammophila-Festuca* vegetation with changes in dominance of the grasses. Commonly, however, neighbouring *Ammophila-Festuca* swards on more fixed sands have a greater variety and abundance of smaller herbs and bryophytes, with species such as *Hypochoeris radicata*, *Taraxacum officinale* agg.; *Leontodon taraxacoides*, *Hieracium pilosella*, *Brachythecium albicans*, *Tortula ruralis* ssp. *ruraliformis* and *Hypnum cupressiforme* becoming important. With the move to more heavily grazed stretches of stabilised dunes, some of these same plants make a frequent appearance where the *Ammophila-Arrhenatherum* grassland gives way to *Festuca-Galium* vegetation. In transitions to the *Luzula* sub-community of that kind of grassland, for example, a common vegetation type on Northumbrian dunes, the decline in *Arrhenatherum* is accompanied by a rise in their frequency along with small tussock grasses like

Koeleria, *Agrostis capillaris*, *Anthoxanthum* and *Festuca ovina* among the grazed down *F. rubra*, *P. pratensis*, *C. arenaria* and *L. campestris*. Where more acidic sands occur, this zonation can continue into stands of *Carex-Festuca-Agrostis* grassland.

Alternatively, the *Ammophila-Arrhenatherum* community can pass to Typical *Festuca-Galium* vegetation where the occasional occurrence of *Arrhenatherum*, *Dactylis*, *Heracleum* and *Cirsium arvense*, with frequent *Achillea* and *S. jacobaea* in sometimes rough *F. rubra*-dominated swards, can give a greater measure of continuity to the zonation. Mixtures of these two communities can also be found among patches of scrubby vegetation on less heavily grazed dunes. There, the *Ammophila-Arrhenatherum* grassland can give way to clumps of *Rubus-Holcus* underscrub, the grassy margins of which, with rank *H. lanatus*, *Arrhenatherum* and *Dactylis*, umbellifers and tall weedy herbs, have much in common with the surrounding dune swards (Figure 14). Or, where bracken invades along with the bramble, there may be transitions to *Pteridium-Rubus* underscrub. In other cases, particularly on some of the Humber dune systems, *Hippophae rhamnoides* scrub can be found among *Ammophila-Arrhenatherum* grassland, *Festuca rubra* and *Ammophila* tending to persist most prominently in the Typical sub-community, *Arrhenatherum* remaining common, often among nitrophilous weeds, in the *Urtica-Galium* sub-community. Or there can be zonations to *Ligustrum vulgare*-dominated stands of the *Crataegus-Hedera* scrub or grassy *Prunus-Rubus* scrub.

Distribution

The *Ammophila-Arrhenatherum* grassland can be found locally on suitably stable dunes around many parts of the British coast, but it is much commoner in north-east England, with the *Geranium* sub-community in particular strongly concentrated in Northumberland.

Affinities

Arrhenatherum figures hardly at all in descriptions of British dune vegetation (Tansley 1939, Gimingham 1964a, Birse 1980) and no account of this sort of grassland has previously been given. In phytosociological terms, it is probably best placed among the Ammophilion communities, rather than with the inland Arrhenatherion swards, although plants like *G. sanguineum* and *R. pimpinellifolia* also provide a link with Geranion sanguinei scrubby grasslands. The *Geranium* sub-community thus comes close to the cliff *Geranietum* described by Malloch (1970, 1971) from sunny, sheltered places with base-rich soils on the Lizard peninsula.

Figure 14. Zonation of vegetation types on a dune hinterland with withdrawal of grazing and management for golf.

In the foreground, landscaping, fertilising and frequent mowing have transformed the grassland of the fixed dune sands to some type of MG6 *Lolio-Cynosuretum*, with periodically cut MG5 *Centaureo-Cynosuretum* forming stretches of rough and SD8 *Festuca-Galium* grassland surviving on the remnants of the original dune ridges. Beyond the fence, elimination of grazing by stock and rabbits has encouraged the development of SD9 *Ammophila-Arrhenatherum* grassland, W24 *Rubus-Holcus* underscrub and W21 *Craetaegus-Hedera* scrub.

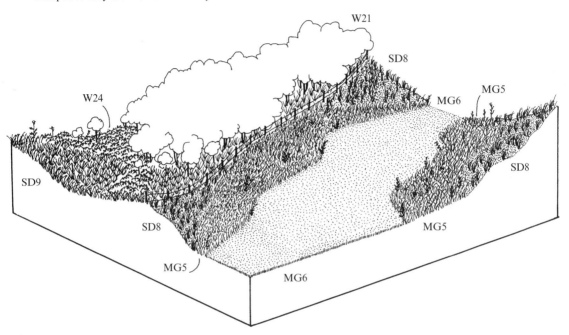

Floristic table SD9

	a	b	9
Festuca rubra	V (3–9)	V (4–9)	V (3–9)
Poa pratensis	V (1–7)	V (1–7)	V (1–7)
Ammophila arenaria	V (2–9)	IV (1–8)	V (1–9)
Achillea millefolium	IV (1–5)	IV (1–4)	IV (1–5)
Arrhenatherum elatius	V (3–9)	III (1–8)	IV (1–9)
Cirsium arvense	III (1–3)	I (1–3)	II (1–3)
Senecio jacobaea	II (1–3)	I (1–3)	I (1–3)
Elymus pycnanthus	I (3–7)		I (3–7)
Myosotis arvensis	I (1–3)		I (1–3)
Crepis capillaris	I (2–3)		I (2–3)
Leymus arenarius	I (1)		I (1)
Dactylis glomerata	III (1–7)	IV (1–8)	III (1–8)
Plantago lanceolata	II (1–3)	IV (1–4)	III (1–4)
Geranium sanguineum	I (1–3)	V (2–8)	III (1–8)
Lotus corniculatus	I (1–3)	III (1–4)	II (1–4)
Trifolium repens	I (3)	II (1–7)	II (1–7)
Trisetum flavescens	I (3–4)	II (1–5)	II (1–5)
Primula veris	I (1–3)	II (1–4)	II (1–4)

Floristic table SD9 (*cont.*)

	a	b	9
Rosa pimpinellifolia	I (3)	II (1–7)	I (1–7)
Lathyrus pratensis	I (1–5)	II (1–4)	I (1–5)
Carex flacca		I (1–4)	I (1–4)
Koeleria macrantha		I (3–5)	I (3–5)
Astragalus danicus		I (1–4)	I (1–4)
Tortula ruralis ruraliformis		I (2–7)	I (2–7)
Thymus praecox		I (3–5)	I (3–5)
Rumex acetosa		I (1–3)	I (1–3)
Pimpinella saxifraga		I (1–3)	I (1–3)
Ranunculus bulbosus		I (1–2)	I (1–2)
Sanguisorba minor		I (4–6)	I (4–6)
Bellis perennis		I (1–4)	I (1–4)
Veronica chamaedrys	II (1–4)	III (1–3)	III (1–4)
Heracleum sphondylium	III (1–4)	II (1–4)	II (1–4)
Galium verum	II (1–5)	III (1–6)	II (1–6)
Ononis repens	II (1–6)	III (1–7)	II (1–7)
Holcus lanatus	II (1–8)	II (2–7)	II (2–7)
Carex arenaria	II (2–6)	II (1–4)	II (1–6)
Pseudoscleropodium purum	II (3–7)	II (2–7)	II (2–7)
Cerastium fontanum	I (1–3)	II (1–3)	II (1–3)
Taraxacum officinale agg.	I (1–3)	II (1–3)	II (1–3)
Campanula rotundifolia	I (2–3)	II (1–3)	II (1–3)
Thalictrum minus	I (1–7)	II (1–7)	II (1–7)
Luzula campestris	I (2–4)	II (1–4)	I (1–4)
Hypochoeris radicata	I (1–3)	I (1–3)	I (1–3)
Brachythecium rutabulum	I (1–4)	I (2–4)	I (1–4)
Centaurea nigra	I (2–7)	I (1–6)	I (1–7)
Vicia sativa	I (1–3)	I (1–3)	I (1–3)
Cruciata laevipes	I (1–5)	I (1–3)	I (1–5)
Hieracium pilosella	I (2–3)	I (2–6)	I (2–6)
Equisetum arvense	I (1–3)	I (2–4)	I (1–4)
Viola riviniana	I (2–3)	I (1–3)	I (1–3)
Silene alba	I (1–3)	I (1–3)	I (1–3)
Rhytidiadelphus triquetrus	I (5–8)	I (1)	I (1–8)
Potentilla reptans	I (2–4)	I (3)	I (2–4)
Elymus farctus	I (1–4)	I (1)	I (1–4)
Leontodon hispidus	I (3–4)	I (1–4)	I (1–4)
Anthoxanthum odoratum	I (4)	I (2–7)	I (2–7)
Rhytidiadelphus squarrosus	I (6)	I (1–5)	I (1–6)
Agrostis capillaris	I (5)	I (1–5)	I (1–5)
Elymus repens	I (2)	I (2–5)	I (2–5)
Hypnum cupressiforme	I (1)	I (2–7)	I (1–7)
Pteridium aquilinum	I (3–6)	I (1–3)	I (1–6)
Fragaria vesca	I (1)	I (1–3)	I (1–3)
Torilis japonica	I (1)	I (1–4)	I (1–4)

	a	b	9
Number of samples	42	52	94
Number of species/sample	12 (7–26)	18 (9–29)	15 (7–29)
Slope (°)	3 (0–20)	11 (0–45)	8 (0–45)
Soil pH	7.6 (5.2–8.8)	7.7 (5.4–9.0)	7.7 (5.2–9.0)

a   Typical sub-community

b   *Geranium sanguineum* sub-community

9   *Ammophila arenaria-Arrhenatherum elatius* dune grassland (total)

SD10
Carex arenaria dune community

Synonymy
Carex arenaria community   West 1936, 1937; *Coryne-phorus canescens* localities   Marshall 1967 *p.p.*; *Carex arenaria* vegetation   Noble 1982 *p.p.*

Constant species
Carex arenaria.

Rare species
Astragalus danicus, Corynephorus canescens.

Physiognomy
The *Carex arenaria* community includes very open to more or less closed swards in which the sand-sedge is the most abundant plant, where vascular associates are few in number and usually sparsely distributed and where there is hardly any contribution from mosses or lichens on what is often still a somewhat mobile sand surface. Where the sedge is invading freshly-deposited material, its initial cover may be very low, the shoots emerging, spaced out and generally single, in striking straight lines from the far-creeping rhizomes. If accretion continues at a fairly modest rate, well-established plants can keep pace but, as the sand starts to become stabilised, clones can thicken up very considerably with densely-packed shoots growing tiller-like from short, closely-spaced branches. Up to several hundred shoots per m^2 have been recorded in this vegetation, but vigour is affected by soil conditions in a particular stand and by grazing which can reduce the sedge to very squat proportions (Tidmarsh 1939, Noble 1982).

In younger stands, or where *C. arenaria* has pre-empted a site and remained strongly dominant, there may be very little else growing among it. Often, though, the sedge is accompanied by some *Festuca rubra* or *F. ovina*, the former tending to favour coastal situations, the latter the more common of the two in those few stretches of inland sands where this vegetation can be found. In coastal stands, too, there may be plants of *Elymus farctus* or *Ammophila arenaria*, though any abundance of these generally presages a spatial or successional shift to other dune communities. Around the East Anglian coast, the patches of bare, shifting sand where this vegetation develops may also be invaded by the rare grass *Corynephorus canescens*, and its dense tussocks of glaucous shoots may be locally abundant. Then, there is occasionally some *Holcus lanatus* and *Dactylis glomerata* while, among smaller grasses, *Koeleria macrantha* and *Agrostis capillaris* are sometimes found, particularly in transitions to grazed swards on more stable ground, with *Aira paraecox* sometimes making an appearance on sheltered open areas.

No dicotyledons occur frequently throughout the community but coarser weedy plants such as *Senecio jacobaea*, *S. vulgaris*, *Matricaria maritima* and *Epilobium angustifolium* are found in some stands, with *Plantago lanceolata*, *Achillea millefolium*, *Cerastium fontanum*, *Galium verum*, *Taraxacum officinale* agg., *Potentilla anserina* and *Rumex acetosella* also recorded occasionally. The community can also provide a locus for the rare *Astragalus danicus*, though this plant is not so characteristic here as in shorter, more closed swards with the sedge.

Cryptogams are rare, but very occasionally there may be small patches of *Hypnum cupressiforme s.l.* or *Ceratodon purpureus.*

Sub-communities

***Festuca rubra* sub-community.** *F. rubra* is a frequent associate in this kind of *Carex* vegetation or sometimes there is *Ammophila* or *E. farctus* at low to moderate cover, with *Corynephorus* locally prominent in some East Anglian stands. *Potentilla anserina*, *Plantago lanceolata* and *Matricaria maritima* occur occasionally and, where the community develops among foredunes, *Honkenya peploides.*

***Festuca ovina* sub-community.** *F. ovina* usually replaces *F. rubra* here and there can be frequent *H. lanatus*, with scattered *S. jacobaea* giving a weedy appearance to the vegetation. In other stands, small tussocks of *K. macrantha* and patches of *A. capillaris* give some continuity with closed grassy swards, and there is occasional *C. fontanum, G. verum, R. acetosella* and *P. lanceolata*.

Habitat

The *Carex arenaria* community is a pioneer vegetation type of freshly-deposited calcareous and acid sands in more sheltered places among dunes all around the British coast and at a few inland sites. It is especially common around the foot of lee slopes of mobile dunes and among blow-outs, or marking the track of recently-disturbed sand over more stable ground, usually forming a minor and relatively short-lived element of dune vegetation. Some stands are grazed and this may help perpetuate the dominance of the sedge, but animal activity can precipitate erosion and destruction of the community.

C. arenaria is ubiquitous around the seaboard of north-west Europe, its northern limit, coinciding roughly with the 0 °C isotherm for mean January air temperature, falling just beyond the latitude of Orkney (Noble 1982). Throughout this range, it is most characteristic of deep pure sands out of reach of tidal inundation and salt-spray, although it exceptionally invades mixtures of sand and shingle (Marshall 1967). In mainland Europe, it is common inland, extending far into the Baltic lowlands on the sandy plains of Sweden, Denmark, Poland and Germany and into The Netherlands whereas, with us, suitable substrates are virtually confined to coastal dune systems. British records away from such habitats are few (Perring & Walters 1962) and strongly concentrated on the small tracts of unreclaimed sand and sandy brown soils (Avery 1980) that remain in Breckland and Lincolnshire (Gibbons 1975, Trist 1979, Noble 1982).

In such situations in Britain, *C. arenaria* is a common and sometimes abundant member of a range of vegetation types but the dominance characteristic of this community is very much associated with the fresh deposition of wind-blown sand in relatively sheltered places. Establishment of the sedge from seed on such surfaces demands the maintenance of moist conditions, a certain temperature fluctuation, and freedom from burial or the predation of herbivores (Tidmarsh 1939, Noble 1982), conditions most likely to be met where sand movement impinges on damper, ungrazed hollows among the dunes. Elsewhere, colonisation often takes place by vegetative extension from individuals already present in the surrounding swards: the rhizomes of *C. arenaria* are able to grow for many metres, and branching and shoot production show renewed vigour as the underground stems penetrate into the loose bare sand, so that clone margins are able to extend and thicken up rapidly to form virtually monospecific stands as opportunity arises. More-

over, if accretion continues, the plants can keep pace to some extent by the production of series of tiered lateral rhizomes at successively higher levels, though deep and sudden burial may overwhelm the leading shoots (Tidmarsh 1939). Likewise, where young rhizomes grow into the face of a dune, apical dominance is lost, new laterals developing further back and consolidating the hold in the shallower sand (Noble 1982). If extensive invasion of this kind allows the sedge to pre-empt a site, big stands of the community can remain prominent even though the sand becomes fixed, as happened over parts of the Blakeney dunes in Norfolk after myxomatosis (White 1961). And, in Breckland, Watt (1937) considered that the abundance of *C. arenaria* in a variety of vegetation types in the 1930s marked the trail of a series of severe sand-storms that had occurred as long as three centuries before. For the most part, however, stands of this *Carex* community are a local and more short-lived feature of the dune environment, being replaced by other more species-rich assemblages as the surface becomes stabilised but recurring where and when bouts of wind erosion or other disturbance favour the renewed deposition of sand.

In this kind of dune habitat, shortage of water and nutrients are important factors influencing the nature of the vegetation. Although *C. arenaria* itself is very tolerant of widely differing ground moisture regimes and often establishes among developing slack vegetation on damp sand, or even where there is some winter flooding in dune hollows (Willis *et al.* 1959*b*, Willis 1985*b*) beneath this community, the sand surface is usually dry. At moderate elevations, the sedge may still be able to benefit from ground water and any nutrients dissolved in it, because it has robust sinker roots which can grow rapidly and penetrate deep, as far as 2–3 m (Tidmarsh 1939, Robards *et al.* 1979, Noble 1982) but, on dunes raised far above the water-table, it must rely much of the time on rain for its moisture supply, a mass of fine superficial roots absorbing the water as it quickly percolates away. A marked resistance to drought in *C. arenaria* equips well-established clones to survive the frequent periods of water shortage in such situations. Episodes of rain may also be important for flushing such nutrients as there are in the sand down the dune slopes into just those places where the sedge can thrive in this community with minimal competition from other species. In general, however, it seems likely that shortage of nitrogen and phosphorus in the raw sands where this vegetation gets a hold imposes the major constraint on the vigour of *C. arenaria* and its associates (Willis & Yemm 1961, Pemadasa & Lovell 1974*b*, Noble 1982).

Variation in soil reaction, on the other hand, seems to affect the sedge little. Where this community develops among younger dunes around our coasts, the sands are usually very lime-rich, often with a high proportion of shell fragments and not yet subject to much leaching, such that the superficial pH is often 8 or more (Noble 1982). On sands derived from blow-outs among older, fixed dunes, the substrate may be more depleted of calcium and, rather locally, there is a plentiful supply of acid sands for the earlier stages of dune building, as at Winterton in Norfolk, where this vegetation is found on ground with a surface pH as low as 4 (Marshall 1967). Inland stands can also occur on highly acidic sands, as in Breckland (Watt 1936, 1957), but both here and around the coast, differences in soil reaction tend to make themselves felt among the flora in the later stages of sward development which succeed this community.

Although older stands of *Carex* vegetation can be found where there is little else growing among densely-packed ageing sedge shoots, it is usually the younger stages of colonisation that exhibit the more impoverished associated floras here, with other species appearing as the *C. arenaria* consolidates its hold and enters a mature phase of growth (Noble 1982). Such companions are often of rather sporadic and chancy occurrence at first, but prominent among them are plants well-adapted in one way or another to the inhospitable dune environment, particularly in the coastal stands of the *F. rubra* sub-community, where extensive tracts of vegetation with other potential invaders of freshly-turned sand generally surround the community. Some of these plants, like the rhizomatous grasses *Ammophila*, *F. rubra* and *E. farctus*, may get an early hold along with the sedge by vegetative spread from established individuals nearby. For the rare associate *Corynephorus*, invasion is a more precarious affair, being dependent on seed germination which requires dampening of the surface by rain, and but a modest amount of accretion if the seedlings are not to be overwhelmed. Even where these conditions are satisfied, survival among established *C. arenaria* is especially poor, competition for moisture tending to favour the sedge (Marshall 1967). More ephemeral plants, too, like *Matricaria maritima*, *Senecio vulgaris* and *Aira praecox*, may only be able to get a hold where there is more open, but stable, ground.

Mixtures of perennial and more short-lived associates also characterise developing stands of the *F. ovina* sub-community seen on inland sands, though in the non-maritime environment, there is little overlap with the flora of the *F. rubra* sub-community, the companions being mostly early invaders from the grassy swards and heaths established on the surrounding stable sandy soils.

Many of the dune systems among which the *Carex* community can be found are subject to grazing, and cattle, rabbits and hares commonly eat the young aerial shoots, particularly during late winter and early spring when the evergreen foliage offers a quite tender bite and little alternative herbage is available (Bhadresa 1977). Such predation may result in stunted growth and a severe curtailment of flowering if the grazing extends into the late spring (Tidmarsh 1939, Noble 1976), although it can help maintain species-poor stands of the community where more palatable invaders are selectively grazed out: in Breckland in the 1930s, for example, *Carex* vegetation commonly spread around rabbit burrows replacing the preferentially grazed *Calluna* heath (Tansley 1939). Heavy grazing, though, can be very destructive, even where such small herbivores as voles are responsible: these tunnel through the herbage and gnaw off the shoots at ground level (Tidmarsh 1939). Burrowing or scuffing of the sand and trampling may also damage the vegetation cover by direct injury to the rhizomes and shoots or by precipitating erosion. Human visitors to dunes can also encourage the destruction of the *Carex* community, though all these kinds of disturbance may release new supplies of sand for deposition elsewhere, with the possibility of further stands developing.

Zonation and succession

The *Carex arenaria* community is generally found as a minor element among zonations and mosaics of other vegetation types of coastal and inland sands, interrupting the patterns wherever the sedge has been able to capitalise on local deposition in sheltered places. It is replaced by other communities as more aggressive plants are able to dominate on still mobile sand or where, with increased stability of the surface, the vigour of the sedge wanes. Grazing may influence these successional changes or help bring about a renewed round of invasion on newly-disturbed ground.

On coastal sites, the *F. rubra* sub-community is sometimes found invading areas of sand among foredunes where the *Elymus farctus* community is the predominant colonising vegetation, but more often it appears among *Ammophila* and *Ammophila-Festuca* stands on somewhat older mobile dunes, where marram has not yet asserted, or reasserted, its dominance. Boundaries between the vegetation types in these patterns are marked by shifts in the proportions of the monocotyledons, although *Ammophila* can be quite common and abundant among developing *Carex* vegetation, while the sedge persists frequently, though never as more than a sub-dominant, in the *Carex* sub-community of *Ammophila* vegetation and in certain kinds of *Ammophila-Festuca* sward. The scattered occurrence throughout of associates like *F. rubra* and *S. jacobaea* can also accentuate the continuity among the vegetation cover.

Where accretion is rapid, *Ammophila* readily assumes dominance over the sedge with a succession to these other communities in a resumption of the main trend of development, although local deposition of sand in sheltered spots among the dunes and around blow-outs which subsequently form can favour a resurgence of

Carex vegetation. Indeed, such a return to this pioneer community is possible among stretches of fixed dunes where the *Festuca-Galium* vegetation is often the successor to the *Ammophila-Festuca* community where the sands remain calcareous. *C. arenaria* is a common associate in the sward and may be quick to take advantage of accretion, especially where *Ammophila* has become very debilitated. Boundaries are generally sharp in such situations, though the margins of the *Festuca-Galium* turf may become fretted away by wind erosion or animal disturbance or blurred by a thin overlay of sand and, if the surface becomes stabilised once again, the richer sward can re-establish itself around the margins or among the *Carex* vegetation. In certain situations, however, it seems as if the sedge community can remain in long occupation of fixed dunes, as on Blakeney where, with the disappearance of rabbits in the myxomatosis epidemic, *C. arenaria* attained a dominance that has not been readily challenged by *F. rubra* (White 1961).

Among coastal dunes built of more acidic sands, or where the surface has become strongly leached with time, the patterns are somewhat different. If accretion proceeds at a fairly modest rate, and especially where the sand is sufficiently acidic from the outset to inhibit the vigour of marram (Huiskes 1979), the sedge may play a more prominent role throughout the succession. Then, especially where the vegetation is grazed, the *Carex* community can give way to the *Carex-Festuca-Agrostis* or *Carex-Cornicularia* swards. In the former, mixtures of *C. arenaria* and smaller grasses usually dominate, with such *Ammophila* as is present often sparse and puny, and the early appearance among the developing sedge cover of plants like *Agrostis capillaris, H. lanatus, P. lanceolata* and *Achillea millefolium*, may make the boundaries between the vegetation types indistinct. Generally, however, high frequencies of *Anthoxanthum odoratum, Luzula campestris* and *Galium saxatile* characterise the later stages of sward development, together with the appearance of a variety of calcifuge bryophytes that hardly ever find a place on the still mobile sand. In other situations, perhaps where heavy rabbit grazing has played an important role in influencing the succession, the *Carex-Cornicularia* community can occupy much of the more stable sand surface. Here, again, the sedge retains a frequent and often abundant place in the vegetation, commonly with some *F. rubra* and sparse marram, but *Rumex acetosella* and a variety of diminutive ephemerals now appear and there is typically an extensive carpet of lichens, with *Cornicularia aculeata* and a variety of *Cladonia* spp. particularly prominent. Where either of these swards becomes disrupted with degeneration and wind erosion, the *Carex* community can reappear on drifts of accumulated sand. A striking feature of some sites with this kind of pattern, as at Winterton, is the prominence, early in the succession

and re-seeding for a few years within the more close turf, of *Corynephorus* (Marshall 1967).

It is mixtures of grassy and lichen-rich swards that make up much of the vegetation context for the development of inland stands of the *Carex* community. Away from the coast, tracts of mobile sand are very localised though, in regions like Breckland, wind erosion of sandy soils has played an important part in the evolution of the landscape. Here, blow-outs have been observed forming (Watt 1937) as a result of the action of local cyclonic winds on degenerating *Calluna-Festuca* heath, *Carex-Cornicularia* or *Carex-Festuca-Agrostis* grasslands on sand-smeared podzols. Where these have become reduced to a mat of humus or a scabby carpet of lichens among decaying grass shoots, the surface is readily disrupted by the impact of sun, rain and wind, the sand accumulating in low dunes around the developing hollow or being blown away in shallow drifts by frontal erosion. It is on this newly-accreted material that the *F. ovina* sub-community of the *Carex* vegetation is able to get a hold, sometimes persisting for a considerable time in its impoverished form, in other places grading back into grazed swards. Then, the *Carex-Festuca-Agrostis* or *Carex-Cornicularia* communities may form a transition to the *Festuca-Agrostis-Rumex* grassland, the contribution of the sedge waning, and dominance passing to small tussock grasses, ephemerals and patches of lichens and bryophytes. Very occasionally in these fine mosaics, where there is a transition to less sandy soils of higher pH, the *Carex* community can be closely juxtaposed with the more basiphilous of the Breckland grassheaths of the *Festuca-Hieracium-Thymus* type (Watt 1940, Noble 1982).

On shallower, stabilising drifts of sand where there is little or no grazing, an alternative successional development is the reappearance of heath, because in such conditions *Calluna* is an important competitor to *C. arenaria* (Watt 1936, 1937). On Breckland sands, it is the *Calluna-Festuca* community that is the usual kind of sub-shrub vegetation and its *Carex* sub-community, with co-dominant mixtures of heather and sedge, can form a transitional zone in shifts from more to less mobile substrates (Tidmarsh 1939, Gibbons 1975). *C. arenaria* also remains a constant feature of many stands of heath which develop on the more stable, acid sands of ungrazed or lightly-grazed coastal dunes. Around parts of the East Anglian coast, *Calluna-Festuca* heath indistinguishable from that inland can be found in such situations but, elsewhere, *Calluna-Carex* heath is the usual community, with *Erica cinerea* or, much more locally, *Empetrum nigrum* spp. *nigrum*, invading along with the heather. Such heaths are sometimes subject to burning to regenerate the sub-shrub cover, and such treatment may help stimulate flowering among ageing *C. arenaria* (Tidmarsh 1939).

A further complication of such patterns is the appearance of *Pteridium aquilinum*, because the sedge and bracken are both able to spread rapidly on loose and deep, well-aerated substrates and, in both inland and coastal sites, the *Carex* community can be found in intimate association with bracken vegetation, the balance between the dominants along the boundaries being related to the cyclical upbuilding and ageing of the *Pteridium* along its front (Watt 1936, 1940; Tidmarsh 1939). More mature stands of bracken in such patterns on acid sands tend to be of the *Pteridium-Galium* type, with *Pteridium-Rubus* underscrub developing on less base-poor and impoverished substrates.

Distribution
The community occurs in suitable situations on dunes all around the British coast and inland in Breckland and Lincolnshire.

Affinities
The aggressive, gregarious character of sand sedge was early recognised, but accounts of British coastal dune vegetation have usually treated the species as little more than a locally prominent element in communities dominated by *Ammophila* (Tansley 1911, 1939), although Watt (1936, 1937), in his account of the vegetation of inland Breck sands, gave prominence to *Carex arenaria* stands, and Noble (1982) recognised a variety of assemblages with the sedge.

In phytosociological schemes, *C. arenaria* is seen as a characteristic plant of the Corynephorion, an alliance of pioneer swards on dry, acid sands which has been described from The Netherlands (Westhoff & den Held 1969), Germany (Ellenberg 1978, Oberdorfer 1978) and Poland (Matuszkiewicz 1981). Generally, however, the sedge is less prominent in the associations in the process of sand fixation, than *Corynephorus* and only in a very few sites in Britain can the sequence of vegetation types typical of this kind of succession be seen. Thus, although swards like those of the *Carex-Cornicularia* and *Carex-Festuca-Agrostis* communities would find a ready place among the Sedo-Scleranthetea, there might be some argument for retaining the *Carex* community itself in the Ammophilion.

Floristic table SD10

	a	b	10
Carex arenaria	V (5–10)	V (4–10)	V (4–10)
Festuca rubra	III (2–5)		II (2–5)
Ammophila arenaria	II (2–3)		I (2–3)
Elymus farctus	II (5–6)		I (5–6)
Potentilla anserina	II (3–5)		I (3–5)
Matricaria maritima	II (2–4)		I (2–4)
Honkenya peploides	I (2–4)		I (2–4)
Festuca ovina		IV (3–7)	II (3–7)
Senecio jacobaea		III (3)	II (3)
Holcus lanatus		III (3–7)	II (3–7)
Cerastium fontanum		II (3–4)	I (3–4)
Rumex acetosella		II (3–4)	I (3–4)
Koeleria macrantha		II (4)	I (4)
Galium verum		II (2–3)	I (2–3)
Astragalus danicus		I (3)	I (3)
Ceratodon purpureus		I (3)	I (3)
Plantago lanceolata	II (1–3)	II (3–4)	II (1–4)
Achillea millefolium	I (2)	I (3)	I (2–3)
Taraxacum officinale agg.	I (1–3)	I (1–3)	I (1–3)
Aira praecox	I (4)	I (3–4)	I (3–4)
Epilobium angustifolium	I (4)	I (3–5)	I (3–5)
Senecio vulgaris	I (1–2)	I (3)	I (1–3)
Agrostis capillaris	I (3)	I (3–5)	I (3–5)
Hypnum cupressiforme	I (4)	I (3–4)	I (3–4)
Dactylis glomerata	I (1)	I (3–5)	I (1–5)
Number of samples	18	18	36
Number of species/sample	8 (3–14)	8 (2–19)	8 (2–19)
Vegetation height (cm)	14 (1–50)	42 (5–50)	35 (1–50)
Vegetation cover (%)	62 (20–100)	93 (50–100)	77 (20–100)

a   *Festuca rubra* sub-community
b   *Festuca ovina* sub-community
10 *Carex arenaria* dune community (total)

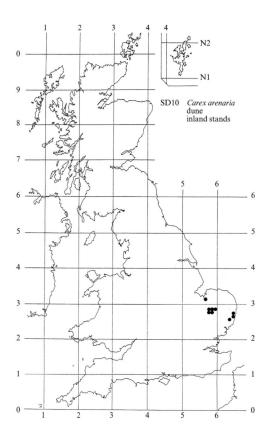

SD10 *Carex arenaria*
dune
inland stands

SD11
Carex arenaria-Cornicularia aculeata dune community

Synonymy
Caricetum arenariae Tansley 1939 *p.p.*

Constant species
Carex arenaria, Cornicularia aculeata.

Rare species
Astragalus danicus, Corynephorus canescens.

Physiognomy
The *Carex arenaria-Cornicularia aculeata* community includes lichen-rich swards in which sand-sedge is the only constant vascular plant. It is, though, never a vigorous dominant here, only occasionally having high cover and more usually occurring as scattered tufts of shoots, sometimes distinctly moribund. Neither are any other herbs consistently abundant. *Festuca ovina* and *F. rubra* are quite commonly found, for example, but usually as small scattered tussocks and, though *Agrostis capillaris* occasionally forms patches in the turf, these are rarely dense. Then, although *Luzula campestris* is fairly frequent, it generally occurs as rather sparse individuals. Sometimes more abundant, though only very locally around the East Anglian coast, is the rare grass *Corynephorus canescens* which, while being essentially a pioneer plant of more open sands, can continue to seed into these swards for several years (Marshall 1967). Certain annual grasses like *Aira praecox, A. caryophyllea* and *Phleum arenarium* may also occur with considerable local abundance where bare patches develop. Finally, among the monocotyledons, *Ammophila arenaria* is a frequent feature of coastal stands of the community though, even more than *Carex arenaria*, it is characteristically of very debilitated growth, persisting as sparse, scattered tussocks.

Perennial dicotyledons are few and usually of low cover, though some can be locally prominent. Most common is *Rumex acetosella*, with occasional *Hypochoeris radicata, Galium saxatile* and *Sedum acre*, patches of this last sometimes quite plentiful and eye-catching with their yellow flowers. Then, there can be some *Thymus praecox, Galium verum, Plantago lanceolata* and *Viola canina* while, at certain localities, this vegetation provides a locus for the rare *Astragalus danicus*. More ephemeral herbs may be found, too, with *Senecio jacobaea, Viola tricolor, Filago minima, Erodium cicutarium* and *Teesdalia nudicaulis* coming and going on areas of open ground. Other stands can have a somewhat heathy appearance with small bushes of *Calluna vulgaris*, but the cover of this is never extensive.

Typically, all these plants are of low growth, and often grazed back into very stunted individuals, and it is the extent and diversity of the lichen carpet that generally forms the most impressive feature of the vegetation, its colour suggesting to some the epithet 'grey' for the kind of dunes on which it is found (Tansley 1939). Most frequent and plentiful of the lichens are *Cornicularia aculeata, Cladonia arbuscula, C. foliacea* and *C. impexa* with *C. furcata, C. tenuis, C. pyxidata, C. uncialis, C. gracilis, C. squamosa* and *Hypogymnia physodes* occasional to common in the different sub-communities, *C. fimbriata, C. rangiformis* and *C. rangiferina* more rarely represented.

Bryophytes may also contribute to the ground cover, though as a group they are nowhere near as abundant as the lichens. However, *Dicranum scoparium, Polytrichum piliferum, P. juniperinum, Ceratodon purpureus* and *Hypnum cupressiforme* occur quite frequently sometimes as extensive patches, with *Ptilidium ciliare, Brachythecium albicans, Racomitrium canescens, Tortula ruralis* ssp. *ruraliformis* and *Rhytidium rugosum* being found less commonly.

Sub-communities

***Ammophila arenaria* sub-community.** The constancy of small amounts of *Ammophila* is the most distinctive feature of the vascular element in this sub-community, with both *F. rubra* and *F. ovina* occurring occasionally. *A. praecox* is common, with *A. caryophyllea* and *P. arenarium* being recorded less frequently and it is in this kind of *Carex-Cornicularia* vegetation that *Corynephorus* can be found. *R. acetosella, H. radicata, S. acre* and *V. tricolor* are the commonest associated herbs, with *G. verum, T. praecox* and *V. canina* more occasional and, among the annuals, *E. cicutarium, L. minima* and *Teesdalia*.

The most frequent and abundant lichens are usually *Cornicularia aculeata, C. arbuscula* and *C. furcata* with *C. foliacea* and *Hypogymnia physodes* also very common, *C. impexa, C. fimbriata, C. tenuis* and *C. rangiformis* more occasional but sometimes extensive. *D. scoparium, P. piliferum, Ceratodon* and *H. cupressiforme* are fairly frequent and can be of locally high cover.

Festuca ovina sub-community. The general appearance of this kind of *Carex-Cornicularia* vegetation is similar to the above but *Ammophila* is absent and the fescue in the sward is almost always *F. ovina*, mixtures of this with *C. arenaria, L. campestris* and *A. capillaris* generally forming the bulk of the rather sparse vascular cover. *R. acetosella* is frequent and *G. saxatile* occasional, with *Astragalus* occurring at some sites but associated herbs typically occur again as scattered individuals.

Cornicularia aculeata and *Cladonia arbuscula* remain very frequent and abundant in the lichen carpet, but *C. furcata* is scarce and the other most common species are *C. foliacea* and *C. impexa*, with *C. pyxidata, C. uncialis, C. gracilis* and *C. squamosa* preferentially frequent. Among the bryophytes, *D. scoparium, P. piliferum* and *H. cupressiforme* are occasional to common and locally plentiful.

Habitat

The *Carex-Cornicularia* community is characteristic of fixed and rather acid sands, or compacted mixtures of sand and shingle, where the ground remains very drought-prone and impoverished. It is a rather local community, better developed in the drier east of Britain, but it can be found on both coastal and inland sands, being most typical of flat areas that lie out of reach of the ground water-table between or behind old dune ridges or on stable sand plains. Heavy grazing by rabbits has probably been of considerable importance in maintaining this vegetation in the past, but trampling by stock and humans or disturbance by vehicles is very destructive of the lichen carpet and may help initiate erosion of the surface.

This community cannot become established on wind-blown sand until accretion has virtually ceased and, though the ground may become subject to renewed disturbance, stands are generally found in places beyond the reach of freshly-deposited material from beach sources, in areas of stable and subdued relief in the older parts of dune systems, or where old erosion surfaces have become compacted (Figure 15). Active invaders of

mobile sand do not therefore have a vigorous role in the control of the physical environment here and play a minor part in the composition and structure of the sward. Where this vegetation occurs among coastal dunes, for example, *Ammophila* often persists, though it is as a debilitated relic, with the clones reduced to the typically senile clusters of shoots and scarcely ever flowering (Gimingham 1964a, Huiskes 1979). Edaphic changes initiated with increased stability of the sand surface probably play some part in this (Salisbury 1952, Willis *et al.* 1959a, b; Huiskes 1977a, 1979) because, compared with the immature sands beneath the *Ammophila-Festuca* community, and even the more fixed sediments of the *Festuca-Galium* vegetation, the superficial sand layers here are typically poor in lime, with a pH usually less than 5.5. Where the wind-blown material is initially calcareous, as in dune systems where shall fragments make up a considerable proportion of the beach sediments, such surface acidity can come about only through prolonged leaching (Ranwell 1972, Willis 1985b) though, with sands that are more acid from the start, younger surfaces may offer a congenial substrate, provided they are not actively accreting. Apart from leached or siliceous sands of beach origin, suitable sediments have been derived inland or along the coast from Pliocene and Pleistocene Crag on the Suffolk Sandlings, Greensand and glacial sands on the Norfolk Commons and the more acidic of the Pleistocene and aeolian deposits in Breckland (Watt 1940, Hodge & Seale 1966, Corbett 1973, *Soil Survey* 1983, Hodge *et al.* 1984).

Even where the soils are of considerable age, however, with the surface layers darkened by the accumulation of some organic matter and humic staining, their water-retentive capacity is generally very limited. Absent from those stretches of flat ground among dunes which are subject to winter flooding, indeed typically out of reach of any influence of ground water, this vegetation is thus dependent on rain for the bulk of its moisture. Characteristically here, such a source is limited because the *Carex-Cornicularia* community is largely confined to the drier east of Britain where annual precipitation is less than 800 mm (*Climatological Atlas* 1952) with often fewer than 120 wet days yr^{-1} (Ratcliffe 1968), and mean annual maxima frequently in excess of 27 °C (Conolly & Dahl 1970). The tendency to parching is thus very strong, particularly around the East Anglian coast and in Breckland where conditions are most markedly continental, with rainfall quite often below 500 mm annually and a high likelihood of a water deficit in late spring (Gregory 1957, Chandler & Gregory 1976, Smith 1976).

The other important feature of the sands is their nutrient-poor character, something that is inherently typical of siliceous sediments, but perhaps accentuated over many generations by particular kinds of biotic activity. Grazing by rabbits, for example, which defaecate their re-ingested pellets in latrines, can continually remove nutrients from the system, perhaps appreciable amounts of phosphorus and nitrogen (Watt 1981*a*), and long histories of rabbit-rearing, and of sheep-grazing with folding on arable, have been characteristic of places like the Suffolk Sandlings and Breckland (Crompton & Sheail 1975, Sheail 1979, Chadwick 1982, Webb 1986). This, together with past generations of shifting cultivation on inland sands, has helped preserve stretches of suitably impoverished ground where the community has been able to persist.

These climatic and edaphic conditions affect the vegetation in a variety of ways. First, there is the generally open character of the vascular cover. *Carex arenaria*, for example, though better able than *Ammophila* to tolerate the base-poor soil environment (Huiskes 1979, Noble 1982) and strongly resistant to drought (Noble

Figure 15. Sand dune and transitions to salt-marsh at Scolt Head Island, Norfolk.
The main spit comprises a shingle ridge with various kinds of SD6 *Ammophila* vegetation fronted, at the far point, by a very fragmentary SD2 *Honkenya-Cakile* strandline. Behind, on semi-fixed dune sand, is a zone of SD7 *Ammophila-Festuca* grassland with areas of the *Ammophila* sub-community of SD11 *Carex-Cornicularia* vegetation patchily colonised by W24 *Rubus-Holcus* underscrub. Shingle lows have stretches of vegetation resembling the SM21 *Suaeda-Limonium* community with SM25 *Suaeda vera* vegetation fringing the transition to the salt-marsh proper. (Redrawn from Hadley *et al.* 1990, by permission of the Joint Nature Conservation Committee.)

1982), is usually in a senile phase of growth in this community. The fresh deposition of sand that favours active clone extension has ceased here and the vigour of the plants is severely constrained by the shortage of major nutrients (Willis & Yemm 1961, Noble 1982). Where rabbits remain numerous, they can graze off inflorescences in the spring (Ranwell 1960*a*, *b*) and perhaps completely prevent the establishment of new seedlings (White 1961, Noble 1982). Perennial grasses also fare badly. In fact, it has been suggested that colonisation by other species among senile *C. arenaria* is markedly inhibited by allelopathic compounds secreted by the sedge (Symonides 1979) but, even if this is ineffectual, the parched, oligotrophic character of the sands is likely to maintain the sparse and tussocky aspect of the swards, with the herbage often looking crisp and brown by midsummer. In this respect, the vegetation can contrast quite sharply with the *Carex-Festuca-Agrostis* grassland, which is typical of fixed acid sands in cooler and wetter parts of Britain. There, the swards are generally closed, with *C. arenaria* growing more vigorously, grasses like *Poa pratensis*, *Anthoxanthum* or even *Holcus lanatus* contributing to the cover, and dicotyledonous herbs such as *Viola riviniana* and *Cerastium fontanum* occurring more often.

In the *Carex-Cornicularia* community, such species are scarce or absent and, though the poor competitive ability of the grasses allows for a potentially large contribution from herbaceous associates, these are few in number and of particular kinds, the perennials being light-demanding and drought-tolerant, species like *Rumex acetosella*, *Hypochoeris radicata* and *Sedum acre* providing the most distinctive element. Patches of bare

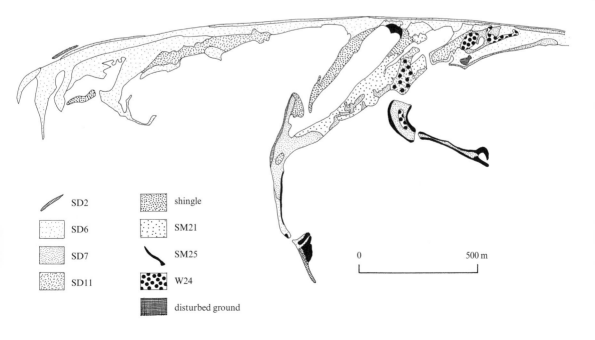

SD2

SD6

SD7

SD11

shingle

SM21

SM25

W24

disturbed ground

0   500 m

ground, occasionally renewed by the demise of such perennials in severe droughts (Watt 1971*b*), also offer opportunity for the continued seeding-in of *Corynephorus*, and colonisation by ephemerals, sometimes coarse weedy plants like *Senecio jacobaea*, more usually diminutive annuals such as *Aira praecox, Phleum arenarium, Viola tricolor, Erodium cicutarium, Logfia minima* and *Teesdalia nudicaulis*. In the data available, such plants have been more frequently recorded in the coastal stands, but this is probably fortuitous.

The other component of the vegetation able to capitalise on the soil and climatic conditions here are the lichens, particularly the bulkier fruticose species like *Cornicularia aculeata, Cladonia arbuscula, C. furcata* and *C. foliacea*, some of these perhaps making an early appearance in the colonisation of compact erosion surfaces (Watt 1938). Bryophytes tend to be less prominent, though the most abundant are generally acrocarpous invaders of bare ground, with a noticeable scarcity of the pleurocarps common in *Carex-Festuca-Agrostis* swards, like *Rhytidiadelphus squarrosus, Pseudoscleropodium purum, Hylocomium splendens* and *Pleurozium schreberi*.

Natural environmental conditions may be sufficient to maintain the essential character of the *Carex-Cornicularia* swards but, quite apart from its influence on soils, grazing by rabbits and stock probably has other important influences on the vegetation. Close cropping can keep the herbage short, helping maintain the balance between perennial grasses and dicotyledons, and the annuals and cryptogams able to take advantage of open ground, while scuffing creates new bare patches for invasion. Grazing is also often very selective, with certain vascular plants like *Sedum acre* remaining uneaten. As far as rabbits are concerned, such choosiness may be especially beneficial to the lichens which they seem to avoid. Sheep, too, may have some effect on the composition of this element of the vegetation: in similar swards to these around the Baltic, selective grazing appears to modify the distribution of *Cladonia* spp. (Sjögren 1971) and such influences may contribute to the contrasts seen between the sub-communities here.

Trampling by stock, however, or by humans is detrimental to the lichen carpet and may help open up the surface to renewed sand movement. Although military manoeuvres can contribute to maintaining the sort of open landscape in which the *Carex-Cornicularia* community can survive (Sheail 1979), tracked army vehicles are very destructive of the vegetation.

Zonation and succession

The *Carex-Cornicularia* community is commonly found as a minor element in mosaics with calcifuge grasslands and heaths on the subdued topography of older dunes and sand plains. On siliceous sediments, these vegetation types may represent late stages in a distinctive line of succession, but they also develop where calcareous sands have been long-leached and may then occur among less mature and more basiphilous dune communities, particularly in coastal sites. This vegetation can also be part of secondary cycles of succession related to renewed erosion and is probably often dependent for its continuing survival on heavy grazing, particularly by rabbits. Where this is relieved, the community probably gives way to heath or, where the ground is less impoverished, to scrub and woodland.

In the heart of its distribution in East Anglia, the *Carex-Cornicularia* vegetation often occurs, at both coastal and inland sites, with the *Festuca-Agrostis-Rumex* grassland which, apart from the scarcity of *C. arenaria* can be very similar in its physiognomy and composition. In the *Cornicularia-Cladonia* sub-community of the grassland, for example, *Cornicularia aculeata, Cladonia arbuscula, C. tenuis, C. impexa, C. foliacea, C. uncialis* and *C. furcata* all remain frequent and abundant, with *D. scoparium, P. piliferum* and *R. acetosella*, among an open tussocky turf of *F. ovina* and *A. capillaris*. Transitions of this kind may reflect increasing stability of the surface, with the final extinction of the sand sedge, while in more scuffed and disturbed places, as along path edges, the *Carex-Cornicularia* community may give way to the *Erodium-Teesdalia* sub-community of the *Festuca-Agrostis-Rumex* grassland with its rich and varied suite of ephemerals. Gross disruption of the swards in such mosaics may result in erosion of the sand, with the development of blow-outs and tracts of freshly-deposited sediment (Watt 1938) and then there is an opportunity for renewed invasion by the sand sedge from rhizome fragments or by seed (Noble 1982), remnant stretches of sward occurring among young stands of the *Carex* community. Where the surface of such siliceous sands becomes quickly stable once again, the sedge decreasing in vigour and the spread of other perennials being restricted by inhospitable soil conditions, it is possible that the *Carex-Cornicularia* community re-establishes itself fairly quickly (Watt 1938) and this may represent a natural sequence on such base-poor sediments. In contrast to such situations in The Netherlands (Westhoff & den Held 1969), Germany (Oberdorfer 1978, Ellenberg 1978) and Poland (Matuszkiewicz 1981), however, *Corynephorus* plays a very restricted role with us in the earlier stages, being an important colonist at just a few coastal sites (Marshall 1967) and remaining now at only a single inland station (Trist 1979).

On sands which were initially more calcareous, as in many coastal dune systems, the persistence of *Ammophila* is a reminder that the *Carex-Cornicularia* community may represent a late stage in a rather different successional sequence, with marram playing the major role in fixing the sediments before they become stable and leached. In such situations, the *Carex-Cornicularia* community can be found among older dunes, sometimes again with *Festuca-Agrostis-Rumex* swards or among stretches of *Festuca-Galium* grassland, particularly where the sands have not been so strongly decalcified. In the *Luzula* sub-community of such grassland, plants like *F. ovina*, *L. campestris* and *H. radicata* remain frequent in a turf with senile *Ammophila* and sometimes sparse *C. arenaria*, and where the *Carex-Cornicularia* community extends into eastern Scotland, *Astragalus danicus* can occur throughout the zonation. Increasingly, however, with the shift to the wetter and cooler part of Britain, the *Carex-Festuca-Agrostis* grassland becomes important as a transition between parched calcifuge swards of the *Carex-Cornicularia* type and the more closed grasslands of fixed dune sands. Then, such fragments of the former as remain, grade almost imperceptibly through a thickening turf into *Festuca-Galium* vegetation, species such as *Poa pratensis*, *A. capillaris*, *F. rubra*, *Galium verum* and *Lotus corniculatus* becoming increasingly important, and bryophytes occluding lichens as the major cryptogamic element.

Grazing is an important factor in maintaining mosaics of these different grasslands and some sites now betray ample evidence of the demise of rabbits or the cessation of sheep pasturing with an increase in size of the grass tussocks and a disappearance of the species demanding of open ground. A spread of *Calluna* is also very likely in such situations, the vegetation becoming increasingly heathy as the bushes multiply and enlarge. In East Anglia, such succession is usually to the *Calluna-Festuca* heath, among certain types of which *Carex arenaria* and lichens remain very frequent and patchily abundant. Where the climate is not so strongly continental, *Calluna-Carex* heath often develops from calcifuge dune grasslands, *Erica cinerea* being a common associate among lichen-rich vegetation with frequent *C. arenaria*, *F. ovina*, *L. campestris* and *A. praecox*. Disturbed and enriched ground may have patches of *Ulex-*

Rubus scrub with fragments of calcifuge sward persisting among the gorse bushes, while deep, loose sands can support dense *Pteridium-Galium* vegetation. Invasion of birch or pine may presage a development to some kind of Quercion woodland, typically, in the heart of the range of the *Carex-Cornicularia* community, of the *Quercus-Betula-Deschampsia* type.

Distribution

The *Carex-Cornicularia* community is found mainly on the east coast of Britain, with the *Ammophila* sub-community concentrated on the coastal dunes of Norfolk and Suffolk, with some stands in eastern Scotland. The *Festuca* sub-community extends the range inland, on to the sands of Breckland.

Affinities

Although the cryptogamic component of this kind of vegetation was described from old fixed dunes at Blakeney by Richards (1929) and included within a *Caricetum arenariae* by Tansley (1939), the community figured little in early accounts of British dune vegetation. More recently, stands were listed in ecological studies of *Corynephorus* (Marshall 1967) and *Carex arenaria* (Noble 1982), where attention was drawn to the similarity between this sort of assemblage and Corynephorion communities described from coastal and inland sands from The Netherlands (Westhoff & den Held 1969), through Germany (Oberdorfer 1978, Ellenberg 1978) into Poland (Matuszkiewicz 1981). Closest among these to our own *Carex-Cornicularia* community is the *Spergulo morisonii-Corynephoretum canescentis* R.Tx. (1928) 1955 which has virtually identical suites of lichens and mosses, together with frequent records for *R. acetosella*, *Teesdalia* and *Filago minima*. Although *Corynephorus* is generally more important in sand fixation in such Continental vegetation than *Carex arenaria*, it seems sensible to locate British stands within the Corynephorion, and thus emphasise the genesis of at least some of them from a distinctive succession on initially acidic sands.

Floristic table SD11

	a	b	11
Carex arenaria	V (1–9)	V (4–8)	V (1–9)
Cornicularia aculeata	IV (1–6)	III (2–5)	IV (1–6)
Ammophila arenaria	IV (1–6)		III (1–6)
Cladonia furcata	III (1–7)	I (3)	III (1–7)
Aira praecox	III (1–5)	I (3)	II (1–5)
Hypogymnia physodes	III (1–5)		II (1–5)
Festuca rubra	II (1–3)		I (1–3)
Sedum acre	II (2–5)		I (2–5)
Hypochoeris radicata	II (2–5)		I (2–5)
Corynephorus canescens	II (1–7)		I (1–7)
Viola tricolor	II (1–4)		I (1–4)
Polytrichum juniperinum	II (2–7)		I (2–7)
Cladonia tenuis	II (2–5)		I (2–5)
Galium verum	I (1–3)		I (1–3)
Thymus praecox	I (3–7)		I (3–7)
Phleum arenarium	I (2–4)		I (2–4)
Cladonia coccifera	I (1–6)		I (1–6)
Viola canina	I (1–3)		I (1–3)
Racomitrium canescens	I (4–9)		I (4–9)
Cladonia rangiformis	I (2–8)		I (2–8)
Anthoxanthum odoratum	I (1–3)		I (1–3)
Erodium cicutarium	I (1–3)		I (1–3)
Logfia minima	I (1–5)		I (1–5)
Tortula ruralis ruraliformis	I (3–4)		I (3–4)
Cladonia rangiferina	I (1–6)		I (1–6)
Festuca ovina	I (2–5)	V (3–8)	I (2–8)
Cladonia pyxidata	I (2–4)	III (3)	I (2–4)
Cladonia uncialis	I (2–4)	III (1–5)	I (1–5)
Cladonia gracilis	I (2–3)	III (3–4)	I (2–4)
Calluna vulgaris	I (2)	II (3)	I (2–3)
Cladonia squamosa	I (3)	II (3)	I (3)
Astragalus danicus		I (2)	I (2)
Rhytidium rugosum		I (6)	I (6)
Cladonia arbuscula	III (1–9)	III (2–4)	III (1–9)
Cladonia foliacea	III (1–6)	III (2–3)	III (1–6)
Rumex acetosella	III (1–4)	III (2–4)	III (1–4)
Cladonia impexa	II (1–8)	III (2–4)	II (1–8)
Dicranum scoparium	II (1–5)	III (3–5)	II (1–5)
Luzula campestris	II (1–3)	II (4–5)	II (1–5)
Polytrichum piliferum	II (2–8)	II (3–4)	II (2–8)
Hypnum cupressiforme	II (2–8)	II (2–3)	II (2–8)
Agrostis capillaris	II (1–7)	I (4)	II (1–7)
Cladonia fimbriata	II (2–3)	I (3)	II (2–3)
Ceratodon purpureus	II (1–6)	I (3)	II (1–6)
Galium saxatile	I (2)	II (3)	I (2–3)

Ptilidium ciliare	I (1)	II (4–5)	I (1–5)
Brachythecium albicans	I (2–3)	I (4)	I (2–4)
Teesdalia nudicaulis	I (2)	I (4)	I (2–4)
Senecio jacobaea	I (1–2)	I (1)	I (1–2)
Plantago lanceolata	I (4)	I (3)	I (3–4)
Campanula rotundifolia	I (2)	I (3)	I (2–3)
Number of samples	46	7	53
Number of species/sample	12 (8–19)	17 (7–30)	16 (7–30)

a　*Ammophila arenaria* sub-community

b　*Festuca ovina* sub-community

11　*Carex arenaria-Cornicularia aculeata* dune community (total)

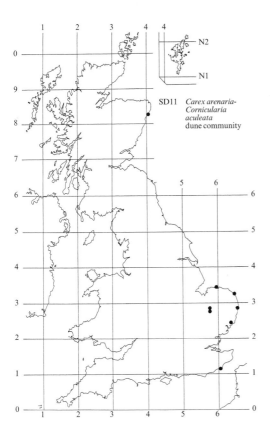

SD11　*Carex arenaria-*
Cornicularia
aculeata
dune community

SD12

Carex arenaria-Festuca ovina-Agrostis capillaris dune grassland

Synonymy

Dune grassland Pearsall 1934; Dune pasture Gimingham 1964*a p.p.*; *Astragalo-Festucetum arenariae* Birse 1980 *p.p.*

Constant species

Agrostis capillaris, Ammophila arenaria, Carex arenaria, Festuca ovina, Poa pratensis.

Rare species

Astragalus danicus.

Physiognomy

The *Carex arenaria-Festuca ovina-Agrostis capillaris* community comprises short, generally closed swards dominated by various mixtures of *Carex arenaria* and a number of grasses, of which *Festuca ovina*, *Agrostis capillaris* and *Poa pratensis s.l.* are the most frequent and abundant. *F. rubra* also occurs quite commonly, and *Anthoxanthum odoratum*, *Luzula campestris* and *Holcus lanatus* are variously represented in the different subcommunities, with *Koeleria macrantha* occasional. *Ammophila arenaria* persists at high frequency, too, though it hardly ever shows any vigour, being usually reduced to sparse and debilitated shoots.

Dicotyledonous herbs are rather few in number, but the assemblage is distinctive, with *Galium saxatile*, *Lotus corniculatus*, *Galium verum*, *Cerastium fontanum* and *Rumex acetosella* occasional to frequent throughout. In some stands, *Hypochoeris radicata*, *Thymus praecox* and *Hieracium pilosella* become more common, with *Campanula rotundifolia*, *Viola riviniana* and *Achillea millefolium* occurring more often in others. Compared with more open dune swards, however, ephemeral plants are noticeably scarce. There are very occasionally some scattered bushes of *Calluna vulgaris*, frequently nibbled down, but any increase in the cover of ericoid sub-shrubs usually marks a transition to heath.

Bryophytes are fairly numerous and sometimes abundant in the sward and, though *Dicranum scoparium* is very common, it is bigger pleurocarps that are generally more obvious, with frequent records for *Rhytidiadelphus squarrosus*, *Pseudoscleropodium purum*, *Hylocomium splendens* and *Pleurozium schreberi* and more occasional occurrences of *Rhytidiadelphus triquetrus* and *Hypnum cupressiforme*. *Brachythecium albicans*, *Polytrichum juniperinum* and *Ceratodon purpureus* also occur in some stands.

Lichens are generally not so numerous or extensive in their cover, and certainly nothing like so prominent as in the *Carex-Cornicularia* community. However, *Peltigera canina* is occasionally found and there can be patches of *Cladonia arbuscula*, *C. rangiformis* and *Cornicularia aculeata*.

Sub-communities

***Anthoxanthum odoratum* sub-community.** *F. ovina*, occasionally accompanied or sometimes replaced by *F. rubra*, is the usual dominant in this sub-community, with mixtures of varying amounts of *C. arenaria*, *A. capillaris*, *P. pratensis* and, preferentially frequent here, *Anthoxanthum*. *Ammophila* is constant but rarely of even moderate cover and usually distinctly moribund. *L. campestris* is also preferential at low cover, with occasional *K. macrantha* and, of more local occurrence, *Danthonia decumbens* and *Deschampsia flexuosa*. In contrast to the other sub-community, *H. lanatus* is not very common.

Among the dicotyledons of the community, *G. saxatile*, *L. corniculatus*, *C. fontanum* and *R. acetosella* remain frequent, with *G. verum*, *Trifolium repens* and *Veronica officinalis* more occasional while, among the preferentials, *Hypochoeris radicata* is often found. With the appearance of plants such as *Thymus praecox*, *Hieracium pilosella* and *Plantago lanceolata*, the sward may take on something of the look of a calcicolous grassland, while other stands can have *Potentilla erecta* and scattered clumps of *Calluna*, *Erica cinerea* or even *E. tetralix* in transitions to damper heath. In some sites, the

Anthoxanthum sub-community also provides a locus for *Astragalus danicus*.

The community mosses *D. scoparium*, *R. squarrosus*, *P. purum*, *H. splendens* and *Pleurozium schreberi* all remain common and each is able to show considerable local abundance, while *Rhytidiadelphus triquetrus* and *Hypnum cupressiforme* occur more occasionally, with *Polytrichum junipericum* preferential at low frequency and cover.

***Holcus lanatus* sub-community.** *C. arenaria* is generally the most abundant plant here, with *A. capillaris*, *F. ovina* (and occasional *F. rubra*) and *P. pratensis* making up much of the rest of the sward. However, in contrast to the previous sub-community, *H. lanatus* is very common and it can be moderately abundant, while *Anthoxanthum* and *L. campestris* are only occasional. *Ammophila* is also rather patchy in its occurrence here, though locally it may retain some vigour and show strongly tussocky growth.

G. saxatile and *R. acetosella* remain frequent in the sward, together with *L. corniculatus* and *C. fontanum*, but *H. radicata* is much less common and there is preferential *C. rotundifolia*, *V. riviniana* and *A. millefolium* with *Prunella vulgaris* and *Ranunculus repens* in some stands. Most of the bryophytes of the community remain frequent and *Ceratodon purpureus* is preferentially common, but these plants are not usually so abundant as in the *Anthoxanthum* sub-community.

Habitat

The *Carex-Festuca-Agrostis* grassland is characteristic of grazed stretches of fixed, acidic sands around the coasts of Britain, developing where quartz sand has become stabilised or more calcareous wind-blown sediments leached with the passage of time. It is thus most often found on areas of subdued relief among older dunes and on long-established sand plains, and is commoner towards the north and west where the wetter and cooler climate enhances leaching and providing some protection against drought.

Even more so than the *Festuca-Galium* grassland, this is a community of fixed sand surfaces, attaining its best development only where accretion has fallen to negligible levels and where erosion is limited to areas of disturbed ground. On more sheltered coasts, stands may be set back just a short way inland but this kind of vegetation is usually found on surfaces which are now far removed from sources of mobile material. Thus, although *Ammophila* persists at high frequency and may show a measure of renewed vigour locally when there is the occasional bout of fresh deposition of sand, it generally survives as a puny relic exerting next to no influence on the character of the substrate or the structure of the vegetation (Gimingham 1964*a*; Huiskes 1979). Also limiting to its growth here is the fact that the sands are base-poor, at least in the upper layers, the surface pH being usually below 5 (Salisbury 1952, Willis *et al.* 1959*a, b*; Huiskes 1977*a*, 1979). In places where the wind-blown sediments are lime-rich, as where shell fragments make up a considerable proportion of the sands, such superficial acidity may develop only after very prolonged leaching (Ranwell 1972, Willis 1985*b*), which is one reason why this community is still of only local occurrence on many stretches of machair: there, despite the considerable age of the surfaces and the very wet climate, the ground often remains markedly calcareous. By contrast, on those dune systems where the sediments are largely made up of quartz sand, younger stable surfaces can offer a congenial substrate for this vegetation, even in regions where the climate is much drier.

The occurrence of suitably siliceous sands of Pliocene and Pleistocene origin around the coast of East Anglia gives the *Carex-Festuca-Agrostis* community some important outposts in the drier and warmer south-east of Britain, but most stands of this vegetation are found further north and west. There, annual rainfall of more than 800 mm (*Climatological Atlas* 1952), with usually over 140 wet days yr^{-1} (Ratcliffe 1968), and mean annual maxima often below 25 °C (Conolly & Dahl 1970), help maintain the distinctive character of these swards on sands which were initially more calcareous. Also, the greater consistency of the rainfall and the cloudier character of the climate help protect the vegetation against the tendency of the ground to parching. Older soils here, with some accumulation of organic matter in their upper layers, can be more moisture-retentive than the raw sands of younger dunes, but the plants are still largely dependent on regular rainfall for sustained growth.

This combination of edaphic and climatic features gives the *Carex-Festuca-Agrostis* grassland much of its distinctive floristic and structural character. As in the *Festuca-Galium* swards, the waning of the dominance of *Ammophila* on the more fixed sand surface allows the smaller rhizomatous plants like *C. arenaria*, *P. pratensis* and *F. rubra* to increase their contribution, though here, perhaps with the reduction of any influence of salt-laden winds, *F. ovina* is often the more prominent fescue. More obvious than this, however, is the better representation of calcifuges. On these more acidic soils, *C. arenaria* itself thrives better than *Ammophila* (Huiskes 1979, Noble 1982), but more diagnostic against the *Festuca-Galium* grassland are *A. capillaris*, *Anthoxanthum* and *G. saxatile* which usually exceed in frequency plants like *L. corniculatus*, *G. verum*, *T. repens* and *P. lanceolata*. Then, among the bryophytes, there is the common occurrence of such species as *D. scoparium*, *H. splendens* and *P. schreberi*. When all of these are well represented, as in the *Anthoxanthum* sub-community, the impression is of a Nardo-Galion grassland with little of the calcicolous or mesophytic aspect seen in many *Festuca-Galium* swards.

However, compared with the *Carex-Cornicularia* community, which is the other herbaceous vegetation of fixed, acidic sands, the *Carex-Festuca-Agrostis* grassland generally has a more closed and less tussocky cover, something which reflects the less droughty character of the soils here. In the *Carex-Cornicularia* swards, most of the perennials perform rather poorly, with even *C. arenaria* often in a senile phase (Noble 1982) and the grasses frequently patchy and prone to parching in summer. Here, by contrast, the sedge and grass matrix is more or less continuous and usually retains its fresh green colour throughout the year. *Poa pratensis* becomes constant, too, a good marker of better moisture supply within the sands and then, in the *Holcus* sub-community, there is the frequent appearance of *H. lanatus*, *V. riviniana* and *A. millefolium*. The development of this kind of *Carex-Festuca-Agrostis* grassland may reflect treatment differences, but it could also extend the community on to damper and less impoverished ground around dune hollows. Certainly, it has an aspect never seen in the *Carex-Cornicularia* community.

Even in the *Anthoxanthum* sub-community, however, there are other obvious floristic distinctions. Thus, while light-demanding perennial herbs such as *H. radicata*, *R. acetosella* and *T. praecox* remain quite common in the short swards, ephemerals able to capitalise rapidly on the appearance of bare patches are decidedly scarce. Then, among the cryptogams, it is not drought-tolerant lichens that dominate, with acrocarpous mosses colonising open areas, but moisture-demanding pleurocarpous mosses that form a sometimes abundant weft among the turf.

In addition to these environmental effects, however, there is the influence of grazing because this is almost always a plagioclimax vegetation maintained by the predation of herbivores. More particularly, it is probably stock, as much as or more than rabbits, which are important here. With the *Carex-Cornicularia* community, rabbit grazing is (or has been) a very likely influence in maintaining the impoverished condition of the soils and favouring the abundance of lichens in the sward, as well as keeping the herbage very short. With the *Carex-Festuca-Agrostis* grassland, sheep and cattle can crop the vegetation close, helping to maintain some diversity among the vascular plants, but their heavier trampling is destructive of lichens. Further, because their faeces and urine are distributed over the sward, rather than concentrated into latrines, they can have a considerable manurial effect, enhancing the nutrient status of the sands and increasing moisture retention with the accumulation of an organic fraction. Such effects are perhaps especially important in the development of the more mesophytic

Holcus sub-community which may be associated with the greater enrichment and less assiduous cropping that cattle, rather than sheep, bring. But, whatever the herbivores, they are ultimately important in holding in check the invasion or spread of any ericoids or shrubs and trees and, without their predation, most stands would probably progress quickly to heath or woodland.

Zonation and succession
The *Carex-Festuca-Agrostis* community is typically found in zonations and mosaics with other grasslands and heath on the fixed sands of the hinterland of coastal dunes and sand plains, where the vegetation patterns are primarily determined by edaphic variation and differences in treatment. Relaxation of grazing mediates a succession to heath or woodland, but most stands are maintained by continuing use as pasture for stock. Where stretches of fixed dunes have been reclaimed for more intensive agricultural use or for forestry or amenity purposes, the *Carex-Festuca-Agrostis* grassland may survive more fragmentarily among improved swards or arable land, plantations or golf-course greens.

Among the other kinds of grassland found on fixed sands, the community is most often seen with some type of *Festuca-Galium* vegetation, grading to this where the ground is more base-rich, with substrates of shell-sand that is not so strongly leached or periodically renewed by very modest accretion. Species like *C. arenaria*, fescues and *P. pratensis* continue to provide much of the cover in a fairly short sward with just scattered tufts of *Ammophila*, but *F. rubra* generally exceeds *F. ovina*, and there is usually a more consistent occurrence of *G. verum*, *P. lanceolata*, *T. repens*, *L. corniculatus* and *A. millefolium*. In the *Luzula* sub-community, however, *L. campestris*, *A. capillaris*, *F. ovina* and *Anthoxanthum* become quite common and this sort of *Festuca-Galium* sward may pass imperceptibly to the *Anthoxanthum* sub-community of the *Carex-Festuca-Agrostis* grassland where there are gentle shifts in the base-richness of the sands. On somewhat moister and less impoverished ground, the *Holcus* sub-community of the *Carex-Festuca-Agrostis* grassland may grade to the *Prunella* type of *Festuca-Galium* sward.

Where the sand retains a greater degree of mobility with the move on to younger dunes, the *Festuca-Galium* grassland passes in turn to *Ammophila-Festuca* vegetation, where marram is still vigorous on the upbuilding surface and where the associates make up an often patchy cover between the tussocks. Sometimes, the *Carex-Festuca-Agrostis* grassland itself gives way to the *Ammophila-Festuca* community in places where the sand has become mobile again with disturbance of the surface, but more often, and particularly where the sands are acidic, it is the *Carex arenaria* community that is the secondary colonising vegetation. In that vegetation, the cover can be much more open at first, with rejuvenated *C. arenaria* extending its rhizomes out through the loose sand, but the sedge may thicken up in time, and other associates appear, with tussocks of the fescues and scattered *R. acetosella* and *P. lanceolata* providing a measure of continuity with the more intact swards around. Indeed, on base-poor sands, this kind of invasion may be the natural line of primary succession that leads eventually to the development of *Carex-Festuca-Agrostis* grassland.

On coasts in warmer and drier parts of Britain, and more locally elsewhere, the community can also be found on acidic sands with the *Carex-Cornicularia* vegetation. In that assemblage, the sedge cover thins and the perennial grasses are reduced to scattered tussocks, with just occasional rosette herbs and patches of chamaephytes and, very strikingly, a switch to lichens rather than pleurocarps as the dominant cryptogams. Such a change can be associated with concentrated rabbit activity, but exposure to parching on sunnier dune slopes may also favour the development of the *Carex-Cornicularia* community on old erosion surfaces that have compacted and impoverished sands.

Both the *Carex-Cornicularia* and especially the *Carex-Festuca-Agrostis* community are prone to invasion by *Calluna* and other ericoids where grazing is lax. In drier parts of the country, where both these vegetation types can contribute to the swards on fixed acid sands, colonisation by heather usually leads to the development of the *Calluna-Festuca* heath, where *C. arenaria* can remain locally prominent with *F. ovina* and occasional *A. capillaris*, *G. saxatile* and *R. acetosella*, along with *D. scoparium* and *H. cupressiforme* in the ground carpet. *Erica cinerea* may also occur in such heath in East Anglia, but this plant becomes much more common in northern and western localities where the *Calluna-Carex* heath is the usual kind of dune ericoid vegetation. Here again, mixtures of sub-shrubs, *C. arenaria* and *F. ovina* provide a diagnostic element, but *L. campestris*, *Anthoxanthum* and various bryophytes give greater floristic continuity with the *Carex-Festuca-Agrostis* grassland. In this type of dune heath in eastern Scotland, *Empetrum nigrum* tends to replace *E. cinerea* as the associated sub-shrub.

Quite often, where grazing has fluctuated in intensity complex mosaics of heath and grassland occur on pastured dunes, the expansion of the sub-shrubs within the *Carex-Festuca-Agrostis* community being held or pushed back by renewed predation by stock. Elsewhere, neglect has allowed the development of leggy heather which excludes virtually all associates until it begins to degenerate. Or, where seedling shrubs and trees get away, there may be transitions to scrub or woodland, with birch and pine, more locally oak, leading a succession to Quercion vegetation. Then, there may be patches of the *Pteridium-Galium* community where bracken has preempted the spread of heather in the grassland or shaded it out, with *Ulex-Rubus* scrub marking places that have been disturbed or enriched. Fragments of all these may survive, along with patches of the *Carex-Festuca-Agrostis* vegetation, around plantations on dunes, in field corners of improved pastures and on golf-course rough.

Distribution
The community has scattered localities around the coasts of south-east England, but it is commoner towards the north and west in those places where older, leached sands have been grazed. Were suitable substrates more widely distributed through north-west Scotland, it would probably be extensive there.

Affinities
More calcifuge dune vegetation of this kind has been described from a few sites around the British coast (Pearsall 1934, Gimingham 1964a), and it was included by Birse (1980, 1984) in the *Astragalo-Festucetum arenariae*, his grassland of fixed dunes on the less oceanic coasts of eastern Scotland. In this scheme, regional phytogeographic distinctions among the swards of less mobile dunes seem less obvious than those which reflect variation in the base-richness of the substrate, and throughout its range the *Carex-Festuca-Agrostis* grassland retains its integrity as the nearest approach to Nardo-Galion vegetation among dune communities. Because of the prevailingly calcicole character of much *Astragalo-Festucetum*, Birse (1980, 1984) grouped it within the Koelerion alliance of the Sedo-Scleranthetea, whereas the *Carex-Festuca-Agrostis* grassland as defined here would be best located in the Corynephorion of the same class.

Floristic table SD12

	a	b	12
Agrostis capillaris	V (1–9)	V (1–9)	V (1–9)
Carex arenaria	V (2–10)	IV (1–6)	V (1–10)
Festuca ovina	IV (2–6)	V (2–9)	IV (2–9)
Poa pratensis	IV (2–5)	IV (1–7)	IV (1–7)
Ammophila arenaria	V (2–8)	III (1–6)	IV (1–8)
Anthoxanthum odoratum	V (2–7)	II (2–4)	III (2–7)
Luzula campestris	V (1–4)	II (2–4)	III (1–4)
Hypochoeris radicata	III (1–5)	I (1)	II (1–5)
Koeleria macrantha	II (1–5)	I (1–2)	I (1–5)
Hieracium pilosella	II (1–6)		I (1–6)
Polytrichum juniperinum	II (1–3)		I (1–3)
Plantago lanceolata	II (1–4)		I (1–4)
Thymus praecox	II (1–6)		I (1–6)
Potentilla erecta	I (1–6)		I (1–6)
Calluna vulgaris	I (1–6)		I (1–6)
Danthonia decumbens	I (1–4)		I (1–4)
Deschampsia flexuosa	I (1–5)		I (1–5)
Erica cinerea	I (1–5)		I (1–5)
Cladonia furcata	I (1–3)		I (1–3)
Carex pilulifera	I (2–3)		I (2–3)
Ononis repens	I (1–3)		I (1–3)
Astragalus danicus	I (1)		I (1)
Holcus lanatus	II (1–6)	IV (2–8)	III (1–8)
Campanula rotundifolia	I (1–4)	III (1–4)	II (1–4)
Viola riviniana	I (1)	III (2–4)	II (1–4)
Ceratodon purpureus	I (1–4)	III (2–4)	II (1–4)
Achillea millefolium	I (2–4)	II (1–4)	I (1–4)
Ranunculus repens		I (2–4)	I (2–4)
Prunella vulgaris		I (2–4)	I (2–4)
Lathyrus pratensis		I (2–4)	I (2–4)
Urtica dioica		I (2–4)	I (2–4)
Galium saxatile	III (1–7)	III (2–5)	III (1–7)
Dicranum scoparium	III (1–7)	III (2–5)	III (1–7)
Rhytidiadelphus squarrosus	III (1–6)	III (2–5)	III (1–6)
Lotus corniculatus	III (1–6)	II (2–5)	III (1–6)
Pseudoscleropodium purum	III (1–8)	II (1–5)	III (1–8)
Festuca rubra	III (1–9)	II (2–4)	II (1–9)
Hylocomium splendens	II (1–9)	III (2–4)	II (1–9)
Pleurozium schreberi	II (1–7)	III (2–4)	II (1–7)
Cerastium fontanum	II (1–2)	III (2–4)	II (1–4)
Rumex acetosella	II (1–5)	II (2–4)	II (1–5)
Galium verum	II (1–6)	II (2–4)	II (1–6)
Hypnum cupressiforme	II (1–7)	II (2)	II (1–7)
Rhytidiadelphus triquetrus	II (2–5)	II (2–3)	II (2–5)
Trifolium repens	II (1–4)	I (2–4)	II (1–4)

Veronica officinalis	II (1–6)	I (2–4)	II (1–6)
Brachythecium albicans	I (1)	II (2–4)	I (1–4)
Peltigera canina	I (1–6)	I (2)	I (1–6)
Senecio jacobaea	I (1)	I (2–4)	I (1–4)
Cerastium diffusum diffusum	I (1–3)	I (2)	I (1–3)
Cirsium arvense	I (1)	I (2–4)	I (1–4)
Cladonia arbuscula	I (1–6)	I (2)	I (1–6)
Rumex acetosa	I (1–3)	I (2–4)	I (1–4)
Vicia lathyroides	I (1–2)	I (1)	I (1–2)
Geranium molle	I (2)	I (2)	I (2)
Number of samples	37	30	67
Number of species/sample	18 (10–27)	16 (5–28)	17 (5–28)

a   *Anthoxanthum odoratum* sub-community

b   *Holcus lanatus* sub-community

12   *Carex arenaria-Festuca ovina-Agrostis capillaris* dune grassland (total)

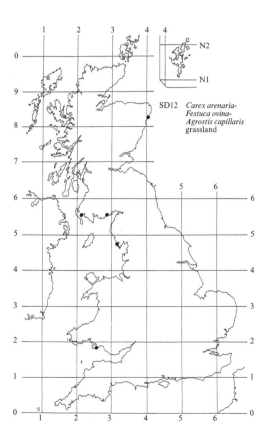

SD12   *Carex arenaria-Festuca ovina-Agrostis capillaris* grassland

SD13
Sagina nodosa-Bryum pseudotriquetrum dune-slack community

Synonymy
Calliergon cuspidatum-Salix repens nodum   Jones 1992
p.p.; Young dry slack nodum   Jones 1992

Constant species
Carex arenaria, Juncus articulatus, Leontodon hispidus, Sagina nodosa, Salix repens, Aneura pinguis, Bryum pseudotriquetrum.

Rare species
Equisetum variegatum, Pyrola rotundifolia, Moerckia hibernica, Petalophyllum ralfsii.

Physiognomy
The *Sagina nodosa-Bryum pseudotriquetrum* dune-slack community comprises short and often rather open swards dominated by a low and patchy cover of *Salix repens*, a patchy turf of grasses and diminutive herbs and a sometimes extensive contingent of mosses and thalloid liverworts. The vegetation is typically submerged briefly and to shallow depths in the winter and it is on the bare damp patches of sand and shell debris exposed as the water level falls that the various ephemerals and less competitive bryophytes of the community can gain a hold.

Apart from *S. repens*, the commonest vascular perennials are *Carex arenaria, Juncus articulatus* and *Agrostis stolonifera*, all typically represented by sparse shoots, with small scattered rosettes of *Leontodon hispidus*. Among these, by late spring, there are often numerous tiny plants of *Sagina nodosa* with the more prominent *Centaurium eythraea*. Seedlings of *Rubus caesius* occasionally figure and there are sparse plants of *Samolus valerandi, Mentha aquatica, Carex serotina, Equisetum palustre* and *Taraxacum officinale* agg. *Ammophila arenaria* is typically very scarce in this vegetation and of insignificant cover.

Among the bryophytes, *Bryum pseudotriquetrum* is usually the most prominent, its robust dark green shoots, often choked with sand, occurring patchily in the turf, but small thalli of *Aneura pinguis* are also very common with *Pellia endiviifolium* a distinctive occasional. In one sub-community, this cryptogam contingent shows a distinctive enrichment.

Sub-communities

***Poa annua-Moerckia hibernica* sub-community:** *Centaurio-Saginetum moniliformis* Diemont, Sissingh & Westhoff 1940 *p.p.* In this kind of *Sagina-Bryum* vegetation, the cover is usually more open, with *Salix repens* more patchy, and less grassy, though *Poa annua* is a distinctive preferential at low abundance. Also frequent are *Hydrocotyle vulgaris* and *Blackstonia perfoliata* with *Prunella vulgaris, Leontodon autumnalis, Epilobium palustre* and *Senecio jacobaea* occasional. Striking cryoptogam associates are *Moerckia hibernica* and, less commonly, *Petalophyllum ralfsii*.

***Holcus lanatus-Festuca rubra* sub-community.** Here, *S. repens* is more extensive and the sward grassier with *Agrostis stolonifera, Holcus lanatus, Festuca rubra* and *Poa pratensis* all very frequent, though each usually of only moderate cover. Common vascular associates are *Lotus corniculatus, Anthyllis vulneraria, Euphrasia officinalis* agg. with *Carex flacca* and *Hieracium pilosella* occasional and this kind of *Sagina-Bryum* vegetation also provides a locus for the nationally rare *Equisetum variegatum* and *Pyrola rotundifolia*.

Habitat
The *Sagina-Bryum* community includes pioneer and early stages of vegetation in slacks that are damp in winter but dry on the surface in summer among stabilised dunes of calcareous shell-sand. It is a very local community kept immature by periodic, brief and shallow submergence but probably also dependent for its open character on grazing, trampling and scuffing by stock and rabbits.

Studies in The Netherlands (Westhoff 1947, Schat

1982) where similar vegetation to that included here as the *Poa-Moerckia* sub-community has been described in the *Centaurio-Saginetum moniliformis* Diemont, Sissingh & Westhoff 1940 suggest that this assemblage is characteristic of dune slacks with quite marked seasonal fluctuations in the water-table, flooded perhaps briefly and to only shallow depth in the winter and with a summer water-table 20–50 cm below the surface. Hydrological studies at Newborough (Ranwell 1959) showed that, between 1950 and 1953, the more open dune-slack vegetation with *Sagina nodosa* and *Centaurium littorale* had a water-table between about 60 and 160 cm down and winter flooding of less than 2 cm depth. In the Voorne dunes in The Netherlands, too, van der Laan (1979) found that neither of these two distinctive species of the *Sagina-Bryum* community occurred in slacks where the summer water was less than 52 cm below the surface. In fact, Schat (1982) and co-workers showed that both the germination and growth of these plants are reduced or actually prevented by flooding. At Braunton, most of the young slacks of this general kind studied by Willis *et al.* (1959*b*) and Hope-Simpson & Yemm (1979) had a flood duration of much less than 160 days.

The other type of *Sagina-Bryum* vegetation, in the *Holcus-Festuca* sub-community, seems characteristic of even drier situations but in slacks that have been stabilised for just a short time – perhaps only 20 years or so at Kenfig on the South Wales coast, for example (Jones 1992). The less open grassy sward with more abundant *Salix repens* marks such a shift in the hydrological regime though the biomass of the vascular perennials is not such as to totally exclude ephemerals or the distinctive bryophytes of the community.

Though relatively young and often seasonally wet, the ground waters beneath this kind of vegetation do not seem to have much or any lingering saline influence. In contrast to some Dutch stands of the *Centaurio-Saginetum*, for example, species like *Glaux maritima* and *Juncus gerardi* are very scarce in the *Sagina-Bryum* community. Typically, however, the superficial pH is high, maintained by a combination of the calcium carbonate of the shell sand substrate and periodic flooding with base-rich ground waters.

Fluctuating ground water-tables are clearly a key factor in maintaining the open, immature character of this vegetation, particularly in the *Poa-Moerckia* sub-community. Redeposition of modest amounts of sand, blown about in the drier conditions of summer, may also create new barer places. Interestingly, *Petalophyllum ralfsii* can withstand some degree of burial by producing new thalli from the sub-surface tissues and some surface mobility may help its survival. Disturbance and cropping of the herbage by rabbits and stock could be important, too. Jones (1992) also suggested that the prominence of *Poa annua* in this kind of *Sagina-Bryum* vegetation might be due to import of seed by trekking ponies.

Zonation and succession

The *Sagina-Bryum* community is a very local vegetation type representing early or repeatedly-renewed stages in the colonisation of young, only briefly-wetted slacks among stabilised dune ridges. It can be found in systems with more mature and drier slacks and with more summer-damp slacks according to the pattern of ground-water fluctuation and gives way on drier dune ridges to a range of fixed grasslands. A combination of drying and freedom from grazing can facilitate eventual development of wet scrubby woodland.

Damper slacks on similar base-rich dunes to those with the *Sagina-Bryum* community can have stands of the *Salix-Campylium* community or, where older slacks are kept very wet and have a more vigorous shading canopy of *Salix repens*, stretches of *Salix-Calliergon* vegetation. Generally speaking, such slacks are separated by ridges carrying *Festuca-Galium* grassland or where there is still some measure of sand movement, the more stable types of *Ammophila-Festuca* grassland. To these, the *Sagina-Bryum* vegetation can grade through the *Holcus-Festuca* sub-community, the grass cover thickening up, *Ammophila* becoming somewhat more vigorous and ephemerals and less competitive bryophytes being crowded out. Drier slack vegetation of the *Salix-Holcus* type may occur as a transitional zone in such sequences.

As slacks with the *Sagina-Bryum* community age and dry, a process assisted by trapping of sand among the *Salix repens*, this vegetation is probably succeeded by the *Salix-Holcus* community which in turn may be invaded by shrubs and trees, as in its *Ononis* sub-community. *Hippophae rhamnoides* always has the potential to supervene in such seral progressions.

Distribution

Young drier slacks of the *Sagina-Bryum* type have been described from only very few sites around the British coast (Dargie 1993, 1995; Radley 1994), most notably from Kenfig in South Wales (Jones 1992) but also from Lindisfarne and Sefton in England and Torrs Warren in Scotland.

Affinities

The *Sagina-Bryum* community is very similar to the young dune-slack vegetation described from The Netherlands as the *Centaurio-Saginetum moniliformis* Diemont, Sissingh & Westhoff 1940, though that assemblage has a modest contingent of halophytes like *Glaux maritima*, *Juncus gerardi* and *Sagina maritima* which reflects its occurrence in slacks that have a lingering saline influence. In our stands, too, *Centaurium littorale* and *C. pulchellum* are replaced by *C. erythraea*. Traditionally, such vegetation has been assigned to the Nano-cyperion, though a recent review of this alliance in The Netherlands (Lemaire & Weeda 1994) suggests that the Saginion maritimae might provide a more sensible locus because of the halophytic character of the vegetation.

Floristic table SD13

	a	b	13
Salix repens	V (1–8)	V (2–9)	V (1–9)
Bryum pseudotriquetrum	V (1–4)	V (2–7)	V (1–7)
Aneura pinguis	V (1–5)	V (1–4)	V (1–5)
Sagina nodosa	V (1–4)	IV (1–4)	V (1–4)
Leontodon hispidus	IV (1–3)	V (1–4)	IV (1–4)
Carex arenaria	IV (1–8)	V (1–5)	IV (1–8)
Juncus articulatus	IV (3–4)	IV (1–4)	IV (1–4)
Poa annua	IV (1–3)		III (1–3)
Hydrocotyle vulgaris	III (1–4)	I (1)	II (1–4)
Moerckia hibernica	III (1–4)	I (1–3)	II (1–4)
Blackstonia perfoliata	III (1–3)	I (1)	II (1–3)
Prunella vulgaris	II (1–4)	I (1)	II (1–4)
Leontodon autumnalis	II (1–3)	I (1)	I (1–3)
Campylium stellatum	II (1–4)	I (3)	I (1–4)
Epilobium palustre	II (1–3)		I (1–3)
Senecio jacobaea	II (1–3)		I (1–3)
Hippophae rhamnoides seedling	II (1)		I (1)
Petalophyllum ralfsii	II (1–3)		I (1–3)
Pulicaria dysenterica	I (1–3)		I (1–3)
Arenaria serpyllifolia	I (1–3)		I (1–3)
Agrostis stolonifera	III (1–4)	IV (1–4)	III (1–4)
Holcus lanatus	II (1–3)	V (1–6)	III (1–6)
Poa pratensis	I (1)	V (1–3)	III (1–3)
Lotus corniculatus		IV (1–5)	II (1–5)
Festuca rubra		IV (3–5)	II (3–5)
Anthyllis vulneraria		IV (1–6)	II (1–6)
Equisetum variegatum	I (1–6)	III (2–5)	II (1–6)
Euphrasia officinalis agg.	I (1–2)	III (1–3)	II (1–3)
Pyrola rotundifolia	I (1–3)	III (1–4)	II (1–4)
Carex flacca	I (1–5)	II (3–7)	I (1–7)
Hieracium pilosella		II (1–2)	I (1–2)
Epipactis palustris		II (1–4)	I (1–4)
Ononis repens		I (1–4)	I (1–4)
Polygala vulgaris		I (1–2)	I (1–2)
Galium verum		I (1–2)	I (1–2)
Dactylorhiza incarnata		I (1)	I (1)
Centaurium erythraea	III (1–3)	III (1–3)	III (1–3)
Rubus caesius	II (1–3)	II (1)	II (1–3)
Pellia endiviifolia	I (3)	I (4)	I (3–4)
Samolus valerandi	I (2–3)	I (2–3)	I (2–3)

Equisetum palustre	I (2)	I (3–5)	I (2–5)
Mentha aquatica	I (1)	I (1)	I (1)
Carex serotina	I (1–4)	I (1–4)	I (1–4)
Amblystegium serpens	I (1)	I (1–3)	I (1–3)
Taraxacum officinale agg.	I (1)	I (1)	I (1)
Sonchus arvensis	I (1)	I (1)	I (1)
Ammophila arenaria	I (1)	I (1)	I (1)
Cerastium fontanum	I (1)	I (1–2)	I (1–2)
Phragmites australis	I (4)	I (3–4)	I (3–4)
Barbula tophacea	I (1)	I (1–3)	I (1–3)
Number of samples	28	20	48
Number of species/sample	15 (11–24)	18 (13–26)	16 (11–26)

a   *Poa annua-Moerckia hibernica* sub-community
b   *Holcus lanatus-Festuca rubra* sub-community
13   *Sagina nodosa-Bryum pseudotriquetrum* dune-slack (total)

SD14

Salix repens-Campylium stellatum dune-slack community

Synonymy
Sandscale *Salix repens* dunes   Pearsall 1934; *Campylium stellatum-Salix repens* nodum   Jones 1992; *Carex flacca*-Thalloid Liverwort nodum   Jones 1992 *p.p.*

Constant species
Agrostis stolonifera, Carex flacca, Epipactis palustris, Equisetum variegatum, Hydrocotyle vulgaris, Mentha aquatica, Salix repens, Calliergon cuspidatum, Campylium stellatum.

Rare species
Dactylorhiza majalis ssp. *praetermissa* and ssp. *purpurella, Juncus acutus, Liparis loeselii, Pyrola rotundifolia.*

Physiognomy
The *Salix repens-Campylium stellatum* community comprises generally closed vegetation of dune slacks with an often extensive low bushy carpet of *S. repens*, frequently species-rich assemblages of vascular associates and a usually extensive carpet of bryophytes among which both *Campylium stellatum* and *Calliergon cuspidatum* figure very commonly, the former generally especially abundant.

The constant combination of *Carex flacca, Equisetum variegatum* and, in most sub-communities, *Epipactis palustris* with more widely occurring dune-slack plants such as *Hydrocotyle vulgaris, Agrostis stolonifera* and *Mentha aquatica*, is what gives this vegetation its particular character. Among these species, *E. variegatum* can be especially abundant and it sometimes assumes a striking tussock growth-form, perhaps because of rapid shoot proliferation in the early stages of colonisation (Jones 1992). Also, as with the *Salix-Calliergon* community, the only occasional occurrence of *Potentilla anserina* and *Carex nigra* here, and the generally low cover of the latter, is an important contrast with the *Potentilla-Carex* slack vegetation.

Among the other commoner associates, *Carex arenaria, Leontodon autumnalis, Ranunculus flammula,* *Rubus caesius, Lotus corniculatus* and *Juncus articulatus* are the most frequent with more occasional *Taraxacum officinale* agg. and *Equisetum palustre. Glaux maritima, Juncus gerardii* and *J. maritimus* figure at low frequencies and locally an abundance of *Phragmites australis, Cladium mariscus* and *Juncus acutus* can give stands a striking individuality. Though not represented among our data, *Schoenus nigricans* can also figure prominently in this vegetation. *Samolus valerandi* is a nationally scarce plant that occurs here and the community also provides a locus at a few sites for the rare *Liparis loeselii.*

Along with *Campylium stellatum* and *Calliergon cuspidatum*, other distinctive bryophytes occurring occasionally throughout are *Drepanocladus sendtneri, D. lycopodioides* and *Riccardia chamaedryfolia.*

Sub-communities

***Carex serotina-Drepanocladus sendtneri* sub-community.** *C. flacca* is joined here by *C. arenaria* and, especially distinctive, *C. serotina*, with occasional *Eleocharis palustris* and *Dactylorhiza majalis* ssp. *purpurella*. Also frequent, though less strongly preferential, are *Ranunculus flammula* and *Juncus articulatus. Drepanocladus sendtneri* is a little more frequent here than elsewhere in the community.

***Rubus caesius-Galium palustre* sub-community:** *Campylium stellatum-Salix repens* nodum, Typical and *Carex nigra* sub-types   Jones 1992. *Carex arenaria, Ranunculus flammula* and *Juncus articulatus* remain very common here and *Leontodon autumnalis* also appears among the constants. More striking, though, is the preferential frequency of *Rubus caesius* and *Galium palustre* and, more occasionally, *Carex nigra.*

***Bryum pseudotriquetrum-Aneura pinguis* sub-community:** *Carex flacca*-Thalloid Liverwort nodum   Jones 1992 *p.p.* In this, the most particular kind of *Salix-Campylium* vegetation, the vascular contingent has preferentially frequent records for *Juncus articulatus* and *Lotus corniculatus* and is additionally enriched by *Poa pratensis*, *Prunella vulgaris*, *Leontodon hispidus*, *Euphrasia officinalis* agg., *Linum catharticum* and *Anagallis tenella*. Equally distinctive among the extensive bryophyte carpet is the constancy of *Bryum pseudotriquetrum* and a range of thalloid liverworts including *Pellia endiviifolia*, *Aneura pinguis* and, more occasionally, *Moerckia hibernica* and *Preissia quadrata*.

Festuca rubra **sub-community:** *Campylium stellatum-Salix repens* species-rich nodum, dry sub-type Jones 1992 *p.p. Leontodon autumnalis*, *Lotus corniculatus*, *Poa pratensis*, *Prunella vulgaris* and *Rubus caesius* all remain frequent here but the sward has a grassier look, with *Festuca rubra* and *Holcus lanatus* constant at sometimes moderate to high cover and *Trifolium repens* and *T. pratense* are also very common. *Pulicaria dysenterica* is especially distinctive among the preferentials and there is occasional *Ranunculus repens*, *R. acris*, *Lotus uliginosus* and *Sonchus arvensis*. *Brachythecium rutabulum* is frequent in the bryophyte mat.

Habitat

The *Salix-Campylium* community is a scarce vegetation type found in slacks of young to moderate age and kept moist by fluctuations in quite base-rich ground waters.

As slacks go, this type includes much of the more calcicolous vegetation found on British dune systems and values for calcium concentration of well in excess of 40 mg l^{-1} (comparable with some inland Caricion davallianae fens) have been recorded in this kind of situation (Gorham 1958, Jones 1967). Yet, as Jones (1992) points out, the contingent of more base-demanding plants here is relatively modest. It is possible, then, that some other hydrological factor inhibits the expression of this aspect of the vegetation, perhaps considerable range in fluctuation of the water-table. In comparable systems in The Netherlands, for example, van der Laan (1979) showed that the summer water-table was between 10 and 60 cm below the surface with a possibility of superficial drought in warmer summers (Jones 1992). Winter flooding in this kind of slack can attain from 10 to 50 cm in depth, but typically it is not so long-lasting as in the *Salix-Calliergon* community. One other important factor is the deficiency in major nutrients in slacks (Willis & Yemm 1961, Willis 1963), perhaps even more marked than in inland Caricion davallianae fens (Jones 1992).

The youngest type of *Salix-Campylium* vegetation is the *Bryum-Aneura* sub-community (Jones 1992) where the *Campylium* carpet is not yet too extensive or thick, thus allowing the thalloid liverworts to thrive. In fact, it seems that the moss mat can develop here relatively quickly – in 35 years or so in some Kenfig slacks, for example (Jones 1992) – and as soon as it attains 5 cm, the thalloid liverworts typically of the *Bryum-Aneura* sub-community are virtually extinguished.

The differences between the habitats of the three other sub-communities are unclear, though the *Festuca-rubra* sub-community is perhaps characteristic of drier situations. As with other dune-slack communities, grazing by stock and rabbits may play some part in determining the floristics and physiognomy of the vegetation.

Zonation and succession

The *Salix-Campylium* community typically occupies slacks of moderate wetness in systems with a range of ground-water conditions, generally between ridges with various *Festuca-Galium* or *Ammophila-Festuca* swards. It is probably maintained by continuing winter flooding but any tendency to drying is marked by eventual progression to woodland.

Within individual slacks, depressions of natural or artificial origin where the moss carpet is opened up, can have local stands of the *Bryum-Aneura* sub-community among other kinds of *Salix-Campylium* vegetation. Drier stands can pass to the *Salix-Holcus* community or directly to *Festuca-Galium* swards on the dune-ridges around, a generally sharp shift with the change in ground moisture conditions.

Successional change within the community sees the *Bryum-Aneura* type giving way to other sub-communities and, in this process, the development of a tussock habit in *Equisetum variegatum* may play a distinctive part in the two or three decades it can take for *Salix repens* to form a dominant canopy (Hope-Simpson & Yemm 1979). As tussocks grow and fuse, creeping and stoloniferous species infilling any remaining spaces, it seems that the surface of the vegetation mat elevates perhaps several decimetres (Jones 1992).

Distribution

This is an uncommon community, occurring locally around the English and Welsh coasts, more sparsely still in Scotland.

Affinities

Vegetation similar to this has been described from calcareous dune slacks in The Netherlands (Westhoff 1947, Duvigneaud 1947, van der Maarel & Westhoff 1964, Van der Laan 1979), some of it included in the *Pyrolo-Salicetum* Meltzer 1941, some in the *Junco baltici-Schoenetum nigricantis* Westhoff 1946. More recently, Schaminée *et al.* (1995) have provided a new description of the *Junco baltici-Schoenetum* Westhoff ex Westhoff & van Oosten 1991 with a floristic profile very similar to that of the *Salix-Campylium* community, apart from the consistent prominence there of *Schoenus nigricans* and infrequent occurrence of *Juncus balticus*.

Floristic table SD14

	a	b	c	d	14
Salix repens	V (1–10)	V (3–9)	V (3–9)	V (3–9)	V (1–10)
Campylium stellatum	V (2–10)	V (3–10)	V (3–10)	IV (1–10)	V (1–10)
Hydrocotyle vulgaris	V (3–8)	V (3–10)	IV (1–7)	V (1–10)	V (1–10)
Equisetum variegatum	IV (3–9)	V (2–10)	IV (3–10)	V (3–8)	V (2–10)
Carex flacca	III (1–8)	V (1–5)	V (3–8)	V (1–5)	V (1–8)
Agrostis stolonifera	III (1–5)	V (1–6)	V (1–8)	IV (1–5)	IV (1–8)
Mentha aquatica	IV (1–5)	V (1–7)	III (1–4)	IV (1–6)	IV (1–7)
Calliergon cuspidatum	III (1–9)	IV (1–10)	IV (1–6)	V (1–10)	IV (1–10)
Epipactis palustris	II (1–5)	V (1–6)	IV (1–5)	IV (1–5)	IV (1–6)
Carex arenaria	IV (1–5)	V (1–5)	III (1–4)	II (1–4)	III (1–5)
Drepanocladus sendtneri	III (1–10)	II (1–10)	II (1–4)	I (2–6)	II (1–10)
Carex serotina	III (1–5)	I (1–3)	II (2–4)	I (1–3)	II (1–5)
Eleocharis palustris	II (1–4)	I (2–3)		I (1)	I (1–4)
Dactylorhiza majalis purpurella	I (1–2)				I (1–2)
Centaurium pulchellum	I (1)				I (1)
Leontodon autumnalis	I (1–2)	IV (1–5)	IV (1–5)	IV (1–5)	III (1–5)
Ranunculus flammula	III (1–4)	IV (1–4)	III (1–4)	II (1–4)	III (1–4)
Rubus caesius	II (1–5)	IV (1–8)	II (1–5)	III (1–5)	III (1–8)
Galium palustre	I (1–3)	IV (1–4)	I (2)	I (1–3)	II (1–4)
Carex nigra	I (1–9)	II (1–8)	I (5)	I (1–5)	I (1–9)
Lotus corniculatus	I (1–4)	III (1–5)	V (1–5)	IV (1–5)	III (1–5)
Juncus articulatus	III (1–5)	III (1–9)	IV (1–5)	II (1–4)	III (1–9)
Poa pratensis		I (1–3)	IV (1–5)	V (1–6)	II (1–6)
Prunella vulgaris	I (1–3)	I (1–3)	IV (1–5)	III (1–6)	II (1–6)
Leontodon hispidus	II (1–6)	I (1–5)	IV (1–5)	II (2–5)	II (1–6)
Pellia endviifolia	I (1–5)	I (3)	IV (1–6)	II (1–6)	II (1–6)
Bryum pseudotriquetrum	I (1–5)	I (1–2)	V (1–8)	I (1–4)	II (1–8)
Aneura pinguis	I (1–4)	I (1–3)	IV (1–6)	I (1–3)	I (1–6)
Linum catharticum	I (1–4)	I (1–3)	III (1–3)	I (1–3)	I (1–6)
Euphrasia officinalis agg.	I (1–3)	I (1–3)	II (1–4)	I (1–4)	I (1–4)
Anagallis tenella	I (1–5)	I (1–4)	II (1–4)	I (1–4)	I (1–5)

Species				
Moerckia hibernica	I (1–3)	II (1–3)	I (3)	I (1–3)
Preissia quadrata	I (1–4)	II (1–6)		I (1–6)
Leiocolea badensis		II (3–10)		I (3–10)
Carex hirta		I (1–4)		I (1–4)
Equisetum arvense		I (1–3)		I (1–3)
Festuca rubra	I (1–3)	II (1–6)	V (1–9)	II (1–9)
Holcus lanatus	I (3)	III (1–6)	IV (1–5)	II (1–6)
Pulicaria dysenterica	I (1–2)	II (1–8)	IV (1–8)	II (1–8)
Trifolium pratense	I (3)	I (1–5)	IV (2–8)	II (1–8)
Trifolium repens	II (1–5)	III (2–5)	III (1–4)	II (1–5)
Brachythecium rutabulum	I (4)	II (1–7)	III (3–9)	I (1–9)
Ranunculus repens	I (1–3)	I (1–3)	III (1–4)	I (1–4)
Danthonia decumbens	I (2–3)	I (1–3)	II (1–5)	I (1–5)
Ranunculus acris	I (1)	I (1)	II (1–4)	I (1–4)
Sonchus arvensis	I (2)	I (1)	II (1–5)	I (1–5)
Lotus uliginosus	I (2–5)		II (1–6)	I (1–6)
Plantago lanceolata		I (2–3)	II (1–5)	I (1–5)
Lophocolea bidentata		I (3)	II (1–4)	I (1–4)
Eupatorium cannabinum			I (1–4)	I (1–4)
Eurhynchium praelongum			I (1–4)	I (1–4)
Dactylorhiza incarnata			I (1–3)	I (1–3)
Angelica sylvestris			I (1–3)	I (1–3)
Briza media			I (2–4)	I (2–4)
Potentilla anserina	II (1–5)	I (1–5)	II (1–5)	II (1–5)
Equisetum palustre	I (1–4)	I (3–5)	II (1–4)	II (1–5)
Taraxacum officinale agg.	I (3)	II (1–3)	I (1–3)	I (1–3)
Dactylorhiza majalis praetermissa	I (1–3)	I (1–3)	I (1–3)	I (1–3)
Glaux maritima	I (1–4)	I (1–4)	I (1–3)	I (1–4)
Lycopus europaeus	I (1–3)	I (3)	I (1–5)	I (1–5)
Phragmites australis	I (1–5)	I (8–10)	I (2–10)	I (1–10)
Carex panicea	I (3–6)	I (7)	I (4–8)	I (3–8)
Riccardia chamaedryfolia	I (1–3)	I (3)	I (1–3)	I (1–3)
Ophioglossum vulgatum	I (1–2)		I (1–3)	I (1–4)
Eleocharis quinqueflora	I (1–4)	I (1–3)		I (1–4)
Senecio jacobaea	I (1–2)	I (1–2)	I (1–4)	I (1–4)

Floristic table SD14 (cont.)

	a	b	c	d	14
Bellis perennis	I (2–4)		I (1–3)	I (1–3)	I (1–4)
Drepanocladus lycopodiodes	I (1–10)	I (10)		I (5)	I (1–10)
Oenanthe lachenalii	I (1)	I (2)		I (1–3)	I (1–3)
Liparis loeselii	I (1–4)	I (1–3)	I (3)		I (1–4)
Plantago major	I (1)		I (1)	I (1)	I (1)
Polygala vulgaris	I (1)		I (1–2)	I (2)	I (1–2)
Juncus gerardii	I (4)	I (4)		I (1–3)	I (1–4)
Juncus maritimus	I (3)	I (1–3)	I (5)		I (1–5)
Equisetum fluviatile	I (1–2)	I (3)			I (1–3)
Salix caprea	I (2)	I (1–3)		I (2)	I (1–3)
Pseudoscleropodium purum	I (2)		I (5)	I (1)	I (1–5)
Juncus acutus	I (1)	I (1)		I (5)	I (1–5)
Plantago coronopus	I (3–4)		I (4–6)		I (3–6)
Cladium mariscus	I (8–9)	I (4–8)			I (4–9)
Epilobium parviflorum	I (1)	I (2–5)			I (1–5)
Samolus valerandi	I (1–3)		I (1)		I (1–3)
Sagina nodosa	I (1–2)		I (1–3)		I (1–3)
Blackstonia perfoliata	I (1–3)		I (1)		I (1–3)
Drepanocladus aduncus	I (1–4)		I (8)		I (1–8)
Carex demissa	I (6)		I (4)		I (4–6)
Leontodon taraxacoides	I (3–4)			I (1–2)	I (1–4)
Trifolium fragiferum	I (3)			I (3)	I (3)
Anthoxanthum odoratum	I (2)			I (2)	I (2)
Carex hostiana	I (2)			I (5)	I (2–5)
Senecio erucifolius		I (1)		I (3–4)	I (1–4)
Carex caryophyllea		I (2)		I (5)	I (2–5)
Juncus inflexus		I (2)		I (3)	I (2–3)
Pyrola rotundifolia			I (1–5)	I (1–6)	I (1–6)
Gymnadenia conopsea			I (1)	I (1–3)	I (1–3)
Hippophae rhamnoides			I (1–4)	I (1)	I (1–4)
Fragaria vesca			I (1)	I (1–4)	I (1–4)
Gentianella amarella			I (1–2)	I (3)	I (1–3)

	a	b	c	d	14
Anthyllis vulneraria			I (1)	I (1)	I (1)
Parnassia palustris			I (2–4)	I (3)	I (2–4)
Rhinanthus minor			I (2)	I (1–3)	I (1–3)
Succisa pratensis			I (2)	I (4–5)	I (2–5)
Viola riviniana			I (1)	I (1–2)	I (1–2)
Number of samples	70	85	34	57	246
Number of species/sample	14 (8–27)	17 (10–24)	24 (14–42)	23 (12–31)	19 (8–42)

a   *Carex serotina-Drepanocladus sendtneri* sub-community

b   *Rubus caesius-Galium palustre* sub-community

c   *Bryum pseudotriquetrum-Aneura pinguis* sub-community

d   *Festuca rubra* sub-community

14 *Salix repens-Campylium stellatum* dune-slack (total)

SD15
Salix repens-Calliergon cuspidatum dune-slack community

Synonymy
Calliergon cuspidatum-Salix repens noda   Jones 1992.

Constant species
Hydrocotyle vulgaris, Mentha aquatica, Salix repens, Calliergon cuspidatum.

Physiognomy
The *Salix repens-Calliergon cuspidatum* dune-slack community shares, with the *Salix-Campylium* vegetation, the high frequency of *Salix repens, Hydrocotyle vulgaris, Mentha aquatica* and *Calliergon cuspidatum* but the other vascular constants of that assemblage, *Carex flacca, Agrostis stolonifera, Equisetum variegatum* and *Epipactis palustris*, are of more restricted occurrence here and *Campylium stellatum* is extremely scarce.

Also, with more frequent records throughout for *Galium palustre, Epilobium palustre, Equisetum palustre* and *Lotus uliginosus*, the overall stamp of the vegetation tends to be like a poor fen rather than a rich fen. Also occasional in the community are *Festuca rubra* and *Poa pratensis* with local enrichment from tall herbs such as *Iris pseudacorus, Filipendula ulmaria* and *Phragmites australis*.

Apart from *Calliergon cuspidatum*, which is usually extremely abundant in a thick carpet, there are no frequent bryophytes and the striking contingent of calcicolous thalloid liverworts characteristic of some kinds of *Salix-Campylium* dune-slack is never present.

Sub-communities

Carex nigra sub-community: *Calliergon cuspidatum-Salix repens* noda, species-poor sub-type   Jones 1992. Both *Rubus caesius* and *Galium palustre* are constant here but more strongly preferential are *Carex nigra* and *Scutellaria galericulata* with occasional *Lysimachia vulgaris* and *Equisetum fluviatile*.

Equisetum variegatum sub-community: *Campylium stellatum-Salix repens* nodum, *Equisetum variegatum* sub-type

Jones 1992 *p.p.* It is here and in the next sub-community that the *Salix-Calliergon* community comes closest to the *Salix-Campylium* type with frequent records for *Agrostis stolonifera, Equisetum variegatum, Carex flacca* and *Epipactis palustris* but *Campylium stellatum* itself is still only very scarce. However, *Carex nigra* and *Galium palustre* remain very common and there is often some *Carex arenaria, Potentilla anserina* and *Ranunculus repens*. *Cladium mariscus* is a scarce but sometimes locally abundant associate.

Carex flacca-Pulicaria dysenterica sub-community: *Calliergon cuspidatum-Salix repens* noda, Herb-rich sub-type   Jones 1992 *p.p. Agrostis stolonifera, Equisetum variegatum, Carex flacca* and *Epipactis palustris* all remain very frequent in this kind of *Salix-Calliergon* vegetation but *Carex flacca* becomes more common and *Pulicaria dysenterica, Eupatorium cannabinum* and *Ranunculus flammula* and, more occasionally, *Oenanthe lachenalii* are preferential.

Holcus lanatus-Angelica sylvestris sub-community. *Rubus caesius, Carex flacca, Pulicaria dysenterica* and *Eupatorium cannabinum* remain frequent here but more diagnostic are *Holcus lanatus, Angelica sylvestris* and *Succisa pratensis* and, less commonly, *Molinia caerulea, Cirsium palustre* and *Vicia sativa* ssp. *nigra*. Also, *Phragmites australis* is quite frequent and locally abundant giving a fen-like stamp to the vegetation. In a few localities, a local abundance of *Juncus acutus* is distinctive.

Habitat
The *Salix-Calliergon* community is characteristic of older dune slacks kept very wet by prolonged flooding with circumneutral ground-waters.

Lengthy inundation through the year is essential for the development of this kind of vegetation: at Braunton Burrows, for example, Willis *et al.* (1959b) encountered it in slacks flooded for up to 8 months and Jones (1992) drew a parallel with the *Carex-Calliergon* vegetation of inland fens where similar assemblages are sustained by fluctuations of small amplitude, at most from 5 cm above the surface to 40 cm below. Critically, the rooting zone is only rarely out of contact with the capillary fringe of the water table (Jones 1992). Such prolonged wetness and the shade cast by the often dense *Salix repens* cover encourage the luxuriant development of the shade-tolerant *Calliergon cuspidatum*, the thick mat of which is itself inimical to the invasion of less competitive plants. This is often a more species-poor assemblage than the *Salix-Campylium* community, for example and, more particularly, it lacks the thalloid liverworts which depend on the more open conditions typical of that kind of slack vegetation.

It is also somewhat less calcicolous than the *Salix-Campylium* community which suggests that the ground waters here are less base-rich, tending perhaps to values below pH 6 which are characteristic of the *Potentilla-Carex* community. At Braunton Burrows, for example, Willis (1985*a*) found calcium levels substantially lower beneath this kind of vegetation than in the sand around which carried *Ammophila*: less than 50 mg g^{-1}, compared to over 70 mg g^{-1}. Sodium and potassium levels, though, were quite high. Similarities to poor-fen vegetation are best seen here in the *Carex nigra* sub-community whereas, in the *Equisetum* and *Carex flacca-Pulicaria* sub-communities, the presence of plants such as *Equisetum variegatum* and *Epipactis palustris* suggests slightly more base-rich conditions. The differences between these last two types of *Salix-Calliergon* vegetation may also relate to the frequency of ground-water fluctuations. The *Equisetum* sub-community, for example, has something of the look of an inundation grassland that experiences more frequent variation in surface wetness.

Conditions akin to those in tall-herb fens, with some moderate enrichment with major nutrients, are perhaps most typical of the *Holcus-Angelica* sub-community, here, particularly where *Phragmites* is present in abundance. However, an additional factor of importance in this kind of *Salix-Calliergon* vegetation may be grazing. Stock or rabbits may actually be effective in hindering the development of a thick and extensive carpet of *Calliergon* in this community (Jones 1992) but, once free of such predation, the maintenance of wet conditions could encourage the spread of helophytes, bulky grasses and tall dicotyledons.

Zonation and succession
The *Salix-Calliergon* community typically occupies older and wetter slacks among stabilised dune systems, quite commonly with other types of slack vegetation in younger and drier hollows disposed according to the age of the dune ridges and the variation in the water-table (Figure 16). Transitions to surrounding dune grasslands depend on the configuration of the slacks and ridges. The high water-table and extensive moss carpet of this

vegetation inhibit colonisation by shrubs and trees but, in drier conditions, grazing by stock and rabbits may also be important in setting back succession.

Where there is variation in hydrological conditions within individual large slacks, different types of *Salix-Calliergon* vegetation can be found in close proximity, grading the one into the other. Sometimes, too, where conditions become somewhat drier, this kind of slack assemblage can grade through the *Equisetum* sub-community to the *Carex flacca* sub-community of the *Potentilla-Carex* slack.

In the opposite direction, where slacks contain stretches of permanent open water, the *Holcus-Angelica* sub-community can pass to some kind of *Phragmites-Eupatorium* fen, *Phragmites* swamp or the *Eleocharitetum palustris*, more particularly the *Littorella* sub-community.

The *Salix-Calliergon* community seems characteristic of late stages in succession (van der Laan 1979): at Kenfig Jones (1992) noted that stands appeared to have taken many years to develop and had not changed noticeably in 8 years. Where the ground continues to be flooded for considerable periods, it seems likely that this kind of vegetation might have some stability. However, where woody plants do get a hold, it is generally *Salix cinerea*, *S. caprea* and *Betula pubescens* that colonise first, giving rise to some type of *Salix-Betula-Phragmites* or *Salix-Galium* woodland. In some sites, patches of these can be seen among stands of *Salix-Calliergon* vegetation.

Distribution
The *Salix-Calliergon* community is one of the more widely distributed kinds of slack vegetation found, for example, on most Welsh dune systems which have slacks, scattered around the coast of England and occurring locally in Scotland.

Affinities
Vegetation of this general type with *Salix repens*, *Carex nigra* and *Calliergon cuspidatum* has been described previously from various sites in Britain (Blanchard 1952, Willis *et al.* 1959*b*, Ranwell 1960*a*) and from the Voorne dunes in The Netherlands by van der Maarel & Westhoff (1964). The latter authors provisionally considered the Dutch vegetation to be part of the *Acrocladio-Salicetum* Braun-Blanquet & de Leeuw 1936, an association of the Caricion davallianae. Following Westhoff & den Held (1969), Schaminée *et al.* (1995) incorporate this assemblage into the *Junco baltici-Schoenetum* Westhoff 1943 and describe a new association, the *Equiseto variegati-Salicetum repentis* Westhoff & Schaminée 1995, which seems more like some kinds of *Salix-Calliergon* slack. Despite its only moderately calcicolous character, the Caricion davallianae seems to be the most appropriate alliance in which to include the community.

Figure 16. Slacks and swamps in the dune system at
Crymlyn Burrows, South Wales.
Well-developed SD2 *Honkenya-Cakile* vegetation is not
present at this site, though there is a persistent kind of
strandline interface around parts of the extensive
embayments of salt-marsh, some of which is clearly
SM24 *Atriplici-Elymetum pycnanthi*. Most of the more
mobile dunes are occupied by various kinds of SD6
Ammophila vegetation, with a small area of SD4
Elymus farctus foredune at the seaward point of the
main spit. Behind are extensive stretches of the SD7
Ammophila-Festuca community on semi-fixed sand
with SD8 *Festuca-Galium* grassland on some areas of
fixed dunes, SD12 *Carex-Festuca-Agrostis* grassland on

more acid sands where there is grazing. Much of the
less heavily grazed dune hinterland has SD9
Ammophila-Arrhenatherum vegetation widely colonised
by various kinds of scrub. Low-lying areas with a high
water table have SD15 *Salix-Calliergon*, SD16 *Salix-
Holcus* and SD17 *Potentilla-Carex* vegetation
distributed according to the degree of wetness and the
base status of the flooding waters. The slacks are
backed by extensive areas of S4 *Phragmites* swamp
which pass to salt-marsh through the *Oenanthe* sub-
community of SM18 *Juncus maritimus* vegetation.
(Redrawn from Dargie 1990, by permission of the Joint
Nature Conservation Committee)

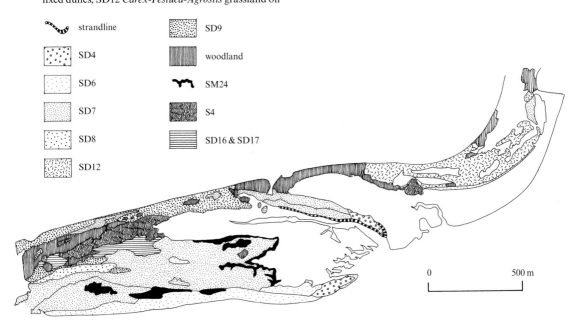

Floristic table SD15

	a	b	c	d	15
Salix repens	V (1–10)	V (3–10)	V (3–10)	V (1–10)	V (1–10)
Calliergon cuspidatum	V (1–10)	V (3–10)	V (4–10)	IV (4–10)	V (1–10)
Hydrocotyle vulgaris	V (1–9)	V (3–9)	V (1–10)	III (1–6)	V (1–10)
Mentha aquatica	IV (1–7)	V (1–7)	IV (1–7)	IV (2–6)	IV (1–7)
Rubus caesius	V (1–8)	V (1–8)	I (1–5)	III (1–8)	III (1–8)
Galium palustre	IV (1–5)	V (1–5)	III (1–4)	I (1–2)	III (1–5)
Carex nigra	IV (1–10)	III (1–7)	II (3–5)	II (2–6)	III (1–10)
Scutellaria galericulata	III (1–4)	II (1–4)		I (2)	II (1–4)
Lysimachia vulgaris	II (3–7)				I (3–7)
Equisetum fluviatile	II (2–5)				I (2–5)
Agrostis stolonifera	II (1–4)	IV (1–8)	IV (1–4)	II (3–6)	III (1–8)
Carex arenaria	II (1–4)	IV (1–5)	III (1–7)	II (1–5)	III (1–7)
Equisetum variegatum	I (1–7)	IV (2–9)	V (2–9)	I (1–2)	III (1–9)
Potentilla anserina	II (1–5)	III (1–6)	II (1–4)	I (2–5)	II (1–6)
Ranunculus repens	I (1–5)	III (1–6)	II (1–4)	I (1–3)	II (1–6)
Trifolium repens		II (1–7)	I (3–5)	I (2–4)	I (1–7)
Parnassia palustris		I (3–6)			I (3–6)
Linum catharticum		I (2)			I (2)
Homalothecium lutescens		I (4–5)			I (4–5)
Cladium mariscus		I (7–9)			I (7–9)
Carex flacca	I (1–4)	III (1–5)	V (1–5)	III (2–4)	III (1–5)
Epipactis palustris	I (3–4)	III (1–6)	IV (1–8)	I (1–4)	II (1–8)
Pulicaria dysenterica	I (1–4)	I (1–3)	IV (1–6)	IV (1–6)	II (1–6)
Eupatorium cannabinum	I (1–8)	I (1–2)	III (1–4)	III (1–6)	II (1–8)
Ranunculus flammula	I (1–4)	II (1–3)	III (1–3)	I (1–3)	II (1–4)
Oenanthe lachenalii			II (1–3)	I (1)	I (1–3)
Glaux maritima			I (2–4)		I (2–4)
Holcus lanatus	I (1–4)	I (1–3)	II (1–4)	IV (1–7)	II (1–7)
Angelica sylvestris	I (1–4)	I (4)	I (1–4)	IV (1–4)	II (1–4)
Phragmites australis	I (1–5)	I (4)	I (1–5)	III (2–10)	II (1–10)

Floristic table SD15 (cont.)

	a	b	c	d	15
Succisa pratensis			II (1–5)	III (1–6)	II (1–6)
Molinia caerulea		I (1)	I (1–2)	III (3–8)	II (1–8)
Eurhynchium praelongum	I (3)	I (3–5)	I (3)	II (3–8)	I (3–8)
Juncus acutus	I (8)	I (1)	I (3–5)	II (1–5)	I (1–8)
Cirsium palustre	I (1–5)		I (1–4)	II (1–5)	I (1–5)
Salix caprea		I (3)	I (1–4)	II (1–7)	I (1–7)
Lotus corniculatus		I (1–5)	I (2–4)	II (3–7)	I (1–7)
Vicia sativa nigra			I (3)	II (1–2)	I (1–3)
Vicia cracca				I (4)	I (4)
Arrhenatherum elatius				I (3–5)	I (3–5)
Equisetum palustre	IV (1–9)	II (1–4)	II (1–5)	IV (1–6)	III (1–9)
Lotus uliginosus	II (1–5)	II (1–5)	II (1–5)	III (1–4)	II (1–5)
Epilobium palustre	I (1–3)	II (1–3)	I (1–3)	II (1–4)	II (1–4)
Festuca rubra	I (1–5)	I (2–4)	II (1–8)	II (1–5)	I (1–8)
Poa pratensis	I (1–5)	I (1–3)	II (1–8)	II (1–8)	I (1–8)
Juncus maritimus		I (3–4)	II (4–8)	II (3–7)	I (3–8)
Juncus articulatus	I (1–3)	I (2–4)	I (1–3)	I (1–2)	I (1–4)
Lycopus europaeus	I (1–4)	I (2–4)	I (1–4)	I (4–6)	I (1–6)
Prunella vulgaris	I (3)	I (3–4)	I (1–4)	I (1–3)	I (1–4)
Juncus inflexus	I (1–5)	I (1–4)	I (1–2)	I (1–6)	I (1–6)
Iris pseudacorus	I (1–5)	I (1)	I (1–2)	I (4)	I (1–5)
Filipendula ulmaria	I (1)	I (2–3)	I (1)	I (1–8)	I (1–8)
Cardamine pratensis	I (1–4)	I (2–3)	I (3)		I (1–4)
Ranunculus acris	I (1–2)	I (2–3)	I (1–4)		I (1–4)
Potentilla reptans	I (1–5)	I (1–5)		I (1–3)	I (1–5)
Lychnis flos-cuculi	I (1–5)	I (1–4)		I (3–5)	I (1–5)
Drepanocladus sendtneri	I (3–10)	I (1–5)	I (3)		I (1–10)
Ophioglossum vulgatum	I (1–5)	I (1–4)		I (2–3)	I (1–5)
Dactylorhiza incarnata	I (1)	I (1–3)	I (1)		I (1–3)
Campylium stellatum	I (10)	I (3–10)	I (6–8)		I (3–10)
Carex panicea	I (2–4)	I (2)		I (2)	I (2–4)
Equisetum arvense	I (1–4)	I (3)		I (2–3)	I (1–4)

	a	b	c	d	15
Dactylorhiza majalis praetermissa	I (1)	I (1–3)	I (1–3)		I (1–3)
Eleocharis palustris	I (1–3)	I (1–2)	I (1)		I (1–3)
Danthonia decumbens	I (4)	I (4)	I (2–3)		I (2–4)
Anagallis tenella		I (3)	I (1–3)	I (1)	I (1–3)
Lathyrus pratensis	I (1)	I (2–3)		I (3)	I (1–3)
Pseudoscleropodium purum		I (3)	I (6)	I (4–5)	I (3–6)
Senecio jacobaea		I (2)	I (1–2)	I (1)	I (1–2)
Cirsium arvense	I (2–3)		I (3)	I (1–3)	I (1–3)
Juncus subnodulosus	I (6)		I (3)	I (7–8)	I (3–8)
Solanum dulcamara	I (1)	I (1)		I (5)	I (1–5)
Agrostis capillaris	I (3)	I (2–4)			I (2–4)
Epilobium parviflorum	I (2–5)	I (1–2)			I (1–5)
Carex hirta	I (3–4)	I (1–4)			I (1–4)
Caltha palustris	I (3)	I (4–5)			I (3–5)
Drepanocladus lycopodiodes	I (1)	I (5–7)			I (1–7)
Lophocolea bidentata	I (3–4)	I (3–4)		I (3–4)	I (3–4)
Agrimonia eupatoria	I (4)	I (4)		I (1)	I (1–4)
Eriophorum angustifolium	I (3–5)	I (3–5)	I (3)		I (3–5)
Scirpus maritimus	I (5–10)	I (5–10)	I (6)		I (5–10)
Trifolium pratense		I (1–4)	I (1–4)		I (1–4)
Leontodon autumnalis		I (1–4)	I (1–2)		I (1–4)
Glechoma hederacea		I (1–3)		I (4)	I (1–4)
Agrostis canina	I (5)	I (5)		I (4)	I (4–5)
Brachythecium mildeanum	I (4)	I (4)		I (4)	I (4)
Plantago lanceolata		I (1–3)	I (1–3)	I (1–2)	I (1–3)
Centaurea nigra		1 (1–4)	I (3–4)	I (1)	I (1–4)
Lythrum salicaria		I (1–3)	I (1–3)	I (3)	I (1–3)
Samolus valerandi		I (1–2)	I (1–2)	I (1–3)	I (1–3)
Number of samples	81	57	48	33	229
Number of species/sample	12 (6–19)	14 (8–23)	16 (9–27)	17 (10–25)	14 (6–27)

a   *Carex nigra* sub-community

b   *Equisetum variegatum* sub-community

c   *Carex flacca-Pulicaria dysenterica* sub-community

d   *Holcus lanatus-Angelica sylvestris* sub-community

15   *Salix repens-Calliergon cuspidatum* dune-slack (total)

SD16

Salix repens-Holcus lanatus dune-slack community

Synonymy
Braunton Damp Pasture Willis *et al.* 1959; *Salix repens-Holcus lanatus* nodum Jones 1992; *Festuca rubra-Brachythecium rutabulum* nodum Jones 1992.

Constant species
Carex flacca, Festuca rubra, Holcus lanatus, Lotus corniculatus, Salix repens.

Physiognomy
In the *Salix repens-Holcus lanatus* dune-slack community, *Salix repens* is generally dominant in a bushy canopy that can be several decimetres tall. Its most frequent associates are *Holcus lanatus* and *Festuca rubra*, the abundance of which often give a rank grassy aspect to the sward among the willow. *Agrostis stolonifera* and *Poa pratensis* are at least occasional throughout but also common among the associates are *Carex flacca* and a range of dicotyledonous herbs including *Lotus corniculatus, Euphrasia officinalis* agg., *Hieracium pilosella, Senecio jacobaea, Prunella vulgaris, Leontodon autumnalis* and *Epipactis palustris. Ononis repens* is also quite frequent overall but rather patchy in its representation among the various sub-communities and not consistently abundant.

Bryophytes are not typically a prominent element in the vegetation but *Bryum pseudotriquetrum* is occasional and there is sometimes patchily prominent *Calliergon cuspidatum* and *Campylium stellatum*.

Sub-communities

Ononis repens sub-community. *Ononis repens* is at its most frequent and abundant here with preferentially common *Carex arenaria, Hypochoeris radicata* and *Galium verum.* Young shrubs and trees are quite frequent too, with bushes or saplings of *Salix caprea, Betula pubescens* and *Quercus robur.*

Rubus caesius sub-community. *O. repens* and saplings of the above woody species remain occasional to sparse

here but *Rubus caesius* is constant with occasional *Potentilla anserina.*

Prunella vulgaris-Equisetum variegatum sub-community. *Campylium stellatum-Salix repens* species-rich nodum, dry sub-type Jones 1992 *p.p.* This is the most species-rich type of *Salix-Holcus* vegetation in which *Poa pratensis* and *Prunella vulgaris* have a peak in their frequency but where especially distinctive is the common occurrence of *Equisetum variegatum, Pyrola rotundifolia, Trifolium pratense, Fragaria vesca* and the bryophytes *Brachythecium rutabulum, Amblystegium serpens, Eurhynchium praelongum* and *Lophocolea bidentata s.l.* Thalloid liverworts like *Pellia endiviifolia, Riccardia chamaedryfolia* and *Moerckia hibernica* occur occasionally.

Agrostis stolonifera sub-community. Here, *F. rubra* is replaced by *A. stolonifera* as the commonest grass in the vegetation and preferentials of the other sub-communities are almost all very sparse. Distinctive here are *Hydrocotyle vulgaris, Juncus articulatus, Leontodon taraxacoides* and, less frequently, *Carex serotina, Anagallis tenella, Ranunculus flammula* and *Dactylorhiza incarnata.* Particularly scarce but striking are *Samolus valerandi, Parnassia palustris* and *Petalophyllum ralfsii.*

Habitat
The *Salix-Holcus* community is characteristic of older and drier dune slacks, rarely flooded to any great extent, even in wetter winters, and often accessible to grazing stock and rabbits throughout the year.

Little systematic information is available about the flooding regime which helps sustain this kind of vegetation but data from Ranwell (1972) and Jones (1992) suggest that the water table is from 50 cm down to 2 m below the surface in the summer months, that is, beyond capillary contact with the rooting zone in the growing season. Winter flooding is rare and generally brief so the soil profile is almost never in a reducing state. Such con-

ditions are what give more mesophytic species an opportunity to contribute substantially to the sward, even
where *S. repens* remains abundant, and limit the extent
of a mat of moisture-demanding bryophytes and many
of the herbs typical of wetter slacks. Only in the *Agrostis
stolonifera* sub-community does this last aspect of the
vegetation become more conspicuous.

The community occurs commonly on dune systems
where grazing occurs and stock, or rabbits where they
exist, can play some part in keeping this vegetation more
open and diverse. Where grazing is reduced, this kind of
dune slack is always susceptible to invasion of shrubs
and trees, a process already in train in the *Ononis* sub-
community.

Zonation and succession

The *Salix-Holcus* community is a widespread type of
slack vegetation occupying the older dry slacks in large
and complex dune systems and sometimes comprising
the bulk of cover between stable dune ridges. Through
the *Agrostis* and *Poa* sub-communities, it can grade to
wetter slack vegetation in areas where the water-table
breaches the surface in winter but often it occupies the
entire area of individual slacks, the most obvious zonations being to the *Festuca-Galium* grassland on grazed
dune ridges. With the shift to drier ground, the dominance of *Salix repens* declines and a mixture of grasses
and smaller dicotyledons assumes dominance. Where
the *Prunella-Equisetum* sub-community of the *Salix-
Holcus* slack gives way to the *Prunella* type of *Festuca-
Holcus* sward or the *Ononis* and *Rubus* sub-communities
of the former pass to Typical *Festuca-Holcus* grassland,
the shift can be quite gradual apart from the loss of *Salix
repens*. Where grazing is less intensive, the *Salix-Holcus*
slack can give way to *Ammophila-Arrhenatherum* grassland.

The *Salix-Holcus* community is a later stage in the
development of slack vegetation, probably succeeding
the *Sagina-Bryum* vegetation quite quickly where inundation ceases, replacing wetter slack communities more
slowly as the surface dries. The lowering of the ground
water table can play an obvious part in such a process
but trapping of sand by *S. repens* and upbuilding of a
mat upon which grasses can root may also be important.
Where the canopy is a little more open, invasion of
bushy *Salix* spp. and *Betula pubescens* can initiate the
development of woodland. Where *Hippophae* occurs
locally, it can supervene in such successions.

Distribution

The *Salix-Holcus* community is widespread and
common and appears to be the most extensive kind of
slack vegetation around the Welsh and English coasts
and in south-east Scotland.

Affinities

Van der Maarel & Westhoff (1964), and London (1971)
characterised a variety of communities from southern
Dutch dunes where *S. repens*, *H. lanatus* and *F. rubra* are
the dominants whose affinities seem partly with the
Caricion davallianae, partly with the Elymo-Rumicion.
As Jones (1992) points out, there are also strong links
with the Festuco-Plantaginion swards of the *Festuca-
Galium* grassland. However, the prominence of *Pyrola
rotundifolia* in Welsh stands of the *Salix-Holcus* slack
community led Jones (1992) to equate at least some of
this vegetation with the *Pyrolo-Salicetum*, a very heterogeneous syntaxon in Continental descriptions. On
balance, it seems sensible to locate this community
among the scrubs of sandy substrates in the Salicion
repentis arenariae Tüxen 1952.

Floristic table SD16

	a	b	c	d	16
Salix repens	V (1–10)	V (2–10)	V (4–10)	V (4–9)	V (1–10)
Holcus lanatus	IV (1–7)	V (1–8)	V (3–8)	III (1–6)	IV (1–8)
Lotus corniculatus	IV (2–9)	III (1–7)	IV (1–5)	V (1–5)	IV (1–9)
Festuca rubra	IV (1–7)	IV (2–9)	IV (1–9)	I (3–5)	IV (1–9)
Carex flacca	II (2–5)	III (1–5)	V (1–6)	V (1–6)	IV (1–6)
Ononis repens	IV (1–8)	II (1–5)	III (1–6)	I (1)	III (1–8)
Carex arenaria	IV (1–8)	II (2–9)	II (1–4)	II (2–5)	III (1–9)
Hypochoeris radicata	III (1–4)	II (1–4)	I (3)	I (2–3)	II (1–4)
Salix caprea	III (1–7)	I (1–5)	I (1–4)		I (1–7)
Betula pubescens sapling	II (1–8)	I (6)	I (1)		I (1–8)
Galium verum	II (1–5)	I (3–5)	I (3)		I (1–5)
Quercus robur sapling	II (1–4)	I (1)			I (1–4)
Rubus caesius	II (1–6)	IV (1–8)	II (1–5)	I (3–5)	II (1–8)
Potentilla anserina	I (4)	II (1–7)	I (3)	I (1–5)	I (1–7)
Juncus maritimus		I (1–8)			I (1–8)
Juncus inflexus		I (1)			I (1)
Poa pratensis	III (1–8)	III (2–7)	V (1–6)	I (1–4)	III (1–8)
Prunella vulgaris	II (2–5)	II (2–4)	IV (1–5)	III (1–4)	III (1–5)
Leontodon hispidus	II (1–7)	I (1)	IV (1–7)		II (1–7)
Equisetum variegatum	I (1–4)	I (3–4)	IV (1–5)		II (1–6)
Brachythecium rutabulum	I (2)	I (1–3)	IV (2–10)	II (2–6)	I (1–10)
Pyrola rotundifolia	I (1–8)	I (1–4)	III (1–7)	I (1–4)	I (1–8)
Trifolium pratense	I (4)	I (3–5)	III (2–7)	I (4)	I (2–7)
Amblystegium serpens		I (1)	III (1–4)	I (4–5)	I (1–5)
Eurhynchium praelongum		I (2)	III (1–7)		I (1–7)
Fragaria vesca	I (3–4)		III (1–6)		I (1–6)
Lophocolea bidentata s.l.			III (1–6)		I (1–6)
Pulicaria dysenterica		I (1–3)	II (1–8)	I (2)	I (1–8)
Rhinanthus minor	I (3)	I (1–4)	II (1–4)	I (1–2)	I (1–4)
Sonchus arvensis	I (1)		II (1–5)		I (1–5)
Riccardia chamaedryfolia			II (1–3)	I (4)	I (1–4)

Species	1	2	3	4
Luzula campestris				I (1–4)
Pellia endiviifolia				I (1–4)
Moerckia hibernica				I (1–3)
Agrostis stolonifera	II (2–7)	II (1–4)	V (3–7)	II (1–7)
Hydrocotyle vulgaris	II (1–5)	II (1–6)	III (1–6)	II (1–6)
Juncus articulatus	I (1–4)	I (2–4)	III (2–6)	I (1–6)
Leontodon taraxacoides	I (1–2)		III (1–5)	I (1–5)
Carex serotina	I (1)	I (1–3)	II (2–5)	I (1–5)
Dactylorhiza incarnata	I (3)	I (1)	II (1–4)	I (1–4)
Anagallis tenella	I (7)	I (3)	II (2–6)	I (2–7)
Ranunculus flammula		I (1)	II (1–3)	I (1–3)
Samolus valerandi			I (1–3)	II (1–3)
Petalophyllum ralfsii			I (2–3)	I (2–3)
Parnassia palustris			I (1–4)	I (1–4)
Euphrasia officinalis agg.	III (1–6)	III (1–4)	I (1–4)	II (1–6)
Hieracium pilosella	III (2–5)	III (1–4)	I (1–3)	II (1–5)
Bryum pseudotriquetrum	II (2–5)	II (1–7)	II (1–9)	III (1–9)
Epipactis palustris	II (1–8)	I (1–4)	II (1–4)	II (1–8)
Leontodon autumnalis	II (3–4)	II (1–4)	I (1–4)	I (1–4)
Senecio jacobaea	II (1–4)	II (1–4)		I (1–4)
Anthyllis vulneraria	I (5)	I (1–9)	I (2)	I (1–9)
Mentha aquatica	II (2–5)	II (1–4)	I (1–3)	I (1–5)
Ranunculus repens	I (3)	II (1–3)	I (1)	I (1–8)
Plantago lanceolata	I (2–5)	II (1–6)	I (2)	I (1–6)
Trifolium repens	I (3)	II (1–4)	I (1–4)	I (1–5)
Linum catharticum	II (1–3)	II (1–3)	I (2)	I (1–3)
Calliergon cuspidatum	I (2–3)	II (1–10)	II (1–8)	I (1–10)
Campylium stellatum	I (4–9)	II (1–5)	II (1–7)	I (1–7)
Sagina nodosa	I (4)	II (1–3)	I (1–4)	I (1–4)
Bellis perennis	I (1–3)	I (1–4)	I (3)	I (1–4)
Carlina vulgaris	I (1–4)	I (1)	I (2)	I (1–3)
Juncus acutus	I (2)	I (4–5)	I (2)	I (1–5)
Ranunculus acris	I (4–5)	I (1–3)	I (1)	I (1–4)
Erigeron acer	I (1–4)	I (1)	I (1–2)	I (1–3)
Trifolium dubium	I (3)	I (4)	I (1)	I (1–4)

Floristic table SD16 (*cont.*)

	a	b	c	d	16
Carex nigra	I (3–4)	I (3)		I (1–5)	I (1–5)
Pseudoscleropodium purum	I (3–8)	I (3–5)	I (3–10)		I (3–10)
Cynosurus cristatus	I (3–5)	I (3)	I (3–4)		I (3–5)
Epilobium palustre	I (1–3)	I (2–3)	I (3)		I (1–3)
Ammophila arenaria	I (1–3)	I (2–3)	I (4)		I (1–4)
Cerastium fontanum	I (1–3)	I (1–3)	I (1–3)		I (1–3)
Epipactis helleborine	I (1–3)	I (2)	I (1–2)		I (1–3)
Viola riviniana	I (3)	I (3)	I (1–5)		I (1–5)
Taraxacum officinale agg.	I (1)		I (1–2)	I (3)	I (1–3)
Hypnum cupressiforme	I (3–8)	I (1–8)		I (1)	I (1–8)
Brachythecium albicans	I (4–8)		I (4)	I (1)	I (1–8)
Gentianella amarella	I (2–4)		I (1–4)	I (1)	I (1–4)
Equisetum palustre		I (1–7)	I (2–4)	I (1)	I (1–7)
Equisetum arvense		I (1–3)	I (3)	I (1–3)	I (1–3)
Agrostis capillaris		I (2–4)	I (3–4)	I (1–3)	I (1–4)
Thymus praecox	I (3–5)	I (5)			I (3–5)
Poa annua		I (2–3)	I (2)		I (2–3)
Eupatorium cannabinum		I (1–6)	I (2–4)		I (1–6)
Phragmites australis	I (1–4)	I (4)	I (1–10)		I (1–10)
Tortula ruralis ruraliformis	I (2–6)		I (3)		I (2–6)
Cirsium arvense		I (1–4)		I (3)	I (1–4)
Crepis capillaris	I (3–4)	I (3)			I (3–4)
Eleocharis quinqueflora			I (4)	I (3)	I (3–4)
Number of samples	67	47	58	31	203
Number of species/sample	16 (8–28)	15 (6–36)	22 (14–32)	15 (7–29)	17 (6–36)

a *Ononis repens* sub-community
b *Rubus caesius* sub-community
c *Prunella vulgaris-Equisetum variegatum* sub-community
d *Agrostis stolonifera* sub-community
16 *Salix repens-Holcus lanatus* dune-slack (total)

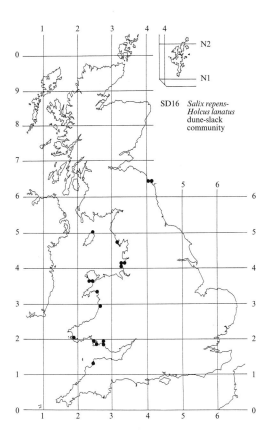

SD16 *Salix repens-*
Holcus lanatus
dune-slack
community

SD17

Potentilla anserina-Carex nigra dune-slack community

Constant species
Agrostis stolonifera, Carex nigra, Potentilla anserina, Calliergon cuspidatum.

Physiognomy
The *Potentilla anserina-Carex nigra* dune-slack community comprises generally closed swards up to a decimetre or so high dominated by mixtures of grasses, sedges and rosette or mat-forming herbs. Especially distinctive is the combination of *Agrostis stolonifera* and *Carex nigra* with *Potentilla anserina*, each of these sometimes present in abundance and the last particularly striking when dominating in a ground carpet. *Festuca rubra* and *Holcus lanatus* are also very frequent in certain of the sub-communities and there is occasionally some *Juncus articulatus*. *Salix repens* is typically of limited occurrence here, though it is frequent and locally abundant in one type of *Potentilla-Carex* slack.

The other most frequent vascular associates overall are *Ranunculus repens, Trifolium repens, Cardamine pratensis, Equisetum palustre* and *Euphrasia officinalis* agg. Tall herbs like *Iris pseudacorus, Angelica sylvestris* and *Filipendula ulmaria* can give a locally distinctive stamp to the vegetation while species such as *Elymus repens, Triglochin maritima* and *Carex disticha* may lend a hint of the upper salt-marsh.

The only bryophyte which is at all frequent in the community is *Calliergon cuspidatum* and it can be locally abundant. Other species like *Plagiomnium rostratum, Brachythecium rutabulum* and *Eurhynchium praelongum* occur very sparsely.

Sub-communities

***Festuca-rubra-Ranunculus repens* sub-community.** In this grassy kind of *Potentilla-Carex* vegetation, *Festuca rubra* is very frequent and of quite high cover with common records too for *Ranunculus repens, Bellis perennis* and *Trifolium repens*. More occasional are *Holcus lanatus, Carex panicea, Rumex crispus, Glaux maritima*

and *Taraxacum officinale* agg. with *Sagina procumbens* sometimes finding a hold in more open patches.

***Carex flacca* sub-community.** *F. rubra, T. repens* and *Holcus lanatus* all remain frequent here and *Poa pratensis* is also weakly preferential, but most particularly diagnostic is the common occurrence of *Carex flacca* and *Prunella vulgaris* with occasional records for *Plantago maritima, Carex arenaria, Lathyrus pratensis* and *Lotus corniculatus*. There are also sometimes young *Salix caprea* and shoots of *Rubus caesius. Parnassia palustris* can be found here, too, at low frequency.

***Caltha palustris* sub-community.** Among the various grasses of the community, *Holcus lanatus* is the commonest here but *Cynosurus cristatus* also becomes frequent with *Poa trivialis* and *Anthoxanthum odoratum* occasional. More striking, however, is the high frequency of *Caltha palustris, Rhinanthus minor* and *Lychnis flos-cuculi* with *Epilobium palustre, Rumex acetosa, Pedicularis palustris, Leontodon autumnalis, Plantago lanceolata, Vicia cracca, Cerastium fontanum, Dactylorhiza fuchsii* and *Equisetum fluviatile* occasional.

***Hydrocotyle vulgaris-Ranunculus flammula* sub-community.** Almost all the preferentials of the other types of *Potentilla-Carex* vegetation are very scarce in this sub-community which is characterised by high frequencies of *Hydrocotyle vulgaris, Ranunculus flammula, Eleocharis palustris* and *Galium palustre*. It is here, too, that *Salix repens* has its best representation in the community as a sometimes patchily extensive cover of low bushes. *Mentha aquatica* and *Equisetum variegatum* are occasional.

Habitat
The *Potentilla-Carex* community is characteristic of damp or wet dune slacks kept moist by the fluctuation of less base-rich ground waters, particularly in the moister climate of northern Britain.

No qualitative data are available on the flooding regime associated with this kind of slack vegetation but it seems clear that the inundations can be substantial, perhaps up to 50 cm depth, lengthy and of considerable amplitude. However, exposure of the vegetation to water shortage when the ground water level falls is offset by the high rainfall experienced in northern Britain which keeps the swards damp and replenishes the underlying supplies.

Among the various sub-communities, the *Caltha* and *Hydrocotyle-Ranunculus* types seem to be associated with wetter ground, the former perhaps where the waters and sands are a little more enriched. The *Carex flacca* sub-community is from drier situations and the *Festuca-Ranunculus* type can have some lingering saline influence.

Very commonly, this kind of slack vegetation is accessible to grazing stock, forming part of the machair pastures in western Scotland. This contributes to the floristics of the community, particularly in drier situations, and helps set back any invasion of shrubs and trees.

Zonation and succession

Among complexes of dune slacks, the *Potentilla-Carex* community occupies wetter and less base-rich hollows, where there can be considerable fluctuations in ground water levels, generally among stable dune ridges with fixed grasslands. Repeated substantial inundation help maintain the community but where drying conditions combine with a reduction in grazing, this vegetation can be readily invaded by shrubs and trees.

Shifts from the *Potentilla-Carex* community to the *Festuca-Galium* grassland, which provides its usual context, can be quite sharp where there is a marked change in slope and ground water levels on the surrounding dune ridges. Often, however, particularly on the sand-plains of north-west Scotland where this kind of slack vegetation is especially common, the shifts are gradual. Then, the *Carex flacca* sub-community of the

Potentilla-Carex vegetation can pass gently to the *Prunella* sub-community of the *Festuca-Galium* grassland over undulating machair whose whole patchwork of swards is used for extensive pasturing.

Very commonly, too, in Scotland, the *Potentilla-Carex* community forms part of a continuum with wetter grasslands to the vegetation of swamps and mires (Dargie 1993). Then it is particularly the *Caltha* and *Hydrocotyle-Ranunculus* sub-communities that play a prominent role, the latter often passing to the *Eleocharitetum* swamp in open waters with *Phragmitetum* more locally. The *Festuca-Agrostis-Potentilla* grassland and something like the *Caltha-Cynosurus* pasture can figure with the former, though data accumulated since the NVC suggests the existence of undescribed assemblages of this general type in which *Carex nigra* continues to play a prominent role (Dargie 1993).

Where stagnation of less base-rich waters occurs in the wetter hollows of dune systems in northern Britain, peat can accumulate and sustain wet heath of the *Scirpus-Erica* or *Erica-Sphagnum* types. With these vegetation types, the *Potentilla-Carex* community can form complex mosaics and transitions.

Distribution

The *Potentilla-Carex* community is widespread in dune systems around Britain but especially important to the north where it comprises the commonest and most extensive vegetation type in slacks and damp machair depressions.

Affinities

Among British dune slacks, this vegetation type is distinct from other communities in its poor representation of *Salix repens* and its overall similarity to inundation grasslands of the Elymo-Rumicion and, on balance, it is much better placed in the latter alliance than either the Caricion davallianae or Caricion nigrae. Further sampling and analysis are needed to clarify relationships among these swards and the Calthion grasslands.

Floristic table SD17

	a	b	c	d	17
Potentilla anserina	V (3–8)	IV (2–8)	IV (4–8)	IV (1–9)	IV (1–9)
Calliergon cuspidatum	IV (2–8)	V (2–9)	IV (2–8)	IV (3–10)	IV (2–10)
Carex nigra	III (4–8)	IV (2–9)	V (2–9)	V (1–10)	IV (1–10)
Agrostis stolonifera	V (4–8)	III (3–6)	III (3–6)	IV (1–9)	IV (1–9)
Festuca rubra	IV (3–8)	IV (2–8)	II (3–5)	I (1)	III (1–8)
Ranunculus repens	IV (2–9)	III (2–8)	III (2–7)	II (1–6)	III (1–9)
Bellis perennis	III (3–8)	II (1–3)	I (2–3)	I (1–3)	II (1–8)
Carex panicea	II (2–6)	I (1–5)	I (2–3)	I (2–6)	I (1–6)
Sagina procumbens	II (2–5)	I (1–3)	I (3–5)		I (1–5)
Rumex crispus	II (2–4)		I (2–3)	I (1–3)	I (1–4)
Glaux maritima	II (3–4)		I (2)		I (2–4)
Taraxacum officinale agg.	II (3–4)	I (1–5)	I (2)	I (2)	I (1–5)
Trifolium repens	III (3–7)	IV (1–7)	III (2–7)	I (2)	III (1–7)
Carex flacca	I (3–8)	IV (1–8)	I (2–5)	I (2–4)	II (1–8)
Poa pratensis	II (2–5)	III (1–6)	I (2–5)	I (1)	II (1–6)
Prunella vulgaris	I (2)	III (1–9)	I (3–4)	I (1–3)	I (1–9)
Plantago maritima	I (3)	II (1–7)	I (3)		I (1–7)
Lathyrus pratensis	I (4–5)	II (2–6)	I (2)		I (2–6)
Carex arenaria	I (3)	II (1–8)		I (1–5)	I (1–8)
Rubus caesius		II (4–8)		I (1–5)	I (1–8)
Epipactis palustris		II (1–4)		I (3–4)	I (1–4)
Lotus corniculatus		II (1–4)			I (1–4)
Salix caprea		II (1–5)			I (1–5)
Parnassia palustris	I (3)	I (1–5)			I (1–5)
Juncus inflexus		I (1–4)		I (3)	I (1–4)
Linum catharticum	I (3–5)	I (2–5)	I (3)		I (2–5)
Galium verum		I (2–5)			I (2–5)
Holcus lanatus	II (2–6)	III (2–7)	IV (2–5)	I (1–3)	III (1–7)
Caltha palustris	II (2–7)	I (1–3)	IV (2–6)	I (1–3)	II (1–7)
Cynosurus cristatus	I (3)	I (3–8)	III (2–7)	I (1)	II (1–8)
Rhinanthus minor	I (2–3)	II (2–5)	III (3–5)		II (2–5)

Species					
Lychnis flos-cuculi		I (2–3)	III (1–4)	I (3)	I (1–4)
Poa trivialis	I (3–4)	I (2)	II (2–6)	I (1)	I (1–6)
Epilobium palustre	I (3)	I (1–3)	II (2–4)	I (1–6)	I (1–6)
Leontodon autumnalis	I (1–2)	I (1–4)	II (2–3)	I (2)	I (1–4)
Plantago lanceolata	I (2–3)	I (2–3)	II (2–6)	I (3)	I (2–6)
Vicia cracca	I (2–8)	I (2–6)	II (2–8)		I (2–8)
Cerastium fontanum	I (2–3)	I (1)	II (2–3)		I (1–3)
Anthoxanthum odoratum	I (3)		II (2–9)	I (1)	I (1–9)
Rumex acetosa		I (1–3)	II (2–4)		I (1–4)
Dactylorhiza fuchsii			II (1–3)	I (3)	I (1–3)
Pedicularis palustris			II (2–5)	I (1–3)	I (1–5)
Equisetum fluviatile			II (2–7)	I (1–5)	I (1–7)
Hydrocotyle vulgaris	I (3)	III (3–5)	I (3–7)	V (2–10)	III (2–10)
Ranunculus flammula	I (4)	I (1–3)	I (2)	IV (1–4)	II (1–4)
Eleocharis palustris	I (3–8)	I (3)	II (2–7)	III (1–8)	II (1–8)
Galium palustre	I (3)	I (4)	I (3)	III (1–6)	II (1–6)
Salix repens		I (1–8)		III (1–9)	I (1–9)
Mentha aquatica	I (2–4)	I (3–5)		II (1–4)	I (1–5)
Equisetum variegatum		I (1–4)		II (3–9)	I (1–9)
Drepanocladus sendtneri				I (4–10)	I (4–10)
Campylium stellatum				I (3–9)	I (3–9)
Juncus articulatus	II (2–9)	II (1–6)	II (2–7)	II (1–4)	II (1–9)
Cardamine pratensis	II (2–6)	II (1–3)	II (2–4)	II (1–5)	II (1–6)
Equisetum palustre	I (3–8)	II (1–3)	II (2–7)	II (1–5)	II (1–8)
Euphrasia officinalis agg.	I (2–3)	II (1–5)	II (2–4)		I (1–5)
Angelica sylvestris	I (2)	I (2–3)	I (3–5)	I (1)	I (1–5)
Plagiomnium rostratum	I (3)	I (3)	I (3–5)	I (3)	I (3–5)
Myosotis laxa caespitosa	I (3)	I (2)	I (2–3)	I (1–4)	I (1–4)
Rhytidiadelphus squarrosus	I (3–5)	I (3–5)	I (2–3)		I (2–5)
Elymus repens	I (7)	I (3)	I (5)		I (3–7)
Equisetum arvense	I (2–4)	I (1–6)	I (4)		I (1–6)
Brachythecium rutabulum	I (2–4)	I (7)	I (3–5)		I (2–7)
Triglochin maritima	I (2)	I (4)	I (3–4)		I (2–4)
Lolium perenne	I (3)	I (4)	I (3–5)		I (3–5)
Poa annua	I (3)	I (3)	I (2)		I (2–3)

Floristic table SD17 (*cont.*)

	a	b	c	d	17
Centaurea nigra	I (3)	I (5)	I (3–4)		I (3–5)
Eurhynchium praelongum	I (4)	I (4)	I (3)		I (3–4)
Agrostis capillaris	I (2)	I (2–5)		I (5)	I (2–5)
Iris pseudacorus	I (3–8)		I (1–5)	I (1–7)	I (1–8)
Filipendula ulmaria	I (4)		I (3)	I (2–4)	I (2–4)
Cratoneuron filicinum	I (3–4)		I (3–4)	I (4)	I (3–4)
Dactylorhiza incarnata		I (1–2)	I (1–2)	I (1)	I (1–2)
Lophocolea bidentata s.l.		I (3)	I (2–5)	I (3)	I (2–5)
Dactylorhiza majalis purpurella		I (1–2)	I (2–3)	I (1)	I (1–3)
Polygonum persicaria	I (3)	I (2)	I (2–5)	I (3–4)	I (2–5)
Juncus effusus		I (3–4)	I (2–5)	I (3)	I (2–5)
Carex disticha		I (5–7)	I (3–4)	I (5)	I (3–7)
Homalothecium lutescens	I (2–3)	I (3–5)			I (2–5)
Odontites verna	I (2–3)	I (3–4)			I (2–4)
Juncus bufonius	I (4–6)	I (2–3)			I (2–6)
Eriophorum angustifolium	I (4)		I (2–3)		I (2–4)
Alopecurus geniculatus	I (3)		I (5–8)		I (3–8)
Juncus acutiflorus	I (3–5)			I (5)	I (3–5)
Drepanocladus aduncus	I (2–3)			I (3–5)	I (2–5)
Potentilla palustris			I (2)	I (1–6)	I (1–6)
Number of samples	24	32	40	46	142
Number of species/sample	12 (8–24)	19 (11–39)	17 (9–26)	12 (5–29)	15 (5–39)

a Festuca rubra-Ranunculus repens sub-community
b Carex flacca sub-community
c Caltha palustris sub-community
d Hydrocotyle vulgaris-Ranunculus flammula sub-community
17 Potentilla anserina-Carex nigra dune-slack (total)

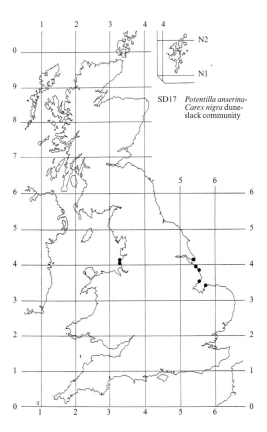

SD17 *Potentilla anserina-*
Carex nigra dune-
slack community

SD18

Hippophae rhamnoides dune scrub

Synonymy
Hippophae rhamnoides scrub Tansley 1911; *Hippophaëtum* Tansley 1939; *Hippophaë* communities Pearson & Rogers 1962; Dune scrub Gimingham 1964a *p.p.*

Constant species
Hippophae rhamnoides.

Rare species
Hippophae rhamnoides.

Physiognomy
Hippophae rhamnoides occurs sparsely and at low cover in a variety of vegetation types of less mobile dune sands, but the *Hippophae* scrub includes stands where this thorny, deciduous shrub is becoming a consistent and more prominent feature. The density and height of the bushes is, however, very variable. At one extreme, the community includes grassier vegetation where the buckthorn is scattered and of only moderate cover overall, whereas other stands have a very thickly-set mass of suckering stems with a virtually impenetrable canopy of the stiff branches. Exceptionally, the bushes can attain over 3 m in height but 1–2 m is more usual and, in more exposed situations, the cover may be severely wind-trimmed.

The extent and character of the associated flora is very much dependent on the degree of *Hippophae* dominance, and no other species are constant throughout the community. Interestingly, the most frequent dune grass overall is *Elymus pycnanthus*, but there is quite commonly some *Poa pratensis* and *Holcus lanatus*, with more occasional *Agrostis stolonifera*, *Dactylis glomerata* and *Carex arenaria*. The most widely distributed dicotyledon is *Senecio jacobaea*, with *Cirsium arvense*, *Sonchus arvensis*, *Epilobium angustifolium* and *Heracleum sphondylium* occurring more rarely but with local abundance. *Rubus fruticosus* agg. is also fairly frequent and can thicken up in places among and around the buckthorn.

Grasses such as *Ammophila arenaria*, *Festuca rubra* and *Leymus arenarius* tend to persist only in the earlier stages of *Hippophae* invasion, though there they can be very common and abundant and, along with a variety of more occasional dune herbs, provide strong continuity with the vegetation being colonised. In denser stands, however, these plants become much scarcer, the associated plants thinning out beneath the canopy but providing a distinctively weedy and nitrophilous aspect around the bushes.

Sub-communities

Festuca rubra sub-community: Open *Hippophae* community Pearson & Rogers 1962. Here, the *Hippophae* bushes are usually more scattered and small, and not yet densely suckering, providing a cover that is generally less than 50%. Between the buckthorn, there is grassy vegetation, often rank and sometimes strongly tussocky, typically dominated by mixtures of *F. rubra* and *Ammophila*, with locally abundant *E. pycnanthus* and, preferential at low frequencies here, *Leymus* and *E. farctus*. Along with scattered *S. jacobaea* and *C. arvense*, there are occasional plants of *Ononis repens* with, in more open places, *Hypochoeris radicata*, *Plantago lanceolata*, *Taraxacum officinale* and *Leontodon taraxacoides*. Even some strandline plants, such as *Honkenya peploides* or *Cakile maritima*, can occur where there are still patches of mobile sand and, on more stable bare areas, there can be ephemerals like *Cerastium diffusum* ssp. *diffusum*, *C. semidecandrum*, *Aira praecox* and *Phleum arenarium*, patches of *Brachythecium albicans* and *Ceratodon purpureus* or thalli of various *Cladonia* spp.

Urtica dioica-Arrhenatherum elatius sub-community: Dense *Hippophae* community Pearson & Rogers 1962. In this kind of *Hippophae* scrub, the cover of the bushes is rarely less than 50% and usually much denser, with patches of continuous canopy and small areas of open ground between. In these, most of the characteristic dune plants of the *Festuca* sub-community are very poorly represented and in their place are patches of *Urtica dioica* and *Arrhenatherum elatius* with trailing masses of *Galium aparine* and *Solanum dulcamara*. *Cirsium vulgare, Stellaria media, Montia perfoliata* and *Poa trivialis* occur along with *S. jacobaea, C. arvense, E. angustifolium* and *Heracleum* though, among the densest buckthorn, only very frail individuals of the most shade-tolerant species survive among sparse wefts of *Brachythecium rutabulum* and *Eurhynchium praelongum*. In places, the canopy is enriched by bushes of *Sambucus nigra* or, more rarely, *Crataegus monogyna*.

Habitat
The *Hippophae* scrub is a local feature of less mobile dune sand at scattered localities around the British coast, also occurring very occasionally on soft sea-cliffs. It is a natural vegetation type along the east and south-east coasts, but has developed elsewhere from planted stock, and could probably establish on all but the most acidic dunes.

H. rhamnoides ssp. *rhamnoides* has an extensive distribution through the mountains of southern Europe and central Asia, but occurs as a coastal plant in north-west Europe from Brittany to north Norway and the Baltic (Pearson & Rogers 1962). It has been native with us since the Late Glacial (Godwin 1975) but its natural localities appear to be concentrated between east Sussex and Humberside with a few stations northwards to the Lothians (Perring & Walters 1962). Elsewhere, it has become widely naturalised and there is probably no climatic hindrance to its survival anywhere around our coasts or indeed in the kind of inland river gravel habitat that it favours through its Eurasian range.

Hippophae is able to colonise dune sand as it becomes more fixed in the later stages of *Ammophila* dominance. Initial invasion may be by seed because, though the species is dioecious, fruiting can be prolific and bird-dispersal could transport the seed some distance. Germination requires a cold pre-treatment, but such a need is generally met in early winter and viability can be very high. Once established, however, vegetative reproduction is the more important means of local spread, an extensive network of horizontal roots putting up aerial shoots, irregularly spaced but often clustered, which can grow rapidly and produce a densely branching canopy.

Young plants can withstand some accretion, so long as they are not completely buried, and will respond to deposition by putting out new adventitious roots into the drifting sand. *Hippophae* is also resistant to drought, the bushes eventually producing roots which penetrate deeply and often make contact with moist layers of sand (Pearson & Rogers 1962). On the other hand, periodic flooding does not seem to damage established colonies and this kind of scrub can be found in some winter-wet

dune hollows. The bushes are resistant to wind-cut and mature foliage is not damaged by sea-spray, though growth is often noticeably better in sheltered situations and colonies are often centred in depressions between ridges and over the lee faces of dunes.

The floristic character of younger and more open stands is largely inherited from the invaded vegetation, generally some form of the *Ammophila-Festuca* community, where marram is still vigorous but often accompanied by *F. rubra* and *P. pratensis* and, locally down the east coast, by *Leymus*. It is thus these grasses, together with their associated dicotyledons from the dune swards, that give the distinctive stamp to the *Festuca* sub-community.

The buckthorn is easily able to compete with the rank grassy cover of the invaded vegetation by putting up its vigorous suckers and, as the bushes thicken up, the dune flora becomes increasingly confined to enclaves and is then largely shaded out. The flora of the *Urtica-Arrhenatherum* sub-community thus consists of the rather sparse survivors from this earlier phase, together with new plants tolerant of the shady scrub margins and able to benefit from the protection from grazing that the bushes provide. Among these, the prominence of plants like *U. dioica, G. aparine, Arrhenatherum* and *Sambucus* reflects the more eutrophic soil conditions that seem to be characteristic of denser *Hippophae* stands. Nitrogen is significantly greater in the substrates of such vegetation than in the surrounding sands (Pearson & Rogers 1962) and, though some of this may derive from steady accumulation of humus with the ageing of the fixed sediments, much may come from fixation of atmospheric nitrogen in root nodules on the buckthorn (Bond *et al*. 1954, 1956). Such nodules develop within a matter of weeks when seedlings are grown in soil with a suitable inoculum, though the exact nature of the mycorrhizal organism is uncertain (Pearson & Rogers 1962). Beneath the denser stretches of *Hippophae* canopy, of course, the beneficial effects of soil enrichment on the associated flora are largely offset by the deep shade, but the elements of a sparse woodland flora may appear with time.

Seedlings of *Hippophae* are eaten by rabbits and such grazing may be effective in preventing the establishment of the shrub from seed: after myxomatosis, buckthorn rapidly got a hold in some sites (Dargie in Shimwell 1971c) and the demise of rabbits may have played a major role in the recent spread of this vegetation. Older bushes, though, are well-armed with thorns and their leaves appear bitter to stock, so stands often show few signs of predation and may be difficult to keep in check. However, mowing young shoots can reduce them to a prostrate mat of contorted shoots and ploughing or ditching can prevent the spread of the roots. The stabilising effect of this scrub on dunes, the protection from wind that it can afford to tree saplings and the decorative character of the foliage and fruit have favoured the wide-spread use of the shrub for coastal defence, shelter and screening, but such plantings quickly get a natural look and may become troublesome.

Zonation and succession
The *Hippophae* community is usually found in mosaics with dune grasslands, weedy tall-herb vegetation and other kinds of scrub developed over less heavily grazed stretches of fixed sands. It develops as a part of natural successions in its east coast localities, elsewhere as an introduction, and, where it is not overtopped, it could perhaps have a measure of stability in more exposed habitats. However, many of our stands are too young for us to know what the end product of this succession might be.

Although *Hippophae* scrub can extend on to quite mobile sands carrying the *Ammophila* community, it is commonly seen among *Ammophila-Festuca* vegetation on somewhat more stable sediments. Then, there can be a continuous gradation between the *Festuca* sub-community of the scrub and the dune swards around, mixtures of *Ammophila*, *F. rubra* and *P. pratensis* assuming dominance away from the buckthorn bushes. *Leymus* can be prominent in the dune vegetation, too, and at some localities along the Lincolnshire and Norfolk coasts, there is a local abundance of *Elymus pycnanthus* throughout the zonation. Where the *Hippophae* cover thickens, the *Festuca* sub-community can then give way to the *Urtica-Arrhenatherum* sub-community.

In certain sites, this basic pattern is complicated by the occurrence of *Ammophila-Arrhenatherum* grassland, the *Arrhenatherum* sub-community of which can grade into both the *Ammophila-Festuca* dune vegetation with the disappearance of *Arrhenatherum*, *Dactylis glomerata*, *Heracleum* and *Cirsium arvense* or into denser patches of scrub with the appearance of clumps of *Urtica* and tangles of *G. aparine* and *Solanum*. The Arrhenatherion plants may also provide continuity with stands of *Rubus-Holcus* underscrub where brambles dominate among mixtures of rank grasses and weedy herbs, or there may be stands of tall-herb vegetation

dominated by *Urtica* or *Epilobium angustifolium*. Then, locally, the occasional bushes of *Sambucus* seen among the *Hippophae* may thicken up with transitions to *Crataegus-Hedera* or *Prunus-Rubus* scrub from among which the buckthorn is eventually shaded out. With time and shelter from winds, it is possible that such scrub on dunes progresses to Carpinion woodland but even our *Hippophae* scrub still seems to be in an immature state compared with, for example, Dutch stands and we have little information about what the natural successions beyond buckthorn stands might be.

Distribution
The community is well established on the east coast between Kent and Fife, especially in Lincolnshire and north Norfolk, and has become firmly naturalised elsewhere, as at Formby-Ainsdale in Lancashire, at scattered sites between Devon and Cromarty and, more recently, in Wales.

Affinities
The distinctive role of *Hippophae* as a colonist of dunes was early recognised (Tansley 1911, 1939) and the study of Pearson & Rogers (1962) characterised the two kinds of scrub distinguished here. Compared with the buckthorn vegetation described from the Dutch coast, however, our own stands seem immature. The *Festuca* sub-community is similar to a *Hippophae-Ammophila* assemblage which van der Maarel & Westhoff (1964) thought a precursor to the *Hippophaeo-Ligustretum* Meltzer 1941 *emend.* Boerboom 1960, a scrub with mixtures of *Hippophae*, *Ligustrum vulgare*, *Berberis vulgaris*, *Rhamnus catharticus*, *Euonymus europaeus*, *Crataegus monogyna* and *Rosa* spp. The *Urtica-Arrhenatherum* sub-community, on the other hand is more similar to the *Hippophao-Sambucetum* Boerboom 1960 where canopies of *Hippophae* and *Sambucus* are associated with a more nitrophilous suite of herbs. These kinds of vegetation have traditionally been placed among the scrubs of the Berberidion alliance in the Prunetalia (Westhoff & den Held 1969, Ellenberg 1978).

Floristic table SD18

	a	b	18
Hippophae rhamnoides	V (4–10)	V (3–10)	V (3–10)
Festuca rubra	V (3–9)	I (3–6)	III (3–9)
Ammophila arenaria	IV (3–9)	I (1–6)	III (1–9)
Leymus arenarius	II (1–4)	I (1–4)	II (1–4)
Elymus farctus	II (2–7)	I (5)	I (2–7)
Hypochoeris radicata	II (1–7)	I (1–2)	I (1–7)
Taraxacum officinale agg.	II (1–4)	I (1–2)	I (1–4)
Plantago lanceolata	II (1–4)	I (3)	I (1–4)
Ononis repens	II (2–7)		I (2–7)
Leontodon taraxacoides	I (1–5)		I (1–5)
Cerastium diffusum diffusum	I (1–2)		I (1–2)
Daucus carota	I (1–3)		I (1–3)
Honkenya peploides	I (1–5)		I (1–5)
Brachythecium albicans	I (3–6)		I (3–6)
Salix repens	I (3)		I (3)
Aira praecox	I (2–3)		I (2–3)
Calystegia soldanella	I (3–4)		I (3–4)
Cerastium semidecandrum	I (1–3)		I (1–3)
Cerastium fimbriata	I (1–3)		I (1–3)
Ceratodon purpureus	I (1–6)		I (1–6)
Cladonia furcata	I (1–3)		I (1–3)
Achillea millefolium	I (2–3)		I (2–3)
Anacamptis pyramidalis	I (2–3)		I (2–3)
Phleum arenarium	I (1–3)		I (1–3)
Diplotaxis tenuifolia	I (1–4)		I (1–4)
Sagina procumbens	I (3–5)		I (3–5)
Urtica dioica		IV (1–8)	III (1–8)
Arrhenatherum elatius	I (3–7)	III (3–9)	II (3–9)
Galium aparine	I (1)	III (1–4)	II (1–4)
Solanum dulcamara	I (1)	III (1–6)	II (1–6)
Sambucus nigra	I (3)	II (1–8)	I (1–8)
Brachythecium rutabulum	I (3–5)	II (2–6)	I (2–6)
Cirsium vulgare	I (1)	II (1–3)	I (1–3)
Rubus caesius		I (3–6)	I (3–6)
Stellaria media		I (2–5)	I (2–5)
Montia perfoliata		I (1–3)	I (1–3)
Poa trivialis		I (3–4)	I (3–4)
Cynoglossum officinale		I (1–2)	I (1–2)
Crataegus monogyna		I (3)	I (3)
Glechoma hederacea		I (3)	I (3)
Senecio jacobaea	III (1–6)	III (1–4)	III (1–6)
Elymus pycnanthus	II (1–7)	II (1–8)	II (1–8)
Poa pratensis	II (1–6)	II (1–6)	II (1–6)
Cirsium arvense	II (1–2)	II (1–6)	II (1–6)
Rubus fruticosus agg.	II (1–7)	II (1–7)	II (1–7)

Floristic table SD18 (*cont.*)

	a	b	18
Holcus lanatus	II (2–4)	II (2–5)	II (2–5)
Epilobium angustifolium	I (2–4)	I (2–6)	I (2–6)
Sonchus arvensis	I (1–5)	I (1–3)	I (1–5)
Agrostis stolonifera	I (5)	I (2–6)	I (2–6)
Heracleum sphondylium	I (1–3)	I (1–3)	I (1–3)
Carex arenaria	I (1–4)	I (5)	I (1–5)
Dactylis glomerata	I (1–4)	I (2–4)	I (1–4)
Galium verum	I (1–3)	I (2)	I (1–3)
Eurhynchium praelongum	I (3–4)	I (4–7)	I (3–7)
Lotus corniculatus	I (1–4)	I (3)	I (1–4)
Phragmites australis	I (5)	I (1–5)	I (1–5)
Convolvulus arvensis	I (3–5)	I (5)	I (3–5)
Number of samples	36	37	73
Number of species/sample	11 (3–25)	10 (4–25)	10 (3–25)

a   *Festuca rubra* sub-community
b   *Urtica dioica-Arrhenatherum elatius* sub-community
18 *Hippophae rhamnoides* dune scrub (total)

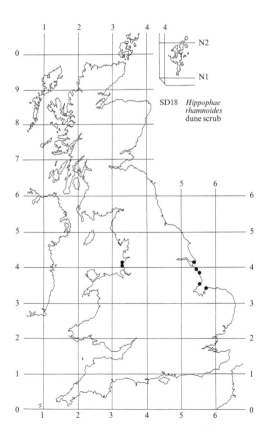

SD18 *Hippophae rhamnoides* dune scrub

SD19

Phleum arenarium-Arenaria serpyllifolia dune annual community
Tortulo-Phleetum arenariae (Massart 1908) Br.-Bl & De Leeuw 1936

Constant species
Ammophila arenaria, Arenaria serpyllifolia, Carex arenaria, Festuca rubra, Phleum arenarium.

Rare species
Mibora minima, Vulpia membranacea.

Physiognomy
The *Tortulo-Phleetum* comprises generally open assemblages of ephemerals which make a brief appearance in the dampness of spring and early summer among gaps in a cover of perennial grasses and dicotyledonous herbs of semi-fixed to fixed dunes.

The most distinctive species of the community are the annual grass *Phleum arenarium* with *Arenaria serpyllifola, Cerastium diffusum* ssp. *diffusum, Aira praecox* and *Viola tricolor* ssp. *curtisii* and the moss *Tortula ruralis* ssp. *ruraliformis. Ammophila arenaria, Festuca rubra* and *Carex arenaria* are consistently represented in the grassland matrix in which this vegetation makes its appearance, together with *Ononis repens, Lotus corniculatus, Senecio jacobaea* and, more occasionally, *Poa pratensis.*

Light-demanding mat plants like *Thymus praecox* and *Sedum acre* figure quite commonly and there are occasional records for *Cerastium semidecandrum, Desmazeria marina, Euphorbia paralias, Logfia minima, Erophila verna, Trifolium dubium, T. campestre, Erodium cicutarium, Geranium molle* and *Centaurium erythraea.* This community also provides a locus for the nationally rare *Mibora minima* and *Vulpia membranacea.*

In addition to *T. ruralis* ssp. *ruraliformis, Hypnum cupressiforme* is frequent, *Homalothecium lutescens, Brachythecium albicans* and *Tortella flavovirens* occasional. *Peltigera canina* can also occur with some local abundance.

Habitat
The *Tortulo-Phleetum* is a community that capitalises on the appearance of gaps with essentially stable sand that develop where semi-fixed and fixed dune swards are opened up by drought or locally disturbed, often by rabbits or modest anthropogenic activity.

The distinctive annuals of the community, being small, are unable to establish themselves on shifting sand or in closed swards of perennials, so the patchwork of open but stabilising ground characteristic of maturing stands of the *Ammophila-Festuca* community offers a very congenial habitat. Wind can play some part in the dispersal of seed or fruits in the plants, though usually only over very short distances because of their low flowering stems (Pemadasa et al. 1974, Pemadasa & Lovell 1974a, Watkinson 1978b, c), but burial of seed under just a few centimetres of sand has been shown to be highly inimical to germination in *Aira praecox, Cerastium diffusum* ssp. *diffusum, Erophila verna* and *Mibora minima* (Pemadasa & Lovell 1975), and wind-drag is very destructive of seedling establishment in *Vulpia fasciculata* (Watkinson 1978a). This grass can actually grow quite bulky and will tolerate a modest amount of accretion, putting out short underground stems if its tussocks bases get covered (Watkinson 1978c), but this is unusual and most of the annuals remain vulnerable to being overwhelmed or uprooted throughout their life cycle. They thus play little part in this vegetation until the dune surface has become more or less fixed with the drop in sand supply or the provision of local shelter.

The establishment of patches of the mosses upon which the annuals can germinate in the *Turtulo-Phleetum* may be of very considerable importance. At the same time, however, conditions must remain such as to hinder the development of a continuous cover of taller perennial plants among the marram tussocks if the annuals are not to be crowded out. This is especially true of potentially vigorous associates like *Festuca rubra* which is a constant feature of the vegetation of semi-fixed and fixed dunes: indeed, in the study of Pemadasa *et al.* (1974), the distribution of annuals in this community was essentially the inverse of the patterning in this grass in the dune sward around. At this stage of dune growth, mobility of the sand is of much reduced consequence in keeping the cover open, and it is shortage of water and nutrients that now play the leading role. The amounts of moisture, nitrogen and phosphorus may be little higher here than in the raw sands of more mobile dunes and remain strongly limiting to plant growth (Willis & Yemm 1961, Pemadasa & Lovell 1974*b*, Noble 1982), there being little return as yet of decaying organic matter to the soils, and probably only a small contribution from nitrogen-fixing plants like *Ononis repens* and *Lotus corniculatus* which can improve their local edaphic environment and eventually benefit the surrounding vegetation (Willis 1985*a*, Jones & Turkington 1986). Death of perennials in more severe summer droughts, and the annual die-back of deciduous plants like *O. repens* and *Anthyllis*, can also make room for the establishment of annuals (Watkinson 1978*c*), and trampling and grazing may assist by scuffing the sand surface and keeping down the height of the herbage. Grazing, though, can be highly deleterious to the annuals themselves because, when flowering or fruiting shoots are devoured, this effectively prevents re-establishment of a population in the following season unless there are some seed stores in the sand (Watkinson & Harper 1978, Watkinson 1978*c*). With *V. fasciculata*, for example, which is selectively grazed by rabbits, particularly in its reproductive phase, there was a very dramatic spread on some dune systems after myxomatosis (Willis 1967, Watkinson 1978*c*).

Nutrient shortage is also limiting to the growth of the annuals themselves (Willis & Yemm 1961, Willis 1963, Pemadasa & Lovell 1974*b*) though, with the sub-optimal light and temperatures of autumn, when germination in these plants characteristically takes place (Ratcliffe 1961, Pemadasa & Lovell 1975), this may be less critical: even with the addition of balanced nutrients, these climatic factors continue to control growth in the short period before the onset of virtual dormancy in the winter months (Pemadasa & Lovell 1974*a*, 1976). However, this timing of the life-cycle is of crucial importance to the success of these plants because it enables them to complete most of their growth while the perennials are least active (Pemadasa & Lovell 1974*a*, 1975) and, with flowering complete by summer, to avoid the worst effects of water shortage by passing the driest part of the year as seed (Ratcliffe 1961, Pemadasa & Lovell 1976, Rozijn & van der Werf 1986).

Germination in these winter annuals is, in fact, controlled by complex interactions between innate dormancy and the need for after-ripening, characteristics which vary considerably from species to species, and climatic and soil conditions, which can be very variable from one year to the next (Ratcliffe 1961, Pemadasa & Lovell 1975). Generally speaking, however, germination is timed to take advantage of rains which moisten the dune surface in autumn. The overall distribution of some of these plants has been correlated with a minimum level of precipitation through the autumn and winter months (Watkinson 1978*c*) and for unimpeded germination an adequate and continuous supply of moisture is needed. Both viability and germination rate have been shown to decline markedly in a number of species with decreasing soil moisture, desiccation often arresting establishment (Pemadasa & Lovell 1975).

Seedlings of winter annuals and maturing plants are often resistant to the short periods of drought which can characterise these dunes through the winter and particularly in spring (Pemadasa & Lovell 1975, Watkinson 1978*c*) and this may be critical for their survival. With the two *Aira* spp., for example, which are both drought-avoiders in their life-cycles, it may be the poorer drought-resistance of *A. caryophyllea* as compared with *A. praecox*, that largely excludes it from this community (Pemadasa & Lovell 1974*a*, Rozijn & van der Werf 1986). In general, however, supplies of soil moisture continue to be important for growth and reproduction (Newman 1967, Pemadasa & Lovell 1974*a*, Ernst 1981, Watkinson 1982) though, despite the coming of more favourable light and temperature regimes in spring, vegetative activity often shows a sharp decline with the initiation of flower primordia as the season advances (Pemadasa & Lovell 1976). Survival through the driest months is then assured by adequate seed-set, high summer temperatures imposing dormancy and a need for prolonged hydration before germination can take place (Newman 1963, Pemadasa & Lovell 1975).

Translocation of nitrogen and phosphorus from ageing vegetative tissues to generative organs during this final phase of growth is sometimes very striking and interactions between water and nutrient supplies as the time for reproduction approaches can be sharply contrasting. With *Erodium cicutarium* for example, Ernst (1983) showed that shortage of major nutrients delayed fruiting, whereas in *Phleum arenarium* the process was accelerated, perhaps because the shallower rooting system of this grass rendered it additionally more susceptible to drought (Ernst 1981). In this species, too, fewer nutrients made for larger numbers of lighter fruits, propagules which have a higher degree of dormancy and offered a better chance of contributing to a modest seed-bank and so enhancing the possibility of survival into future seasons (Ernst 1981). Differences such as these among the annuals make a further contribution to their varied representation from one year to the next.

Zonation and succession
The *Tortulo-Phleetum* is characteristically a minor and local element within stretches of more stable dune grasslands. Its distinctive contingent of ephemerals comes and goes as particular locations become congenial and it can be rapidly overwhelmed in any one place

by renewed vigour among the perennials in the surrounding swards.

Zonations to the *Ammophila-Festuca* and *Festuca-Galium* grasslands, the two communities that generally provide the context for this assemblage, involve a thickening up of the perennial grasses and recolonisation by dicotyledonous herbs. Where local disturbance becomes more acute, however, for example where rabbit activity loosens the sand surface, there can be renewed erosion and a spread of *Carex arenaria* or *Ammophila* vegetation which rapidly overwhelms the *Tortulo-Phleetum*.

Distribution
The community is widespread but local on dune systems in England and Wales, more scarce in Scotland.

Affinities
Phytosociological investigations have characterised a number of syntaxa from among assemblages of dune annuals, locating them in the Sedo-Scleranthetea, the class of largely ephemeral vegetation of dry, sandy soils (or the Koelerio-Cornephoretea as Westhoff & den Held (1969) have it). The commonest association of this kind along warmer coasts in north-west Europe is this *Tortulo-Phleetum arenarii*, which Moore (1977) and Schouten & Nooren (1977) recorded from Eire. From cooler, oceanic seaboards, Braun-Blanquet & Tüxen proposed recognising a *Viola curtisii-Tortuletum ruraliformis* and this was subsequently recorded from Eire by Ivimey-Cook & Proctor (1966), Beckers *et al.* (1976) and Ni Lamhna (1982), and from Scottish dunes by Birse (1980, 1984). However, the distinction between this and the *Tortulo-Phleetum* is often slim (White & Doyle 1982) and the available data do not reveal much consistent variety among the assemblages of annuals encountered. Irish workers and Birse (1980, 1984) locate these associations in the Koelerion (or Galio-Koelerion as it was renamed by Westhoff).

Floristic table SD19

Phleum arenarium	V (1–5)	*Euphorbia paralias*	I (1–4)
Arenaria serpyllifolia	IV (2–4)	*Aira caryophyllea*	I (2–4)
Festuca rubra	IV (1–7)	*Tortella flavovirens*	I (1–3)
Ammophila arenaria	IV (1–8)	*Cirsium arvense*	I (1–2)
Carex arenaria	IV (2–5)	*Logfia minima*	I (1–2)
		Mibora minima	I (1–3)
Tortula ruralis ruraliformis	III (2-7)	*Erophila verna*	I (1–2)
Cerastium diffusum diffusum	III (1–4)	*Trifolium dubium*	I (1–3)
Aira praecox	III (1–4)	*Trifolium campestre*	I (1–3)
Ononis repens	III (1–5)	*Plantago lanceolata*	I (1–3)
Viola tricolor curtisii	III (1–7)	*Peltigera canina*	I (1–5)
Hypnum cupressiforme	III (2–5)	*Anthoxanthum odoratum*	I (1–5)
Lotus corniculatus	III (1–5)	*Erodium cicutarium*	I (1–3)
Sedum acre	III (1–5)	*Geranium molle*	I (1–4)
Senecio jacobaea	III (1–5)	*Plantago coronopus*	I (2–4)
Thymus praecox	II (2–8)	*Trifolium repens*	I (1–6)
Galium verum	II (1–5)	*Centaurium erythraea*	I (3–4)
Trifolium arvense	II (1–3)	*Luzula campestris*	I (2–4)
Anthyllis vulneraria	II (1–3)	*Tortella tortuossa*	I (2–6)
Homalothecium lutescens	II (1–5)		
Cerastium semidecandrum	II (2–6)	Number of samples	28
Crepis capillaris	II (1–3)	Number of species/sample	19 (8–33)
Poa pratensis	II (2–4)		
Vulpia membranacea	II (1–5)	Herb height (cm)	15 (2–50)
Brachythecium albicans	II (1–4)	Herb cover (%)	76 (40–100)
Desmazeria marina	I (1–5)	Ground height (mm)	14 (3–30)
Hypochoeris glabra	I (1–3)	Ground cover (%)	23 (1–60)

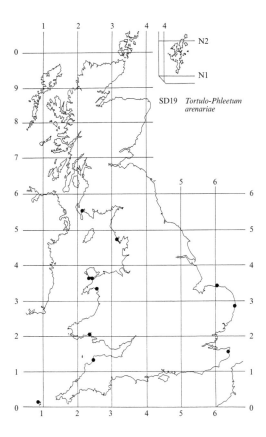

SD19 *Tortulo-Phleetum
arenariae*

MARITIME CLIFF COMMUNITIES

INTRODUCTION TO MARITIME CLIFF COMMUNITIES

The sampling of maritime cliff vegetation

Apart from along the eastern shores of England, from Flamborough Head south to the Thames estuary, steep rocky cliffs figure prominently around much of our coastline, splashed by spray to varying degrees, home to colonial sea-birds and often providing grazing for coastal farms. Yet, until relatively recently, the distinctive plant communities they support, often disposed in striking zonations related to maritime influence, the geology and soils and the influence of stock and wild herbivores, had attracted little attention, at least on a large scale. Thus, they have usually figured only in local studies, often as part of descriptions of both coastal and inland vegetation of particular sites (e.g. Petch 1933, Poore & Robertson 1949, Goodman & Gillham 1954, Gillham 1953, McVean 1961a, Coombe 1961, Birks 1973) or, when surveyed more widely, have been described in relation to broadly-defined habitat groups (Goldsmith 1975).

We were therefore especially fortunate in this project to have the benefit of a detailed phytosociological study of sea-cliff vegetation from all around the coast of Britain carried out by Andrew Malloch of Lancaster University and building on his earlier survey of the Lizard and other parts of Cornwall (Malloch 1970, 1971, 1972). As with the general approach we adopted for this project, Malloch had located samples only on the basis of floristic and structural homogeneity of the vegetation, had used an identical range of quadrat sizes and had recorded species composition using the Domin scale of cover/abundance. In so far as possible in such dangerous terrain, he had covered the full range of vegetation right down to the splash-zone, where vascular assemblages were replaced by lichens and then algae in the truly marine environment, and had included grasslands, heath and scrub on the unenclosed cliff-tops. Within the wider framework of the whole project, we were able to supplement the latter with samples of the complete range of vegetation abutting on to rocky coasts in farmland, unimproved inland grasslands, mires and non-maritime heaths, ensuring there were no gaps in coverage. We were less successful in sampling extensively on earth cliffs, locally important along the south and particularly the east coast of England and, though less strikingly maritime in the latter region where the winds often blow offshore, still supporting distinctive vegetation on their unstable sand and clay substrates. In this range of situations, few vascular plants give cause for uncertainty in identification but, as usual, we recorded *Rubus fruticosus* and *Taraxacum officinale* to the aggregate. Bryophytes and macrolichens (*sensu* Dahl 1968) were included in the records.

In addition to the floristic data, we had from Malloch and our own data the usual notes on physiognomy to assist in describing and interpreting the vegetation with, for example, details on the organisation of crevice communities and the pattern of dominance in cliff-top grasslands and heaths. Particular note was taken of the context of the vegetation being sampled, its place in any zonational sequence and any signs of successional changes related to natural processes of colonisation or shifts in management.

The usual environmental data were recorded and supplemented with information on features of particular significance on sea-cliffs. To notes on altitude, aspect, slope and geology, for example, were added details of the configuration and profile of cliffs, whether the faces were exposed to or sheltered from prevailing winds and spray, whether the cliffs were vertical, gently sloping or bevelled, and just how the lithology and stratigraphy of the various weathered bedrocks had influenced the extent and disposition of crevices and ledges.

Especially valuable were data on maritime influence from salt-spray expressed in terms of soil sodium/loss-on-ignition, analyses for which Malloch had carried out on many soils underlying his samples. His survey also provided data on other physico-chemical features of the soil environment including pH, water content, calcium, magnesium, potassium and phosphorus, giving us an unusually detailed opportunity to explore edaphic influences among these vegetation types. Also of importance was information on the extent of any grazing of the cliff vegetation and the type of stock used and notes on the occurrence of nesting or roosting sea-birds.

A total of just over 1500 samples was available for analysis and the geographical distribution of these is shown in Figure 17.

Data analysis and the description of maritime cliff communities

For the analysis, only the floristic data from the samples were used, the environmental information being employed afterwards for interpreting the vegetation types characterised. The quantitative scores for all vascular plants, bryophytes and macrolichens were entered into the analysis, no particular weighting being given to maritime species or any other taxa of supposed significance.

Twelve communities of sea-cliff crevice vegetation, maritime grasslands and bird colony vegetation were characterised from these data (Figure 18). Heath and scrub, though important on some cliffs and often referred to in this volume in discussions of zonations, are included elsewhere in the scheme together with their inland counterparts (Rodwell 1991, 1992). Among the

vegetation types described here, the major influence on floristic composition and the extent and disposition of the communities on particular stretches of cliff are the amount of input from salt-spray, geology and soils, climate, grazing and disturbance and enrichment from sea-birds. The production of spray depends largely on the impact of the sea with the shoreline in the breaking of waves, the height and frequency of which are determined by the speed and duration of the winds and the fetch of sea over which they blow. The general prevalence of south-westerly winds across the British Isles (e.g. Shellard 1976) is a very important reason, along with the softer character of many coastal rocks there, for the poorer development of maritime cliff vegetation down the eastern seaboard of England. Even on the east coast of Scotland, where north-westerlies blow down from the Arctic Ocean on to more substantial rocky cliffs, the zone of truly maritime cliff vegetation is relatively narrow compared with the west coast.

Sea-spray has more or less the same ionic content and balance as sea-water (Junge 1963, Malloch 1970, 1972) and, of its components, sodium chloride is the most influential on vegetation. Apart from observable tissue

Figure 17. Distribution of samples available from sea-cliffs.

Figure 18. Distribution of vegetation types characterised from sea-cliffs.

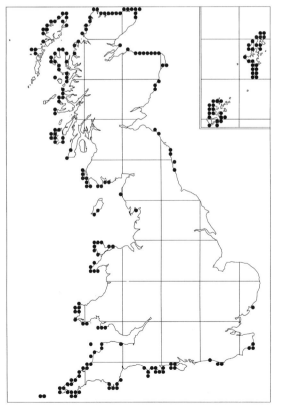

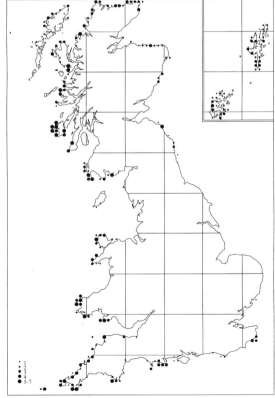

damage, well seen after summer gales when delicate shoots can blacken and wither, salt or chloride can affect the water balance in soil and cells and upset metabolism (e.g. Goldsmith 1975). Species which fare best on sea-cliffs are those able to survive, to a greater or lesser degree, higher concentrations of salt than inland plants and to avoid such damage: they thus have an absolute or competitive advantage over these plants on sites which receive much sea-spray (Malloch & Okusanya 1979, Okusanya 1979a, b). Among the communities characterised here, there is a highly significant correlation between the percentage of plants more or less confined to the coastal fringe or Britain and the soil sodium loss on ignition value. It is the prominence of such maritime species which gives coastal cliff vegetation its distinctive stamp: most frequent throughout the communities characterised here are *Festuca rubra*, often the densely caespitose ecotype sometimes referred to ssp. *pruinosa*, *Armeria maritima* and *Silene vulgaris* ssp. *maritima*.

Spray deposition falls away rapidly with distance from the point of production even in very exposed and windy situations where more spray is generated and blown further inland (Fujiwara & Umejima 1962, Junge 1963, Edwards & Claxton 1964, Malloch 1970, 1972). Such reduction is initially very great and then more gradual: on The Lizard, for example, Malloch (1970, 1972) showed that only 100 m inland, deposition rates were less than 20% of those at the cliff edge and, after 500 m, deposition was very low. Such gradations are largely responsible for the often very pronounced zoning of vegetation types on sea-cliffs from the more maritime crevice communities through grasslands to heath, scrub and inland vegetation (Figure 19).

Maritime cliff crevice and ledge communities

In the most exposed situations, the effect of salt-spray often overrides any influence of lithological differences between bedrocks and derived variations in soils but lithology and stratigraphy can still have important physical impacts on the habitat and vegetation. Coastal cliffs are found in many types of rock and a great variety of coastal geology in Britain is reflected in a wide range of cliff scenery (see Steers 1953). Differential weathering and erosion of these rocks provides complexes of exposed and sheltered surfaces and variations in cliff profile. Vertical cliffs, which tend to generate more spray and receive more at their tops, are commoner in softer rocks. Bevelled cliffs, with a quieter initial interception of waves and cliff-tops set back beyond the zone of heavy spray deposition, occur more widely in harder deposits of igneous or metamorphic origin.

For crevice communities, the bedding and jointing of rocks is also important in providing niches of varying width, depth and disposition, for colonisation by the

most salt-tolerant plants. Two such communities are of widespread occurrence in such exposed habitats, one in southern Britain, one in the north. The former, the *Crithmum maritimum-Spergularia rupicola* community (MC1, *Crithmo-Spergularietum rupicolae* Géhu 1964), represents an extension around our spray-splashed rocky coasts of crevice vegetation typical of south-west and southern Europe, placed in the alliance of Crithmo-Armerion maritimae Géhu 1968. Its northern counterpart, replacing it beyond the Mull of Galloway, is the *Armeria maritima-Ligusticum scoticum* community

Figure 19. The influence of salt spray on sea-cliff zonations.
The figures show the disposition of some common sea-cliff communities in relation to salt-spray on (a) an exposed cliff, (b) a sheltered cliff, (c) a cove and headland with variation in exposure and with (d) complexities related to overwash at X and eddying at Y.
MC1 *Crithmo-Spergularietum* maritime rock-crevice community
MC8 *Festuca-Armeria* maritime grassland
MC9 *Festuca-Holcus* maritime grassland
H7 *Calluna-Scilla* maritime heath

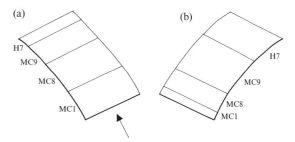

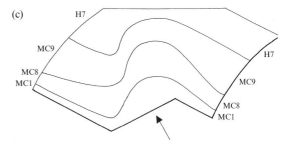

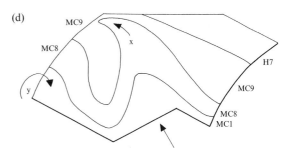

(MC2) which can be seen as an extension southwards of the spray-zone vegetation of the sub-Arctic (Birks 1973) and perhaps best located in a different alliance. The loss to the north of *C. maritimum*, *S. rupicola* and their companion in the *Crithmo-Spergularietum*, *Inula crithmoides* is related in some way to the greater sensitivity of these plants to cold than *Ligusticum* (Okusanya 1979c), perhaps through a limitation on their vegetation growth or seed set and maturation.

Much more local, and replacing the *Crithmo-Spergularietum* on crumbling ledges and cliff edges on south-facing sea-cliffs of chalk and other coastal limestones in southern Britain, is the *Brassica oleracea* community (MC4). In its more maritime expression, where *Beta vulgaris* ssp. *maritima* is a distinctive preferential on more spray-splashed limey cliffs, the vegetation is virtually identical with the *Brassicetum oleraceae* Géhu 1962, another Crithmo-Armerion community, but, in Britain the *Brassica* assemblage also shows strong links with cliff-top Mesobromion grasslands.

In the generally warmer and dried south of Britain, it is also possible that lower precipitation, especially in summer, accentuates the impact of spray, allowing these more maritime assemblages to extend further up cliffs. In this part of Britain, too, there is generally a more continuous shift from maritime crevice vegetation to therophyte-rich vegetation in shallow soils in crevices and around outcrops which are prone to summer drought but less splashed by spray. Here, the *Armeria maritima-Cerastium diffusum* ssp. *diffusum* community (MC5) is characteristic, an assemblage with a conspicuous contingent of winter annuals and one which provides a locus for numerous rare coastal plants, that are intolerant of competition. It can be seen as representing the maritime extreme among British communities of the alliance Thero-Airion Tüxen 1951.

On the west Scottish coast, by contrast, where the impact of spray is lessened by the wetter climate and shallow soils in crevices and on ledges do not experience such drought, the *Armeria-Cerastium* community is not so common and less strictly maritime vegetation can occupy positions quite close to the sea. Here, ungrazed ledges above the *Armeria-Ligusticum* zone often have the *Rhodiola rosea-Armeria maritima* community (MC3). In this vegetation, *Ligusticum* remains occasional but *Rhodiola* is a constant feature and, with *Holcus lanatus*, *Rumex acetosa* and other scattered tall herbs making up luxuriant herbage, the vegetation looks like a maritime equivalent of the alliance Adenostylion alliariae Br.-Bl. 1926 (Birks 1973).

Maritime grasslands of sea-cliffs

Beyond the most spray-splashed zone on sea-cliffs, the influence of salt input dies away and other factors play an increasing role in determining the character and dis-

position of the vegetation. Prime among these are the nature of the soils and the impact of grazing.

In less exposed situations, and as the steepness of the cliff profile lessens, deeper and more mature soils accumulate. The balance of minerals in the profile is strongly influenced by the lithology of the bedrock and, among maritime cliff communities as a whole, contrasts between vegetation types on rendzinas at one extreme of soil variation and rankers at the other are very clear. A second edaphic trend can be seen among communities of, on the one hand, very shallow, excessively draining and summer-parched soils and, on the other, deeper loams which are protected against drought by the gentler topography on which they occur and by their texture. In general, the shift up the cliff is marked in the vegetation by an increasing prominence of *Festuca rubra* and associates of the Arrhenatherion – *Holcus lanatus*, *Dactylis glomerata*, *Agrostis stolonifera*, *Plantago lanceolata* and *Trifolium repens*.

It is a moderate maritime influence and these edaphic factors which govern much of the floristics in and extent of the *Festuca rubra-Armeria maritima* (MC8), *Festuca rubra-Holcus lanatus* (MC9) and *Festuca rubra-Daucus carota* ssp. *gummifer* (MC11) grasslands. The first community is the most maritime of these swards, extending down to the more spray-splashed zone of cliffs where it shows transitions to crevice communities and usually occupying the bulk of the slopes above, at least on less heavily grazed cliffs. Where *F. rubra* is able to dominate, the herbage can be mattress-like and very species-poor but, on shallower soils in sunnier situations, richer assemblages of associates, with plants like *Anthyllis vulneraria*, *Lotus corniculatus*, *Silene vulgaris* ssp. *maritima* and *Sedum anglicum*, can join *Armeria maritima*.

On more sheltered slopes, usually gentler and with deeper, though generally free-draining, soils, the *Festuca-Armeria* grassland gives way to the *Festuca-Holcus* grassland. *F. rubra* often remains the most prominent grass here but the sward has an altogether more mesophytic look with frequent *H. lanatus*, *Dactylis*, *Trifolium repens*, *Rumex acetosa*, *Plantago lanceolata* and *Cerastium fontanum*. Where spray input is becoming negligible, this community shows floristic transitions to Cynosurion grasslands on more fertile soils and to Nardion swards on more acidic substrates.

With the shift to sea-cliffs cut into calcareous rocks, where rendzinas deepen to brown calcareous soils on gentler slopes with less spray deposition, the *Festuca-Holcus* community tends to be replaced in the grassland zonation by the *Festuca-Daucus* grassland. Here *F. rubra* and *Dactylis glomerata* are usually the most abundant grasses, but *D. carota* ssp. *gummifer* is the really distinctive preferential, giving the sward a striking appearance when flowering in abundance. The maritime element in the vegetation is small and the floristic variation is related most obviously to differences in soil depth and the degree of droughtiness. On cliff-tops with somewhat deeper soils, the grassland grades to Mesobromion swards.

Superimposed on these contrasts are influences related to the intensity of grazing. Many sea-cliffs are open to stock and, in some regions, they provide an important element of unenclosed pasture, though difficult topography generally limits access to the naturally less maritime grasslands and these, in any case, provide more luxuriant herbage for the animals. The overall effects of grazing are a shortening and tightening of the sward and a selection in favour of more resistant or less palatable species: among the commoner plants of cliff grasslands, it is usually *Festuca rubra*, *Holcus lanatus* and *Dactylis glomerata* that decline in favour of *Plantago lanceolata*, *P. maritima* and *P. coronopus* but the extent of this shift varies according to the numbers and kinds of stock. The trend towards the *Festuca rubra-Plantago* spp. grassland (MC10) can already be seen in the *Festuca-Armeria* and *Festuca-Holcus* swards but it is on Scottish cliff-tops, where grazing by sheep is especially intensive, that this community reaches its most extensive development. Typically, there is a closed, very tight sward, in which short-cropped *F. rubra* and grazing-resistant rosette plants predominate though, because of the wetter climate in many areas where grazing is heavy, there is often some *Carex nigra* and *C. panicea* in the vegetation and more strongly flushed swards show transitions to poor-fen and wet heath of the Ericion tetralicis type. Drier *Festuca-Plantago* swards also occur locally on cliff-tops in south-western Britain but here stock are less frequently given free range of cliffs and there is often a sharp boundary between ungrazed maritime grasslands (or heath) and the enclosed pasture of farms. Even where cliffs are grazed, it is usually by beef cattle rather than sheep and their less assiduous cropping does not have such a marked effect on the sward.

All of these maritime grasslands can be readily accommodated in a new alliance, the Silenion maritimae, proposed by Malloch (1972, 1976) as part of the order Glauco-Puccinellietalia Beeftink & Westhoff 1962. A further vegetation type of gentler cliff-top slopes with relatively little spray deposition, the *Festuca rubra-Hyacinthoides non-scripta* community (MC12),

could also probably be located there, though it has clear affinities with Arrhenatherion grasslands and Rubion sub-scrub. In this community, *F. rubra* remains a prominent component of the vegetation along with *H. lanatus*, *Rumex acetosa* and also often *Dactylis*, but *Hyacinthoides* is constant and frequently abundant in lush herbage. Usually this vegetation replaces the *Festuca-Holcus* grassland with the shift to deep, moist and fertile brown earths, often on shadier north-facing slopes or in gullies, and always where there is no grazing. Despite close floristic relationships to scrub, it is hard to see how some stands of the *Festuca-Hyacinthoides* community could ever have had a woody canopy or, with even the moderate spray input, could develop one in the future.

Vegetation of sea-bird cliffs

Coastal cliffs provide locations for the resting, roosting and nesting activities of various kinds of sea-birds, particularly gulls and auks, especially in places where human disturbance or predators are scarce, where there are ample ledges and where good supplies of food are accessible. Sea-birds, particularly when gathered in colonies, have striking effects on vegetation which often override the influence of other environmental factors. Particularly important are the repeated disturbance of the soil and herbage by nesting activities and display and the deposition of faeces rich in phosphorus and nitrogen. Two vegetation types associated with such effects have been characterised in this scheme. The *Atriplex prostata-Beta vulgaris* ssp. *maritima* community (MC6, *Atriplici-Betetum maritimae* J.-M. & J. Géhu 1969), variously dominated by luxuriant growth of *Atriplex* spp., *Beta* or *Lavatera arborea*, is typical of sea-bird colonies on cliffs in southern Britain, replacing crevice vegetation where there is abundant guano or occurring locally within the *Festuca-Armeria* zone. The *Stellaria media* community (MC7) is more widely distributed and tends to replace the *Atriplici-Betetum* on more northerly cliffs. Its maritime character is less pronounced but both vegetation types are probably best located with certain shingle strandlines in the Honkenyo-Crambion J.-M. & J. Géhu 1969, an alliance which brings together communities of enriched coastal habitats, mostly boreal in distribution. Indeed, vegetation often indistinguishable from the *Atriplici-Betetum* can be found on standline detritus around southern coasts.

KEY TO MARITIME CLIFF COMMUNITIES

With something as complex and variable as vegetation, no key can pretend to offer an infallible short cut to diagnosis. The following should thus be seen as simply a crude guide to identifying the types of vegetation found on maritime cliffs and must always be used in conjunction with the data tables and community descriptions. It relies on floristic (and, to a lesser extent, physiognomic) features of the vegetation and demands a knowledge of the British vascular flora and, in only a few cases here, of bryophytes and lichens. It does not make primary use of any habitat features, though these may provide a valuable confirmation of a diagnosis.

Because the major distinctions between the vegetation types in the classification are based on inter-stand frequency, the key works best when sufficient samples of similar composition are available to construct a constancy table. It is the frequency values in this (and, in some cases, the ranges of abundance) which are then subject to interrogation with the key.

Samples should always be taken from homogeneous stands and be 2×2 m or 4×4 m according to the scale of the vegetation or, where stands are irregular, of identical size but different shape.

---

1   Open, generally species-poor, vegetation of rock crevices with usually 25% of *Festuca rubra* and one or more of *Crithmum maritimum, Spergularia rupicola, Inula crithmoides*, rayed *Aster tripolium, Ligusticum scoticum* and *Schistidium maritimum*; *Sedum* spp. usually absent   2

Not as above   5

2   *Ligusticum scoticum* present, *Crithmum maritimum* absent

    MC2 *Armeria maritima-Ligusticum scoticum* maritime rock-crevice community

Ligusticum scoticum absent, one or more of *Crithmum maritimum, Spergularia rupicola, Inula crithmoides* and rayed *Aster tripolium* present

    MC1 *Crithmum maritimum-Spergularia rupicola* maritime rock-crevice community
    Crithmo-Spergularietum rupicolae Géhu 1964   3

In a small area of south-west Scotland, notably around the Mull of Galloway, *Ligusticum scoticum* may occur together with *Crithmum maritimum* and/or *Spergularia rupicola* in crevice vegetation on sea-cliffs.

---

3   Rayed *Aster tripolium* present without *Inula crithmoides*

    MC1 *Crithmo-Spergularietum*
    Aster tripolium sub-community

Rayed *Aster tripolium* absent or, if present, then *Inula crithmoides* also present   4

4   One or more of *Inula crithmoides, Limonium binervosum* and *Parapholis incurva* present

    MC1 *Crithmo-Spergularietum*
    Inula crithmoides sub-community

Crithmum maritimum and *Spergularia rupicola* present without the above species

    MC1 *Crithmo-Spergularietum*
    Typical sub-community

5   Generally small stands of open vegetation with perennial grasses and *Plantago maritima* having less than 50% cover and/or with *Cerastium diffusum* ssp. *diffusum* and two of *Sedum anglicum, S. acre, Bromus hordeaceus* ssp. *ferronii, Aira praecox, Desmazeria marina* and *Arenaria serpyllifolia*   6

Generally closed vegetation with perennial grasses having more than 50% cover and/or without the above combination of species 10

6 *Sedum* spp. absent 7

Sedum anglicum and/or *S. acre* present 8

7 *Armeria maritima* dominant with some *Festuca rubra* in a very species-poor vegetation; *Desmazeria marina* and *Bromus hordeaceus* ssp. *ferronii* absent

MC8 *Festuca rubra-Armeria maritima* maritime grassland
Armeria maritima-dominated sub-community

Armeria maritima, *Festuca rubra* and *Plantago coronopus* constant and sometimes abundant with *Desmazeria marina* and *Bromus hordeaceus* ssp. *ferronii*

MC5 *Armeria maritima-Cerastium diffusum* ssp. *diffusum* maritime therophyte community
Desmazeria marina sub-community

8 *Sedum acre* present with *Arenaria serpyllifolia, Bromus hordeaceus* ssp. *ferronii, Dactylis glomerata* and *Thymus praecox*

MC5 *Armeria maritima-Cerastium diffusum* ssp. *diffusum* maritime therophyte community
Arenaria serpyllifolia sub-community

Sedum anglicum present or *S. acre* without *Arenaria serpyllifolia* 9

9 *Sedum anglicum* (or rarely *S. acre*) present with frequent *Desmazeria marina, Spergularia rupicola* and *Anthyllis vulneraria*

MC5 *Armeria maritima-Cerastium diffusum* ssp. *diffusum* maritime therophyte community
Anthyllis vulneraria sub-community

Sedum anglicum (or rarely *S. acre*) present with infrequent records for the above associates but with *Aira praecox* and *Festuca ovina* constant

MC5 *Armeria maritima-Cerastium diffusum* ssp. *diffusum* maritime therophyte community
Aira praecox sub-community

10 Luxuriant ledge vegetation with prominent *Rhodiola rosea*

MC3 *Rhodiola rosea-Armeria maritima* maritime cliff-ledge community

Not as above 11

11 Scruffy vegetation with one or more of *Atriplex prostrata* agg., *Beta vulgaris* ssp. *maritima, Matricaria maritima, Lavatera arborea* and *Stellaria media* conspicuous 12

Closed grassy swards without the above species 13

12 *Stellaria media* present usually with *Rumex acetosa* and *Holcus lanatus*

MC7 *Stellaria media-Rumex acetosa* sea-bird cliff community

Atriplex prostrata agg., *Beta vulgaris* ssp. *maritima* and/or *Lavatera arborea* dominant

MC6 *Atriplex prostrata-Beta vulgaris* ssp. *maritima* sea-bird cliff community
Atriplici-Betetum maritimae J.-M. & H. Géhu 1969

13 Vegetation dominated by *Festuca rubra* with a little *Crithmum maritimum* or *Ligusticum scoticum* 14

Crithmum maritimum and *Ligusticum scoticum* absent 15

14 *Crithmum maritimum* and *Daucus carota* ssp. *gummifer* present

MC8 *Festuca rubra-Armeria maritima* maritime grassland
Crithmum maritimum sub-community

Ligusticum scoticum present

MC8 *Festuca rubra-Armeria maritima* maritime grassland
Ligusticum scoticum sub-community

In a small area of south-west Scotland, notably around the Mull of Galloway, *Ligusticum scoticum* and *Crithmum maritimum* may occur together in these transitions between maritime grassland and crevice vegetation.

15 Vegetation overwhelmingly dominated by *Festuca rubra* which forms a thick mattress-like sward; *Holcus lanatus* and *Anthyllis vulneraria* infrequent and never abundant

MC8 *Festuca rubra-Armeria maritima* maritime grassland
Typical sub-community

Festuca rubra generally dominant but *Holcus lanatus* constant and sometimes abundant and *Rumex acetosa* frequent but *Plantago lanceolata* infrequent

MC8 *Festuca rubra-Armeria maritima* maritime grassland
Holcus lanatus sub-community

It may sometimes be difficult to partition samples between the *Holcus lanatus* sub-community of the *Festuca rubra-Armeria maritima* maritime grassland and the *Festuca rubra-Holcus lanatus* maritime grassland in gradual transitions between the two communities: see 28.

Festuca rubra generally dominant but with constant and sometimes abundant *Anthyllis vulneraria* and *Silene vulgaris* ssp. *maritima*

    MC8 *Festuca rubra-Armeria maritima* maritime grassland
    Anthyllis vulneraria sub-community

Festuca rubra may be prominent but not with the above combinations of species   16

---

16   *Festuca rubra*, *Dactylis glomerata* and *Daucus carota* ssp. *gummifer* present with one or more of *Brassica oleracea*, *Ononis repens*, *Bromus hordeaceus* ssp. *ferronii* and *Sanguisorba minor*   17

Not as above   20

---

17   *Brassica oleracea* and *Beta vulgaris* ssp. *maritima* present

    MC4 *Brassica oleracea* maritime cliff-ledge community
    Beta vulgaris ssp. *maritima* sub-community

Brassica oleracea absent or present without *Beta vulgaris* ssp. *maritima*   18

---

18   *Ononis repens*, *Centaurea scabiosa* and *Rumex acetosa* present with *Brassica oleracea* and *Silene nutans*

    MC4 *Brassica oleracea* maritime cliff-ledge community
    Ononis repens sub-community

Ononis repens present, sometimes with *Centaurea scabiosa* and *Rumex acetosa* but with little or no *Brassica oleracea* or *Silene nutans*

    MC11 *Festuca rubra-Daucus carota* ssp. *gummifer* maritime grassland
    Ononis repens sub-community

It may sometimes be difficult to partition samples between these two vegetation types in gradual transitions on the tops of limestone cliffs.

Not as above   19

---

19   *Sanguisorba minor* present, often with *Brachypodium pinnatum*

    MC11 *Festuca rubra-Daucus carota* ssp. *gummifer* maritime grassland
    Sanguisorba minor sub-community

Sanguisorba minor and *Brachypodium pinnatum* absent; *Armeria maritima* and *Bromus hordeaceus* ssp. *ferronii* frequent

    MC11 *Festuca rubra-Daucus carota* ssp. *gummifer* maritime grassland
    Bromus hordeaceus ssp. *ferronii* sub-community

---

20   *Festuca rubra*, *Plantago lanceolata*, *P. maritima* and *P. coronopus* all abundant in a short tight turf   21

Not as above   24

---

21   *Carex panicea* present   23

Carex panicea absent   22

---

22   *Euphrasia* spp. and *Plantago lanceolata* infrequent

    MC8 *Festuca rubra-Armeria maritima* maritime grassland
    Plantago coronopus sub-community

Euphrasia spp. and *Plantago lanceolata* frequent

    MC10 *Festuca rubra-Plantago* spp. maritime grassland
    Armeria maritima sub-community

It may sometimes be difficult to partition samples between these two vegetation types where grazing varies in intensity on cliff-tops.

---

23   *Schoenus nigricans*, *Carex serotina*, *Danthonia decumbens*, *Potentilla erecta* and *Molinia caerulea* constant; *S. nigricans* or *M. caerulea* may dominate

    MC10 *Festuca rubra-Plantago* spp. maritime grassland
    Schoenus nigricans sub-community

Above species infrequent and rarely abundant; *Lotus corniculatus* constant

    MC10 *Festuca rubra-Plantago* spp. maritime grassland
    Carex panicea sub-community

24   *Festuca rubra* generally dominant with abundant *Holcus lanatus* in a rank tussocky sward; *Plantago lanceolata* and *P. maritima* may occur but *P. coronopus* is absent   27

Not as above   25

25   *Hyacinthoides non-scripta* present

MC12 *Festuca rubra-Hyacinthoides non-scripta* maritime grassland   26

Hyacinthoides non-scripta absent   27

26   *Armeria maritima* and *Silene vulgaris* ssp. *maritima* constant

MC12 *Festuca rubra-Hyacinthoides non-scripta* maritime grassland
Armeria maritima sub-community

Ranunculus ficaria and *Heracleum sphondylium* frequent in small amounts

MC12 *Festuca rubra-Hyacinthoides non-scripta* maritime grassland
Ranunculus ficaria sub-community

27   *Heracleum sphondylium* constant in small amounts though conspicuous when flowering; *Plantago maritima* absent

MG1 *Arrhenatheretum elatioris*

Plantago maritima frequent and *Heracleum sphondylium* rare

MC9 *Festuca rubra-Holcus lanatus* maritime grassland   28

It may sometimes be difficult to partition samples between these two communities in less maritime situations with deep, moist neutral soils on ungrazed cliffs.

28   *Achillea millefolium* and *Galium verum* constant often with occasionals of rich neutral or calcareous grasslands

MC9 *Festuca rubra-Holcus lanatus* maritime grassland
Achillea millefolium sub-community

Achillea millefolium and *Galium verum* infrequent in a more species-poor sward   29

29   *Dactylis glomerata* constant and often abundant with frequent *Scilla verna* and *Daucus carota* ssp. *gummifer*

MC9 *Festuca rubra-Holcus lanatus* maritime grassland
Dactylis glomerata sub-community

Samples of *Festuca rubra-Daucus carota* ssp. *gummifer* maritime grassland may key out here but *Holcus lanatus* is much less frequent in that community and the various sub-community differentials are almost always present: see 18 in the key.

Not as above   30

30   *Anthoxanthum odoratum*, *Agrostis capillaris*, *Poa subcaerulea* and *Potentilla erecta* frequent

MC9 *Festuca rubra-Holcus lanatus* maritime grassland
Anthoxanthum odoratum sub-community

Not as above   31

Samples of cliff-top *Festuca-Agrostis-Galium* grassland may key out here, especially if *Festuca rubra* and *F. ovina* have been confused, but that vegetation is generally a short tight turf and *F. ovina* is very infrequent in the *Festuca rubra-Holcus lanatus* maritime grassland.

31   *Plantago maritima* constant with infrequent *Achillea millefolium*, *Dactylis glomerata* and *Anthoxanthum odoratum*

MC9 *Festuca rubra-Holcus lanatus* maritime grassland
Plantago maritima sub-community

Lightly-grazed samples of *Festuca rubra-Plantago* spp. maritime grassland may key out here, especially in the north where *Plantago coronopus*, a generally diagnostic species of that community, is less frequent.

Primula vulgaris constant with frequent *Geranium sanguineum*

MC9 *Festuca rubra-Holcus lanatus* maritime grassland
Primula vulgaris sub-community

COMMUNITY DESCRIPTIONS

MC1

Crithmum maritimum-Spergularia rupicola maritime rock-crevice community
Crithmo-Spergularietum rupicolae   Géhu 1964

Synonymy
Armeria maritima-Aster tripolium provisional nodum Ivimey-Cook & Proctor 1966 *p.p.*; *Crithmum maritimum* rock-crevice community Proctor 1975; Crithmion communities Shimwell 1976 ms.

Constant species
Crithmum maritimum, Spergularia rupicola, Festuca rubra, Armeria maritima.

Rare species
Parapholis incurva, Limonium recurvum (and probably other apomicts of the *L. binervosum* group, e.g. *L. paradoxum, L. transwallianum*; see also Ingrouille 1981).

Physiognomy
The *Crithmo-Spergularietum* has a low-growing, very open cover of scattered vascular perennials rooted in rock crevices. The distribution of the plants and the overall appearance of the vegetation are strongly influenced by the nature of the substrate and, in general, none of the association constants can be said to be truly dominant, though all but *Spergularia rupicola* may be particularly abundant in individual stands. *Plantago maritima* is the most frequent associate throughout the association. Bryophytes are rarely conspicuous with only *Schistidium maritimum* and *Tortella flavovirens* recorded very occasionally. The only lichens are epilithic.

Sub-communities

Typical sub-community: *Crithmo-Spergularietum rupicolae typicum*   Géhu 1964. The four association constants here account for the bulk of the vegetation cover and, of the associates, only *Plantago maritima* and *P. coronopus* attain a constancy above I.

***Iula crithmoides* sub-community:** *Crithmo-Spergularietum rupicolae plantaginetosum coronopi*   Géhu 1964 *p.p. Spergularia rupicola* is rather less frequent in this sub-community but *Inula crithmoides* and *Plantago coronopus* are both constant and the former may be abundant. *Limonium binervosum* (including its apomicts), *Parapholis incurva* and *Desmazeria marina* are all preferentially frequent.

Rayed *Aster tripolium* sub-community: *Armeria maritima-Aster tripolium* provisional nodum   Ivimey-Cook & Proctor 1966. Apart from *Armeria maritima*, the frequency of the association constants is reduced in this, the most open of the sub-communities. Rayed *Aster tripolium* is constant and *Cochlearia officinalis, Atriplex hastata, Cerastium tetrandum* ssp. *diffusum, Matricaria maritima* and *Silene vulgaris* ssp. *maritima* are all preferentially frequent.

Habitat
Within its geographical range, the *Crithmo-Spergularietum* occupies the most maritime zone of vascular plant vegetation on rocky cliffs, where there is the largest input of salt from sea-spray and onshore winds and the highest salt content in the soil (Malloch & Okusanya 1979). *Crithmum maritimum, Spergularia rupicola* and *Inula crithmoides* have all been shown to germinate and grow adequately at high salinities (Okusanya 1979*a, b*).

Within this extreme maritime zone, the occurrence of the association is limited mainly by the availability of crevices, narrow ledges and friable rock surfaces: it is, for example, rare on massively-jointed granites even where these are very exposed as on much of the Land's End peninsula. The chemical composition of the rock and of the soils, which are generally simply skeletal accumulations of mineral and organic debris, seems to be of minor importance, though there is a tendency for the *Inula* sub-community to be associated with more calcareous situations, especially where the bedrock surface is friable, as on thinly-bedded limestones. Higher soil calcium content may also derive from accumulation of wind-blown shell-sand or shells dropped by sea-birds or from input of calcium-rich drainage waters from above. Soil pH is consistently high (mean = 7.5) irrespective of rock and soil type because of the high input of sodium ions from spray. The soils are usually free-draining and may experience summer parching. The association also occurs in the crevices of harbour walls and moles.

The *Crithmo-Spergularietum* is rarely grazed, being usually inaccessible to stock but the vegetation is some-times disturbed and enriched manurially by sea-birds.

Zonation and succession
Seawards, the *Crithmo-Spergularietum* overlaps with and grades into the grey xeric-supralittoral lichen zone, generally occupied by the *Ramalinetum scopularis* (DR 1925) Klem. 1955 (Fletcher 1973a, b; James *et al.* 1977). Above, there is usually an abrupt transition to the *Festuca-Armeria* maritime grassland on deeper, moister soils in less maritime conditions. In some cases, the zonation occurs gradually over progressively deeper and less saline soils through the *Crithmum* sub-community of the grassland. Where excessively-drained soils occur above the most maritime zone on cliffs, the *Crithmo-Spergularietum* gives way to the *Armeria-Cerastium* community.

On sea-walls, the association may be replaced by crevice vegetation which has a mixture of maritime and inland mural species like the *Cymbalarietum*.

There is no evidence of any successional progression from *Crithmo-Spergularietum*.

Distribution
The association can be seen as the northernmost exten-sion of a complex of maritime communities centred around the Mediterranean. It is restricted in Britain to the south and west coasts, terminating northwards at the Mull of Galloway where it is replaced by the *Armeria-Ligusticum* community in an abrupt transition that is probably governed by temperature acting directly on *Crithmum maritimum*, *Spergularia rupicola* and *Inula crithmoides* or retarding their flowering and fruiting (Okusanya 1979c).

The typical and *Inula* sub-communities occur throughout the range of the association, though the latter is commonest on the limestones of south Wales and southern England. By contrast, the *Aster* sub-com-munity appears to be more Atlantic in its distribution, being confined in Britain to the western extremities of Wales and Cornwall, with the very similar community described by Ivimey-Cook & Proctor (1966) occurring widely on the western coast of Eire. This distribution may be a response to the higher precipitation/evapora-tion ratios on this extreme seaboard.

Affinities
The Crithmo-Armerietalia was defined by Géhu (1964) within the Crithmo-Limonietea as containing maritime crevice communities of the Atlantic coast of Europe and the vegetation described here is essentially identical to the *Crithmo-Spergularietum* of Brittany. A similar com-munity has been described from Alderney (Proctor 1975).

Géhu's *typicum* is identical to the typical sub-commu-nity described here and the *Inula* sub-community is similar to his *Plantago coronopus* sub-association. Although *Aster tripolium* is absent from the Brittany stands, Géhu has a variant of *Cochlearia officinalis* which bears some resemblance to the *Aster* sub-commu-nity.

Floristic table MC1

	a	b	c	1
Armeria maritima	V (2–7)	V (2–7)	V (1–7)	V (1–7)
Festuca rubra	V (1–5)	V (1–8)	III (1–7)	IV (1–8)
Crithmum maritimum	V (2–7)	V (1–8)	III (1–4)	IV (1–8)
Spergularia rupicola	V (1–5)	III (1–5)	III (1–4)	IV (1–5)
Inula crithmoides		V (1–8)	I (1)	III (1–8)
Plantago coronopus	II (1–3)	IV (1–4)	I (1–4)	II (1–4)
Limonium binervosum		III (1–7)		II (1–7)
Parapholis incurva	I (2–3)	II (2–4)		I (2–4)
Desmazeria marina	I (1–2)	II (1–3)	I (1–3)	I (1–3)
Aster tripolium (rayed)		I (1–4)	V (1–4)	II (1–4)
Cochlearia officinalis	I (1–2)	I (1–4)	III (1–4)	II (1–4)
Atriplex prostrata	I (1–3)	I (2)	II (1–3)	I (1–3)
Cerastium diffusum diffusum	I (3)	I (2)	II (1–4)	I (1–4)
Matricaria maritima	I (1)	I (1–3)	II (1–3)	I (1–3)
Silene vulgaris maritima	I (1–4)	I (2–3)	II (1–5)	I (1–5)
Plantago maritima	III (1–4)	II (1–5)	II (1–3)	II (1–5)
Asplenium marinum	I (1)	I (1–2)	I (1–3)	I (1–3)
Beta vulgaris maritima	I (1–4)	I (1–4)	I (3–4)	I (1–4)
Cochlearia danica	I (1–3)	I (1–4)	I (2)	I (1–4)
Daucus carota gummifer		I (1–4)	I (2)	I (1–4)
Tortella flavovirens	I (2–3)	I (2–4)	I (2–5)	I (2–5)
Number of samples	26	60	37	123
Number of species/sample	6 (3–12)	8 (3–12)	6 (3–13)	7 (3–13)
Vegetation height (cm)	8 (3–20)	13 (3–50)	10 (3–50)	11 (3–50)
Total cover (%)	26 (5–8)	46 (5–98)	20 (5–80)	34 (5–98)
Altitude (m)	10 (3–40)	16 (3–30)	13 (3–29)	14 (3–40)
Slope (°)	35 (4–80)	30 (0–90)	36 (5–90)	33 (0–90)
Soil depth (cm)	3 (1–5)	11 (2–40)	9 (3–20)	9 (1–40)
Number of soil samples	6	15	7	28
Superficial pH	6.8 ±0.3	7.8 ±0.2	7.3 ±0.2	7.4 ±0.1
Water content (% soil dry weight)	73 ±24	31 ±6	47 ±16	46 ±7
Loss on ignition (% soil dry weight)	21 ±6	11 ±2	19 ±9	15 ±3
Sodium (mole g^{-1})	89 ±32	82 ±23	80 ±54	83 ±18
Potassium (mole g^{-1})	14 ±3	13 ±2	8 ±3	12 ±1
Magnesium (mole g^{-1})	101 ±25	55 ±8	50 ±11	64 ±7
Calcium (mole g^{-1})	49 ±9	73 ±9	65 ±9	66 ±6
Phosphorus (mole g^{-1})	6.2 ±2.5	0.8 ±0.4	1.1 ±0.5	2.0 ±0.7
Sodium/loss on ignition (mole g^{-1})	411 ±44	830 ±190	593 ±194	681 ±115

a   Typical sub-community
b   *Inula crithmoides* sub-community
c   *Aster tripolium* sub-community
1   *Crithmo-Spergularietum rupicolae* maritime rock-crevice community (total)

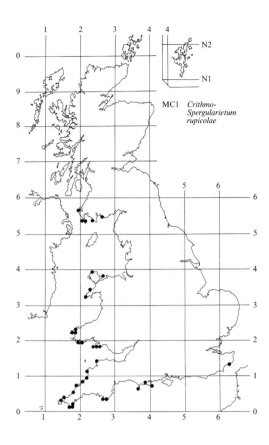

MC1   *Crithmo-*
Spergularietum
rupicolae

MC2
Armeria maritima-Ligusticum scoticum maritime rock-crevice community

Synonymy

Armeria maritima-Grimmia maritima rock crevice community Ostenfeld 1908; *Armeria maritima-Ligusticum scoticum* low cliff vegetation Petch 1933; *Armeria maritima-Grimmia maritima* and *Asplenium marinum-Grimmia maritima* Associations Birks 1973; Habitat Groups II & III Goldsmith 1975; *Armeria maritima-Grimmia maritima* association Malloch & Okusanya 1979.

Constant species

Armeria maritima, Festuca rubra, Ligusticum scoticum, Schistidium maritimum.

Physiognomy

As in the *Crithmo-Spergularietum*, the vegetation comprises a low-growing, very open cover, mainly of vascular perennials whose precise arrangement is strongly influenced by the availability and pattern of rock-crevices. There is no single dominant but *Armeria maritima, Festuca rubra* and *Ligusticum scoticum* may each be abundant in particular stands. *Plantago maritima, Rhodiola rosea* and *Silene vulgaris* ssp. *maritima* are frequent. Although *Schistidium maritimum* is occasionally present in the *Crithmo-Spergularietum*, it is here constant, though always in small amounts. Other bryophytes are rare. Birks (1973) recorded *Anaptychia fusca* and *Ramalina siliquosa* as frequently occurring lichens exclusive to this vegetation on Skye.

Asplenium marinum is an occasional in the community but it may become locally abundant in sheltered rock-crevices with *Trichostomum brachydontium* (Birks 1973).

Habitat

The *Armeria-Ligusticum* community replaces *Crithmo-Spergularietum* as the most maritime vascular plant community north of Galloway. It occurs on all rock types, being limited mainly by the physical structure of the deposits. On softer sandstones, such as parts of the Old Red Sandstone which comprises much of the Caithness cliffs, the vegetation cover tends to be greater than usual. The community also occurs among moderately large pebbles on some spray-drenched shingle beaches.

The soils are always simply skeletal accumulations of rock fragments, blown sand and organic debris and, though moist, they are free-draining. Superficial pH is always high, around 7.

The local abundance of *Asplenium marinum* in sheltered situations is probably due to its susceptibility to air-frosting in the open.

The inaccessibility of the stands normally precludes grazing, though where sheep gain access *Ligusticum scoticum* is readily eaten (Goldsmith 1973, Tutin 1980*b*).

Zonation and succession

Towards high-water mark, the community grades into the *Ramalinetum scopularis* lichen zone which, in sheltered sites in western Scotland, may have some of the larger foliose lichens characteristic of Lobarion communities (James *et al*. 1977). Above, there is often a transition to the *Festuca-Armeria* maritime grassland, sometimes abrupt, in other cases more gradual through the *Ligusticum* variant of the typical sub-community of the grassland. On some very tall cliffs, where there are more sheltered ledges above, there may be a switch from the *Armeria-Ligusticum* community to the *Rhodiola-Armeria* community before an intact grassland develops. Where waterlogged saline soils occur on ledges or abut onto low cliff-tops, the community is replaced by perched salt-marsh vegetation of the *Leontodon* sub-community of the *Juncetum gerardi*.

Distribution

The *Armeria-Ligusticum* community is the northern equivalent of the *Crithmo-Spergularietum* extending from the Mull of Galloway round to Shetland with a few east coast occurrences as far south as St Abb's Head.

The switch from one community to the other is probably climatically controlled: the growth of *Ligusticum scoticum* is less sensitive to cold than *Crithmum maritimum*, *Spergularia rupicola* or *Inula crithmoides* and maximum germination requires cold, wet conditions. Drought sensitivity may also restrict its extension southwards (Okusanya 1979c).

Similar vegetation has been described form the Faeroes (Ostenfeld 1908) and from Norway (Nordhagen 1922, Störmer 1938, Skogen 1965).

Affinities
The vegetation included here belongs to the Arctic counterpart of the predominantly Atlantic maritime crevice communities grouped by Géhu (1964) in the Crithmo-Armerietalia. An alternative treatment of the *Asplenium marinum*-rich component would be to regard it as a separate maritime Asplenietea community, perhaps part of the *Asplenietum marinae* Br.-Bl. & R.Tx. 1952 (e.g. Birks 1973).

Floristic table MC2

Armeria maritima	V (2–7)
Festuca rubra	IV (2–8)
Ligusticum scoticum	IV (1–6)
Schistidium maritimum	IV (2–4)
Plantago maritima	III (2–4)
Rhodiola rosea	III (2–5)
Silene vulgaris maritima	III (2–4)
Agrostis stolonifera	II (1–4)
Cochlearia officinalis	II (1–4)
Matricaria maritima	II (1–4)
Plantago coronopus	I (1–4)
Leontodon autumnalis	I (1–2)
Cerastium fontanum	I (1–3)
Rumex crispus	I (1–3)
Atriplex hastata	I (2–4)
Spergularia rupicola	I (3–4)
Asplenium marinum	I (1–2)
Number of samples	41
Number of species/sample	7 (3–11)
Vegetation height (cm)	9 (2–20)
Total cover (%)	20 (5–100)
Altitude (m)	12 (2–50)
Slope (°)	29 (0–80)

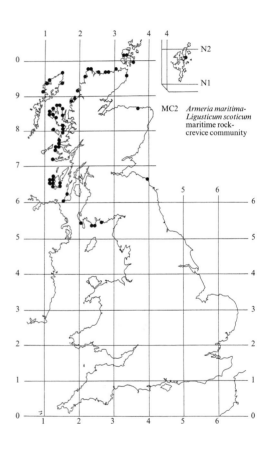

MC2 *Armeria maritima-Ligusticum scoticum maritime rock-crevice community*

MC3
Rhodiola rosea-Armeria maritima maritime cliff-ledge community

Synonymy
Rhodioletum roseae Nordhagen 1922 *p.p.*

Constant species
Festuca rubra, Armeria maritima, Rhodiola rosea, Rumex acetosa.

Physiognomy
The luxuriant herbaceous vegetation of this community has no single dominant. Usually, robust plants of *Rhodiola rosea* and *Rumex acetosa* and large cushions of *Armeria maritima* grow from a matrix of *Festuca rubra* in irregular and often fragmentary stands on cliff ledges. *Plantago maritima, P. lanceolata* and *Holcus lanatus* are frequent, the last sometimes occurring in abundance. Tall herbs such as *Angelica sylvestris* and *Silene dioica* may be prominent. No bryophytes were recorded.

Habitat
The *Rhodiola-Armeria* community occurs on ledges, mostly north-facing, above the most maritime zone on sea-cliffs. It is found on a variety of rock types where suitable ledges are developed: particularly fine stands may occur on prominently-bedded rocks such as the Old Red Sandstone of Caithness. The soils are irrigated rankers. Stands are quite inaccessible to grazing animals.

Zonation and succession
Stands of the community are generally discrete but may form part of a fragmented zonation up cliffs from the *Armeria-Ligusticum* maritime rock-crevice community to a more intact *Festuca-Armeria* maritime grassland or maritime heath above. On very tall cliffs, there may be a transition to less maritime ledge vegetation such as that described by Birks (1973) from Skye as the *Luzula sylvatica-Silene dioica* Association.

Distribution
The *Rhodiola-Armeria* community is a northern vegetation type, occurring around the coasts of Scotland from Islay round to Aberdeenshire. Similar vegetation has been described from the Faeroes (Ostenfeld 1908) and from Norway (Nordagen 1922, Skogen 1965, Engelskjön 1970).

Affinities
Birks (1973) placed this sea-cliff ledge vegetation with the inland tall-herb communities of the Mulgedion alpini. The more distinctly maritime nature of the *Rhodiola-Armeria* community suggests that it is probably best seen within the compass of the major maritime vegetation types.

Floristic table MC3

Festuca rubra	V (3–8)
Armeria maritima	V (3–7)
Rhodiola rosea	V (4–6)
Rumex acetosa	IV (3–8)
Holcus lanatus	III (3–7)
Plantago lanceolata	III (2–3)
Plantago maritima	III (2–4)
Silene vulgaris maritima	II (3–5)
Agrostis stolonifera	II (3–4)
Angelica sylvestris	II (2–5)
Ligusticum scoticum	II (2–4)
Silene dioica	II (3–6)
Lotus corniculatus	I (3–4)
Matricaria maritima	I (3–4)
Primula vulgaris	I (2–5)
Leontodon autumnalis	I (2)
Number of samples	13
Number of species/sample	8 (5–14)
Vegetation height (cm)	16 (10–30)
Total cover (%)	82 (20–100)
Altitude (m)	65 (5–210)
Slope (°)	59 (0–80)

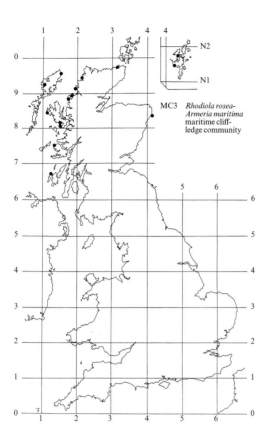

MC3   *Rhodiola rosea-
Armeria maritima*
maritime cliff-
ledge community

MC4
Brassica oleracea maritime cliff-ledge community

Synonymy
Includes *Brassicetum oleraceae* Géhu 1962.

Constant species
Festuca rubra, Brassica oleracea, Dactylis glomerata, Daucus carota ssp. *gummifer.*

Rare species
Brassica oleracea, Ophrys sphegodes, Silene nutans.

Physiognomy
The *Brassica oleracea* community generally has an irregular grassy cover of *Festuca rubra* and some *Dactylis glomerata* with prominent erect or decumbent plants of *B. oleracea* and a little *Daucus carota* spp. *gummifer. Plantago lanceolata* is the most frequent associate throughout, but it is never abundant. *Cheiranthus cheiri* and *Sonchus oleraceus*, though infrequent, may be conspicuous when flowering.

Sub-communities

***Beta vulgaris* ssp. *maritima* sub-community:** *Brassicetum oleraceae* Géhu 1962. *B. oleracea* is more abundant in this species-poor sub-community and *Beta vulgaris* ssp. *maritima* is an additional constant which may be co-dominant with *B. oleracea* and *Festuca rubra*. Maritime species, such as *Armeria maritima, Silene vulgaris* ssp. *maritima* and *Bromus hordeaceus* ssp. *ferronii* are confined to this sub-community though none is ever abundant.

***Ononis repens* sub-community.** Additional constants here are *O. repens, Silene nutans, Centaurea scabiosa* and *Rumex acetosa. B. oleracea* is less abundant here than in the *Beta* sub-community. Also frequent are *Brachypodium pinnatum* (which often co-dominates with *F. rubra*), *Hieracium pilosella* and *Teucrium scorodonia* and among the occasional species are some characteristic of calcicolous grasslands including *Ophys sphegodes* (Summerhayes 1968).

Habitat
The community is most characteristic of the crumbling edges and sloping ledges of south-facing cliffs in calcareous rocks. Soils are rendziniform, usually very shallow and dry, often fragmentary and maintained in a state of immaturity by substrate instability. Mitchell & Richards (1979) have suggested that *B. oleracea* is associated with phosphate-rich soil systems and that its distribution may be partly related to manurial enrichment by sea-birds.

The community occurs from the splash-zone to cliff-tops but, though it is more generally characteristic of sheltered coasts, the *Beta* sub-community seems to favour more maritime conditions.

Zonation and succession
The *Brassica* community can constitute the most maritime vegetation on relatively sheltered, dry calcareous cliffs though it occurs occasionally above a zone of *Crithmo-Spergularietum*. Inland it passes to the *Festuca-Armeria* or *Festuca-Daucus* maritime grasslands or to non-maritime calcareous grasslands or *Ligusticum* scrub.

Distribution
Samples were available only from the south coast of England and further investigation is needed to identify the floristic context of *B. oleracea* in its more northern stations (Mitchell 1976, Mitchell & Richards 1979).

Affinities
The *Beta* sub-community is virtually identical with the *Brassicetum oleraceae* described by Géhu (1962) from the French Channel coast and placed with other splash-zone communities in the Crithmo-Limonietea. The *Ononis* sub-community can be seen as a transition to Mesobromion calcareous grasslands and it may be peculiar to Britain.

Floristic table MC4

	a	b	4
Festuca rubra	V (3–9)	IV (2–8)	V (2–9)
Brassica oleracea	V (3–6)	V (2–4)	V (2–6)
Dactylis glomerata	IV (2–4)	V (3–4)	IV (2–4)
Daucus carota gummifer	V (2–3)	V (2–4)	V (2–4)
Beta vulgaris maritima	IV (3–7)		II (3–7)
Armeria maritima	III (2–4)		II (2–4)
Galium aparine	III (3–4)		II (3–4)
Potentilla reptans	II (4)		I (4)
Brassica nigra	II (4)		I (4)
Bromus hordeaceus ferronii	II (1–3)		I (1–3)
Sedum acre	II (1–2)		I (1–2)
Senecio vulgaris	II (2–3)		I (2–3)
Cirsium arvense	II (2–4)		I (2–4)
Silene vulgaris maritima	II (1)		I (1)
Ononis repens	I (2)	V (2–4)	III (2–4)
Silene nutans		IV (2–5)	III (2–5)
Centaurea scabiosa		IV (2–3)	II (2–3)
Rumex acetosa		IV (2–3)	II (2–3)
Brachypodium pinnatum		III (3–9)	II (3–9)
Hieracium pilosella		III (3–5)	II (3–5)
Teucrium scorodonia		III (3–4)	II (3–4)
Centaurea nigra		II (2–4)	II (2–4)
Festuca arundinacea		II (1–5)	II (1–5)
Senecio jacobaea		II (1–2)	II (1–2)
Echium vulgare	I (1)	II (1–2)	II (1–2)
Agrostis stolonifera		II (3–6)	I (3–6)
Anthyllis vulneraria		II (2–3)	I (2–3)
Taraxacum sp.		II (2–3)	I (2–3)
Tragopogon pratensis		II (2)	I (2)
Plantago lanceolata	II (2–3)	III (1–3)	III (1–3)
Sonchus oleraceus	II (2–4)	II (1–2)	II (1–4)
Cheiranthus cheiri	I (1)	II (1–4)	II (1–4)
Leucanthemum vulgare	I (2)	II (3)	II (2–3)
Plantago coronopus	II (1–3)	I (2)	II (1–3)
Number of samples	8	10	18
Number of species/sample	10 (6–12)	16 (10–28)	13 (6–28)
Vegetation height (cm)	52 (15–80)	19 (10–50)	34 (10–80)
Total vegetation cover (%)	86 (20–100)	88 (50–100)	87 (20–100)
Altitude (m)	23 (3–40)	23 (4–65)	23 (3–65)
Slope (°)	26 (0–80)	33 (10–60)	29 (0–80)

a   *Beta vulgaris* ssp. *maritima* sub-community

b   *Ononis repens* sub-community

4   *Brassica oleracea* maritime cliff-ledge community (total)

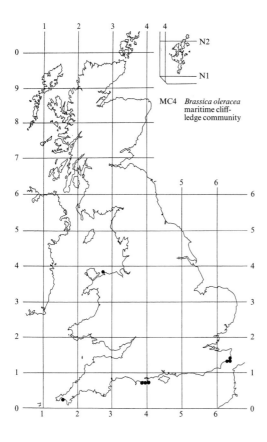

MC4 *Brassica oleracea*
maritime cliff-
ledge community

MC5
Armeria maritima-Cerastium diffusum ssp. *diffusum* maritime therophyte community

Synonymy

Includes *Trifolium occidentale-Herniaria ciliolata-Catapodium marinum* and *Trifolium occidentale-Scilla autumnalis-Jasione montana* noda Coombe 1961; *Sagino-Catapodietum marinae* R.Tx. & Westhoff 1963; *Thero-Sedetum anglici* Malloch 1971; Sedion anglici communities Proctor 1975.

Constant species

Armeria maritima, Plantago coronopus, Festuca rubra, Cerastium diffusum ssp. *diffusum, Sedum* spp. (*S. anglicum* and *S. acre*).

Rare species

Allium schoenoprasum, Astragalus danicus, Brassica oleracea, Centaurium capitatum, Genista pilosa, Herniaria ciliolata, Lotus hispidus, Mibora minima, Minuartia verna, Ononis reclinata, Ornithopus pinnatus, Poa bulbosa, P. infirma, Polycarpon tetraphyllum, Romulea columnae, Scilla autumnalis, S. verna, Senecio integrifolius ssp. *maritimus, Trifolium occidentale, T. suffocatum.*

Physiognomy

The *Armeria-Cerastium* community has a very short open turf in which cushions of *A. maritima*, tussocks of rather poorly-growing *F. rubra, P. coronopus* or *Sedum* spp. may dominate. Sprawling plants of *C. diffusum* ssp. *diffusum* occur throughout and the annual grasses *Desmazeria marina* and *Bromus hordeaceus* ssp. *ferronii* are frequent. Numerous winter annuals flourish on the patches of bare ground and many of these are national rarities. *Trifolium* spp. of restricted distribution are also characteristic of some sub-communities. Bryophytes occur at low frequencies throughout but in some sub-communities they and lichens may attain up to 20% cover.

Sub-communities

Desmazeria marina sub-community: *Sagino-Catapodietum marinae* R.Tx. & Westhoff 1963 *p.p.* In this, the most species-poor of the sub-communities, *A. maritima* and either *F. rubra* or *D. marina* tend to co-dominate with scattered plants of *C. diffusum* ssp. *diffusum, B. hordeaceus* ssp. *ferronii* and *Spergularia rupicola*. Other species are relatively few but *Sagina maritima, Cochlearia officinalis* and slimy pads of *Collema* spp. are distinctive.

Anthyllis vulneraria sub-community: *Trifolium occidentale-Herniaria ciliolatia-Catapodium marinum* nodum Coombe 1961; *Sagino-Catapodietum marinae* R.Tx. & Westhoff 1963 *p.p.*; Sedion anglici releves 1–10 Proctor 1975. Although this sub-community shares with the last the constancy of *Desmazeria marina* and frequent occurrence of *Spergularia rupicola*, they are here joined by *Sedum anglicum* which may occasionally be abundant and sometimes intermixed with a little *S. acre*. This and the next sub-community share a large number of associates but the positive features here are the high frequency and occasional abundance of *A. vulneraria* and the slightly preferential occurrence of *Herniaria ciliolata, Trifolium arvense* and *T. occidentale.*

Aira praecox sub-community: *Trifolium occidentale-Scilla autumnalis-Jasione montana* nodum Coombe 1961; Sedion anglici releves 11–15 Proctor 1975. *S. anglicum* is here joined by *A. praecox* and *Festuca ovina* as constants and maritime therophyte and crevice species are much less prominent than in the two previous sub-communities. Of the species shared with *Anthyllis* sub-community, most are more frequent here, notably *Jasione montana, Aira caryophyllea, Hypnum cupressiforme* and the lichens *Cladonia foliacea, C. rangiformis* and *C. chlorophaea.*

Arenaria serpyllifolia sub-community. Although *Sedum acre* occurs in the two previous sub-communities, it here totally replaces *S. anglicum* and is occasionally co-dominant with *F. rubra. A. serpyllifolia* is a differential constant and *B. hordaeceus* ssp. *ferronii, Thymus praecox* and *Dactylis glomerata* (perhaps in the form ssp. *hispanica*) are also constant. Many low frequency differentials occur in this sub-community, notably *Desmazeria rigida*, *Echium vulgare*, *Hieracium pilosella*, *Salvia horminoides*, *Taraxacum* sp. and species characteristic of inland calcareous grasslands as well as the national rarities *Ononis reclinata* and *Poa bulbosa*.

Habitat

The *Armeria-Cerastium* community is characteristic of excessively-draining, often very shallow soils at all levels of rocky cliffs, occurring most often in crevices and hollows which accumulate skeletal mixtures of mineral and organic matter or around rock outcrops where deeper soils thin out. It is generally ungrazed: even where there is no cliff-top enclosure, the vegetation is usually out of the reach of stock.

Floristic variation between the sub-communities can be related to differences in maritime influence and in bedrock and soil type. The *Desmazeria* sub-community occurs in the most maritime situations on all rock types and here the effects of salt-spray are probably responsible for the total exclusion of *Sedum* spp. The *Anthyllis* sub-community is the next most maritime and it is found on all rock types except chalk and the more friable limestones. On harder limestones, as on the south Wales coast, *S. acre* tends to replace *S. anglicum* in this sub-community. Both the *Aira* and *Arenaria* sub-communities are characteristic of the least maritime situations in which this community occurs. The *Aira* sub-community is more typical of rankers over non-calcareous rocks, the *Arenaria* sub-community of rendziniform soils on chalk and the softer limestones.

Zonation and succession

The community usually occurs in mosaics with other maritime vegetation types which vary according to the degree of maritime influence and the rock type. The *Desmazeria* sub-community is found at the level of the *Crithmo-Spergularietum* and typical *Festuca-Armeria* grassland and the *Anthyllis* sub-community within the *Festuca-Armeria* and *Festuca-Holcus* grasslands. The *Aira* sub-community occurs in the zone of the *Festuca-Holcus* grassland and maritime heaths and the *Arenaria* sub-community in comparable positions within the *Brassica* and *Festuca-Daucus* communities. Gradations to each of the surrounding communities may be gentle or sharp according to the change in soil depth.

There is some evidence that the *Armeria-Cerastium* community may initiate colonisation of disturbed and eroding rock surfaces on cliffs, being succeeded by the vegetation characteristic of the particular more stable combination of exposure and lithology.

Distribution

The *Armeria-Cerastium* community is predominantly southern since, in the cooler and damper climate of more northern cliffs, even the shallowest soils are able to carry more extensive crevice vegetation or maritime grassland. The *Anthyllis* sub-community is the most widespread with scattered occurrences up the west coast and also in north-west Scotland. The *Aira* sub-community has a similar distribution though it is much more common around the Mull of Galloway. The *Desmazeria* sub-community extends from Dorset to north Wales and the *Arenaria* sub-community is more frequent along the Channel coast.

Affinities

The community has floristic affinities with a variety of maritime vegetation types among the developing or eroding cover of which its distinctive therophyte element is able to gain a temporary or recurrent hold. The occurrence of halophyte ephemerals among this component has led some to place such vegetation with the communities of disturbed places on salt-marshes in the Saginetea maritimae (e.g. Tüxen & Westhoff 1963) but it is more appropriate to set them alongside similar inland vegetation in the Thero-Airion of the Sedo-Scleranthetea.

Floristic table MC5

	a	b	c	d	5
Armeria maritima	V (2–8)	V (2–8)	V (1–6)	II (2–5)	V (1–8)
Plantago coronopus	V (1–8)	V (1–8)	IV (2–8)	III (2–4)	V (1–8)
Festuca rubra	III (3–7)	V (2–8)	III (2–7)	V (3–8)	IV (2–8)
Cerastium diffusum diffusum	III (3–4)	V (1–4)	III (1–4)	IV (2–5)	IV (1–5)
Desmazeria marina	V (1–8)	IV (1–6)	I (3–4)	III (2–4)	III (1–8)
Spergularia rupicola	III (2–6)	III (1–5)	I (1–4)	I (2)	II (1–6)
Sagina maritima	II (2–3)	I (2–3)			I (2–3)
Radiola linoides	I (1–2)	I (2)			I (1–2)
Matricaria maritima	I (2)	I (1–3)			I (1–3)
Collema spp.	I (2)				I (2)
Cochlearia officinalis	I (1–2)				I (1–2)
Sedum anglicum		V (1–8)	V (2–9)		III (1–9)
Anthyllis vulneraria	I (3)	III (1–8)	I (1–4)		II (1–8)
Herniaria ciliolata		I (1–5)	I (1–2)		I (1–5)
Trifolium arvense		I (3–4)	I (1–4)		I (1–4)
Trifolium occidentale		I (2–4)	I (3)		I (2–4)
Aira praecox		II (2–5)	V (1–7)	I (4)	III (1–7)
Festuca ovina		I (4–9)	IV (3–9)	I (5–7)	II (3–9)
Holcus lanatus	I (2–3)	II (1–8)	II (1–6)		II (1–8)
Cladonia foliacea		II (2–5)	II (2–6)		II (2–6)
Jasione montana		I (2–4)	II (1–4)		I (1–4)
Aira caryophyllea		I (1–4)	II (1–4)		I (1–4)
Cladonia rangiformis		I (1–4)	II (2–6)		I (1–6)
Hypnum cupressiforme		I (3–5)	II (2–4)		I (2–5)
Agrostis capillaris		I (3)	I (1–4)		I (1–4)
Scilla autumnalis		I (2–5)	I (2–5)		I (2–5)
Cladonia chlorophaea		I (2)	I (2–4)		I (2–4)
Sagina apetala			I (1–4)		I (1–4)
Arenaria serpyllifolia				V (1–4)	I (1–4)
Sedum acre			I (5)	V (2–7)	I (2–7)
Bromus hordeaceus ferronii	III (2–5)	III (1–6)	I (1–3)	V (2–5)	III (1–6)

Floristic table MC5 (*cont.*)

	a	b	c	d	5
Dactylis glomerata	I (2)	II (1–5)	II (1–5)	V (2–6)	II (1–6)
Thymus praecox		I (3–5)	II (1–8)	IV (2–5)	II (1–8)
Desmazeria rigida				II (2–4)	I (2–4)
Echium vulgare				II (1–2)	I (1–2)
Hieracium pilosella				II (2–4)	I (2–4)
Salvia horminoides				II (1–5)	I (1–5)
Taraxacum sp.				II (1–3)	I (1–3)
Myosotis ramosissima			I (3–4)	I (2–3)	I (2–4)
Euphorbia portlandica			I (2)	I (1–3)	I (1–3)
Festuca arundinacea				I (2–3)	I (2–3)
Hippocrepis comosa				I (2–4)	I (2–4)
Diplotaxis muralis				I (2–3)	I (2–3)
Cirsium acaule				I (1–2)	I (1–2)
Filipendula vulgaris				I (2–3)	I (2–3)
Senecio jacobaea				I (2–3)	I (2–3)
Ranunculus acris				I (2)	I (2)
Vicia sativa				I (1–2)	I (1–2)
Brassica oleracea				I (1–4)	I (1–4)
Veronica arvensis				I (2–4)	I (2–4)
Achillea millefolium				I (1–3)	I (1–3)
Brachypodium pinnatum				I (3–4)	I (3–4)
Sanguisorba minor				I (3–5)	I (3–5)
Senecio vulgaris				I (1–3)	I (1–3)
Reseda lutea				I (1–2)	I (1–2)
Carex flacca				I (2)	I (2)
Atriplex littoralis				I (2–4)	I (2–4)
Daucus carota gummifer	I (1–3)	II (1–4)	II (1–4)	III (2–4)	II (1–4)
Koeleria macrantha		I (2–6)	II (2–6)	III (2–6)	II (2–6)
Plantago lanceolata	I (1)	II (1–5)	II (1–4)	III (1–5)	II (1–5)
Lotus corniculatus		II (2–5)	II (1–4)	III (1–5)	II (1–5)
Leontodon taraxacoides		II (1–4)	II (1–4)	II (2–3)	II (1–4)
Plantago maritima	I (2–4)	I (1–5)	II (1–5)	I (3)	I (1–5)
Scilla verna	I (1)	I (1–4)	II (1–4)	I (2)	I (1–4)
Silene vulgaris maritima	II (3–7)	II (2–7)	I (3–4)	II (3–7)	II (2–7)

	a	b	c	d	5
Agrostis stolonifera		II (3–5)	I (2–4)	I (1–4)	I (1–5)
Bellis perennis		I (2–3)	I (2–3)	II (2–3)	I (2–3)
Galium verum		I (1–5)	I (2–4)	II (1–5)	I (1–5)
Beta vulgaris maritima	I (4–5)	I (1–4)		I (4)	I (1–5)
Cochlearia danica	I (1–4)	I (1–5)		I (4)	I (1–5)
Tortella flavovirens	I (2–3)	I (3–4)	I (3)		I (2–4)
Trifolium repens	I (2–3)	I (2)	I (3)		I (2–3)
Lolium perenne			I (2)	I (2)	I (2)
Centaurium erythraea		I (1–3)	I (1–3)	I (2)	I (1–3)
Cerastium fontanum		I (2)	I (1–2)	I (2)	I (1–2)
Sonchus oleraceus		I (1–3)		I (1–2)	I (1–3)
Trifolium scabrum		I (2–4)		I (2–4)	I (2–4)
Erodium cicutarium		I (3–4)	I (1–2)	I (2–3)	I (2–4)
Number of samples	19	78	75	21	193
Number of species/sample	8 (5–13)	14 (7–23)	14 (7–25)	17 (12–23)	14 (5–25)
Vegetation height (cm)	6 (2–30)	4 (1–20)	3 (1–10)	4 (1–15)	4 (1–30)
Total vegetation cover (%)	77 (10–100)	79 (10–100)	77 (20–100)	85 (70–100)	79 (10–100)
Altitude (m)	18 (3–60)	28 (3–70)	32 (3–215)	65 (35–150)	33 (3–215)
Slope (°)	10 (0–35)	14 (0–45)	14 (0–60)	9 (0–30)	13 (0–60)
Soil depth (cm)	14 (1–40)	5 (1–41)	4 (1–15)	10 (3–30)	6 (1–41)
Number of soil samples	5	19	20	4	48
Superficial pH	5.9 ±0.8	6.2 ±0.2	5.0 ±0.1	7.5 ±0.3	5.8 ±0.2
Water content (% soil dry weight)	46 ±13	51 ±9	59 ±9	35 ±22	53 ±6
Loss on ignition (% soil dry weight)	21 ±5	21 ±3	32 ±4	21 ±7	26 ±2
Sodium (mole g^{-1})	49 ±7	33 ±5	48 ±10	20 ±4	40 ±5
Potassium (mole g^{-1})	10 ±1	12 ±1	12 ±1	12 ±1	12 ±1
Magnesium (mole g^{-1})	61 ±18	43 ±6	36 ±5	33 ±10	41 ±4
Calcium (mole g^{-1})	82 ±43	47 ±10	32 ±8	179 ±7	55 ±9
Phosphorus (mole g^{-1})	3.3 ±0.8	7.1 ±1.6	4.3 ±1.3	0.9 ±0.9	5.0 ±0.9
Sodium/loss on ignition (mole g^{-1})	304 ±69	169 ±21	156 ±20	103 ±13	172 ±15

a   *Desmazeria marina* sub-community

b   *Anthyllis vulneraria* sub-community

c   *Aira praecox* sub-community

d   *Arenaria serpyllifolia* sub-community

5   *Armeria maritima-Cerastium diffusum* ssp. *diffusum* maritime therophyte community (total)

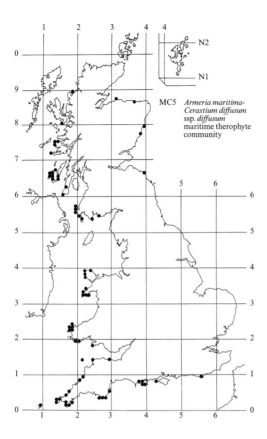

MC5 *Armeria maritima-Cerastium diffusum* ssp. *diffusum* maritime therophyte community

MC6
Atriplex prostrata-Beta vulgaris ssp. *maritima* sea-bird cliff community
Atriplici-Betetum maritimae   J.-M. & J. Géhu 1969

Synonymy

Atriplicetum Gillham 1953; Ornithocoprophilous vegetation Gillham 1956b *p.p.*; *Lavateretum arboreae* J.-M. & J. Géhu 1961; *Beta maritima* – sociatie Beeftink 1962; *Atriplici-Betetum perennis* J.-M. & J. Géhu 1969; *Beto-Tripleurospermetum maritimi* Malloch 1970; Herring gull colony vegetation Sobey & Kenworthy 1979 *p.p.*

Constant species

Atriplex prostrata agg., *Beta vulgaris* ssp. *maritima*, *Festuca rubra*, *Matricaria maritima*.

Physiognomy

The *Atriplici-Betetum* is very variable in its floristics and appearance. Usually one or more of *Atriplex prostata* agg. (including *A. glabriuscula* and *A. babingtonii*: Aellen 1964), *Beta vulgaris* ssp. *maritima* and *Lavatera arborea* dominate in an open or closed, often scruffy, cover with sometimes abundant *Matricaria maritima*. *Festuca rubra*, *Spergularia rupicola*, *Armeria maritima* and *Dactylis glomerata* occur frequently in isolated but often vigorous fragments of crevice vegetation or maritime sward. Various species characteristic of open disturbed places, such as *Polygonum aviculare* agg. and *Rumex crispus*, may be prominent on the sometimes extensive areas of bare ground. There may be a marked seasonal variation in the vegetation with a dramatic spring growth of overwintered *Cochlearia officinalis* (or *C. danica*: Gillham 1953) being succeeded by *A. prostrata*.

Habitat

The community is most characteristic of rocky coastal sites where there is a combination of high maritime influence and intense disturbance by sea-birds, notably gulls (*Larus* spp.), razorbill (*Alca torda*) and guillemot (*Uria aalge*). Essentially similar mixtures of *Atriplex* spp. and *Beta* can also be found on strandline debris in sandy and shingle foreshores.

It has been generally assumed that the major influence of the sea-birds is to enrich the soils with nutrients through their guano but, in a study of herring gull (*Larus argentatus*) colonies on the east coast of Scotland, Sobey & Kenworthy (1979) have demonstrated that physical disturbance in treading, nest-building and particularly in boundary clashes, is also of considerable importance. Such disturbance fragments, and may eventually destroy, the existing vegetation, both directly and by making it more susceptible to wind erosion. It also creates a suitable habitat for colonisation by species, some of them non-maritime, characteristic of open situations. Some, once established, may show a resistance to further disturbance by virtue of a stout rooting system (e.g. *Rumex crispus*, *Matricaria maritima*, *Lavatera arborea*) but many are ephemerals well able to exploit the repeatedly-disturbed patchwork of open ground, especially when there is also some nutrient enrichment. Along foreshores, turbulent wave action repeatedly creates a congenial open environment.

Sea-bird guano is rich in a variety of potential nutrients and, though the soils carrying the community are of varying depth and pH (generally acid to neutral), they are all characterised by high levels of cations and particularly large amounts of available phosphorus and nitrogen during the period of occupation by the birds. During the winter, the levels of these nutrients decline (Sobey & Kenworthy 1979) and the extreme maritime microclimate is the distinctive feature of the habitat.

Zonation and succession

The *Atriplici-Betetum* usually represents the most maritime vegetation where it occurs, replacing the *Crithmo-Spergularietum* or *Armeria-Ligusticum* community though it also occurs higher up cliffs in the zone normally occupied by the *Festuca-Armeria* maritime grassland (Figure 20). Fragments of each of these communities may occur in mosaics with the *Atriplici-Betetum* where the effect of sea-birds is not too severe and may be able to expand if sites are abandoned by the birds. Very intense or prolonged activity may result in the total degeneration of the normal maritime vegetation (see Sobey 1976 on the Isle of May). Foreshore stands occur with other strandline vegetation like the *Honkenya-Cakile* and *Matricaria-Galium* communities.

Distribution
The community occurs patchily around the cliffed coasts and foreshores of the south and west extending north into Scotland. Particularly fine examples are found in sites less subject to human disturbance of nesting sea-birds as on islands. The distribution of *Lavatera arborea* within the community is probably limited by climatic factors: Okusanya (1979c) has shown this species to be injured by low temperatures and destroyed by slight frosts ($-5\,°C$).

Affinities
As described here, the community includes some of the vegetation noted in accounts of sea-bird colonies

Figure 20. Vegetation of sea-bird cliffs at St Govan's Chapel, Stackpole.
MC1 *Crithmo-Spergularietum* vegetation occurs in crevices on the lower spray-splashed stretches of the cliffs at this site but is replaced by luxuriant stands of MC6 *Atriplici-Betetum* vegetation beneath ledges occupied by nesting guillemots and gulls. Higher up the cliffs, as the influence of spray becomes less intense, there are ledges with, first, the *Crithmum* sub-community of MC8 *Festuca-Armeria* vegetation, then the *Holcus* sub-community. Where sightseers trample the cliff-top sward, this is replaced by the *Plantago coronopus* sub-community.
(Redrawn from Cooper 1987, by permission of the Joint Nature Conservation Committee.)

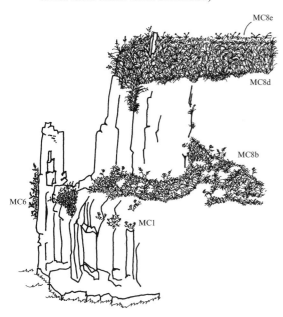

(Gillham 1953, 1956b; Sobey & Kenworthy 1979) and fouled foreshores (Beeftink 1962). In Britain there seems no justification for separating a *Lavatera arborea* community from one more generally dominated by *A. prostrata* and *B. maritima* (cf. Géhu & Géhu 1969). Géhu & Géhu (1969) placed relatively short-lived communities of organically-enriched maritime environments in a new alliance, the Honckenyo-Crambion maritimae, within the Elymetea pycnanthi.

Floristic table MC6

Atriplex prostrata	V (2–8)
Festuca rubra	V (2–5)
Beta vulgaris maritima	IV (3–9)
Matricaria maritima	IV (3–5)
Spergularia rupicola	IV (2–4)
Dactylis glomerata	III (2–5)
Armeria maritima	III (2–4)
Lavatera arborea	II (2–9)
Desmazeria marina	II (2–6)
Rumex crispus	II (2–4)
Polygonum aviculare	II (1–4)
Cochlearia officinalis	II (2–4)
Silene vulgaris maritima	I (4)
Plantago coronopus	I (6)
Daucus carota	I (1)
Bromus hordeaceus ferronii	I (1)
Sonchus oleraceus	I (1)
Taraxacum sp.	I (1)
Number of samples	8
Number of species/sample	7 (5–10)
Vegetation height (cm)	23 (3–50)
Total vegetation cover (%)	83 (50–100)
Altitude (m)	26 (5–33)
Slope (°)	11 (0–20)
Soil depth (cm)	11 (4–24)

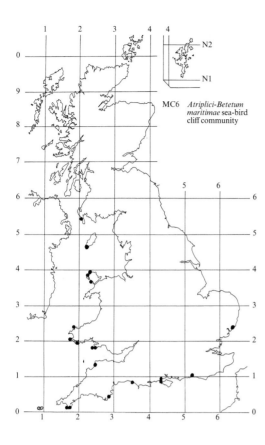

MC6 *Atriplici-Betetum maritimae* sea-bird cliff community

MC7
Stellaria media-Rumex acetosa sea-bird cliff community

Synonymy
Bird cliff vegetation   Petch 1933; Zooplethismic vegetation   Poore   &   Robertson   1949;   *Cochlearietum*
Goodman & Gillham 1954 *p.p.*; Herring gull colony vegetation   Sobey & Kenworthy 1979 *p.p.*

Constant species
Stellaria media.

Physiognomy
The *Stellaria-Rumex* community has a scruffy but generally closed cover of *S. media* with some *Festuca rubra*,
Agrostis stolonifera and *Holcus lanatus* (each of which
may be abundant) and a little *R. acetosa* and *Armeria
maritima*. There is a variety of low frequency occasionals of maritime and inland grasslands and of disturbed
habitats. *Cochlearia officinalis* may be much more abundant early in spring than later.

Habitat
The community is always associated with disturbance
and manuring by colonial sea-birds. On St Kilda, it is
particularly frequent around the nest burrows of puffin
(*Fratercula arctica*) where there is considerable excavation of soil.

Zonation and succession
As with the *Atriplici-Betetum* community, this vegetation forms a mosaic with surrounding maritime crevice
communities and grasslands and its spatial and temporal relationships with these are likely to be governed by
the intensity of sea-bird activity and the high maritime
influence.

Distribution
The community has been recorded from scattered sites
from Pembrokeshire round to the Firth of Forth.

Affinities
Among the apparently rather variable assemblages of
species able to take advantage of the seasonal enrichment of soils exposed to high maritime influence, this is
distinct from the latter in the replacement of *B. vulgaris*
ssp. *maritima* and *A. prostrata* by *S. media* and *R.
acetosa*. The abundance of grasses characteristic of less
maritime situations suggests an affinity with the vegetation of enriched and disturbed inland habitats placed in
the Chenopodietea.

Floristic table MC7

Stellaria media	V (2–8)
Rumex acetosa	III (4–5)
Holcus lanatus	III (3–7)
Agrostis stolonifera	III (4–8)
Festuca rubra	III (7–9)
Armeria maritima	III (2–3)
Plantago maritima	II (4)
Leontodon autumnalis	II (2)
Poa subcaerulea	II (4–6)
Cochlearia officinalis	II (1–2)
Rumex crispus	II (1)
Atriplex prostrata	II (2)
Silene vulgaris maritima	II (1–3)
Matricaria maritima	I (4)
Polygonum aviculare agg.	I (2)
Cerastium glomeratum	I (5)
Poa annua	I (3)
Sedum anglicum	I (4)
Senecio vulgaris	I (2)
Cerastium fontanum	I (2)
Potentilla anserina	I (5)
Trifolium repens	I (4)
Juncus articulatus	I (4)
Spergula arvensis	I (2)
Number of samples	7
Number of species/sample	7 (3–13)
Vegetation height (cm)	9 (2–20)
Total vegetation cover (%)	98 (90–100)
Altitude (m)	31 (10–59)
Slope (°)	7 (0–15)
Soil depth (cm)	34 (23–64)

MC8
Festuca rubra-Armeria maritima maritime grassland

Synonymy
Festuca rubra community McLean 1935; *Festucetum rubrae* Gillham 1953; *Armerietum* Goodman & Gillham 1954, McVean 1961; *Sileno maritimae-Festucetum pruinosae* R.Tx. 1963 *p.p.*, *Armerieto maritimae-Daucetum gummiferi* Géhu 1964 *p.p.*, *Festuco-Armerietum rupestris* Malloch 1971.

Constant species
Festuca rubra, Armeria maritima.

Rare species
Asparagus officinalis ssp. *prostratus, Astragalus danicus, Brassica oleracea, Carpobrotus edulis, Parapholis incurva, Scilla verna, Senecio integrifolius* spp. *maritimus, Silene nutans, Trifolium occidentale.*

Physiognomy
The *Festuca rubra-Armeria maritima* community is a grassland with a generally closed sward, usually dominated by *F. rubra* which often forms a thick mattress. *A. maritima* may be abundant as scattered bulky cushions but it is not usually a dominant species. Frequent throughout the community are *Agrostis stolonifera, Plantago maritima, Daucus carota* ssp. *gummifer* and *Silene vulgaris* ssp. *maritima.* Bryophytes are generally sparse.

Sub-communities

Typical sub-community: *Festuca rubra* community McLean 1935; Coastal *Armerietum* Goodman & Gillham 1954 *p.p.*; Hirta mixed grassland *p.p.* & Rona *Armerietum* McVean 1961; *Festuco-Armerietum rupestris,* typical sub-association Malloch 1971. In this rather species-poor sub-community, *F. rubra* is overwhelmingly dominant as a mattress which may attain a thickness of 40 cm. There are scattered cushions of *Armeria maritima* but *Agrostis stolonifera* is the only other species that is at all frequent. No species is prefe-

rential here but the prostrate maritime form of *Asparagus officinalis* is a notable rarity. Bryophytes are rarely able to find a place in the thick turf.

Crithmum maritimum sub-community: Coastal *Armerietum* Goodman & Gillham 1954 *p.p.*, *Armerieto-Daucetum gummiferi crithmetosum* Géhu 1964; *Festuco-Armerietum rupestris, Crithmum maritimum* sub-association Malloch 1971. *F. rubra* is again dominant as a mattress with scattered *A. maritima* cushions. *Daucus carota* ssp. *gummifer* attains constancy here though it is rarely abundant and the really distinctive feature of the vegetation is the occurrence in the more open areas of the sward of species characteristic of maritime crevices and ledges, notably *Crithmum maritimum, Inula crithmoides, Brassica oleracea* and *Beta vulgaris* ssp. *maritima. Brachypodium pinnatum* is a sometimes abundant preferential species. Vegetation similar to this sub-community was described by Coombe (1961) as one of the contexts for the rare *Trifolium occidentale* (see also Géhu 1973*b*).

Ligusticum scoticum sub-community. This is a more closed sub-community than the latter with a generally intact sward of *F. rubra* with some *Agrostis stolonifera* and occasionally abundant *Holcus lanatus. Armeria maritima* and *Ligusticum scoticum* are both constant though usually in small amounts.

Holcus lanatus sub-community: Inland *Armerietum* Goodman & Gillham 1954; *Festuco-Armerietum rupestris, Holcus-Dactylis* variant Malloch 1971. A thick mattress of *F. rubra* with scattered *A. maritima* is again characteristic but here there is a prominent contribution by species characteristic of inland neutral grasslands. *H. lanatus* is constant and may make a substantial contribution to the grassy cover with smaller amounts of *Dactylis glomerata* and *Agrostis stolonifera. Rumex acetosa, Hypochoeris radicata* and *Plantago lanceolata* are frequent and there are occasional records for

Achillea millefolium, Cirsium arvense and *C. vulgare. Senecio integrifolius* ssp. *maritimus* occurs in this community at its Anglesey locality (see also Smith 1979).

***Plantago coronopus* sub-community:** *Plantago coronopus-Cerastium tetrandum* Association Br.-Bl. & R.Tx. 1952; Grazed *Festucetum rubrae* Gillham 1953; Hirta *Plantago* sward McVean 1961; *Cerastium atrovirens-Plantago coronpus* Association and *Carex distans-Plantago maritima* Association *p.p.* Ivimey-Cook & Proctor 1966; *Festuco-Armerietum rupestris, Plantago coronopus* sub-association Malloch 1971, Habitat Group III Goldsmith 1975 *p.p.* Physiognomically, this sub-community presents a marked contrast to those described above. There is a short tight sward, still generally dominated by *F. rubra* with scattered *Armeria maritima* but here with constant *Agrostis stolonifera, Plantago coronopus* and *P. maritima*. Occasionally, the two *Plantago* spp. may dominate: *P. maritima* tends to be more prominent in the north and *P. coronopus* in the south. A number of smaller herbaceous species are preferential to this sub-community including some sedges, notably *Carex distans* and *C. caryophyllea*, and various annuals such as *Cerastium diffusum* ssp. *diffusum* (= *C. tetrandum* = *C. atrovirens*), *Desmazeria marina, Sagina maritima* and *S. apetala* which may be missed in late sampling. Bryophytes, though never very abundant, are more conspicuous here than in any other sub-community with occasional records for *Eurhynchium praelongum, Tortella flavovirens* and *Trichostomum brachydontium*.

***Anthyllis vulneraria* sub-community.** *F. rubra* and *Anthyllis vulneraria* are co-dominant here with constant *Armeria maritima, Silene vulgaris* ssp. *maritima* and *Agrostis stolonifera* in what is one of the most colourful of maritime communities in the flowering season. *Holcus lanatus, Plantago maritima* and *Lotus corniculatus* are frequent with *Sedum anglicum* and *Sonchus oleraceus* as preferential occasionals.

***Armeria maritima*-dominated sub-community:** Coastal *Armerietum* Goodman & Gillham 1954 *p.p. F. rubra* and *A. maritima* are the sole constants in this, the most open and species-poor of the sub-communities. Cushions of *A. maritima* dominate the vegetation with *F. rubra* only rarely attaining over 10% cover. *Spergularia rupicola* is the only other frequent species and scattered plants of this and the few occasionals are rooted in patches of bare soil between exposed rock.

Habitat
Of the grasslands proper occurring on coastal cliffs, the *Festuca-Armeria* community occupies the most maritime position. It generally occurs on steep to moderate slopes up to about 50 m above sea-level and receives large amounts of sea-spray. The community is found on a wide range of rock types and the soils are generally brown rankers, moderately deep, rich in rock fragments and organic matter (much of it derived from decay of the bulky grass mattress) and of neutral pH.

Some of the floristic variation within the community can be understood in relation to a gradient of maritime influence running from the most maritime situations with (in the south) the *Crithmum* sub-community or (in the north) the *Ligusticum* sub-community, through the typical sub-community to much less maritime situations with the *Holcus* sub-community. Often, this variation is related to a topographic zonation from low-situated steep slopes with shallow soils to high gentle slopes with deep soils but this general pattern is complicated by aspect and the particular configuration of the cliff profile. On exposed, south-facing sites in southern England, for example, high maritime influence and parching of soils may act together and here the *Crithmum* sub-community may extend high up the cliffs on to gentle slopes with quite deep soils. By contrast, on sheltered cliffs, the *Holcus* sub-community may run further downslope than in exposed situations, almost eclipsing the more maritime sub-communities.

Where site drainage becomes excessive, the dominance of *F. rubra* in these grasslands seems to suffer and both the *Anthyllis* and *Armeria* sub-communities seem to be related in part to this effect. The *Anthyllis* sub-community is generally associated with south- or west-facing slopes often with very shallow soils. The *Armeria*-dominated sub-community is especially characteristic of more maritime situations where there is some degree of erosion, as, for example, on fractured cliff edges where *A. maritima* is able to maintain its position by rooting deep into crevices.

The *Plantago coronopus* sub-community is unusual among the various types of *Festuca-Armeria* grassland in that it is grazed: normally the community is naturally inaccessible to stock or beyond the limit of cliff-top enclosure. Grazing, generally by sheep, produces and maintains the close, varied sward of the *P. coronopus* sub-community and stands may occur alongside ungrazed areas of the typical sub-community. The predominance in this sub-community of *P. coronopus* as against *P. maritima* towards the south is perhaps related to the greater ability of the former to withstand higher summer temperatures and moderately low rainfall in a maritime environment (see Dodds 1953).

Zonation and succession
On ungrazed cliffs, the *Festuca-Armeria* community generally forms a zone above the *Crithmo-Spergularietum* or the *Armeria-Ligusticum* crevice communities into which it may grade through its *Crithmum* or *Ligusticum* sub-communities. Above it may pass into the

Festuca-Holcus maritime grassland through its *Holcus* sub-community (Figure 21). This is the general zonation on the cliffs of much of south-west England, Wales and southern Scotland, though on chalk and limestones there is a tendency for the *Crithmum* sub-community to pass directly into cliff-top *Festuca-Daucus* maritime grassland.

On grazed cliffs, the *Plantago coronopus* sub-community may replace all other sub-communities and pass above into the *Festuca-Plantago* maritime grassland. This zonation is characteristic of much of the cliffed coast of north-west Scotland.

Throughout its range, the *Festuca-Armeria* grassland may give way on very shallow dry soils to various types of *Armeria-Cerastium* community, particularly the *Anthyllis* sub-community.

Figure 21. Vegetation pattern at Gurnard's Head, Cornwall.

The basic zonation around Gurnard's Head is from the MC1 *Crithmo-Spergularietum* through the MC8 *Festuca-Armeria* and MC9 *Festuca-Holcus* grasslands to the MC12 *Festuca-Hyacinthoides* community or H7 *Calluna-Scilla* maritime heath. Towards the end of the headland, the proportions of these vegetation types show characteristic variation with shifts in exposure to the prevailing spray-laden winds and there are differences in the kinds of MC8 *Festuca-Armeria* grassland represented: the Typical sub-community occupies the more maritime situations, the *Holcus* sub-community the less.

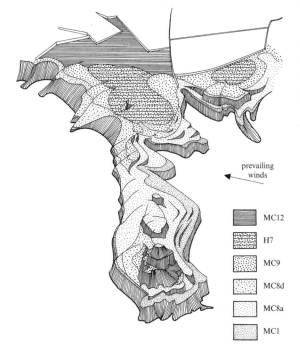

prevailing winds

MC12

H7

MC9

MC8d

MC8a

MC1

Distribution

The community occurs around the whole of the cliffed coastline of Britain, though it is rather rare on the Channel coast. The *Crithmum* sub-community is restricted to the area south of the Mull of Galloway and the typical and *Holcus* sub-communities tend to be commoner in the south. The *Ligusticum* sub-community occurs only north of the Mull of Galloway and the *Plantago coronopus* sub-community is commoner in the north. The *Anthyllis* sub-community seems to be confined to south-west England and Wales.

Affinities

Maritime grasslands dominated by *F. rubra* have been described from salt-marshes and variously allocated to the *Juncetum gerardi* or the *Festucetum littoralis* Corillion 1953. The cliff grasslands of the kind included here are, however, distinct in the general absence of salt-marsh species and have been long, if vaguely, characterised in descriptive accounts as various forms of *Festucetum rubrae* or *Armerietum maritimae*.

There is no exact phytosociological counterpart to the *Festuca-Armeria* grassland: both the *Sileno-Festucetum pruinosae* of Tüxen (1963) and the *Armerieto-Daucetum gummiferi* of Géhu (1964) are more broadly defined and include parts of other maritime grasslands and even some maritime forms of heath. There is, however, fairly general agreement that this kind of community belongs alongside other halophyte swards in the Glauceto-Puccinellietalia of the Asteretea and Malloch (1970, 1971) proposed a new alliance, the Silenion maritimae, to contain the distinctive cliff forms of this vegetation.

Various of the sub-communities of the *Festuca-Armeria* grassland represent clear floristic transitions to other maritime communities. The *Crithmum* and *Ligusticum* sub-communities form a link with the crevice vegetation of the Crithmo-Limonietea and the *Holcus* and *Plantago coronopus* have affinities with less maritime and grazed forms of cliff sward.

Floristic table MC8

	a	b	c	d	e	f	g	8
Festuca rubra	V (7–10)	V (2–10)	V (4–10)	V (5–10)	V (3–10)	V (2–10)	V (2–5)	V (2–10)
Armeria maritima	V (1–5)	IV (1–5)	V (2–7)	V (1–8)	V (2–10)	V (2–5)	V (4–10)	V (1–10)
Agrostis stolonifera	III (1–5)	I (1–5)	IV (2–5)	III (1–5)	IV (1–7)	IV (2–5)	II (4)	III (1–7)
Crithmum maritimum		V (1–8)		I (2–3)	I (4)	I (3)		I (1–8)
Daucus carota gummifer	II (1–5)	IV (1–5)	I (1)	III (1–5)	I (1–5)	II (1–5)	I (2)	II (1–5)
Inula crithmoides		II (1–3)			I (2–3)			I (1–3)
Beta vulgaris maritima	I (1–4)	II (1–6)		I (1)	I (3)			I (1–6)
Brassica oleracea		II (1–5)						I (1–5)
Brachypodium pinnatum		I (4–5)		I (4)				I (4–5)
Ligusticum scoticum			V (3–5)	I (2)	I (3–4)			I (2–5)
Holcus lanatus	I (1–3)	I (3)	III (2–8)	V (1–8)	I (1–5)	III (2–4)	II (1–4)	II (1–8)
Rumex acetosa	I (1–3)	I (2)	I (1–4)	III (1–5)	I (1–4)	I (2)	II (3–5)	I (1–5)
Achillea millefolium				I (2–3)		I (1–3)		I (1–3)
Cirsium arvense	I (1)			I (1–3)				I (1–3)
Plantago coronopus	I (1–4)	II (2–5)	I (2)	I (2–3)	V (1–8)	II (1–3)	II (1–4)	II (1–8)
Plantago maritima	II (1–7)	I (1–4)	III (3–5)	I (1–5)	IV (1–10)	III (3–5)	I (4)	III (1–10)
Cerastium diffusum diffusum	I (1)				II (1–5)		I (2)	I (1–5)
Carex distans		I (2)		I (1)	I (1–4)			I (1–4)
Carex caryophyllea					I (1–4)			I (1–4)
Desmazeria marina		I (1–4)			I (2–3)	I (1)		I (1–4)
Centaurium erythraea				I (1)	I (1–3)	I (1–2)		I (1–3)
Sagina maritima					I (2–5)			I (2–5)
Sagina apetala					I (1–5)			I (1–5)
Tortella flavovirens	I (3)	I (3)			I (1–5)		I (2)	I (1–5)
Trichostomum brachydontium		I (1)			I (2–4)			I (1–4)
Eurhynchium praelongum	I (2)	I (2)			I (3–5)			I (2–5)
Anthyllis vulneraria	I (1–5)	II (1–5)		I (1–4)	I (1–4)	V (4–10)		I (1–10)
Silene vulgaris maritima	II (1–5)	I (2–5)	II (3–4)	II (1–6)	I (2–6)	V (2–4)	II (1–4)	I (1–6)
Sedum anglicum		I (2)			I (1–4)	II (2–3)		I (1–4)

Floristic table MC8 (cont.)

	a	b	c	d	e	f	g	8
Sonchus oleraceus	I (1–3)	I (1–3)		I (1–2)	I (1–2)	II (1–3)		I (1–3)
Lotus corniculatus	II (1–4)	II (1–4)	II (2–4)	III (1–5)	II (2–6)	III (1–4)		II (1–6)
Trifolium repens	I (2–4)	I (2–3)	II (2–4)	II (2–5)	II (2–5)	I (2–3)	II (3–4)	II (2–5)
Cochlearia officinalis	II (1–4)		II (1–3)	I (1–3)	II (1–4)	II (1–3)	I (1–2)	II (1–4)
Matricaria maritima	I (1–2)		II (2–4)	I (1–2)	I (1–2)	I (2)	I (2)	I (1–4)
Plantago lanceolata	I (1–3)	I (1–3)	I (4)	II (1–4)	I (2)	II (1–3)		I (1–4)
Spergularia rupicola	I (1–2)	I (2–3)		I (2–3)	I (1–4)	I (2–3)	III (3–4)	I (1–4)
Cerastium fontanum	I (1–3)	I (1–2)	I (3)	I (1–3)	II (1–3)	I (1–3)		I (1–3)
Hypochoeris radicata	I (1–3)	I (1–3)		II (1–4)	I (1–3)	II (2–3)		I (1–4)
Scilla verna	I (1–3)		I (2–3)	II (1–6)	I (2–6)	II (2–3)		I (1–6)
Dactylis glomerata	I (1–3)	II (2–4)	I (2)	II (1–7)	I (2)			I (1–7)
Leontodon autumnalis	I (2–4)		II (2–3)	I (2–3)	II (1–4)			I (1–4)
Poa subcaerulea	I (3–5)		I (3–4)	I (1–7)	I (3–8)			I (1–8)
Cochlearia danica	I (1)	I (1–3)			I (1–4)		II (1–4)	I (1–4)
Angelica sylvestris	I (4)			I (1–4)				I (1–5)
Leontodon taraxacoides	I (1–3)		II (2–5)	I (1–3)	I (1–4)	I (1–2)		I (1–4)
Potentilla erecta				I (2–4)		I (1–2)		I (1–4)
Number of samples	120	51	16	100	160	18	12	477
Number of species/sample	6 (2–10)	8 (3–15)	8 (4–11)	9 (4–14)	9 (4–16)	10 (5–13)	5 (3–7)	8 (2–15)
Vegetation height (cm)	16 (5–40)	15 (3–50)	15 (5–30)	17 (5–40)	4 (1–15)	14 (7–26)	7 (3–20)	11 (3–50)
Total vegetation cover (%)	99 (60–100)	83 (10–100)	96 (80–100)	100 (95–100)	95 (20–100)	98 (80–100)	75 (5–100)	95 (5–100)
Altitude (m)	21 (4–55)	17 (4–47)	26 (3–57)	35 (8–210)	23 (2–150)	21 (6–55)	55 (10–150)	25 (2–210)
Slope (°)	14 (0–50)	28 (0–90)	27 (5–70)	18 (0–15)	9 (0–40)	26 (0–28)	7 (0–23)	23 (0–90)
Soil depth (cm)	27 (4–80)	22 (4–48)	31 (8–70)	34 (3–75)	23 (2–75)	19 (3–58)	16 (8–45)	26 (2–80)
Number of soil samples	29	15	3	28	63	3	no data	78
Superficial pH	6.5 ±0.2	7.5 ±0.2	5.5	5.7 ±0.2	6.3 ±0.1	5.9		6.3 ±0.1
Water content (% soil dry weight)	93 ±11	45 ±7	132	81 ±11	100 ±12	60		89 ±6
Loss on ignition (% soil dry weight)	24 ±2	12 ±2	43	26 ±3	28 ±3	16		25 ±2

	a	b	c	d	e	f	g	8
Sodium (mole g^{-1})	104 ±15	58 ±8	132	51 ±5	92 ±11	29		82 ±6
Potassium (mole g^{-1})	13 ±1	13 ±1	15	12 ±1	12 ±1	9		12 ±1
Magnesium (mole g^{-1})	66 ±7	57 ±1	47	44 ±4	59 ±5	35		57 ±3
Calcium (mole g^{-1})	49 ±8	73 ±10	49	40 ±9	42 ±4	17		46 ±3
Phosphorus (mole g^{-1})	2.0 ±0.4	0.4 ±0.1	2.3	2.8 ±0.9	1.9 ±0.4	0.8		1.9 ±0.3
Sodium/loss on ignition (mole g^{-1})	469 ±58	569 ±91	336	220 ±27	336 ±21	179		363 ±21

a   Typical sub-community
b   *Crithmum maritimum* sub-community
c   *Ligusticum scoticum* sub-community
d   *Holcus lanatus* sub-community
e   *Plantago coronopus* sub-community
f   *Anthyllis vulneraria* sub-community
g   *Armeria maritima*-dominated sub-community
8   *Festuca rubra-Armeria maritima* maritime grassland (total)

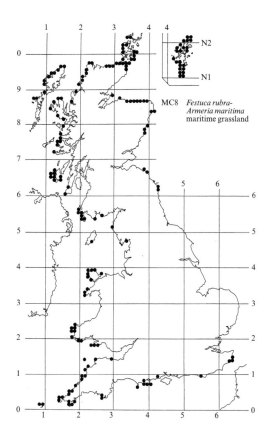

MC8 *Festuca rubra-*
Armeria maritima
maritime grassland

MC9
Festuca rubra-Holcus lanatus maritime grassland

Synonymy

Holcetum lanati Gillham 1953 *p.p.*, Goodman & Gillham 1954 *p.p.*; *Sileno maritimae-Festucetum pruinosae* R.Tx. 1963 *p.p.*, *Armerieto maritimae-Daucetum gummiferi* Géhu 1964 *p.p.*, includes *Festuco-Dactyletum maritimae* Malloch 1971.

Constant species

Festuca rubra, Holcus lanatus, Plantago lanceolata, Armeria maritima.

Rare species

Allium schoenoprasum, Astragalus danicus, Oxytropis halleri, Primula scotica, Scilla verna, Senecio integrifolius ssp. *maritimus, Trifolium occidentale.*

Physiognomy

The *Festuca-Holcus* maritime grassland generally has a closed fairly low-growing but rather rank and often tussocky sward. It is almost always dominated by grasses of which *F. rubra* is usually the most prominent, though *H. lanatus* and, to a lesser extent, *Dactylis glomerata*, are often abundant. Herbaceous dicotyledons are generally an important component of the vegetation and many of these are non-maritime species. *Armeria maritima* and *Plantago lanceolata* are both constant and *Plantago maritima, Rumex acetosa* and *Trifolium repens* are frequent throughout. Bryophytes and lichens are rare.

Sub-communities

Plantago maritima **sub-community.** *F. rubra* is the usual dominant in the thick sward of this sub-community, though *H. lanatus* and, less frequently, *Agrostis stolonifera* and, in the north, *Poa subcaerulea*, may be abundant in particular stands. *Plantago maritima, Trifolium repens* and *Lotus corniculatus* are additional constants here and *P. maritima* is sometimes so abundant as to be a co-dominant, especially where there is a shorter grazed turf. *Scilla verna* is frequent, though never abundant, and

Astragalus danicus, Primula scotica and *Senecio integrifolius* ssp. *maritimus* occur occasionally within their ranges of distribution. *Parnassia palustris* is a low frequency differential.

Dactylis glomerata **sub-community:** *Festuco-Dactyletum maritimae* Malloch 1971. *F. rubra* is again generally dominant though both *H. lanatus* and *Dactylis glomerata* which is constant here, are occasionally very abundant producing a thick, rather luxuriant, sward. *Rumex acetosa, Scilla verna* and *Daucus carota* ssp. *gummifer* are also constant in this sub-community. *Lotus corniculatus* and *Hypochoeris radicata* are frequent. *Silene vulgaris* ssp. *maritima* and *Anthyllis vulneraria* are preferential occasional species.

Achillea millefolium **sub-community.** *F. rubra* is generally dominant, though *Agrostis capillaris*, which is frequent in this sub-community, is sometimes abundant. *Trifolium repens, Achillea millefolium, Galium verum* and *Lotus corniculatus* are additional constants here and *Plantago maritima, Dactylis glomerata, Rumex acetosa, Potentilla erecta* and *Hypochoeris radicata* are frequent. This is the most species-rich of the sub-communities and there are many occasional species characteristic of the richer neutral and calcareous grasslands, notably *Centaurea nigra, Campanula rotundifolia, Helianthemum nummularium, Festuca ovina, Carex caryophyllea, Hieracium pilosella* and *Conopodium majus*. The maritime *Genista tinctoria* ssp. *littoralis* occurs occasionally.

Primula vulgaris **sub-community.** *F. rubra* generally dominates but *H. lanatus* may be abundant. *Lotus corniculatus, Rumex acetosa* and *Primula vulgaris* are additional constants with *Dactylis glomerata, Agrostis stolonifera* and, particularly in the west, *Geranium sanguineum* frequent. *Brachypodium sylvaticum, Ranunculus ficaria* and *Viola riviniana* are low frequency preferential species in the often luxuriant sward.

***Anthoxanthum odoratum* sub-community.** This is the most distictive of the sub-communities. *F. rubra* is much less abundant here and dominance is usually shared by *A. odoratum*, *H. lanatus* and *Agrostis capillaris* all of which are constant. The frequency of *Armeria maritima* is much reduced but additional constants are *Rumex acetosa*, *Poa subcaerulea* and *Potentilla erecta* with *Plantago maritima*, *Trifolium repens*, *Scilla verna*, *Lotus corniculatus* and *Ranunculus acris* frequent, the last being a good preferential species. *Empetrum nigrum* is occasionally present and it may dominate. There are also various low frequency species characteristic of heaths and acid grasslands: *Deschampsia flexuosa*, *Luzula campestris*, *Hypnum cupressiforme*, *Cladonia chlorophaea* and *Peltigera canina*.

Habitat

The *Festuca-Holcus* community is one of a number of grasslands occupying a less maritime position on sea-cliffs, being characteristic of somewhat sheltered situations, either towards the top of cliffs or on lee slopes (Figure 22). A sub-set of soil samples showed a mean sodium/loss-on-ignition ratio almost half that of the *Festuca-Armeria* grassland and approximately the same as the values for the *Festuca-Daucus* and *Festuca-Hyacinthoides* communities. The generally gentle slopes on which the *Festuca-Holcus* grassland occurs usually have deep ranker soils, often moist though always free-draining.

Some of the floristic variation between the sub-communities seems to be related to maritime influence but soil moisture and nutrient status are also important. The *Plantago maritima* and *Dactylis* sub-communities are the more maritime, the former on rather moister soils, the latter on better drained sites, though rarely on chalk or limestones. Of the other three sub-communities, the *Achillea* sub-community is characteristic of drier shallower soils and the *Primula* sub-community of moister soils, often in sheltered gullies where the slopes may be steeper; in both cases the soils are rich in calcium. The *Anthoxanthum* sub-community is found predominantly on north-facing slopes, often on sandstones, where the soils have a low superficial pH and are generally nutrient-poor.

The *Festuca-Holcus* grassland is usually ungrazed. Light grazing has relatively little effect other than a reduction in sward height and an encouragement of the growth of *Plantago maritima* at the expense of the grasses.

Zonation and succession

The *Festuca-Holcus* grassland usually occupies a zone between the *Festuca-Armeria* grassland, into which it passes through the *Holcus* sub-community of the latter, and maritime types of heath. In especially sheltered situations, it may be the most maritime community on cliffs.

Grazing probably mediates a successional relationship with the *Festuca-Plantago* maritime grassland.

Distribution

The community occurs widely on British sea-cliffs except along the south coast. The *Dactylis* sub-community is most common in the south-west with isolated occurrences in Scotland. The *Plantago maritima* sub-community is most common in the north and the *Anthoxanthum* sub-community has been recorded only in Scotland where it is especially common in Caithness and on Orkney, though reaching south to the Mull of Galloway. The *Achillea* and *Primula* sub-communities are less common; the former is particularly abundant around the Solway Firth and the latter is scattered in its distribution.

Figure 22. Sea-cliff zonations with *Festuca-Holcus* grassland.

The figures show stylised zonations of vegetation types from (a) the MC1 *Crithmo-Spergularietum*, through the MC8 *Festuca-Armeria* and MC9 *Festuca-Holcus* grasslands to H7 *Calluna-Scilla* heath, with (b) fragmentation of the pattern on a stepped cliff and (c) reversal of the zonation where there is interception and downwash of spray.

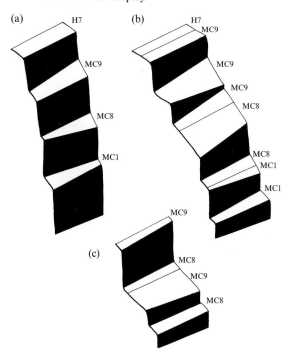

Affinities

The *Festuca-Holcus* grassland is somewhat intermediate between the more maritime grasslands and the neutral swards of the Arrhenatheretalia but it is sufficiently maritime to include within the Glauceto-Puccinellietalia of the Asteretea and Malloch (1970, 1971) placed Cornish stands of this vegetation in his new alliance, the Silenion maritimae. There is no exact phytosociological counterpart to the community, though the *Dactylis* sub-community falls within the range of Tüxen's *Sileno-Festucetum pruinosae* and perhaps the typical sub-association of Géhu's *Armerieto Daucetum gummiferi*.

The sub-communities show floristic affinities with a variety of other maritime vegetation types: the *Dactylis* sub-community with the *Festuca-Daucus* maritime grassland, the *Primula* sub-community with the *Festuca-Hyacinthoides* maritime grassland and the *Anthoxanthum* sub-community with maritime heath.

Floristic table MC9

	a	b	c	d	e	9
Festuca rubra	V (4–10)	V (4–10)	V (4–10)	V (3–10)	V (3–9)	V (3–10)
Holcus lanatus	V (1–8)	V (2–10)	V (2–5)	IV (2–7)	V (3–8)	V (1–10)
Plantago lanceolata	V (1–7)	V (1–4)	V (1–5)	V (1–5)	IV (2–5)	V (1–7)
Armeria maritima	IV (1–6)	IV (1–5)	III (1–4)	III (2–6)	II (2–4)	IV (1–6)
Lotus corniculatus	IV (1–5)	III (1–4)	V (1–5)	IV (2–5)	III (2–4)	III (1–5)
Plantago maritima	V (2–10)	I (2–5)	III (1–6)	II (3–5)	III (2–4)	III (1–10)
Trifolium repens	V (1–5)	I (2–4)	IV (2–4)	II (3–4)	III (2–4)	III (1–5)
Rumex acetosa	II (1–4)	V (1–5)	III (1–5)	V (1–5)	IV (2–5)	III (1–5)
Dactylis glomerata	I (2–7)	V (1–8)	III (1–6)	III (2–7)		II (1–8)
Scilla verna	III (1–4)	V (1–4)	II (1–3)	I (2–4)	III (2–4)	III (1–4)
Daucus carota gummifer	I (1–4)	IV (1–4)	II (1–3)	II (1–2)		II (1–4)
Silene vulgaris maritima	I (2–4)	II (1–5)	I (2–4)	I (2–3)	I (4)	I (1–5)
Anthyllis vulneraria	I (1–4)	II (1–6)	I (1–4)			I (1–6)
Achillea millefolium	I (1–3)	I (2–4)	V (1–4)	II (1–3)	I (3)	II (1–4)
Galium verum	I (2–5)	II (2–5)	V (2–4)	II (2–4)		II (1–4)
Centaurea nigra	I (1–3)	I (2)	II (1–4)			I (1–4)
Campanula rotundifolia			II (1–3)	I (2)		I (1–3)
Genista tinctoria littoralis			I (3–6)			I (3–6)
Carex caryophyllea			I (1–4)			I (1–4)
Cerastium diffusum diffusum			I (1–3)			I (1–3)
Festuca ovina	I (4)	I (5)	I (3–8)	I (5)		I (3–8)
Helianthemum nummularium			I (1–3)			I (1–3)
Conopodium majus	I (1)		I (1–3)			I (1–3)
Hieracium pilosella			I (1–2)			I (1–2)
Trichostomum brachydontium			I (2)			I (2)
Primula vulgaris	I (1–4)	I (2–4)		V (2–6)	I (4)	I (1–6)
Geranium sanguineum			I (1–6)	III (2–5)		I (1–6)
Ranunculus ficaria	I (2–3)	I (3)	I (2)	II (2–5)		I (2–5)
Viola riviniana	I (2–3)	I (1–3)	I (2–5)	II (1–4)	I (2–3)	I (1–5)
Brachypodium sylvaticum		I (4–8)	I (1–7)	II (2–5)		I (1–7)

Species	1	2	3	4	5
Anthoxanthum odoratum	I (4)		I (3)	V (3–8)	I (3–8)
Agrostis capillaris	II (2–6)	I (3–4)	III (1–9)	V (2–8)	II (1–9)
Poa subcaerulea	II (3–8)		I (4)	IV (2–5)	II (2–8)
Potentilla erecta	II (2–4)	I (2–4)	III (2–4)	IV (2–4)	II (2–4)
Ranunculus acris		I (2)	I (1–2)	III (2–4)	I (1–4)
Empetrum nigrum	I (3–6)		I (3–6)	II (2–8)	I (2–8)
Deschampsia flexuosa				I (3–4)	I (3–4)
Luzula campestris	I (1–2)		I (1–3)	I (2–4)	I (1–4)
Hypnum cupressiforme	I (3–4)		I (2–3)	I (3–4)	I (2–4)
Cladonia chlorophaea				I (2)	I (2)
Peltigera canina			I (2–3)	I (2–3)	I (2–3)
Agrostis stolonifera	III (2–8)	II (2–4)	II (2–5)	III (2–6)	III (2–8)
Hypochoeris radicata	II (1–4)	III (1–4)	III (1–4)	I (1–2)	II (1–4)
Cerastium fontanum	II (1–3)	II (1–3)	II (1–2)	II (1–2)	II (1–3)
Cochlearia officinalis	II (1–3)	I (2)	II (2–4)	I (1)	II (1–6)
Carex flacca	I (1–4)	I (2)	II (1–4)	I (4)	I (1–4)
Leontodon autumnalis	II (1–4)	I (3)	I (1–2)	II (1–3)	II (1–4)
Thymus praecox	I (2–4)	I (1–4)	II (2–4)	II (1–3)	I (1–4)
Koeleria macrantha	I (2–4)	I (1–3)	II (2–4)		I (1–4)
Angelica sylvestris	I (1–5)	I (1–4)	I (1–5)		I (1–4)
Carex nigra	I (1–5)	I (1–4)	I (3)	II (2–4)	I (1–5)
Danthonia decumbens	I (3–4)	I (3)	I (3)	II (4–6)	I (1–5)
Bellis perennis	I (2–3)	I (1–4)	I (1–2)	II (2–3)	I (1–4)
Rumex crispus	I (2–4)	I (1)	I (3)		I (1–4)
Heracleum sphondylium	I (1–3)	I (1–4)	I (1–5)		I (1–4)
Senecio jacobaea	I (1–2)	I (1)	I (2)		I (1–2)
Sagina apetala	I (2–3)		I (2–3)	I (3)	I (2–3)
Eurhynchium praelongum	I (2)		I (2–3)	I (5)	I (2–5)
Rhytidiadelphus squarrosus	I (2–4)		I (2–4)	I (3)	I (2–4)
Carex panicea	I (2–4)		I (3–4)	I (2–3)	I (2–4)
Euphorbia portlandica	I (2)		I (2)		I (2)
Serratula tinctoria	I (5)		I (6)		I (2–6)
Sedum anglicum	I (1–2)	I (2–3)	I (2–4)		I (1–4)
Aira praecox	I (2)			I (3)	I (2–3)
Leontodon taraxacoides	I (2–3)	I (1–3)	I (2)		I (1–3)

Floristic table MC9 (*cont.*)

	a	b	c	d	e	9
Leucanthemum vulgare		I (1–2)	I (2–4,	I (1)		I (1–2)
Euphrasia tetraquetra		I (1–3)	I (4)			I (1–4)
Solidago virgaurea				I (1)		I (1–4)
Pseudoscleropodium purum			I (3–4)	I (3)		I (3–4)
Number of samples	123	54	39	20	21	257
Number of species/sample	13 (8–21)	12 (6–18)	18 (9–27)	15 (11–22)	14 (9–20)	14 (6–27)
Vegetation height (cm)	10 (2–30)	15 (3–30)	13 (2–50)	14 (5–35)	13 (4–25)	12 (2–50)
Total vegetation cover (%)	100 (90–100)	100	99 (90–100)	100	100 (95–100)	100 (90–100)
Altitude (m)	27 (3–215)	31 (3–80)	20 (3–80)	20 (2–55)	43 (9–100)	28 (3–215)
Slope (°)	11 (0–45)	18 (0–55)	13 (0–60)	21 (0–50)	12 (0–45)	14 (0–60)
Soil depth (cm)	35 (5–82)	27 (4–81)	22 (3–75)	24 (4–52)	44 (8–70)	31 (3–82)
Number of soil samples	32	10	11	7	6	66
Superficial pH	5.6 ±0.1	5.5 ±0.2	5.7 ±0.3	6.1 ±0.3	5.3 ±0.3	5.6 ±0.1
Water content (% soil dry weight)	96 ±16	77 ±13	71 ±17	130 ±21	136 ±33	96 ±9
Loss on ignition (% soil dry weight)	25 ±3	24 ±2	34 ±6	38 ±5	33 ±6	28 ±2
Sodium (mole g^{-1})	66 ±13	47 ±7	44 ±11	57 ±9	42 ±10	56 ±7
Potassium (mole g^{-1})	11 ±1	13 ±1	10 ±2	11 ±2	11 ±4	11 ±1
Magnesium (mole g^{-1})	42 ±5	49 ±11	51 ±18	65 ±15	35 ±8	46 ±5
Calcium (mole g^{-1})	26 ±5	17 ±4	53 ±16	69 ±20	31 ±11	34 ±5
Phosphorus (mole g^{-1})	1.3 ±0.4	0.6 ±0.2	2.5 ±0.8	1.3 ±0.4	0.9 ±0.4	1.4 ±0.2
Sodium/loss on ignition (mole g^{-1})	255 ±23	205 ±27	123 ±17	146 ±9	131 ±15	203 ±14

a *Plantago maritima* sub-community
b *Dactylis glomerata* sub-community
c *Achillea millefolium* sub-community
d *Primula vulgaris* sub-community
e *Anthoxanthum odoratum* sub-community
9 *Festuca rubra–Holcus lanatus* maritime grassland (total)

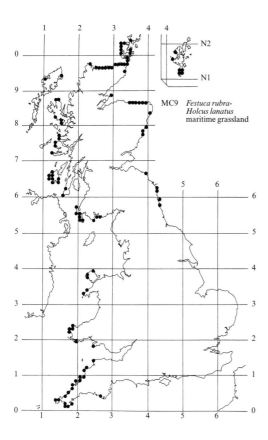

MC9   *Festuca rubra-*
Holcus lanatus
maritime grassland

MC10
Festuca rubra-Plantago spp. maritime grassland

Synonymy
Plantago sward   Praeger 1911, Tansley, 1939, Asprey 1946, Poore & Robertson 1949, McVean 1961 *p.p.*; *Plantaginetum coronopi*   Gillham 1953 *p.p.*; *Plantago maritima* nodum   Malloch 1971; Habitat Group IV Goldsmith 1975 *p.p.*

Constant species
Festuca rubra, Plantago coronopus, P. lanceolata, P. maritima, Agrostis stolonifera.

Rare species
Astragalus danicus, Oxytropis halleri, Primula scotica, Scilla verna, Trifolium occidentale.

Physiognomy
The *Festuca-Plantago* maritime grassland has a closed, very short and tight sward which is generally dominated by *F. rubra* and *Plantago* spp., especially *P. maritima*, with some *Agrostis stolonifera* and a variety of diminutive herbaceous associates of which *Euphrasia* spp. (usually grazed beyond recognition to the species), *Lotus corniculatus, Leontodon autumnalis, Cerastium fontanum* and *Scilla verna* are the most frequent throughout. Bryophytes make a small contribution to the sward with occasional records for *Hypnum cupressiforme* and *Mnium hornum.*

Sub-communities

Armeria maritima **sub-community:** *Plantago* sward *auct. angl. p.p. A. maritima* is an additional constant in the rather species-poor sward of this sub-community, though it is only rarely abundant. *Holcus lanatus, Cochlearia officinalis, Poa subcaerulea* (especially in the north), *Silene vulgaris* ssp. *maritima, Leontodon taraxacoides* and *Cerastium diffusum* ssp. *diffusum* are preferential or differential at low frequency. *Oxytropis halleri, Primula scotica* and *Trifolium occidentale* occur occasionally within their ranges of distribution.

Carex panicea **sub-community.** *Euphrasia* spp., *Lotus corniculatus* and *Leontodon autumnalis* attain constancy in this more species-rich and rather variable sward but the really distinctive feature is the constancy and occasional abundance of *Carex panicea. Thymus praecox* is preferentially frequent here and there are occasional records for a variety of poor-fen species such as *Carex demissa, C. serotina, C. nigra, Anagallis tenella, Molinia caerulea* and *Ranunculus flammula*, though the dominance of *F. rubra* and *P. maritima* remains characteristic. Rare species which occur occasionally are *Festuca vivipara* (indistinguishable from *F. ovina* when heavily-grazed), *Polygonum viviparum, Thalictrum alpinum, Trollius europaeus* and *Coeloglossum viride. Primula scotica* is not as frequent here as in the *Armeria* sub-community. *Salix repens, Nardus stricta* and *Calluna vulgaris* are differential at low frequencies.

Schoenus nigricans **sub-community.** *Carex panicea* and *Euphrasia* spp. are as frequent here as in the *C. panicea* sub-community but additional constants are *Schoenus nigricans, Carex serotina, Danthonia decumbens, Potentilla erecta* and *Molinia caerulea. Festuca rubra* is much less abundant in this sub-community and *Plantago maritima* or *S. nigricans* (as its prostrate maritime ecotype) generally dominate. Wet heath and poor-fen species which are occasional in the *C. panicea* sub-community are more frequent here: *Ranunculus flammula, Erica tetralix, Anagallis tenella, Hydrocotyle vulgaris* and *Succisa pratensis. Polygala serpyllifolia* is differential at low frequency.

Habitat
The *Festuca-Plantago* community is one of the sea-cliff grasslands characteristic of less maritime situations. A sub-set of soil samples had a mean sodium/loss on ignition ratio between the mean values for the *Festuca-Armeria* and *Festuca-Holcus* communities. The most obvious environmental characteristic separating the *Festuca-Plantago* community from these other maritime

grasslands is that it is consistently and heavily grazed, generally by sheep, which maintain the short sward as an often very extensive community of gently-sloping cliff tops.

The floristic differences between the sub-communities can be related mainly to soil variation, though the *Armeria* sub-community also tends to be characteristic of more maritime situations. Under this sub-community, which occurs on generally steeper slopes, the more shallow soils are predominantly mineral, usually brown rankers with a neutral to slightly acid pH. The *C. panicea* sub-community is found in less maritime situations but also usually in regions of wetter climate. Under such conditions, the deeper soils on more gentle slopes are damper and less well drained and often develop a superficial peaty layer (partly derived from the decay of *P. maritima*) over a gleyed mineral horizon. The *Schoenus* sub-community is characteristic of even wetter conditions and its small stands typically occur in depressions or gently-sloping valleys on cliffs where flushing maintains a slow water-flow through deep organic soils.

Zonation and succession

This maritime grassland generally occurs as a zone on grazed sea-cliffs above the *Festuca-Armeria* community (into which it may grade through the *Armeria* sub-community) and at roughly the same level as the *Festuca-Holcus* community. As such, it forms part of the sequence of cliff grasslands developed in relation to decreasing maritime influence on moving inland. Above it usually passes to maritime forms of heath or acid grassland.

It may also form zonations with the *Festuca-Armeria* and *Festuca-Holcus* grasslands in response to variations in grazing pressure and with increased inaccessibility to stock it can grade to these communities through their sub-association of *Plantago* spp. (Figure 23).

The zonation of the various sub-communities reflects the influence of both salt-spray deposition and soil patterns. The *Armeria* sub-community usually forms a zone seaward of the *C. panicea* sub-community and the *Schoenus* sub-community occurs patchily within the latter where there is pronounced flushing.

Grazing probably mediates a successional relationship between the *Festuca-Plantago* community and both the *Festuca-Armeria* and *Festuca-Holcus* communities, particularly the latter. It is possible that the *C. panicea* sub-community develops from wet maritime heath where this is heavily grazed.

Distribution

The *Festuca-Plantago* maritime grassland is predominantly a northern community reaching its most pronounced development on the cliffs of north-west Scotland, the Hebrides, Orkney and Shetland where heavy sheep-grazing occurs in a region of high rainfall. The *C. panicea* sub-community occurs widely as far

south as Galloway but the *Schoenus* sub-community is much more restricted: it has been encountered only on Islay, Skye, Harris, Lewis and the Ardnamurchan peninsula. The *Armeria* sub-community is widely distributed in Scotland but it also occurs patchily in Wales and south-west England.

Affinities

This community includes much of the very distinctive cliff-top vegetation described as *Plantago* sward by numerous authors from Britain (Tansley 1939, Asprey 1946, Poore & Robertson 1949, McVean 1961) and Ireland (Praeger 1911, 1934). It has no existing phytosociological counterpart but can be seen as the grazed equivalent of the *Festuca-Holcus* community (and perhaps also partly the *Festuca-Armeria* community). As such, it could be accommodated among the cliff grasslands of the Asteretea which Malloch (1970, 1971) placed in the new alliance Silenion maritimae.

Figure 23. Zonations showing the impact of grazing on sea-cliff vegetation in north-west Britain.
The upper diagram (a) shows typical patterns on an ungrazed sea-cliff on acidic bedrocks in north-west Britain with (b) the kinds of changes that can be seen where such a cliff is grazed.

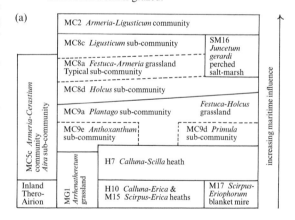

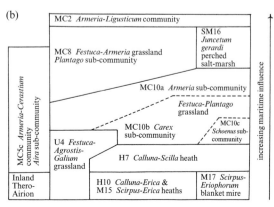

The *C. panicea* sub-community and particularly the *Schoenus* sub-community can be seen as part of a floristic transition from maritime grassland to wet heath. *S. nigricans* occurs in a variety of maritime and sub-maritime vegetation types of salt-marshes and sand-dunes (Sparling 1968) but here the maintenance of high frequencies of *F. rubra* and *Plantago* spp., as well as the absence of a prominent bryophyte component with such species as *Campylium stellatum* and *Scorpidium scorpioides*, argue for retaining the vegetation within this maritime grassland rather than considering it as a maritime variant of *Schoenetum nigricantis*.

Floristic table MC10

	a	b	c	10
Festuca rubra	V (3–10)	V (2–8)	V (3–5)	V (2–10)
Plantago coronopus	V (1–7)	V (2–7)	V (2–5)	V (1–7)
Plantago lanceolata	V (1–6)	V (1–5)	V (3–4)	V (1–6)
Plantago maritima	V (2–10)	V (3–10)	V (2–8)	V (2–10)
Agrostis stolonifera	IV (2–7)	IV (2–5)	IV (2–5)	IV (2–7)
Armeria maritima	V (1–8)	II (2–5)	I (2–4)	III (1–8)
Holcus lanatus	II (2–6)			I (2–6)
Cochlearia officinalis	II (1–4)	I (2–3)		I (1–4)
Poa subcaerulea	II (2–8)	I (2–6)		I (2–8)
Silene vulgaris maritima	I (2–5)			I (2–5)
Leontodon taraxacoides	I (1–4)			I (1–4)
Cerastium diffusum diffusum	I (1–3)			I (1–3)
Carex panicea		IV (1–7)	V (3–5)	II (1–7)
Euphrasia spp.	III (1–4)	IV (1–4)	IV (2–3)	III (1–4)
Lotus corniculatus	II (1–7)	IV (1–5)	I (1–3)	II (1–7)
Leontodon autumnalis	II (1–5)	IV (1–5)	III (2–3)	II (1–5)
Thymus praecox	I (2–6)	III (1–5)	I (3)	II (1–6)
Aira praecox	I (1–5)	II (1–3)		I (1–5)
Anthoxanthum odoratum	I (2–4)	II (1–6)		I (1–6)
Bellis perennis	I (2–4)	II (1–4)		I (1–4)
Luzula campestris	I (1–3)	II (1–4)	I (1)	I (1–4)
Ranunculus acris	I (1–8)	II (1–3)	I (2)	I (1–8)
Sagina procumbens	I (2–3)	II (2–3)	I (2)	I (2–3)
Viola riviniana		II (1–3)	I (1–3)	I (1–3)
Salix repens		I (1–8)		I (1–8)
Nardus stricta		I (2–6)		I (2–6)
Calluna vulgaris		I (1–4)		I (1–4)
Schoenus nigricans		I (1)	V (2–8)	I (1–8)
Carex serotina	I (2–4)	I (2–4)	V (2–4)	I (2–4)
Danthonia decumbens	I (2–3)	II (2–5)	V (2–4)	I (2–5)
Potentilla erecta	I (2–3)	II (1–4)	V (2–3)	I (1–4)
Molinia caerulea		I (2–8)	IV (2–9)	I (2–9)
Ranunculus flammula		I (1–3)	III (1–3)	I (1–3)
Polygala serpyllifolia			II (1–2)	I (1–2)
Anagallis tenella	I (3)	I (2–3)	II (2–4)	I (2–4)
Erica tetralix	I (3)	I (4)	II (2–5)	I (2–5)
Hydrocotyle vulgaris	I (3)	I (1)	II (2–3)	I (1–3)
Succisa pratensis	I (1)	I (1–6)	II (2–3)	I (1–6)
Pinguicula vulgaris			I (1)	I (1)

	a	b	c	10
Cerastium fontanum	III (1–4)	III (1–3)	II (2)	III (1–4)
Scilla verna	III (1–4)	III (1–4)	III (2–3)	III (1–4)
Carex nigra	I (2–4)	II (2–5)	II (3)	I (2–5)
Prunella vulgaris	I (1–2)	II (1–4)	III (1–3)	I (1–4)
Trifolium repens	II (1–5)	III (2–4)	I (3)	II (1–5)
Silene acaulis	I (2–7)	I (4–5)	I (3)	I (2–7)
Agrostis capillaris	I (2–5)	I (3–5)	I (4)	I (2–5)
Parnassia palustris	I (2)	I (3)	I (3)	I (2–3)
Sedum anglicum	I (1–4)	I (2–3)		I (1–4)
Sagina apetala	I (1–3)	I (1–3)		I (1–3)
Koeleria macrantha	I (2–5)	I (2–5)		I (2–5)
Hypochoeris radicata	I (1–4)	I (1)		I (1–4)
Hypnum cupressiforme	I (2–4)	I (2–5)		I (2–5)
Daucus carota gummifer	I (1–5)	I (2–3)		I (1–5)
Anthyllis vulneraria	I (1–5)	I (1–6)		I (1–6)
Rumex acetosa	I (1–3)	I (2–4)		I (1–4)
Mnium hornum	I (2–3)	I (2–4)		I (2–4)
Festuca ovina	I (5)	I (3–6)		I (3–6)
Empetrum nigrum	I (2–4)	I (1–8)		I (1–8)
Selaginella selaginoides		I (2–3)	I (2–3)	I (2–3)
Gentianella campestris		I (1–3)	I (1–2)	I (1–3)
Angelica sylvestris		I (1–3)	I (1)	I (1–3)
Linum catharticum		I (2–3)	I (3)	I (2–3)
Juncus acutiflorus		I (2–4)	I (3)	I (2–4)
Dactylorchis majalis purpurella		I (3)	I (1)	I (1–3)
Number of samples	138	113	11	262
Number of species/sample	12 (8–17)	17 (9–26)	17 (11–23)	14 (8–26)
Vegetation height (cm)	4 (1–15)	4 (1–15)	5 (2–19)	4 (1–19)
Total vegetation cover (%)	99 (80–100)	99 (80–100)	99 (90–100)	99 (80–100)
Altitude (m)	22 (2–100)	24 (4–80)	28 (15–45)	23 (2–100)
Slope (°)	11 (0–39)	8 (0–35)	6 (0–15)	10 (0–39)
Soil depth (cm)	25 (5–81)	32 (5–75)	35 (21–57)	28 (5–81)
Number of soil samples	48	38	3	89
Superficial pH	5.5 ±0.1	5.5 ±0.1	5.5	5.5 ±0.1
Water content (% soil dry weight)	94 ±13	139 ±21	261	119 ±12
Loss on ignition (% soil dry weight)	28 ±3	35 ±4	57	32 ±2
Sodium (mole g^{-1})	77 ±10	67 ±8	127	74 ±7
Potassium (mole g^{-1})	13 ±1	12 ±1	16	13 ±1
Magnesium (mole g^{-1})	56 ±5	43 ±4	66	51 ±4
Calcium (mole g^{-1})	28 ±3	26 ±3	32	27 ±2
Phosphorus (mole g^{-1})	1.6 ±0.3	1.2 ±0.3	0.9	1.4 ±0.2
Sodium/loss on ignition (mole g^{-1})	279 ±23	207 ±12	222	246 ±14

a *Armeria maritima* sub-community
b *Carex panicea* sub-community
c *Schoenus nigricans* sub-community
10 *Festuca-Plantago* spp. maritime grassland (total)

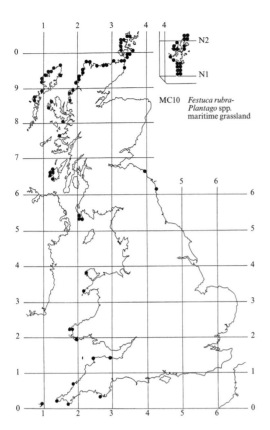

MC10   *Festuca rubra-*
       Plantago spp.
       maritime grassland

MC11

Festuca rubra-Daucus carota ssp. *gummifer* maritime grassland

Synonymy
Armerieto maritimae-Daucetum gummiferi Géhu 1964 *p.p.*

Constant species
Festuca rubra, Dactylis glomerata, Daucus carota ssp. *gummifer.*

Rare species
Brassica oleracea, Scilla verna, Silene nutans.

Physiognomy
The *Festuca-Daucus* maritime grassland has a fairly short, rather tussocky sward generally dominated by grasses of which *F. rubra* is usually the most abundant. *Dactylis glomerata* is constant and, though rarely abundant, may be in the form variously described as var. *abbreviata*, var. *maritima* or ssp. *hispanica*, which Tutin (1980*a*) considered to be one of the tetraploid populations within the species. *Daucus carota* ssp. *gummifer* is also constant in generally small amounts and, when flowering, it gives the vegetation a distinctive stamp. The maritime element in the community is small and the only other frequent species throughout are *Plantago lanceolata* and *Lotus corniculatus*. Bryophytes are rare.

Sub-communities

***Bromus hordeaceus* ssp. *ferronii* sub-community:** *Armerieto-Daucetum gummiferi typicum* Géhu 1964 *p.p. B. hordeaceus* ssp. *ferronii* and *Armeria maritima* are additional constants here and *Plantago coronopus, P. lanceolata* and *Lotus corniculatus* occur frequently. Although *F. rubra* is often abundant, there is usually no single dominant and the sward is sufficiently open to permit repeated colonisation by *B. hordeaceus* ssp. *ferronii* and other therophytes such as *Desmazeria marina, Senecio vulgaris* and *Vicia sativa.*

***Ononis repens* sub-community.** *O. repens* is an additional constant in this sub-community and occasionally it is co-dominant with *F. rubra*. The only other frequent species is *Plantago lanceolata* but the vegetation is distinctive in the occasional occurrence of a variety of species characteristic of open and/or calcareous situations: *Carlina vulgaris, Crambe maritima, Blackstonia perfoliata, Echium vulgare* and *Glaucium flavum.*

***Sanguisorba minor* sub-community.** The generally taller and lusher sward of this vegetation is characterised by the additional constancy of *S. minor* (differential to this sub-community), *Plantago lanceolata* and *Lotus corniculatus. Centaurea scabiosa, Galium verum* and *Brachypodium pinnatum* are frequent and the last may be co-dominant with *F. rubra*. There are numerous occasional species characteristic of inland calcicolous grasslands, notably *Festuca ovina, Helianthemum nummularium* and *Hieracium pilosella.*

Habitat
The *Festuca-Daucus* community is one of the sea-cliff grasslands characteristic of less maritime situations: it generally receives similar amounts of salt-spray an as the *Festuca-Holcus* community. It is, however, virtually confined to cliffs of calcareous rocks with rendziniform soils of high pH and calcium status. The *Bromus* sub-community is the most maritime and both it and the *Ononis* sub-community are especially characteristic of dry south-facing slopes and cliff edges. Where there are excessively-drained soils in such situations, species like *Crambe maritima, Plantago coronopus, Crithmum maritimum* and *Echium vulgare* are most common and abundant within the *Ononis* sub-community.

The *Sanguisorba* sub-community is the least maritime of the sub-communities and it occurs in more stable situations where the soils are somewhat moister. Unlike the other sub-communities it is occasionally grazed.

Zonation and succession
The community usually occurs as a fairly narrow zone inland of the *Brassica* cliff-edge community or sometimes the *Crithmo-Spergularietum* crevice vegetation.

On particularly sheltered cliffs, it may be the most maritime vegetation. The three sub-communities may themselves be zoned in relation to maritime influence with the *Bromus*, *Ononis* and *Sanguisorba* sub-communities succeeding one another on moving inland (Figure 24). Above, the *Festuca-Daucus* community usually passes to calcareous grassland or scrub.

The community may occur at the same level on cliffs as the *Festuca-Holcus* grassland and grade to it through the *Dactylis* sub-community of the latter with an increase in soil moisture.

Figure 24. Sequence of vegetation types on a limestone cliff in southern England.

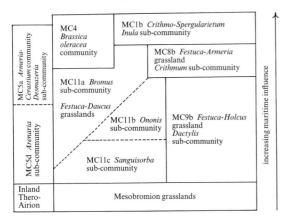

Floristic table MC11

	a	b	c	11
Festuca rubra	V (4–10)	V (3–10)	V (4–9)	V (3–10)
Dactylis glomerata	V (1–5)	V (2–7)	V (2–5)	V (1–7)
Daucus carota gummifer	V (2–5)	IV (1–4)	V (2–4)	V (1–5)
Armeria maritima	V (1–7)	II (2–4)	I (2–5)	III (1–7)
Bromus hordeaceus ferronii	V (2–5)	I (3)	II (2–4)	II (2–5)
Plantago coronopus	III (2–4)	II (1–4)	I (1–3)	II (1–4)
Desmazeria marina	II (2–4)	I (1)	I (1–3)	I (1–4)
Senecio vulgaris	II (1–4)	I (3)		I (1–4)
Silene vulgaris maritima	II (1–7)	I (6)	I (3)	I (1–7)
Vicia sativa	II (2–5)		I (3)	I (2–5)
Ononis repens		V (1–8)	I (2–4)	II (1–8)
Carlina vulgaris		II (1–2)	I (1–3)	I (1–3)
Rumex acetosa		II (2–3)	I (1–4)	I (1–4)
Blackstonia perfoliata		I (1–3)		I (1–3)
Crambe maritima		I (1–5)		I (1–5)
Crithmum maritimum		I (1–6)		I (1–6)
Echium vulgare		I (1–5)	I (1)	I (1–5)

Distribution
The community is most common the chalk and limestone cliffs of the south coast west to Dorset and in south Wales with isolated occurrences in the extreme south-west of England, in north Wales and in Cumbria. Okusanya (1979c) suggested that susceptibility to frost may be one of the main factors limiting *D. carota* ssp. *gummifer* to cliffs in the southern part of Britain.

Though the *Ononis* sub-community tends to be the most widespread, there is little difference in the distributions of the sub-communities.

Affinities
The *Festuca-Daucus* grassland forms part of a floristic sequence among the more calcicolous cliff communities running from the maritime *Brassica* community to inland calcareous grasslands. Apart from passing references (e.g. Tansley 1939, Mitchell & Richards 1979) there has been no previous description of the community from Britain. The typical sub-association of Géhu's *Armerieto-Daucetum gummiferi* described from the French Channel coast (1964) is similar to the *Bromus* sub-community described here but his association is much broader than the *Festuca-Daucus* grassland and includes vegetation which is to be better placed within the *Festuca-Holcus* community.

Glaucium flavum		I (1–4)		I (1–4)
Senecio jacobaea		I (1–3)		I (1–3)
Sanguisorba minor			V (1–7)	III (1–7)
Plantago lanceolata	III (1–4)	III (2–4)	V (2–4)	III (1–4)
Lotus corniculatus	III (3–7)	II (3–6)	IV (2–4)	III (2–7)
Brachypodium pinnatum		I (1–4)	III (3–10)	II (1–10)
Centaurea scabiosa		I (1–3)	III (1–5)	II (1–5)
Galium verum	I (2–4)	I (2)	III (2–4)	II (2–4)
Festuca ovina			II (5–7)	I (5–7)
Helianthemum nummularium			II (2–5)	I (2–5)
Hieracium pilosella			II (2–3)	I (2–3)
Ranunculus bulbosus			II (1–4)	I (1–4)
Carex flacca		I (2–3)	II (1–3)	I (1–3)
Brassica oleracea	I (4–5)	I (1)	II (2–7)	I (1–7)
Achillea millefolium		I (2)	II (2–3)	I (2–3)
Hippocrepis comosa		I (3)	II (1–8)	I (1–8)
Koeleria macrantha	I (3)	I (3)	II (2–5)	I (2–5)
Thymus praecox		I (2–4)	II (2–6)	I (2–6)
Sedum acre			I (2–3)	I (2–3)
Silene nutans			I (1–6)	I (1–6)
Carex caryophyllea			I (1–4)	I (1–4)
Centaurium erythraea			I (2–3)	I (2–3)
Cirsium acaule			I (1–2)	I (1–2)
Cynosurus cristatus			I (2–3)	I (2–3)
Avenula pratensis			I (2–4)	I (2–4)
Ranunculus acris			I (1–2)	I (1–2)
Scilla verna			I (2–3)	I (2–3)
Stachys officinalis			I (2–4)	I (2–4)
Teucrium scorodonia			I (2–3)	I (2–3)
Anthyllis vulneraria	I (1–3)	II (1–4)	II (2–5)	II (1–5)
Festuca arundinacea	I (1)	II (1–4)	I (2–3)	II (1–4)
Taraxacum sp.	I (2)	I (1)	II (1–2)	I (1–2)
Leontodon taraxacoides	I (1–3)	I (1–3)	I (2–4)	I (1–4)
Trifolium repens	I (2–4)	I (3)	I (2)	I (2–4)
Agrostis stolonifera	I (3)	I (2–3)	I (3)	I (2–3)
Cirsium vulgare	I (1)	I (1–2)	I (1)	I (1–2)
Holcus lanatus	I (3)	I (3–5)	I (3–4)	I (3–5)
Sonchus oleraceus	I (1)	I (2–3)	I (2)	I (1–3)
Agrostis capillaris	I (3–4)		I (1–5)	I (1–5)
Convolvulus arvensis	I (2)		I (3)	I (2–3)
Lolium perenne	I (2–3)		I (2)	I (2–3)
Arenaria serpyllifolia	I (3)		I (2–3)	I (2–3)
Potentilla reptans	I (3–4)	I (3)		I (3–4)
Medicago lupulina	I (2)	I (2–4)		I (2–4)
Bellis perennis		I (1–2)	I (1–4)	I (1–4)
Euphorbia portlandica		I (2–3)	I (2–3)	I (2–3)
Hypochoeris radicata		I (1–3)	I (2)	I (1–3)

Floristic table MC11 (*cont.*)

	a	b	c	11
Centaurea nigra		I (1)	I (1)	I (1)
Leucanthemum vulgare		I (2–3)	I (2–3)	I (2–3)
Number of samples	15	25	23	63
Number of species/sample	10 (6–16)	11 (5–17)	18 (11–27)	13 (5–27)
Vegetation height (cm)	10 (2–50)	14 (3–50)	16 (2–50)	14 (2–50)
Total vegetation cover (%)	94 (60–100)	86 (40–100)	98 (80–100)	92 (40–100)
Altitude (m)	31 (6–45)	40 (3–150)	48 (3–100)	41 (3–150)
Slope (°)	16 (0–30)	25 (5–70)	15 (2–40)	19 (0–70)
Soil depth (cm)	19 (6–47)	36 (6–51)	14 (3–30)	23 (3–51)
Number of soil samples	2	5	3	10
Superficial pH	7.5	7.0 ±0.5	7.2	7.2 ±0.3
Water content (% soil dry weight)	19	20 ±5	49	29 ±8
Loss on ignition (% soil dry weight)	17	8 ±1	27	16 ±3
Sodium (mole g^{-1})	27	21 ±5	27	24 ±5
Potassium (mole g^{-1})	15	8 ±2	9	10 ±2
Magnesium (mole g^{-1})	30	20 ±4	27	24 ±4
Calcium (mole g^{-1})	197	102 ±23	148	135 ±17
Phosphorus (mole g^{-1})	2.1	1.0 ±0.9	0.5	1.1 ±0.6
Sodium/loss on ignition (mole g^{-1})	138	260 ±59	96	186 ±41

a   *Bromus hordeaceus* ssp. *ferronii* sub-community
b   *Ononis repens* sub-community
c   *Sanguisorba minor* sub-community
11   *Festuca rubra-Daucus carota* ssp. *gummifer* maritime grassland (total)

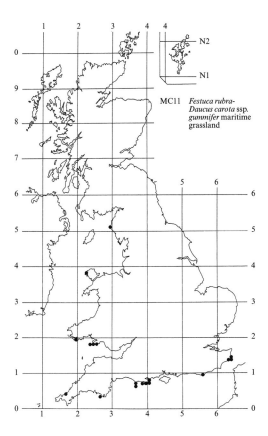

N2

N1

MC11   *Festuca rubra-*
Daucus carota ssp.
gummifer maritime
grassland

MC12

Festuca rubra-Hyacinthoides non-scripta maritime bluebell community

Synonymy
Pteridietum aquilini Goodman & Gillham 1954 *p.p.*;
Endymio-Armerietum maritimae Malloch 1971.

Constant species
Festuca rubra, Hyacinthoides non-scripta, Rumex acetosa, Holcus lanatus.

Rare species
Scilla verna.

Physiognomy
The *Festuca-Hyacinthoides* community comprises a lush carpet of *F. rubra* and *H. non-scripta* with *Holcus lanatus* (sometimes abundant) and scattered plants of *Rumex acetosa*. Few other species are uniformly frequent throughout but *Daucus carota* ssp. *gummifer* occurs occasionally. *Scilla verna* and *Sedum telephium* are distinctive at low frequencies.

Sub-communities

Armeria maritima sub-community: *Endymio-Armerietum maritimae, Cochlearia officinalis* sub-association Malloch 1971. *A. maritima* and *Silene vulgaris* ssp. *maritima* attain constancy in the grassier sward of this sub-community where *H. lanatus* and (especially) *F. rubra* tend to rival *H. non-scripta* in abundance. *Cochlearia officinalis* is differential and there are small amounts of other maritime species: *C. danica, Beta vulgaris* spp. *maritima* and *Spergularia rupicola*.

Ranunculus ficaria sub-community: *Endymio-Armerietum maritimae, Ranunculus ficaria* sub-association Malloch 1971. *H. non-scripta* is much more consistently dominant here though *F. rubra, H. lanatus* and *Dactylis glomerata* are all constant contributors to the sward. The most distinctive feature of the vegetation, however, is the constancy in small amounts of

R. ficaria and the preferential occurrence of *Heracleum sphondylium, Galium verum* and *Pteridium aquilinum* and a variety of species characteristic of scrub and woodland: *Brachypodium sylvaticum, Primula vulgaris, Viola riviniana, Hedera helix* and *Rubus fruticosus* agg.

Habitat
The *Festuca-Hyacinthoides* community is most characteristic of situations where deep, moist and fertile brown soils develop under moderate maritime influence and it is usually found on gentle to moderate north-facing slopes or in deep gullies fairly high on cliffs. It occurs on most rock types apart from chalk and limestones and is especially well developed on more easily weathered materials like the hornblende schists and diabases of the Lizard and Land's End where stands of the community present a very striking appearance. It is always ungrazed.

The conditions under which the community is best developed match well those which Blackman & Rutter (1950, 1954) considered optimal for the growth of *H. non-scripta* apart from the soil mineral status which is here fairly high. They suggested that a low mineral status helps prevent the growth of luxuriant grasses which might effectively compete with the *H. non-scripta*. It is possible that here even the moderate maritime influence prohibits the spread of such species as *Arrhenatherum elatius* which might otherwise be encouraged.

Zonation and succession
The community occurs at about the same level on cliffs as maritime heaths, grading below to *Festuca-Holcus* maritime grassland. Despite the representation of scrub and woodland species within the community, there is little evidence of any succession from or to such communities. It is hard to see how some stands of the *Festuca-Hyacinthoides* community could have ever carried a shrub or tree canopy; nor are they ever likely to

develop one in even the moderate exposure to salt-spray. The spread of *Pteridium aquilinum* may likewise be inhibited by maritime influence but it is possible that high levels of soil moisture also play a part in preventing its spread.

Distribution

The community is sparsely distributed from south-west England northwards to Skye.

Affinities

Apart from incidental references (e.g. Blackman & Rutter 1954, Goodman & Gillham 1954, Géhu 1960), there is no account of this kind of vegetation in the descriptive or phytosociological literature. Malloch (1970, 1971) proposed that it could be accommodated with other maritime swards within his new alliance, the Silenion maritimae, of the Asteretea.

Floristic table MC12

	a	b	12
Festuca rubra	IV (2–6)	V (4–9)	V (2–9)
Hyacinthoides non-scripta	V (3–9)	V (2–9)	V (2–9)
Rumex acetosa	IV (1–4)	V (1–5)	V (1–5)
Holcus lanatus	III (1–8)	IV (2–6)	IV (1–8)
Ranunculus ficaria	V (1–5)		II (1–5)
Dactylis glomerata	V (2–6)	II (1–7)	III (1–7)
Heracleum sphondylium	III (1–3)	II (1–4)	II (1–4)
Galium verum	II (1–3)	I (2)	II (1–3)
Pteridium aquilinum	II (1–5)	I (2–5)	II (1–5)
Brachypodium sylvaticum	II (2–6)	I (4–5)	II (2–6)
Primula vulgaris	II (2–5)	I (2–5)	II (2–5)
Viola riviniana	II (2–4)	I (2–3)	I (2–4)
Hedera helix	II (1–8)		I (1–8)
Rubus fruticosus agg.	II (1–4)		I (1–4)
Vicia sativa	I (1–2)		I (1–2)
Silene vulgaris maritima	II (1–5)	IV (1–6)	III (1–6)
Armeria maritima	I (1–5)	V (1–5)	III (1–5)
Cochlearia officinalis		III (2–5)	II (2–5)
Lotus corniculatus	I (1–3)	II (2–3)	II (1–3)
Beta vulgaris maritima		I (1–4)	I (1–4)
Arrhenatherum elatius		I (5–6)	I (5–6)
Cochlearia danica		I (1)	I (1)
Spergularia rupicola		I (3)	I (3)
Lotus uliginosus		I (3)	I (3)
Daucus carota gummifer	II (1–3)	II (1–3)	II (1–3)
Achillea millefolium	I (2–4)	I (2–3)	I (2–4)
Anthoxanthum odoratum	I (4–6)	I (4)	I (4–6)
Hypochoeris radicata	I (1–2)	I (1–3)	I (1–3)
Plantago lanceolata	I (1–3)	I (1–3)	I (1–3)
Potentilla erecta	I (3)	I (2)	I (2–3)
Scilla verna	I (1–2)	I (2–3)	I (1–3)
Sedum telephium	I (3)	I (1–4)	I (1–4)
Sonchus oleraceus	I (1)	I (1)	I (1)
Cirsium vulgare	I (1)	I (1)	I (1)
Ononis repens	I (3)	I (5)	I (3–5)

Floristic table MC12 (*cont.*)

	a	b	12
Number of samples	17	21	38
Number of species/sample	11 (8–17)	10 (6–18)	10 (6–18)
Vegetation height (cm)	25 (6–40)	27 (10–50)	25 (6–50)
Total vegetation cover (%)	99 (85–100)	99 (80–100)	99 (80–100)
Altitude (m)	28 (3–50)	31 (3–48)	29 (3–50)
Slope (°)	13 (7–35)	17 (0–60)	15 (0–60)
Soil depth (cm)	29 (6–50)	30 (10–75)	30 (6–75)
Number of soil samples	5	9	14
Superficial pH	5.3 ±0.3	5.6 ±0.2	5.5 ±0.2
Water content (% soil dry weight)	157 ±45	104 ±13	123 ±18
Loss on ignition (% soil dry weight)	34 ±5	31 ±3	32 ±3
Sodium (mole g^{-1})	43 ±7	69 ±15	60 ±11
Potassium (mole g^{-1})	14 ±2	15 ±2	15 ±2
Magnesium (mole g^{-1})	59 ±18	69 ±15	65 ±11
Calcium (mole g^{-1})	15 ±3	22 ±5	20 ±3
Phosphorus (mole g^{-1})	0.65 ±0.2	4.19 ±1.7	2.9 ±1.1
Sodium/loss on ignition (mole g^{-1})	138 ±27	222 ±29	192 ±23

a   *Ranunculus ficaria* sub-community
b   *Armeria maritima* sub-community
12   *Festuca rubra-Hyacinthoides non-scripta*
    maritime bluebell community (total)

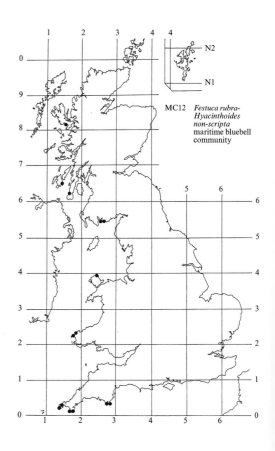

MC12   *Festuca rubra-Hyacinthoides non-scripta* maritime bluebell community

VEGETATION OF OPEN HABITATS

INTRODUCTION TO VEGETATION OF OPEN HABITATS

The sampling of weed vegetation and other assemblages of open habitats

In the days before political correctness was an issue, the editors of a renowned book on weeds made the memorable observation that, as a class, weeds had 'much in common with criminals': that is, when not engaged in nefarious activities, they were often admirable. Certainly, that book (Salisbury 1964), then and still the most accessible account of these plants, did much to inform and engage, and it grouped weeds into broad ecological categories according to habitat – cornfield and arable weeds, grassland weeds, weeds of sandy soils, of chalk, of roadsides and wastes, and of gardens. This was scarcely a classification, of course, but it was a considerable advance on Tansley (1939) which, like its predecessor (Tansley 1911), made only fleeting references to weed vegetation of any kind, and then usually within the context of successions to other plant communities. In fact, Salisbury himself (1964) was building on previous work, like the survey of Buckman (1856), who described arable weeds of different soil types, and the studies of Brenchley (1911, 1912, 1913), who made more comprehensive field lists in various localities, noting the geology, soils and crops with which the plants were associated. However, for a long time there developed a widely held notion that, like criminals perhaps, these vegetation types cannot readily be brought to book and understood within the stricter confines of classification, that the behaviour of these plants is too wayward for them to be considered as belonging to communities like other species.

Meanwhile, however, the distribution and behaviour of many individual weed species, particularly rare and declining plants or aliens, have remained an abiding interest and their occurrence has featured in local floras, sometimes with 'habitat studies', or with a focus on distinctive groups of species like the bulbfield weeds of The Scillies (Lousley 1971). Details of decline, with information on ecology and informal lists of associated plants, were also given for a number of scarcer weeds in Stewart *et al.* (1994).

In sharp contrast to such neglect of their community ecology in Britain, elsewhere in Europe there has long been a deep interest in the phytosociology of weed vegetation, with a clear conviction that assemblages could be characterised and classified in just the same way as with other communities (e.g. Braun-Blanquet 1928, Tüxen 1937, Kruseman & Vlieger 1939, Westhoff & den Held 1969, Oberdorfer 1983, Ellenberg 1988, Pott 1992, Mucina *et al.* 1993). Such attention also included Ireland, from where a number of weed communities were described by Braun-Blanquet & Tüxen (1952) and summarised in White & Doyle (1982). This common interest has not meant that these schemes were uncontentious: indeed, reorganisation of the hierarchies of weed associations has been an abiding preoccupation of many Continental phytosociologists. However, such argument does not itself undermine the value of characterising associations and trying to understand the relationship of these assemblages to the various environmental factors influencing their composition and distribution.

It was not until the study of Silverside (1977), however, that such an approach was applied in Britain, concentrating on arable weed vegetation, assembling large numbers of phytosociological samples, describing communities and relating them to syntaxa defined elsewhere in Europe. It was greatly to our benefit that we were able to make use of the unpublished data from this doctoral research to put alongside our own accumulating samples, particularly where Silverside focused on scarcer kinds of weed vegetation. We were then able ourselves to concentrate on collecting data from assemblages in less striking but very widespread habitats like intensive arable land, weedy pastures and leys, gateways, paths and verges, wasteland and urban habitats like pavements and courtyards.

We have also included in this section of the work samples of vegetation from other kinds of open, disturbed or colonising habitats – assemblages from periodically inundated ground on river banks and shoals, around pools and reservoirs and in the bottom of ephemeral ponds, and vegetation of talus, spoil, rock outcrops and wall crevices. This was partly a practical decision, to ensure that all remaining data were incorporated into the scheme although, of course, there is some ecological sense in this grouping. However, the complexity of our task as a whole meant that the organisation of the communities across the entire scheme was not wholly felicitous: samples from fern communities of siliceous screes, for example, have already been included with upland vegetation (Rodwell 1992) and crevice assemblages of coastal rocks are described among the maritime cliff communities in this volume. As throughout the classification, vegetation dominated by cryptogams, especially important in certain rocky habitats, was omitted altogether from our survey.

As always, in all these different situations, we located our samples only on the basis of floristic and structural homogeneity of the vegetation, and usually found that samples of 2×2 m were adequate for recording or, in taller vegetation dominated by, say, *Urtica dioica*, *Epilobium hirsutum* or *E. angustifolium*, 4×4 m. On verges or tracks, or among crops in marked, well-spaced rows, it was necessary to use rectangular samples of equivalent area. Complex systems of crevices in rock outcrops, on talus, walls and pavements, could pose a more particular challenge for sampling but, so far as possible, we aimed to include an equivalent area of homogeneous vegetation to 2×2 m, even if of very irregular shape.

On some arable land, it was possible to sample only very early or after harvesting, so as to avoid damage to the crops and, as Silverside (1977) warned, weed vegetation can show phenological change so some plants were probably missed then or shifts in their proportions unremarked upon. For example, certain weeds are autumn germinators, so can dominate the winter or early aspect of the vegetation, being replaced by other plants later. In other cases, less ephemeral species can produce a flush of growth in stubble, having survived the harvest by virtue of their low habit and then benefiting from the sudden influx of light.

In recording the floristic data, all vascular plants, bryophytes and lichens were noted, as much care as possible being taken with taxonomically difficult plants. In combining our data with those of Silverside (1977), some discrepancy in identifying *Matricaria perforata* and *M. maritima* seemed likely in that we were probably not as scrupulous as he had been. Also, it is possible that some fumitories were not accurately determined by our team. As usual, *Rubus fruticosus* was recorded as the aggregate.

In arable fields, the crop plants themselves were not recorded as part of the vegetation, though some garden species, like *Oxalis pes-caprae* and *Gladiolus byzantinus*, figure prominently as escapes in certain situations and were included. However, the type of crop and its cover were noted as part of the environment of the vegetation sampled from arable fields. A fair range of crops was included in our own sampling but one of the particular virtues of Silverside's (1977) study was the very wide range covered there – many types of cereal, root, vegetable and salad crops, as well as bulbs, were sampled from farms, smallholdings and gardens. Detailed information on management – for example, the pattern of cultivation, timing and intensity of fertiliser or herbicide applications – was not available but general observations on the style of agriculture were made. Similarly, where weed vegetation was sampled in pastures and leys, along tracks and verges, in gateways and yards, as much information as possible was included about the evident impacts of grazing, mowing, trampling and neglect. The character of the soil was noted and an attempt was made to set the vegetation sampled in its wider landscape context.

To our own data and those from Silverside (1977), we also had the benefit of adding samples from Scotland provided by workers then at the Macaulay Institute (Birse & Robertson 1976, Birse 1980, 1984) and from a few literature sources which are noted in the relevant community descriptions. In total, just over 1500 samples were available. Geographical coverage was well spread but thin, although a wide range of habitats, both anthropogenic or more natural, was included. However, it is important to acknowledge that certain kinds of vegetation in disturbed or colonising habitats are poorly covered: the diminutive ephemeral assemblages of damp tracks and muddy pools, for example, some inundation communities and vegetation of eroding earth banks, certain fern and crevice communities, and tall-herb or scrubby vegetation with *Impatiens glandulifera*, *Reynoutria japonica* and *Buddleja* spp. Such omissions are noted elsewhere in the text and summarised, along with other deficiencies in the scheme, in the Phytosociological Conspectus included in this volume.

A further advantage of the Conspectus is that it highlights the various relationships, in terms of floristic composition and habitat, between the vegetation types characterised in this section of the work and communities defined elsewhere in the scheme. These are summarised in the next section and detailed in particular community descriptions.

Data analysis and the description of weed and other communities

In the usual fashion, only floristic data were used to characterise the vegetation types, any environmental information being used after the sample groups had been defined to aid ecological interpretation. Quantitative scores for cover/abundance for all the vascular plants, bryophytes and lichens were employed, no special weight being given to rare plants or presumed indicator species of any kind and the vegetation types were defined on the basis of the frequency and abundance of the constituent plants.

In all, 42 communities are included in this section. Although among some weed assemblages, there seems to be an element of chance as to which particular species become(s) prominent in what are often more or less uniform habitats, there was in fact no great difficulty in characterising regular groupings that are distinctive of broadly recognisable situations. In this country, however, where we are at the limit of the warm, dry continental climate that prevails over much of central Europe, vegetation types that are well-defined there tend to lose their integrity with us. In particular, our winters are often mild enough for certain important arable crops – and the weeds which grow among them – to be able to germinate in the autumn and survive to complete their life cycle in the following year. This rather confounds the traditional grouping of associations into neat higher level syntaxa like alliances that are based on an agricultural routine dominated by the production of summer cereals.

Climatic contrasts with other parts of Europe are complicated by the intensive style of arable cultivation that now prevails in this country with the use of cleaner seed, less cultivation, heavy applications of fertilisers and the use of herbicides. These practices all tend to favour a certain group of rather generalist ephemerals which, in various combinations, dominate several communities and which, even where they are less frequent, can readily overwhelm more striking weeds characteristic of particular climatic or edaphic conditions. The greater uniformity of arable agriculture means that, in most localities, high fertility dominates the soil environment much more than the reaction or texture.

With other vegetation types, like the inundation communities, where one of a number of opportunists can quickly monopolise a particular stretch of ground, stands can be composed virtually of single species or of various proportions of a few possible dominants. This can pose some problems for classification, but as with swamps, we have tried to take a common-sense approach to such situations.

The 42 communities are described in outline below under the following broad headings: arable weed and trackside communities of light, less fertile, acid soils (6 communities), arable weed and wasteland communities of fertile loams and clays (8), arable weed communities of light, limey soils (3), gateways, tracksides and courtyard vegetation (6), tall-herb weed communities (4), inundation communities (5), dwarf-rush communities of ephemeral ponds (4) and crevice, scree and spoil communities (6).

Arable weed and track-side communities of light, less-fertile acid soils

Six communities can be sensibly considered together because they have various ephemerals typical of open ground on light, often sandy soils, usually quite acidic and inherently infertile, in the warmer and drier south of Britain – plants like *Spergula arvensis*, *Chrysanthemum segetum*, *Erodium cicutarium*, *Anagallis arvensis*, *Briza minor*, *Silene gallica* and *Aphanes microcarpa*. These are species of Continental or Lusitanian affinities, dependent on warmer temperatures and able to complete their life cycle before any prospect of summer drought. Necessarily, in landscapes where lime-poor bedrocks or superficials are scarce, the edaphic conditions needed to sustain assemblages of such plants are very localised. Moreover, even where such soils are repeatedly disturbed for cultivation, they have often been improved by fertilising, when more nutrient-demanding and competitive weeds overwhelm these more diminutive plants, or have been treated with herbicides which can keep most ephemerals at bay. Only in places where more traditional arable agriculture or market-gardening has persisted, where distinctive cropping routines still leave some opportunity for these plants to flourish or along scuffed tracks in heathy landscapes, can these assemblages find opportunities to come and go as disturbance occurs.

The way in which more Continental weed assemblages lose their integrity in highly improved landscapes towards their climatic limit is well seen in the *Viola arvensis-Aphanes microcarpa* community (OV1), a very local vegetation type of arable crops on impoverished sands in the warmer and drier east of Britain. While providing a locus for such Continental species as *Scleranthus annuus*, *Anthoxanthum aristatum* and at one time for the now extinct *Arnoseris minima*, it is more common plants like *Aphanes microcarpa*, *Spergula arvensis* and *Rumex acetosella* which are the more frequent distinctive species here. Similarity to the *Teesdalio-Arnoseridetum minimae* (Malcuit 1929) R.Tx. (1937) 1950 or *Sclerantho-Arnoseridetum* R.Tx. 1937 is therefore attained only in fragmentary fashion.

Somewhat more distinctive floristically is the *Briza minor-Silene gallica* community (OV2) where, among *Anagallis arvensis*, *Aphanes microcarpa*, *Bromus hordeaceus* ssp. *thominii*, *Vulpia bromoides* and *V. myuros*, the nationally rare *Silene gallica* and introduced *Briza minor* show high frequency. This occurs on similar soils to the *Viola-Aphanes* community but in the more oceanic climate of the far south-west of England where bulb cultivation in The Scillies now usually provides the necessary disturbance for its repeated appearance. Again, it can be seen as a fragmentary equivalent of a Continental association, the *Airo multiculmis-Arnoseridetum minimae* (Allorge 1932) R.Tx. 1950, an assemblage which replaces the *Teesdalio-Arnoseridetum* in the south-west of Europe.

A further type of weed vegetation of infertile sands which has just a finger-hold in Britain is the *Digitaria ischaemum-Erodium cicutarium* community (OV5). Of extremely localised occurrence among various arable crops in Surrey, it has frequent *Erodium cicutarium* and *Spergula arvensis* and occasional *Ornithopus perpusillus* among commoner weeds like *Stellaria media*, *Senecio vulgaris* and *Poa annua*. It also provides a locus for the introduced Eurasian grass *D. ischaemum* and probably also *Echinochloa crus-galli* and *Setaria viridis*. It can be seen as representing the *Echinochloa-Setarietum* (Kruseman & Vlieger 1939) *emend.* Kruseman & Vlieger *apud* Sissingh, Vlieger & Westhoff 1940, at its north-western European limit.

Also favouring light, acidic soils but more widespread than any of these assemblages in rather more fertile arable fields throughout the British lowlands is the *Chrysanthemum segetum-Spergula arvensis* community (OV4, *Spergulo-Chrysanthemetum segetum* (Br.-Bl. & de Leeuw 1936) R.Tx. 1937). Quite common in both cereal and root crops, it is *Spergula* and *Chrysanthemum* which give a distinctive character here, along with *Stellaria media*, *Capsella bursa-pastoris*, *Poa annua*, *Polygonum aviculare* and *Elymus repens*. The resistance of *C. segetum*, together with some of the commoner associates of the community to some herbicide treatments, means that this community can persist in more intensive arable landscapes, although it is still most strikingly seen in traditional settings like the arable land of the Hebridean machair.

Towards the far south-west of England, the *Spergulo-Chrysanthemetum* takes on an oceanic character and is replaced on relatively fertile sands of bulb and vegetable fields in the moist climate of Cornwall and The Scillies by the *Cerastium glomeratum-Fumaria muralis* ssp. *boraei* community (OV6). Here *Chrysanthemum* is accompanied by the distinctive constants *Cerastium glomeratum*, *Juncus bufonius*, *F. muralis* ssp. *boraei* and frequent *Briza minor*, *Ranunculus parviflorus*, *Coronopus didymus*, *Silene gallica* and *Medicago arabica*. One type of this *Cerastium-Fumaria* vegetation provides an important locus for *Fumaria bastardii* and the association with bulb cultivation provides an opportunity for *Oxalis pes-caprae* and *Gladiolus byzantinus* to figure frequently as escapes.

Finally, among this group and on light, infertile soils that are less acidic, there is the *Papaver rhoeas-Viola arvensis* community (OV3, *Papaveretum argemones* (Libbert 1933) Kruseman & Vlieger 1939). With its startling display of poppies, growing among *Viola arvensis*, *Veronica persica*, *V. arvensis* and *Anagallis arvensis*, this assemblage is a very distinctive sight among cereal fields that have escaped herbicide treatment throughout the warmer and drier lowlands of Britain, although the community is seen in its full expression only in the more Continental south-east where *Papaver argemone* joins the somewhat more widespread *P. dubium* and common *P. rhoeas* as a frequent element. The *Papaveretum argemones* is widespread through Europe, south to Italy and north to Scandinavia.

Arable weed and wasteland communities of fertile loams and clays

A further six weed communities are characterised by high frequencies and often abundant luxurious growth of ephemerals such as *Stellaria media*, *Matricaria perforata*, *Chenopodium album*, *Capsella bursa-pastoris*, *Poa annua*, *Polygonum aviculare*, *Bilderdykia convolvulus* and *Veronica persica*, together with *Elymus repens*. These are assemblages of more fertile, loamy and clayey soils and, since many of these plants are somewhat resistant to certain herbicides, they will persist in fields under quite intensive arable agriculture. Some of the communities in this group show a climatic limitation to certain parts of the country but others are among the commonest weed assemblages of our cereal, root and vegetable crops of the alliance Polygono-Chenopodion polyspermi W. Koch 1926 *emend.* Sissingh 1946.

Of these vegetation types, the most familiar is probably the *Matricaria perforata-Stellaria media* community (OV9), where, together with these two species, *Polygonum aviculare* is constant, and plants like *Poa annua*, *Bilderdykia*, *Elymus repens* and *Sonchus asper* frequent. In fact, this is the most widespread kind of weed vegetation in which *M. perforata* plays a prominent role in this country, showing no climatic limitation through the British lowlands but quite strongly confined to the free-draining soils which this plant prefers. Occurring in both cereals and other arable crops, many of its constituents respond markedly to additions of nitrogen, so heavy fertilising can greatly stimulate the development of such vegetation.

Also having frequent *M. perforata*, but restricted to lighter, fertile, circumneutral loams in the warmer and drier south-east of Britain, is the *Veronica persica-V. polita* community (OV7, *Veronico-Lamietum hybridi* Kruseman & Vlieger 1939). *V. polita*, which is constant here, *Lamium hybridum*, which is the most distinctive occasional, and also to some degree the associates *Euphorbia helioscopia* and *Solanum nigrum*, all show Continental affinities in the limitation of their ranges in this country. Cereals and other field crops in the east Midlands, East Anglia and southern England provide the most favoured habitat for this community which is essentially the same with us as on the Continental mainland.

V. persica, *M. perforata*, *E. helioscopia* and *S. nigrum* all remain frequent to occasional in the *V. persica-*

Alopecurus myosuroides community (OV8, *Alopecuro-Matricarietum chamomillae* Wasscher 1941), but *A. myosuroides* is the distinctive constant here, *Elymus repens*, *Avena fatua* and *Sinapis arvensis* are frequent companions and *Chamomilla recutita* is a notable occasional. This assemblage is again common only in the warmer parts of Britain but typically on soils whose loamy or clayey texture affords some protection against drought in drier summers. It is especially associated with winter-sown cereals and other field crops in the south-east Midlands and East Anglia. It occurs elsewhere in north-west Europe but has often been subsumed into the *Veronico-Lamietum*.

On loamy and clayey soils in the less continental climate of western Britain, mostly in cereal crops but occasionally among vegetables, these assemblages are replaced by the *Poa annua-Stachys arvensis* community (OV11). *Spergula arvensis* shows a resurgence of frequency here, but the really distinctive feature is the constancy of *Stachys arvensis*.

Also best included among this group of weed communities is the *Poa annua-Senecio vulgaris* community (OV10). This shares with the above frequent records for plants like *Poa annua*, *Stellaria media*, *Polygonum aviculare* and *Capsella* but *M. perforata*, *Bilderdykia* and *Elymus repens* are more limited in occurrence and *Senecio vulgaris* and *Lolium perenne* become common. This is a pioneer assemblage of open cultivated or trampled ground, particularly where fertile soils have become moist. It is very widespread throughout the lowlands in arable fields and gardens, badly managed leys, waysides and on dumped earth. Also common in leys and pastures, particularly where trampling and dunging have opened up swards in wet winter weather, is the *Poa annua-Myosotis arvensis* community (OV12). *P. annua* is consistently accompanied here by *P. trivialis* and *Agrostis solonifera*, *Polygonum aviculare* is often joined by *P. persicaria* and *Myosotis arvensis* is a distinctive feature.

Two further communities of fertile loams show strong floristic affinities with the Polygono-Chenopodion alliance although they comprise vegetation which has traditionally been grouped in the Fumario-Euphorbion Th. Müller ex Görs 1966 alliance. Again, in the widely improved landscapes in the Atlantic climate of Britain, it is difficult to maintain certain traditional phytosociological distinctions: often, in stands of these vegetation types, treated with fertilisers and herbicides, it is more widespread generalist weeds that strike the observer as much as the distinctive ephemerals which they threaten to overwhelm.

In the *Stellaria media-Capsella bursa-pastoris* community (OV13), for example, plants like *Stellaria*, *Capsella*, *Senecio vulgaris*, *Polygonum aviculare* and *Chenopodium album* can be very frequent and abundant, with the more distinctive *Fumaria officinalis*, *F. muralis* ssp. *boraei*, *F.*

bastardii and *Urtica urens* often playing a subordinate role. So, this assemblage subsumes what has sometimes been characterised as a *Fumarietum officinalis* R.Tx. 1950 and a *Fumarietum bastardii* Br.-Bl. in Br.-Bl. & Tx. 1952. It is widespread through the lowlands of this country occurring especially among root, vegetable and salad crops, but also in cereals and on waste ground where loamy soils are naturally rich or have been fertilised.

The *Urtica urens-Lamium amplexicaule* community (OV14) can show a similar general character with these two distinctive species often rather crowded out by commoner ephemerals. This is a more localised community than the above, largely confined to the warmer and drier south-east of Britain and typical of light and somewhat more base-poor soils where it grades to the *Spergulo-Chrysanthemetum*. In fact, *S. arvensis* is a common associate here, though it is plants like *Galinsoga parviflora* and *Solanum nigrum* that provide a more distinctive element. And, very strikingly in some places, this kind of weed vegetation provides a locus for a diversity of wool-aliens originating from shoddy waste used as a fertiliser.

Arable weed communities of light, limey soils

Two communities are very distinct from all the above in their floristics and environmental preferences. *Polygonum aviculare*, *Stellaria media*, *Bilderdykia convolvulus*, *Elymus repens* and *Veronica persica* all remain frequent here but *Capsella* and *Senecio vulgaris* are absent and a variety of calcicolous ephemerals with Continental affinities are preferential. These, then, are weed assemblages of lime-rich soils, mostly among cereal crops, in the warmer and drier south-east of Britain.

The more widespread, though still very local vegetation type is the *Anagallis arvensis-Veronica persica* community (OV15, *Kickxietum spuriae* Kruseman & Vlieger 1939) where, if it were not for the fertilising of arable crops, *Kickxia elatine*, *K. spuria* and *Euphorbia exigua* would probably be constants along with more widespread weeds like *Anagallis arvensis*, *Veronica persica*, *Polygonum aviculare* and *Bilderdykia convolvulus*. Other distinctive species in the richest type of *Kickxietum* are *Legousia hybrida*, *Chaenorhinum minus*, *Sherardia arvensis*, *Reseda lutea*, *Valerianella dentata*, *Ajuga chamaepitys* and *Scandix pecten-veneris*. The community occurs on calcareous soils derived from Chalk and other limestones or from limey superficials, the contingent of summer annuals often surviving harvesting because of their low habit and therefore providing a striking display when flowering later among stubble.

The other assemblage of this group, the *Papaver rhoeas-Silene noctiflora* community (OV16, *Papaveri-Sileneetum noctiflori* Wasscher 1941) shares the widespread weed species listed above but is additionally characterised by *Silene noctiflora* and *Papaver rhoeas*, the former in particular giving a distinct geographical restriction to this kind of vegetation. It occurs on light, calcareous soils, usually among cereals and their stubble, from Dorset north-east to Lincolnshire.

Both these communities have been assigned in Continental schemes to the Caucalidion platycarpi R.Tx. 1950, the alliance of cereal weeds on base-rich soils, and part of the order which is now usually termed the Centaureetalia cyani R.Tx., Lohmeyer & Preising in R.Tx. 1950. One further British weed assemblage could also be considered among this group though Silverside (1977), who first described it, placed it in the Polygono-Chenopodion. This is the *Reseda lutea-Polygonum aviculare* community (OV17, *Descurainio-Anchusetum arvensis* Silverside 1977), an assemblage which appears to have no direct equivalent from elsewhere in Europe but which he considered an analogue of the *Anchusetum arvensis* (Raabe 1944) Passarge 1964. Its distinctive features are constant *Reseda*, *Anchusa* and *Descurainia*, this last plant perhaps not a native to Britain but long established on waste ground and now primarily found among root crops and cereals where it occurs in this community on reasonably calcareous soils derived from superficials over chalk in East Anglia.

Gateways, tracksides and courtyard vegetation

Six assemblages are characterised by various mixtures of *Poa annua*, *Plantago major* and *Chamomilla suaveolens*, with *Lolium perenne* playing a frequent but generally subordinate role in the vegetation. These are open weedy or grassy communities with prominent contingents of ephemerals typical of silty, loamy or clay soils, often wet in winter, retentive of moisture in summer and frequently trampled by stock and people. Virtually ubiquitous throughout the lowlands, they are a feature of almost every earth path and farm gateway, road verge fringe, unweeded wall bottom, snikket, ginnel and entryside, pavement edge, cobbled and flagged courtyard, ill-managed ley, recreational sward and lonesome goal mouth. Together with more intact swards, sown and more thoughtfully managed as leys or amenity grasslands (and treated in this scheme among the mesotropic grasslands), they comprise the order Polygono-Poetalia R.Tx. in Géhu *et al.* 1974 corr. Rivas-Martinez *et al.* 1991.

Grassier vegetation types, closer to leys in their composition and appearance, less trampled, with *Lolium* usually more important and *Taraxacum officinale* agg. often prominent, especially in spring when the dandelions flower in abundance, form part of the Lolio-Plantaginion Sissingh 1960. Two communities of this alliance have been characterised in this volume. The *Poa annua-Taraxacum officinale* agg. community (OV22) is the more open of the two, typical of tracksides, wall-bottoms and the edges of road verges where there is

some disturbance but little actual treading, or of ill-sown leys where poaching is not too severe. The *Lolium perenne-Dactylis glomerata* community (OV23) is a more closed, weedy but essentially a perennial grassy community characteristic of unmown verges, lightly trampled gateway edges and rough places among recreational swards. It can be seen as transitional to the rank grasslands of the Arrhenatherion.

Four communities are additionally distinguished by *Polygonum aviculare*, more particularly *P. arenastrum*, and by higher frequencies for *Capsella bursa-pastoris*, *Sagina procumbens* and *Bryum argenteum*. These have traditionally been grouped in the Polygonion avicularis Br.-Bl. ex Aichinger 1933, the alliance of vegetation of trampled places along earth paths, in gateways and in crevices between courtyard cobbles and paving which escape weeding. In fact, the assemblage of these latter habitats, the *Poa annua-Sagina procumbens* community (OV20, *Sagino-Bryetum argentii* Diemont, Sissingh & Westhoff 1940) is so particular in its composition and structure as sometimes to have been allocated to a separate alliance, the Saginion procumbentis R.Tx. & Obha in Géhu *et al.* 1972. However, seen alongside the other Polygonion communities, the *Sagino-Bryetum* can be considered as an attenuated extreme of this sort of vegetation with species like *Poa annua*, *Chamomilla suaveolens*, *Plantago major*, *Polygonum aviculare* and *Capsella* reduced to often tiny scattered individuals rooted among the crevices of cobbles and paving slabs with strips of *Bryum argenteum* and rosettes of *Sagina procumbens*.

Along pathways and in gateways, the *Poa annua-Plantago major* community (OV21) usually represents the most open and trampled assemblage, the typical plants of the alliance increasing in frequency and cover somewhat compared to the *Sagino-Bryetum* but still often open and dominated by species both resistant to treading but able to take ready advantage of the open moist soil left after a winter of trampling and showers. Somewhat more closed and with *Chamomilla suaveolens* and *Capsella* becoming more prominent is the *Polygonum aviculare-Chamomilla* community (OV18, *Chamomilla-Polygonetum arenastri* T. Müller in Oberdorfer 1971), characteristic of moderately trampled pathsides and gateways. Finally, in this group is the *Poa annua-Matricaria perforata* community (OV19) where *Chamomilla suaveolens* tends to be less frequent and prominent than the more trample-sensitive *M. perforata* along disturbed road verge fringes and the side of farm tracks.

Tall-herb weed communities

Four of the communities are each characterised by the strong dominance of tall perennial herbs whose colonial growth and dense annual crops of shoots eliminates many potential associates or limits them to patchy occurrence. There is an element of naturalness in the way in which each of these plants can take advantage of, for example, fire, tree-fall, the drying of fen margins or accumulation of strandline detritus but, in the present British landscape, the vegetation types they create are generally dependent on anthropogenic events like disturbance, forest clearance and neglect for their development.

Two of them, both dominated by *Urtica dioica*, are especially widespread through the improved agricultural lowlands, around settlements and on wasteland, wherever fertile loamy soils are disturbed or nutrient-rich waste dumped. These are communities of the alliance Galio-Alliarion (Oberdorfer 1957) Lohmeyer & Oberdorfer in Oberdorfer *et al.* 1967. The generally more species-poor *Urtica dioica-Galium aparine* community (OV24) with its tangles of goosegrass and patchy carpet of puny *Poa trivialis* is the more ubiquitous in its range of habitats. The *Urtica dioica-Cirsium arvense* community (OV25) is often more open in the nettle cover with scattered *Cirsium arvense* and *C. vulgare* as the commonest associates. A frequent vegetation type in the recolonisation of ploughed ground and disturbed waysides, this is a usually grassier community with species such as *Elymus repens*, *Dactylis glomerata* and *Lolium perenne* figuring often. Both communities show transitions to ranker Arrhenatherion swards.

U. dioica remains a constant feature, as a subordinate but patchily prominent associate, in the *Epilobium hirsutum* community (OV26). This, the major vegetation type in Britain of the alliance Convolvulion sepium R.Tx. 1947, is also geographically widespread through the lowlands though more confined than the nettle communities to fertile soils that are kept moist by ground water around fens, ponds and ditches but which remain well aerated and free-draining. Floristic transitions seen in the various sub-communities thus include increasing representation of helophytes and tall-fen herbs in the one direction and of rank grasses and dicotyledons of the Arrhenatherion in the other.

These three communities and their two alliances are now grouped together in the class Galio-Urticetea Passarge ex Kopecky 1969. A further vegetation type, dominated by *Epilobium angustifolium*, is the only community characterised in this scheme from among the assemblages of cleared or burned forests and heaths included in the Epilobietea angustifolii Tüxen & Preising in Tüxen 1950. With *E. angustifolium* as the only constant, supremely well adapted to exploit open ground on neutral or acidic soils in such situations, this community has diverse groups of associates typical of the various situations in which it can arise.

Inundation communities

Five of the communities characterised are typical of periodically inundated habitats, usually with fine-textured substrates like silts or sands, in situations where flooding by fresh waters often destroys any existing vegetation by submergence or shifting of sediments and creates a new open and moist habitat available for colonisation as the waters fall or evaporate. Such habitats can be found around the margins of fluctuating ponds, on reservoir drawn-down zones, on river islands and banks and more fragmentarily along paths and trackways, in poached pastures and on dumped soil and made ground where drainage is poor and rainwater accumulates. Very often, such situations are naturally quite eutrophic because the sediments are derived from weathered shales, clays or superficials or periodically enriched by deposited alluvium. Commonly, though, there is further enrichment from dung of watering stock or dissolved nutrients derived from fertilisers in ground waters. The combination of eutrophic substrates and a mild, moist Atlantic climate ensures very congenial conditions for luxuriant growth of a range of ephemerals and nutrient-demanding perennials as temperatures rise in spring and early summer.

Three of the communities among this group are distinguished by high frequencies and often an abundance of knotweeds like *Polygonum hydropiper*, *P. persicaria* and *P. lapathifolium*, with *Bidens tripartita*, *Rorippa palustris* and *Ranunculus sceleratus*, and *Rumex maritimus* as a rare plant which finds an occasional locus here. These are the British representatives of the alliance Bidention Nordhagen 1940 *emend.* R.Tx. *apud* Poli & J.Tx. 1960. Central to the group are the two closely related syntaxa, the *Bidens tripartita-Polygonum amphibium* community (OV30, *Polygono-Bidentetum tripartitae* Lohmeyer in R.Tx. 1950) and the *Polygonum lapathifolium-Poa annua* community (OV33), the latter lacking *B. tripartita* and showing high frequency of various Polygono-Chenopodion plants. Such a feature reflects the common occurrence of such vegetation in wetter situations among made and disturbed ground, where conditions are transitional to damp and weedy arable land. A further Bidention assemblage, the *Myosotis scorpioides-Ranunculus sceleratus* community (OV32, *Ranunculetum scelerati* R.Tx. 1950 ex Passarge 1959) occurs in similar situations where there is heavy enrichment from dung, fertiliser run-off or exposure of ditch-dredgings.

Unfortunately, no samples were available of vegetation representative of the other alliance of the class Bidentetea R.Tx. *et al.* in R.Tx. 1950, the Chenopodion glauci Hejny 1974, although Williams (1969) and Silverside (1977) noted the possible occurrence in Britain of the *Chenopodietum glauco-rubri* Lohmeyer 1950 *apud* Oberdorfer 1957, in which more nutrient-demanding species of goosefoot occurred around sodden dung piles, compost heaps and manure-polluted streamsides (see also Ellenberg 1988).

The two other inundation communities can have quite frequent records for *Polygonum hydropiper*, *P. persicaria* and *P. lapathifolium* and also *Rorippa palustris*, but they are more obviously characterised by mats of *Agrostis stolonifera* or *Alopecurus geniculatus*, often with conspicuous *Ranunculus repens*. These are the *Agrostis stolonifera-Ranunculus repens* community (OV28, *Agrostio-Ranunculetum repentis* Oberdorfer et al. 1967) and the *Alopecurus geniculatus-Rorippa palustris* community (OV29, *Ranunculo-Alopecuretum geniculati* R.Tx. 1937), characteristic of silts, clays and sands in seasonally-flooded pastures, lowland streams and ditches, watering places and river shoals and banks. They comprise part of a diverse group of vegetation types, some more ephemeral and often quite open, others essentially like perennial grasslands, the floristic diversity of which depends on sediment texture and the length and frequency of flooding.

The syntaxomic treatment of this group whose various other members in this scheme are described in the sections on mesotrophic grasslands, sand-dunes and salt-marsh, has been about as unstable as the habitats of which they are characteristic. Originally, they were included in the Elymo-Rumicion crispi Nordhagen 1940 *emend.* R.Tx. 1950, an alliance now renamed the Lolio-Potentillion R.Tx. 1947 or Potentillion anserinae R.Tx. 1947. To simplify cross-referencing to other volumes, we have retained the first and older name in this scheme and placed this alliance along with the Lolio-Plantaginion and Polygonion, in the Agrostietalia stoloniferae Oberdorfer in Oberdorfer et al. 1967 of the class Polygono-Poetea Rivas-Martinez 1975 corr. Rivas-Martinez et al. 1991.

Dwarf-rush communities of ephemeral ponds

Four communities characterised from the data are British representatives of the alliance Nanocyperion Koch 1926, ephemeral vegetation with dwarf rushes and other diminutive herbs that occurs on drying pond muds and other briefly-exposed mineral soils along tracks and streamsides across Western and Central Europe. Of fleeting occurrence and sporadic reappearance, these assemblages of generally inconspicuous plants have largely escaped notice in Britain, except as providing a locus for a number of species which have become increasingly rare in our ever more tidy landscapes. *Limosella aquatica*, *Juncus capitatus*, *J. pygmaeus*, *Cicendia filiformis*, *Ranunculus tripartitus* and *Illecebrum verticillatum* are some of the scarce plants which occur in these habitats (Perring & Farrell 1977, Stewart et al. 1994) but whose vegetation context has rarely been systematically described. In this survey, too, very few data were available and further sampling is essential to ensure better coverage. Elsewhere in Europe, the Nanocyperion has long proved a bewitching attraction (e.g. Weeda 1994 and other papers in that collection).

Commoner plants characteristic of this group include *Juncus bufonius*, *Isolepis setacea*, *Filaginella uliginosa*, *Lythrum portula*, *Radiola linoides*, *Plantago major* ssp. *intermedia*, *Sagina apetala* and *S. nodosa*, and a further distinctive feature is the frequent presence of *Riccia* spp., like *R. glauca*, *R. cavernosa*, *R. beyrichiana*, and other bryophytes typical of bare arable land: *Bryum klinggraeffii*, *Barbula unguiculata*, *Pottia starkeana*, *Physcomitriella patens*. Then, there can be geophytes which are able to take advantage of temporarily favourable conditions. Characteristically, too, samples of such miniature vegetation types often include elements of the perennial matrix of grassland, heath, woodland or aquatic communities among which they have developed.

Of the four assemblages defined in this scheme, the *Allium schoenoprasum-Plantago maritima* community (OV34) was first noted in Britain by Coombe & Frost (1956a), although it was not systematically characterised until a study by Hopkins (1983). It is extremely local in occurrence, even for a Nanocyperion community, being confined to winter-wet erosion pans below seepage lines over serpentine rocks on The Lizard in Cornwall. With distinctive rarities like *Allium schoenoprasum*, *Isoetes hystrix*, *Scilla verna*, *S. autumnalis*, *Herniaria ciliolata*, *Sagina subulata* and *Trifolium bocconei*, the sward has the look of an open maritime heath, stunted by the extreme habitat.

A rather different but equally striking situation is provided for another Nanocyperion assemblage in shallow depressions developed from periglacial ground-ice hollows in Cambridgeshire. Now taken into arable fields, ploughed each autumn and sown with cereals, these hollows are flooded in winter. The standing water kills the young crop and the bare moist ground exposed as the water recedes is colonised by the *Lythrum hyssopifolia-Juncus bufonius* community (OV36), first described by Preston & Whitehouse (1986) and Preston (1989). By harvest time, a dense vegetation cover is established with *L. hyssopifolia*, *J. bufonius*, *Plantago major* var. *intermedia*, some Caucalidion and Bidention herbs and a striking suite of ephemeral bryophytes.

The third vegetation type in this group, the *Lythrum portula-Ranunculus flammula* community (OV35) is less distinctive though more widespread on muddy pond margins and periodically wetted trackways across southern Britain with sporadic occurrence in lowland areas further north. Characterised by an abundance of *L. portula* and occasional *Juncus bufonius* and *Filaginella uliginosa*, it also has a strong contingent of aquatic and amphibious plants including *Ranunculus flammula*, *Eleocharis palustris*, *Alisma plantago-aquatica*, *Callitriche stagnalis* and *Littorella uniflora*. Locally, despite its more mundane character, this vegetation provides a locus for *Limosella aquatica*, *Alopecurus aequalis* and *Pilularia globulifera*.

In man-made habitats not subject to summer drought nor subsequent disturbance, Nanocyperion assemblages are often overtaken by Bidention communities, not strictly as a successional replacement but in the form of bulkier plants which colonise shortly after the diminutive pioneers as the drying sediments release nutrients, and which go on to overwhelm them. In the equable and moist climate of Britain, the line between the Nanocyperion and Bidention is especially hard to draw (Rodwell 1994*b*) and this is well seen in the *Rorippa palustris-Filaginella uliginosa* community (OV31) first described by Birse (1984) and typical of dried-out ponds, pool margins and muddy tracks. There, typical Nanocyperion plants like *F. uliginosa*, *J. bufonius* and *Lythrum portula* are usually outnumbered by a developing abundance of *Polygonum persicaria*, *P. hydropiper* and *Bidens tripartita*.

Finally, it is worth noting here that the colonising vegetation of dune slacks characterised elsewhere in this volume as the *Sagina nodosa-Bryum pseudotriquetrum* community (SD13) shows a strong resemblance to the *Centaurio-Saginetum nodosae* Diemont, Sissingh & Westhoff 1940, a Nanocyperion assemblage described from The Netherlands and Germany.

Crevice, scree and spoil vegetation

Five communities of sometimes very open and fragmentary vegetation with ferns, *Parietaria diffusa* or *Cymbalaria muralis* have been characterised from crevices in rock outcrops or walls or from among scree and it is also sensible to include here an additional vegetation type characteristic of heavy-metal spoil. Two further assemblages from talus, with *Cryptogramma crispa* and/or *Athyrium distentifolium*, have, in fact, already been described among the calcifuge grasslands and montane communities (Rodwell 1992).

Two communities with frequent records for smaller asplenoid ferns, *Cystopteris fragilis* and *Ceterach officinarum*, together with brophytes such as *Fissidens cristatus*, *Tortella tortuosa* and *Weissia controversa*, have been distinguished from smaller crevices in outcrops of calcareous bedrocks or limestone and mortared walls. The *Asplenium trichomanes-A. ruta-muraria* community (OV39, *Asplenietum trichomano-rutae-murariae* R.Tx. 1937), in which the two named ferns are the characteristic constants, is typical of sunny situations at lower altitudes, providing a locus in western Britain for the oceanic *Ceterach officinarum* and more widely, but only locally, for *A. adiantum-nigrum*. At higher altitudes in western and northern Britain, and on shaded aspects, where cooler and more humid conditions prevail, this assemblage is replaced by the *Asplenium viride-Cystopteris fragilis* community (OV40, *Asplenio-Cystopteridetum fragilis* (Kuhn 1939) Oberdorfer 1949) where *A. trichomanes*

and *A. ruta-muraria* remain very frequent but where *A. viride* and *C. fragilis* are the distinctive constants and *Woodsia alpina* and *Polystichum lonchitis* notable rarities.

Traditionally, this latter community is placed among the fern assemblages of shaded habitats in the Cystopteridion fragilis Richard 1972. Following Segal's (1969) treatment of crevice vegetation, the *Asplenietum* is now often located in his Cymbalario-Asplenion, an alliance of communities from sunny situations in more temperate parts of Europe as distinct from the north alpine assemblages of the Potentillion caulescentis Br.-Bl. in Br.-Bl. & Jenny 1926.

This alliance is also now used to contain the kind of sunny wall crevice vegetation in which *A. ruta-muraria* and *A. trichomanes* figure as subordinate elements to *Cymbalaria muralis* (OV42, *Cymbalarietum muralis* Görs 1966). Vegetation of shaded wall crevices, where asplenoioid ferns are subordinate to *Parietaria diffusa* and a weak contingent of nitrophilous associates (OV41, *Parietarietum judaicea* (Arènes 1928) Oberdorfer 1977) has traditionally been located in a Centrantho-Parietarion Rivas-Martinez 1960 or Parietarion Segal 1969. In fact, Segal (1969) recognised a whole series of *Parietaria* communities of different climatic provinces across Europe, again emphasising the distinct character of crevice vegetation in the equable Atlantic realm as distinct from the Mediterranean zone.

The other fern assemblage recognised in this section is the *Gymnocarpium robertianum-Arrhenatherum elatius* community (OV38), vegetation dominated by bulkier plants, the two eponymous species light-demanding, but many of the associates being herbs of scrub or woodland. This is a community of local distribution, almost entirely in England and Wales, in open sunny situations on ungrazed coarse limestone talus. It is the British representative of the Stipion calamagrostis Jenny-Lips ex Br.-Bl. *et al.* 1952, the alliance of sub-montane and montane, base-rich scree vegetation also from Central European and Scandinavian mountains. It has an equivalent on siliceous screes in the British sub-montane zone in the *Cryptogramma crispa-Deschampsia flexuosa* community (U21 *Cryptogrammetum crispae*), traditionally placed in the Androsacion alpinae Br.-Bl. in Br.-Bl. & Jenny 1926.

Finally, best considered among this group, there is the distinctive vegetation tolerant of the extreme environment presented by soils rich in heavy metals developed on lead-mine spoil and over mineral veins. Here, in the *Festuca ovina-Minuartia verna* community (OV37, *Minuartio-Thlaspietum alpestris* Koch 1932), the characteristic feature, among an open turf of tolerant ecotypes of *F. ovina* and *Agrostis capillaris* with *Thymus praecox*, is the frequent occurrence of *M. verna* and *Thlaspi alpestre*. Despite the absence of *Viola calaminaria* from British stands of this vegetation, the community is essentially identical to an association described from the Continent and placed in the alliance Thlaspion calaminaris Ernst 1964 comprising western European heavy-metal assemblages. Discussion continues as to whether such communities deserve a separate class, the Violetea calaminariae R.Tx. 1961, or should be considered among the calcicolous grasslands of the Festuco-Brometea.

KEY TO VEGETATION OF OPEN HABITATS

With something as complex and variable as vegetation, no key can pretend to offer an infallible short cut to diagnosis. The following should thus be seen as simply a crude guide to identifying the types of vegetation found in open habitats and must always be used in conjunction with the data tables and community descriptions. It relies on floristic (and, to a lesser extent, physiognomic) features of the vegetation and demands a knowledge of the British vascular flora and, in only a few cases here, of bryophytes and lichens. It does not make primary use of any habitat features, though these may provide a valuable confirmation of a diagnosis.

Because the major distinctions between the vegetation types in the classification are based on inter-stand frequency, the key works best when sufficient samples of similar composition are available to construct a constancy table. It is the frequency values in this (and, in some cases, the ranges of abundance) which are then subject to interrogation with the key.

Samples should always be taken from homogeneous stands and be 2 × 2 m or 4 × 4 m according to the scale of the vegetation or, where stands are irregular, of identical size but different shape.

1　Vegetation dominated, at least patchily, by tall perennial herbs *Urtica dioica, Epilobium angustifolium* or *E. hirsutum*, often in locally dense stands　　2

Low, herbaceous vegetation, sometimes closed and quite luxuriant, with some of *Polygonum hydropiper, P. lapathifolium, P. persicaria, P. amphibium, Agrostis stolonifera, Alopecurus geniculatus, Bidens tripartita, Rorippa palustris, Ranunculus repens* and *R. sceleratus*　　14

Grassy or weedy vegetation, often open or trampled, with some of *Lolium perenne, Chamomilla suaveolens, Poa annua* and *Plantago major*　　20

Open vegetation of rock crevices, screes or spoil with some of *Gymnocarpium robertianum, Asplenium trichomanes,*

A. ruta-muraria, A. viride, Cystopteris fragilis, Cymbalaria muralis or *Parietaria diffusa*　　36

More or less open vegetation with occasional to frequent scattered *Juncus bufonius* and some of *Allium schoenoprasum, Lythrum portula* or *L. hyssopifolia*　　42

Open tussocky sward of *Festuca ovina* and *Agrostis capillaris* with constant *Minuartia verna, Thymus praecox* and *Campanula rotundifolia* and frequent *Thlaspi alpestre*

　　OV37 *Festuca ovina-Minuartia verna* community
　　Minuartio-Thlaspietum alpestris　Koch 1932　43

Not as above　　44

2　*Urtica dioica* a constant and abundant feature of the vegetation, with some of *Galium aparine, Cirsium arvense, C. vulgare* and *Arrhenatherum elatius* frequent but *Epilobium angustifolium* and *E. hirsutum* scarce and never abundant　　3

U. dioica can be frequent and locally abundant but only with constant and plentiful *E. angustifolium* or *E. hirsutum*　　7

3　*G. aparine* and *Poa trivialis* both common with *C. arvense* of irregular occurrence

　　OV24 *Urtica dioica-Galium aparine* community
　　　　　　　　　　　　　　　　　　　　　　　　4

G. aparine occasionally found but *P. trivialis* scarce and *C. arvense* constant, *C. vulgare* occasional

　　OV25 *Urtica dioica-Cirsium arvense* community
　　　　　　　　　　　　　　　　　　　　　　　　5

4　*Arrhenatherum elatius, Heracleum sphondylium, Rubus fruticosus* agg. and *Taraxacum officinale* agg. frequent

　　OV24 *Urtica dioica-Galium aparine* community
　　Arrhenatherum elatius-Rubus fruticosus agg. sub-community

Above species all scarce but *Cirsium arvense* frequent and *Bromus sterilis* occasional

> **OV24** *Urtica dioica-Galium aparine* community
> Typical sub-community

5   *Papaver rhoeas, Arrhenatherum elatius* and *Lolium perenne* frequent without *Elymus repens*

> **OV25** *Urtica dioica-Cirsium arvense* community
> *Lolium perenne-Papaver rhoeas* sub-community

E. repens constant with occasional *L. perenne* and *A. elatius* but without *P. rhoeas*   6

6   *Dactylis glomerata, Rumex obtusifolius, Artemisia vulgaris* and *Heracleum sphondylium* frequent

> **OV25** *Urtica dioica-Cirsium arvense* community
> *Rumex obtusifolius-Artemisia vulgaris* sub-community

Above species occasional at most but *Holcus lanatus, Sonchus asper, Agrostis stolonifera, Poa annua* and *Cerastium fontanum* all common

> **OV25** *Urtica dioica-Cirsium arvense* community
> *Holcus lanatus-Poa annua* sub-community

7   *Epilobium hirsutum* constant and usually abundant, *E. angustifolium* infrequent and never prominent

> **OV26** *Epilobium hirsutum* community   8

E. angustifolium constant and usually abundant, *E. hirsutum* scarce and never prominent

> **OV27** *Epilobium angustifolium* community   11

8   *Phragmites australis, Iris pseudacorus, Solanum dulcamara, Eupatorum cannabinum* and *Calystegia sepium* frequent

> **OV26** *Epilobium hirsutum* community
> *Phragmites australis-Iris pseudacorus* sub-community

Above species all infrequent   9

9   *Filipendula ulmaria, Angelica sylvestris, Cirsium palustre, Mentha aquatica, Galium aparine* and *Lathyrus palustris* frequent

> **OV26** *Epilobium hirsutum* community
> *Filipendula ulmaria-Angelica sylvestris* sub-community

Above species not occurring together with any frequency   10

10   *Arrhenatherum elatius, Heracleum sphondylium* and *Cirsium arvense* frequent with occasional *Rumex crispus* and *R. obtusifolius*

> **OV26** *Epilobium hirsutum* community
> *Arrhenatherum elatius-Heracleum sphondylium* sub-community

Above species all uncommon but *Juncus effusus* and *Ranunculus repens* frequent

> **OV26** *Epilobium hirsutum* community
> *Juncus effusus-Ranunculus repens* sub-community

Apart from *E. hirsutum, U. dioica* and *Cirsium arvense* are the only common associates

> **OV26** *Epilobium hirsutum* community
> *Urtica dioica-Cirsium arvense* sub-community

11   *Ammophila arenaria, Festuca rubra, Poa pratensis* and *Senecio jacobaea* constant

> **OV27** *Epilobium angustifolium* community
> *Ammophila arenaria* sub-community

Above species scarce at most   12

12   Saplings of *Acer pseudoplatanus, Fraxinus excelsior, Fagus sylvatica* and *Ulmus glabra* frequent with *Sambucus nigra* and *Circaea lutetiana*

> **OV27** *Epilobium angustifolium* community
> *Acer pseudoplatanus-Sambucus nigra* sub-community

Above species all usually absent   13

13   *Urtica dioica* and *Cirsium arvense* frequent with occasional *Galium aparine, Arrhenatherum elatius, Dactylis glomerata* and *Heracleum sphondylium*

> **OV27** *Epilobium angustifolium* community
> *Urtica dioica-Cirsium arvense* sub-community

Holcus lanatus and *Festuca ovina* frequent, *Anthoxanthum odoratum, Potentilla erecta* and *Teucrium scorodonia* occasional in the absence of the above

> **OV27** *Epilobium angustifolium* community
> *Holcus lanatus-Festuca ovina* sub-community

Rubus fruticosus agg. and *Dryopteris dilatata* constant with all above species scarce

> **OV27** *Epilobium angustifolium* community
> *Rubus fruticosus* agg.-*Dryopteris dilatata* sub-community

14   *Ranunculus sceleratus* and *Myosotis scorpioides* constant with frequent *Veronica catenata* and *Nasturtium officinale* agg.

OV32 *Myosotis scorpioides-Ranunculus sceleratus* community
Ranunculetum scelerati R.Tx. 1950 ex Passarge 1959

Above species absent or scarce 15

15 Some of *Polygonum hydropiper, P. lapathifolium, P. persicaria, P. amphibium, Bidens tripartita* and *Filaginella uliginosa* frequent and often prominent with grasses generally subordinate 16

Agrostis stolonifera and/or *Alopecurus geniculatus* constant and often extensive with above species occasional and only patchily prominent 18

16 *B. tripartita* and *F. uliginosa* frequent with occasional *Juncus bufonius* 17

Above species scarce or totally absent among often dense stands of *P. persicaria, P. lapathifolium P. aviculare*

OV33 *Polygonum lapathifolium-Poa annua* community

17 *Phalaris arundinacea* and *Polygonum amphibium* constant

OV30 *Bidens tripartita-Polygonum amphibium* community
Polygono-Bidentetum tripartitae Lohmeyer in R.Tx. 1950

Above species occasional at most but *Rorippa palustris* common

OV31 *Rorippa palustris-Filaginella uliginosa* community

18 *Alopecurus geniculatus* and *Rorippa palustris* constant with *Agrostis stolonifera* scarce

OV29 *Alopecurus geniculatus-Rorippa palustris* community
Ranunculo-Alopecuretum geniculati R.Tx. (1937) 1950

Agrostis stolonifera and *Ranunculus repens* constant with *Alopecurus geniculatus* scarce

OV28 *Agrostis stolonifera-Ranunculus repens* community
Agrostio-Ranunculetum repentis Oberdorfer *et al.* 1967 19

19 *Stellaria media, Poa annua* and *Plantago major* constant

OV28 *Agrostio-Ranunculetum*
Polygonum hydropiper-Rorippa sylvestris sub-community

Not as above

OV28 *Agrostio-Ranunculetum*
Poa annua-Polygonum aviculare sub-community

20 *Dactylis glomerata* constant and often abundant with *Lolium perenne, Plantago lanceolata* and *Taraxacum officinale* agg. but little or no *Chamomilla suaveolens*, in usually closed, somewhat rank grassy vegetation

OV23 *Lolium perenne-Dactylis glomerata* community 21

D. glomerata scarce and *C. suaveolens* generally frequent 24

21 *Arrhenatherum elatius* constant with frequent *Medicago lupulina* and *Agrostis capillaris*

OV23 *Lolium perenne-Dactylis glomerata* community
Arrhenatherum elatius-Medicago lupulina sub-community

Above species scarce 22

22 *Trifolium repens, Plantago major* and *Ranunculus repens* frequent with *Crepis vesicaria, Rumex obtusifolius, Trifolium dubium* and *Hordeum murinum* scarce

OV23 *Lolium perenne-Dactylis glomerata* community
Plantago major-Trifolium repens sub-community

T. repens, P. major and *R. repens* occasionally present but then with some of the other species listed above 23

23 *Trifolium dubium* and *Hordeum murinum* frequent without *Rumex obtusifolius* or *Poa annua*

OV23 *Lolium perenne-Dactylis glomerata* community
Typical sub-community

Crepis vesicaria, R. obtusifolius and *Poa annua* constant

OV23 *Lolium perenne-Dactylis glomerata* community
Crepis vesicaria-Rumex obtusifolius sub-community

24 *Sagina procumbens* and *Bryum argenteum* frequent with *Poa annua* and *Chamomilla suaveolens* in often fragmentary vegetation of pavement and cobble cracks

OV20 *Poa annua-Sagina procumbens* community
Sagino-Bryetum argentii Diemont, Sissingh &
Westhoff 1940   25

S. procumbens and *B. argenteum* at most occasional   26

On moderately trampled pavements and courtyards, it is difficult to separate the *Sagino-Bryetum*
from the Typical sub-community of the *Poa-Plantago* community but the cover is generally more
closed there.

25   *Lolium perenne, Plantago major* and *Agrostis stolonifera* frequent

OV20 *Sagino-Bryetum argentii*
Lolium perenne-Chamomilla suaveolens sub-community

Above species scarce

OV20 *Sagino-Bryetum argentii*
Typical sub-community

26   *Matricaria perforata* constant, usually with
Elymus repens and *Agrostis stolonifera* in grassy vegetation along road and track margins

OV19 *Poa annua-Matricaria perforata* commu-
nity   27
M. perforata occasional at most and usually without *E.
repens*   31

27   *Lolium perenne* and *Capsella bursa-pastoris* frequent with occasional *Holcus lanatus*   28

Above species scarce or absent   30

28   *Atriplex prostrata, A. patula, Chenopodium
album, Sonchus oleraceus* and *Medicago lupulina* constant

OV19 *Poa annua-Matricaria perforata* community
Atriplex prostrata-Chenopodium album sub-community

Above species occasional at most and not in consistent
combination   29

29   *Polygonum persicaria* and *Papaver rhoeas* frequent

OV19 *Poa annua-Matricaria perforata* community
Lolium perenne-Capsella bursa-pastoris sub-community

Cirsium arvense, Rumex obtusifolius and *Sonchus asper*
occasional to frequent in absence of *Polygonum peri-
caria* and *Papaver rhoeas*

OV19 *Poa annua-Matricaria perforata* community
Chamomilla suaveolens-Plantago major sub-community

30   *Senecio squalidus* and *Epilobium angustifolium*
frequent

OV19 *Poa annua-Matricaria perforata* community
Senecio squalidus-Epilobium angustifolium sub-
community

Above species absent

OV19 *Poa annua-Matricaria perforata* community
Elymus repens sub-community

31   *Taraxacum officinale* constant with *Poa annua* but
without *Polygonum aviculare*

OV22 *Poa annua-Taraxacum officinale* commu-
nity   32
T. officinale occasional but generally with *P. aviculare* 33

32   *Crepis vesicaria* and *Epilobium adenocaulon* frequent

OV22 *Poa annua-Taraxacum officinale* community
Crepis vesicaria-Epilobium adenocaulon sub-community

Cirsium vulgare and *C. arvense* constant, *Rumex crispus*
and *R. obtusifolius* occasional to common

OV22 *Poa annua-Taraxacum officinale* community
Cirsium vulgare-C. arvense sub-community

Senecio vulgaris and *S. squalidus* frequent with above
species scarce or absent

OV22 *Poa annua-Taraxacum officinale* community
Senecio vulgaris sub-community

33   *Capsella bursa-pastoris* constant

OV18 *Polygonum aviculare-Chamomilla suaveo-
lens* community   34

C. bursa-pastoris occasional at most

OV21 *Poa annua-Plantago major* community 35

34 *Plantago major* constant usually without *Sisymbrium officinale* and *Polygonum aequale*

OV18 *Polygonum aviculare-Chamomilla suaveolens* community
Plantago major sub-community

Plantago major scarce but *S. officinale* and *Polygonum aequale* frequent

OV18 *Polygonum aviculare-Chamomilla suaveolens* community
Sisymbrium officinale-Polygonum arenastrum sub-community

35 *Polygonum aviculare* and *Ranunculus repens* constant

OV21 *Poa annua-Plantago major* community
Polygonum aviculare-Ranunculus repens sub-community

Lolium perenne constant with little or no *P. aviculare* and *R. repens*

OV21 *Poa annua-Plantago major* community
Lolium perenne sub-community

Above species generally scarce but *Sagina procumbens* and *Bryum argenteum* occasional

OV21 *Poa annua-Plantago major* community
Typical sub-community

36 *Gymnocarpium robertianum* constant and often abundant with *Arrhenatherum elatius*, *Geranium robertianum*, *Teucrium scorodonia* and *Mercurialis perennis* on screes and around rocky outcrops

OV38 *Gymnocarpium robertianum-Arrhenatherum elatius* community
Gymnocarpietum robertianae (Kuhn 1937) R.Tx. 1937

Gymnocarpium robertianum a scarce and local plant at most 37

37 *Asplenium trichomanes* and *A. ruta-muraria* constant without *Cymbalaria muralis* or *Parietaria diffusa* 38

Asplenoid ferns absent or, if present, then subordinate in cover to *C. muralis* or *P. diffusa* 40

38 *A. viride* and *Cystopteris fragilis* constant

OV40 *Asplenium viride-Cystopteris fragilis* community
Asplenio viridis-Cystopteridetum fragilis (Kuhn 1939) Oberdorfer 1949

A. viride and *C. fragilis* generally absent

OV39 *Asplenium trichomanes-A. ruta-muraria* community
Asplenietum trichomano-rutae-murariae R.Tx. 1937 39

39 *Sedum acre*, *Arenaria serpyllifolia*, *Koeleria macrantha* and *Helianthemum nummularium* constant

OV39 *Asplenietum trichomano-rutae-murariae*
Sedum acre-Arenaria serpyllifolia sub-community

Above species scarce or absent but bryophyte element further enriched by frequent *Tortula intermedia* and *Trichostomum crispulum*

OV39 *Asplenietum trichomano-rutae-murariae*
Trichostomum crispulum-Tortula intermedia sub-community

40 *Cymbalaria muralis* constant and plentiful in open crevice vegetation

OV42 *Cymbalaria muralis* community
Cymbalarietum muralis Görs 1966

Parietaria diffusa constant and plentiful with little or no *C. muralis*

OV41 *Parietaria diffusa* community
Parietarietum judaicae (Arènes 1928) Oberdorfer 1977 41

41 *Festuca rubra*, *Dactylis glomerata*, *Daucus carota* and *Centaurea scabiosa* constant with frequent *Brassica oleracea*, *Euphorbia portlandica* and *Sanguisorba minor*

OV41 *Parietarietum judaicae*
Daucus carota sub-community

Above species scarce or absent but bryophyte element enriched by frequent *Homalothecium sericeum* and *Tortula muralis*

OV41 *Parietarietum judaicae*
Homalothecium sericeum-Tortula muralis sub-community

42 *Allium schoenoprasum*, *Scilla verna* and *Sagina subulata* constant in an open sward of *Festuca ovina*, *Plantago maritima*, *Agrostis stolonifera*, *Koeleria macrantha* and *Thymus praecox* with sparse sprigs of *Calluna vulgaris*

OV34 *Allium schoenoprasum-Plantago maritima* community

Lythrum portula and *Ranunculus flammula* constant with occasional *Callitriche stagnalis* and *C. hamulata*, *Eleocharis palustris*, *Littorella uniflora*, *Galium palustre* and *Alisma plantago-aquatica*

OV35 *Lythrum portula-Ranunculus flammula* community

Lythrum hyssopifolia and *Plantago major* constant with *Bryum klinggraeffii* and *Riccia glauca*

OV36 *Lythrum hyssopifolia-Juncus bufonius* community

As the growing season advances, the more diminutive members of the last two communities tend to become overwhelmed by a luxuriant growth of *Polygonum hydropiper*, *P. aviculare* or *P. persicaria* and then the vegetation may resemble the *Polygonum-Poa* community.

---

43   *Cladonia rangiformis*, *C. chlorophaea*, *C. pyxidata* and *Cornicularia aculeata* constant with frequent scattered sprigs of *Calluna vulgaris*

OV37 *Minuartio-Thlaspietum alpestris* *Cladonia* spp. sub-community

Achillea millefolium, *Euphrasia officinalis* agg., *Plantago lanceolata* and *Anthoxanthum odoratum* constant

OV37 *Minuartio-Thlaspietum* *Achillea millefolium-Euphrasia officinalis* sub-community

All above species scarce or absent

OV37 *Minuartio-Thlaspietum* Typical sub-community

44   Open or closed, often rank, sometimes luxuriant, weedy vegetation with frequent records for at least some of *Polygonum aviculare*, *Stellaria media*, *Matricaria perforata*, *Chenopodium album*, *Poa annua*, *Elymus repens* and *Bilderdykia convolvulus* but usually without *Spergula arvensis*, *Chrysanthemum segetum*, *Rumex acetosella*, *Aphanes microcarpa*, *Briza minor* or *Silene gallica*   45

At least some of *S. arvensis*, *C. segetum*, *R. acetosella* and *A. microcarpa* frequent in open or closed weed or footpath vegetation   60

45   *Veronica persica*, *V. polita* and *Anagallis arvensis* occasional to frequent with some of *Euphorbia exigua*,

Kickxia elatine, *K. spuria*, *Silene noctiflora*, *Papaver rhoeas*, *Reseda lutea*, *Descurania sophia* and *Lycopsis arvensis*   46

Capsella bursa-pastoris and *Urtica urens* frequent with some of *Fumaria officinalis*, *F. muralis* ssp. *boraei*, *F. bastardii*, *Lamium amplexicaule* and *Galinsoga parviflora* 48

Above combinations of species not present   49

46   *P. rhoeas* and *S. noctiflora* constant with *Polygonum aviculare*, *Elymus repens*, *Matricaria perforata*, *Stellaria media* and *Bilderdykia convolvulus*

OV16 *Papaver rhoeas-Silene noctiflora* community *Papaveri-Sileneetum noctiflori*   Wasscher 1941

Reseda lutea, *Descurainia sophia* and *Anchusa arvensis* constant with occasional *P. rhoeas* but no *S. noctiflora*

OV17 *Reseda lutea-Polygonum aviculare* community *Descurainio-Anchusetum arvensis*   Silverside 1977

Above combinations of species absent

OV15 *Anagallis arvensis-Veronica persica* community *Kickxietum spuriae*   Kruseman & Vlieger 1939
                                                        47

47   *Euphorbia exigua* and *Kickxia* spp. very frequent with *Legousia hybrida*, *Chaenorhinum minus*, *Sherardia arvensis* and *Valerianella dentata* common

OV15 *Kickxietum spuriae* *Legousia hybrida-Chaenorhinum minus* sub-community

Agrostis stolonifera and *Ranunculus repens* both constant with frequent records for some of *Phascum cuspidatum*, *Barbula unguiculata*, *Bryum rubens*, *B. klinggraeffii*, *B. microerythrocarpum*, *Pottia truncata*, *Dicranella staphylina*, *D. schreberana* and *D. varia*

OV15 *Kickxietum spuriae* *Agrostis stolonifera-Phascum cuspidatum* sub-community

Stellaria media and *Convolvulus arvensis* frequent with *Legousia hybrida*, *Chaenorhinum minus*, *Ranunculus repens* or acrocarpous mosses only scarce at most

OV15 *Kickxietum spuriae* *Stellaria media-Convolvulus arvensis* sub-community

48 *Lamium amplexicaule, Galinsoga parviflora, Solanum nigrum* and *S. sarrachoides* occasional to frequent

> **OV14** *Urtica urens-Lamium amplexicaule* community

At least one of *Fumaria officinalis, F. muralis* ssp. *boraei* or *F. bastardii* frequent with *Euphorbia helioscopa* without the above species

> **OV13** *Stellaria media-Capsella bursa-pastoris* community
> *Fumaria officinalis-Euphorbia helioscopia* sub-community

49 *Veronica persica, Polygonum aviculare, Bilderdykia convolvulus, Stellaria media, Matricaria perforata* and *Chenopodium album* frequent with one or more of *V. polita, Laminium hybridum* and *Alopecurus myosuroides* 50

Above combinations of species absent 51

50 *Alopecurus myosuroides* constant and often abundant

> **OV8** *Veronica persica-Alopecurus myosuroides* community
> *Alopecuro-Chamomilletum recutitae* Wasscher 1941

Veronica polita constant with *Lamium hybridum* occasional

> **OV7** *Veronica persica-V. polita* community
> *Veronico-Lamietum hybridi* Kruseman & Vlieger 1939

51 *Stachys arvensis* constant with frequent *Agrostis stolonifera* and *Ranunculus repens*

> **OV11** *Poa annua-Stachys arvensis* community

Myosotis arvensis and *Poa trivialis* constant with frequent *A. stolonifera* and *R. repens*

> **OV12** *Poa annua-Myosotis arvensis* community 52

Stachys arvensis and *Myosotis arvensis* not present in combination with *A. stolonifera* and *R. repens* 53

52 *Aphanes arvensis* and *Veronica arvensis* frequent with some of *Bryum rubens, B. violaceum, B. klinggraeffii, B. erythrocarpum* and *B. microerythrocarpum* and *Dicranella staphylina*

> **OV12** *Poa trivialis-Myosotis arvensis* community
> *Dicranella staphylina-Bryum* spp. sub-community

Sonchus asper and *Polygonum persicaria* frequent in absence of the above species

> **OV12** *Poa trivialis-Myosotis arvensis* community
> Typical sub-community

53 *Senecio vulgaris* constant 54

S. vulgaris scarce

> **OV9** *Matricaria perforata-Stellaria media* community 57

54 *Lolium perenne* frequent

> **OV10** *Poa annua-Senecio vulgaris* community 55

L. perenne scarce 56

55 *Ranunculus repens, Sonchus asper, Viola arvensis, Anagallis arvensis, Anthemis cotula* and *Polygonum persicaria* constant

> **OV10** *Poa annua-Senecio vulgaris* community
> *Polygonum persicaria-Ranunculus repens* sub-community

Agrostis stolonifera constant with some of *Holcus lanatus, Rumex crispus, Senecio squalidus* and *Taraxacum officinale* agg.

> **OV10** *Poa annua-Senecio vulgaris* community
> *Agrostis stolonifera-Rumex crispus* sub-community

Dactylis glomerata, Plantago lanceolata and *Agrostis capillaris* frequent with above-listed species scarce

> **OV10** *Poa annua-Senecio vulgaris* community
> *Dactylis glomerata-Agrostis capillaris* sub-community

Chamomilla suaveolens, Polygonum aviculare, Chenopodium album and *Matricaria perforata* common with above-listed species scarce

> **OV10** *Poa annua-Senecio vulgaris* community
> *Polygonum aviculare-Matricaria perforata* sub-community

56 *Sonchus asper, Cirsium arvense, Urtica dioica* and *Galium aparine* constant

> **OV13** *Stellaria media-Capsella bursa-pastoris* community
> *Urtica dioica-Galium aparine* sub-community

Matricaria perforata, Poa annua and *Agrostis stolonifera* frequent with above-listed species scarce

OV13 *Stellaria media-Capsella bursa-pastoris* community
Matricaria perforata-Poa annua sub-community

Above-listed species scarce but *Sinapis arvensis* frequent

OV13 *Stellaria media-Capsella bursa-pastoris* community
Typical sub-community

57   *Elymus repens, Agrostis stolonifera, Chenopodium album, Bilderdykia convolvulus, Capsella bursa-pastoris, Chamomilla suaveolens* and *Trifolium repens* all frequent
58

Above-listed species not consistently frequent as a group
59

58   *Plantago major, Rumex obtusifolius, Veronica arvensis* and *Senecio vulgaris* all common

OV9 *Matricaria perforata-Stellaria media* community
Bilderdykia convolvulus-Veronica persica sub-community

Potentilla anserina and *Equisetum arvense* common, *Lamium purpureum, Bromus sterilis, Tussilago farfara* and *Ranunculus arvensis* occasional

OV9 *Matricaria perforata-Stellaria media* community
Elymus repens-Potentilla anserina sub-community

59   *Poa annua, Galeopsis tetrahit* and *Cerastium fontanum* frequent

OV9 *Matricaria perforata-Stellaria media* community
Poa annua-Galeopsis tetrahit sub-community

Anagallis arvensis and *Viola arvensis* constant with *P. annua* occasional at most

OV9 *Matricaria perforata-Stellaria media* community
Anagallis arvensis-Viola arvensis sub-community

60   *Cerastium glomeratum, Juncus bufonius* and *Fumaria muralis* ssp. *boraei* constant with frequent *Chrysanthemum segetum, Briza minor, Ranunculus parviflorus* and *Silene gallica* occasional to frequent

OV6 *Cerastium glomeratum-Fumaria muralis* ssp. *boraei* community
61

C. segetum, B. minor and *S. gallica* can be present but not in combination with the other above-listed species
62

61   *Urtica urens, Valerianella locusta, Polycarpon tetraphyllum* and *Rumex obtusifolius* frequent with *Barbula convoluta, Riccia sorocarpa, Dicranella staphylina* and *Pottia truncata* among an abundant and diverse moss component

OV6 *Cerastium glomeratum-Fumaria muralis* ssp. *boraei* community
Valerianella locusta-Barbula convoluta sub-community

Stellaria media, Plantago lanceolata, Spergula arvensis and *Aphanes microcarpa* all frequent with some of *Ranunculus muricatus, Myosotis discolor, Cerastium fontanum, Trifolium repens* and *Rumex crispus*

OV6 *Cerastium glomeratum-Fumaria muralis* ssp. *boraei* community
Aphanes microcarpa-Ranunculus muricatus sub-community

Vicia hirsuta, Papaver dubium and *Trifolium dubium* constant with frequent *Elymus repens* and *Vicia sativa* ssp. *nigra*

OV6 *Cerastium glomeratum-Fumaria muralis* ssp. *boraei* community
Vicia hirsuta-Papaver dubium sub-community

62   *Digitaria ischaemum, Erodium cicutarium* and *Spergula arvensis* constant with *Crepis capillaris, Geranium molle* and *Rumex acetosella*

OV5 *Digitaria ischaemum-Erodium cicutarium* community

S. arvensis can occur but not in combination with the above-listed species
63

63   *Chrysanthemum segetum* and *S. arvensis* constant with occasional *Misopates orontium* and *Euphorbia helioscopa*

OV4 *Chrysanthemum segetum-Spergula arvensis* community
Spergulo-Chrysanthemetum segetum (Br.-Bl. & de Leeuw 1936) R.Tx. 1937
64

Above combination of species absent
65

64   *Ranunculus repens* and *Sonchus asper* constant with frequent *Polygonum persicaria* and *Potentilla anserina*

OV4 *Spergulo-Chrysanthemetum segetum*
Ranunculus repens-Sonchus asper sub-community

Matricaria perforata frequent with above-listed species scarce

OV4 *Spergulo-Chrysanthemetum segetum*
Typical sub-community

65   *Briza minor* and *Silene gallica* constant with *Anagallis arvensis*, *Trifolium dubium* and *Rumex acetosella* and frequent *Vulpia bromoides*, *V. myuros*, *Aphanes microcarpa* and *Aira caryophyllea*

OV2 *Briza minor-Silene gallica* community

R. acetosella, *A. microcarpa* and *T. dubium* can be occasional to frequent but *S. gallica* and *B. minor* are scarce or absent

OV1 *Viola arvensis-Aphanes microcarpa* community

V. arvensis and *A. arvensis* are constant but also with *Papaver rhoeas* and often also *P. argemone* and *P. dubium*

OV3 *Papaver rhoeas-Viola arvensis* community
Papaveretum argemones (Libbert 1933) Kruseman & Vlieger 1939

COMMUNITY DESCRIPTIONS

OV1
Viola arvensis-Aphanes microcarpa community

Synonymy
Teesdalio-Arnoseridetum minimae (Malcuit 1929) R.Tx. (1937) 1950 *sensu* Silverside 1977

Constant species
Aphanes microcarpa, Poa annua, Rumex acetosella, Viola arvensis.

Rare species
Anthoxanthum aristatum, Briza minor.

Physiognomy
Both *Viola arvensis* and *Aphanes microcarpa* are constants in the open annual vegetation of the *Viola-Aphanes* community, although *Poa annua* often contributes much of the cover by early summer. The diminutive perennial herb *Rumex acetosella* is very frequent too, producing its annual tufts of shoots in the gaps among the other plants.

Frequent associates, usually at low cover, include *Matricaria perforata, Stellaria media, Veronica arvensis* and *Polygonum aviculare* with *Bilderdykia convolvulus, Trifolium dubium, Spergula arvensis, Ornithopus perpusillus, Viola tricolor, Chrysanthemum segetum, Alopecurus myosuroides, Scleranthus annuus* and *Anagallis arvensis* among the more common occasionals. This community also provides a locus for the introduced grass, *Anthoxanthum aristatum*, originally from the Mediterranean and spreading widely in southern England in the nineteenth century but now extremely rare (Salisbury 1964, Perring & Farrell (1977). *Briza minor*, another national rarity, has also been recorded here but it is more usually found now in other communities.

Areas of bare ground can have scattered plants of acrocarpous mosses such as *Bryum microerythrocarpum, B. rubens, Dicranella staphylina, Ceratodon purpureus, Phascum cuspidatum* and *Pleuridium subulatum*.

Habitat
The *Viola-Aphanes* community is typically found among arable crops on impoverished base-poor sandy soils in the more Continental eastern parts of Britain. The local occurrence of suitable soils and the prevalence of intensive cereal production make this a very scarce vegetation type now.

Most of the more frequent species of this assemblage have a widespread distribution through the British lowlands and are able to take advantage of any kind of disturbance on lighter soils. The more distinctive plants, however, like *Aphanes arvensis, Spergula arvensis, Scleranthus annuus* and *Anthoxanthum aristatum* are particularly associated with the sort of impoverished sands, especially acid sands, that are decidedly local in the lowlands of this country and usually now under woodland, heath or settlements. *Arnoseris minima*, a plant that has been extinct with us since 1971, but which was previously characteristic of this kind of vegetation, also showed such a habitat preference (Salisbury 1964). Moreover, species such as *S. annua, A. aristata* and *Arnoseris* are plants with a distinctly Continental distribution, associated with the hot droughty summers typical of the more south-easterly parts of Britain.

The uncertainty of the British climate, the local occurrence of suitable habitats and the intensive character of cereal cultivation in recent decades, using lime, fertilisers and herbicides to produce a congenial growing environment for the crops, has greatly reduced the extent of the habitat for the community. It was encountered by Silverside (1977) mostly in barley and fallow arable fields.

Zonation and succession
The community occurs patchily in arable crops or on bare sandy ground. In more enriched situations, it can be replaced by the *Matricaria perforata-Stellaria* community. Less disturbed sandy tracksides and gateways may have some kind of *Festuca-Agrostis-Rumex* grassland. Repeated disturbance for sowing crops effectively regenerates the community.

Distribution
The *Viola-Aphanes* community occurs very locally in southern and eastern Britain north to Angus (Silverside 1977).

Affinities

The *Viola-Aphanes* community can be seen as the fragmentary equivalent towards the north-west limit of its range of an assemblage variously described as the *Teesdalio-Arnoseridetum* (Malcuit 1929) R.Tx. (1937) 1950 or the *Sclerantho-Arnoseridetum* R.Tx. 1937. This is the most widespread association of base-poor sands in Continental parts of Europe, and has been described from The Netherlands (Westhoff & den Held 1969), Germany (Oberdorfer 1983, Pott 1992) and Austria (Mucina *et al.* 1993).

Floristic table OV1

Poa annua	V (1–8)
Aphanes microcarpa	V (1–4)
Viola arvensis	IV (1–4)
Rumex acetosella	IV (1–3)
Matricaria perforata	III (1–3)
Stellaria media	III (1–4)
Veronica arvensis	III (1–3)
Polygonum aviculare	III (1–5)
Bilderdykia convolvulus	II (1–3)
Trifolium dubium	II (1–3)
Chamomilla suaveolens	II (1–3)
Spergula arvensis	II (1–3)
Ornithopus perpusillus	II (1–2)
Anthoxanthum aristatum	II (1–6)
Viola tricolor	II (1–3)
Scleranthus annuus	II (1–3)
Chrysanthemum segetum	II (1–3)
Alopecurus myosuroides	II (1–3)
Anagallis arvensis	II (3)
Bryum microerythrocarpum	II (1–3)
Dicranella staphylina	II (1–6)
Trifolium repens	II (1–4)
Bryum rubens	II (1–5)
Agrostis stolonifera	II (3–4)
Bromus sterilis	II (1)
Ceratodon purpureus	II (1–5)
Pleuridium subulatum	II (1–3)
Phascum cuspidatum	II (1–3)
Holcus mollis	I (5)
Trifolium arvense	I (3)
Briza minor	I (4)
Myosotis arvensis	I (1)
Sinapis arvensis	I (1)
Anthemis cotula	I (1)
Senecio vulgaris	I (1)
Veronica persica	I (1)
Riccia sorocarpa	I (1)
Number of samples	8
Number of species/sample	19 (13–22)
Herb cover (%)	94 (30–100)
Bryophyte cover (%)	20 (0–80)

OV2

Briza minor-Silene gallica community

Synonymy

Airo multiculmis-Arnoseridetum minimae (Allorge 1922) R.Tx. 1950 *sensu* Silverside 1977.

Constant species

Anagallis arvensis, Briza minor, Rumex acetosella, Silene gallica, Trifolium dubium.

Rare species

Briza minor, Silene gallica, Trifolium suffocatum.

Physiognomy

In the *Briza minor-Silene gallica* community, common annuals like *Anagallis arvensis, Trifolium dubium, Aphanes microcarpa* and *Vicia sativa* ssp. *nigra*, together with *Rumex acetosella*, are very common constituents of the open cover. However, the more striking feature here is the constancy of the two nationally scarce annuals *Silene gallica* and *Briza minor*.

Other diminutive ephemeral grasses also figure with some frequency: *Bromus hordeaceus* ssp. *thominei, Vulpia bromoides*, the scarcer *V. myuros, Aira caryophyllea* (including plants sometimes distinguished as ssp. *multiculmis*) and *A. praecox*. There are also typically sporadic elements of the swards within which this assemblage develops, like sparse fronds of *Pteridium aquilinum* pushing up into open ground, and occasional *Anthoxanthum odoratum, Hypochoeris radicata* and *Plantago lanceolata*.

Other occasionals of the community include *Geranium molle, Chrysanthemum segetum, Sherardia arvensis, Veronica arvensis, Myosotis ramosissima, Ornithopus perpusillus* and sometimes naturalised representatives of the bulb crops among which this vegetation often occurs, like *Allium triquetrum* and *Gladiolus byzantinus*. The nationally rare *Trifolium suffocatum* has been recorded here and three naturalised introductions that can occur are *Briza maxima, Phalaris minor* and *Bromus diandrus*.

Habitat

The *Briza-Silene* community is confined to disturbed sandy soils in the extreme oceanic climate of the far

south-west of Britain where it now occurs most characteristically among the bulb fields of The Scillies.

Like the *Viola-Aphanes* community, this assemblage is typical of disturbed acid sands and includes some species like *Vulpia bromoides, V. myuros* and *Aphanes microcarpa* which show a widespread distribution through the warmer southern parts of Britain. More distinctive is the constancy of *Silene gallica*, a plant which was once also found widely on sandy soils across this part of the country but which shows some limitation by low winter temperatures: it is a winter annual, germinating mainly in autumn, but its seedlings are killed by temperatures of $-10\,°C$ (Stewart *et al.* 1994), so it tends to favour most equable climates. Much more confined by the cold season, perhaps again because of frost sensitivity, is *Briza minor*, an introduction from the Mediterranean and western Europe which has always been confined to the more oceanic fringes of south-west England in its extent here.

Occurring together in this very striking community, these species benefit from the virtually frost-free climate typical of The Scillies (*Climatological Atlas* 1952) where the growing season is effectively year-long (Smith 1976) and where summer and autumn rains provide sufficient moisture to encourage good germination before the cooler season. The other characteristic of this area is that the particular form of arable agriculture there still provides congenial situations for these species to reappear year by year. Many of the annuals of this community have shown shrinking ranges as more traditional agricultural practices have given way to intensive arable or abandonment to heath and woodland: the present limitation of *Silene gallica* to the sub-maritime zone is probably due to such shifts (Stewart *et al.* 1994).

In the past, such vegetation as this was more widespread because of considerable cereal cultivation in the south-west. Now, although its species survive fragmentarily along tracks and roadsides, it is in bulb fields that the community is most often seen. The bulbs are typically cultivated in widely-spaced rows so there is ample light for the numerous poor competitors of this assemblage.

Also, Silverside (1977) pointed out that, because weed control is usually delayed until the bulbs are lifted in late May, there is ample time for many of the annuals to flower, fruit and disperse their seed before their herbage is killed by the herbicides: *B. minor*, for example, flowers from March onwards in The Scillies (Stewart *et al*. 1994).

Zonation and succession
On somewhat more fertile soils, like sandy loams, and where the ground is a little moister, the *Briza-Silene* community can give way to the *Cerastium-Fumaria* vegetation. Both *B. minor* and *S. gallica* remain frequent there but *Cerastium glomeratum*, *Fumaria muralis* ssp. *boraei*, *Juncus bufonius* and *Ranunculus parvifloris* become more common.

This assemblage depends on repeated disturbance for an opportunity to re-establish itself on areas of open ground and any possibility of succession is prevented by renewed cultivation.

Distribution
The *Briza-Silene* community occurs only on The Scillies.

Affinities
Vegetation of this kind was first described as the *Airo multiculmis-Arnoseridetum* (Allorge 1922) R.Tx. 1950 and seen as a replacement for the *Teesdalio-Arnoseridetum* (or *Sclerantho-Arnoseridetum*) in the south-west of Europe. Certainly, this community shows some floristic links with this part of the Continent.

Floristic table OV2

Silene gallica	V (2–5)	Holcus lanatus	I (3)
Anagallis arvensis	V (2–3)	Juncus bufonius	I (2)
Trifolium dubium	V (3–5)	Briza maxima	I (5)
Briza minor	V (2–3)	Ornithogalum umbellatum	I (2)
Rumex acetosella	IV (2–7)	Taraxacum officinale agg.	I (2)
		Montia perfoliata	I (2)
Bromus hordeaceus thominei	III (2–8)	Oxalis articulata	I (1)
Pteridium aquilinum	III (1–3)	Phalaris minor	I (2)
Vulpia bromoides	III (2–9)	Eurhynchium praelongum	I (5)
Vulpia myuros	III (5–8)	Geranium dissectum	I (2)
Aphanes microcarpa	III (2–5)	Myosotis discolor	I (4)
Aira caryophyllea	III (3–5)	Ranunculus ficaria	I (2)
Vicia sativa nigra	III (2–3)	Rumex obtusifolius	I (2)
Plantago lanceolata	III (2–5)	Brachythecium velutinum	I (2)
Anthoxanthum odoratum	II (2–5)	Ranunculus muricatus	I (3)
Aira praecox	II (2–5)	Ranunculus repens	I (2)
Hypochoeris radicata	II (2)	Trifolium arvense	I (3)
Chrysanthemum segetum	II (3–5)	Anthriscus caucalis	I (5)
Geranium molle	II (2–3)	Carex arenaria	I (2)
Aira multiculmis	II (3–5)	Desmazeria marina	I (3)
Sherardia arvensis	II (2)	Papaver dubium	I (2)
Veronica arvensis	II (2–3)	Ranunculus parviflorus	I (2)
Myosotis ramosissima	II (2)	Senecio vulgaris	I (2)
Ornithopus perpusillus	II (3)	Sonchus oleraceus	I (2)
Allium triquetrum	II (2–5)	Valerianella locusta	I (3)
Gladiolus byzantinus	II (3)	Bromus diandra	I (5)
Sagina apetala	I (5)	Polycarpon diphyllum	I (3)
Plantago coronopus	I (2)	Montia fontana	I (2)
Cerastium fontanum	I (2)	Trifolium subterraneum	I (3)
Trifolium repens	I (5)	Trifolium suffocatum	I (2)
Cerastium glomeratum	I (2)		
Spergula arvensis	I (2)	Number of samples	8
Daucus carota	I (2)	Number of species/sample	22 (14–33)

OV3

Papaver rhoeas-Viola arvensis community
Papaveretum argemones (Libbert 1933) Kruseman & Vlieger 1939

Constant species

Anagallis arvensis, Medicago lupulina, Papaver rhoeas, Poa annua, Polygonum aviculare, Veronica persica, Viola arvensis.

Rare species

Veronica triphyllos.

Physiognomy

The *Papaveretum argemones* comprises annual weed vegetation in which the most striking feature is the abundance of various species of red poppy. *Papaver rhoeas* is the most frequent and usually the most abundant of these, its bright scarlet flowers appearing from mid-June and continuing intermittently until October, but also common, particularly in the south and east of Britain and beginning to flower earlier, is *P. argemone*. A third species, like *P. argemone* in its smaller and paler scarlet flowers, though more widespread, is *P. dubium*.

Other frequent contributors to the more or less open cover of this vegetation are *Viola arvensis, Veronica persica, V. arvensis, Anagallis arvensis, Poa annua, Medicago lupulina, Polygonum aviculare, Bilderdykia convolvulus, Capsella bursa-pastoris* and *Trifolium repens. Elymus repens* and *Agrostis stolonifera* are common, too, and can be locally abundant.

Occasionals in the *Papaveretum* include *Arenaria serpyllifolia* (with some plants distinguished as ssp. *leptoclados*), *Galium aparine, Myosotis arvensis, Chenopodium album, Matricaria perforata, Chamomilla suaveolens, Sonchus asper, S. oleraceus* and *Senecio vulgaris*. The nationally rare *Veronica triphyllos* has also been recorded in this vegetation.

Habitat

The *Papaveretum* is characteristic of disturbed, light and friable soils that are not too calcareous, throughout the warmer and drier lowlands of Britain and it is especially frequent in cereal fields that have escaped herbicide treatment.

P. rhoeas has a wide distribution on disturbed sands and loams throughout the British lowlands; *P. dubium* extends somewhat further north and is commoner in certain parts of Wales; *P. argemone* is more confined to the Continental south and east (Perring & Walters 1962, McNaughton & Harper 1964). In its full expression, then, with all these represented, this is a community of the warmer and drier parts of the country and is there limited by the distribution of suitable soils and a pattern of repeated disturbance that is not characterised by heavy additions of fertilisers or herbicides. *P. argemone* in particular is susceptible to many weedkillers, including some of the earliest developed and it cannot tolerate much competition from crops on greatly enriched soils. All the species of poppy here also have poor dispersal so, though seed may be viable for some time and produce a stunning display of flowering among highly gregarious offspring (McNaughton & Harper 1964), persistence in any particular location necessitates some relief from the current intensive style of arable agriculture.

Both the more widespread *P. rhoeas* and the more confined *P. argemone* can germinate in autumn and spring but the latter at least seems better represented in autumn-sown crops. The community is most often seen among cereals which usually receive less nitrogenous fertiliser than roots and vegetables.

Zonation and succession

The *Papaveretum* is typically seen as a patchy or marginal assemblage in parts of cereal fields that have escaped herbicide treatment. On more calcareous soils, it is replaced by the *Papaveri-Sileneetum* and it can pass to other weed communities like the *Veronica* or *Stellaria-Capsella* assemblages where crops have been more heavily fertilised or where more herbicide-resistant species prevail. Where the community occurs on disturbed waste ground, it can give way to *Urtica-Cirsium* vegetation or the *Arrhenatheretum* where *Papaver rhoeas* can remain locally frequent.

The community depends on repeated disturbance for

its reappearance and cultivation effectively prevents any further succession in arable fields. On disturbed ground or dumped soil heaps, abandonment may see subsequent colonisation by tall herbs or rank mesotrophic swards.

Distribution
The *Papaveretum* is widespread in the southern part of Britain and was found by Silverside (1977) north to Angus.

Affinities
British stands of this kind of vegetation clearly belong to the *Papaveretum argemones* which occurs throughout Europe, south to Italy and north as far as Scandinavia (Westhoff & den Held 1969, Oberdorfer 1983, Pott 1992, Mucina *et al.* 1993). Early workers tended to include this association in the Aphanion, later ones in the Aperion, an alliance of weed assemblages on loamy soils, often with autumn-sown cereals.

Floristic table OV3

Papaver rhoeas	V (2–7)	*Aphanes microcarpa*	II (2)
Viola arvensis	IV (1–3)	*Geranium dissectum*	II (1–3)
Veronica persica	IV (2–5)	*Vicia sativa nigra*	II (2–3)
Anagallis arvensis	IV (2–5)	*Cirsium arvense*	II (1–5)
Poa annua	IV (2–5)	*Poa trivialis*	II (1–5)
Polygonum aviculare	IV (3–5)	*Taraxacum officinale* agg.	II (2–3)
Medicago lupulina	IV (1–5)	*Rumex crispus*	II (1–2)
		Achillea millefolium	II (1–2)
Papaver argemone	III (2–5)	*Urtica dioica*	II (1–2)
Veronica arvensis	III (2–3)	*Convolvulus arvensis*	II (2–7)
Bilderdykia convolvulus	III (2–3)	*Anchusa arvensis*	II (1–3)
Capsella bursa-pastoris	III (2–5)	*Anthemis arvensis*	I (1–5)
Stellaria media	III (2–7)	*Arabidopsis thaliana*	I (1–3)
Elymus repens	III (1–8)	*Veronica triphyllos*	I (1–3)
Agrostis stolonifera	III (2–7)	*Pulicaria dysenterica*	I (1–3)
Trifolium repens	III (1–7)	*Bryum rubens*	I (1–3)
Papaver dubium	II (1–5)	*Heracleum sphondylium*	I (1–3)
Arenaria serpyllifolia	II (2–5)	*Valerianella locusta*	I (1–3)
Arenaria leptoclados	II (1–3)	*Bryum sauteri*	I (1–5)
Galium aparine	II (1–2)	*Phascum cuspidatum*	I (1–3)
Myosotis arvensis	II (2–5)	*Eupatorium cannabinum*	I (3)
Chenopodium album	II (1–2)	*Epilobium hirsutum*	I (4)
Matricaria perforata	II (2–5)	*Spergula arvensis*	I (1–3)
Senecio vulgaris	II (2–3)	*Descurainia sophia*	I (4)
Sonchus asper	II (1–3)		
Sonchus oleraceus	II (1–3)	Number of samples	14
Silene alba	II (2–5)	Number of species/sample	30 (13–42)
Chamomilla suaveolens	II (1–2)		
Artemisia vulgaris	II (1–5)	Herb cover (%)	79 (50–100)
Sisymbrium officinale	II (1–2)		

OV4
Chrysanthemum segetum-Spergula arvensis community
Spergulo-Chrysanthemetum segetum (Br.-Bl. & De Leeuw 1936) R.Tx. 1937

Constant species
Chrysanthemum segetum, Poa annua, Polygonum aviculare, Spergula arvensis.

Physiognomy
The distinctive feature of the annual vegetation of the *Spergulo-Chrysanthemetum* is the constancy of *Spergula arvensis* and *Chrysanthemum segetum*, the latter often abundant and especially striking by summer with its big, bright yellow flower heads. *Poa annua* and *Polygonum aviculare* are also constant in the community, *Elymus repens, Capsella bursa-pastoris* and *Stellaria media* frequent. Occasionals occurring throughout are *Bilderdykia convolvulus, Misopates orontium, Euphorbia helioscopa* and *Chamomilla suaveolens*. The perennial grasses *Elymus repens* and, somewhat less commonly *Lolium perenne* and *Agrostis stolonifera* can get a hold each year, though renewed disturbance curtails their continuing growth.

Sub-communities

Typical sub-community. In this sub-community, the only additional distinguishing features are frequent records for *Matricaria perforata* and occasional *Poa trivialis, Polygonum lapathifolium, Fumaria muralis* ssp. *boraei* and *Lamium purpureum.*

Ranunculus repens-Sonchus asper sub-community. Here, *Ranunculus repens* and *Sonchus asper* are strongly preferential with *Polygonum persicaria* and *Potentilla anserina* also frequent and occasional records for *Anagallis arvensis, Viola arvensis, Myosotis arvensis, Crepis capillaris, Vicia sativa* ssp. *nigra* and the perennials *Bellis perennis, Trifolium repens, Holcus lanatus, Cerastium fontanum* and *Plantago major.*

Habitat
The *Spergulo-Chrysanthemetum* is a widespread community of disturbed, light, fertile acidic soils throughout the British lowlands, and a common feature of both cereal and root crops.

Both *Chrysanthemum segetum* (Howarth & Williams 1972) and *Spergula arvensis* (New 1961) are intolerant of non-acidic soils and are most characteristic of sands and light loams, with an optimum pH for *S. arvensis* of 4.5–5. *C. segetum* may also have a requirement for a well-aerated substrate: although this community can extend on to moister soils, and overall has a concentration in the wetter west of Britain, the substrates are generally free-draining. In the drier east of the country, it extends somewhat on to heavier soils which may give protection against drought. Both of these species germinate in spring from seed that can survive lengthy burial. *S. arvensis* may have a requirement for temperature fluctuation (New 1961) and seed needs to be at or near the soil surface for germination. Cultivation in May can destroy a first generation of seedlings but stimulate a subsequent flush.

Also, though *S. arvensis* is often found on infertile sands, *C. segetum* grows best on more eutrophic soils so this community is excluded from very impoverished sands. Indeed, many of the associates are characteristic of fertilised fields, so this is not an assemblage confined to low-input arable agriculture. *C. segetum*, with its waxy foliage, is also somewhat resistant to herbicide treatment (Silverside 1977) though it is certainly a weed that farmers try to exclude from cereal crops: its semi-succulent fruits hinder the drying of grain (Howarth & Williams 1972). In its distribution, then, this community is better represented on arable land on light soils that are naturally quite fertile or which have been lightly manured but not limed: indeed, liming is a treatment that can help control both *C. segetum* and *S. arvensis.*

The differences between the two sub-communities are related to soil moisture. The *Ranunculus-Sonchus* type is more characteristic of wetter substrates and tends to prevail in areas of rainier climate, as on the machair of the Outer Hebrides, or on somewhat more retentive soils.

Zonation and succession

The *Spergulo-Chrysanthemetum* occurs patchily within and around the edges of arable fields, sometimes giving way to other weed assemblages where there are local shifts in soil conditions, or differences in treatments and crops. In fields on clayey soils in the south-west and in Wales, for example, it tends to be replaced by the *Poa annua-Stachys arvensis* community and among summer cereals by the *Papaveretum*. More heavily fertilised or herbicide-treated stretches of crop or fields can see a switch to communities like *Matriciaria-Stellaria* assemblage. With the move to the more extreme oceanic climate of the far south-west of England, the *Spergulo-Chrysanthemetum* is replaced by the *Cerastium-Fumaria* community on similar soils.

Continuing cultivation of arable crops repeatedly creates congenial conditions for the community but traditional alternations of cereals or roots and fallow grassland, as on the machair of the Outer Hebrides, leads to a temporary successional replacement of the *Ranunculus repens* sub-community by swards of the *Festuca-Agrostis-Potentilla* or *Festuca-Galium* types (I. Crawford, personal communication).

Distribution

The *Spergulo-Chrysanthemetum* occurs widely throughout Britain on suitable soils, with highest frequency on less intensive arable land in the west.

Affinities

This kind of weed vegetation is clearly identical to the *Spergulo-Chrysanthemetum* (Br.-Bl. & de Leeuw 1936) R.Tx. 1937 that has been widely described from across Europe, south into France and east to Poland (Westhoff & den Held 1969, Oberdorfer 1983, Pott 1992). It is the major association among the weed assemblages of more base-poor soils in the Atlantic and sub-Atlantic zones of Europe, vegetation usually grouped in the Polygono-Chenopodion. Much of Silverside's (1977) *Briza minor* variant of the *Spergulo-Chrysanthemetum* is here included among the *Cerastium-Fumaria* community.

Floristic table OV4

	a	b	4
Chrysanthemum segetum	V (1–8)	V (1–6)	V (1–8)
Spergula arvensis	V (1–6)	IV (1–3)	V (1–6)
Polygonum aviculare	V (1–3)	IV (1–6)	V (1–6)
Poa annua	IV (1–5)	IV (1–4)	IV (1–5)
Matricaria perforata	III (1–6)	I (6)	II (1–6)
Poa trivialis	II (1–3)	I (3)	I (1–3)
Polygonum lapathifolium	II (1–3)	I (1)	I (1–3)
Fumaria muralis ssp. *boraei*	II (1–3)	I (5)	I (1–5)
Lamium purpureum	II (1–3)		I (1–3)
Ranunculus repens		V (1–4)	III (1–4)
Sonchus asper	I (3)	IV (1–3)	II (1–3)
Polygonum persicaria	I (1)	III (1–5)	II (1–3)
Potentilla anserina	I (3)	III (1–3)	II (1–3)
Bellis perennis		II (1–3)	II (1–3)
Anagallis arvensis	I (1–3)	II (1–3)	II (1–3)
Viola arvensis	I (1–3)	II (1–2)	II (1–3)
Trifolium repens	I (1)	II (1–3)	II (1–3)
Myosotis arvensis		II (1–3)	I (1–3)
Cerastium fontanum		II (1–3)	I (1–3)
Holcus lanatus		II (1–5)	I (1–5)
Crepis capillaris		II (1–3)	I (1–3)
Vicia sativa nigra		II (1–3)	I (1–3)
Plantago major	I (1)	II (1–3)	I (1–3)
Daucus carota		II (1–3)	I (1–3)

Floristic table OV4 (*cont.*)

	a	b	4
Cirsium vulgare		I (1–2)	I (1–2)
Trifolium dubium		I (1–3)	I (1–3)
Rumex acetosella		I (1–3)	I (1–3)
Veronica arvensis		I (1–3)	I (1–3)
Cerastium glomeratum		I (1–2)	I (1–2)
Brassica rapa		I (1–3)	I (1–3)
Veronica agrestis		I (1–2)	I (1–2)
Galeopsis tetrahit agg.		I (3–4)	I (3–4)
Taraxacum officinale agg.		I (1–3)	I (1–3)
Elymus repens	III (1–5)	III (1–3)	III (1–5)
Stellaria media	III (1–6)	III (1–6)	III (1–6)
Capsella bursa-pastoris	III (1–3)	III (1–5)	III (1–5)
Bilderdykia convolvulus	II (1–3)	II (1–3)	II (1–3)
Lolium perenne	II (1–3)	II (1–3)	II (1–3)
Misopates orontium	II (1–3)	II (1–3)	II (1–3)
Euphorbia helioscopa	II (1–3)	II (1–3)	II (1–3)
Chamomilla suaveolens	II (1–3)	II (1–8)	II (1–8)
Agrostis stolonifera	II (1–5)	II (1–4)	II (1–5)
Anchusa arvensis	I (1–3)	I (1–3)	I (1–3)
Raphanus raphanistrum	I (1–3)	I (1–5)	I (1–5)
Stachys arvensis	I (1)	I (1–3)	I (1–3)
Filaginella uliginosa	I (1–3)	I (1–3)	I (1–3)
Bryum rubens	I (1–3)	I (1–3)	I (1–3)
Agrostis capillaris	I (1–3)	I (1–3)	I (1–3)
Solanum nigrum	I (1–3)	I (1–3)	I (1–3)
Coronopus didymus	I (8)	I (1)	I (1–8)
Senecio vulgaris	1 (1–2)	I (3)	I (1–3)
Lolium multiflorum	I (1–3)	I (1)	I (1–3)
Cirsium arvense	I (1–3)	I (1)	I (1–3)
Rumex crispus	I (1)	I (1–3)	I (1–3)
Dactylis glomerata	I (1–3)	I (3)	I (1–3)
Geranium dissectum	I (1)	I (1)	I (1)
Artemisia vulgaris	I (1)	I (1)	I (1)
Number of samples	13	12	25
Number of species/sample	17 (7–22)	24 (17–41)	19 (7–41)
Vegetation cover (%)	82 (50–100)	77 (20–100)	80 (20–100)

a   Typical sub-community
b   *Ranunculus repens-Sonchus asper* sub-community
4   *Spergulo-Chrysanthemetum segetum* (total)

OV5
Digitaria ischaemum-Erodium cicutarium community

Synonymy
Echinochloo-Setarietum (Kruseman & Vlieger 1939) *emend.* Kruseman & Vlieger *apud* Sissingh, Vlieger & Westhoff 1940 *sensu* Silverdale 1977

Constant species
Crepis capillaris, Digitaria ischaemum, Elymus repens, Erodium cicutarium, Geranium molle, Poa annua, Senecio vulgaris, Spergula arvensis, Stellaria media.

Rare species
Apera spica-venti.

Physiognomy
The rare annual introduced grass *Digitaria ischaemum*, together with *Erodium cicutarium* and *Spergula arvensis*, provide a distinctive group of constants in this *Digitaria-Erodium* annual weed community. Also very frequent are *Stellaria media, Poa annua, Senecio vulgaris, Geranium molle, Crepis capillaris* and *Elymus repens*.

Other common associates are *Bilderdykia convolvulus, Capsella bursa-pastoris, Rumex acetosella, Papaver dubium, Chenopodium album* and *Taraxacum officinale* agg. Occasionals of the community include *Polygonum aviculare, Medicago sativa, Ornithopus perpusillus, Scleranthus annuus, Raphanus raphanistrum* and *Urtica urens* with *Holcus mollis* and *Achillea millefolium* sometimes figuring. Bryophytes are occasionally found with *Bryum rubens* and *Pleuridium subulatum* most common.

Habitat
The *Digitaria-Erodium* community is confined to fertilised sandy soils disturbed by the cultivation of root crops and cereals in a very localised part of south-east England.

D. ischaemum is a native of warm-temperate parts of Europe and Asia first recorded in East Anglia in about 1805 and since then locally established among arable crops in sandy fields in southern and south-east England (Hubbard 1984), having probably spread in contaminated seed (Salisbury 1964). It has declined markedly in its occurrences with the shift to more intensive kinds of arable agriculture, although it can persist in situations with quite considerable soil enrichment. In this community, it survives with some species characteristic of sandy soils like *Erodium cicutarium, Scleranthus annuus* and *Ornithopus perpusillus*, as well as with more widely distributed weeds of more fertilised soils.

Although no samples were available with such plants, it seems clear that a number of other introduced grasses like *Echinochloa crus-galli*, a warm-temperate and tropical species which became especially frequent as a weed during World War II when contaminated seed came from North America, and *Setaria viridis*, a Eurasian warm-temperate plant, have often been recorded in this kind of vegetation among root crops like carrots, turnips and mangolds on sandy soils in southern England (Salisbury 1964, Hubbard 1984). These plants, too, have greatly declined in frequency in recent decades (Perring & Walters 1962). Silverside (1977) noted that *Echinochloa* does not germinate until the soil temperature rises to 15 °C with an optimum at 20–30 °C, so is likely to persist only sporadically, even if suitable soils were available.

Zonation and succession
The *Digitaria-Erodium* community has been found within a variety of crops in its single location. Cultivation repeatedly sets back any successional development and encourages a return of the assemblage.

Distribution
This vegetation has been recorded only from one locality on the Bagshot Sands of Surrey.

Affinities
Silverside (1977), from whose study these samples originate, grouped them in the *Echinochloa-Setarietum* (Kruseman & Vlieger 1939) *emend.* Kruseman & Vlieger *apud* Sissingh, Vlieger & Westhoff 1940, an association

characterised by *Echinochloa, S. viridis, S. glauca, Digitaria ischaemum, Galinsoga parviflora* and *G. ciliata* and widely described from The Netherlands (Westhoff & den Held 1969), through Germany (Oberdorfer 1985) and Austria (Mucina *et al.* 1993) to Poland (Matuszkie-wicz 1984). It is seen by most authorities as subsuming a *Digitarietum ischaemum* R.Tx. & Preising (1942) 1950. British stands are clearly towards the geographical limit of such a range and with us the syntaxon lacks any real integrity.

Floristic table OV5

Digitaria ischaemum	V (1–4)	*Viola arvensis*	II (1–3)
Stellaria media	V (1–8)	*Anagallis arvensis*	II (1–3)
Elymus repens	V (1–6)	*Anchusa arvensis*	II (1–3)
Poa annua	V (1–6)	*Digitaria sanguinalis*	I (1)
Erodium cicutarium	V (1–5)	*Phascum cuspidatum*	I (1)
Senecio vulgaris	V (1–4)	*Bryum microerythrocarpum*	I (1)
Spergula arvensis	V (1–6)	*Zygogonium ericetorum*	I (1)
Crepis capillaris	IV (1–3)	*Ranunculus repens*	I (1)
Geranium molle	IV (1–3)	*Brachythecium rutabulum*	I (1)
		Dicranella staphylina	I (1)
Bilderdykia convolvulus	III (1–3)	*Amaranthus retroflexus*	I (1)
Capsella bursa-pastoris	III (1–4)	*Polygonum persicaria*	I (1)
Rumex acetosella	III (1–3)	*Chenopodium ficifolium*	I (1)
Chenopodium album	III (1–3)	*Apera spica-venti*	I (1)
Papaver dubium	III (1–3)	*Matricaria perforata*	I (1)
Bryum rubens	III (1–3)	*Trifolium repens*	I (1)
Taraxacum officinale agg.	III (1–3)	*Plantago lanceolata*	I (1)
Holcus mollis	II (1–4)	*Rumex crispus*	I (1)
Polygonum aviculare	II (1–3)	*Rumex obtusifolius*	I (1)
Achillea millefolium	II (1–3)	*Cirsium arvense*	I (1)
Medicago sativa	II (1–3)	*Aphanes microcarpa*	I (1)
Ornithopus perpusillus	II (1–3)	*Dactylis glomerata*	I (1)
Scleranthus annuus	II (1–3)		
Raphanus raphanistrum	II (1–3)	Number of samples	6
Equisetum arvense	II (1–3)	Number of species/sample	16 (9–23)
Pleuridium subulatum	II (1–3)		
Urtica urens	II (1–3)	Vegetation cover (%)	68 (30–90)
Solanum nigrum	II (1–3)		

OV6
Cerastium glomeratum-Fumaria muralis ssp. *boraei* community

Synonymy

Spergulo-Chrysanthemetum (Br.-Bl. de Leeuw 1936) R.Tx. 1937 *sensu* Silverside 1977 *p.p.*; *Medicagi-Ranunculetum parviflori* Silverside 1977 *p.p.*

Constant species

Anagallis arvensis, Cerastium glomeratum, Fumaria muralis ssp. *boraei, Juncus bufonius, Poa annua, Senecio vulgaris, Sonchus asper.*

Rare species

Allium babingtonii, Briza minor, Fumaria bastardii, Silene gallica.

Physiognomy

The *Cerastium glomeratum-Fumaria muralis* ssp. *boraei* community is one of the British weed assemblages in which fumitories figure with some prominence. *F. muralis* ssp. *boraei* is the most common taxon here and it can be abundant. Less frequent in the community, though locally prominent, are *F. officinalis* (a plant with a wider distribution nationally than *F. muralis* ssp. *boraei*), and the nationally scarce *F. bastardii* and *F. occidentalis.*

Other distinctive constants of this kind of vegetation are *Cerastium glomeratum* and *Juncus bufonius* with *Anagallis arvensis, Poa annua, Senecio vulgaris* and *Sonchus asper. P. trivialis* and *Coronopus didymus* are among the frequent associates but more striking is the common occurrence of *Chrysanthemum segetum* and the nationally rare or scarce *Ranunculus parviflorus, Briza minor* and *Silene gallica. Gladiolus byzantinus* is a common escape from the bulb crops among which this assemblage is often found.

Occasionals include *Veronica persica, Medicago arabica* and *Rumex acetosella* with the South African introduction *Oxalis pes-caprae* naturalised here, though with varying frequency in the different sub-communities, and another introduced sorrel *O. corniculata* occurring much more rarely.

Sub-communities

***Aphanes microcarpa-Ranunculus muricatus* sub-community:** *Spergulo-Chrysanthemetum* (Br.-Bl. & de Leeuw 1936) R.Tx. 1937, *Briza minor* variant Silverside 1977 *p.p. Plantago lanceolata* and *Stellaria media* are more frequent here than elsewhere in the community but more distinctive preferentials are *Aphanes microcarpa* and the Mediterranean introduction *Ranunculus muricatus.* Also frequent are *Spergula arvensis, Cerastium fontanum, Myosotis discolor, Trifolium repens* and *Rumex crispus* with occasionals *Bellis perennis, Ranunculus repens, R. ficaria, Myosotis arvensis* and *Medicago polymorpha.*

***Valerianella locusta-Barbula convoluta* sub-community:** *Medicagi-Ranunculetum parviflori* Silverside 1977 *p.p. Veronica arvensis, Sagina procumbens, Agrostis stolonifera* and, among mosses of bare ground, *Bryum rubens* are all weakly preferential to this sub-community but among the vascular plants *Valerianella locusta, Urtica urens, Polycarpon tetraphyllum* and *Fumaria officinalis* are more striking, with *Barbula convoluta* and *Riccia sorocarpa* among the distinctive bryophytes. Other frequent plants are *Capsella bursa-pastoris, Rumex obtusifolius, Heracleum sphondylium* and *Chenopodium album* with *Fumaria bastardii, Allium babingtonii* and *Arum italicum* ssp. *neglectum* occasional.

***Vicia hirsuta-Papaver dubium* sub-community.** *Oxalis pes-caprae, Sonchus oleraceus* and *Galium aparine* are somewhat preferential in this kind of *Cerastium-Fumaria* vegetation but better distinguishing features are high frequencies of *Vicia hirsuta, Papaver dubium* and *Trifolium dubium* with *V. sativa* ssp. *nigra* and *Elymus repens* also common, *Erodium moschatum, Mercurialis annua, Anthriscus caucalis, Convolvulus arvensis* and *Ranunculus marginatus* var. *trachycarpus* occasional.

Habitat

The *Cerastium-Fumaria* community is confined to fertile, light, non-calcareous soils in the extreme oceanic climate of the far south-west of Britain where bulb and vegetable cultivation provides the regular disturbance necessary for its development.

In some ways, this kind of weed vegetation can be seen as a replacement for the *Spergulo-Chrysanthemetum* in the more equable and moist oceanic climate of the South-West Peninsula. It is similarly characteristic of more acidic sandy and loamy soils, sometimes even extending on to peats, that are more free-draining and relatively fertile, either inherently so or because of modest manuring. *C. segetum* itself remains very common here, while *S. arvensis* is somewhat more restricted but the particular assemblage of constants, with species like *Cerastium glomeratum*, *Anagallis arvensis* and *Fumaria muralis* ssp. *boraei* occurring along with *Juncus bufonius*, indicates a distinctive combination of a light texture with moistness that is often met on the cultivated soils of The Scillies and parts of Cornwall.

Other species in this community reflect the more distinctively oceanic character of the climate in this part of Britain. *Ranunculus parviflorus*, for example, is a native plant typical of damper habitats in regions of south-west Europe and the Mediterranean with a moister climate. Like *Medicago arabica* and *Erodium cicutarium*, less common but still distinctive here, it is a winter annual but frost-sensitive, so limited to areas with mild temperatures in the cold season. *Silene gallica* and the introduced *Briza minor*, for which this vegetation provides their second major locus in Britain, are also restricted to more equable climates.

Bulb and vegetable cultivation provides the necessary disturbance that encourage seed germination and establishment of this assemblage.

Zonation and succession

On the Scillies, the *Cerastium-Fumaria* community can be seen in close association with the *Briza-Silene* assemblage which tends to replace it on drier soils in bulb fields. More heavily fertilised areas may see a shift to the *Stellaria-Capsella* community. The annual cycle of bulb and cereal cultivation renews the conditions suitable for the community and effectively prevents any successional developments.

Distribution

The community occurs only on The Scillies and in south-west Cornwall around Gulval, Trenance and Gweek (Silverside 1977).

Affinities

The *Cerastium-Fumaria* community brings together the more oceanic samples which Silverside (1977) included in his *Spergulo-Chrysanthemetum* as the *Briza minor* variant and what he defined as a new syntaxon, the *Medicagi-Ranunculetum*, characteristic of even more Atlantic conditions. The difference between these associations is preserved here at sub-community level.

Floristic table OV6

	a	b	c	6
Cerastium glomeratum	V (2–7)	V (2–5)	V (2–5)	V (2–7)
Anagallis arvensis	V (2–5)	V (2–7)	V (1–3)	V (1–7)
Fumaria muralis boraei	IV (2–8)	V (2–7)	V (3–7)	V (2–8)
Poa annua	V (2–8)	V (2–7)	V (2–5)	V (2–8)
Juncus bufonius	IV (2–7)	V (2–5)	IV (2–5)	IV (2–7)
Senecio vulgaris	IV (2–7)	V (2–7)	IV (2–3)	IV (2–7)
Sonchus asper	IV (2–3)	V (2–3)	III (2–3)	IV (2–3)
Plantago lanceolata	IV (2–5)	II (1–4)	II (2–3)	III (1–5)
Stellaria media	IV (2–5)		II (2–3)	II (2–5)
Aphanes microcarpa	IV (2–7)		I (3)	II (2–7)
Spergula arvensis	III (2–8)	I (2)		II (2–8)
Cerastium fontanum	III (2–5)			II (2–5)
Myosotis discolor	III (2–5)	I (3)	I (1)	II (1–5)
Trifolium repens	III (2–3)	II (2–5)		II (2–5)
Ranunculus muricatus	III (2–8)			II (2–8)

Rumex crispus	III (1–2)	I (2)	I (2)	II (1–2)
Bellis perennis	II (2)		I (2)	I (2)
Ranunculus repens	II (1–8)	I (1–3)		I (1–8)
Myosotis arvensis	II (2–3)			I (2–3)
Medicago polymorpha	II (2–5)		I (3)	I (2–5)
Ranunculus ficaria	II (2–3)	I (3)		I (2–3)
Veronica arvensis	III (1–4)	IV (2–5)	I (3)	III (1–5)
Sagina procumbens	II (2–3)	IV (2–3)	III (2–3)	III (2–3)
Bryum rubens	I (1–6)	IV (1–3)	III (2–5)	III (1–6)
Agrostis stolonifera	I (1–5)	IV (2–8)	III (2–7)	III (1–8)
Barbula convoluta		IV (2–3)	I (2)	III (2–3)
Valerianella locusta	I (2–5)	IV (2–5)	I (2–5)	III (2–5)
Urtica urens		IV (2–3)	I (2)	II (2–3)
Polycarpon tetraphyllum	I (5)	IV (2–5)	I (2)	II (2–5)
Capsella bursa-pastoris	I (2–3)	III (2)	II (2–3)	II (2–3)
Rumex obtusifolius	II (2–3)	III (2–5)	I (1)	II (1–5)
Riccia sorocarpa	I (2)	III (1–2)	II (2–3)	II (1–3)
Heracleum sphondylium	I (3)	III (2–5)	II (2)	II (2–5)
Dicranella staphylina	I (1–2)	III (2–3)	I (2)	II (1–3)
Pottia truncata	I (1)	III (1–2)	I (2)	II (1–2)
Phascum cuspidatum	I (2)	III (2–3)	I (2)	II (2–3)
Fumaria officinalis		III (2–7)		II (2–7)
Eurhynchium praelongum	I (2–3)	II (2)	I (3)	II (2–3)
Fumaria bastardii	I (3)	II (2–5)	I (2–3)	I (2–5)
Pleuridium subulatum	I (2–3)	II (1–2)	I (1–2)	I (1–3)
Veronica hederifolia	I (2)	II (2–3)	I (2)	I (2–3)
Chenopodium album	I (3)	II (2–3)		I (2–3)
Allium babingtonii	I (3)	II (2)		I (2–3)
Aphanes arvensis	I (2–3)	II (2)		I (2–3)
Ceratodon purpureus	I (1–3)	II (1–3)	I (2–3)	I (1–3)
Viola arvensis		II (2–5)	I (2)	I (2–5)
Arum italicum neglectum		II (2–3)		I (2–3)
Galium aparine	II (1–4)	III (2–3)	IV (2–3)	III (1–4)
Sonchus oleraceus	II (1–5)	III (1–2)	IV (2)	III (1–5)
Oxalis pes-caprae	II (1–8)	III (2–7)	V (2–10)	III (1–10)
Vicia hirsuta	I (3)		V (2–7)	II (2–7)
Papaver dubium	I (2)	I (3)	IV (2–3)	II (2–3)
Trifolium dubium	I (1–4)		IV (2–5)	II (1–5)
Elymus repens	I (5)		III (3–7)	II (3–7)
Vicia sativa nigra		II (2)	III (2–3)	II (2–3)
Erodium moschatum	I (2–3)	I (3)	II (2–3)	I (2–3)
Mercurialis annua		I (2)	II (1–2)	I (1–2)
Anthriscus caucalis			II (2)	I (2)
Fumaria occidentalis	I (7)		II (5–10)	I (5–10)
Convolvulus arvensis			II (2)	I (2)
Ranunculus marginatus trachycarpus			I (5)	I (5)
Poa trivialis	III (2–8)	III (2–5)	III (2–5)	III (2–8)
Chrysanthemum segetum	III (1–10)	III (2–3)	III (2–7)	III (1–10)
Briza minor	III (1–4)	II (2–3)	III (2–3)	III (1–4)

Floristic table OV6 (*cont.*)

	a	b	c	6
Ranunculus parviflorus	III (2–7)	III (2–3)	II (2–3)	III (2–7)
Gladiolus byzantinus	II (1–2)	III (2–5)	III (2–4)	III (1–5)
Coronopus didymus	III (2–5)	II (2–5)	II (2–3)	III (2–5)
Silene gallica	III (2–7)	II (2)	III (2–7)	III (2–7)
Veronica persica	II (1–5)	III (2–5)	II (3)	II (1–5)
Medicago arabica		III (2–8)	III (2–3)	II (2–8)
Montia perfoliata	II (2–5)	I (3)	II (3)	II (2–5)
Rumex acetosella	II (3–7)		II (2–5)	I (2–7)
Holcus lanatus	I (2–8)	I (3)	I (2–3)	I (2–8)
Allium triquetrum	I (2)	I (3)	I (2)	I (2–3)
Taraxacum officinale agg.	I (2)	I (2)	I (2)	I (2)
Bryum bicolor	I (1)	I (3)	I (2)	I (1–3)
Bromus diandrus	I (2)	I (1–2)	I (2–3)	I (1–3)
Sherardia arvensis	I (2)	I (2)	I (3)	I (2–3)
Lamium purpureum	I (2–3)	I (2)		I (2–3)
Oxalis corniculata	I (5)	I (2)		I (2–5)
Papaver rhoeas	I (2)	I (2)		I (2)
Dactylis glomerata	I (2–5)	I (2)		I (2–5)
Geranium dissectum	I (2–5)	I (2)		I (2–5)
Polygonum aviculare	I (1)	I (3)		I (1–3)
Phalaris minor	I (2–3)		I (2–5)	I (2–5)
Crepis capillaris	I (2)		I (2)	I (2)
Thlaspi arvensis		I (2)	I (2)	I (2)
Atriplex prostrata		I (2)	I (2)	I (2)
Allium roseum bulbiferum		I (2)	I (3)	I (2–3)
Ditrichum cylindricum		I (2)	I (2)	I (2)
Number of samples	24	27	11	62
Number of species/sample	28 (18–41)	29 (20–37)	29 (12–39)	29 (12–41)

a   *Aphanes microcarpa-Ranunculus muricatus* sub-community
b   *Valerianella locusta-Barbula convoluta* sub-community
c   *Vicia hirsuta-Papaver dubium* sub-community
6   *Cerastium glomeratum-Fumaria muralis* ssp. *boraei* community (total)

OV7

Veronica persica-Veronica polita community
Veronico-Lamietum hybridi Kruseman & Vlieger 1939

Synonymy
Setario-Veronicetum politae Oberdorfer 1957 *sensu* Silverside 1977; *Tripleurospermum maritimum* stands Kay 1994 *p.p.*

Constant species
Bilderdykia convolvulus, Chenopodium album, Matricaria perforata, Polygonum aviculare, Stellaria media, Veronica persica, Veronica polita.

Physiognomy
In the *Veronico-Lamietum*, both *Veronica persica* and the more geographically restricted *V. polita* are constant, the former often in some abundance by summer. *Polygonum aviculare* also frequently has high cover and *Stellaria media* and *Bilderdykia convolvulus* are constant contributors among the smaller herbs. Often more conspicuous, however, are the mealy shoots of *Chenopodium album* and the big daisy-like inflorescences of *Matricaria perforata*, especially where this shows a second flush of flowering in unploughed stubble remaining after harvest into the autumn.

Lamium hybridum is a particularly diagnostic associate of the above constants but it is not universally present in this community. More frequent are *Poa annua, Elymus repens, Senecio vulgaris, Capsella bursa-pastoris* and *Chamomilla suaveolens* with *Euphorbia helioscopia, Sonchus oleraceus, Solanum nigrum* and *Urtica urens* among the occasionals.

Habitat
The *Veronico-Lamietum* is characteristically a weed community of cereals and other annual field crops on lighter, well-drained, highly fertile circumneutral soils in the warmer and drier lowlands of southern and eastern Britain.

Suitable soil conditions for most of the common species of this assemblage are very widespread through the lowland agricultural landscape of this country but both *Veronica polita* and, more strikingly, *Lamium*

hybridum are more restricted to areas with a Continental climate (Perring & Walters 1962). Within this zone, it is the better-draining soils of the east Midlands, East Anglia and southern England, widely cultivated for cereals, roots and vegetable crops, that regularly provide the kind of disturbance and enrichment that favours the rapid establishment of this community.

Most of the cover of *M. perforata* establishes in spring, though, having frost-hardy seedlings (Kay 1994), this species may get a head start in the vegetation among autumn-sown crops. Being able to produce vigorous laterals from undamaged lower parts after upper shoots have been cut when cereals are harvested, it is also one of the more long-lasting species in this assemblage, sometimes flowering on into October where there is no early autumn cultivation. This is, in fact, one of the major assemblages for this troublesome weed (Kay 1994), a plant that has been shown in field trials to be more competitive in winter wheat than many of its usual companions (Wilson & Wright 1990). Efficient ploughing with complete sod reversal kills existing plants, but it is a prolific seeder with a long potential for dormancy (Kay 1994).

Like most of the common species here, *M. perforata* has no particular method of seed dispersal but long local survival in the soil, and dispersal in mud and through dung, is enough to ensure the reappearance of the assemblage from year to year. For *M. perforata* a requirement for diurnal temperature fluctuations, for *Chenopodium album* (Williams 1963) low temperatures, and for many of the other species the need for light and high nutrient content, are readily met in the disturbance provided by autumn or spring cultivation for arable crops. Occasionally, vegetable gardens, farmyards and disturbed waysides provide suitable habitat conditions for the *Veronicio-Lamietum* within its general geographical range.

It should be noted that one of the most frequent plants of this assemblage, *Veronica persica*, is not a native species. It originates from western Asia (Salisbury

1964) but spread rapidly through much of England, probably through contaminated seed and dung, in the decades after its first appearance in Berkshire in 1825 in the kind of situations increasingly provided by intensifying arable cultivation.

Zonation and succession

Typically, the community occurs patchily within or around the margins of arable fields sown with cereals, roots of other crops. Where banks or hedgerows remain, it can pass to weedy *Arrhenatheretum* or in gateways to some kind of Polygonion or Lolio-Plantaginion vegetation. Renewed cultivation effectively prevents any succession.

Distribution

The *Veronico-Lamietum* occurs widely on suitable soils through the southern and eastern parts of Britain and scattered on arable land around coasts elsewhere (Silverside 1977).

Affinities

Although *Lamium hybridum* is less characteristic of this kind of vegetation with us than in other parts of Europe, this community is essentially the same as the *Veronico-Lamietum* first described from The Netherlands by Kruseman & Vlieger (1939). Similar vegetation has been noted in Ireland under this association name (White & Doyle 1982), though Lambe (1971) considered *Veronica persica* assemblages of this general kind in to be part of what here is called the *Stellaria-Capsella* community. Silverside (1977) also characterised a syntaxon equivalent to the *Setario-Veronicetum politae* of Oberdorfer (1957) from a single locality in Breckland but this South Germany and Swiss association, with its distinct continental character, cannot really be separated among the available data from the *Veronico-Lamietum*. Silverside (1977) also placed both syntaxa in the Fumario-Euphorbion whereas Westhoff & den Held (1969) located the *Veronico-Lamietum* in the Polygono-Chenopodion.

Floristic table OV7

Veronica persica	IV (1–3)
Veronica polita	IV (1–3)
Polygonum aviculare	IV (1–8)
Bilderdykia convolvulus	IV (1–5)
Stellaria media	IV (1–6)
Matricaria perforata	IV (1–3)
Chenopodium album	IV (1–3)
Poa annua	III (1–3)
Elymus repens	III (1–3)
Senecio vulgaris	III (1–3)
Capsella bursa-pastoris	III (1–4)
Chamomilla suaveolens	III (1–3)
Euphorbia helioscopa	II (1–3)
Sonchus oleraceus	II (1–3)
Lamium hybridum	II (1–4)
Solanum nigrum	II (1–8)
Urtica urens	II (1–3)
Lamium purpureum	II (1–4)
Anagallis arvensis	II (1–3)
Viola arvensis	II (1–3)
Atriplex patula	II (1–3)
Avena fatua	II (1–3)
Papaver rhoeas	II (1–3)
Diplotaxis muralis	II (1–3)
Sonchus asper	II (1–3)
Plantago major	II (1–3)
Galium aparine	II (1–3)
Ranunculus repens	II (1–4)
Agrostis stolonifera	II (1–6)
Sisymbrium officinale	II (1–3)
Coronopus squamatus	I (1–5)
Euphorbia peplus	I (1–3)
Thlaspi arvense	I (1–3)
Lamium amplexicaule	I (1–4)
Polygonum lapathifolium	I (1–3)
Erucastrum gallicum	I (1)
Aethusa cynapium	I (1)
Atriplex prostrata	I (1)
Plantago lanceolata	I (1–3)
Cirsium arvense	I (1–3)
Conyza candensis	I (1)
Aphanes arvensis	I (1)
Number of samples	22
Number of species/sample	18 (10–34)
Vegetation cover (%)	19 (5–90)

OV8

Veronica persica-Alopecurus myosuroides community
Alopecuro-Chamomilletum recutitae Wascher 1941

Constant species

Alopecurus myosuroides, Bilderykia convolvolus, Elymus repens, Polygonum aviculare, Stellaria media, Veronica persica.

Physiognomy

The *Alopecuro-Chamomilletum* is the most distinctive weed assemblage in which the annual grass *Alopecurus myosuroides* is found. Its tufts can be very abundant here, along with *Elymus repens, Polygonum aviculare* and *Stellaria media*, each of which can also show locally high cover. *Veronica persica* and *Bilderdykia convolvulus* are also constant, though generally in smaller amounts.

The characteristic mayweed of this vegetation, *Chamomilla recutita*, is only occasional overall but *Matricaria perforata* is frequent, along with *Chenopodium album, Capsella bursa-pastoris, Sinapis arvensis, Sonchus asper, Senecio vulgaris, Avena fatua, Poa annua, Plantago major, Polygonum persicaria* and *Cirsium arvense*. Among the more common occasionals are *Convolvulus arvensis, Euphorbia helioscopa, Solanum nigrum, Urtica urens* and the scarce goosefoot *Chenopodium hybridum*.

Habitat

This assemblage is typical of winter-sown cereals and other field crops on loamy and clayey soils that occur in the warmer and drier south-east of England but which are protected from drought by their heavy texture.

A. myosuroides is native across much of Europe but is only common in Britain in the more Continental parts of the country where the July mean is above 15 °C (Naylor 1972), being especially abundant in the south-east Midlands and East Anglia. It can get a head start among arable crops in this region by autumn germination and survive through the winter as a rosette, flowering from May to August of the following year. Even on very wet or waterlogged soils, it does not seem to suffer in the coldest months and a high water requirement

when growth recommences may account for the scarcity of the plant on very sharply-draining soils (Barallis 1968).

Other species in the assemblage appear most prolifically in the spring but spring-sown cereals can suppress the growth of *A. myosuroides* itself. Closely-spaced crops can also overwhelm what is apparently a somewhat puny grass, whose problematic character probably depends on its ability to compete with cereal crops for nitrogen (Wellbank 1963, Naylor 1972). It is also notoriously resistant to many herbicides and this weed assemblage has shown a striking increase in infestation with the concentration on cereal monocultures in recent decades. Disturbance and compaction can check seedlings so cultivation after harvest or the hoeing of vegetable crops can help eliminate the plant but the seeds can remain dormant for some years and light stimulates germination, so tilling can expose seed and encourage establishment of a new crop of the grass.

Areas of waste ground on suitable soils within the range of *A. myosuroides* can also carry this community.

Zonation and succession

Characteristically, the *Alopecuro-Chamomilletum* occurs within or around the edges of arable fields, sometimes with other weed assemblages typical of heavier soils like the *Polygonum-Ranunculus* sub-community of the *Poa-Senecio* community where *Anthemis cotula* is a distinctive associate. Repeated disturbance from year to year effectively prevents any succession.

Distribution

The community occurs widely on suitable soils within the range of *A. myosuroides*, concentrated around the Fenland fringes and scattered west to Dorset.

Affinities

The *Alopecuro-Matricarietum* was first described from The Netherlands by Wasscher (1941), though it has been

subsumed by various authors since, including Westhoff
& den Held (1969), in the *Veronico-Lamietum*. If that
course were adopted with British vegetation of this kind,
it would constitute a distinctive sub-community with
preferentially frequent *A. myosuroides*, like the *alopecu-
retosum* of Westhoff & den Held (1969). These authors
placed this syntaxon in the Polygono-Chenopodion
while Silverside (1977) located it among the more distinctly basiphilous associations of the Fumario-
Euphorbion alliance.

Floristic table OV8

Veronica persica	V (1–2)
Polygonum aviculare	V (1–9)
Elymus repens	V (1–6)
Alopecurus myosuroides	IV (1–8)
Stellaria media	IV (1–6)
Bilderdykia convolvulus	IV (1–6)
Chenopodium album	III (1–3)
Capsella bursa-pastoris	III (1–5)
Sonchus asper	III (1–3)
Senecio vulgaris	III (1–3)
Sinapis arvensis	III (1–3)
Avena fatua	III (1–3)
Plantago major	III (1–3)
Polygonum persicaria	III (1–3)
Matricaria perforata	III (1–6)
Poa annua	III (1–4)
Cirsium arvense	III (1–4)
Convolvulus arvensis	II (1–6)
Euphorbia helioscopa	II (1–3)
Chenopodium hybridum	II (1–3)
Solanum nigrum	II (1–5)
Urtica urens	II (1–3)
Sisymbrium officinale	II (1–3)
Galium aparine	II (1–3)
Chamomilla recutita	II (1–8)
Geranium dissectum	II (1)
Coronopus squamatus	I (1–3)
Rumex crispus	I (1–3)
Trifolium repens	I (1–5)
Chamomilla suaveolens	I (1–3)
Sonchus oleraceus	I (1–3)
Euphorbia peplus	I (1–3)
Lamium purpureum	I (1–2)
Papaver rhoeas	I (1–2)
Sonchus arvensis	I (1–2)
Aethusa cynapium	I (1–3)
Atriplex patula	I (1–3)
Atriplex prostrata	I (1)
Lapsana communis	I (1–3)
Anthemis cotula	I (1–2)
Urtica dioica	I (1)
Number of samples	13
Number of species/sample	16 (10–28)
Vegetation cover (%)	74 (50–100)

OV9
Matricaria perforata-Stellaria media community

Synonymy
Tripleurospermum inodorum stands   Kay 1994 *p.p.*

Constant species
Matricaria perforata, Polygonum aviculare, Stellaria media.

Physiognomy
The *Matricaria perforata-Stellaria media* community comprises annual vegetation dominated by mixtures of *Stellaria media, Polygonum aviculare* and *Matricaria perforata*. Also quite frequent overall but variously represented in the different sub-communities are *Poa annua, Elymus repens, Bilderdykia convolvulus* and *Sonchus asper*. Occasionals of the community include *Cirsium arvense, Atriplex patula, Raphanus raphanistrum* and *Urtica dioica*. Acrocapous mosses are sometimes conspicuous on bare ground, *Bryum rubens, Barbula unguiculata* and *Dicranella staphylina* being recorded most commonly.

Sub-communities

***Anagallis arvensis-Viola arvensis* sub-community.** The vegetation here tends to be quite species-poor but enlivened by the pretty flowers of *Viola arvensis* and *Anagallis arvensis*.

***Poa annua-Galeopsis tetrahit* sub-community.** *P. annua* shows its peak of frequency here with *Galeopsis tetrahit, Cerastium fontanum* and *Elymus repens* also frequent, *Aphanes arvensis* and *Juncus bufonius* occasional.

***Elymus repens-Potentilla anserina* sub-community.** *E. repens* and *Poa annua* both remain frequent in this sub-community but *Agrostis stolonifera* also becomes common, along with *Potentilla anserina* and *Equisetum arvense*. Also more frequent, though showing their peak of occurrence in the next sub-community, are *Bilderdykia convolvolus, Chenopodium album, Capsella bursa-*

pastoris, Chamomilla suaveolens, Trifolium repens and *Veronica persica. Lamium purpureum, Bromus sterilis, Tussilago farfara* and *Ranunculus arvensis* are more distinctive occasionals.

***Bilderdykia convolvulus-Veronica persica* sub-community:** *Lycopsietum arvensis*   (Raabe 1944) Passarge 1964 *sensu* Silverside 1977; *Alchemillo-Matricarietum chamomillae* R.Tx. 1937 *emend.* Passarge 1957 *sensu* Silverside 1977. The vegetation here has much in common with the last sub-communiity but *B. convolvulus* and *V. persica* become especially common and there is also frequent *V. arvensis, Rumex obtusifolius, Plantago major, Senecio vulgaris, Solanum nigrum* and *Taraxacum officinale* agg. *Matricaria recutita* and *Anchusa arvensis* are distinctive preferentials that are quite common in this sub-community.

Habitat
The *Matricaria-Stellaria* community is found widely throughout the British lowlands on disturbed and nutrient-rich, well-drained, circumneutral soils in arable fields, market gardens, farmyards and waysides.

This is the most widely distributed type of weed vegetation in which *M. perforata* is prominent, showing no climatically-related confinement to the warmer and drier parts of the country. What is limiting to its development throughout the lowlands is the texture of the soil because *M. perforata* (Kay 1994) and another important species here, *Stellaria media* (Sobey 1981), are both intolerant of poorly-drained profiles. These annuals can germinate in both autumn and spring, so early establishment of this community in autumn-sown cereals and other crops is common.

In fact, *S. media* can germinate throughout the growing season and seems especially well-adapted to take advantage of congenial conditions for its establishment here. Its seeds have been shown to be especially widespread in arable fields (Roberts & Stokes 1966) and display highly variable germination requirements.

Reserves of seed with uniform needs for temperature and light conditions can germinate prolifically more or less simultaneously while more variable elements of the seed bank are able to respond sporadically to particular different situations (van der Vegte 1978). Both *S. media* and *M. perforata* are also especially responsive to nitrogen additions, so the heavy fertilising associated with modern arable cultivation is very encouraging of luxuriant growth here (Sobey 1981, Kay 1994).

Neither of these species, nor the other constant, *P. aviculare*, which usually germinates in the spring, has any special dispersal mechanism but seeding is prolific and re-establishment very ready. As in other communities where it is well represented, *M. perforata* can show a second flush of flowering here in autumn where stubble is left unploughed because the bases of plants remaining after cereals have been cut can produce vigorous new shoots. Elements of this community can therefore last a full year.

The particular habitat relationships of the various sub-communities are unclear but the *Bilderdykia-Veronica* type only gains its full complement of species in spring-sown crops or disturbed places on sandier or light chalky soils where there is space for establishment at the start of the growing season. The *Elymus-Potentilla* sub-community extends the occurrence of the *Matricaria-Stellaria* vegetation to damper soils in gateways to arable fields, farm tracks and waysides.

Zonation and succession

In arable fields and market gardens, the *Matricaria-Stellaria* community can occur patchily among the crops or around the field margins, alone or in patchworks with other weed communities characteristic of lighter, fertile soils. Along tracks through fields and in gateways, the community often gives way to the *Polygonum-Chamomilla* assemblage which is more tolerant of trampling.

Stands of the *Matricaria-Stellaria* community on disturbed waysides can have a fringe of the *Poa-Matricaria* vegetation, occur with other eutrophic tall-herb communities like the *Urtica-Cirsium* assemblage or give way to closed Lolio-Plantaginion swards or the *Arrhenatheretum* on verges.

Regular disturbance in arable fields prevents any successional developments but where the community occurs on abandoned ground, it can progress to Lolio-Plantaginion grassland and Rubion sub-scrub.

Distribution

The community occurs widely throughout the lowlands of Britain wherever suitable soils and disturbance occur. The *Bilderdykia-Veronica* sub-community is more common in eastern England.

Affinities

This community corresponds closely to some of the arable weed vegetation tabulated in Kay's (1994) study of *Matricaria perforata*, termed there *Tripleurospermum inodorum*. It includes the *Lycopsietum arvensis* (Raabe 1944) Passarge 1964 as characterised by Silverside (1977), a syntaxon which, as described by Oberdorfer (1983) and Pott (1992), has more ephemerals of acid sands than is usual in the community described here. In this scheme, too, the *Matricaria-Stellaria* vegetation also includes much of what Silverside (1977) grouped in the *Alchemillo-Matricarietum* R.Tx. 1937 *emend.* Passarge 1957, an association recognised by Oberdorfer (1983) as including widely occurring weed vegetation in central and eastern Germany (see also Heinrich & Weber 1979) but which Westhoff & den Held (1969) subsumed in the *Papaveretum argemones* Libbert 1933 *emend.* Kruseman & Vlieger 1939. In this somewhat broader conception, the *Matricaria-Stellaria* community is best accommodated in the Polygono-Chenopodion alliance.

Floristic table OV9

	a	b	c	d	9
Polygonum aviculare	V (2–4)	V (1–5)	IV (1–10)	V (1–5)	V (1–10)
Stellaria media	IV (2–5)	III (1–5)	IV (1–5)	IV (1–7)	IV (1–7)
Matricaria perforata	IV (2–4)	IV (1–7)	III (1–7)	IV (2–5)	IV (1–7)
Anagallis arvensis	IV (2–5)	III (1–3)	I (1)	II (2–3)	II (1–5)
Viola arvensis	IV (1–3)	II (1–3)	II (1)	III (1–3)	II (1–3)
Hordeum vulgare	I (2–3)				I (2–3)
Epilobium angustifolium	I (2–3)				I (2–3)
Symphytum officinale	I (2–3)				I (2–3)
Poa annua	II (1–3)	V (2–8)	III (1–3)	IV (1–3)	III (1–8)
Galeopsis tetrahit		III (1–5)			I (1–5)
Cerastium fontanum		III (1–2)		I (1–2)	I (1–2)
Aphanes arvensis	I (2–4)	II (1–5)			I (1–5)
Juncus bufonius		II (1–3)			I (1–3)
Elymus repens		III (1–5)	IV (1–7)	IV (1–8)	III (1–8)
Agrostis stolonifera		II (2–3)	IV (2–5)	IV (2–10)	III (2–10)
Potentilla anserina		II (2–3)	III (1–4)		II (1–4)
Equisetum arvense	I (3)	I (1)	III (1–3)		I (1–3)
Lamium purpureum			II (3)	I (3)	I (3)
Bromus sterilis			II (3–4)		I (3–4)
Tussilago farfara			II (1–3)		I (1–3)
Ranunculus arvensis			II (1–3)		I (1–3)
Bilderdykia convolvulus		II (1–2)	III (1)	V (2–5)	III (1–5)
Chenopodium album	I (8)	I (1)	III (1–3)	IV (1–5)	III (1–8)
Veronica persica	II (2–8)	II (1–3)	II (1–5)	IV (1–5)	III (1–8)
Capsella bursa-pastoris	II (2–5)		III (1–3)	IV (1–5)	II (1–5)
Chamomilla suaveolens	I (2–4)	II (2–5)	III (1–3)	IV (1–5)	II (1–5)
Trifolium repens			III (2–5)	IV (1–5)	II (1–5)
Plantago major	I (2)	II (1–3)	I (3)	III (1–7)	II (1–7)
Veronica arvensis	I (2)	II (1–2)	I (3)	III (1–4)	II (1–4)
Rumex obtusifolius		II (1–3)	I (1)	III (1–3)	II (1–3)

Floristic table OV9 (*cont.*)

	a	b	c	d	9
Senecio vulgaris	I (3)	I (1–2)	II (1)	III (1–3)	I (1–3)
Taraxacum officinale agg.		I (1)		III (1–3)	I (1–3)
Solanum nigrum				III (1–5)	I (1–5)
Crepis capillaris		I (1–2)		III (1–3)	I (1–3)
Anchusa arvensis	I (3)			III (1–4)	I (1–4)
Holcus lanatus		I (2)		II (2–5)	I (2–5)
Chamomilla recutita				II (1–3)	I (1–3)
Sisymbrium officinale				II (1–3)	I (1–3)
Rumex acetosella				I (1–2)	I (1–2)
Mercurialis annua				I (1–5)	I (1–5)
Papaver dubium				I (2–3)	I (2–3)
Misopates orontium				I (1–2)	I (1–2)
Sonchus asper	II (3)	II (1–2)	III (1–2)	III (1–5)	III (1–5)
Cirsium arvense	I (2)	III (1–3)		III (1–3)	II (1–3)
Atriplex patula	I (5)	I (1)	III (1–7)	III (1–2)	II (1–7)
Raphanus raphanistrum	II (1–2)		II (1)	II (1)	II (1–2)
Urtica dioica		I (1)	II (1–2)	II (1–3)	II (1–3)
Bryum rubens		II (1–3)	I (1)	II (2–5)	II (1–5)
Fumaria officinalis	I (3–5)	I (3)	I (3)	I (8)	I (3–8)
Rumex crispus		I (1–3)	I (1)	I (1)	I (1–3)
Agrostis capillaris	I (3)	I (3)	I (4)		I (3–4)
Poa trivialis		I (2)	I (1)	I (3)	I (1–3)
Artemisia vulgaris	I (2)	I (2)	I (3)	I (1–3)	I (1–3)
Sonchus arvensis		I (1)	I (1)	I (1)	I (1)
Barbula unguiculata		I (1–2)	I (1)	I (1–2)	I (1–2)
Dicranella staphylina		I (3–5)	I (2)	I (1–5)	I (1–5)
Lolium perenne		I (2)	I (3)	I (1–3)	I (1–3)
Papaver rhoeas	I (2)		I (3)	I (2)	I (2–3)
Filaginella uliginosa	I (1)	I (2–3)	I (1)	I (1)	I (1–3)
Myosotis arvensis	I (2)	I (1–2)	I (1)	II (1)	I (1–2)
Lamium hybridum	I (1)			I (3)	I (1–3)
Polygonum arenastrum	I (2)			I (2–5)	I (2–5)

Species	a	b	c	d	9
Polygonum persicaria		I (1–5)		I (1)	I (1–5)
Atriplex prostrata	I (3)			I (2)	I (2–3)
Sherardia arvensis	I (3)			I (2)	I (2–3)
Chrysanthemum segetum	I (5)			I (7)	I (5–7)
Holcus mollis		I (2)	I (3)		I (2–3)
Arrhenatherum elatius			I (1)	I (1–2)	I (1–2)
Plantago lanceolata	I (2)			I (1–3)	I (1–3)
Geranium molle	I (3)			I (2–3)	I (2–3)
Epilobium montanum		I (1)		I (1)	I (1)
Lapsana communis		I (1–5)		I (1)	I (1–5)
Dactylis glomerata		I (2)		I (2)	I (2)
Trifolium pratense		I (2)	I (1)	I (2–3)	I (2–3)
Eurhynchium praelongum		I (2)		I (1–3)	I (1–3)
Pottia intermedia		I (2)		I (1–2)	I (1–2)
Riccia sorocarpa		I (1)		I (3)	I (1–3)
Phascum cuspidatum		I (1)		I (1–2)	I (1–2)
Number of samples	11	12	9	25	57
Number of species/sample	10 (6–15)	18 (6–32)	14 (8–18)	26 (9–51)	19 (6–51)

a   *Anagallis arvensis-Viola arvensis* sub-community
b   *Poa annua-Galeopsis tetrahit* sub-community
c   *Elymus repens-Potentilla anserina* sub-community
d   *Bilderdykia convolvulus-Veronica persica* sub-community
9   *Matricaria perforata-Stellaria media* community (total)

OV10
Poa annua-Senecio vulgaris community

Constant species
Poa annua, Senecio vulgaris.

Physiognomy
The *Poa annua-Senecio vulgaris* community brings together a variety of weed assemblages which are distinctive in their combinations of common species rather than by the presence of striking differentials. *Poa annua* and *Senecio vulgaris* are the only constants throughout but *Lolium perenne* is very frequent and *Capsella bursa-pastoris* and *Cerastium fontanum* occur commonly in various sub-communities. Occasionals include *Cirsium arvense, Plantago major, Poa trivialis, Veronica persica* and *Urtica dioica*. The assemblages vary in their total cover and, though most of the more frequent species here are emphemeral, a perennial grassy element can be seen establishing in some sub-communities.

Sub-communities

Polygonum persicaria-Ranunculus repens **sub-community.** The abundance of knotweeds and spurges is often the most striking feature here, with *Polygonum aviculare* and, more strongly preferential, *P. persicaria* both constant, *Euphorbia helioscopa* and *E. exigua* frequent. *Stellaria media* is also more common in this sub-community than elsewhere with *Ranunculus repens, Anagallis arvensis, Viola arvensis, Bilderdykia convolvulus, Anthemis cotula, Trifolium repens* and *Lapsana communis*. Preferential occasionals include *Sinapis arvensis, Raphanus raphanistrum, Spergularia arvensis* and *Fumaria officinalis*.

Polygonum aviculare-Matricaria perforata **sub-community.** *P. aviculare* remains constant here with *P. persicaria* and *P. arenastrum* occasional, but more striking is the high frequency and often abundance of mayweeds: *Chamomilla suaveolens,C. recutita* and *M. perforata* are all common in this sub-community. *Chenopodium album* and *Coronopus squarratus* are also weakly preferential.

Agrostis stolonifera-Rumex crispus **sub-community.** *Agrostis stolonifera* and *Holcus lanatus* are both frequent here in a patchy grassy cover with knotweeds and mayweeds figuring occasionally. More obvious, though, are docks and thistles, with *Rumex crispus* and *R. obtusifolius, Cirsium vulgare* and *C. arvense* all common. *Senecio squalidus* often accompanies *S. vulgaris* and *Taraxacum officinale* agg. is frequent. *Epilobium angustifolium* occurs occasionally, though not in abundance, along with *Tussilago farfara, Poa pratensis* and *Sonchus oleraceus. Bryum argenteum* is sometimes seen on bare earth.

Dactylis glomerata-Agrostis capillaris **sub-community.** *Dactylis glomerata* and *Agrostis capillaris* are both frequent in this sub-community, sometimes at quite high cover, giving a grassier look to the vegetation. *Plantago lanceolata, P. media, Achillea millefolium, Medicago lupulina, Vicia sativa, Bromus sterilis, Rumex acetosa* and *Erodium cicutarium* are all quite common.

Habitat
The *Poa-Senecio* community is characteristically a pioneer weed assemblage of open cultivated or trampled ground, especially where fertile soils have become moist. It is ubiquitous through the British lowlands, being particularly frequent in arable land, gardens, ill-sown and badly-poached leys and recreational grasslands, waysides, gateways and freshly-dumped earth on building sites and roadworks.

This is one of a range of weed communities in which the success of *Poa annua* as a colonist of open, moist soils is very evident (Hutchinson & Seymour 1982). It is an extremely widespread species, successful on all but very acid, basic, impoverished or saline soils but it performs especially well on moister loams and clays where, through cultivation, trampling or the delayed establishment of perennials, the ground remains open. It shows peaks of germination in spring and autumn (Law 1981), light, alternating temperatures and high nitrate (Roberts

& Benjamin 1979) all enhancing germination conditions likely to prevail in spring- or autumn-sown cereals or other field crops under intensive arable cultivation. However, seeds will also germinate in the dark (Thompson *et al.* 1977) and among established swards where open ground appears. Though establishment is best in loose soil (Roberts & Stokes 1965), germination can occur at low oxygen concentration (Müllverstadt 1963) and the plant is noticeably tolerant of soil compaction, temporary flooding and waterlogging. All these, in fact, can help create opportunities for this assemblage to establish by hindering the development of perennials or destroying them where they occur in existing swards. Plants can survive warm summer conditions and *P. annua* often capitalises on droughty periods by appearing in gaps created in grasslands after the soil has been subsequently wetted.

Like *P. annua*, the seed of *Senecio vulgaris*, the other characteristic constant of the community, is able to germinate very quickly and other common species of this assemblage are all ready colonisers of open, fertile ground. The high frequency of *Lolium perenne* reflects the widespread occurrence of this plant along disturbed waysides and in gateways but also the common appearance of the *Poa-Senecio* assemblage within badly-managed pastures and leys.

Of the various sub-communities, the *Polygonum-Ranunculus* type is especially characteristic of arable and garden crops, poorly-sown leys and disturbed ground on heavier clay and clay-loam soils in the warmer and drier south-east of Britain. *Anthemis cotula*, a distinctive preferential of this kind of *Poa-Senecio* vegetation, is a plant with a more or less Continental distribution in Britain (Perring & Walters 1962, Kay 1971), most abundant where the July mean is above 15.6 °C and annual precipitation less than 880 mm or where, as in south-west England, high summer temperatures offset the impact of higher rainfall. It flowers from mid-June onwards but can show a second flush among unploughed stubble because the bases of shoots cut during harvest are able to produce vigorous new growth (Kay 1971).

Among the other sub-communities, all of which are more widespread in their occurrence, the *Polygonum-Matricaria* type is characteristic of lighter sands and loams, the *Dactylis-Agrostis* type of disturbed, somewhat improved pastures and waysides on slightly more acidic soils and the *Agrostis-Rumex* type of poorly-

managed leys, pastures and recreational swards on neutral loams.

Zonation and succession

In arable fields, the *Poa-Senecio* community can occur patchily within or around the crop, alone or with other Polygono-Chenopodion assemblages typical of cereals, roots or vegetables, the *Polygonum-Matricaria* or, mostly in the south-east, the *Polygonum-Ranunculus* sub-community being the typical forms here.

In weedy leys or pastures, the *Agrostis-Rumex* sub-community often occurs among some Lolio-Plantaginion sward like the *Lolio-Plantaginetum*, sometimes with patches of *Urtica-Cirsium* vegetation. In very badly poached areas, these may give way to some sort of Bidention assemblage like the *Polygonium-Poa* community. Around drier gateways, there is often a sequence of *Polygonum-Chamomilla* and *Poa-Plantago* assemblages. On slighter more acidic and less eutrophic soils, the *Dactylis-Agrostis* sub-community replaced the *Agrostis-Rumex* type. This sort of *Poa-Senecio* vegetation can also be seen with *Lolium-Dactylis* grassland on disturbed waysides and verges.

Both *P. annua* (Hutchinson & Seymour 1982) and *S. vulgaris* (Salisbury 1964) are able to complete their life cycle very quickly and, where conditions do not remain congenial, this community can have but a fleeting existence. In arable fields or seasonally-poached leys, it may return year after year but, where swards close, it is typically replaced by some form of Lolio-Plantaginion vegetation like the *Lolium-Dactylis* community. This in turn may pass to the *Arrhenatheretum*.

Distribution

The *Poa-Senecio* community occurs throughout Britain, except for the *Polygonum-Ranunculus* sub-community where is more confined to the south and east of the country.

Affinities

This kind of impoverished weedy vegetation has attracted little attention and it is very difficult to define using the sorts of character species developed elsewhere in Europe. It should be seen as an extremely generalised assemblage transitional in floristics and habitat between the Polygono-Chenopodion and the Lolio-Plantaginion.

Floristic table OV10

	a	b	c	d	10
Poa annua	V (1–4)	V (2–5)	V (1–5)	V (2–4)	V (1–5)
Senecio vulgaris	V (1–3)	V (1–4)	V (1–6)	V (1–5)	V (1–6)
Polygonum aviculare	V (1–4)	IV (2–7)	II (3–4)	I (4)	II (1–7)
Stellaria media	V (1–4)	III (3–4)	III (1–5)	I (2)	II (1–5)
Ranunculus repens	V (1–4)		II (1–4)	II (1–2)	I (1–4)
Sonchus asper	V (1–4)	I (3)	II (1–5)	I (2)	I (1–5)
Anagallis arvensis	V (1–3)	I (1)	I (3)	I (3)	I (1–3)
Viola arvensis	V (1–4)	I (1–2)	I (4)	I (3)	I (1–4)
Polygonum persicaria	IV (2–3)	II (3–7)	I (1)		I (1–7)
Bilderdykia convolvulus	IV (1–3)	II (4–5)			I (1–5)
Anthemis cotula	IV (3–4)	I (3)		I (2–3)	I (2–4)
Trifolium repens	III (1–7)	I (1–2)	II (3–7)	I (2–3)	I (1–7)
Euphorbia exigua	III (1–3)				I (1–3)
Euphorbia helioscopa	III (1–4)				I (1–4)
Lapsana communis	III (1–3)				I (1–3)
Sinapis arvensis	II (1–2)	I (2)	I (1)		I (1–2)
Raphanus raphanistrum	II (1–2)	I (2)			I (1–2)
Spergularia arvensis	II (3–8)				I (3–8)
Fumaria officinalis	II (1–3)				I (1–3)
Chamomilla suaveolens	I (2)	III (3–6)	III (3–4)	II (2–4)	II (2–6)
Chenopodium album	II (1–3)	III (2–4)	II (1–3)	I (1–4)	I (1–4)
Matricaria perforata	I (1–4)	III (2–7)	II (3–5)	I (1–2)	I (1–7)
Chamomilla recutita		II (1–3)	I (1–3)		I (1–3)
Polygonum arenastrum		II (3–4)	I (3)		I (3–4)
Coronopus squamatus		II (2)			I (2)
Agrostis stolonifera	II (1–5)		IV (1–4)	I (3–4)	II (1–5)
Rumex crispus	II (2–4)	I (1)	III (1–3)	I (2)	I (1–4)
Holcus lanatus			III (1–5)	II (2–5)	I (1–5)
Senecio squalidus			III (3–5)		I (3–5)
Taraxacum officinale agg.			III (1–5)		I (1–5)
Rumex obtusifolius		I (2–4)	II (3–5)	I (1)	I (1–5)

Cirsium vulgare		I (1)	I (3–4)	I (1–4)
Epilobium angustifolium		I (3)		I (2–3)
Tussilago farfara	I (1)			I (1–5)
Poa pratensis			I (1–3)	I (1–3)
Bryum argenteum			I (2)	I (1–3)
Sonchus oleraceus				I (1–3)
Dactylis glomerata	I (3)	I (1–3)	III (2–6)	II (1–6)
Plantago lanceolata		I (1)	III (2–4)	II (1–4)
Agrostis capillaris			III (3–7)	I (3–7)
Achillea millefolium	I (2)		II (1–4)	I (1–4)
Medicago lupulina	I (1)		II (2–4)	I (1–4)
Vicia sativa	I (1)	I (3)	II (2–3)	I (1–3)
Bromus sterilis		I (1)	II (2–5)	I (1–5)
Rumex acetosa		I (1)	II (1)	I (1)
Plantago media			II (2–3)	I (2–3)
Erodium cicutarium			II (3–7)	I (3–7)
Alliaria petiolata			I (2–6)	I (2–6)
Festuca ovina			I (2–3)	I (2–3)
Erophila verna			I (1–3)	I (1–3)
Lolium perenne	IV (1–6)	II (2–6)	III (1–8)	III (1–8)
Capsella bursa-pastoris		III (2–3)	II (2–3)	III (1–5)
Cerastium fontanum	I (4)		III (2–3)	III (1–4)
Cirsium arvense	II (1–4)	I (1–4)	III (1–6)	II (1–6)
Plantago major	III (1–2)	II (2)		II (1–5)
Poa trivialis	II (2–4)	I (4)	I (1–3)	II (1–5)
Veronica persica	II (3–4)	II (3–4)		II (1–4)
Urtica dioica		II (1–5)	II (3)	II (1–5)
Myosotis arvensis	II (1)			I (1–3)
Lamium purpureum		II (1–2)	I (2)	I (1–3)
Elymus repens	I (1–2)	I (1)	I (6)	I (1–6)
Galium aparine	I (1)	I (1)	I (4–5)	I (1–5)
Heracleum sphondylium	I (1)	I (2)	I (1)	I (1–2)
Trifolium pratense	I (1–2)	I (1)	I (1)	I (1–3)
Urtica urens	I (2)	I (8)	I (2–3)	I (1–8)
Senecio jacobaea	I (1)		I (2–3)	I (1–3)

Floristic table OV10 (*cont.*)

	a	b	c	d	10
Veronica polita	I (1)	I (2–3)		I (3)	I (1–3)
Aphanes arvensis	I (3)		I (1)	I (4)	I (1–4)
Veronica arvensis	I (1)		I (2)	I (2)	I (1–2)
Leucanthemum vulgare	I (2)	I (3–6)	I (1)		I (1–6)
Lamium hybridum	I (1)		I (1)		I (1)
Arctium minus agg.		I (1)		I (1)	I (1)
Veronica chamaedrys		I (3)	I (2)		I (2–3)
Atriplex prostrata		I (1)	I (3)		I (1–3)
Number of samples	7	10	18	12	47
Number of species/sample	29 (19–40)	12 (8–24)	20 (6–35)	19 (10–30)	22 (6–40)

a   *Polygonum persicaria-Ranunculus repens* sub-community
b   *Polygonum aviculare-Matricaria perforata* sub-community
c   *Agrostis stolonifera-Rumex crispus* sub-community
d   *Dactylis glomerata-Agrostis capillaris* sub-community
10 *Poa annua-Senecio vulgaris* community (total)

OV11

Poa annua-Stachys arvensis community

Synonymy
Stachys arvensis community   Silverside 1977.

Constant species
Anagallis arvensis, Poa annua, Polygonum aviculare, Stachys arvensis.

Rare species
Fumaria bastardii.

Physiognomy
The constancy of *Stachys arvensis* is the most striking feature of the *Poa annua-Stachys* community, along with *P. annua, Polygonum aviculare* and *Anagallis arvensis.* Also very common throughout the assemblage are *Spergula arvensis, Stellaria media, Plantago major, Matricaria perforata, Chamomilla suaveolens, Ranunculus repens, Elymus repens* and *Agrostis stolonifera.* Occasionals include *Capsella bursa-pastoris, Bilderdykia convolvulus, Fumaria muralis* ssp. *boraei, Senecio vulgaris, Taraxacum officinale* agg., *Trifolium repens, Holcus lanatus* and *Lolium perenne.* The total cover of vascular plants is usually high and some stands have a distinctly grassy appearance. In one sub-community, acrocarpous mosses can be varied and quite abundant.

Sub-communities

Chenopodium album-Euphorbia helioscopa sub-community. *Chamomilla suaveolens, Sonchus asper* and *Veronica persica* show somewhat higher frequency than usual here but more striking preferentials are *Chenopodium album* and *Euphorbia helioscopa.* Knotweeds are quite often prominent with *Polygonum lapathifolium, P. nodosum* and *P. persicaria* occasionally joining *P. aviculare,* and *Viola arvensis, Atriplex patula, Sinapis arvensis, Sonchus arvensis, Sherardia arvensis* and *Agrostis capillaris* are all preferential at low frequency. Among nationally-scarce plants, this sub-community occasionally provides a locus for *Kickxia elatine* and *Misopates orontium.*

Cerastium fontanum-Bryum rubens sub-community. *Polygonum persicaria* increases in frequency in his sub-community but more exclusive preferentials are *Cerastium fontanum* and a variety of acrocarpous mosses of which *Bryum rubens, Pottia truncata* and *Dicranella staphylina* are the most common with *B. klinggraeffii* and *B. violaceum* less frequent. *Brachythecium rutabulum* and *Eurhynchium praelongum* also occasionally form sparse wefts. Other vascular associates here are *Poa trivialis, Lamium purpureum, Trifolium dubium, Leontodon autumnalis* and *Rumex crispus.*

Habitat
The *Poa-Stachys* community is mostly associated with cereal crops on less limey loam and clay-loam soils in the western parts of England and Wales.

S. arvensis is a plant with a somewhat western distribution in Britain (Perring & Walters 1962) and characteristic of soils that are not so dry and acidic as those favoured by the *Spergulo-Chrysanthemetum.* This assemblage occurs typically on loamy or clayey soils such as those derived from the Old Red Sandstone or boulder clay. It is found largely west of a line from Dorset to Cheshire, being commonest in Pembrokeshire and Anglesey (Silverside 1977). It has been encountered mostly in oats and barley, occasionally in vegetable crops. The *Cerastium-Bryum* sub-community is characteristic of damper ground that has been undisturbed for some time, as among stubble that has not been burned or ploughed in.

Zonation and succession
On sandier and more acidic soils, the *Poa-Stachys* community tends to be replaced by the *Spergulo-Chrysanthemetum* and, on more fertilised areas within crops, by the *Stellaria-Capsella* or *Matricaria-Stellaria* communities. Continuing cultivation for growing arable crops creates suitable conditions for re-establishment of the community each year and sets back any tendency for succession.

Distribution

The community is largely confined to south-west England and Wales.

Affinities

This assemblage was first characterised by Silverside (1977) as distinct from the *Spergulo-Chrysanthemetum* on the basis of a shift in the balance of constants: *Spergula arvensis* and especially *C. segetum* are less common here, *Stachys arvensis* much more frequent. He recognised analogus trends in the Dutch data of Westhoff & den Held (1969) and among the communities described by Oberdorfer (1957, 1983) who separated off a *Setario-Stachyetum* from the *Lycopsietum*.

Floristic table OV11

	a	b	11
Stachys arvensis	V (1–6)	V (1–6)	V (1–6)
Poa annua	V (1–6)	V (1–8)	V (1–8)
Anagallis arvensis	IV (1–8)	IV (1–4)	IV (1–8)
Polygonum aviculare	IV (1–8)	IV (1–4)	IV (1–8)
Chamomilla suaveolens	IV (1–6)	III (1–4)	III (1–6)
Sonchus asper	IV (1–4)	III (1–6)	III (1–6)
Chenopodium album	IV (1–8)	II (1–3)	III (1–8)
Veronica persica	III (1–8)	II (1–3)	II (1–8)
Euphorbia helioscopa	III (1–3)	I (1–4)	II (1–4)
Agrostis capillaris	II (1–4)	I (1–3)	I (1–4)
Kickxia elatine	II (1–4)	I (1)	I (1–4)
Polygonum lapathifolium	II (1–6)		I (1–6)
Polygonum nodosum	II (1–4)		I (1–4)
Viola arvensis	II (1–3)		I (1–3)
Atriplex patula	II (1–3)		I (1–3)
Sinapis arvensis	II (1–6)		I (1–6)
Sonchus arvensis	II (1–3)		I (1–3)
Sherardia arvensis	II (1–4)		I (1–4)
Misopates orontium	II (1–3)		I (1–3)
Euphorbia exigua	I (1–3)		I (1–3)
Polygonum persicaria	II (1–8)	IV (1–4)	III (1–8)
Cerastium fontanum		III (1–4)	II (1–4)
Bryum rubens		III (1–3)	II (1–3)
Pottia truncata		III (1–4)	I (1–4)
Dicranella staphylina		III (1–3)	I (1–3)
Poa trivialis	I (1)	II (1–4)	I (1–4)
Lamium pupureum	I (1–3)	II (1–3)	I (1–3)
Brachythecium rutabulum		II (1–3)	I (1–3)
Trifolium dubium		II (1–3)	I (1–3)
Leontodon autumnalis		II (1–3)	I (1–3)
Rumex crispus		II (1–3)	I (1–3)
Eurhynchium praelongum		I (1–3)	I (1–3)
Pleuridium subulatum		I (1–3)	I (1–3)
Aphanes arvensis		I (1–3)	I (1–3)
Bryum klinggraeffii		I (1–4)	I (1–4)
Bryum violaceum		I (1–4)	I (1–4)

Plantago major	III (1–3)	III (1–4)	III (1–4)
Spergula arvensis	III (1–6)	III (1–4)	III (1–6)
Stellaria media	III (1–8)	III (1–6)	III (1–8)
Matricaria perforata	III (1–8)	III (1–3)	III (1–8)
Ranunculus repens	III (1–3)	III (1–3)	III (1–3)
Agrostis stolonifera	III (1–8)	III (1–3)	III (1–8)
Elymus repens	III (1–6)	III (1–4)	III (1–6)
Capsella bursa-pastoris	II (1–3)	III (1–3)	II (1–3)
Bilderdykia convolvulus	II (1–3)	III (1–3)	II (1–3)
Holcus lanatus	II (1–3)	III (1–3)	II (1–3)
Fumaria muralis ssp. *boraei*	II (1–6)	II (1–3)	II (1–6)
Senecio vulgaris	II (1–8)	II (1–3)	II (1–8)
Trifolium repens	II (1–3)	II (1–3)	II (1–3)
Taraxacum officinale agg.	II (1–3)	II (1–3)	II (1–3)
Lolium perenne	II (1–3)	II (1–3)	II (1–3)
Plantago lanceolata	II (1–3)	II (1–4)	II (1–4)
Rumex obtusifolius	II (1–3)	II (1–3)	II (1–3)
Geranium dissectum	II (1–3)	II (1–3)	II (1–3)
Potentilla anserina	II (1–3)	II (1–3)	II (1–3)
Fumaria bastardii	I (1–3)	I (1–3)	I (1–3)
Veronica arvensis	I (1)	I (1–3)	I (1–3)
Chrysanthemum segetum	I (1–3)	I (1–3)	I (1–3)
Daucus carota	I (1)	I (1–2)	I (1–2)
Aphanes microcarpa	I (1)	I (1)	I (1)
Coronopus didymus	I (1)	I (1)	I (1)
Cerastium glomeratum	I (1)	I (1)	I (1)
Senecio sylvaticus	I (1)	I (1)	I (1)
Solanum nigrum	I (1)	I (1)	I (1)
Number of samples	23	16	39
Number of species/sample	22 (14–33)	26 (12–48)	23 (14–48)
Herb cover (%)	75 (25–100)	80 (55–100)	77 (25–100)
Bryophyte cover (%)		6 (1–30)	1 (0–30)

a *Chenopodium album-Euphorbia helioscopa* sub-community
b *Cerastium fontanum-Bryum rubens* sub-community
11 *Poa annua-Stachys arvensis* community (total)

OV12
Poa annua-Myosotis arvensis community

Constant species
Agrostis stolonifera, Myosotis arvensis, Poa annua, Poa trivialis, Polygonum aviculare.

Physiognomy
Poa annua and *P. trivialis* are constant in the *Poa-Myosotis arvensis* community and the former in particular can have high cover. *Agrostis stolonifera* and *Elymus repens* are also very frequent overall and, among these grasses, are scattered plants of *Polygonum aviculare* and *Myosotis arvensis*. Other common associates include *Stellaria media, Chamomilla suaveolens, Ranunculus repens, Veronica persica* and *Anagallis arvensis*. More occasional are *Lamium purpureum, Bilderdykia convolvolus* and *Lolium perenne*.

Sub-communities

Typical sub-community. *Polygonum persicaria* and *Sonchus asper* are rather more frequent here with occasional *Urtica dioica, Dactylis glomerata, Lolium multiflorum, Chenopodium album* and *Atriplex patula*.

Dicranella staphylina-Bryum sub-community. Among vascular plants, *Aphanes arvensis* and *Veronica arvensis* become frequent in this sub-community with occasional *Lapsana communis, Viola arvensis, Heracleum sphondylium* and *Trifolium repens*. However, the more striking feature over the surface of the soil is the variety and abundance of diminutive acrocarpous mosses. *Dicranella staphylina, Phascum cuspidatum*, various *Bryum* spp. (including *B. rubens, B. erythrocarpum, B. microerythrocarpum, B. violaceum* and *B. klinggraeffii*) are frequent with occasional *Pottia truncata, P. intermedia, Barbula convoluta* and *B. unguiculata*.

Habitat
The *Poa-Myosotis* community is characteristic of trampled and dunged areas within damp leys, pastures and recreational swards. In such situations, *P. annua* germinates very readily where seeding has been poor or where trampling creates gaps. Its seed remains viable in the dung of cows and horses and can germinate in cattle dung (Hutchinson 1979). Compaction, periodic flooding and waterlogging also present no real hindrance to establishment (Hutchinson & Seymour 1982). Such moist conditions also favour the spread of associates like *Agrostis stolonifera, Poa trivialis* and *Ranunculus repens*. *Myosotis arvensis*, too, finds such situations very congenial for its germination, most of which occurs in autumn (Salisbury 1964). The *Dicranella-Bryum* sub-community is especially characteristic of early stages of colonisation or where bare soil patches persist.

Zonation and succession
Typically, the *Poa-Myosotis* community occurs as patches with Lolio-Plantaginion swards, often with the *Polygonum-Chamomilla* assemblage in moderately trampled places, the *Agrostis-Ranunculus* community where less disturbed swards are kept moist or the *Polygonum-Poa* community where water stands long in winter. Patches of *Urtica-Cirsium* vegetation may also figure.

Where the community occurs along paths, it can pass via patchy zones of the *Polygonum-Chamomilla* vegetation to the *Lolium-Dactylis* grassland or mown recreational swards of the Lolio-Plantaginion.

Renewed care of damaged leys or reseeding of bare and poached areas can lead to replacement of the community with Lolio-Plantaginion swards but continuing neglect can lead to a run-down of this vegetation into ranker weedy assemblages.

Distribution
The community occurs widely throughout the country.

Affinities
Vegetation of this type has not been described before from Britain and has no analogue in published accounts of weed communities from other parts of Europe. It can be placed in the Polygono-Chenopodion alliance or perhaps the Polygonion. Certainly, it is transitional between the Stellarietea and Plantaginetea.

Floristic table OV12

	a	b	12
Poa trivialis	V (1–5)	V (2–5)	V (1–5)
Poa annua	V (1–10)	V (2–10)	V (1–10)
Polygonum aviculare	IV (1–8)	V (1–3)	V (1–8)
Myosotis arvensis	IV (2–5)	IV (1–5)	IV (1–5)
Agrostis stolonifera	III (3–8)	IV (2–10)	IV (2–10)
Sonchus asper	III (1–3)	II (1–5)	II (1–5)
Polygonum persicaria	III (1–3)	II (2–3)	II (1–3)
Urtica dioica	II (1–2)	I (1)	I (1–2)
Dactylis glomerata	II (1–2)	I (2)	I (1–2)
Lolium multiflorum	II (2–3)	I (2)	I (2–3)
Chenopodium album	II (1–8)	I (2–3)	I (1–8)
Atriplex patula	II (1–3)	I (1)	I (1–3)
Bryum spp.*		IV (1–3)	III (1–3)
Dicranella staphylina		III (1–5)	II (1–5)
Phascum cuspidatum		III (1–5)	II (1–5)
Aphanes arvensis	I (1–3)	III (1–5)	II (1–5)
Veronica arvensis	I (2)	III (1–5)	II (1–5)
Lapsana communis	I (2)	II (1–7)	I (1–7)
Viola arvensis	I (2)	II (1–3)	I (1–3)
Pottia truncata		II (1–3)	I (1–3)
Pottia intermedia		II (1–3)	I (1–3)
Barbula convoluta		II (1–2)	I (1–2)
Barbula unguiculata		II (1–2)	I (1–2)
Trifolium repens		II (1–2)	I (1–2)
Heracleum sphondylium		II (1–2)	I (1–2)
Stellaria media	III (1–7)	III (1–7)	III (1–7)
Chamomilla suaveolens	III (1–4)	III (2–5)	III (1–5)
Elymus repens	III (2–6)	III (1–7)	III (1–7)
Ranunculus repens	III (1–3)	III (1–3)	III (1–3)
Veronica persica	II (1–5)	III (1–3)	III (1–5)
Anagallis arvensis	II (2–5)	III (1–3)	III (1–5)
Matricaria perforata	II (1–7)	III (2–7)	III (1–7)
Lamium purpureum	II (1–3)	II (2–3)	II (1–3)
Bilderdykia convolvulus	II (2–3)	II (1–2)	II (1–3)
Lolium perenne	II (1–5)	II (2–5)	II (1–5)
Plantago major	I (1–3)	II (1–3)	II (1–3)
Convolvulus arvensis	I (2–3)	II (2–5)	I (2–5)
Cerastium fontanum	I (1–2)	II (1–3)	I (1–3)
Number of samples	13	21	34
Number of species/sample	18	20	20

* Includes records for *Bryum rubens, B. erythrocarpum, B. microerythrocarpum, B. violaceum* and *B. klinggraeffii*.

a Typical sub-community

b *Dicranella staphylina-Bryum* spp. sub-community

12 *Poa annua-Myosotis arvensis* community (total)

OV13
Stellaria media-Capsella bursa-pastoris community

Synonymy
Includes *Fumarietum officinalis*   R.Tx. 1950 and *Fumarietum bastardii*   Br.-Bl. 1950.

Constant species
Capsella bursa-pastoris, Chenopodium album, Polygonum aviculare, Senecio vulgaris, Stellaria media.

Rare species
Fumaria bastardii.

Physiognomy
The *Stellaria media-Capsella bursa-pastoris* community is an annual vegetation type dominated by mixtures of *Stellaria media, Capsella bursa-pastoris, Senecio vulgaris, Polygonum vulgare* and *Chenopodium album.* Also more or less frequent overall but rather unevenly represented in the various sub-communities are *Poa annua, Elymus repens, Chamomilla suaveolens* and *Urtica urens.* More occasional are *Sonchus asper, Cirsium arvense* and *Polygonum persicaria.* Scarcer associates in the community include *Rumex obtusifolius, Convolvulus arvensis, Solanum nigrum* and *Avena fatua.*

Sub-communities

Typical sub-community. Apart from the species mentioned above, there is little that is distinctive about the vegetation here. Occasionally, *Sinapis arvensis, Sisymbrium officinale* and *Lolium perenne* are seen.

Matricaria perforata-Poa annua sub-community. *Poa annua* and, more particularly, *Matricaria perforata* are preferentially frequent in this sub-community along with common *Agrostis stolonifera.*

Fumaria officinalis-Euphorbia helioscopa sub-community. A number of quite common community associates, like *Elymus repens, Veronica persica* and *Lamium purpureum* are especially frequent here, but more striking is the preferential occurrence of *Fumaria officinalis* and *Euphorbia helioscopa.* More occasional are *E. peplus, Sonchus oleraceus, Veronica agrestis, Mercurialis annua, Polygonum nodosum* and *Geranium dissectum.* Around the coastal lowlands of western Britain, this vegetation provides a locus for the nationally rare *Fumaria bastardii* and, at scattered localities in England and Wales, for *Chenopodium urbicum,* probably an introduced plant.

Urtica dioica-Galium aparene sub-community. *Sonchus asper* and *Cirsium arvense* are somewhat more common here than in other sub-communities but more obviously preferential are *Urtica dioica* and *Galium aparine,* with occasional *Papaver rhoeas, Bromus sterilis, Polygonum convolvulus, P. lapathifolium, Cirsium vulgare, Brassica napus* and *Chenopodium bonus-henricus.*

Habitat
The *Stellaria-Capsella* community occurs widely on fertile loamy soils throughout the British lowlands, as weed vegetation among root, vegetable and salad crops, often even where these have been treated by herbicides, but also among cereals and on dumped topsoil and disturbed ground.

   The most common species of this community all grow best on disturbed ground that is naturally eutrophic or, more commonly, where there has been some enrichment through fertilising, dumping of organic waste or disturbance (e.g. Sobey 1981, Hutchinson & Seymour 1982, Kay 1994). Many are prodigious seeders, able to remain dormant for some years and some, like *Poa annua, Stellaria media* and *Capsella bursa-pastoris,* show an intermittent germination pattern that enables them to take advantage of disturbance and opening up of the ground at any time through the growing season. Moreover, species such as *S. media, Bilderdykia convolvulus, Polygonum* spp., *Chamomilla suaveolens* and *Veronica persica* are all somewhat resistant to many of the herbicides that are in common use (Silverside 1977), so this kind of weed vegetation is one of the commonest assemblages associated with intensive root, vegetable and salad crops on farms and market gardens, as well as in smallholdings and on allotments.

Such situations, and fertilised cereal crops, are most characteristic for the Typical and *Matricaria-Poa* sub-communities, while the *Urtica-Galium* type is more often found on disturbed and dumped soil, around manure piles and in derelict pastures. By contrast, the *Fumaria-Euphorbia* sub-community preserves a little more of the species diversity associated with less intensively cultivated arable crops. Over much of the British lowlands, certainly towards the east, *F. officinalis* is the typical fumitory but it is replaced in essentially the same vegetation in the west by *F. muralis* ssp. *boraei*, the commoner plant in western England and Wales, and down the western seaboard by *F. bastardii*.

Zonation and succession

The *Stellaria-Capsella* community replaces other weed assemblages on loamy soils as cultivation practices are intensified. Where stretches of crop are less effectively fertilised and sprayed, it can be found with the *Papaveretum argemones* and, on more obviously calcareous soils in the warmer and drier south-east, by the *Kickxietum spuriae*. On more sandy and less base-rich soils in eastern Britain it can give way to the *Urtica urens-Lamium* community and, in the far south-west, to the *Cerastium-Fumaria* community. More widely, in intensive arable landscapes, the *Stellaria-Capsella* community is found with the *Veronico-Lamietum*, the *Alopecurio-Chamomilletum* and the *Matricaria-Stellaria* community.

Continued cultivation effectively prevents any succession.

Distribution

The *Stellaria-Capsella* community occurs widely through the British lowlands with the *Fumaria-Euphorbia* type the most local.

Affinities

This is the most widespread weed community in Britain that has recognisable affinities with the Fumario-Euphorbion, the alliance of emphemeral vegetation types characteristic of less acidic loams and clays throughout western Europe. As such, it is equivalent to the *Fumarietum officinalis* R.Tx. 1950 or its various manifestations: the *Mercuriali-Fumarietum* Kruseman & Vlieger 1939 *emend.* J.Tx. 1955 (as in Westhoff & den Held 1969), the *Thlaspio-Fumarietum* Görs in Oberdorfer *et al.* 1967 ex Passarge & Jurko 1975 (as in Pott 1992) or the *Mercurialetum annuae* Kruseman & Vlieger 1939 *emend.* Th. Müller (as in Oberdorfer 1983). The *Fumarietum bastardii* Br.-Bl. in Br.-Bl. & Tx. 1952 was defined from Ireland on the basis of four samples, with Brun-Hool & Wilmans (1982) subsequently assigning some possible new samples to this syntaxon. However, the floristic differences among the British data seem insufficient to recognise this as distinct and even the *Fumarietum officinalis* is so poorly developed as to be hard to distinguish from its now much more widespread impoverished derivative. Such a problem is hardly unexpected when the *Fumarietum* is itself a community of naturally fertile loams, soils which now provide the bulk of the land for intensive vegetable cultivation. Among the Fumario-Euphorbion, these British stands are therefore transitional to the Polygono-Chenopodion. In fact, in The Netherlands, the Fumario-Euphorbion has been subsumed in this latter alliance (Westhoff & den Held 1969).

Floristic table OV13

	a	b	c	d	13
Capsella bursa-pastoris	V (2–7)	IV (1–5)	V (1–8)	V (1–3)	V (1–8)
Stellaria media	V (1–6)	V (1–7)	V (1–7)	III (3–5)	V (1–7)
Senecio vulgaris	IV (3–5)	V (1–3)	V (1–5)	III (2–4)	V (1–5)
Polygonum aviculare	IV (2–3)	IV (1–10)	III (3–5)	IV (3–5)	IV (1–10)
Chenopodium album	III (3–9)	V (1–5)	V (1–5)	IV (2–5)	IV (1–9)
Sinapis arvensis	III (2–4)	I (1–8)	II (1–5)	I (4)	I (1–8)
Sisymbrium officinale	II (1–4)	II (1–3)		I (3)	I (1–4)
Lolium perenne	II (3–4)				I (3–4)
Poa annua	II (2–7)	IV (1–8)	IV (1–3)	II (1–4)	III (1–8)
Matricaria perforata	I (1)	V (1–8)	II (2–3)		II (1–8)
Agrostis stolonifera	II (3–5)	III (1–5)	I (2)	I (4)	II (1–5)
Elymus repens	III (2–4)	II (1–5)	V (1–3)	IV (3–8)	III (1–8)
Veronica persica	III (2–7)	III (1–7)	V (1–5)	I (1)	III (1–7)
Lamium purpureum	III (1–5)	II (1–5)	IV (1–5)	I (3)	III (1–5)
Fumaria officinalis	I (3)	II (1–5)	III (1–5)		II (1–5)
Euphorbia helioscopa	I (1)	I (1–3)	III (1–4)		II (1–4)
Sonchus oleraceus	I (1)	I (1–3)	II (1–3)		II (1–3)
Euphorbia peplus		I (1–2)	II (1–4)		II (1–4)
Veronica agrestis			II (2–3)		I (2–3)
Mercurialis annua			II (1–8)		I (1–8)
Chenopodium urbicum		I (1–3)	II (1–4)	I (3)	I (1–4)
Polygonum nodosum		I (2)	II (3)		I (2–3)
Geranium dissectum	I (2)	I (1)	II (1–3)		I (1–3)
Fumaria muralis boraei			II (1–5)		I (1–5)
Fumaria bastardii			II (1–8)		I (1–8)
Sonchus asper	II (1–3)	I (1)	III (1–3)	V (2–5)	III (1–5)
Cirsium arvense	III (1–7)	II (1–5)	III (1–5)	IV (2–4)	II (1–7)
Galium aparine	I (2)	I (1–2)	I (3)	IV (2–5)	I (1–5)
Urtica dioica	I (4)	I (1)		IV (1–4)	I (1–4)
Papaver rhoeas		I (2)	I (1)	II (3–5)	I (1–5)
Bromus sterilis			I (2)	II (3–4)	I (2–4)
Polygonum lapathifolium		I (1)	I (1)	II (3–9)	I (1–9)

	a	b	c	d	13
Cirsium vulgare	III (1–6)		I (1)	II (2–5)	I (1–5)
Brassica napus	I (2)			II (2–3)	I (2–3)
Chenopodium bonus-henricus	II (2–4)			II (2–4)	I (2–4)
Chamomilla suaveolens		II (1–5)	III (1–5)	II (2–6)	III (1–6)
Urtica urens		III (1–8)	III (1–3)	I (1)	III (1–8)
Polygonum persicaria		II (1–5)	II (1–7)	II (3)	II (1–7)
Rumex obtusifolius		II (1–3)	I (1)	II (2–4)	I (1–4)
Convolvulus arvensis		I (1)	II (3–8)	II (2–4)	I (1–4)
Solanum nigrum		II (1–3)	II (1–2)		I (1–3)
Bilderdykia convolvulus		II (1–3)	II (3–5)	II (3–5)	I (1–5)
Plantago major		II (1–3)	II (1–5)		I (1–5)
Avena fatua		II (1–3)	I (1)	II (3–4)	I (1–4)
Lamium album	I (9)	I (2)	I (2)		I (2–9)
Atriplex patula	I (4–6)	I (3)	I (5)		I (3–5)
Thlaspi arvense	I (2)	I (1)	I (3)		I (1–3)
Chenopodium rubrum	I (3)	I (2–3)	I (1)	I (1)	I (1–3)
Rumex crispus	I (3)	I (1)		I (3)	I (1–3)
Atriplex prostrata	I (2)	I (1)	I (8)		I (1–8)
Coronopus squamatus		I (2)	I (2)		I (2)
Calystegia sepium		I (2)		I (7)	I (2–7)
Poa pratensis		I (1–2)	I (1–4)		I (1–4)
Arctium minus agg.		I (3–5)	I (3)		I (3–5)
Sonchus arvensis		I (2)	I (5–7)		I (2–7)
Lamium hybridum		I (3)		I (2)	I (2–3)
Chamomilla recutita			I (1–2)	I (3)	I (1–3)
Spergula arvensis	I (5)	I (1)		I (3)	I (1–5)
Trifolium repens		I (3)	I (1–2)		I (1–3)
Alopecurus myosuroides		I (3–5)	I (3)		I (3–5)
Veronica polita		I (2–4)		I (3)	I (2–4)
Number of samples	10	19	39	8	76
Number of species/sample	12 (5–20)	17 (11–32)	19 (6–39)	16 (12–20)	17 (5–39)

a   Typical sub-community
b   *Matricaria perforata-Poa annua* sub-community
c   *Fumaria officinalis-Euphorbia helioscopa* sub-community
d   *Urtica dioica-Galium aparine* sub-community
13   *Stellaria media-Capsella bursa-pastoris* community (total)

OV14
Urtica urens-Lamium amplexicaule community

Synonymy
Spergula arvensis-Lamium amplexicaule community Sissingh 1950

Constant species
Capsella bursa-pastoris, Chenopodium album, Poa annua, Senecio vulgaris, Stellaria media, Urtica urens.

Rare species
Erodium moschatum, Medicago polymorpha, Sisymbrium irio.

Physiognomy
In the annual vegetation of the *Urtica urens-Lamium amplexicaule* community, the usual dominants are *Stellaria media, Poa annua* or *Capsella bursa-pastoris*, with *Chenopodium album* making a very frequent but somewhat more variable contribution to the cover, *Senecio vulgaris* and *Urtica urens* constant but generally of less abundance.

Also frequent, sometimes with locally high cover, are *Lamium amplexicaule, Solanum nigrum* and the now widely naturalised *Galinsoga parviflora*, a South American plant which escaped from Kew around 1860 (Salisbury 1964). *Spergula arvensis, Veronica persica, Matricaria perforata, Chamomilla suaveolens, Polygonum aviculare* and *Elymus repens* are common, too, though usually at low cover and another South American introduction, *Solanum sarrachoides*, is occasional. *Polygonum persicaria, Sonchus oleraceus, S. asper* and *Bilderdykia convolvulus* are scarce companions.

More locally, this vegetation can show a very distinctive enrichment from a variety of 'shoddy aliens'. These are plants whose seed was brought in with rags from different parts of Europe, the waste from which, after recovery of wool fibre, was dumped or spread on fields as a fertiliser. In recent years (e.g. Lavin & Wilmore 1994), *Stellaria-Urtica* vegetation in such fields in West Yorkshire has provided a locus for plants like *Sisym-*

brium irio, S. loeselii, S. orientale, Ammi majus, Echinochloa crus-galli, Medicago arabica, M. polymorpha, M. minima, Xanthium spinosum, Erodium chium, E. botrys, E. moschatum and, more rarely *Amaranthus hybridus, A. albus, A. deflexus, Scorpiurus muricatus, Carduus pycnocephalus, Trifolium tomentosum* and *Carthamus lanatus*.

Habitat
The *Urtica urens-Lamium amplexicaule* community is characteristically found among root and vegetable crops on light and more base-poor soils in the warmer south and east of England.

Lamium amplexicaule is a native plant of lighter soils in southern and eastern England and all up the eastern lowlands of Scotland, overlapping in its edaphic preferences with *Spergula arvensis* but extending further on to less acidic loams. Here it occurs also with *Urtica urens*, an annual of lighter soils that is normally poorly represented in vegetation like the *Spergulo-Chrysanthemetum* because of the lower nutrient content of the substrates there.

Along with other widely occurring nitrophilous weeds, *Urtica, Lamium* and *Spergula* are found here in association with *Solanum nigrum*, its introduced relative *S. sarrachoides* and the garden escape *Galinsoga parviflora*, which add a more Continental character to the assemblage: these are plants more strikingly confined to the warmer and drier south-east of the country.

Such plants have found a congenial habitat in the fertilised but not so strongly herbicide-treated arable crops of market gardens and smallholdings: Silverside (1977) recorded this community in a wide variety of root, vegetable and salad crops, often grown in a strip-farming system that allowed plenty of room for the assemblage to develop.

This kind of situation also seems to provide conditions suitable for a variety of aliens that are brought into Britain on imported wool, notably the shoddy used by mills concentrated around Dewsbury, Ossett and Morley (Rodwell 1994a). The waste from shoddy used to be widely distributed as a slow-acting organic manure, and was especially significant in enriching the weed flora of the market-gardening area of the Thames Valley and Bedfordshire (Dony 1953*a*, *b*). Similar entertaining diversity is still seen in stands of this community in West Yorkshire (Lavin & Wilmore 1994).

Zonation and succession
Where fertilising is more intensive among arable crops, the *Urtica-Lamium* community is replaced by other assemblages like the *Stellaria-Capsella* vegetation. Cultivation year after year helps regenerate the community provided there is no great shift in treatment of the crops and continuing additions of shoddy waste can maintain distinctive diversity.

Distribution
The community is widespread in southern and eastern England with local stands further north on the eastern side of the country.

Affinities
A *Spergula arvensis-Lamium amplexicaule* community was referred to briefly by Sissingh (1950) as a replacement in The Netherlands for the *Echinochloo-Setarietum* on sandy alluvium after the harvesting of late-season root crops. This latter community was seen by Silverside (1977) as represented here by what we have termed the *Digitaria-Erodium* assemblage, a much more localised syntaxon of drought-prone sands in the warmer south-east.

Floristic table OV14

Stellaria media	V (1–9)
Urtica urens	V (1–8)
Capsella bursa-pastoris	V (1–8)
Poa annua	V (1–4)
Senecio vulgaris	V (1–8)
Chenopodium album	IV (1–4)
Lamium amplexicaule	III (1–4)
Solanum nigrum	III (1–6)
Galinsoga parviflora	III (1–7)
Spergula arvensis	III (1–4)
Veronica persica	III (1–3)
Matricaria perforata	III (1–3)
Polygonum aviculare	III (1–3)
Elymus repens	III (1–3)
Chamomilla suaveolens	III (1–6)
Solanum sarrachoides	II (1–5)
Erodium cicutarium	I (1)
Polygonum persicaria	I (1–4)
Sonchus oleraceus	I (1–3)
Lamium purpureum	I (6)
Fumaria officinalis	I (1)
Euphorbia peplus	I (1)
Sonchus asper	I (1–2)
Bilderdykia convolvulus	I (1–2)
Papaver rhoeas	I (1)
Agrostis stolonifera	I (1)
Rumex obtusifolius	I (1–3)
Convolvulus arvensis	I (1–3)
Cirsium arvense	I (1)
Plantago lanceolata	I (1)
Artemisia vulgaris	I (1)

Number of samples	18
Number of species/sample	12 (9–18)
Vegetation cover (%)	66 (30–100)

OV15

Anagallis arvensis-Veronica persica community
Kickxietum spuriae Kruseman & Vlieger 1939

Constant species

Anagallis arvensis, Bilderdykia convolvulus, Polygonum aviculare, Veronica persica.

Rare species

Ajuga chamaepitys, Scandix pecten-veneris.

Physiognomy

The *Kickxietum spuriae* comprises annual vegetation that is usually dominated by smaller ephemerals like *Anagallis arvensis, Veronica persica* and *Polygonum aviculare* with twining trails of *Bilderdykia convovlulus.* However, by mid- to late summer, the most distinctive feature in many stands is the presence of one or other, often both, of *Kickxia elatine* and *K. spuria*, their downy shoots spreading among the cereal stubble that usually forms the habitat of this vegetation. *Euphorbia exigua* is another summer annual that occurs more or less frequently throughout.

Less distinctive but still common through the community as a whole are *Poa annua, Matricaria perforata, Sonchus asper, Myosotis arvensis, Elymus repens* and *Plantago major*, with *Veronica polita, Lapsana communis, Papaver rhoeas, Cirsium arvense, Galium aparine* and *Lolium perenne* occasional.

Sub-communities

***Stellaria media-Convolvulus arvensis* sub-community:** *Kickxietum spuriae* Kruseman & Vlieger 1939; unassigned aufnahmen *sensu* Silverside 1977. The two *Kickxia* spp. and *E. exigua* are frequent here but are accompanied by less distinctive weeds like *Stellaria media* and tangles of *Convolvulus arvensis*. The grasses *Avena fatua, Alopecurus myosuroides* and *Agrostis stolonifera* are quite common, the inflorescences of the first two often growing up among the cereal crop. *Aethusa cynapium, Chenopodium album* and *Anthemis cotula* are occasional.

***Legousia hybrida-Chaenorhinum minus* sub-community:** *Kickxietum spuriae* Kruseman & Vlieger 1939, *sherar-*
dietosum sensu Silverside 1977; *Adonis autumnalis-Iberis amara* Association (Allorge 1913) R.Tx. 1950. The two *Kickxia* spp. and, more particularly, *E. exigua* show maximum frequency here and the contingent of summer annuals is further enriched by *Legousia hydrida, Chaenorhinum minus* and *Valerianella dentata*, creating a pretty sight when all are flowering, sometimes well into early autumn. Also quite common in this sub-community are *Sherardia arvensis, Reseda lutea* and *Mentha arvensis* with the nationally scarce *Ajuga chamaepitys* and *Scandix pecten-veneris* sometimes figuring where seedings established in autumn manage to survive damp winters or any tilling that precedes winter cereal sowing.

***Agrostis stolonifera-Phascum cuspidatum* sub-community:** *Kickxia elatine*-Aphanion vegetation and *Ranunculus repens* noda Silverside 1977. Summer annuals, including *K. spuria*, tend to be less common in this sub-community, where *A. stolonifera* and, more noticeably, *Ranunculus repens* are preferentially frequent. Often more striking, however, is the diversity and local abundance of a variety of acrocarpous mosses over the soil surface. Among these *Phascum cuspidatum, P. floerkianum, Barbula unguiculata, B. convoluta, Bryum rubens, B. klinggraeffii, B. microerythrocarpum, Dicranella staphylina* and *D. schreberana* are most frequent with *Eurhynchium praelongum* also common.

Habitat

The *Kickxietum spuriae* is characteristically a weed community of cereal crops on base-rich soils in the warmer and drier south-east of Britain.

It is the two species of *Kickxia* and *Euphorbia exigua* which give this community its distinctive character overall. Among the other frequent plants of this kind of weed vegetation, typical together of light and only moderately fertile soils, these are all of more Continental distribution in Britain, common only in the warmer and drier regions south-east of a line from the Severn to the Humber. They are also all fairly calcicolous, and this community as a whole is characteristic of lime-rich soils derived from Chalk or other limestones like Cornbrash, Lower Purbeck or Oolite, or the limey superficials common through this region.

The most striking type of the *Kickxietum* is the *Legousia-Chaenorhinum* sub-community where a further group of calcicolous plants, *L. hybrida*, *C. minus* and also *Reseda lutea* and *Valerianella dentata*, emphasise the edaphic preference of the community as a whole. Of these, the first and last are also rather strikingly Continental in their climatic preferences. Another of the weaker preferentials, *Ajuga chamaepitys*, a nationally rare plant which finds one of its loci here, also reflects the association with open lime-rich conditions. Two other of the common preferentials of this sub-community, *Sherardia arvensis* and *Mentha arvensis*, have broader edaphic and phytogeographic affinities, being typical of disturbed soils throughout the lowlands.

The characteristic arable crops on these calcareous soils of south-east England are cereals, mostly barley and wheat (Silverside 1977), and the most frequent of the more distinctive plants of the *Kickxietum* are summer annuals (Salisbury 1964), germinating largely in spring and often surviving the harvesting of the crop by virtue of their low habit. Indeed, the late flowering of the *Kickxia* species among stubble when late summer and autumn are warm and sunny can be an especially striking feature here.

On more clayey soils, the frequency of most of these plants is reduced, but then the preferentials of the *Agrostis-Phascum* sub-community increase reflecting the moist and open ground conditions, particularly when summer and early autumn rains fall on cut cereal fields. On these heavier soils, Silverside (1977) found wheat as often as barley to be the crop, whereas generally the latter prevailed as the context of this vegetation.

The less distinctive assemblage of the *Stellaria-Convolvulus* sub-community, where *Euphorbia exigua* and the *Kickxia* species remain quite frequent but are often overwhelmed in cover by *S. media* and other nitrophilous weeds, is probably associated with more intensively fertilised crops.

Zonation and succession

The *Kickxietum* is typically found within and around cereal crops and their stubble and can occur with the *Papaveri-Sileneetum* or less distinctive weed assemblages of lighter soils like the *Matricaria perforata-Stellaria* community where crops are more heavily fertilised. Repeated cultivation enables this kind of vegetation to reappear each year and sets back any tendency to succession.

Distribution

The *Kickxietum* occurs widely but locally on suitable soils across south-east England.

Affinities

The *Kickxietum* has been described from The Netherlands (Westhof & den Held 1969), Germany (Oberdorfer 1983, Pott 1992) and Austria (Mucina *et al.* 1993). It is one of the central associations of the Caucalidion, the alliance of calcicolous weed assemblages, usually from cereal crops, in the more Continental parts of Europe. As described here, the community could include the vegetation separated off by Silverside (1977) as the *Adonis autumnalis-Iberis amara* Association (Allorge 1913) R.Tx. 1950.

Floristic table OV15

	a	b	c	15
Anagallis arvensis	V (1–5)	V (2–5)	V (2–5)	V (1–5)
Veronica persica	V (2–7)	V (1–5)	IV (2–5)	V (1–7)
Polygonum aviculare	IV (2–7)	IV (2–8)	IV (2–7)	IV (2–8)
Bilderdykia convolvulus	IV (2–5)	IV (2–5)	III (1–5)	IV (1–5)
Stellaria media	IV (1–7)	II (1–3)	II (2–7)	III (1–7)
Convolvulus arvensis	III (2–3)	II (1–5)	II (2–3)	II (1–5)
Avena fatua	II (2–7)	I (2–7)	I (2–3)	I (2–7)
Aethusa cynapium	II (2–5)	I (2–5)	I (2–3)	I (2–5)
Alopecurus myosuroides	II (2–7)	I (2)	I (3)	I (2–7)
Chenopodium album	II (1–2)	I (2–3)	I (1–3)	I (1–3)
Anthemis cotula	II (1–5)	I (4)	I (1–3)	I (1–5)
Euphorbia exigua	III (1–5)	V (1–4)	II (2–3)	III (1–5)
Kickxia elatine	III (1–3)	IV (2–5)	III (1–3)	III (1–5)
Kickxia spuria	III (1–3)	IV (2–5)	I (2–3)	III (1–5)
Legousia hybrida	I (1–2)	IV (1–3)	II (2–4)	II (1–4)
Chaenorhinum minus	I (1)	IV (1–3)	I (2–3)	II (1–3)
Sherardia arvensis		III (1–5)	I (1–2)	II (1–5)
Reseda lutea		III (1–5)	I (2)	II (1–5)
Valerianella dentata		II (1–3)	I (3)	I (1–3)
Mentha arvensis		III (1–5)	I (2)	I (1–5)
Ajuga chamaepitys		II (1–3)		I (1–3)
Silene alba		I (1–3)		I (1–3)
Filago pyramidata		I (1–3)		I (1–3)
Adonis annua		I (2–3)		I (2–3)
Scandix pecten-veneris		I (4)		I (4)
Agrostis stolonifera	III (1–5)	II (1–5)	IV (1–10)	III (1–10)
Ranunculus repens	I (1–3)	II (1–3)	IV (1–3)	III (1–3)
Phascum cuspidatum			III (1–3)	II (1–3)
Barbula unguiculata			III (1–5)	II (1–5)
Eurhynchium praelongum			III (1–3)	II (1–3)
Bryum rubens			III (1–3)	II (1–3)
Bryum klinggraeffii			II (1–3)	I (1–3)
Bryum microerythrocarpum			II (1–3)	I (1–3)
Pottia truncata			II (1–3)	I (1–3)
Dicranella staphylina			II (1–5)	I (1–5)
Dicranella schreberana			II (1–3)	I (1–3)
Barbula convoluta			II (1–3)	I (1–3)
Phascum floerkianum			II (1–3)	I (1–3)
Dicranella varia			II (1–3)	I (1–3)
Phascum curvicollum			I (1)	I (1)
Barbula fallax			I (1)	I (1)
Pottia starkeana conica			I (1–2)	I (1–2)
Poa annua	III (2–7)	III (2–5)	III (2–7)	III (2–7)
Myosotis arvensis	III (1–3)	III (1–3)	III (1–7)	III (1–7)
Matricaria perforata	III (2–5)	II (2–5)	III (2–5)	III (2–5)

Sonchus asper	III (1–3)	III (1–3)	II (1–2)	III (1–3)
Elymus repens	III (2–5)	II (1–7)	III (2–5)	III (1–7)
Plantago major	III (1–3)	II (2–5)	III (2–3)	III (1–5)
Veronica polita	II (2–3)	III (2–5)	II (1–3)	II (1–5)
Lapsana communis	II (1–5)	III (1–5)	II (1–5)	II (1–5)
Papaver rhoeas	II (1–5)	III (1–5)	II (2–3)	II (1–5)
Cirsium arvense	II (1–3)	II (1–2)	I (1–3)	II (1–3)
Galium aparine	II (1–3)	II (2–5)	I (2–3)	II (1–5)
Lolium perenne	I (1–3)	II (2–5)	II (2–7)	II (1–7)
Medicago lupulina	I (1–3)	I (1–5)	I (1–5)	I (1–5)
Poa trivialis	I (1–3)	I (1–3)	I (1–3)	I (1–3)
Trifolium repens	I (1–3)	I (1–3)	I (1–3)	I (1–3)
Odontites verna serotina	I (1–3)	I (1)	I (1)	I (1–3)
Cerastium fontanum	I (1)	I (1)	I (1–3)	I (1–3)
Euphorbia helioscopa	I (3)	I (3)	I (2–3)	I (2–3)
Fumaria officinalis wirtgenii	I (1)	I (2–3)	I (1–3)	I (1–3)
Atriplex patula	I (1–3)	I (1–5)	I (2–3)	I (1–5)
Dactylis glomerata	I (1–3)	I (1–3)	I (1)	I (1–3)
Trifolium pratense	I (1–2)	I (1)	I (1)	I (1–2)
Heracleum sphondylium	I (1–3)		I (1)	I (1–3)
Urtica dioica	I (1–3)		I (1)	I (1–3)
Geranium dissectum	I (1)		I (1–3)	I (1–3)
Cirsium vulgare	I (1)		I (1)	I (1)
Number of samples	20	18	24	62
Number of species/sample	20 (8–32)	24 (17–33)	27 (13–50)	24 (8–50)
Herb cover (%)	55 (25–90)	57 (30–100)	72 (45–100)	62 (25–100)
Bryophyte cover (%)	–	–	9 (1–50)	3 (1–50)

a   *Stellaria media-Convolvulus arvensis* sub-community

b   *Legousia hybrida-Chaenorhinum minus* sub-community

c   *Agrostis stolonifera-Phascum cuspidatum* sub-community

15   *Kickxietum spuriae* (total)

OV16

Papaver rhoeas-Silene noctiflora community
Papaveri-Sileneetum noctiflori Wasscher 1941

Constant species

Bilderdykia convolvulus, Elymus repens, Matricaria perforata, Papaver rhoeas, Polygonum aviculare, Silene noctiflora, Stellaria media, Veronica persica.

Rare species

Silene noctiflora.

Physiognomy

The *Papaveri-Sileneetum* is an annual community in which *Stellaria media*, *Matricaria perforata* and *Polygonum aviculare* usually provide the bulk of the herbage, along with small ephemerals like *Veronica persica*, *V. polita*, *Anagallis arvensis*, twines of *Bilderdykia convolvulus* and young shoots of *Elymus repens*. The most striking feature, however, is the constancy of *Papaver rhoeas* and the nationally scarce *Silene noctiflora*, a plant whose peak of flowering is in mid-summer when the distinctive blooms, their yellow-backed petals inrolled until the cool of the evening, are seen most prolifically among cereal stubble or autumn-harvested crops.

Other frequent plants of this kind of vegetation are *Chenopodium album*, *Agrostis stolonifera* and *Galium aparine* with many occasionals including *Fumaria officinalis* ssp. *wirtgenii*, *Viola arvensis*, *Aethusa cynapium* and *Linaria vulgaris*.

Habitat

The *Papaveri-Sileneetum* is confined to light, well-drained calcareous soils, mostly among cereals, across the warmer and drier south-east of England.

The most distinctive plant of this assemblage, *S. noctiflora*, has a more or less Continental distribution in Europe, rare in both the Mediterranean and Scandinavia, preferring better-drained soils and reproducing poorly in wet years (Stewart *et al.* 1994). With us, it has a marked south-easterly range, extending from Dorset to the Scottish border, where the climate is more congenial for its survival and where calcareous bedrocks are especially extensive. There it occurs in this community among arable crops, particularly cereals, occasionally in root crops, not too heavily fertilised nor treated with herbicides and seems particularly to favour rotations which include spring-sown crops. It germinates primarily in spring and was found by Wilson (1990) to develop best among crops sown towards late March.

S. noctiflora has declined in its extent quite considerably since the 1950s, though it and this assemblage have benefited more recently from the creation of conservation headlands around arable fields.

Zonation and succession

Other weed assemblages of light, calcareous soils in south-east England where arable crops have not been too heavily fertilised or sprayed with herbicides include the *Kickxietum spuriae* and richer types of *Stellaria-Capsella* vegetation. Where soils are less base-rich, the *Papaveri-Sileneetum* tends to be replaced by the *Papaveretum argemones* and the *Urtica-Lamium* community. With increased intensification of arable agriculture, the *Papaveri-Sileneetum* gives way to the *Veronico-Lamietum* or *Matricaria-Stellaria* community or more species-poor forms of *Stellaria-Capsella* vegetation.

Repeated ploughing for arable crops prevents any prospect of succession.

Distribution

The community occurs on suitable soils from Dorset and Wiltshire, north-east to Lincolnshire.

Affinities

Although this syntaxon, first described from The Netherlands by Wasscher (1941), was regarded as of dubious status by Sissingh (1950), it was recognised by Westhoff & den Held (1969) and similar vegetation has also been described from Germany (Passarge 1964, Oberdorfer 1957, 1983). Pott (1992) in Germany and also Mucina *et al.* (1993) in Austria, characterise a similar assemblage as the *Euphorbio exiguae-Melandrietum* G. Müller 1964. Most authorities agree on placing the syntaxon in the Caucalidion, the alliance of calcicolous assemblages on lime-rich sands and loams in Continental Europe reaching its north-west limit in Britain.

Floristic table OV16

Veronica persica	V (1–4)
Polygonum aviculare	V (1–8)
Elymus repens	V (1–8)
Bilderdykia convolvulus	V (1–4)
Silene noctiflora	V (1–4)
Matricaria perforata	IV (1–6)
Stellaria media	IV (1–5)
Papaver rhoeas	IV (1–6)
Chenopodium album	III (1–4)
Anagallis arvensis	III (1–3)
Veronica polita	III (1–3)
Agrostis stolonifera	III (1–4)
Galium aparine	III (1–4)
Fumaria officinalis wirtgenii	II (1–4)
Viola arvensis	II (1–3)
Lapsana communis	II (1–8)
Capsella bursa-pastoris	II (1–3)
Senecio vulgaris	II (1–3)
Sonchus asper	II (1–3)
Chamomilla suaveolens	II (1–4)
Poa annua	II (1–4)
Cirsium arvense	II (1–4)
Aethusa cynapium	II (1–4)
Linaria vulgaris	II (1–3)
Plantago major	II (1–3)
Trifolium repens	II (1–3)
Lolium perenne	II (1–3)
Sisymbrium officinale	II (1–3)
Bromus sterilis	II (1)
Silene alba	I (1–4)
Nepeta cataria	I (1)
Valerianella dentata	I (1)
Arenaria leptoclados	I (1)
Euphorbia exigua	I (1)
Kickxia spuria	I (1)
Avena fatua	I (1–3)
Sinapis arvensis	I (1)
Sonchus arvensis	I (1)
Silene vulgaris	I (1–4)
Medicago lupulina	I (1–3)
Malva sylvestris	I (1–2)
Number of samples	14
Number of species/sample	21 (14–27)
Vegetation cover (%)	70 (20–95)

OV17

Reseda lutea-Polygonum aviculare community
Descurainio-Anchusetum arvensis   Silverside 1977

Constant species

Anchusa arvensis, Bilderdykia convolvulus, Chenopodium album, Descurainia sophia, Elymus repens, Polygonum aviculare, Reseda lutea.

Physiognomy

The *Descurainio-Anchusetum* is an ephemeral community in which *Descurainia sophia*, probably a long-established introduction (Rich 1991), *Anchusa arvensis* and *Reseda lutea* comprise a distinctive group of constants, along with very frequent *Elymus repens*, *Chenopodium album*, *Polygonum aviculare* and *Bilderdykia convolvulus*.

Also very common are *Veronica persica*, *V. polita*, *Chamomilla suaveolens*, *Matricaria perforata*, *Stellaria media*, *Senecio vulgaris*, *Solanum nigrum*, *Silene alba* and *Colvolvulus arvensis*. More distinctive occasionals include *Erodium cicutarium*, *Conyza canadensis*, *Urtica urens*, *Papaver rhoeas* and *Linaria vulgaris* with the grasses *Poa annua*, *Agrostis capillaris*, *A. stolonifera* and *Dactylis glomerata*.

Habitat

The *Descurainio-Anchusetum* is characteristic of disturbed, dry, sandy soils among arable crops in the Continental climate of East Anglia.

D. sophia is perhaps not native to Britain (Rich 1991, Stace 1995) but it is long established and was formerly, according to Salisbury (1964), much more common than now as a plant of waste ground. It remains frequent among arable crops in East Anglia from where Silverside (1977) characterised this assemblage on soils derived from superficials over chalk, reasonably calcareous, often sandy, though not always rapidly draining. Even in the very dry climate of Breckland, where the community was especially distinctive, the soils could be moist, particularly where irrigation was frequent. *Anchusa arvensis*, *Reseda lutea* and *Veronica polita* are three other species here which reflect the combination of light soils in a more Continental climate typical of the community.

This kind of weed vegetation was encountered by Silverside (1977) among a variety of root crops, in barley and in fallow fields. He recognised some tentative sub-associations on soils of varying texture and dryness but in this scheme those samples are included in other different communities.

Zonation and succession

Where soils are somewhat more clayey and calcareous in arable fields, the *Descurainio-Anchusetum* can give way to the *Kickxietum* with the appearance of *Euphorbia exigua*, *Kickxia elatine* and *Chaenorhinum minus*. More intensively fertilised fields usually see a transition to the *Matricaria-Stellaria* community where *Lycopsis arvensis* can persist with some frequency.

Distribution

The community was found by Silverside (1977) only in East Anglia.

Affinities

The *Descurainio-Anchusetum* was first described as an association by Silverside (1977) and has no apparent equivalent anywhere else in Europe. He considered it as an analogue of the *Lycopsietum arvensis* (Raabe 1944) Passarge 1964, an association which is here subsumed within the *Matricaria-Stellaria* community.

Floristic table OV17

Reseda lutea	V (1–8)	*Artemisia vulgaris*	II (1–3)
Polygonum aviculare	V (1–6)	*Dactylis glomerata*	II (1–3)
Elymus repens	V (1–6)	*Medicago lupulina*	II (1–3)
Chenopodium album	V (1–6)	*Anagallis arvensis*	II (1–3)
Bilderdykia convolvulus	IV (1–6)	*Geranium dissectum*	II (1–3)
Descurainia sophia	IV (1–4)	*Linaria vulgaris*	II (1–3)
Anchusa arvensis	IV (1–6)	*Sisymbrium orientale*	I (1–4)
		Diplotaxis muralis	I (1–3)
Veronica persica	III (1–6)	*Bromus sterilis*	I (1–3)
Senecio vulgaris	III (1–3)	*Malva neglecta*	I (1–3)
Stellaria media	III (1–3)	*Echium vulgare*	I (1)
Silene alba	III (1–3)	*Papaver dubium*	I (4)
Matricaria perforata	III (1–8)	*Polygonum persicaria*	I (1–3)
Veronica polita	III (1–3)	*Polygonum nodosum*	I (1–3)
Convolvulus arvensis	III (1–3)	*Lamium amplexicaule*	I (1)
Solanum nigrum	III (1–3)	*Sisymbrium officinale*	I (1–3)
Chamomilla suaveolens	III (1–6)	*Sinapis arvensis*	I (1–3)
Poa annua	II (1–4)	*Medicago sativa*	I (1)
Erodium cicutarium	II (1–4)	*Silene vulgaris*	I (1–3)
Conyza canadensis	II (1–3)	*Rumex crispus*	I (1)
Agrostis capillaris	II (1–3)	*Urtica dioica*	I (1–3)
Spergula arvensis	II (1–6)		
Urtica urens	II (1–6)	Number of samples	17
Papaver rhoeas	II (1–3)	Number of species/sample	17 (7–30)
Capsella bursa-pastoris	II (1–3)		
Agrostis stolonifera	II (1–4)	Vegetation cover (%)	62 (15–90)

OV18
Polygonum aviculare-Chamomilla suaveolens community

Constant species
Capsella bursa-pastoris, Chamomilla suaveolens, Lolium perenne, Poa annua, Polygonum aviculare.

Physiognomy
The *Polygonum aviculare-Chamomilla suaveolens* community comprises open swards characterised in late spring and summer by mixtures of *Polygonum aviculare, Chamomilla suaveolens, Capsella bursa-pastoris, Poa annua* and young, scattered plants of *Lolium perenne. Matricaria perforata* and *Chenopodium album* are occasional to frequent but they do not dominate and typically other associates are relatively few in number and not abundant. Smaller herbs, such as *Urtica urens, Veronica persica, Anagallis arvensis, Lamium purpureum, Plantago lanceolata* and *Medicago lupulina*, or grasses like *Elymus repens, Dactylis glomerata* and *Agrostis stolonifera* are commonest among these companions with just very occasional taller herbs such as *Rumex obtusifolius, R. crispus* and *Artemisia vulgaris*. Bryophytes are absent or very sparse.

Sub-communities

***Sisymbrium officinale-Polygonum arenastrum* sub-community.** *Polygonum arenastrum* is a frequent preferential here, forming denser prostrate mats among the more branched and somewhat ascending *P. aviculare*, and *Bilderdykia convolvulus* also occurs occasionally, its long flexuous stems trailing over the ground and other low herbs. *Sisymbrium officinale* is common, too, with *Sinapis arvensis, Stellaria media* and *Senecio vulgaris* occasional.

***Plantago major* sub-community.** Scattered plants of *Plantago major* and, more occasionally, *Coronopus squamatus* are characteristic here, their rosettes often flattened by trampling.

Habitat
The *Polygonum-Chamomilla* community is an ephem-eral vegetation type of disturbed and moderately trampled loamy and sandy soils throughout the lowlands of Britain. It is a virtually ubiquitous feature of paths and gateways in agricultural landscapes, on waste ground and in recreation areas.

In such situations, some longer-lived plants tolerant of trampling, like *Lolium* and *P. major*, can find a place among more ephemeral species. The disturbed muddy conditions also provide a very congenial habitat for *Chamomilla*, an introduction to Britain, supposedly from Oregon in the US, though probably originally from north-east Asia, and first recorded here only in 1871 (Salisbury 1964). It is now found in almost every part of the country (Perring & Walters 1962), probably having benefited from the early era of road travel before surfaced roads were so universal: the fruits have no pappus and are dispersed in mud and rainwash (Salisbury 1964). Typically, *Matricaria perforata* has a subordinate role here to *Chamomilla*, a reflection of its lower tolerance of trampling (Kay 1994).

Of the two sub-communities, the *Sisymbrium-Polygonum* type is perhaps associated with more disturbed situations.

Zonation and succession
The *Polygonum-Chamomilla* community occurs with other weed assemblages and grasslands around tracks and gateways in sequences of vegetation related to the intensity of disturbance and trampling.

Often, the community occupies an intermediate position between the *Poa-Plantago* vegetation of the most heavily trampled situations and various pasture, ley or verge grasslands where treading is not so severe but where grazing or mowing maintain the sward. A widespread pattern is for this community to pass to the *Lolio-Plantaginetum* on recreational swards and frequently cut verges, or to some *Lolium* ley or the *Lolio-Cynosuretum* in agricultural enclosures. On less frequently mown path edges, the *Lolium-Dactylis* grassland can figure or, on disturbed verge margins, the *Poa-Taraxacum* community.

More disturbed tracksides and gateways can also have the *Poa-Senecio* or *Stellaria-Capsella* communities, where species such as *Senecio vulgaris*, *Stellaria media*, *Sinapis arvensis* and *Chenopodium album* join *Poa annua*, *Chamomilla suaveolens*, *Polygonum aviculare* and *Capsella bursa-pastoris* as prominent elements of the flora.

Some of these transitions also represent successional developments from the *Polygonum-Chamomilla* community with a reduction in trampling pressure. Such shifts can continue through the more grassy *Lolium-Dactylis* community or, with disturbance, through the weed communities of the Polygono-Chenopodion and Fumario-Euphorbion. Where disturbance ceases, progression to sub-scrub vegetation of the *Rubus-Holcus* type and then to Prunetalia scrub, is usual.

Distribution

The community is ubiquitous through the lowlands and upland fringes of Britain.

Affinities

Vegetation of this type has been widely recorded from various parts of Continental Europe, sometimes as the *Chamomillo-Polygonetum arenastri* T. Müller in Oberdorfer 1971 (originally the *Matricario-Polygonetum avicularis*), as in Austria (Mucina *et al.* 1993), the *Polygono-Matricarietum discoidea* Br.-Bl. 1930 *emend.* Lohmeyer 1975, as in Germany (Pott 1992), or the *Coronopo-Matricarietum* Sissingh (1966) 1969, as in The Netherlands (Westhoff & den Held 1969). Oberdorfer (1983) subsumed German examples into the *Lolio-Polygonetum* which, in this scheme, is more like the *Poa-Plantago* community. The assemblage was also recorded in Ireland by Braun-Blanquet & Tüxen (1952) and noted as widespread by White & Doyle (1982). Most authors place it in the Polygonion aviculare Br.-Bl. ex Aichinger 1933, or its replacement the Chamomillo-Polygonion arenasti Rivas-Martinez 1975 corr. Rivas-Martinez *et al.* 1991.

Floristic table OV18

	a	b	18
Chamomilla suaveolens	IV (2–6)	V (3–8)	V (2–8)
Lolium perenne	V (2–7)	V (1–6)	V (1–7)
Polygonum aviculare	III (2–4)	V (2–10)	V (2–10)
Capsella bursa-pastoris	IV (1–8)	IV (1–8)	IV (1–8)
Poa annua	III (2–4)	IV (1–7)	IV (1–7)
Sisymbrium officinale	III (2–4)	I (3)	II (2–4)
Polygonum arenastrum	III (2–6)		II (2–6)
Sinapis arvensis	II (2–5)	I (3)	I (2–5)
Stellaria media	II (2–3)	I (1–5)	I (1–5)
Senecio vulgaris	II (1–3)	I (2)	I (1–3)
Bilderdykia convolvulus	II (2–4)		I (2–4)
Agrostis capillaris	II (2–4)		I (2–4)
Crepis capillaris	I (3–4)		I (3–4)
Silene alba	I (2–4)		I (2–4)
Malva sylvestris	I (1–5)		I (1–5)
Reseda lutea	I (3–4)		I (3–4)
Agrimonia eupatoria	I (2–4)		I (2–4)
Heracleum sphondylium	I (2–3)		I (2–3)
Arrhenatherum elatius	I (2–4)		I (2–4)
Poa pratensis	I (3–4)		I (3–4)
Plantago major	I (3)	IV (1–5)	III (1–5)
Coronopus squamatus	I (2)	II (2–3)	II (2–3)
Matricaria perforata	III (2–6)	II (2–6)	II (2–6)
Chenopodium album	II (1–6)	II (2–3)	II (1–6)
Elymus repens	I (2–6)	I (1–4)	I (1–6)

Floristic table OV18 (*cont.*)

	a	b	18
Urtica urens	I (3)	I (4)	I (3–4)
Veronica persica	I (1)	I (3–8)	I (1–8)
Anagallis arvensis	I (2)	I (3)	I (2–3)
Plantago lanceolata	I (4–5)	I (2–3)	I (2–5)
Dactylis glomerata	I (2–5)	I (3–4)	I (2–5)
Agrostis stolonifera	I (3–4)	I (3–4)	I (3–4)
Taraxacum officinale agg.	I (2–3)	I (1–3)	I (1–3)
Lamium purpureum	I (2–3)	I (2–3)	I (2–3)
Medicago lupulina	I (2–3)	I (2)	I (2–3)
Rumex obtusifolius	I (2)	I (1–2)	I (1–2)
Papaver rhoeas	I (2–4)	I (4)	I (2–4)
Convulvulus arvensis	I (3)	I (2–3)	I (2–3)
Trifolium pratense	I (1–9)	I (2)	I (1–9)
Artemisia vulgaris	I (2–3)	I (2)	I (2–3)
Hordeum murinum	I (6)	I (2)	I (2–6)
Rumex crispus	I (4)	I (1–3)	I (1–4)
Fumaria officinalis	I (3–4)	I (1)	I (1–4)
Coronopus didymus	I (2–5)	I (1–5)	I (1–5)
Sonchus asper	I (3–4)	I (2)	I (2–4)
Plantago media	I (2–3)	I (3)	I (2–3)
Veronica chamaedrys	I (2–3)	I (3)	I (2–3)
Thlaspi arvense	I (3)	I (2)	I (2–3)
Galium aparine	I (4)	I (2)	I (2–4)
Number of samples	16	45	61
Number of species/sample	15 (6–29)	9 (4–22)	10 (4–29)

a   *Sisymbrium officinale-Polygonum arenastrum* sub-community

b   *Plantago major* sub-community

18   *Polygonum aviculare-Chamomilla suaveolens* community (total)

OV19
Poa annua-Matricaria perforata community

Constant species

Elymus repens, Matricaria perforata, Poa annua.

Physiognomy

The *Poa annua-Matricaria perforata* community includes coarse weedy vegetation with a variety of more ephemeral herbs, some small, others more bulky, and some perennial grasses. *Poa annua* is the commonest grass but *Elymus repens* and *Agrostis stolonifera* are frequent in many of the sub-communities and *Lolium perenne* is also often prominent. *Matricaria perforata* and, somewhat less commonly, *Chamomilla suaveolens*, are characteristic, too, with *Polygonum aviculare*. No other associates of the community as a whole are frequent throughout but *Capsella bursa-pastoris, Holcus lanatus, Chenopodium album, Rumex obtusifolius* and *R. crispus* occur commonly in several sub-communities. *Plantago lanceolata, Taraxacum officinale* agg., *Stellaria media, Sinapis arvensis* and *Anagallis arvensis* are scarce throughout. Bryophytes are sparse but *Pottia truncata, Funaria hygrometrica, Bryum argenteum, B. caespiticium* and *B. microerythrocarpon* can occasionally be seen, sometimes in great abundance.

Sub-communities

Senecio squalidus-Epilobium angustifolium **sub-community.** Perennial grasses are very sparse here and mixtures of *Poa annua* and *Matricaria perforata* with *Chamomilla suaveolens* and *Polygonum aviculare* form the bulk of the cover with *Senecio squalidus* and *Epilobium angustifolium* frequent preferentials. Scattered plants of *Tussilago farfara* and *Reseda lutea* can sometimes be seen and carpets of *Funaria hygrometrica* and *Bryum* spp. are occasionally very extensive.

Lolium perenne-Capsella bursa-pastoris **sub-community.** *Lolium perenne, Holcus lanatus* and *Poa trivialis* are all frequent in this sub-community along with *Capsella*

bursa-pastoris and *Polygonum persicaria*. *Papaver rhoeas, Lapsana communis, Chenopodium album, Plantago major, Atriplex prostrata, Rumex obtusifolius* and *Cirsium arvense* occur occasionally.

Atriplex prostrata-Chenopodium album **sub-community.** *C. bursa-pastoris* and the grasses of the above sub-community all remain quite frequent here, but *Atriplex prostrata, A. patula, C. album, Plantago major, Sonchus oleraceus* and *Medicago lupulina* are all additionally preferential. *Rumex crispus, R. obtusifolius, Cirsium arvense, C. vulgare, Polygonum arenastrum, Senecio vulgaris* and *Artemisia vulgaris* are occasional. *Vulpia myuros* and *Lactuca serriola* are scarcer plants sometimes recorded in this vegetation.

Chamomilla suaveolens-Plantago major **sub-community.** Of the characteristic species of the previous sub-community, only *P. major* remains at all frequent here but coarser herbs like *Rumex obtusifolius, R. crispus, Cirsium arvense* and *Sonchus asper* are frequent.

Elymus repens **sub-community.** *E. repens* attains its peak of frequency in this sub-community but there are few other distinguishing features.

Habitat

The *Poa-Matricaria* community is an ephemeral vegetation type characteristic of disturbed verge edges along roads, on farm tracks and around gateways where there is only moderate trampling.

Matricaria perforata is most often found as a weed of arable land, farmyards, hen-runs and pig fields (Kay 1994) in assemblages like the *Stellaria-Polygonum* community, but it can persist in the distinctive *Poa-Matricaria* vegetation where disturbance by traffic and spray-wash from vehicles helps prevent establishment of closed perennial weedy vegetation or grassland. Kay (1994) noted that *M. perforata* seemed to be increasing in frequency in such situations along heavily-used roads. There, smaller ephemerals like *Anagallis arvensis*, *Myosotis arvensis*, *Veronica persica* and *V. arvensis*, characteristic of arable crops, are less able to get a hold. Similar conditions can be found in the disturbed gateways of arable fields.

The particular habitat preferences of each of the subcommunities are imperfectly known, but the *Senecio-Epilobium* type is often found on verges and waste ground that have been burned. *E. angustifolium* is a native plant but, like *S. squalidus*, which is a Sicilian species that appears to have escaped from the Botanic Garden in Oxford in the late eighteenth century, it first came to prominent notice on wartime bomb sites which provided an especially congenial habitat. In fact, both species had become widely distributed before this time: they disperse very efficiently by wind-borne fruits that are produced in prodigious quantities and which readily germinate on open ground in autumn (Salisbury 1964, Myerscough 1980). Although these plants can produce bulky herbage by the following season, the vegetation remains sufficiently open for *M. perforata* to make some consistent contribution at low cover and for carpets of mosses to be conspicuous.

Of the other sub-communities, the *Lolium-Capsella* type is typical of transitions to sown swards on verges and the *Elymus* type of arable crops.

Zonation and succession

Most commonly, the *Poa-Matricaria* community occurs as a fringe to sown verges, with *Lolium-Dactylis* vegetation or the *Lolio-Plantaginetum*. With increased trampling as around gateways, it is replaced by the *Polygonum-Chamomilla* community, then by the *Poa-Plantago* community. In arable fields, it gives way to *Stellaria-Polygonum* and related weed assemblages with the shift from the gateway to the crop.

Where disturbance ceases, the *Poa-Matricaria* community is succeeded by grassy Lolio-Plantaginion swards.

Distribution

The community occurs widely in suitable habitats through lowland Britain, particularly in the south and east.

Affinities

Matricaria perforata figures in various weed assemblages of this general type recognised in previous descriptions of Continental vegetation (e.g. Oberdorfer 1983), although no exact equivalent appears to have been described. As elsewhere in Europe, this species is more usually encountered in arable weed communities of the Polygono-Chenopodietalia.

Floristic table OV19

	a	b	c	d	e	19
Matricaria perforata	V (2–4)	IV (1–7)	V (1–8)	IV (2–8)	V (1–4)	V (1–8)
Poa annua	IV (2–8)	III (2–5)	V (2–5)	IV (1–5)	V (1–10)	V (1–10)
Elymus repens		III (1–5)	IV (1–7)	III (1–4)	V (1–10)	IV (1–10)
Senecio squalidus	III (1–5)	I (3–4)	II (1–3)	I (2)		II (1–5)
Epilobium angustifolium	III (2–5)	I (1–2)	I (1)			I (1–5)
Tussilago farfara	II (2–3)		I (1–2)	I (1)		I (1–3)
Funaria hygrometrica	II (7–9)		I (1)			I (1–9)
Bryum argenteum	II (3–10)		I (3–5)			I (3–10)
Reseda lutea	II (2–5)					I (2–5)
Diplotaxis tenuifolia	I (5–7)					I (5–7)
Lolium perenne		IV (3–6)	IV (1–5)	III (2–5)	I (1–2)	III (1–6)
Capsella bursa-pastoris	I (2)	III (1–5)	III (1–3)	II (1–2)	I (1–3)	III (1–5)
Holcus lanatus	I (3)	III (2–5)	II (1–2)	III (2–5)	I (2)	III (1–5)
Polygonum persicaria		III (1–6)	II (2–3)		I (1–7)	II (1–7)
Papaver rhoeas	I (2)	II (1–4)	II (1–2)		I (5)	I (1–5)
Lapsana communis		II (1–2)	I (2)		I (3)	I (1–3)
Rumex acetosa		I (2–3)				I (2–3)
Potentilla reptans		I (2–3)				I (2–3)
Brassica rapa		I (1–4)				I (1–4)
Plantago major	I (2)	II (1–5)	V (2–3)	IV (1–6)	I (1)	II (1–6)
Chenopodium album	II (2–3)	II (1–8)	IV (2–5)	II (1–2)	II (1–3)	II (1–8)
Atriplex prostrata	II (2–8)	II (1–6)	IV (1–5)			II (1–8)
Poa trivialis		III (2–6)	IV (1–5)	I (2)	I (2)	II (1–6)
Sonchus oleraceus	I (3–5)	I (1–3)	IV (1–3)	II (1–2)	I (1)	II (1–5)
Atriplex patula		I (1–4)	IV (1–3)	I (1–3)	I (1–2)	II (1–4)
Medicago lupulina		I (3–6)	IV (1–7)	II (2–3)		II (1–7)
Rumex crispus		I (1–3)	III (1–3)	II (1–2)	I (1)	II (1–3)
Polygonum arenastrum		I (1–4)	III (2–5)	II (4–7)		II (1–7)
Trifolium repens		I (3)	III (1–5)	II (2–5)	II (1–3)	II (1–5)
Senecio vulgaris	I (2)	II (1–5)	III (1–3)	I (1)		II (1–5)
Artemisia vulgaris	I (5–8)	I (3)	II (1–2)	I (1–2)	I (1–2)	I (1–8)

Floristic table OV19 (*cont.*)

	a	b	c	d	e	19
Cirsium vulgare	I (2)	I (1–7)	II (1–2)	I (1)		I (1–7)
Picris echioides		I (3–4)	II (1–3)	I (3)		I (1–4)
Urtica dioica		I (2)	II (2)	I (2)		I (1–2)
Sisymbrium officinale		I (1–8)	II (2–4)		I (1)	I (1–8)
Ranunculus repens		I (2–3)	II (1–2)	I (2)	I (1)	I (1–3)
Epilobium adenocaulon		I (1)	II (1–2)	I (2)	I (2)	I (1–2)
Conyza canadensis			II (1–2)	I (2–3)		I (1–3)
Vulpia myuros			II (1–2)	I (3)		I (1–3)
Lactuca serriola			II (1–7)	I (1)		I (1–7)
Geranium molle			I (1–5)			I (1–5)
Rumex obtusifolius		II (1–5)	II (1–4)	III (1–2)	I (3)	II (1–5)
Cirsium arvense	I (2)	II (1–3)	II (1)	III (1–5)		II (1–5)
Sonchus asper		I (1–6)	I (2)	II (1–4)		I (1–6)
Veronica persica	I (3)	I (1–4)	I (2)	I (3)	II (1–5)	I (1–5)
Bilderdyckia convolvulus		I (3–4)	I (1–2)		II (1–3)	I (1–4)
Bryum rubens					II (1–3)	I (1–3)
Polygonum aviculare	III (1–8)	III (1–8)	V (2–10)	III (1–3)	III (1–9)	III (1–10)
Agrostis stolonifera	I (5–7)	II (2–7)	IV (1–3)	V (2–10)	III (2–5)	III (1–10)
Chamomilla suaveolens	II (3–5)	III (1–4)	III (2–3)	V (1–6)	V (1–6)	III (1–6)
Plantago lanceolata	I (2)	I (2–3)	I (2–3)	I (1)	I (1)	I (1–3)
Taraxacum officinale agg.	I (2)	I (2–4)	I (2)	I (3)	I (1)	I (1–4)
Geranium dissectum		I (1–4)	I (1)	I (2)	I (1)	I (1–4)
Spergula arvensis	I (4)	I (3–5)		I (3)	I (1)	I (1–5)
Pottia truncata		I (3)	I (2)	I (1–2)	I (1–3)	I (1–3)
Bromus sterilis		I (2)	I (1)	I (1)	I (3)	I (1–3)
Urtica urens	I (2–3)	I (3)	I (5)		I (1)	I (1–5)
Stellaria media		I (3–5)	I (1–4)	I (1)		I (1–4)
Dactylis glomerata		I (2–3)	I (2)		I (1)	I (1–3)
Sinapis arvensis		I (1)	I (1–2)	I (2)		I (1–2)
Anagallis arvensis	I (1–3)	I (1)	I (2)		I (1–2)	I (1–3)
Raphanus raphanistrum	I (4)	I (2)	I (2)			I (2–4)

	a	b	c	d	e	19
Heracleum sphondylium		I (1)	I (1)		I (1)	I (1)
Crepis capillaris		I (2)	I (2)	I (3)		I (2–3)
Myosotis arvensis		I (1–3)		I (3)	I (1)	I (1–3)
Filaginella uliginosa		I (3)	I (3)		I (2)	I (2–3)
Sonchus arvensis	I (3)	I (1–4)	I (1)		I (2)	I (1–4)
Anthemis cotula	I (2)	I (3)			I (1)	I (1–3)
Senecio jacobaea	I (2)	I (1)	I (1)			I (1–2)
Solanum nigrum	I (3–4)	I (1)	I (2)			I (1–4)
Achillea millefolium		I (1–5)		I (2–3)	I (1)	I (1–5)
Veronica polita	I (2)	I (6)			I (2–5)	I (2–6)
Coronopus squamatus		I (3)	I (2)		I (1)	I (1–3)
Silene alba			I (2)	I (5)	I (1)	I (1–5)
Triticum aestivum		I (3)	I (2)			I (2–3)
Agrostis capillaris		I (2–3)	I (3)			I (2–3)
Coronopus didymus		I (1–2)		I (2–5)		I (1–5)
Cerastium fontanum		I (1–2)		I (2)		I (1–2)
Chamomilla recutita	I (3–6)	I (5–9)	I (3)			I (3–9)
Convolvulus arvensis		I (2–3)		I (1–3)		I (1–3)
Euphorbia helioscopa					I (1)	I (1–5)
Descurainia sophia	I (2)				I (1–3)	I (1–3)
Juncus bufonius		I (1)		I (2–5)		I (1–5)
Ballota nigra		I (4)	I (2–3)		I (3)	I (2–4)
Vicia sativa nigra			I (2)	I (1–3)		I (1–3)
Picris hieracioides		I (8)		I (1–2)		I (1–8)
Bryum caespiticium	I (2–7)			I (3–8)		I (2–8)
Bryum microerythrocarpon		I (3–5)		I (3–5)	I (1–2)	I (1–5)
Number of samples	22	10	24	12	15	83
Number of species/sample	16 (9–27)	11 (7–19)	21 (9–46)	16 (5–32)	13 (5–24)	16 (5–46)

a   *Senecio squalidus-Epilobium angustifolium* sub-community
b   *Lolium perenne-Capsella bursa-pastoris* sub-community
c   *Atriplex prostrata-Chenopodium album* sub-community
d   *Chamomilla suaveolens-Plantago major* sub-community
e   *Elymus repens* sub-community
19   *Poa annua-Matricaria perforata* community (total)

OV20

Poa annua-Sagina procumbens community
Sagino-Bryetum argentii   Diemont, Sissingh &
Westhoff 1940

Constant species
Poa annua, Sagina procumbens.

Physiognomy
The *Sagino-Bryetum* is a species-poor but highly distinctive community in which cushions of *Bryum argenteum*
and small scattered individuals of *Poa annua* and *Sagina
procumbens* form the most consistent features, often disposed in striking patterns between cobble stones and
pavement cracks. No other species are frequent throughout but *Capsella bursa-pastoris, Plantago major* and
Agrostis stolonifera are occasional and *Stellaria media*
and *Medicago lupulina* can sometimes be seen. Mosses
like *Bryum bicolor* and *Schistidium apocarpum* occur as
scarce associates of *B. argenteum.*

Sub-communities

Typical sub-community. There are no additional consistent features here but *Arenaria serpyllifolia, Juncus
bufonius* and *Polygonum aviculare* occur occasionally
with *Ceratodon purpureus. Sagina apetala* and *S. subulata* can sometimes be found among the *S. procumbens.*

***Lolium perenne-Chamomilla suaveolens* sub-community.**
The vegetation in this sub-community is richer and more
extensive in cover than above, with *L. perenne* and *C.
suaveolens* both constant, *Plantago major* and *Agrostis
stolonifera* frequent, *Dactylis glomerata, P. major* and
Agrostis stolonifera frequent,and *Ranunculus repens*
occasional.

Habitat
The *Sagino-Bryetum* is a very widespread community of
crevices between cobble stones and paving slabs in streets,
on pavements and in courtyards in urban and suburban
areas and around farm and country dwellings, where
there is trampling by pedestrians or light vehicle traffic.

Such crevices provide a demanding habitat which only
rather particular plants are able to exploit (Diemont *et*

al. 1940, Segal 1969). Soil accumulation is very sparse
and treading compacts the material though, because nitrification is slow with the lack of aeration, organic debris
tends to accumulate. Indeed, the uppermost layer of soil
is often simply the compacted humic remains of the vascular plants and mosses. The moisture regime can also
be rather extreme since, after rain, water seeps away only
slowly while, in prolonged periods of dry weather, the
soils may become baked and dusty in the sun. Small
therophytes therefore prevail among the vascular contingent, rapid growers and prostrate and rosette-forming
plants being especially successful.

The composition and appearance of the vegetation are
determined by the width and configuration of the crevices and the amount of trampling. The Typical sub-
community is characteristic of narrower cracks, the
Lolium-Chamomilla sub-community of wider ones and
situations where there is less trampling. The amount of
moisture can also be influential with shaded or more frequently wetted situations (as around watered gardens and
leaking down-spouts) developing a distinctive appearance. No samples were available, but a form of *Sagino-
Bryetum* with *Marchantia polymorpha* is known to occur
in such places in this country (cf. also Segal 1969).

Zonation and succession
The *Sagino-Bryetum* often occurs isolated from other
vegetation types in urban streets and yards. Among care-
fully kept stonework, only the *Parietaria* and *Cymbalaria* communities or *Asplenium* vegetation among wall
cracks may accompany it to bring some touch of green
to the built environment. With less assiduous street
maintenance, the *Poa-Taraxacum* community is also
often present down pavement edges and gross neglect
brings other Polygonion and Lolio-Plantaginion assemblages, as well as tall-herb weed communities. Among
urban ruins, for example, or ancient monuments where
there is still some trampling, the *Sagino-Bryetum* often
gives way to the *Poa-Plantago* and *Polygonum-
Chamomilla* communities in places where crevices are

larger and soil accumulation more extensive. On resown areas, the *Lolium-Dactylis* community can also figure.

Some of this patterning represents successional developments from the *Sagino-Bryetum* to more complex herbaceous vegetation, though it is only the *Lolium- Chamomilla* sub-community of larger crevices that is readily colonised by bigger herbs. The Typical sub-community, in its very narrow cracks, is more resistant to such invasion, though trees and shrubs may colonise directly and prise open cobbles and paving as their roots grow. *Sambucus nigra*, *Acer pseudoplatanus* and the garden escape *Buddleja davidii* are commonly seen in such situations.

Distribution
The community occurs ubiquitously through the lowlands and upland fringes.

Affinities
As recognisable in the courtyards of Berlin, Paris, Amsterdam, Dublin and Vienna as in London, the *Sagino-Bryetum* has often figured in accounts of vegetation elsewhere in Europe (Westhoff & den Held 1969, Matuszkiewicz 1981, Brun-Hool & Wilmanns 1982, Oberdorfer 1983, Pott 1992, Mucina *et al.* 1993) and was part of Segal's (1969) compendious review of wall, pavement and street vegetation. With Oberdorfer (1983), we have placed this in the Polygonion, though many authors now recognise a distinct alliance, the Saginion procumbentis R.Tx. & Oberdorfer in Géhu *et al.* 1972. This is usually located in the Polygono-Poetea although Westhoff & den Held (1969) place the Sagino-Bryetum in a Polygono-Coronopion alliance in the Stellarietea.

Floristic table OV20

	a	b	20
Poa annua	V (1–8)	V (3–8)	V (1–8)
Sagina procumbens	IV (1–6)	IV (1–4)	IV (1–6)
Arenaria serpyllifolia	II (2–5)	I (1)	I (1–5)
Polygonum aviculare	II (2–5)		I (2–5)
Ceratodon purpureus	II (1–7)		I (1–7)
Juncus bufonius	II (1–5)		I (1–5)
Cardamine hirsuta	I (1–3)		I (1–3)
Erodium cicutarium	I (5)		I (5)
Rumex obtusifolius	I (2–3)		I (2–3)
Honkenya peploides	I (1)		I (1)
Lythrum portula	I (2–3)		I (2–3)
Rorippa islandica	I (2–4)		I (2–4)
Filaginella uliginosa	I (2–6)		I (2–6)
Sagina apetala	I (2–3)		I (2–3)
Sagina subulata	I (1–6)		I (1–6)
Lolium perenne	II (1)	V (1–5)	III (1–5)
Chamomilla suaveolens	II (1–7)	V (1–4)	III (1–7)
Plantago major	II (1–5)	III (2–4)	II (1–5)
Agrostis stolonifera	II (1–3)	III (1–5)	II (1–5)
Dactylis glomerata	I (2)	II (1–3)	I (1–3)
Plantago lanceolata		II (2–3)	I (2–3)
Taraxacum officinale agg.		II (1–2)	I (1–2)
Ranunculus repens		II (1–4)	I (1–4)
Senecio vulgaris		I (1–3)	I (1–3)
Senecio jacobaea		I (1–3)	I (1–3)
Bryum argenteum	III (2–6)	III (1–4)	III (1–6)
Capsella bursa-pastoris	II (2–3)	II (1––3)	II (1–3)

Floristic table OV20 (*cont.*)

	a	b	20
Bryum bicolor	I (1–3)	I (1)	I (1–3)
Stellaria media	I (3)	I (2)	I (2–3)
Medicago lupulina	I (2–3)	I (1–4)	I (1–4)
Atriplex patula	I (3)	I (1)	I (1–3)
Schistidium apocarpum	I (2)	I (2–4)	I (2–4)
Rumex crispus	I (1)	I (2–4)	I (1–4)
Cirsium arvense	I (2)	I (1)	I (1–2)
Holcus lanatus	I (1)	I (2–3)	I (1–3)
Trifolum repens	I (1–2)	I (1)	I (1–2)
Rumex acetosella	I (2)	I (2)	I (2)
Number of samples	24	10	34
Number of species/sample	8 (3–21)	15 (6–42)	10 (3–42)

a   Typical sub-community
b   *Lolium perenne-Chamomilla suaveolens* sub-community
20   *Sagino-Bryetum argentii* (total)

OV21
Poa annua-Plantago major community

Constant species
Chamomilla suaveolens, Plantago major, Poa annua.

Physiognomy
The *Poa annua-Plantago major* community comprises open swards in which rosettes of *Plantago major* and scattered plants of *Poa annua* and *Chamomilla suaveolens* are the only consistent feature. In contrast to the *Polygonum-Chamomilla* community, *P. aviculare* and *Lolium perenne* are of very variable frequency in the different sub-communities and *Capsella bursa-pastoris* is never more than occasional. The scarcity of *Matricaria perforata* and *Elymus repens* provide a distinction from the *Poa-Matricaria* community. *Agrostis stolonifera* and *Taraxacum officinale* agg. are occasional through the community and there are scarce records for *Rumex crispus, R. obtusifolius, Bellis perennis, Filaginella uliginosa, Medicago lupulina, Alopecurus geniculatus, Anagalllis arvensis* and *Senecio vulgaris.*

Sub-communities

Typical sub-community. Few species other than the three community constants occur with any frequency here but occasionally there is some *Lolium perenne, Polygonum aviculare, Agrostis stolonifera* and *Taraxacum officinale.* More distinctive is the presence in some samples of *Sagina procumbens* and *Bryum argenteum.*

***Lolium perenne* sub-community.** *L. perenne* is a constant associate here with occasional *Capsella bursa-pastoris, Polygonum arenastrum, Trifolium repens, Ranunculus repens, Agrostis stolonifera* and *Taraxacum officinale* agg.

***Polygonum aviculare-Ranunculus repens* sub-community.** *L. perenne* remains quite frequent here, along with *Trifolium repens* but more distinctive is the common occurrence of *P. aviculare, R. repens, A. stolonifera* and *T. officinale* agg. with occasional *Potentilla anserina, Matricaria maritima, Cerastium fontanum, Stellaria media,*

Polygonum persicaria, P. lapathifolium, P. hydropiper, Holcus lanatus and *Dactylis glomerata.*

Habitat
The *Poa-Plantago* community is characteristic of more heavily trampled tracks and gateways throughout the lowlands of Britain, and is a ubiquitous feature of urban recreation areas, wasteland, country paths, roadsides and farms.

This kind of vegetation generally requires more substrate than the *Sagino-Bryetum* though, in the Typical sub-community, it extends on to cobbles and paving where the crevices are a little larger and more robust vascular plants can gain ascendancy there over mosses and diminutive ephemerals. This kind of *Poa-Plantago* vegetation is also extremely common in the heavily-trampled centres of dirt paths and gateways where treading and disturbance are often combined with periodic wetting by rain, creating a congenial substrate for colonisation from spring right through to autumn.

The *Lolium* sub-community is also common in this habitat, particularly where paths run through resown recreational grasslands or along verges and where gateways open on to permanent pastures or leys, from which rye-grass can readily spread. The *Polygonum-Ranunculus* sub-community is more typical of wetter habitats, occurring in trampled and poached areas of ill-drained and periodically-flooded pastures and around watering places for stock by rivers and streams.

Zonation and succession
The *Poa-Plantago* community is very commonly found as part of zonations and mosaics with grasslands and other weed communities where the patterning is related to the degree of disturbance and trampling which the vegetation experiences.

A usual situation is for the *Poa-Plantago* community to occupy the most trampled zone of vegetated ground along paths and in gateways, giving way to the *Polygonum-Chamomilla* community where treading and disturbance are less severe. The *Lolium* sub-community represents an intermediate stage in such a zonation. The sequence may then continue to some kind of *Lolium* ley in enclosed pastures, to the *Lolio-Plantaginetum* in recreational swards of regularly mown verges or the *Lolium-Dactylis* community on infrequently mown or recently-neglected ground. The latter may pass in turn to the *Arrhenathetum* on verges which receive one or two annual cuts. More frequently trimmed verge margins may have the *Poa-Taraxacum* community. The *Poa-Plantago* community is also found as a more abrupt intrusion among calcicolous, calcifuge, dune and cliff-top grasslands, wherever heavy trampling along paths and around viewpoints disrupts and transforms the existing swards.

Wetter tracks, ill-drained or periodically-flooded pastures often have the *Polygonum-Ranunculus* sub-community in trampled and poached places, giving way to the *Festuca-Agrostis-Potentilla* grassland where treading and disturbance are less severe. This in turn can pass to the *Lolio-Cynosuretum* or a *Lolium* ley on better-drained ground.

Successional developments from the *Poa-Plantago*

grassland depend on the intensity of trampling. Where tracks and gateways become disused, the community is probably replaced by the *Polygonum-Chamomilla* or *Lolium-Dactylis* community, or, where there is enrichment from dunging, by *Urtica-Galium* or *Urtica-Cirsium* vegetation. Increased flooding of watering places can lead to the appearance of Elymo-Rumicion vegetation like the *Agrostis-Ranunculus* community or Bidention assemblages such as the *Polygono-Bidentetum* or *Polygonum-Poa* community.

Distribution
The *Poa-Plantago* community is universally distributed through the lowlands and upland fringes.

Affinities
Vegetation of this type figures in accounts from The Netherlands (Westhoff & den Held 1969) and Germany (Oberdorfer 1983) and the nearest equivalent syntaxon seems to be the *Lolio-Polygonetum arenastri* Br.-Bl. 1930 *emend.* Lohmeyer 1975, where *L. perenne*, as here, is not always so frequent as the name of the association implies.

Floristic table OV21

	a	b	c	21
Poa annua	V (1–8)	V (2–8)	IV (2–8)	V (1–8)
Plantago major	IV (1–5)	V (1–7)	IV (1–5)	IV (1–7)
Chamomilla suaveolens	IV (1–6)	V (1–6)	IV (1–7)	IV (1–7)
Bryum argenteum	II (1–5)	I (1–2)	I (3)	I (1–5)
Sagina procumbens	II (1–4)	I (1–2)	I (2)	I (1–4)
Schistidium apocarpum	I (4)			I (4)
Puccinellia distans	I (2–5)			I (2–5)
Lolium perenne	II (1–8)	V (1–9)	III (2–5)	III (1–9)
Capsella bursa-pastoris	I (2–5)	II (1–7)	I (1–3)	I (1–7)
Polygonum arenastrum		II (3–7)	I (4)	I (3–7)
Cynosurus cristatus		I (1–5)		I (1–5)
Leontodon autumnalis		I (1–3)		I (1–3)
Polygonum aviculare	II (2–10)	I (1–5)	IV (1–5)	III (1–10)
Trifolium repens	I (1–3)	III (1–5)	IV (1–5)	III (1–5)
Ranunculus repens	I (2–3)	II (1–4)	IV (1–7)	III (1–7)
Agrostis stolonifera	II (2–10)	II (1–5)	III (1–7)	II (1–10)
Taraxacum officinale agg.	II (2–4)	II (1–5)	III (1–7)	II (1–7)
Matricaria maritima	I (2–3)	I (4)	II (1–8)	I (1–8)
Potentilla anserina		I (1–4)	II (1–5)	I (1–5)
Cerastium fontanum	I (2)	I (3)	II (1–5)	I (1–5)
Holcus lanatus	I (3)		II (2–5)	I (2–5)
Dactylis glomerata		I (1–4)	II (2–3)	I (1–4)
Stellaria media	I (1)	I (1)	II (1–5)	I (1–5)
Juncus bufonius	I (4)	I (1)	II (2–6)	I (1–6)
Polygonum persicaria		I (3)	II (1–7)	I (1–7)

	a	b	c	21
Sonchus asper	I (2)		II (1–3)	I (1–3)
Cirsium arvense	I (1–2)	I (1)	II (1–5)	I (1–5)
Polygonum lapathifolium			I (1–3)	I (1–3)
Urtica dioica			I (1–3)	I (1–3)
Phleum pratense			I (1–2)	I (1–2)
Polygonum hydropiper			I (4–5)	I (4–5)
Bromus sterilis			I (3–8)	I (3–8)
Rumex crispus	I (1–2)	I (1–2)	I (1–4)	I (1–4)
Rumex obtusifolius	I (3)	I (2)	I (3–5)	I (2–5)
Bellis perennis	I (1)	I (1–3)	I (1–3)	I (1–3)
Filaginella uliginosa	I (2–4)	I (1)	I (1–9)	I (1–9)
Medicago lupulina	I (1–3)	I (2)	I (4)	I (1–4)
Alopecurus geniculatus	I (3)	I (1)	I (2–6)	I (1–6)
Anagallis arvensis	I (1–3)	I (2)	I (2–7)	I (1–7)
Senecio vulgaris	I (5)	I (1)	I (3–4)	I (1–5)
Coronopus squamatus	I (1)	I (3–4)		I (1–4)
Aphanes arvensis	I (2)		I (1–2)	I (1–2)
Odonitites verna	I (3)		I (4)	I (3–4)
Cardamine hirsuta	I (1)		I (2–3)	I (1–3)
Kickxia elatine	I (3)		I (3)	I (3)
Papaver rhoeas	I (3)		I (1–2)	I (1–2)
Atriplex prostrata	I (2–4)		I (2)	I (2–4)
Myosotis arvensis	I (3)		I (1–2)	I (1–3)
Viola arvensis	I (2)		I (1–3)	I (1–3)
Senecio jacobaea	I (1–2)		I (1–3)	I (1–3)
Elymus repens	I (1)		I (7)	I (1–7)
Geranium molle	I (1)		I (4)	I (1–4)
Heracelum sphondylium	I (1)		I (1–4)	I (1–4)
Artemisia vulgaris	I (1–5)		I (1–4)	I (1–5)
Reseda lutea	I (3–5)		I (3)	I (3–5)
Veronica officinalis	I (3)		I (3)	I (3)
Prunella vulgaris	I (2–3)		I (2)	I (2–3)
Poa trivialis		I (5–6)	I (2–7)	I (2–7)
Phleum bertolonii		I (2–3)	I (4)	I (2–4)
Epilobium angustifolium		I (2)	I (1–3)	I (1–3)
Trifolium pratense		I (3)	I (2)	I (2–3)
Brachythecium rutabulum		I (1–3)	I (2)	I (1–3)
Plantago media		I (1–4)	I (3)	I (1–4)
Poa pratensis		I (4)	I (3–4)	I (3–4)
Plantago lanceolata		I (2–3)	I (2–4)	I (2–4)
Trifolium dubium		I (1–3)	I (2–3)	I (1–3)
Convolvulus arvensis		I (3)	I (2–3)	I (2–3)
Number of samples	29	32	33	94
Number of species/sample	9 (4–16)	8 (4–20)	18 (7–34)	10 (4–34)

a   Typical sub-community

b   *Lolium perenne* sub-community

c   *Polygonum aviculare-Ranunculus repens* sub-community

21   *Poa annua-Plantago major* community (total)

OV22
Poa annua-Taraxacum officinale community

Constant species
Poa annua, Taraxacum officinale agg.

Physiognomy
The *Poa annua-Taraxacum officinale* agg. community comprises more or less open coarse weedy vegetation in which *Poa annua* and various dandelions are the sole constants. *Plantago major* is frequent throughout and *Senecio vulgaris,Chamomilla suaveolens, Lolium perenne* and *Stellaria media* are common in various of the sub-communities but otherwise only *Sagina procumbens* and *Bromus sterilis* are more than scarce.

Sub-communities

Senecio vulgaris sub-community. Apart from the two community constants and *Plantago major*, only *Senecio vulgaris* and *S. squalidus* are frequent here with *Cerastium fontanum, Matricaria maritima, Holcus lanatus* and *Veronica arvensis* occasional.

Cirsium vulgare-C. arvense sub-community. *L. perenne* and *C. suaveolens* join *Poa annua* and *Taraxacum officinale* agg. as constants in this sub-community and *Stellaria media* and *Trifolium repens* are common but more distinctive is the high frequency of the coarse tall herbs *Cirsium vulgare, C. arvense, Rumex obtusifolius, R. crispus* and *Chenopodium album.*

Crepis vesicaria-Epilobium adenocaulon sub-community. *C. suaveolens* and *P. major* remain as frequent associates here but more distinctively preferential are *Crepis vesicaria* and *Epilobium adenocaulon* with *Sonchus oleraceus, Poa trivialis, Trifolium dubium, Cardamine hirsuta, Coronopus didymus* and *Capsella bursa-pastoris* occasional.

Habitat
The *Poa-Taraxacum* community occurs on disturbed, but only lightly trampled, loamy soils along tracksides, pathways, wall-bottoms, pavement edges and road verges, in gardens, farm fields and on waste ground.
Periodic disturbance without too much treading is the typical feature of the habitat here. Such conditions can be found along verge edges where the *Senecio* sub-community is often found or on recently turned and neglected garden soil, ploughed-up grassland, waste ground, and churned up tracks where the *Cirsium* sub-community commonly occurs. This kind of *Poa-Taraxacum* vegetation is also frequently seen in leys and resown pastures where seeding has been poor or subsequent grazing management injudicious.
The *Crepis-Epilobium* sub-community is especially frequent in southern Britain where the two most characteristic preferentials began their spread from late last century – *C. vesicaria* is a south-west European introduction, *E. adenocaulon* is from North America – but it can be found widely further north, especially on waste ground and in the distinctive habitat of wall-bottoms and pavement edges.

Zonation and succession
Where the *Poa-Taraxacum* community occurs in poorly-seeded and badly-managed leys and pastures as the *Cirsium* sub-community, it typically passes to some kind of Lolio-Plantaginion grassland where the sward is more intact or, around trampled areas, to the *Polygonum-Chamomilla* or *Poa-Plantago* communities. On verges, too, away from the edge, the *Poa-Taraxacum* assemblage is replaced by Lolio-Plantaginion grasslands or, where management is not so assiduous, by the *Lolium-Dactylis* community. On waste ground, patches of *Urtica-Cirsium* vegetation are a common feature of these patterns.

Distribution
The community occurs very widely throughout the country.

Affinities
Vegetation of this kind, though extremely widespread and common, has rarely attracted much attention and has sometimes been subsumed in other Lolio-Plantaginion assemblages. Hutchinson & Seymour (1982), in their study of *Poa annua*, noted this sort of community and Oberdorfer (1983) recognised a similar *Poa annua*-Gesellschaft.

Floristic table OV22

	a	b	c	22
Poa annua	V (1–7)	IV (2–9)	V (2–6)	V (1–9)
Taraxacum officinale agg.	IV (1–4)	IV (1–3)	IV (2–7)	IV (1–7)
Senecio vulgaris	IV (1–7)	III (1–4)	I (1–2)	III (1–7)
Senecio squalidus	III (1–4)	I (1–4)	II (1–4)	II (1–4)
Cerastium fontanum	II (2–3)	I (3)		II (2–3)
Matricaria maritima	II (2–5)	I (1–3)	I (1–4)	II (1–5)
Holcus lanatus	II (1–3)	I (2)	I (2)	I (1–3)
Veronica arvensis	II (1–5)	I (2)	I (5)	I (1–5)
Reseda lutea	I (3–7)			I (3–7)
Silene alba	I (3–8)			I (3–8)
Cirsium vulgare	I (1–2)	V (1–3)	II (1–3)	III (1–3)
Chamomilla suaveolens	II (2–3)	IV (1–4)	III (2–4)	III (1–4)
Lolium perenne	II (2)	IV (2–7)	II (2–4)	III (2–7)
Cirsium arvense	I (1–2)	IV (1–6)	I (1–2)	II (1–6)
Stellaria media	II (1–3)	III (1–7)	II (1–3)	II (1–7)
Trifolium repens		III (2–6)	II (1–8)	II (1–8)
Rumex obtusifolius	I (2–4)	III (1–5)	II (1–3)	II (1–5)
Chenopodium album	I (1–2)	II (1–2)	I (1)	I (1–2)
Rumex crispus	I (1–3)	II (1–7)		I (1–7)
Plantago lanceolata		II (2–5)		I (2–5)
Ranunculus repens		II (1–4)		I (1–4)
Spergula arvensis		I (2–3)		I (2–3)
Lolium multiflorum		I (2–5)		I (2–5)
Conium maculatum		I (2–3)		I (2–3)
Elymus repens		I (3–7)		I (3–7)
Crepis vesicaria	I (4)	II (1–4)	IV (2–6)	III (1–6)
Epilobium adenocaulon	I (2)	I (2–3)	IV (1–5)	II (1–5)
Sonchus oleraceus	II (1–2)	I (2)	III (1–3)	II (1–3)
Poa trivialis	I (3)	I (2–3)	II (2–5)	II (2–5)
Trifolium dubium		I (2)	II (1–7)	I (1–7)
Cardamine hirsuta			II (1–3)	I (1–3)
Coronopus didymus		I (2–3)	II (2–6)	I (2–6)
Capsella bursa-pastoris	I (1–3)	I (3)	II (1–7)	I (1–7)
Linaria purpurea			I (2)	I (2)
Cymbalaria muralis			I (1–4)	I (1–4)
Parietaria diffusa			I (2–7)	I (2–7)
Cheiranthus cheiri			I (1–3)	I (1–3)
Plantago major	III (2–4)	III (1–5)	III (1–5)	III (1–5)
Sagina procumbens	II (1–3)		II (2–3)	II (1–3)
Bromus sterilis	I (3)	II (2–6)	II (1–5)	II (1–6)
Medicago lupulina	I (3)	I (1–4)	I (2)	I (1–4)
Agrostis capillaris	I (3)	I (2–3)	I (4)	I (2–4)
Polygonum arenastrum	I (2)	I (7)	I (2–3)	I (2–7)
Heracleum sphondylium	I (4)	I (4)	I (2)	I (2–4)
Achillea millefolium	I (3)	I (1–4)	I (3)	I (1–4)
Malva sylvestris	I (3)	I (2)	I (2–4)	I (2–4)

Floristic table OV22 (*cont.*)

	a	b	c	22
Arrhenatherum elatius	I (3)	I (2–3)	I (3)	I (2–3)
Lactuca serriola	I (2–3)	I (3)	I (1–2)	I (1–3)
Epilobium angustifolium	I (2)	I (2–4)		I (2–4)
Tussilago farfara	I (2)	I (2)		I (2)
Bryum argenteum	I (2)	I (2)		I (2)
Funaria hygrometrica	I (2)	I (5)		I (2–5)
Atriplex prostrata	I (2)	I (3)		I (2–3)
Dactylis glomerata	I (2)	I (2–3)		I (2–3)
Atriplex patula	I (1)	I (2–4)		I (1–4)
Rubus fruticosus agg.	I (2)	I (3–4)		I (2–4)
Convolvulus arvensis	I (2)	I (2–3)		I (2–3)
Lamium purpureum	I (2)	I (2–3)		I (2–3)
Polygonum persicaria	I (1)	I (2–3)		I (1–3)
Potentilla reptans	I (2)	I (1–3)		I (1–3)
Bellis perennis	I (1–6)	I (3)		I (1–6)
Epilobium hirsutum		I (2)	I (2–3)	I (2–3)
Trifolium pratense		I (3)	I (2–6)	I (2–6)
Hordeum murinum		I (2–3)	I (2–5)	I (2–5)
Bromus hordeaceus hordeaceus		I (2–5)	I (1–5)	I (1–5)
Sisymbrium officinale		I (2–3)	I (3–4)	I (2–4)
Cerastium glomeratum		I (3)	I (2–3)	I (2–3)
Number of samples	12	15	18	45
Number of species/sample	14 (9–23)	19 (7–42)	12 (6–23)	14 (6–42)

a   *Senecio vulgaris* sub-community
b   *Cirsium vulgare-Cirsium arvense* sub-community
c   *Crepis vesicaria-Epilobium adenocaulon* sub-community
22 *Poa annua-Taraxacum officinale* community (total)

OV23
Lolium perenne-Dactylis glomerata community

Constant species

Dactylis glomerata, Lolium perenne, Plantago lanceolata, Taraxacum officinale agg.

Physiognomy

The *Lolium perenne-Dactylis glomerata* community comprises coarse weedy grassland vegetation in which *Lolium perenne* and *Dactylis glomerata* usually make up the bulk of the more or less closed cover, along with a variety of perennial associates and scattered ephemerals which find a place in locally disturbed places. *Plantago lanceolata* and *Taraxacum officinale* agg. are the commonest of these companions but *Achillea millefolium*, *Plantago major*, *Trifolium pratense*, *Agrostis stolonifera*, *Urtica dioica*, *Hypochoeris radicata* and *Potentilla reptans* all figure occasionally among the perennials, *Poa annua*, *Bromus hordeaceus* ssp. *hordeaceus* and *B. sterilis* among the annuals.

Sub-communities

Typical sub-community. Apart from the community constants, only frequent records for *Trifolium dubium* and *Hordeum murinum* with occasional *Vicia sativa* and *Senecio squalidus* are distinctive here.

Crepis vesicaria-Rumex obtusifolius sub-community. *Poa annua* becomes constant here but better preferentials are *Crepis vesicaria* and *Rumex obtusifolius* with *Poa trivialis*, *Senecio vulgaris* and *Cirsium arvense* occasional. Seedlings of *Buddleja davidii* are sometimes found.

Plantago major-Trifolium repens sub-community. *Poa annua* and *Holcus lanatus* remain very frequent here but *Plantago major* and *Trifolium repens* are more distinctive with *Ranunculus repens* and *Rumex crispus* occasional.

Arrhenatherum elatius-Medicago lupulina sub-community. The grass contingent of the vegetation is further augmented here by constant *H. lanatus* and, more preferential, *Arrhenatherum elatius* and *Agrostis capillaris*.

Also very frequent are *Achillea millefolium*, *Medicago lupulina* with occasional *Cerastium fontanum*, *Vicia sativa* and taller herbs such as *Artemisia vulgaris*, *Daucus carota*, *Heracleum sphondylium*, *Senecio jacobaea*, *Centaurea nigra* and, on chalky soils in the south-east, *Cichorium intybus*.

Habitat

The *Lolium-Dactylis* community is characteristic of resown recreation areas like verges, playing fields and institutional grounds where there is only occasional summer mowing, continuing disturbance or a measure of neglect.

Reseeding of disturbed ground or made areas around residential buildings, institutions, factories and urban road schemes often involves the use of rye-dominated mixtures (Hubbard 1968). In such situations, with periodic mowing through the growing season but little else by way of management, bulky perennial grasses are able to maintain some ascendancy over smaller and more ephemeral plants though local or periodic disturbance often provides opportunity for weedy plants to continue to figure.

Such disturbance may be very particular. The high frequency of *Hordeum murinum* in the Typical sub-community, for example, is often seen around lamp-posts and trees on suburban verges where dogs urinate. More widely, trampling provides a source of disturbance and the *Plantago-Trifolium* sub-community is most common around paths through such resown swards where treading favours frequent occurrence of *P. major* and provides an opportunity for *Poa annua* to colonise. The *Crepis-Rumex* sub-community experiences more gross disturbance, being typical of churned-up verges and waste ground.

By contrast, the *Arrhenatherum-Medicago* sub-community is found on those resown verges and recreational areas where mowing occurs but once or twice each spring or summer, or where abandonment of management favours the further spread of bulkier grasses and taller dicotyledonous herbs.

Zonation and succession

The *Lolium-Dactylis* community is commonly found in zonations and mosaics with other grasslands and weed communities, on verges, recreation and waste ground, patterns being dependent upon the frequency of disturbance, trampling and mowing.

Paths through stretches of this vegetation usually see a sharp transition through the *Plantago-Trifolium* sub-community to the *Poa-Plantago* community along the trampled strip (Figure 25). Where there is more extensive trampling, an intervening zone of *Polygonum-Chamomilla* vegetation can mark the areas with lighter

Figure 25. Vegetation pattern on an ill-maintained urban street.
The sown and occasionally mown strips of turf on the pavement are Typical OV23a *Lolium-Dactylis* vegetation with an abundance of *Hordeum murinum* around the lamp-post indicating a favourite spot for dogs to urinate. More trampled sections of the verge have the OV23c *Plantago-Trifolium* sub-community, giving way to the OV21b *Lolium* sub-community of *Poa-Plantago* vegetation where there is heavier pedestrian pressure. Between the cobbles of the much-used snicket behind, there is Typical OV20a *Sagino-Bryetum argentii* and, along the crevices at the foot of the wall, the OV22c *Crepis-Epilobium* sub-community of *Poa-Taraxacum* vegetation. In crevices on the wall itself, small stands of OV41 *Parietarietum judaicae* can be seen, with a stand of OV24 *Urtica-Galium* vegetation in a run-down garden.

treading. Verges with the *Lolium-Dactylis* community may have a disturbed fringe along the roadside with *Poa-Matricaria* vegetation.

Where resown grasslands are less frequently mown or disturbed, *Lolium-Dactylis* vegetation can grade through the *Arrhenatheretum-Medicago* sub-community to the *Arrhenatheretum* and this can represent a common successional development where management becomes less intensive. On verges or in recreation areas where this happens, the usual further stage is for *Rubus-Holcus* underscrub and *Crataegus-Hedera* scrub to develop. Similar mixtures of rank grasslands and woody vegetation can be found on wasteland and abandoned building sites where *Lolium-Dactylis* vegetation spreads on to spoil heaps.

Distribution

The community is ubiquitous through the British lowlands.

Affinities

This is a difficult assemblage to place within a phytosociological frame because it lies close to the border between the Lolio-Plantaginion alliance and the Arrhenatherion where coarser grasses like *Dactylis*, *H. lanatus* and *Arrhenatherum* become important. On balance, it seems better to locate it in the former, as a weedier assemblage than the swards included in this scheme among the *Lolium* leys. It has no direct equivalent in the European literature.

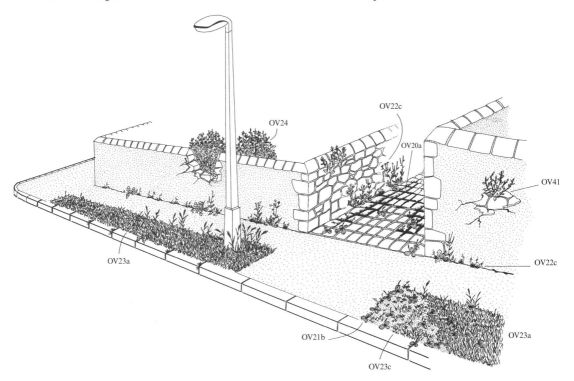

Floristic table OV23

	a	b	c	d	23
Lolium perenne	V (1–8)	IV (3–5)	V (2–7)	IV (2–7)	V (1–8)
Dactylis glomerata	IV (2–6)	IV (2–7)	III (1–4)	V (1–8)	IV (1–8)
Plantago lanceolata	V (1–7)	III (2–4)	V (1–5)	IV (2–4)	IV (1–7)
Taraxacum officinale agg.	V (1–4)	IV (2–5)	V (1–4)	III (2–4)	IV (1–5)
Trifolium dubium	III (2–6)	II (2)	I (2–4)	I (4)	II (2–6)
Hordeum murinum	III (1–9)	I (3)		I (3)	I (1–9)
Senecio squalidus	II (1–2)		I (1)	I (2)	I (1–2)
Stellaria media	I (2–3)				I (2–3)
Crepis vesicaria	II (2)	IV (2–5)	I (1)		II (1–5)
Rumex obtusifolius		IV (1–5)	II (1–4)	I (3)	II (1–5)
Poa trivialis	I (2)	II (4–8)	I (3)	I (4)	I (1–8)
Buddleja davidii seedling		II (2–4)	I (7)		I (2–7)
Senecio vulgaris	I (1–2)	II (1–3)	I (1–3)		I (1–3)
Cirsium arvense	I (1)	II (2–4)	I (2)		I (1–4)
Poa annua	I (3–5)	IV (2–5)	IV (2–6)	II (2–4)	III (2–6)
Plantago major	I (1)	II (1–2)	IV (1–4)	II (3–4)	II (1–4)
Trifolium repens	II (2–5)	I (4–5)	IV (2–6)	II (1–4)	II (1–6)
Ranunculus repens	I (3)	II (2–3)	III (1–3)	I (2–3)	II (1–3)
Rumex crispus	I (3)	I (3)	II (2–4)	I (1–3)	I (1–4)
Spergula arvensis			I (5–7)		I (5–7)
Polygonum arenastrum			I (2–3)		I (2–3)
Phleum bertolonii			I (1–4)		I (1–4)
Plantago coronopus			I (1–3)		I (1–3)
Achillea millefolium	III (2–5)	I (1)	II (2–4)	IV (2–5)	III (1–5)
Holcus lanatus	I (3)	I (2–5)	III (1–5)	IV (2–4)	II (1–5)
Medicago lupulina	I (3–4)		II (3–8)	IV (2–5)	II (2–8)
Arrhenatherum elatius	I (4)	I (4)	I (1)	IV (3–4)	II (1–4)
Agrostis capillaris			I (4)	III (3–5)	I (3–5)
Artemisia vulgaris		I (6)	I (1)	II (2–6)	I (1–6)
Cerastium fontanum	I (2)	I (4)	I (2)	II (1–4)	I (1–4)

Floristic table OV23 (*cont.*)

	a	b	c	d	23
Heracleum sphondylium			I (1)	II (1–3)	I (1–3)
Daucus carota			I (2)	II (2–4)	I (2–4)
Brachythecium rutabulum			I (1–3)	II (1–4)	I (1–4)
Senecio jacobaea				II (2–4)	I (2–4)
Festuca ovina				II (2–4)	I (2–4)
Cichorium intybus				II (4)	I (4)
Centaurea nigra				II (2–4)	I (2–4)
Silene vulgaris				II (2–4)	I (2–4)
Torilis japonica				I (2–4)	I (2–4)
Leucanthemum vulgare				I (3–4)	I (3–4)
Epilobium angustifolium				I (2–6)	I (2–6)
Trifolium pratense	II (2–4)	I (2–4)	II (3–8)	II (3–4)	II (2–8)
Bromus hordeaceus hordeaceus	II (1–4)	II (2–3)	I (1–3)	I (2)	II (1–4)
Urtica dioica	II (1–4)	II (2–5)	I (1)	I (1)	II (1–5)
Vicia sativa	II (3–4)	I (2–3)	I (2)	II (3–6)	II (2–6)
Agrostis stolonifera		I (3)	II (3–8)	II (2–4)	II (2–8)
Hypochoeris radicata	I (1)		II (1–5)	II (2–4)	II (1–5)
Bromus sterilis	II (1–5)	II (2–7)			I (1–7)
Crepis capillaris	II (2–3)		I (2–4)	II (1–4)	I (1–4)
Potentilla reptans	II (2–5)		I (2–4)	II (2–4)	I (2–5)
Chamomilla suaveolens	I (2)	I (1)	I (3)	I (3)	I (2–3)
Sonchus oleraceus	I (4)	I (2)	I (1–2)	I (3)	I (1–4)
Capsella bursa-pastoris	I (2)	I (1)	I (2–5)	I (2)	I (1–5)
Geranium dissectum	I (3)	I (3)	I (1–3)	I (5)	I (2–5)
Sonchus asper	I (1)	I (3)	I (1–3)	I (3)	I (1–3)
Bellis perennis	I (2)	I (2)	I (2–3)	I (3)	I (2–3)
Poa pratensis	I (3–4)	I (2–3)	I (4–5)		I (2–5)
Sisymbrium officinale	I (2–4)	I (3)	I (1)		I (1–4)
Malva sylvestris	I (5–6)	I (3)	I (3)		I (3–6)
Cynosurus cristatus	I (3)	I (2)	I (2)		I (2–3)
Tragopogon pratensis	I (1)	I (4)		I (3)	I (1–4)
Medicago arabica	I (7)	I (3–7)		I (7)	I (3–7)

	a	b	c	d	23
Ceratodon purpureus	I (2)		I (3)	I (1)	I (1–3)
Cirsium vulgare		I (4)	I (1–4)	I (2–3)	I (1–4)
Holcus mollis	I (2)	I (1)	I (1)		I (1–2)
Vulpia myuros	I (4)				I (1–4)
Tussilago farfara	I (7)			I (5)	I (5–7)
Lapsana communis	I (3)		I (2)		I (2–3)
Cymbalaria muralis		I (2)	I (3)		I (2–3)
Ranunculus bulbosus	I (3)	I (2)			I (2–3)
Cerastium glomeratum		I (1)	I (1–2)		I (1–2)
Bryum argenteum			I (1–3)	I (2–5)	I (1–5)
Sagina procumbens			I (1–3)	I (2)	I (1–3)
Festuca rubra			I (1–2)	I (3)	I (1–3)
Galium aparine		I (1)	I (2–3)		I (1–3)
Matricaria maritima			I (4–7)	I (3)	I (3–7)
Veronica arvensis		I (3)	I (1)		I (1–3)
Picris echioides		I (5)	I (1–3)		I (1–5)
Number of samples	13	13	16	14	56
Number of species/sample	12 (7–19)	12 (8–20)	16 (8–35)	19 (8–30)	15 (7–35)

a   Typical sub-community

b   *Crepis vesicaria-Rumex obtusifolius* sub-community

c   *Plantago major-Trifolium repens* sub-community

d   *Arrhenatherum elatius-Medicago lupulina* sub-community

23 *Lolium perenne-Dactylis glomerata* community (total)

OV24

Urtica dioica-Galium aparine community

Constant species

Galium aparine, Urtica dioica.

Rare species

Allium triquetrum.

Physiognomy

The *Urtica dioica-Galium aparine* community comprises generally species-poor tall-herb vegetation dominated by often densely abundant *U. dioica*, frequently growing over 1 m high by mid-summer. *G. aparine* is the only other constant throughout and it typically forms sprawls among the nettles. *Poa trivialis* is also common and locally quite extensive as a thin carpet of shoots over the ground and among the nettle stools. Through the community as a whole, no other species occurs with any frequency and, by late autumn, the bulk of the herbage collapses and is rapidly incorporated through the winter to leave ground that can be virtually bare.

Sub-communities

Typical sub-community. Here, the usual very dense nettle and goosegrass cover has scattered plants of *Cirsium arvense* and occasional *Bromus sterilis*. Seaside stands often have *Smyrnium olusatrum*, sometimes in local abundance.

Arrhenatherum elatius-Rubus fruticosus agg. sub-community. In this vegetation, the cover of nettle is not usually so thick and there are scattered tussocks of *Arrhenatherum elatius*, patches of *Rubus fruticosus* agg. and, by early summer, emergent flowering shoots of first, *Anthriscus sylvestris*, and then *Heracleum sphondylium*. In somewhat more open places, there can be seen *Taraxacum officinale* agg., *Lolium perenne, Dactylis glomerata, Bromus mollis, Achillea millefolium* and *Potentilla reptans* while *Hedera helix* can form a patchy ground carpet. Locally, occasionals such as *Conium maculatum, Artemisia vulgaris, Chenopodium album* or *Malva sylvestris* can give a distinctive stamp.

Habitat

The *Urtica-Galium* community is typical of nutrient-rich, moist but well-aerated soils throughout the lowlands, usually where there has been some kind of disturbance. It occurs very widely around dumps of soil and dung, among rubbish on farmland, in gardens and on suburban or industrial wasteland, on disturbed verges and trackways, along wall-bottoms, around abandoned dwellings, rabbit burrows and rotting carcases and on strandline detritus on beaches and salt-marshes.

U. dioica is a perennial that establishes initially from seed that is dispersed by adhesion to animal fur, skin and clothes, by wind, and in dung (Greig-Smith 1948). Germination, which is mostly in spring, needs light and moisture, so establishment among existing vegetation or on compacted soil requires disturbance to open up or loosen bare ground or the dumping of leaf litter, clippings, brashings, dung or rubbish to smother the existing herbage. Subsequent growth is by vegetative spread, sympodial rhizomes extending out beneath the soil surface, stolons growing and rooting at their nodes above ground and both producing aerial shoots. In loose soil, the rhizomes can ramify down to 30 cm or more but compacted substrates greatly hinder establishment and spread.

The luxuriance of the developing clump is affected by light and nutrients. In full sun, or on dry ground, growth is rather stunted; in deep shade, the shoots can grow tall but produce few flowers. In partial shade and with some shelter from drying winds, they often attain 1 m or more, packed densely and flowering profusely. Growth is more pronounced in nutrient-rich habitats, whether this eutrophic aspect is a natural feature of the soil or is related to disturbance of the substrate and nitrification of organic matter or inputs in ground water or rubbish.

In late summer and autumn, a new generation of rhizomes is produced from the base of the aerial shoots and these continue growing until the aerial shoots die. They then turn up and form a new aerial shoots of their own which grow to 15 cm or so and survive the winter before resuming growth in spring. Meanwhile, the foliage and old shoots disintegrate and are often rapidly incorporated into the soil so just sparse dead shoots survive above the new herbage.

In the relatively dense shade of the nettle clumps, and among the closely-packed mass of shoot bases, stolons and rhizomes, there can sometimes be little opportunity for other plants to gain a hold. The puny shoots of the shade-tolerant *Poa trivialis* and trails of *Galium aparine* are thus the only consistent feature here. Where stands are a little more open or where other species have established coincidentally, there can be more diversity. The Typical sub-community includes various more impoverished nettle beds, the *Arrhenatherum-Rubus* type somewhat richer stands where *Arrhenathum*, *R. fruticosus* and *H. sphondylium* occur in situations where disturbance coincides with reduction in grazing or among established rank swards and sub-scrub on verges and waysides.

Zonation and succession
The *Urtica-Galium* community is found in very diverse patchworks and zonations with weed vegetation, mesotrophic grasslands, fens, scrub and woodland. It may give way to woody vegetation in time but invasion of shrubs and trees can be greatly hindered by the dense habit of the dominant nettle.

A common pattern seen on verges, waysides, neglected pastures and ill-managed recreational areas is for the *Urtica-Galium* community to occur patchily around disturbed places in stretches of *Lolium-Dactylis* grassland or among the *Arrhenatheretum* with stretches of *Rubus-Holcus* underscrub marking the progression to woody vegetation. In such situations, the more open *Urtica-Cirsium* community can also occur and there may be whole suites of weedy assemblages along pathways and around gateways.

In wetter habitats, where the *Urtica-Galium* community has developed along streamsides or around ponds,

lakes and fen systems, it can give way on drier ground to the *Arrhenatheretum* and towards open water to the *Phragmites-Urtica* fen with its characteristic patchy dominance of *Urtica*, *Phragmites* and *Epilobium hirsutum*. Stands of the *Epilobium* community or eutrophic swamps like the *Typhetum latifoliae* or *Glycerietum maximae* can also figure in such zonations, or on drier ground, the *Alnus-Urtica* woodland. This kind of forest is probably the ultimate successor of the *Urtica-Galium* community in this sort of habitat.

Disturbance
The *Urtica-Galium* community is ubiquitous through the lowlands, extending into the upland fringes wherever disturbance creates suitable conditions.

Affinities
More species-poor stands of tall-herb vegetation have sometimes created problems for adequate description and integration into phytosociological schemes, so certain authorities recognise so-called 'basal' syntaxa like the *Urtica dioica*-Gesellschaft of Mucina *et al.* (1993) which are allocated, not to any particular alliances, but more generally to a class. Of more precisely defined assemblages, the *Urtica-Galium* community comes closest, in the *Arrhenatherum-Rubus* sub-community, to the *Urtico-Cruciatetum laevipedis* Dierschke 1973 described from Germany (Oberdorfer 1983, Pott 1992) and Austria (Mucina *et al.* 1993). Authorities differ as to whether this is better placed in the Galio-Alliarion (Oberdorfer 1957) Lohmeyer & Oberdorfer in Oberdorfer *et al.* 1967 or the Aegopodion podagrariae R.Tx. 1967. Certainly, with us, this is not usually a woodland fringe community, vegetation types from which are grouped in the latter alliance. Whatever the choice, these two alliances are now separated off from the Artemisietea into a new class, the Galio-Urticetea Passarge & Kopecký 1967 which contains more eutrophic tall-herb and woodland fringe vegetation.

Floristic table OV24

	a	b	24
Urtica dioica	V (1–10)	V (2–9)	V (1–10)
Galium aparine	IV (2–4)	IV (2–5)	IV (2–5)
Cirsium arvense	III (1–4)	II (2–4)	II (1–4)
Bromus sterilis	II (1–7)	I (3–6)	I (1–7)
Smyrnium olusatrum	II (2–7)		I (2–7)
Veronica hederifolia	I (2–3)		I (2–3)
Fumaria officinalis	I (3–4)		I (3–4)
Allium triquetrum	I (3–4)		I (3–4)
Myosotis arvensis	I (1–4)		I (1–4)
Rumex conglomeratus	I (1–3)		I (1–3)
Heracleum sphondylium	II (1–4)	IV (2–7)	III (1–7)
Arrhenatherum elatius	I (1)	IV (2–8)	III (1–8)
Taraxacum officinale agg.	I (3)	III (1–3)	II (1–3)
Rubus fruticosus agg.	I (2–3)	III (1–5)	II (1–5)
Anthriscus sylvestris	I (3–5)	II (4–6)	II (3–6)
Hedera helix	I (7)	II (3–4)	II (3–7)
Lolium perenne	I (2–7)	II (1–5)	II (1–7)
Dactylis glomerata	I (1)	II (1–4)	II (1–4)
Achillea millefolium		II (1–4)	I (1–4)
Potentilla reptans		II (3–5)	I (3–5)
Bromus hordeaceus hordeaceus		II (3–9)	I (3–9)
Artemisia vulgaris		I (4–7)	I (4–7)
Conium maculatum		I (3–4)	I (3–4)
Glechoma hederacea		I (1–3)	I (1–3)
Trifolium pratense		I (2–3)	I (2–3)
Centaurea nigra		I (1–4)	I (1–4)
Chenopodium album		I (2–4)	I (2–4)
Papaver rhoeas		I (3–5)	I (3–5)
Bromus mollis		I (2–4)	I (2–4)
Hordeum murinum		I (3–8)	I (3–8)
Malva sylvestris		I (3–4)	I (3–4)
Veronica chamaedrys		I (2–3)	I (2–3)
Bilderdykia convolvulus		I (3–6)	I (3–6)
Poa trivialis	III (2–3)	III (1–5)	III (1–5)
Convolvulus arvensis	I (3–7)	I (3–4)	I (3–7)
Holcus lanatus	I (3–5)	I (2–3)	I (2–5)
Geranium robertianum	I (1)	I (3)	I (1–3)
Lamium album	I (2–4)	I (3–4)	I (2–4)
Tussilago farfara	I (2)	I (1–4)	I (1–4)
Arctium minus	I (2–4)	I (4–9)	I (2–9)
Phleum pratense	I (5)	I (4)	I (4–5)
Agrostis stolonifera	I (4–7)	I (1)	I (1–7)
Plantago lanceolata	I (3)	I (3–4)	I (3–4)
Plantago major	I (1)	I (1–2)	I (1–2)
Brachythecium rutabulum	I (2–6)	I (8)	I (2–8)

Stellaria media	I (1–5)	I (3)	I (1–5)
Eurhynchium praeolongum	I (5–6)	I (1)	I (1–6)
Geranium dissectum	I (1)	I (4)	I (1–4)
Sisymbrium officinale	I (2)	I (2–3)	I (2–3)
Senecio vulgaris	I (2)	I (3)	I (2–3)
Silene alba	I (8)	I (3–4)	I (3–8)
Lamium purpureum	I (2–3)	I (3)	I (2–3)
Barbarea vulgaris	I (4)	I (3)	I (3–4)
Cerastium fontanum	I (2)	I (3)	I (2–3)
Poa annua	I (2–3)	I (1)	I (1–3)
Veronica arvensis	I (2–3)	I (1)	I (1–3)
Number of samples	15	43	58
Number of species/samples	10 (1–22)	12 (5–38)	11 (1–38)

a   Typical sub-community
b   *Arrhenatherum elatius-Rubus fruticosus* agg. sub-community
24   *Urtica dioica-Galium aparine* community (total)

OV25
Urtica dioica-Cirsium arvense community

Constant species

Cirsium arvense, Urtica dioica.

Physiognomy

The *Urtica dioica-Cirsium arvense* community typically has a rather open or patchy cover of *U. dioica*, usually grown tall by mid-summer but not so densely developed as to exclude other associates. Among these companions, large thistles and coarse grasses are the most conspicuous, with *Cirsium arvense* and *C. vulgare* both occurring frequently, and often in some abundance, and *Dactylis glomerata, Elymus repens, Holcus lanatus* and *Arrhenatherum elatius* very common in various of the sub-communities. *Galium aparine* is occasional, its shoots sprawling over the taller herbs, and there can be scattered plants of *Lamium purpureum, Leucanthemum vulgare, Epilobium hirsutum* and *Carduus acanthoides.*

Sub-communities

Holcus lantus-Poa annua **sub-community.** *Elymus repens* and *Holcus lanatus* with, somewhat less frequently, *Poa annua* and *Agrostis stolonifera* give a distinctly grassy look to this vegetation and there is often some *Sonchus asper, S. oleraceus, Senecio vulgaris, Daucus carota, Rumex obtusifolius* and *Echium vulgare.* Locally, *Pteridium aquilinum* can figure. Smaller associates in more open places among this tall and coarse herb cover include *Cerastium fontanum, Trifolium repens, T. pratense* and *Myosotis arvensis* and *Vicia sativa* is an occasional climber.

Rumex obtusifolius-Artemisia vulgaris **sub-community.** *Elymus repens, Dactylis glomerata* and *Arrhenatherum elatius* are frequent here, along with the nettle and thistles, but the most distinctive feature is the common occurrence of *Artemisia vulgaris* and *Heracleum sphondylium* with occasional *Calystegia sepium, Malva sylvestris* and *Conium maculatum.*

Lolium perenne-Papaver rhoeas **sub-community.** *Dactylis* and *Arrhenatherum* remain common in this sub-community but more striking is the frequent occurrence of *Lolium perenne* and *Papaver rhoeas* along with occasional *Bromus mollis, B. sterilis, Matricaria maritima, Silene vulgaris, Sisymbrium officinale* and *Anthriscus sylvestris. Rubus fruticosus* agg. can be patchily abundant and stands among fens may have *Phragmites australis.*

Habitat

The *Urtica-Cirsium* community is characteristic of disturbed areas of nutrient-rich loamy soils within badly-managed pastures and leys, on abandoned arable land, waysides, verges and waste ground, and woodland clearings.

The most obvious floristic difference between this kind of nettle vegetation and the *Urtica-Galium* community is the more consistent frequency and abundance here of the two thistles, *Cirsium arvense* and *C. vulgare*, and this reflects some contrast in the characteristic habitats of the two assemblages. These thistles both get a hold and thrive best on open areas of bare soil as on molehills in pastures, places where long-lying dung pats have smothered grassy swards, on newly-seeded verges, on abandoned ground and dumped soil, and on derelict land.

C. arvense produces seed where can germinate in the autumn after flowering, although the warming and fluctuating temperatures of spring are especially favourable to its establishment. The first tap root quickly puts out laterals on which buds produce new vertical vegetative or flowering shoots: these can thus give rise to a patch of thistles and horizontal spread can occur with formidable speed, up to 12 m or more in a single season. With such a successful form of vegetative establishment, it matters less that the flowering heads of single individuals or clones are usually functionally dioecious and the inflorescences sterile. When both male and female plants occur in close proximity, however, viable fruits are freely produced and readily dispersed by wind (Salisbury 1964).

C. vulgare is a biennial when its life cycle proceeds unhindered, and it reproduces entirely by seed, though predation by herbivores can thicken up clumps where stock or rabbits devour flowering stems and stimulate the production of secondary shoots (Klinkhamer & de Jong 1993). In fact, the prickly vegetative shoots are unpalatable to most stock and grazing of surrounding vegetation can greatly help establishment on small areas of bare ground by keeping potential competitors to the thistle rosettes in check.

Both *Cirsium* spp. can colonise soils of varying texture and moisture content although they perform best on circumneutral loams that are free-draining but not prone to drought. Growing together here with *Urtica dioica*, the favoured situations are nutrient-rich, such as fertilised pastures and arable land and disturbed top soil.

The various sub-communities are characteristic to some extent of different habitats although the details of the particular conditions of each are uncertain. The *Lolium-Papaver* type occurs on verges and derelict pastures which have been disturbed, on dumped soil and waste ground and in the fairly early stages of colonisation of abandoned arable land. The *Elymus-Artemisia* sub-community is seen on waste ground and in disturbed woodland clearances or young plantations. The *Holcus-Poa* type is characteristic of dumped soil and waste ground.

Zonation and succession
The community occurs typically with other weed vegetation or among various grasslands and scrub communities. Invasion by brambles, shrubs and trees can continue or restore a succession to woodland.

Among neglected, disturbed or ill-managed pastures and leys, this kind of vegetation can occur very fragmentarily around molehills and patches of broken ground, passing sharply to the *Lolio-Cynosuretum* or Lolio-Plantaginion leys. On grassy waysides and verges, stands can be larger and the context is usually communities like the *Lolium-Dactylis* assemblage or the *Arrhenatheretum*. Along road edges, it can give way to the *Poa-Matricaria perforata* community.

On wasteland, the *Urtica-Cirsium* community can occur with a variety of other weed vegetation like the *Matricaria perforata-Stellaria* or *Stellaria-Capsella* assemblages, in various sub-communities of which both stinging nettle and thistles can remain prominent. Such vegetation types can precede the development of the *Urtica-Cirsium* community on abandoned arable land or waste ground. Subsequent stages in succession can be seen where ground remains undisturbed or unmanaged. Then, *Rubus fruticosus* agg. can become more patchily prominent among the *Urtica-Cirsium* vegetation, thicken up into *Rubus-Holcus* underscrub and this in turn progress to *Crataegus-Hedera* scrub. Mosaics of such communities are a common feature of abandoned waste ground and grossly-disturbed woodland clearings and margins. Eventual development of *Quercus-Pteridium-Rubus* or *Fagus-Rubus* woodland is the likely culmination of such succession on the kind of soils where the *Urtica-Cirsium* community occurs.

Distribution
The community is ubiquitous through the British lowlands.

Affinities
This kind of vegetation presents various general features characteristic of a number of associations recognised in other parts of Europe where *U. dioica* plays a prominent role: the *Urtico-Convolvuletum sepium* Görs & Müller 1969, the *Urtico-Aegopodietum podagrariae* (R.Tx. 1963) Oberdorfer 1964 in Görs 1968 and the *Urtico-Cruciatetum laevipedis* Dierschke 1974. It comes closest in some ways to the last type, though lacks many of the woodland fringe taxa characteristic there. These associations are variously grouped in the Galio-Alliarion (Oberdorfer 1957) Lohmeyer & Oberdorfer in Oberdorfer et al. 1967 or the Aegopodion podagrariae R.Tx. 1967, alliances now placed in the Galio-Urticetea Passarge ex Kopecký 1969.

Floristic table OV25

	a	b	c	25
Urtica dioica	IV (1–4)	IV (3–6)	IV (2–4)	IV (1–6)
Cirsium arvense	V (1–6)	IV (2–6)	IV (3–6)	IV (1–6)
Elymus repens	IV (3–5)	V (3–8)		III (3–8)
Dactylis glomerata	II (2–4)	IV (3–5)	IV (3–4)	III (2–5)
Holcus lanatus	IV (2–5)	I (4)	I (4)	II (2–5)
Sonchus asper	III (1–3)	II (3)	II (2–4)	II (1–4)
Agrostis stolonifera	III (4–7)		II (3–4)	II (3–7)
Poa annua	III (2–5)			II (2–5)
Cerastium fontanum	III (2–3)			I (2–3)
Vicia sativa	II (1–3)		I (2)	I (1–3)
Daucus carota	II (1–5)		I (1)	I (1–5)
Trifolium repens	II (3)		I (3)	I (3)
Echium vulgare	II (4–5)			I (4–5)
Pteridium aquilinum	II (5–6)			I (5–6)
Trifolium pratense	II (3–5)			I (3–5)
Sonchus oleraceus	II (1–3)			I (1–3)
Myosotis arvensis	II (1–5)			I (1–5)
Senecio vulgaris	II (2–4)			I (2–4)
Rumex obtusifolius	II (3)	III (1–4)	I (3)	II (1–4)
Artemisia vulgaris		III (4–8)	I (2)	II (2–8)
Heracleum sphondylium		III (3–4)		II (3–4)
Calystegia sepium	I (3)	II (2–5)		II (2–5)
Conium maculatum		II (3–5)	I (7)	I (3–7)
Malva sylvestris		II (2–7)	I (3)	II (2–7)
Arctium minus		I (3–10)		I (3–10)
Beta vulgaris		I (3–5)		I (3–5)
Lolium perenne	II (4–5)	II (3–4)	V (3–6)	III (3–6)
Arrhenatherum elatius	I (4)	III (2–4)	IV (2–6)	III (2–6)
Papaver rhoeas	I (3)		IV (2–6)	II (2–6)
Bromus hordeaceus hordeaceus		I (4)	II (3–7)	II (3–7)
Matricaria maritima	I (3)		II (3–4)	II (3–4)
Silene vulgaris		I (3)	II (3–4)	II (3–4)
Sisymbrium officinale		I (2–3)	II (3–4)	II (2–4)
Bromus sterilis		I (3)	II (2–6)	II (2–6)
Hordeum murinum		I (3)	II (3–5)	I (3–5)
Rubus fruticosus agg.		I (5)	II (2–3)	I (2–5)
Capsella bursa-pastoris	I (3)		II (2–3)	I (2–3)
Anthriscus sylvestris		I (2–4)	II (4)	I (2–4)
Silene dioica			II (2–3)	I (2–3)
Ranunculus repens			II (3–5)	I (3–5)
Geranium dissectum			II (3)	I (3)
Potentilla reptans			II (3–8)	I (3–8)
Phragmites australis			II (3–4)	I (3–4)
Achillea millefolium			II (2–3)	I (2–3)

	a	b	c	25
Dipsacus fullonum			I (2)	I (2)
Amaranthus albus			I (5)	I (5)
Carduus nutans			I (3)	I (3)
Malva neglecta			I (4)	I (4)
Cirsium vulgare	II (3–4)	II (3–4)	II (3–4)	II (3–4)
Lamium purpureum	II (1)	II (3–4)	I (4)	II (1–4)
Galium aparine	II (3–4)	I (3)	II (2–3)	II (2–4)
Leucanthemum vulgare	I (3)	I (3)	I (3)	I (3)
Centaurea scabiosa	I (4)	I (3)	I (3)	I (3–4)
Atriplex prostrata	I (3)	I (5)		I (3–5)
Epilobium hirsutum	I (2)	I (4–5)		I (2–5)
Carduus acanthoides		I (2–4)	I (5)	I (2–5)
Plantago lanceolata		I (3)	I (4)	I (3–4)
Geranium molle	I (3)		I (3)	I (3)
Senecio jacobaea	I (1)		I (3)	I (1–3)
Aster tripolium	I (2)		I (4)	I (2–4)
Avena fatua	I (3)		I (4)	I (3–4)
Glechoma hederacea	I (3)		I (2)	I (2–3)
Number of samples	8	11	8	27
Number of species/sample	16 (9–31)	11 (8–19)	17 (10–31)	14 (8–31)

a   *Holcus lanatus-Poa annua* sub-community
b   *Rumex obtusifolius-Artemisia vulgaris* sub-community
c   *Lolium perenne-Papaver rhoeas* sub-community
25   *Urtica dioica-Cirsium arvense* community (total)

OV26
Epilobium hirsutum community

Synonymy
Epilobium hirsutum-Filipendula ulmaria community
Wheeler 1980 *p.p.*

Constant species
Epilobium hirsutum, Urtica dioica.

Physiognomy
The *Epilobium hirsutum* community comprises often
species-poor tall-herb vegetation in which *E. hirsutum* is a
constant and generally very abundant feature. It is usually
accompanied by *Urtica dioica*, overall a subordinate
element in the cover but locally abundant in a patchy canopy
of the two species that is often over 1 m tall. No other plants
are frequent throughout but *Filipendula ulmaria, Arrhenatherum elatius* and *Cirsium arvense* are common in
various of the sub-communities and occasional are *Carex
acutiformis, Phalaris arundinacea, Deschampsia cespitosa,
Galium aparine* and *Rumex crispus*. Patches of *Calliergon
cuspidatum* and *Brachythecium rutabulum* are sometimes to
be seen over the soil and stools.

Sub-communities

Juncus effusus-Ranunculus repens **sub-community.** *E. hirsutum* is overwhelmingly dominant here with only occasional and usually not very abundant *U. dioica.
Ranunculus repens* is frequent, creeping among the
willow-herb stools, and tussocks of *Juncus effusus* can
be prominent. Occasionally, there is some *Cardamine
flexuosa, Angelica sylvestris, Mentha aquatica, Holcus
lanatus, Poa trivialis* and *Deschampsia cespitosa*.

Phragmites australis-Iris pseudacorus **sub-community.** *E.
hirsutum* remains dominant in this sub-community but it
is typically accompanied by patches of *U. dioica* and
Phragmites, with scattered individuals of *Iris pseudacorus, Filipendula ulmaria, Eupatorium cannabinum,
Lythrum salicaria, Lycopus europaeus, Lysimachia vulgaris* and sprawling *Solanum dulcamara, Calystegia
sepium* and *Galium palustre*. More locally, *Glyceria
maxima, Carex riparia* or *C. paniculata* can occur with
some abundance.

Filipendula ulmaria-Angelica sylvestris **sub-community.** *F.
ulmaria* becomes constant here, along with *Urtica* and *E.
hirsutum* and it often almost rivals the latter in cover,
forming a patchy canopy among which there are commonly scattered individuals of *Angelica sylvestris, Cirsium
palustre, Scrophularia auriculata, Mentha aquatica, Equisetum palustre, Arrhenatherum elatius* and *Holcus lanatus*
with sprawling and climbing *Galium aparine, Lathyrus
pratensis, Lotus uliginosus* and *Vicia cracca*. More occasionally can be seen *Caltha palustris, Cardamine pratensis,
Ranunculus acris, Rumex sanguinens, R. acetosa, Lychnis
flos-cuculi* and *Hypericum tetrapterum*.

Arrhenatherum elatius-Heracleum sphondylium **sub-
community.** *A. elatius* is more common and abundant
here, among the *E. hirsutum* and *Urtica*, but the fen
herbs of the two previous sub-communities are all occasional at most. Coarse herbs like *Cirsium arvense,
Rumex crispus* and *R. obtusifolius* are conspicuous,
along with *Heracleum sphondylium, Elymus repens, Dactylis glomerata* and *Lolium perenne. Rubus fruticosus*
agg. can be locally abundant.

Urtica dioica-Cirsium arvense **sub-community.** In this
generally most species-poor kind of *Epilobium hirsutum*
vegetation, *Urtica* and *C. arvense* are very frequent with
often just occasional *Rumex crispus* and *Elymus repens*.

Habitat
The *Epilobium hirsutum* community is characteristic of
moist but well-aerated, mesotrophic to eutrophic
mineral soils and fen peats in open-water transition
mires, around ponds, in silting ditches, and along
streamsides throughout the lowlands.

E. hirsutum is a shade-sensitive tall herb that favours
moist soils but which is one of the fen plants that is most

intolerant of ferrous iron, the various toxic effects of which may be very influential in influencing the distribution of the species in such habitats (Snowden & Wheeler 1993). Following observations on the striking contrast in distributions of *E. hirsutum* and *Juncus subnodulosus* in relation to waterlogging around a spring fen (Wheeler *et al*. 1985), Snowden & Wheeler (1993) went on to screen a range of fen plants and showed that those, like *E. hirsutum*, more sensitive to ferrous iron tended to occur in sites with lower iron availability, where sediments were relatively base-rich, fertile and sometimes summer-dry. These species were also less likely to form ochreous precipitates on their roots (Cook 1990), one defence against penetration of the reduced iron. *Filipendula ulmaria*, the most common associate of *E. hirsutum* throughout this community, was also among the most sensitive species in this study. *Urtica dioica*, the other constant here, was not included in the study but is well known to favour aerated sediments and soils.

These kinds of responses probably play some considerable part in the distribution of this community on more nutrient-rich sediments kept moist by ground water but not waterlogged, at least not in summer. Favoured sites therefore include the upper parts of fen sequences around open waters, the edges of pools, ditch and canal sides and places where dredgings have been deposited and become oxidised, releasing nutrients. It can also spread in places where fens are becoming drier and more eutrophicated because of a fall in ground-water level. The other controlling factor in limiting its distribution is grazing: *E. hirsutum* can increase markedly where grazing is reduced around open waters.

The various sub-communities reflect the range of different contexts in which this vegetation can develop. The margins of wetter fens and ditches usually have the *Phragmites-Eupatorium* sub-community; fen-meadows, silted ditches and damp road verges, the *Filipendula-Angelica* type; wet rush-pastures and washlands, the *Juncus-Ranunculus* sub-community; drier pond margins, verges and waste ground, the *Arrhenatherum-Heracleum* sub-community. The *Urtica-Cirsium* type comprises drier stands from a variety of settings.

Zonation and succession
The *Epilobium hirsutum* community occurs in swamps, fens, woodlands and mesotrophic grasslands in zonations and mosaics that are determined by the position of the ground water table and the occurrence of grazing or mowing.

A common sequence around open waters is for the community to occur behind some kind of tall-herb fen, like the *Phragmites-Eupatorium* fen or, in more generally eutrophic systems, the *Phragmites-Urtica* fen. With the latter, it can be difficult to draw the bounds between each community because patchy dominance of *E. hirsutum*

and *Urtica*, along with *Phragmites*, is typical of the fen, and reed may run on with local prominence into the *Epilobium* vegetation.

More abrupt sequences, where there is a sudden shift to wetter ground, as on steep ditch sides, can see the *Epilobium* community giving way directly to the *Phragmitetum*, *Glycerietum*, *Caricetum paniculatae*, *Caricetum ripariae*, or *Typhetum latifoliae* swamps. Along streams where stock or people have access or on canal margins, such patterns may be very fragmentary because of trampling or bank erosion.

Towards drier ground in these systems, the *Epilobium* community can give way to the *Arrhenatheretum* and these two vegetation types can also occur together on verges, the *Filpendula-Angelica* sub-community of the *Epilobium* vegetation being in common in ditches along roadsides.

This sub-community is also widespread on river terraces and in wet meadows where it often occurs with the *Filipendula-Angelica* mire, *E. hirsutum* persisting some distance into the fen as dominance shifts to the meadowsweet. Where such systems are grazed and especially where the soils are not so free-draining, the *Juncus-Cirsium* community or *Holco-Juncetum* tends to replace the *Filipendula-Angelica* fen and, where the *Epilobium* vegetation persists, it is usually of the *Juncus-Ranunculus* type.

Woodlands of various kinds can be found as part of these sequences, particularly the *Salix-Betula-Phragmites* woodland and *Alnus-Carex* woodland in more intact fens and streamsides, the *Alnus-Urtica* woodland in eutrophicated fens and in mature river valleys with alluvial terraces. However, the shade sensitivity of *E. hirsutum* means transitions to the *Epilobium* community are typically very abrupt.

The community is probably expanding its cover in fens with increased eutrophication of ground waters and sediments and, once established, it is probably fairly stable: it is not really an element in active hydroseres. With continued drying, however, it might be expected to progress eventually to some kind of *Alnus-Urtica* woodland.

Distribution
The *Epilobium* community is widespread and common throughout the British lowlands.

Affinities
This broadly-defined type of *Epilobium* community takes in more species-rich vegetation, like that included in the *Epilobio hirsuti-Filipenduletum* Sougnez 1957 (Mucina *et al*. 1993) and simpler assemblages like the Sociatie van *Epilobium hirsutum* of Westhoff & den Held 1969. Such syntaxa have sometimes been placed in the Galio-Urticetea, sometimes in the Molinio-Arrhenatheretea.

Floristic table OV26

	a	b	c	d	e	26
Epilobium hirsutum	V (5–10)	V (5–9)	V (5–9)	V (5–10)	V (5–10)	V (5–10)
Urtica dioica	II (1–4)	V (3–5)	IV (3–5)	IV (3–7)	IV (1–7)	IV (1–7)
Ranunculus repens	III (1–5)	I (3)	I (3)	I (3)	I (1–3)	I (1–5)
Juncus effusus	III (1–7)	I (2–3)	II (3–7)	I (1–4)		I (1–7)
Cardamine flexuosa	II (1–3)				I (3–5)	I (1–5)
Phragmites australis		IV (3–5)	I (7)	I (1–4)	I (7)	II (1–7)
Iris pseudacorus	I (2–6)	III (1–8)	I (3)	I (2–6)		I (1–8)
Solanum dulcamara	I (1–4)	III (2–4)	I (3)	I (3–5)	I (3)	I (1–5)
Eupatorium cannabinum		III (3–5)	I (3–5)	I (2–6)		I (2–6)
Calystegia sepium	I (2)	III (2–5)	I (5)	I (2–4)	I (1)	I (1–5)
Lythrum salicaria		II (2–3)		I (3–5)	I (3)	I (2–5)
Glyceria maxima	I (3)	II (3–5)	I (3–4)	I (5)	I (7)	I (3–7)
Lycopus europaeus	I (3)	II (3–5)	I (3)	I (3)	I (2–4)	I (2–5)
Carex riparia	I (7)	II (3–5)	I (5)	I (3)		I (3–7)
Galium palustre	I (2–4)	II (2–3)	I (3–4)		I (1)	I (1–4)
Carex paniculata	I (3–4)	II (3–5)	I (3–5)			I (3–5)
Scutellaria galericulata	I (3)	II (3)				I (3)
Lysimachia vulgaris		I (3–5)				I (3–5)
Menyanthes trifoliata		I (3)				I (3)
Thelypteris palustris		I (3–5)				I (3–5)
Filipendula ulmaria	I (6–7)	IV (2–7)	V (3–7)	II (2–7)		III (2–7)
Angelica sylvestris	II (1–5)	II (1–4)	IV (3–5)	I (2–4)		II (1–5)
Cirsium palustre	I (1–3)	II (3)	IV (3–5)			II (1–5)
Galium aparine	II (1–2)	II (3)	III (3–5)	II (2–4)	I (1–4)	II (1–5)
Mentha aquatica	II (1–5)	II (3–6)	III (3–4)	I (4–5)		II (1–6)
Equisetum palustre	I (1)	II (1–4)	III (3–6)	I (4–7)		I (1–7)
Lathyrus pratensis	I (2)	I (3)	III (3)	I (2–4)	I (3)	I (2–4)
Lotus uliginosus	I (2–4)	I (3)	III (3–4)	I (3–4)		I (2–4)
Vicia cracca		I (3–4)	III (3–6)	I (2–4)		I (2–6)
Holcus lanatus	II (1–5)	I (3–4)	III (3–5)	I (3–6)	I (4)	I (1–6)
Galium uliginosum		I (3)	III (3)			I (3)
Poa trivialis	II (1–5)	I (1–3)	III (3–5)	I (2–4)	I (3)	I (1–5)

Scrophularia auriculata	I (3)	I (3)		II (3–5)		I (2–5)
Juncus inflexus	I (6–9)			II (3–8)		I (3–9)
Caltha palustris	I (3–6)	I (3)		II (3–4)		I (3–6)
Cardamine pratensis	I (1–3)	I (3)		II (1–3)		I (1–3)
Equisetum fluviatile	I (3–5)	I (3)		II (1–3)	I (2)	I (1–5)
Ranunculus acris	I (5)	I (2)		II (3)	I (3)	I (2–5)
Rumex sanguineus		I (3)		II (3–4)	I (3)	I (2–4)
Rumex acetosa	I (1–4)	I (3)		II (2–3)	I (2)	I (1–4)
Hypericum tetrapterum	I (1)	I (3)		II (3)		I (1–3)
Lychnis flos-cuculi		I (3–4)		II (3)		I (3–4)
Agrostis stolonifera	I (1–4)	I (3)		II (3)	I (2–6)	I (1–6)
Cerastium fontanum				I (3)		I (3)
Carex disticha		I (3)		I (3–7)		I (3–7)
Arrhenatherum elatius	I (3)	I (3–5)	III (3–7)	IV (4–7)	I (2)	II (2–7)
Cirsium arvense	I (2)	I (3)	I (3–6)	III (1–5)	III (2–4)	II (1–6)
Heracleum sphondylium	I (1–3)	I (3)	I (3)	III (2–5)	I (4)	I (1–5)
Rumex crispus	I (2)	I (2–3)		II (2–4)	II (1–3)	I (1–4)
Elymus repens	I (5)		I (2)	II (2–6)	II (1–6)	I (1–6)
Dactylis glomerata	I (1–5)	I (2–4)	I (3)	II (2–7)	I (1–7)	I (1–7)
Rumex obtusifolius	I (2–4)	I (2)		II (2–4)		I (2–4)
Rubus fruticosus agg.	I (4)	I (2)	I (3)	II (3)	I (3)	I (2–4)
Lolium perenne	I (5)			II (3–5)		I (3–5)
Convolvulus arvensis			I (4)	II (4–6)		I (4–6)
Anthriscus sylvestris	I (1)			II (3–5)		I (1–5)
Alliaria petiolata				I (3)		I (3)
Achillea millefolium				I (3–4)		I (3–4)
Polygonum lapathifolium				I (4–6)		I (4–6)
Calliergon cuspidatum	II (1–4)	II (3)		I (3–4)	I (2)	I (1–4)
Carex acutiformis	II (1–5)	II (3–6)		II (3–8)	I (3–5)	I (3–8)
Brachythecium rutabulum	II (1–5)	I (3)	I (3)	II (3)		I (1–5)
Deschampsia cespitosa	I (4)			II (3–5)	I (3–4)	I (1–5)
Phalaris arundinacea	I (1–6)	II (3–7)		II (2–5)	I (2–7)	I (2–7)
Sparganium erectum	I (1–6)	I (4–7)		I (4)	I (2–7)	I (1–7)
Rumex conglomeratus	I (1–3)	I (3)		I (2–3)	I (2)	I (2–6)
Stachys palustris	I (1)	I (2)		I (3)	I (8)	I (1–8)
Myosotis laxa cespitosa	I (2)	I (3)		I (3–4)	I (3–4)	I (2–4)

Floristic table OV26 (*cont.*)

	a	b	c	d	e	26
Plagiomnium undulatum	I (1–4)	I (3)	I (3)	I (1)		I (1–4)
Plantago lanceolata	I (3)	I (4)	I (3)	I (4)		I (3–4)
Stachys sylvatica	I (3)	I (3)	I (3)	I (1–3)		I (1–3)
Potentilla anserina	I (1)	I (3)	I (3)	I (3–5)		I (1–5)
Lemna minor	I (5)	I (3)	I (3)			I (3–5)
Valeriana officinalis	I (6)	I (3–5)	I (3)			I (3–6)
Lophocolea bidentata	I (1–2)	I (3)	I (3)			I (1–3)
Ajuga reptans	I (1)	I (3)	I (3)			I (1–3)
Festuca arundinacea	I (1)		I (3)	I (2)		I (1–3)
Veronica beccabunga	I (2–3)		I (1–3)		I (3)	I (1–3)
Eurhynchium praelongum	I (3–5)	I (3)			I (4)	I (3–5)
Typha latifolia	I (2–3)	I (3–5)			I (2–4)	I (2–7)
Symphytum officinale		I (3)	I (3)	I (2–7)		I (3–7)
Berula erecta		I (4)	I (3)	I (4–7)		I (3–4)
Humulus lupulus		I (3–5)	I (3)			I (3–5)
Salix cinerea sapling		I (3–5)	I (3)			I (3–5)
Epilobium parviflorum		I (1–3)	I (3)			I (1–3)
Juncus subnodulosus		I (3–7)	I (3–8)			I (3–8)
Carex hirta	I (4)		I (3)			I (3–4)
Dactylorhiza fuchsii	I (1)		I (3)			I (1–3)
Juncus articulatus		I (3)	I (3)			I (3)
Phleum pratense			I (3–5)	I (1–3)		I (1–5)
Rumex hydrolapathum		I (3)		I (3)		I (3)
Typha angustifolia		I (3–5)			I (5)	I (3–5)
Nasturium officinale agg.	I (4–5)			I (3–5)	I (4)	I (3–5)
Number of samples	20	15	30	32	19	116
Number of species/sample	16 (6–31)	16 (5–37)	21 (9–42)	13 (5–35)	7 (2–17)	15 (2–42)

a   *Juncus effusus-Ranunculus repens* sub-community
b   *Phragmites australis-Iris pseudacorus* sub-community
c   *Filipendula ulmaria-Angelica sylvestris* sub-community
d   *Arrhenatherum elatius-Heracleum sphondylium* sub-community
e   *Urtica dioica-Cirsium arvense* sub-community
26   *Epilobium hirsutum* community (total)

Constant species
Epilobium angustifolium.

Physiognomy
The *Epilobium angustifolium* community is overwhelmingly dominated by *E. angustifolium*, the tall shoots of which can reach well over 1 m by summer. No other species is frequent throughout but the commonest associates overall are *Rubus fruticosus* agg., *Holcus lanatus*, *Pteridium aquilinum* and *Urtica dioica* with various other species reflecting the different situations in which this kind of vegetation can develop.

Sub-communities

Holcus lanatus-Festuca ovina **sub-community.** In this grassy or heathy vegetation, *E. angustifolium* is accompanied by frequent *Holcus lanatus* and *Festuca ovina*, occasional *Anthoxanthum odoratum*, *Agrostis capillaris*, *Potentilla erecta*, *Galium saxatile* and *Teucrium scorodonia*, and sometimes by scattered *Calluna vulgaris*, *Erica cinerea*, *Ulex europaeus* and *U. gallii* and sparse or patchy *Pteridium*.

Urtica dioica-Cirsium arvense **sub-community.** *U. dioica* is a frequent associate here, though not rivalling *E. angustifolium* in cover, and there are often scattered individuals of *Cirsium arvense* and *C. vulgare*. *Holcus lanatus* remains common but is occasionally accompanied here by *Arrhenatherum elatius*, *Dactylis glomerata*, *Holcus mollis*, *Deschampsia cespitosa* and *Poa trivialis*, with *Heracleum sphondylium* and *Galium aparine*. Patches of *Rubus fruticosus* agg. can sometimes be seen.

Rubus fruticosus **agg.-*Dryopteris dilatata* sub-community.** *Rubus fruticosus* agg. is constant here, along with *Dryopteris dilatata*, and occasional *Pteridium aquilinum* and saplings of *Betula pendula*, *B. pubescens*, *Pinus sylvestris*, *P. nigra*, *Quercus robur* and *Q. petraea*. Bryophytes such as *Mnium hornum*, *Aulacomnium*

androgynum and *Lophocolea bidentata s.l.* are sometimes prominent on decaying wood.

Acer pseudoplatanus-Sambucus nigra **sub-community.** Saplings of *Acer pseudoplatanus*, *Fraxinus excelsior* and, less commonly, *Fagus sylvatica* and *Ulmus glabra* feature prominently in this sub-community along with bushes of *Sambucus nigra* and clumps of *Rubus fruticosus* agg. Herbaceous associates include occasional *Urtica dioica*, *Circaea lutetiana*, *Brachypodium sylvaticum*, *Mercurialis perennis* and ferns like *Dryopteris filix-mas*, *D. borreri* and *D. dilatata*. In spring, there can be a patchy show of *Hyacinthoides non-scripta* or *Allium ursinum*. *Eurhynchium praelongum*, *E. striatum* and *Brachythecium rutabulum* are occasional over the soil surface and litter.

Ammophila arenaria **sub-community.** *Ammophila arenaria*, *Festuca rubra* and *Poa pratensis* are frequent associates of *E. angustifolium* here, sometimes in abundance, and there are frequent scattered individuals of *Senecio jacobaea*, *Lotus corniculatus* and *Hypochoeris radicata*. More occasionally, *Ononis repens*, *Crepis capillaris*, *Luzula campestris* and *Sedum acre* can be seen with ephemerals like *Myosotis ramosissima*, *Carlina vulgaris*, *Valerianella locusta* and *Viola tricolor* appearing on open areas of bare sand.

Habitat
The *Epilobium angustifolium* community is characteristic of damp, fertile soils on disturbed, often burned, ground in woodlands, on heaths, road verges, tracksides, recreation areas and wasteland throughout Britain.

E. angustifolium is a circumpolar plant, widespread through Eurasia and North America and native in Britain (Hultén 1971). In its natural habitats, it is characteristic of two main groups of communities: the tall-herb and scrub vegetation of snow-protected mountain slopes and ledges (e.g. Nordhagen 1943, Dahl 1956, Ellenberg 1988) and the secondary vegetation of forests disturbed by wind-throw, fire or clearance (Tüxen 1950,

Ellenberg 1988). The *Epilobium angustifolium* community includes stands in this second kind of situation but also reflects the success of the plant in colonising more artificial habitats where a congenial combination of conditions can be found. Though the plant was recorded widely in Britain in the eighteenth and nineteenth centuries (Myerscough 1980), it was not until this century that it began to be seen abundantly, an increased prominence attributed by Salisbury (1964) to the greater availability of burned and derelict habitats.

E. angustifolium produces seed in phenomenal quantities – perhaps over 50 000 per flowering shoot (Salisbury 1964, Myerscough 1980) – and this is very widely dispersed by wind. The seed remains viable for only about 18 months or so but readily germinates and establishes up to this age if light, moisture and at least moderate amounts of major nutrients are available (Reinikainen 1964, Myerscough & Whitehead 1967). Cleared and burned areas in woodlands thus provide a very suitable habitat. Felling or fire opens up the ground to high light levels and disturbance or burning encourages mineralisation of humus or provides nutrient-rich ash. Removal of the trees which would otherwise draw water from the soil, shelter from drying winds by surrounding vegetation and the opening up of the soil surface to wetting rains also all help create a substrate that is moist, at least in the early stages of recolonisation.

In this process, *E. angustifolium* also gains a considerable advantage from its rapid growth following establishment. Most seedlings probably appear initially in late summer and autumn and they can gain weight rapidly before overwintering as small rosettes. Subsequent growth is by horizontally spreading roots which bear shoot buds (Moss 1936) even on very young plants (Myerscough & Whitehead 1967). Flowering can take place within the first year and occurs every year thereafter, but vegetative reproduction takes priority for consolidating established colonies and in terms of resource allocation (van Andel & Vera 1977). This strategy is also important because the favourable conditions in clearings and burned areas persist only for a short time, with the initial mobilised supply of nutrients being used up within two or three years (Ellenberg 1988). Essentially, the mature plant is a geophyte.

Once well established, plants of *E. angustifolium* often overtop any lower growing associates by late spring or early summer (Myerscough 1980) and go on increasing in above-ground biomass for a number of years. Associates through the community are therefore generally few, comprising bulkier potential competitors that sprang up with the willow-herb or subsequently overtook it or which are survivors from pre-existing vegetation that persist patchily in more open places. The diversity of these companions reflects the variety of situations in which the community can develop.

The commonest associates overall are *Pteridium aquilinum* and *Rubus fruticosus* agg., especially in woodland habitats, with mosses like *Brachythecium rutabulum* and *Eurhynchium praelongum* occasional in various sub-communities. Very frequently, as in the *Rubus-Dryopteris* type, sprawls of bramble grow up among and around the *E. angustifolium* with scattered fronds or patches of bracken. Both these associates can spring up again after burning, putting up new shoots from underground stems. Scattered *D. dilatata* and young saplings of birch and pine are also typical of this vegetation which is widespread on acidic soils in cleared or burned oak-birch and coniferous woodland (e.g. Hill & Jones 1978). The *Acer-Sambucus* sub-community where bramble and bracken remain quite frequent, but where the usual young woody species are *Acer pseudoplatanus*, *Fraxinus excelsior*, *Fagus sylvatica*, *Ulmus glabra* and *Sambucus nigra*, is a more local type of *E. angustifolium* vegetation developing around old bonfire sites in more mesophytic and calcicolous woodlands. It is also common on railway embankments and derelict land, particularly where trackside fires have occurred or old buildings have been gutted and burned.

Also widespread on road verges, railway embankments, in recreation areas and on wasteland, is the *Urtica-Cirsium* sub-community. Gross disturbance and fires from picnics or carelessly discarded cigarettes commonly destroy the vegetation of such habitats providing an opportunity for *E. angustifolium* to establish, along with other nutrient-demanding weeds and coarse ephemerals. Perennial grasses like *Arrhenatherum elatius*, *Dactylis*, *Deschampsia cespitosa* and *Holcus mollis* can survive light burns and also seed in from sources nearby.

Fires are also a common feature of heaths and the *Holcus-Festuca* sub-community includes stands of *E. angustifolium* vegetation developed among burned areas, with a patchy persistence, regrowth from stools or seeding in of sub-shrubs, and a discontinuous sward of grasses and herbs characteristic of acidic soils or, as with *H. lanatus*, reflecting temporary nutrient enrichment.

Finally, the *Ammophila* sub-community comprises *E. angustifolium* vegetation from burned stable dune systems.

Zonation and succession

Very commonly, stands of *E. angustifolium* vegetation are sharply marked off from their surrounds by the highly localised disturbance that has resulted in their development.

In oak-birch woodlands, for example, stands of the *Rubus-Dryopteris* sub-community often occupy much of small open areas created by clearance of fire, or those places where burning of brashings or coppice waste has been concentrated, surrounded by surviving areas of

Quercus-Pteridium-Rubus, Quercus-Betula-Oxalis, or *Quercus-Betula-Deschampsia* woodlands or coniferised replacements. Sometimes, *Rubus-Pteridium* or *Rubus-Holcus* underscrubs also occur with the willow-herb where bramble or bracken have gained ascendancy or else patches of *Festuca-Agrostis-Rumex* grassland on parched or grazed areas. Heathy woodlands where burning has occurred sometimes have both the *Rubus-Dryopteris* and the *Holcus-Festuca* sub-communities. On unwooded heaths, the latter type of *E. angustifolium* vegetation usually marks out old burns among *Calluna-Festuca, Calluna-Ulex minor, Calluna-Ulex gallii* and *Calluna-Erica cinerea* heaths. *Ulex-Rubus* scrub can also develop around disturbed and burned areas and, on grazed heaths, *Festuca-Agrostis-Rumex* and *Festuca-Agrostis-Galium* grasslands can also form part of the patchwork. The *Acer-Sambucus* sub-community is also commonly seen in woodlands, in this case the *Fraxinus-Acer-Mercurialis* or *Fagus-Mercurialis* types, marking out old bonfire sites. In larger clearings where trees and shrubs are slower to re-establish themselves, *Deschampsia-Holcus* grassland is a common associate, with *Crataegus-Hedera* scrub where shrub colonisation is occurring.

The *Urtica-Cirsium* sub-community is often seen among disturbed or burned stretches of *Arrhenatheretum* or *Deschampsia-Holcus* grassland, often with stands of *Urtica-Galium* and *Urtica-Cirsium* vegetation. The *Ammophila* sub-community occurs locally among *Ammophila-Festuca, Festuca-Galium* and *Ammophila-Arrhenatherum* grasslands where dunes have been burned and disturbed. Sometimes, *Hippophae* scrub also occurs in such situations.

E. angustifolium is a formidable competitor to smaller herbs and seedling trees and shrubs but woody plants can eventually overtop the willow-herb where they have established in more open places or where they resprout from substantial cut stools. The particular kind of woodland succeeding the *E. angustifolium* vegetation will depend on the local soil and climatic conditions and the availability of colonisers, but the range of communities noted above may all eventually replace stands of willow-herb according to particular circumstances. On more mesotrophic or calcareous soils, *Crataegus-Hedera* or *Prunus* scrub may supervene. On sandy soils, resump-

tion of judicious burning or grazing may reinstate some kind of heath rather than acidophilous oak-birch woodland.

Distribution

The *Epilobium angustifolium* community is very widely distributed in suitable habitats throughout the British lowlands.

Affinities

Elsewhere in Europe, can be seen vegetation dominated by *Epilobium angustifolium* in association with such herbs as *Fragaria vesca, Senecio sylvaticus, Galeopsis speciosa* and *Myosotis sylvatica* and colonising shrubs like *Rubus idaeus, Sambucus nigra* and *Salix caprea* (e.g. Westhoff & den Held 1969, Oberdorfer 1978, White & Doyle 1982, Matuszkiewicz 1981, Pott 1992, Mucina *et al.* 1993). The associations most commonly recognised have been the *Digitali-Epilobietum* Schwickerath 1944, where *Digitalis purpurea, Holcus mollis* and *Teucrium scorodonia* are characteristic and the *Senecioni sylvatici-Epilobietum* R.Tx. 1937 with *S. sylvaticus* and *Deschampsia flexuosa.* Both communities can be found in clearances and burned areas within acidophilous Quercion, Fagion and conferous woodlands and the ecological differences between them are not always clear: the former is perhaps more characteristic of better-quality brown soils with moder or mull. Sometimes, other assemblages including *E. angustifolium* have been characterised from scrubby regrowth with *Rubus idaeus.* The scarcity of *Digitalis* among the samples available to us is striking and inexplicable: certainly vegetation resembling the *Digitali-Epilobietum* is widespread in Britain. In Birse's (1984) survey of Scottish vegetation, he characterised the *Senecioni-Epilobietum.* For the moment, it seems better to retain a single, diverse and rather ill-defined assemblage.

Whatever associations have been distinguished, authorities agree on grouping *E. angustifolium* vegetation in a distinct alliance, the Epilobion angustifoli Soó emend. R.Tx. 1950 or the Carici piliferae-Epilobion R.Tx. 1950, characteristic of more base-poor soils, within a special class, the Epilobietea angustifolii R.Tx. & Preising in R.Tx. 1950 of various vegetation types in cleared, thinned, burned and disturbed woodlands.

Floristic table OV27

	a	b	c	d	e	27
Epilobium angustifolium	V (5–10)	V (5–10)	V (5–9)	V (5–8)	V (5–8)	V (5–10)
Holcus lanatus	V (1–5)	III (1–6)	II (4–6)	V (5–8)	III (4–6)	III (1–6)
Festuca ovina	III (3–5)					I (3–5)
Anthoxanthum odoratum	II (2–3)	I (4)	I (2)			I (2–4)
Potentilla erecta	II (2–5)	I (4)				I (1–5)
Teucrium scorodonia	II (3–6)	I (1)		I (5)		I (1–6)
Erica cinerea	II (3–4)					I (3–4)
Calluna vulgaris	II (2–4)					I (2–4)
Galium saxatile	II (2–6)		I (4)			I (2–6)
Ulex europaeus	II (3–7)					I (3–7)
Nardus stricta	I (4)					I (4)
Ulex gallii	I (7)					I (7)
Cytisus scoparius	I (5)					I (5)
Urtica dioica		IV (2–6)	II (3–6)	II (4–5)		II (2–6)
Cirsium arvense	I (1–3)	III (1–5)		I (2)		I (1–5)
Galium aparine		II (1–4)	I (3–4)	I (1–3)	II (3)	I (1–4)
Arrhenatherum elatius	I (2–4)	II (2–8)		I (1)		I (1–8)
Dactylis glomerata	I (4)	II (1–4)				I (1–4)
Heracleum sphondylium		II (1–4)		I (1)		I (1–4)
Deschampsia cespitosa		II (2–5)	I (3)	I (4)		I (2–5)
Holcus mollis		II (1–6)	I (4)			I (1–6)
Poa trivialis		II (1–4)		I (2–4)		I (1–4)
Cirsium vulgare	I (2)	II (1–5)			I (1)	I (1–5)
Solanum dulcamara		II (1–4)				I (1–4)
Elymus repens		I (3–6)				I (3–6)
Anthriscus sylvestris		I (3–4)				I (3–4)
Calystegia sepium		I (3)				I (3)
Epilobium hirsutum		I (2–3)				I (2–3)
Rubus fruticosus agg.	II (3–5)	III (2–7)	IV (2–6)	III (2–7)	I (6)	III (2–7)
Dryopteris dilatata		I (2–3)	IV (1–7)	II (1–7)		I (1–7)
Betula pubescens sapling			II (7)			I (7)

Species	1	2	3	4	5	6
Aulacomnium androgynum			II (1–4)			I (1–4)
Mnium hornum			II (1–3)	I (2)		I (1–3)
Lophocolea cuspidata	I (2)		II (1–5)			I (1–5)
Pinus nigra sapling			II (6)			I (6)
Acer pseudoplatanus sapling		I (6)		V (3–6)		I (3–6)
Fraxinus excelsior sapling		I (1–8)		IV (1–6)		I (1–6)
Sambucus nigra		I (3)		IV (1–4)		I (1–8)
Circaea lutetiana				III (1–3)		I (1–3)
Fagus sylvatica sapling				III (5)		I (5)
Ulmus glabra sapling				III (4–5)		I (4–5)
Eurhynchium striatum				II (1–4)		I (1–4)
Allium ursinum				II (1)		I (1)
Brachypodium sylvaticum				I (1–3)		I (1–3)
Ammophila arenaria					V (4–8)	I (4–8)
Festuca rubra					V (4–9)	I (4–9)
Senecio jacobaea	II (2–3)	I (2–3)	I (2)		IV (1–3)	I (1–3)
Poa pratensis		I (3–6)			IV (2–7)	I (2–7)
Lotus corniculatus			I (2)		III (2–4)	I (2–4)
Hypochoeris radicata	I (3)				III (2–4)	I (2–4)
Ononis repens					II (2–5)	I (2–5)
Crepis capillaris					II (2–3)	I (2–3)
Luzula campestris					II (2)	I (2)
Myosotis ramosissima					II (2)	I (2)
Valerianella locusta					II (1–2)	I (1–2)
Carlina vulgaris					II (1–3)	I (1–3)
Sedum acre					II (1–2)	I (1–2)
Viola tricolor					II (1)	I (1)
Agrostis capillaris	II (3–5)	II (1–6)	I (2)	I (1–4)	I (2)	I (1–6)
Pteridium aquilinum	II (3–5)	II (1–8)	II (5–8)	I (5)		I (1–8)
Brachythecium rutabulum	I (2)	II (2–6)	I (1)	II (2–4)	I (3–4)	I (1–6)
Eurhynchium praelongum		II (3–6)	II (1–3)	II (2–5)		I (1–6)
Epilobium montanum		I (3)	I (2–3)	I (1–3)	I (1)	I (1–3)
Cerastium fontanum	I (2–3)	I (1)	I (1)		I (2–3)	I (1–3)
Juncus effusus	I (5–9)	I (1–4)	I (3)	I (1)		I (1–9)

Floristic table OV27 (*cont.*)

	a	b	c	d	e	27
Dicranum scoparium		I (1)	I (3–4)	I (4)	I (6)	I (1–6)
Rubus caesius		I (3)	I (1–5)	I (2–3)	I (3–5)	I (1–5)
Lathyrus pratensis	I (3)	I (1–3)		I (3)		I (1–3)
Cirsium palustre	I (3)	I (3)		I (1–3)		I (1–3)
Mercurialis perennis		I (4)	I (5)	I (5–6)		I (4–6)
Silene dioica		I (1)	I (3)	I (2–5)		I (1–5)
Rumex obtusifolius		I (3)	I (1)	I (1–3)		I (1–3)
Athyrium filix-femina		I (2)	I (3)	I (1–2)		I (1–3)
Dryopteris filix-mas		I (3–5)	I (3)	I (1–3)		I (1–5)
Hyacinthoides non-scripta		I (8)	I (3–5)	I (3–4)		I (3–8)
Hypnum cupressiforme			I (4–6)	I (2–4)	I (3–5)	I (2–6)
Hedera helix		I (3–4)		I (3–5)		I (3–5)
Achillea millefolium	I (5)	I (4)				I (4–5)
Juncus conglomeratus	I (3)	I (3–5)				I (3–5)
Rumex acetosa	I (3)	I (1–4)				I (1–4)
Digitalis purpurea		I (6)	I (1)			I (1–6)
Rubus idaeus			I (4–7)	I (1–6)		I (1–7)
Rumex sanguineus		I (2)		I (1–2)		I (1–2)
Plagiomnium undulatum		I (1)		I (1–3)		I (1–3)
Glechoma hederacea		I (2–5)		I (1–2)		I (1–5)
Stachys sylvatica		I (2–4)		I (2)		I (2–4)
Number of samples	12	43	14	8	8	85
Number of species/sample	13 (5–45)	11 (4–29)	10 (4–14)	26 (11–41)	15 (5–22)	13 (4–45)

a   *Holcus lanatus-Festuca ovina* sub-community
b   *Urtica dioica-Cirsium arvense* sub-community
c   *Rubus fruticosus* agg.-*Dryopteris dilatata* sub-community
d   *Acer pseudoplatanus-Sambucus nigra* sub-community
e   *Ammophila arenaria* sub-community
27 *Epilobium angustifolium* community (total)

OV28

Agrostis stolonifera-Ranunculus repens community
Agrostio-Ranunculetum repentis   Oberdorfer *et al.* 1967

Constant species

Agrostis stolonifera, Ranunculus repens.

Physiognomy

The *Agrostio-Ranunculetum* comprises open or closed vegetation in which a mat of stolons and runners of *A. stolonifera* and *R. repens* is the characteristic consistent feature. Throughout the community as a whole, no other species is frequent but there is occasionally some *Poa trivialis* and *Trifolium repens* in the ground carpet and scattered shoots or small clumps of *Urtica dioica* and *Cirsium arvense*. *Senecio vulgaris, Atriplex prostrata* and *Taraxacum officinale* agg. are scarce companions.

Sub-communities

Polygonum hydropiper-Rorippa sylvestris sub-community. The cover of the two community constants tends to be higher here and there are occasionally quite conspicuous shoots of *Phalaris arundinacea* and *Juncus effusus* with scattered plants or little patches of *Polygonum hydropiper* and *Rorippa sylvestris* springing up among the perennials. *Galium palustre, Mentha aquatica, Myosotis scorpioides* and *Alopecurus geniculatus* are occasionally seen.

Poa annua-Polygonum aviculare sub-community: *Ranunculetum repentis* Knapp 1946 *sensu* Silverside 1990. *Poa annua* is a constant and sometimes abundant contributor to the ground carpet in this sub-community with frequent records too for *Plantago major, Stellaria media, Polygonum aviculare, P. persicaria* and *Chamomilla suaveolens*. *Lolium perenne, Elymus repens, Potentilla anserina* and *Anagallis arvensis* occur occasionally and there are quite often small patches of mosses on bare areas of damp soil: *Bryum rubens, Pottia truncata* and *Dicranella staphylina* are the most consistent contributors to this element of the vegetation.

Habitat

The *Agrostio-Ranunculetum* is characteristic of damp silts and clays on river islands and banks, in and around sluggish streams, drainage ditches and seasonally-inundated hollows in ill-drained pastures, arable fields and river flood-plains, around waterlogged places in made ground and among dumps of soil and along muddy tracks.

The two sub-communities are typical of rather different situations within this range of habitats. The *Polygonum-Rorippa* sub-community is usually found in wetter places, where water levels fall later in spring or even remain on the surface: for example, on river shoals, around drains and streams and in more or less permanently wet hollows in fields. The *Poa-Plantago* sub-community is more typical of depressions in pastures, among dumped soil and along trackways where the ground is wet in winter but dries somewhat in summer. Poaching by stock or trampling by humans is common.

Zonation and succession

Around wet areas in pastures, on flood-plains and away from river banks, the *Agrostio-Ranunculetum* can give way to the *Festuca-Agrostis-Potentilla* grassland or some kind of *Lolium* ley, with the *Poa-Plantago* community sometimes figuring as an intermediate or, where drier ground is trampled, the *Polygonum-Chamomilla* assemblage. In sluggish streams or watering places, the *Agrostis-Alopecurus* community can figure. In wet arable fields, the *Agrostio-Ranunculetum* can pass to the *Poa-Plantago* or *Matricaria-Stellaria* community.

On river shoals and silty margins of water-courses, the community can be found in mosaics and zonations with the *Ranunculo-Alopecuretum*, the *Polygono-Bidentetum*, the *Polygonum-Poa* community and the *Rorippa-Filaginella* community, sometimes also with patches of *Phalaridetum*.

Repeated inundation sets back any tendency to succession on river shoals, streamsides and pasture hollows and, on drier ground, grazing can play a part in checking any seral change. Where areas are drained and grazed, a likely sequence is for the *Poa-Plantago* sub-community to develop into some kind of *Festuca-Agrostis-Potentilla* sward or with reseeding, which has been a common fate, for it to be replaced by a *Lolium* ley.

Distribution
The *Agrostio-Ranunculetum* occurs widely on suitable substrates throughout the lowlands.

Affinities
Various permutations of *Agrostis stolonifera*, *Alopecurus geniculatus* and *Ranunculus repens*, often with large *Rumex* spp. and *Rorippa* spp. have been characterised in a range of assemblages of this kind: a *Rumici-Agrostietum* Moor 1958 in Mucina *et al.* (1993) from Austria, a *Rorippo-Agrostietum* (Moor 1958) Oberdorfer & T. Müller in T. Müller 1961 from Germany (Pott 1992) and an *Agrostio-Ranunculetum* Oberdorfer *et al.* 1967, also from Germany (Oberdorfer 1983). Very commonly, too,

these assemblages are reduced to very species-poor vegetation dominated by one or other of the species listed, as in the *Ranunculus repens*-Gesellschaft (Oberdorfer 1983, Mucina *et al.* 1993) or the *Agrostis stolonifera* community (Sykora 1983).

The affiliation of these syntaxa to alliances and higher units has been a much debated issue (see, for example, Westhoff & den Held 1969) and many authorities have now abandoned the alliance Elymo-Rumicion Nordhagen 1940 *emend*. R.Tx. 1950 in favour of the Lolio-Potentillion R.Tx. 1947 (Sykora 1983, Pott 1992) or the Potentillion anserinae R.Tx. 1947 (Mucina *et al.* 1993). In this treatment, we have retained the older name to avoid confounding affiliations discussed in earlier volumes. Whatever name is given to this alliance, this community, together with the *Ranunculo-Alopecuretum*, belongs with various mesotrophic grasslands, duneslacks and upper salt-marsh swards, all of which experience somewhat unpredictable seasonal flooding with fresh or brackish waters.

Floristic table OV28

	a	b	28
Agrostis stolonifera	V (2–9)	V (1–8)	V (1–9)
Ranunculus repens	V (2–8)	V (1–4)	V (1–8)
Polygonum hydropiper	III (1–4)		II (1–4)
Phalaris arundinacea	II (1–6)		I (1–6)
Juncus effusus	II (2–7)		I (2–7)
Rorippa sylvestris	II (1–5)		I (1–5)
Galium palustre	II (1–4)		I (1–4)
Mentha aquatica	II (1–5)		I (1–5)
Myosotis scorpioides	II (1–5)		I (1–5)
Alopecurus geniculatus	II (1–6)		I (1–6)
Poa annua	I (1–4)	V (1–8)	II (1–8)
Polygonum aviculare	I (1–2)	V (1–3)	II (1–3)
Plantago major	II (1–5)	V (1–4)	II (1–5)
Stellaria media	I (1–3)	V (1–8)	I (1–8)
Polygonum persicaria	I (1–4)	III (1–3)	I (1–4)
Chamomilla suaveolens		III (1–2)	I (1–2)
Lolium perenne	I (1–4)	II (1–4)	I (1–4)
Potentilla anserina	I (1)	II (1–3)	I (1–3)
Bryum rubens	I (1)	II (1–4)	I (1–4)
Pottia truncata	I (1)	II (1–4)	I (1–4)
Dicranella staphylina		II (1–4)	I (1–4)
Chenopodium album		II (1–2)	I (1–2)
Elymus repens		II (1–8)	I (1–8)
Anagallis arvensis		II (1–3)	I (1–3)
Urtica dioica	II (1–3)	II (1–2)	II (1–3)
Cirsium arvense	II (1–4)	II (1–6)	II (1–6)
Poa trivialis	II (1–4)	II (1–5)	II (1–5)
Trifolium repens	II (1–4)	II (1–3)	II (1–4)
Senecio vulgaris	I (2–6)	I (1–2)	I (1–6)
Taraxacum officinale agg.	I (1)	I (1)	I (1)
Atriplex prostrata	I (2–3)	I (1–4)	I (1–4)
Cerastium fontanum	I (2–3)	I (1–3)	I (1–3)
Filaginella uliginosa	I (2)	I (1)	I (1–2)
Atriplex patula	I (1)	I (1–2)	I (1–2)
Number of samples	34	19	53
Number of species/sample	12 (8–18)	20 (13–29)	14 (8–29)

a   *Polygonum hydropiper-Rorippa sylvestris* sub-community

b   *Poa annua-Polygonum aviculare* sub-community

28   *Agrostio-Ranunculetum repentis* (total)

OV29

Alopecurus geniculatus-Rorippa palustris community
Ranunculo-Alopecuretum geniculati R.Tx. (1937) 1950

Constant species
Alopecurus geniculatus, Rorippa palustris.

Physiognomy
A mat of *Alopecurus geniculatus*, often extensive and lush, is the most distinctive characteristic of the *Ranunculo-Alopecuretum*, with sometimes a small contribution to the carpet from *Potentilla anserina*, *Poa trivialis* and *Ranunculus repens*. There are frequent scattered plants of *Rorippa palustris* and various annual knotweeds, among which *Polygonum lapathifolium*, *P. aviculare* and *P. hydropiper* are the most common. *Rumex crispus* also occurs often and there can be some *Elymus repens* and *Phalaris arundinacea*, though these are generally not very abundant. A wide range of plants of damp, weedy places occur at low frequency.

Habitat
The *Ranunculo-Alopecuretum* is typical of periodically-flooded sills and sands on terraces, bars and islands in mature river valleys, on the edges of seasonal pools and small gentle streams in the lowlands and around fluctuating ponds and lakes with more nutrient-rich waters. Modest trampling by people, stock or wildfowl can be important in maintaining open ground for this vegetation and for dispersal of propagules.

Flooding by river or lake waters or the seasonal accumulation of rainwater in shallow depressions can handicap the growth of pasture vegetation or leave deposits of silt and sand around shores and on the terraces of river valleys. Then the species of this community are often able to take rapid advantage of the areas of bare ground thus created, by germination of seed and spread of vegetative propagules. *Alopecurus geniculatus* can benefit in both these ways: it produces large numbers of seeds but also grows rapidly from broken shoot fragments by stolon production. *Rorippa palustris* and the various knotweeds can also germinate and establish quickly while *Potentilla anserina* and *Elymus repens* show ready vegetative spread. The cover of the sward and its partic-ular character can, however, vary considerably from season to season according to the duration and extent of the inundation.

The flood-waters themselves can be an effective means of transporting seed and shoot fragments but grazing stock or wildfowl can also carry away propagules in mud. Trampling also helps keep such areas open.

Zonation and succession
On expanses of sediment laid totally bare by flooding and then exposed on river islands and margins, the *Ranunculo-Alopecuretum* can develop patchily, either alone or with other inundation communities, their disposition related in part to the texture and wetness of the sediments, though often with an element of chance about which assemblages colonise where (Figure 26). Such patterns also vary considerably from year to year according to the extent and character of the material once more flooded and exposed: Tüxen (in Ellenberg 1988) very aptly compared this variation to the squeezing and release of an accordion.

Very often, the *Agrostio-Ranunculetum* is also involved in such patterns and each of the communities can pass to the *Polygono-Bidentetum* or *Polygonum-Poa* community with an increase in abundance of *Bidens* and *Polygonum* spp. These latter two assemblages are perhaps more common where silts prevail among the sediments. On muds which remain wetter longer, the *Rorippa-Filago* community can figure and, where the water table stays close to the surface all summer, the *Polygonum amphibium* community. Patchily on river islands, sometimes more extensively along banks and towards the limit of fluctuation around reservoirs, the *Phalaridetum arundinaceae* may occur with the *Ranunculo-Alopecuretum*.

Another very common situation for this community is in low-lying stretches of flood-plain pastures which are inundated in winter. Then, the *Ranunculo-Alopecuretum* usually gives way on less flooded ground where there has not been too much agricultural improvement to the *Festuca-Agrostis-Potentilla* grassland. Splashy places in such pastures may have the *Agrostis-Alopecurus* grassland, sometimes with the *Agrostio-Ranunculetum* between these areas and drier ground.

Distribution

The *Ranunculo-Alopecuretum* occurs widely through the British lowlands in suitable situations.

Affinites

Vegetation of this type has been previously allocated to a *Rumici-Alopecuretum geniculati* R.Tx. (1937) 1950 in The Netherlands (Westhoff & den Held 1969) and to the *Ranunculo-Alopecuretum* R.Tx. 1937 in Ireland (Braun-Blanquet & Tüxen 1952) and Germany (Oberdorfer 1983, Pott 1992). In Sykora's (1983) treatment of these inundation communities, he used the latter association for Dutch stands and described a *rorippetosum* which is very similar to the vegetation included here. As explained under the account of the *Agrostio-Ranunculetum*, this assemblage is assigned to the Elymo-Rumicion alliance, rather than its replacements, in many more recent schemes the Lolio-Potentillion or Potentillion.

Figure 26. Inundation communities on the flood banks of the river Lune, north Lancashire.
On the low flood banks around a periodically-flooded pool with A15 *Elodea canadensis* vegetation and fragmentary S10 *Equisetetum fluviatile*, there is a zone of the OV29 *Ranunculo-Alopecuretum geniculati* which gives way over the slightly drier shoal to the OV28a *Polygonum-Rorippa* sub-community of the *Agrostio-Ranunculetum repentis*. This in turn passes to the grazed and periodically-inundated MG11a *Lolium* sub-community of the *Festuca-Agrostis-Potentilla* grassland, then to MG6 *Lolio-Cynosuretum* pastures on the main river terraces.

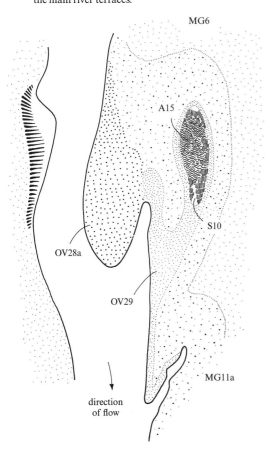

Floristic table OV29

Alopecurus geniculatus	IV (3–9)
Rorippa palustris	IV (1–5)
Potentilla anserina	III (1–4)
Rumex crispus	III (1–5)
Elymus repens	III (2–6)
Poa trivialis	II (3–4)
Polygonum lapathifolium	II (4–6)
Ranunculus repens	II (3–5)
Phalaris arundinacea	II (2–4)
Polygonum aviculare	II (2–3)
Polygonum hydropiper	II (3–7)
Agrostis stolonifera	I (3)
Cirsium arvense	I (3–4)
Glyceria fluitans	I (2–4)
Lolium perenne	I (3)
Plantago major	I (1–4)
Caltha palustris	I (3–4)
Elecocharis palustris	I (3–4)
Juncus articulatus	I (2–4)
Myosotis scorpioides	I (1–4)
Rumex obtusifolius	I (3–4)
Stellaria media	I (1–3)
Lemna minor	I (3–4)
Oenanthe fistulosa	I (2–3)
Bidens tripartita	I (2)
Cardamine flexuosa	I (3)
Carex disticha	I (3)
Cochlearia anglica	I (2)
Equisetum fluviatile	I (4)
Poa annua	I (2)
Polygonum persicaria	I (2)
Number of samples	12
Number of species/sample	9 (3–21)

OV30

Bidens tripartita-Polygonum amphibium community
Polygono-Bidentetum tripartitae Lohmeyer in R.Tx.
1950

Constant speces

Bidens tripartita, Filaginella uliginosa, Phalaris arundi-nacea, Polygonum amphibium, Polygonum hydropiper.

Physiognomy

The *Polygono-Bidentetum* comprises stands of open or closed vegetation variously dominated by *Bidens tripartita, Polygonum amphibium* or *P. hydropiper*. Other knotweeds are typically scarce but *P. persicaria* sometimes occurs. *Filaginella uliginosa* is constant but generally at low cover and there are very often some sparse shoots of *Phalaris arundinacea. Alopecurus geniculatus* and *Agrostis stolonifera* are quite common, occasionally forming dense mats of shoots and *Callitriche hamulata* occurs frequently. Other occasionals include *Alisma plantago-aquatica, Myosotis laxa* ssp. *cespitosa, Juncus bufonius, Plantago major, Potentilla anserina* and *Stellaria alsine.*

Habitat

The *Polygono-Bidentetum* is characteristic of periodically-flooded, eutrophic silts and clays towards the limit of inundation around fluctuating ponds, lakes and reservoirs.

It is the combination of periodic inundation and high levels of nitrogen in waters and/or substrates favoured by the ephemeral species of this assemblage that invade as flooding subsides. Typically, inundation occurs in winter or at least more extensively then, so that exposure of a damp, fertile substrate coincides with the warmer temperatures of spring for germination of the annual plants. In fact, some frequent species in this community are perennials, like *Polygonum amphibium* and *Phalaris arundinacea*, occurring at generally low covers but maintaining themselves towards the upper limits of flooding. *P. amphibium* has far-creeping rhizomes and a truly amphibious habit, good adaptations to the measure of unpredictability in inundation in these more or less unstable substrates. *Phalaris* is one of the very few tall helophytes tolerant of irregular flooding.

The periods of freedom from inundation may allow such species and the carpets of creeping perennial grasses like *Agrostis stolonifera* and *Alopecurus geniculatus* to become a little more extensive for a while but are never lengthy enough to permit complete colonisation of the muds, which would exclude *Bidens* and the other typical ephemerals.

Zonation and succession

The *Polygono-Bidentetum* is typically a patchy element of zonations around fluctuating open waters where shifts in composition and structure of the vegetation are related to duration and depth of inundation.

Quite commonly, this community gives way, on silts and clays subject to longer inundation, to the *Rorippa-Filaginella* community where *Bidens* remains quite common but where dominance usually passes to various annual knotweeds with *Rorippa palustris* and *Filaginella uliginosa* becoming more frequent. Upslope, at and beyond the limits of inundation, the community can give way to a zone of the *Phalaridetum arundinaceae*. Alternatively, around ponds in pastures, there can be a transition through the *Agrostis-Alopecurus* community or *Ranunculetum scelerati* to drier leys or pastures of the Lolio-Plantaginion, Cynosurion or Elymo-Rumicion types.

Distribution

The community occurs widely in suitable habitats through the lowlands of Britain.

Affinities

This community is the central type of Bidention vegetation in Britain and it clearly corresponds to the association variously described as the *Polygono-Bidentetum* Koch 1926 *emend*. Sissingh 1946 from The Netherlands (Westhoff & den Held 1969), Ireland (White & Doyle 1982) and Poland (Matuszkiewicz 1984), or the *Polygono hydropiperis-Bidentetum* Lohmeyer in R.Tx. 1950 from Germany (Oberdorfer 1983, Pott 1982) and Austria (Mucina *et al*. 1993).

Floristic table OV30

Bidens tripartita	V (4–8)
Polygonum amphibium	V (3–7)
Filaginella uliginosa	IV (2–5)
Polygonum hydropiper	IV (4–6)
Phalaris arundinacea	IV (3–5)
Alopecurus geniculatus	III (1–5)
Callitriche hamulata	III (1–3)
Agrostis stolonifera	II (2–4)
Polygonum persicaria	II (2–7)
Alisma plantago-aquatica	II (1–2)
Myosotis laxa cespitosa	II (3)
Juncus bufonius	II (6)
Plantago major	II (1–3)
Potentilla anserina	II (1–3)
Stellaria alsine	II (1–5)
Eleocharis palustris	I (3)
Iris pseudacorus	I (1)
Arrhenatherum elatius	I (2)
Berula erecta	I (4)
Calystegia sepium	I (1)
Carex vesicaria	I (6)
Chenopodium murale	I (2)
Holcus lanatus	I (1)
Juncus effusus	I (4)
Lythrum salicaria	I (4)
Chamomilla recutita	I (1)
Mentha arvensis	I (2)
Poa trivialis	I (4)
Ranunculus flammula	I (4)
Ranunculus repens	I (2)
Salix fragilis sapling	I (4)
Scirpus sylvaticus	I (5)
Solanum dulcamara	I (1)
Sparganium erectum	I (1)
Trifolium repens	I (1)
Salix cinerea sapling	I (5)
Number of samples	7
Number of species/sample	10 (5–17)
Vegetation height (cm)	30 (15–60)
Vegetation cover (%)	95 (80–100)

OV31
Rorippa palustris-Filaginella uliginosa community

Synonymy
Rorippa islandica-Gnaphalium uliginosum community
Birse 1984.

Constant species
Filaginella uliginosa, Rorippa palustris.

Rare species
Limosella aquatica.

Physiognomy
The *Rorippa palustris-Filaginella* community comprises
open or closed vegetation in which *R. palustris* and
Filaginella uliginosa are the most consistent feature and
where either can be abundant over patches of damp
mud, but which may also be variously dominated by the
occasional to frequent annual knotweeds, *Polygonum
persicaria, P. hydropiper* and *P. aviculare. Bidens tripartita* also occurs commonly but not usually with any
abundance and there is usually some *Juncus bufonius*.
Other plants recorded occasionally are *Plantago major,
Lythrum portula, Littorella uniflora* and the grasses
Agrostis stolonifera, Alopecurus geniculatus, Poa annua
and *Phalaris arundinacea*. Scattered individuals of
*Rumex crispus, R. obtusifolius, Chenopodium rubrum,
Cirsium arvense* and *Epilobium obscurum* may catch the
eye but such species are never common. This community
can provide a locus for the nationally rare *Limosella
aquatica*.

Habitat
The *Rorippa-Filaginella* community is characteristically
found in periodically-flooded, eutrophic muds and
sands around the margins of fluctuating pools, lakes and
reservoirs, on islands and banksides in mature river
systems and in ruts along paths and trackways.

The community depends upon the exposure of bare,
damp substrates by lowering of water-tables in rivers
and lakes after the winter or by evaporation from
smaller pools and flooded ruts with the warmth of

spring. The knotweeds which commonly come to dominate this kind of vegetation are especially well adapted
to rapid colonisation of congenial situations after a
period of cold temperatures and can quickly overwhelm
the smaller ephemerals (Justice 1941, Timson 1965,
Courtney 1968, Henson 1969).

High levels of nitrogen are also important in permit-
ting the quick growth of the sorts of lush swards characteristic here. The community is thus especially likely to
develop where nutrient-rich ground waters or eutrophic
substrates like river and lake alluvia occur, or where
there is local enrichment of damp ground by watering
stock or animal traffic along trackways. Even very
limited development of such conditions, as along rutted
paths where cattle, sheep or horses defaecate in largely
impoverished environments like lowland heaths can
permit the temporary development of the community.
In fact, for a rarity like *Limosella aquatica*, such habitats
may be especially important: it too has seeds which germinate rapidly in summer and completes its life cycle
before autumn (Preston in Stewart *et al.* 1994). Typically, like the commoner species of this community, the
populations of *Limosella* can vary greatly in size from
year to year in any one place. The characteristic habitats
here are not only unstable but quite often not precisely
congenial for colonisation by a particular species.

Zonation and succession
The *Rorippa-Filaginella* community can be found in
mosaics and zonations with other kinds of inundation
communities, wet grasslands and water-margin vegetation according to differences in the extent of inundation,
instability of the habitat and amount of nutrients in the
waters and substrates.

On ground that emerges later from inundation and is
not so unstable, the community can be replaced by
amphibious perennial vegetation like the *Polygonum
amphibium* community or by a zone of bare wet muds
too shortly exposed for colonisation. Above, the community can pass to the *Bidens-Phalaris* assemblage or to

the *Polygonum-Poa* community on ground which remains exposed each summer for longer. Where the community marks out periodically-flooded ruts in pasture or heath, it can give way more sharply to Elymo-Rumicion, Lolio-Plantaginion or Cynosurion swards, or to ericoid vegetation.

With an increase in nitrogen enrichment on wetter ground, the *Rorippa-Filaginella* community is replaced by the *Ranunculetum scelerati*.

Distribution
The community occurs widely in suitable habitats throughout the lowlands of Britain.

Affinities
In the high frequency of *F. uliginosa* and *J. bufonius*, and in providing a locus for *Limosella aquatica*, the *Rorippa-Filaginella* community shows strong floristic affinities with the Nanocyperion vegetation of fluctuating pools. However, in the equable and moist Atlantic climate of Britain, such vegetation often supports a variety of bulkier plants, like *Polygonum* spp. and *Bidens*, which can get a hold as the muds dry out and release nutrients but are not too parched in the absence of summer drought (Rodwell 1994b). Then the appearance is more of a Bidention assemblage.

Floristic table OV31

Rorippa palustris	V (1–9)
Filaginella uliginosa	V (2–8)
Polygonum persicaria	III (2–8)
Juncus bufonius	III (2–6)
Polygonum hydropiper	III (3–10)
Polygonum aviculare	III (1–5)
Bidens tripartia	III (1–7)
Plantago major	III (1–8)
Alopecurus geniculatus	II (1–7)
Agrostis stolonifera	II (2–5)
Lythrum portula	II (2–5)
Polygonum amphibium	II (1–5)
Poa annua	II (2–8)
Littorella uniflora	II (1–4)
Phalaris arundinacea	I (2–5)
Polygonum lapathifolium	I (1–5)
Rumex crispus	I (2–4)
Epilobium obscurum	I (2–4)
Ranunculus flammula	I (3–4)
Rumex obtusifolius	I (2–4)
Pohlia carnea	I (3–5)
Ranunculus repens	I (2–4)
Limosella aquatica	I (1–9)
Chenopodium rubrum	I (1–5)
Cirsium arvense	I (2–3)
Glyceria fluitans	I (2–4)
Juncus articulatus	I (2–4)
Callitriche hamulata	I (3–7)
Eleocharis palustris	I (1–4)
Alisma plantago-aquatica	I (2–4)
Callitriche stagnalis	I (3)
Mentha aquatica	I (1–7)
Potentilla anserina	I (1–3)
Physcomitrium pyriforme	I (2–3)
Physcomitriella patens	I (2–4)
Number of samples	49
Number of species/sample	10 (4–18)

OV32
Myosotis scorpioides-Ranunculus sceleratus community
Ranunculetum scelerati   R.Tx. 1950 ex Passarge 1959

Constant species
Agrostis stolonifera, Glyceria maxima, Myosotis scorpioides, Ranunculus sceleratus, Rorippa islandica.

Physiognomy
The *Ranunculetum scelerati* includes open or closed carpets of vegetation, locally lush, in which mixtures of *Myosotis scorpioides, Ranunculus sceleratus, Rorippa islandica* and *Agrostis stolonifera* occur, often among scattered shoots of *Glyceria maxima. Veronica catenata, V. beccabunga, Nasturtium officinale* agg. and *Berula erecta* are occasional to frequent and there are sometimes small tussocks of *Deschampsia cespitosa, Holcus lanatus, Juncus inflexus* and *J. effusus.* Small patches of *Lemna gibba* or *Callitriche stagnalis* can occur scattered on wet mud or shallow pools among the vegetation. Typically, *Bidens tripartita* and knotweeds are, at most, occasional.

Habitat
This is a community of very nitrogen-rich, intermittently wetted and disturbed ground such as the heavily-manured surrounds of ponds and streams where stock water, along nutrient-enriched seasonal watercourses and over wet ditch dredgings.

As far as moisture and disturbance requirements are concerned, this community is probably essentially the same as the *Rorippa-Filaginella* vegetation, being dependent on the exposure of moist soils and clays in the warmer weather of spring when ephemerals are best placed to capitalise on the congenial conditions. Hence, disturbance may often come, not from flooding, but from trampling by stock or by physical operations like the cleaning of silt from streams and ditches. The frequent occurrence among the typical annuals of perennial plants is a reflection of the common occurrence of this assemblage as fragmentary stands among the fringes of water-margin or swamp vegetation. Noticeably, these include plants well able to regenerate from broken rhizome or stem fragments.

Ellenberg (1988) considered the *Ranunculetum scelerati* community to be more nitrophilous than the *Rorippa-Filaginella* vegetation. *R. sceleratus*, for example, can readily grow on the sludge beds of sewage farms, a habitat intolerable for many species.

Zonation and succession
The *Ranunculetum scelerati* is generally found as small, often fragmentary stands in mosaics and zonations with other inundation vegetation, water-margin communities and wet grasslands, the patterns being related to the extent of inundation, disturbance and enrichment.

Where dunging or other forms of eutrophication are absent or very localised around ponds and alongside streams, the community can give way to *Rorippa-Filaginella* vegetation on somewhat less fertile, periodically-inundated ground. At the water's edge, it frequently gives way to Glycerio-Sparganion assemblages like the *Glycerietum fluitantis* or stands dominated by *Nasturtium officinale* agg. or *Veronica beccabunga*, among which plants like *Myosotis scorpioides, Alisma plantago-aquatica* and *Berula erecta* can remain locally prominent. Ragged zonations between these mixtures of plants are very common along small lowland streams and ditches through silty and clay soils.

In other situations, the *Ranunculetum scelerati* can pass more directly to swamps like the *Glycerietum maximae* or to periodically inundated Elymo-Rumicion swards or damp Lolio-Plantaginion leys.

Distribution
The community can be found throughout the lowlands in suitable habitats.

Affinities
Vegetation of this type from elsewhere in Europe has generally been characterised as a *Ranunculetum scelerati* R.Tx. 1950 ex Passarge 1959, as with Oberdorfer (1983) and Pott (1992) in Germany, as a *Ranunculo scelerati-Rumicetum maritimi* Sissingh (1946) 1966, in Westhoff

& den Held (1969) from The Netherlands and White & Doyle (1982) from Ireland or as a *Rumicetum maritimi* Sissingh ex R.Tx. 1950 in Matuszkiewicz (1984) from Poland and Mucina *et al.* (1993) from Austria. *Rumex maritimus* was not in fact recorded in the samples available to us but it does occur as an occasional, scattered

through the lowlands of England, and just into Wales and Scotland, in the kinds of habitats favoured by this community: it prefers nutrient-rich muds kept wet late into spring when it can appear in abundance. However this assemblage is named, it clearly belongs in the Bidention alliance.

Floristic table OV32

Myosotis scorpioides	V (2–5)	*Eleocharis palustris*	I (4)
Ranunculus sceleratus	IV (3–9)	*Epilobium palustre*	I (3)
Agrostis stolonifera	IV (4–7)	*Poa pratensis*	I (2)
Glyceria maxima	IV (3–5)	*Potentilla anserina*	I (1)
Rorippa islandica	IV (1–6)	*Rumex obtusifolius*	I (2)
Veronica catenata	III (2–7)	*Arctium minus* agg.	I (3)
Nasturtium officinale	III (1–3)	*Carex paniculata*	I (1)
Polygonum hydropiper	II (2–3)	*Epilobium angustifolium*	I (3)
Ranunculus repens	II (2–3)	*Epilobium adenocaulon*	I (2)
Berula erecta	II (3–6)	*Iris pseudacorus*	I (2)
Bidens tripartita	II (2–3)	*Juncus bufonius*	I (2)
Veronica beccabunga	II (3–4)	*Lycopus europaeus*	I (5)
Deschampsia cespitosa	II (3–4)	*Phalaris arundinacea*	I (2)
Holcus lanatus	II (2–3)	*Poa annua*	I (2)
Juncus effusus	II (3–5)	*Polygonum persicaria*	I (3)
Juncus inflexus	II (3–5)	*Ranunculus circinatus*	I (2)
Lemna gibba	I (3–4)	*Rorippa sylvestris*	I (3)
Callitriche stagnalis	I (2–4)	*Rumex hydrolapathum*	I (3)
Equisetum fluviatile	I (3)	*Stachys palustris*	I (3)
Glyceria declinata	I (4–6)	*Trifolium repens*	I (1)
Polygonum lapathifolium	I (2)	*Ranunculus peltatus*	I (2)
Mentha aquatica	I (6)	*Azolla filiculoides*	I (3)
Bidens cernua	I (2)	Number of samples	13
Cirsium arvense	I (2)	Number of species/sample	10 (7–14)

OV33
Polygonum lapathifolium-Poa annua community

Constant species
Poa annua, Polygonum aviculare, Polygonum lapathifolium, Polygonum persicaria.

Physiognomy
The *Polygonum-Poa* community includes open or closed stands of vegetation in which various knotweeds are the most characteristic feature. *P. persicaria, P. lapathifolium* and *P. aviculare* are all very common here and each can be abundant, often in locally dense, even more or less monodominant patches. *Poa annua* is also constant but not usually of extensive cover and there is often some *Stellaria media, Chenopodium album, Plantago major* and *Chamomilla suaveolens*. More bulky occasionals include *Elymus repens, Lolium perenne, Urtica dioica* and *Cirsium arvense* but none of these is more than locally prominent. Smaller herbaceous associates found at low frequency are *Viola arvensis, Anagallis arvensis, Ranunculus repens, Trifolium repens, Spergula arvensis* and *Veronica persica*, and there can be occasional scattered individuals of *Rumex crispus, R. obtusifolius, Sonchus asper* and *Euphorbia helioscopa*.

Habitat
This *Polygonum-Poa* community is characteristic of damp eutrophic soils in disturbed places such as gateways, tracks, farmyards and ill-managed leys, on dumped topsoil and made ground around building sites.

Compared with the *Rorippa-Filaginella* community, this vegetation is less dependent on long winter flooding to maintain extensive areas of ground then laid bare by a drop in the water table. Indeed, local impedence of drainage of rainwater, as on heavy or trampled ground, is sufficient to create suitable conditions for the rapid appearance of the annual knotweeds, *Poa annua* and weeds like *Stellaria media, Chenopodium album* and *Chamomilla suaveolens*. With seed sources often nearby, there is also opportunity for the (perhaps temporary) invasion of perennial weeds such as *Elymus repens, Cirsium arvense, Urtica dioica* and big *Rumex* spp., and forage plants like *Lolium perenne* and *Trifolium repens*.

The other feature which all these plants either require or flourish on is an abundance of nitrogen. In the context in which this vegetation develops, general fertility is high anyway, but there may also be local enrichment from defaecation by stock or run-off from local manure piles or organic waste.

Zonation and succession
Typically, the *Polygonum-Poa* community occurs on the wettest ground among patchworks of other kinds of weed vegetation, leys and pasture swards on farmland, recreation areas and along tracks.

Usually, it is the degree of wetness, disturbance and trampling which influences the character and disposition of these assemblages. Where the ground is somewhat drier but still trampled, the *Polygonum-Poa* community often gives way to some kind of Polygonion aviculare vegetation like the *Polygonum-Chamomilla* and *Poa-Plantago major* communities where *P. aviculare* and *Poa annua* remain constant but where the other annual knotweeds fade in importance. This can then pass to Lolio-Plantaginion swards where more trample-resistant plants remain common within a more extensive grassy matrix. Similar sequences to this develop where ill-sown leys are subject to much poaching in wet spring weather.

If irregular disturbance of damp, eutrophic soils remains a feature, as where farm vehicles churn up the ground, Polygono-Chenopodion communities can replace the *Polygonum-Poa* vegetation as with the *Poa-Senecio* or *Poa-Myosotis* communities.

Where disturbance or trampling of such habitats ceases, the *Polygonum-Poa* community is usually replaced by less ephemeral Polygono-Chenopodion weed vegetation, then by eutrophic tall-herb stands of the Artemisietea, usually dominated by *Urtica dioica* and large *Cirsium* spp.

Disturbance
This community is very common in suitable habitats throughout the lowlands.

Affinities

The *Polygonum-Poa* community contains those stands of ephemeral vegetation in which various species of *Polygonum* are prominent in the absence of *Bidens tripartita*. The high frequency and abundance of plants like *Stellaria media*, *Chenopodium album* and *Chamomilla suaveolens* put this assemblage close the Polygono-Chenopodion alliance but it clearly belongs among the Bidention communities. In some existing treatments, vegetation of this type is subsumed with the *Polygono-Bidentetum*.

Floristic table OV33

Polygonum lapathifolium	V (1–9)	*Dactylis glomerata*	I (1–4)
Polygonum persicaria	IV (2–7)	*Fumaria officinalis*	I (1–2)
Polygonum aviculare	IV (1–5)	*Galeopsis tetrahit* agg.	I (1–3)
Poa annua	IV (1–5)	*Galium aparine*	I (1–3)
		Myosotis arvensis	I (1–2)
Stellaria media	III (1–4)	*Poa trivialis*	I (1–4)
Chenopodium album	III (1–6)	*Raphanus raphanistrum*	I (1–2)
Plantago major	III (1–4)	*Senecio vulgaris*	I (2–3)
Chamomilla suaveolens	III (1–5)	*Avena fatua*	I (1)
Elymus repens	II (1–4)	*Cirsium vulgare*	I (1)
Capsella bursa-pastoris	II (1–6)	*Equisetum arvense*	I (2–4)
Cirsium arvense	II (1–5)	*Filaginella uliginosa*	I (1–3)
Lolium perenne	II (1–6)	*Chamomilla recutita*	I (2–5)
Urtica dioica	II (1–3)	*Medicago lupulina*	I (1–2)
Viola arvensis	II (1–3)	*Phleum pratense*	I (1)
Anagallis arvensis	II (1–3)	*Bilderdykia convolvulus*	I (1–4)
Ranunculus repens	II (1–5)	*Sinapis arvensis*	I (1–4)
Rumex crispus	II (1–3)	*Trifolium hybridum*	I (1–3)
Sonchus asper	II (1–2)	*Trifolium pratense*	I (1–4)
Trifolium repens	II (1–3)	*Hordeum vulgare*	I (3–8)
Anthemis cotula	II (3–4)	*Matricaria perforata*	I (1–7)
Euphorbia helioscopa	II (1–4)	*Achillea millefolium*	I (2)
Lapsana communis	II (1–2)	*Atriplex prostrata*	I (3–4)
Rumex obtusifolius	II (2–4)	*Atriplex patula*	I (4–5)
Spergula arvensis	II (2–7)		
Veronica persica	II (3–8)	Number of samples	21
Agrostis stolonifera	I (2–7)	Number of species/sample	17 (7–37)

OV34
Allium schoenoprasum-Plantago maritima community

Synonymy
Allium schoenoprasum-Plantago maritima community Hopkins 1983

Constant species
Allium schoenoprasum, Festuca ovina, Plantago maritima, Scilla verna.

Rare species
Allium schoenoprasum, Isoetes hystrix, Juncus capitatus, Scilla autumnalis, S. verna.

Physiognomy
The *Allium schoenoprasum-Plantago maritima* community comprises usually sparsely scattered individuals of perennial and ephemeral plants occuring as diverse mixtures in very small stands, often less than 1 m² in extent, and with a marked annual pattern of growth. Most of the species are perennial, though few of these are constant or of consistently high cover. However, after the damper weather of winter and with the flushing that is a characteristic feature of the habitat of this vegetation, the scattered usually dimutive tussocks of *Festuca ovina* and *Armeria maritima* and the rosettes of *Plantago maritima* show a flush of green growth in spring and are joined by the emerging shoots of *Scilla verna* and *Allium schoenoprasum*. Twisting leaves of the latter are especially distinctive in the sward and the species is often the most abundant plant with cover values exceeding 25%. It also flowers prolifically here, with a profusion of pretty pink dwarf inflorescences, unlike many of the other species which remain vegetative.

Among the associates, dwarfed plants of *Calluna vulgaris* are frequent and occasionally there can be some *Scilla autumnalis, Thymus praecox, Galium verum, Centaurium erythraea, Polygala vulgaris, Carex flacca, Minuartia verna, Agrostis stolonifera, A. canina* ssp. *montana* and *Koeleria macrantha*. Then, there are some distinctive annuals, such as *Sagina subulata, Aira caryophyllea, Juncus bufonius* and the nationally rare *J. capita-*

tus, which complete their life cycle by late spring and early summer before the drought which typically bakes the sites occupied by this vegetation bone dry. Another very distinctive rarity which has an important locus here, *Isoetes hystrix*, also completes its yearly cycle of growth through late autumn to early spring and then becomes dormant as the drought period sets in.

Other rare vascular plants which can be found here are the Oceanic West European *Herniaria ciliolata* and the Oceanic Southern *Trifolium bocconei* which, with *Juncus capitatus*, seem to be especially common on the edges of stands occurring among stretches of *Erica vagans-Ulex europaeus* heath.

Bryophytes are not common in the community although certain species may have been missed where recording was carried out in dry summer weather. However, *Riccia beyhrichiana* and, less commonly, *R. bifurca* can be found as extensive mats in the winter months and as shrivelled thalli during the droughty summer.

Habitat
The *Allium-Plantago* community occurs only on The Lizard in Cornwall where, in an extreme oceanic climate, it is confined to small shallow pans on serpentine soils, flushed by seepage in winter and baked dry in summer drought.

This vegetation was first noted by Coombe & Frost (1956a) among the striking complex of habitats developed on The Lizard, a peninsula of serpentine, gabbro and schists which experiences the most oceanic climate of mainland Britain (Malloch 1970). Subsequently it was characterised by Hopkins (1983) as occurring where serpentine crops out on the cliff-tops on the west coast of the peninsula, on valley sides and around rock outcrops among the inland heaths of the plateau. Here, it is found locally in places where seepage in the winter months provides a slow but constant flushing and seasonal flooding to depths of a few centimetres. With such seepage, the soils derived from the serpentine consist of but shallow pockets of gravelly bedrock embedded in a clayey matrix, usually less than 4 cm deep. Moreover, in late spring and early summer, the flushing ceases and the pans become droughted as the hot weather ensues. The mean annual maximum temperature of this part of Britain is above 26°C and summer insolation is high. With a distinct minimum in precipitation in spring and early summer, there is thus a marked potential water deficit at this time of year (Malloch 1970, Hopkins 1983).

Some elements of the flora of the *Allium-Plantago* community reflect the generally maritime character of the climate, like *Scilla verna*, *S. autumnalis* and *Plantago maritima*. Others, however, such as the winter annuals or those perennials able to capitalise on the mild, damp winter and avoid the rigours of the summer drought, are more obviously adapted to the very particular conditions of the pan habitat, and give this vegetation its especially distinctive character, thriving among the turf kept open by the relatively poor performance of potential competitors. *Allium schoenoprasum*, for example, is a circumboreal plant widely distributed throughout the Arctic and alpine regions of Europe and finding in these situations on The Lizard one of a number of British habitats which, alternating between very wet and droughted, enable it to thrive. It lies dormant until spring, then produces its leaves and flowers, dying back in early summer. In very extreme situations where flowering becomes sparse, it reproduces vegetatively by the development of new bulbs but, where seed matures, this can germinate in autumn or the following spring (Bougourd in Stewart *et al*. 1994).

Zonation and succession

The *Allium-Plantago* community is found most commonly among stands of the *Erica-Ulex* heath on The Lizard in situations where the serpentine soils become shallow in rocky ground on the plateaus, on valley sides and in transitions to cliff-tops.

In such places the ericoid sub-shrubs and gorse become confined to deeper pockets of soil within crevices among the serpentine, are held in check by the shortage of ground water and occasionally killed by severe drought (Hopkins 1983). The consequent lack of shade and access of stock and rabbits to the sward also help maintain the short and species-rich characteristic of the *Festuca* sub-community of the *Erica-Ulex* heath. Here, *F. ovina*, *Thymus praecox*, *Koeleria macrantha*, *Scilla verna* and *Plantago maritima* become more common among the heath flora and it is such plants which, together with sparse *Calluna*, extend a little way into the pans among the rock exposures where the *Allium-Plantago* community develops.

Closer to cliff-tops, as maritime influence increases, the *Erica-Ulex* heath gives way to the *Calluna-Scilla* heath, the *Viola* sub-community of which shares many associates with the grassy form of the inland heath. The *Allium-Plantago* community can sometimes be found in such transitions but it does not extend far on to the sea-cliffs.

Distribution
The community is confined to The Lizard in Cornwall.

Affinities
This vegetation is the most striking representative in Britain of the associations grouped together in phytosociological schemes as the Nanocyperion alliance – assemblages with a pioneer ephemeral element of therophytes and dwarf cyperaceous plants repeatedly establishing on periodically flooded, then droughted, patches of bare ground with mud, sand or peat. They have been widely described from other parts of Europe like The Netherlands (Westhoff & den Held 1969), Germany (Pott 1992), Austria (Grabherr & Mucina 1993) and Poland (Matuszkiewicz 1984).

Indeed, these communities have sometimes attracted a degree of excitement elsewhere (e.g. Weeda 1994, Lemaire & Weeda 1994) that throws the British neglect of description into sharp relief.

However, it is clear that, with the tendency to wetter summers in Britain, we are on the edge of the distribution of recognisable vegetation of this alliance and that, with us, it grades imperceptibly to the Bidention, the alliance which includes ephemeral vegetation of wetter and more nutrient-rich habitats like silty ponds.

Floristic table OV34

Allium schoenoprasum	V (4–8)
Festuca ovina	V (3–6)
Plantago maritima	V (1–5)
Scilla verna	IV (1–5)
Calluna vulgaris	III (2–5)
Riccia beyhrichiana	III (1–4)
Sagina subulata	III (1–3)
Scilla autumnalis	II (2–5)
Thymus praecox	II (1–2)
Agrostis stolonifera	II (4–5)
Koeleria macrantha	II (2–3)
Agrostis canina montana	II (3–5)
Galium verum	II (1–3)
Centaurium erythraea	II (1–2)
Polygala vulgaris	II (1–2)
Juncus bufonius	II (1–3)
Juncus capitatus	II (1–2)
Carex flacca	II (1–4)
Weissia sp.	II (2–3)
Leontodon taraxacoides	II (2–3)
Minuartia verna	II (1–3)
Ulex sp. seedling	II (1)
Aira caryophyllea	I (1–3)
Herniaria ciliolata	I (1–2)
Juncus bulbosus	I (1–3)
Hypericum pulchrum	I (1–2)
Plantago lanceolata	I (1–4)
Trichostomum brachydontium	I (2–3)
Erica cinerea	I (2)
Danthonia decumbens	I (2)
Filipendula vulgaris	I (1–4)
Juncus articulatus	I (2)
Pedicularis sylvatica	I (1–2)
Isoetes hystrix	I (3)
Sedum anglicum	I (3)
Number of samples	19
Number of species/sample	12 (8–18)
Herb height (cm)	4 (2–8)
Vegetation cover (%)	40 (32–62)

OV35
Lythrum portula-Ranunculus flammula community

Constant species
Lythrum portula, Ranunculus flammula.

Rare species
Alopecurus aequalis, Pilularia globulifera.

Physiognomy

The *Lythrum portula-Ranunculus flammula* community comprises generally open vegetation in which *L. portula* and *R. flammula* are the only constants, the former sometimes occurring with abundance, occasionally in locally monodominant stands. Apart from *Eleocharis palustris*, other associates are recorded rather sporadically and only rarely with high cover values, but can be quite varied: *Agrostis stolonifera, Juncus bufonius, J. effusus, J. articulatus, Galium palustre, Callitriche stagnalis, Littorella uniflora, Mentha aquatica, Alisma plantago-aquatica, Myosotis scorpioides* and *Apium inundatum* have all been found here. The community also provides a locus for the nationally rare *Pilularia globulifera*, a perennial plant but one quick to colonise the damp open muds characteristic of this vegetation (Jermy in Stewart *et al.* 1994), *Limosella aquatica*, an ephemeral herb, and *Alopecurus aequalis*, an annual grass which invades here as water levels drop (Hubbard 1984, Twist in Stewart *et al.* 1994).

Habitat

The *Lythrum-Ranunculus* community is typical of silty or peaty soils, wetted and exposed by fluctuating or temporary waters around pool, lake and reservoir margins and in flooded gravel and brick-earth workings.

Compared with the habitats of Bidention assemblages, the situations colonised by this vegetation are not nutrient-rich and, though the cover of the sward can be extensive, the herbage does not have the lush character of those nitrophilous ephemeral communities. Indeed, the *Lythrum-Ranunculus* community is not dominated by annuals at all nor does it show the striking diversity of dominance typical there. Its cover of perennials is maintained in a somewhat attenuated form by shallow inundation and forms an open matrix in which the more short-lived species can obtain a hold. Most of the plants are diminutive and do not exert a shading effect on their neighbours.

The other important element in helping maintain an open habitat is a modest amount of trampling, sufficient to keep the surface scuffed but not too broken up. Animals may also be important in transporting seed of the species in the community widely across areas where suitable conditions are likely to prevail locally in subsequent years (Ellenberg 1988).

For both *Alopecurus aequalis* and *Pilularia*, both of them declining species (Stewart *et al.* 1994), the habitat characteristic of this community provides widely scattered surviving situations which they can colonise from time to time.

Zonation and succession

The *Lythrum-Ranunculus* community is sometimes found with inundation communities like the *Agrostio-Ranunculetum* or *Polygono-Bidentetum* on periodically-flooded silts around pools and lakes or among the *Holco-Juncetum* in wet pastures on impeded soils. Towards shallow open waters, it can give way to the *Eleocharitetum palustris* or *Callitriche* vegetation. Repeated flooding helps maintain suitable habitats for the community and drainage leads to succession to wet grassland or rush pasture.

Distribution

This is a very local community occurring at scattered sites through the warmer south of Britain.

Affinities

Although the *Lythrum-Ranunculus* community shows some affinities with both Bidention and Littorellion assemblages characterised from Britain, it is probably best accommodated in the Nanocyperion.

Floristic table OV35

Lythrum portula	V (1–8)
Ranunculus flammula	V (1–4)
Eleocharis palustris	III (2–9)
Agrostis stolonifera	II (1–4)
Galium palustre	II (1–3)
Alisma plantago-aquatica	II (2–3)
Callitriche stagnalis	II (3–7)
Juncus bufonius	II (3–5)
Littorella uniflora	II (1–8)
Polygonum hydropiper	II (2–4)
Hydrocotyle vulgaris	II (3–4)
Mentha aquatica	II (2–4)
Juncus effusus	II (3–5)
Myosotis scorpioides	II (1–4)
Poa annua	II (1–4)
Calliergon cuspidatum	II (1–3)
Callitriche hamulata	II (3–4)
Filaginella uliginosa	II (3–5)
Alopecurus aequalis	II (2–5)
Apium inundatum	II (2–5)
Cardamine pratensis	II (2–3)
Juncus articulatus	II (1–2)
Pilularia globulifera	I (4–6)
Rorippa islandica	I (3–5)
Ranunculus hederaceus	I (1–4)
Alopecurus geniculatus	I (1–4)
Ranunculus trichophyllus	I (2–3)
Drepanocladus fluitans	I (2)
Agrostis canina	I (4–5)
Prunella vulgaris	I (1)
Limosella aquatica	I (2–5)
Riccia sp.	I (2)
Salix cinerea seedling	I (1)
Bryum klinggraeffii	I (1)
Number of samples	16
Number of species/sample	11 (4–23)

OV36

Lythrum hyssopifolia-Juncus bufonius community

Synonymy

Lythrum hyssopifolia stands Preston & Whitehouse 1986.

Constant species

Anagallis arvensis, Juncus bufonius, Lythrum hyssopifolia, Matricaria perforata, Plantago major, Polygonum aviculare, Polygonum persicaria, Bryum klinggraeffii, Pottia starkeana ssp. *minutula, Riccia glauca.*

Rare species

Lythrum hyssopifolia.

Physiognomy

The *Lythrum hyssopifolia-Juncus bufonis* community comprises ephemeral vegetation which, by mid-summer, is typically dominated by mixtures of the nationally rare *Lythrum hyssopifolia*, together with *Juncus bufonius, Plantago major* (diagnosed as ssp. *intermedia* by Preston & Whitehouse 1986), *Polygonum persicaria* and *P. aviculare*. Also frequent throughout among the vascular associates are *Mentha arvensis, Rumex crispus* and seedlings of *Salix* spp. *Elymus repens* is occasional and, in sites which have been undisturbed by recent ploughing, *Polygonum amphibium* tends to increase. Bryophytes are an important element in the vegetation with *Bryum klinggraeffii* and *Riccia glauca* common throughout, the former with modest abundance.

Preston & Whitehouse (1986) noted a consistent, but overlapping, contrast between the stands they sampled but not sufficiently sharp to characterise sub-communities. To one extreme, species such as *Anagallis arvensis, Alopecurus myosuroides, Kickxia elatine, K. spuria* and more obviously *Anthemis cotula* and *Aethusa cynapium* ssp. *agrestis*, tended to be associated with *Pottia starkeana* ssp. *minutula, Barbula unguiculata, B. argenteum, Dicranella staphylina* and *D. varia*. To the other, *Matricaria perforata* and *Equisetum arvense* were found with *Physcomitrella patens, Riccia cavernosa, R. subbifurca* and *Pohlia carnea*.

Habitat

The community is confined to winter-flooded hollows in annually-ploughed arable fields in Cambridgeshire where long inundation kills the cereal crop and allows the reappearance of this distinctive combination of opportunist vascular plants and cryptogams.

The most distinctive plant of the community, *Lythrum hyssopifolia*, is characteristic of seasonally-flooded habitats (Coombe *et al*. 1959) and may not be native: at least, its native distribution is greatly confused by its sporadic appearance as an obviously introduced plant through the more Continental parts of south-east England. In this community, first characterised in detail by Preston & Whitehouse (1986), it colonises shallow periglacial ground-ice hollows which persist in the arable landscape of south Cambridgeshire. These experience periodic flooding in winter, of variable duration and depth but sufficient to kill off the cereals sown into the fields after autumn ploughing. As the hollows dry out, the assemblage reappears on the damp, bare mud.

Both flooding and ploughing are essential in maintaining a suitable habitat for the community. *Lythrum* itself is absent from both unflooded areas in the arable fields and in the hollows following winters when there is no flooding (Walters 1978). Hollows which are not ploughed quickly become dominated by dense swards of perennials, some of which, like *Polygonum amphibium, Elymus repens* and *Equisetum arvense* are occasional in this vegetation but whose robust rhizomes are prevented from extensive spread by the physical disruption of cultivation.

In contrast to these scarcer companions, most of the commoner vascular species of the community are ephemerals. All but *Plantago major* among the vascular constants are annuals and, even this hemicryptophyte can behave as an annual in this vegetation: such a strategy (Sagar & Harper 1960) is especially typical of the var. *intermedia* characteristic of unstable habitats.

Lythrum itself, though capable of autumn germination (Salisbury 1968), appears to establish only in spring in this community (Preston & Whitehouse 1986) and *P. aviculare* and *P. persicaria* have seeds which remain dormant until after moist conditions and low temperatures have ensued (Justice 1941, Timson 1965, Courtney 1968, Henson 1969). The seeds of *Plantago major* also have a chill requirement (Sagar & Harper 1960) while *Juncus bufonius* (Cope & Stace 1978) and *Matricaria perforata* (Roberts & Feast 1970) show maximum germination as the weather warms in March. The constants of the community also show considerable seed longevity (Preston & Whitehouse 1986).

The species of the community are well adapted to survival through the disruptions of ploughing and flooding and to rapid reappearance on the bare, damp ground created by these events. Their seeds are also probably easily able to be carried from hollow to hollow or to different fields, by gulls and waders for example (Walters 1972), which would assist local survival.

Almost all species among the very distinctive bryophyte assemblage of this community are also ephemerals with adaptations for rapid vegetative or sexual reproduction. Most, including *Bryum klinggraeffii, B. rubens, B. ruderale, Dicranella staphylina* and *D. varia*, are dioecious species with rhizoid tubers, organs especially frequent in mosses of arable fields and probably able to remain dormant for years. Many of the other bryophytes are monoecious and without means of vegetative reproduction but they regularly produce cleistocarpous capsules with spores that can probably remain dormant for long periods (Preston & Whitehouse 1986).

Zonation and succession

The stands of the *Lythrum-Juncus* community are characteristically clearly marked off from their intensively arable context. By August, as Preston & Whitehouse (1986) vividly describe the sight, they are picked out as pink islands, coloured by the flowers of *Polygonum persicaria*, among the golden ripening cereals.

Without the disturbance necessary for its reappearance, however, the community is rapidly overtaken by Polygono-Chenopodion vegetation with an increasing perennial element (as in the *Elymus-Potentilla* sub-community of the *Matricaria-Stellaria* community) or by tall-herb vegetation of the Artemisietea, where species like *Epilobium hirsutum* dominate. One long-uncultivated hollow noted by Preston & Whitehouse (1986) had progressed to *Salix atrocinerea* scrub.

Distribution

In the review of existing *Lythrum* sites of Preston & Whitehouse (1986), the plant was confined to a very limited part of south Cambridgeshire, all within one 10 × 10 km square of the National Grid.

Affinities

The *Lythrum-Juncus* community is a further British representative of the ephemeral vegetation of the Isoeto-Nanojuncetea Br.-Bl. & Tüxen 1943. However, although in France (Braun-Blanquet 1935), *Lythrum* has been reported from the diminutive open swards characteristic of the Mediterranean Nanocyperion, the vegetation described here clearly shows some Bidention features. Preston & Whitehouse (1986) considered it to belong to the order Cyperetalia fusci (Klika 1935) Muller-Stoll & Pietsch 1961, characteristic of wetter habitats along the Atlantic fringes of Europe, but not all authorities recognise such a separate grouping. Certainly only two of the order character species, *Riccia cavernosa* and *R. glauca*, occur in the community.

Floristic table OV36

Lythrum hyssopifolia	V (4–8)
Juncus bufonius	V (2–7)
Plantago major	V (2–6)
Polygonum persicaria	V (1–7)
Bryum klinggraeffii	V (1–3)
Polygonum aviculare	IV (1–6)
Riccia glauca	IV (1–2)
Matricaria perforata	IV (1–4)
Anagallis arvensis	IV (1–5)
Pottia starkeana ssp. *minutula*	IV (1–3)
Mentha arvensis	III (1–2)
Rumex crispus	III (1–4)
Salix sp. seedling	III (1–3)
Physcomitriella patens	III (1–3)
Alopecurus myosuroides	III (1–5)
Barbula unguiculata	III (1–3)
Elymus repens	III (1–4)
Equisetum arvense	III (1–4)
Riccia cavernosa	III (1–3)
Kickxia elatine	III (1–2)
Sonchus oleraceus	III (1–2)
Bryum argenteum	III (1–3)
Polygonum amphibium	II (1–4)
Ranunculus repens	II (1–3)
Kickxia spuria	II (1–2)
Dicranella staphylina	II (1–3)
Stellaria media	II (1–2)
Bilderdykia convolvulus	II (1–2)
Pohlia carnea	II (2)
Dicranella varia	II (2–3)
Bryum rubens	II (1–3)
Anthemis cotula	II (1–4)
Riccia subbifurca	I (1)
Epilobium hirsutum	I (1)
Medicago lupulina	I (1–2)
Viola arvensis	I (1–2)
Hordeum distichon	I (4–5)
Veronica persica	I (2–3)
Bryum violaceum	I (2)
Taraxacum officinale agg.	I (1–4)
Senecio vulgaris	I (1)
Number of samples	10
Number of species/sample	25 (16–34)
Herb cover (%)	83 (60–100)
Bryophyte cover (%)	16 (0–40)

OV37

Festuca ovina-Minuartia verna community
Minuartio-Thlaspietum alpestris   Koch 1932

Constant species
Agrostis capillaris, Campanula rotundifolia, Festuca ovina, Minuartia verna, Thymus praecox.

Physiognomy
The *Minuartio-Thlaspietum alpestris* characteristically comprises an open turf in which tussocks of *Festuca ovina*, patches of *Agrostis capillaris* and scattered individuals or populations of *Minuartia verna* are the most prominent feature. *Thymus praecox* is also constant and its mats can be locally quite extensive and there is very often a little *Campanula rotundifolia* and *Rumex acetosella*. Among the more frequent associates of the community, *Thlaspi alpestre* is the most distinctive, though it is typically not abundant.

Other species recorded commonly throughout, though not usually in any abundance, are *Rumex acetosa*, *Senecio jacobaea*, *Linum catharticum* and, particularly in transitions to more continuous calcicolous and mesotrophic swards, *Lotus corniculatus* and *Trifolium repens*. Occasionals of the community include *Geranium robertianum*, *Cerastium fontanum*, *Hieracium pilosella*, *Cirsium vulgare*, *Agrostis stolonifera* and *Viola lutea*.

Through the community as a whole, bryophytes are neither numerous nor extensive, though small patches of *Pohlia nutans*, *Weissia controversa* and *Dicranum scoparium* can sometimes be seen. Lichens, too, are absent from many stands but strikingly varied and abundant in one sub-community.

Sub-communities

Typical sub-community: *Minuartio-Thlaspietum typicum* Shimwell 1968a. In this vegetation, *Minuartia verna*, *Festuca ovina* (here occasionally joined by *F. rubra*) and *Agrostis capillaris* dominate the sward with a typical range of community associates. In stands from the Yorkshire Dales, *Galium sterneri* is a distinctive associate.

***Achillea millefolium-Euphrasia officinalis* sub-community:** *Minuartio-Thlaspietum achilletosum* Shimwell 1968a. The cover of the turf in this sub-community is somewhat more extensive than usual and, among the grasses, *Anthoxanthum odoratum* and, less commonly, *Koeleria macrantha* and *Avenula pratensis* join *F. ovina* and *Agrostis capillaris*. Usually more obvious, though, is the increased frequency of *Trifolium repens*, *Lotus corniculatus* and *Linum catharticum* and, especially diagnostic, of *Achillea millefolium*, *Euphrasia officinalis* agg. and *Plantago lanceolata*. *Anthyllis vulneraria* and *Carex flacca* are also preferential at low frequencies. Among bryophytes, *Rhytidiadelphus squarrosus* is very common, though not usually abundant.

***Cladonia* spp. sub-community:** *Minuartio-Thlaspietum cladonietosum* Shimwell 1968a. The vascular contingent here is less extensive than in other types of *Minuartio-Thlaspietum* and there are no distinctive preferentials among them apart from occasional scattered sprigs of *Calluna vulgaris*. Lichens, however, are unusually prominent and sometimes diverse. Most common and abundant among them are *Cladonia rangiformis*, *C. chlorophaea* and *Cornicularia aculeata* with *Cladonia pyxidata*, *C. impexa*, *C. gracilis*, *C. arbuscula* and *C. furcata* less frequent and extensive. *Bryum pallens* is an occasional preferential but bryophytes are generally not important in the sward.

Habitat
The *Minuartio-Thlaspietum* is a local community restricted to the spoil heaps of lead mines or outcrops of veins of heavy metals among calcareous bedrocks around the upland fringes of northern and western Britain. It is usually open to grazing stock but the composition and structure of the vegetation are strongly influenced by the mineralogy of the soil parent material.

Lead-bearing mineral veins are especially associated in Britain with the Carboniferous Limestone deposits of the Mendips, Derbyshire Dales, Yorkshire Dales and north Pennines, and have been exploited in these areas since Roman times. More recent mining activities, particularly from the last century, have left extensive areas of spoil in particular localities in these regions and their mixture of limestone and shale fragments and minerals such as fluorspar and barytes, forms a distinctive soil parent material.

Although data were not available for this survey, vegetation with marked floristic and physiognomic affinities to the *Minuartio-Thlaspietum* is also to be found in a few places on stable river gravels rich in heavy metals in a number of river valleys in northern England, most notably the South Tyne, where washings from mineral processing in the catchments have been redeposited (Sellars & Baker 1988). Serpentine rocks rich in nickel, chromium and cobalt can also support vegetation with close affinities to the *Minuartio-Thlaspietum* in Scotland (Birse 1982).

The community can be found down to altitudes of 150 m, but it is mostly characteristic of somewhat higher ground, extending up to 730 m, and generally the sites experience the more or less cool and wet climate characteristic of the British uplands. Annual precipitation is usually over 1000 mm, up to 2500 mm in the north Pennines, and the mean annual maximum temperature is less than 26 °C (*Climatological Atlas* 1952). Under such circumstances, especially with such a parent material, the development of soil is slow and, particularly on the looser spoil, there is no real profile, simply pockets of gradually weathering mineral debris with local accumulation of humus from decaying vegetation.

Most important in limiting colonisation, however, are the large amounts of heavy metals present in the spoil. Among these, it is not lead, the prize of the mining enterprise, but zinc which if of greatest impact. Ernst (1965) reported zinc contents up to 5060 ppm in soils of such mire spoil near Osnabrück in Germany and concentrations of 2460 and 6790 ppm in the leaves of *Minuartia verna* and *Thlaspi alpestre*. These two species, the most distinctive associates of the community, can accumulate zinc and other plants, like *Festuca ovina* (Ellenberg 1988), *Agrostis capillaris* (Bradshaw 1952) and probably also *Rumex acetosa*, have strains tolerant of heavy metals. For other species, unable to tolerate or reduce such potential toxins, the spoil-heaps provide an extreme environment that can be colonised only where earlier invaders have reduced the levels or where the debris has been intermixed with other adjacent parent material, like head or alluvium at the foot of slopes or along streamsides.

Among the sub-communities, the typical form of the *Minuartio-Thlaspietum* is found on spoil with more immature soils while the *Achillea-Euphrasia* sub-community occurs on more stable material with better-developed profiles and in transitions to rendzinas and brown calcareous earths on less contaminated ground around

the spoil. The main floristic differences between these vegetations types, with an increase in more mesophytic and less zinc-tolerant plants reflects this edaphic shift, though the swards of the *Achillea-Euphrasia* sub-community also tend to be more influenced by grazing as they provide more accessible and palatable herbage.

The distinctive features of the *Cladonia* sub-community reflect the more extreme climatic conditions found at higher altitudes. Stands usually occur between 400 and 600 m in cold and windy situations where the rainfall is considerably higher. Leaching of the surface of the detritus is more obvious than in the drier climate at lower altitudes and, with scattered plants of *Calluna* among the patchy *Cladonia* carpet, the vegetation has something of the appearance of a lichen heath. The maintenance of a lichen carpet is probably also dependent on freedom from trampling by grazing stock.

Zonation and succession

The usual context of the *Minuartio-Thlaspietum* is among grasslands and heaths, any transitions to which are mediated by differences in soils.

In fact, quite often, stands of the community are rather sharply marked off from the grasslands around. At lower altitudes, in the Mendips and Derbyshire, these grasslands are usually of the *Festuca-Avenula* type, most often the *Dicranum* sub-community or, where there has been some improvement for agriculture, the *Holcus-Trifolium* sub-community. Where soils change more gradually, Typical *Minuartio-Thlaspietum* may give way to the Mesobromion pastures through the *Achillea-Euphrasia* sub-community. In other places, the latter vegetation passes to a Cynosurion sward, usually the *Lolio-Cynosuretum*, derived by fertilising and often ploughing and reseeding. Where soils are somewhat less calcareous, as where superficials overlie the limestone, the *Festuca-Agrostis-Galium* grassland can figure among the pastures around the lead-mine spoil or, where grazing has not been so heavy, heath vegetation. In the Mendips, where this situation is more common, this is generally *Calluna-Ulex gallii* heath.

At higher altitudes over most of the Yorkshire Dales, the calcicolous pasture is of the *Sesleria-Galium* type. The Typical sub-community of this grassland can pass sharply to Typical *Minuartio-Thlaspietum* or the *Cladonia* sub-community or, more gradually, through the *Achillea-Euphrasia* sub-community. Elsewhere, on somewhat more acidic soils, the *Sesleria-Galium* grassland can be replaced by the *Festuca-Agrostis-Thymus* grassland or, where grazing has not been so intensive, by grassy forms of *Calluna-Vaccinium* heath.

Where succession has proceeded further around old lead mines, the *Minuartio-Thlaspietum* may be closely hemmed in by *Crataegus-Hedera* scrub and *Fraxinus-Acer-Mercurialis* woodland at lower altitudes or by

Fraxinus-Sorbus-Mercurialis on higher ground to the north. The character of the soils supporting the *Minuartio-Thlaspietum* itself is generally inimical to colonisation by shrubs and trees, but relief from grazing may have some effect on the vegetation in the long term.

Distribution

The community occurs locally in the Mendips, Derbyshire Dales, Yorkshire Dales and north Pennines.

Affinities

First described formally from Britain by Shimwell (1968*a*), this vegetation is clearly synonymous with the *Minuartio-Thlaspeetum* Koch 1932, the core association of the western European vegetation of heavy-metal habitats placed in the alliance Thlaspion calaminaris Ernst 1964. The association has also been described from Germany (Pott 1992) and from Ireland (Doyle 1982, White & Doyle 1982). Following Braun-Blanquet & Tüxen (1943), heavy metal vegetation was placed in a separate class, the Violetea calaminariae (e.g. Westhoff & den Held 1969, White & Doyle 1982) but the favoured view now is that these assemblages should return to their original location in the Festuco-Brometea. Certainly, among British stands of this type, there is a strong similarity to the *Festuca-Hieracium-Thymus* grassland, a Continental sward of rendzina soils which belongs to the Koelerio-Phleion. In his survey of Scottish vegetation, Birse (1980, 1984) described several associations from serpentine soils which he placed in the Violetea but further data are required to clarify the relationship of the various assemblages.

Floristic table OV37

	a	b	c	37
Festuca ovina	V (2–9)	V (1–6)	V (1–8)	V (1–9)
Minuartia verna	V (2–7)	V (1–6)	V (1–6)	V (1–7)
Campanula rotundifolia	IV (1–4)	IV (1–3)	IV (1–3)	IV (1–3)
Thymus praecox	IV (1–5)	IV (1–4)	III (1–3)	IV (1–5)
Agrostis capillaris	III (1–6)	IV (1–3)	IV (1–4)	IV (1–6)
Festuca rubra	II (2–7)		I (1)	I (1–7)
Galium sterneri	II (1–6)		I (5)	I (1–6)
Weissia controversa	II (1–2)		I (1)	I (1–2)
Linum catharticum	III (1–3)	IV (1–3)	III (1–3)	III (1–3)
Trifolium repens	II (1–3)	IV (1–3)	III (1–3)	III (1–3)
Lotus corniculatus	II (2–6)	IV (1–3)	III (1–3)	III (1–3)
Achillea millefolium	I (1)	V (1–3)		III (1–3)
Euphrasia officinalis agg.	I (1)	V (1–3)	I (1)	III (1–3)
Anthoxanthum odoratum	I (1)	IV (1–4)		II (1–4)
Plantago lanceolata		IV (1–3)		II (1–4)
Rhytidiadelphus squarrosus	I (2)	IV (1–4)		II (1–4)
Dicranum scoparium	I (3)	II (1–3)	I (1)	I (1–3)
Koeleria macrantha	I (2)	II (1–3)	I (1)	I (1–3)
Anthyllis vulneraria		II (1–4)		I (1–4)
Avenula pratensis		II (1–3)		I (1–3)
Carex flacca		II (1–3)		I (1–3)
Cladonia rangiformis		I (1)	IV (1–4)	III (1–4)
Cladonia chlorophaea			IV (1–4)	II (1–4)
Cornicularia aculeata			IV (1–2)	II (1–2)
Cladonia pyxidata	I (1)		IV (1–3)	II (1–3)
Calluna vulgaris			III (1–3)	II (1–3)
Cladonia impexa	I (1)		II (3–4)	I (1–4)
Bryum pallens	I (1)		II (1–3)	I (1–3)
Cladonia gracilis			II (2–4)	I (2–4)

	a	b	c	37
Cladonia furcata			I (2)	I (2)
Cladonia arbuscula			I (3)	I (3)
Rumex acetosa	III (1–6)	IV (1–3)	III (1–4)	III (1–6)
Thlaspi alpestre	III (1–4)	III (1–3)	III (1–3)	III (1–4)
Senecio jacobaea	II (1–2)	III (1–3)	II (1–3)	II (1–3)
Geranium robertianum	II (1–3)	II (1–3)	II (1–3)	II (1–3)
Cerastium fontanum	II (1–2)	II (1–3)		II (1–3)
Cirsium vulgare	II (1–3)	II (1–3)	I (1)	II (1–3)
Hieracium pilosella	II (2)	II (1–4)		II (1–4)
Agrostis stolonifera	II (1–3)	I (1)	II (1–3)	II (1–3)
Viola lutea	II (1–4)	I (1)	II (1)	II (1–4)
Galium saxatile	I (1)	I (1)	I (1)	I (1)
Arrhenatherum elatius	I (1)	I (1)	I (1)	I (1)
Dactylis glomerata	I (1)	I (1)		I (1)
Pohlia nutans		I (1)	I (1–3)	I (1–3)
Carex caryophyllea	I (1–2)		I (1)	I (1–2)
Parmelia saxatilis	I (1)		I (1)	I (1)
Number of samples	15	10	13	38
Number of species/sample	14 (9–19)	23 (15–29)	16 (11–27)	16 (9–29)
Herb cover (%)	60 (40–80)	88 (70–100)	51 (40–70)	69 (40–100)
Bryophyte/lichen cover (%)	2 (0–5)	11 (5–25)	56 (30–80)	22 (0–80)

a   Typical sub-community

b   *Achillea millefolium-Euphrasia officinalis* agg. sub-community

c   *Cladonia* spp. sub-community

37   *Minuartio-Thlaspietum alpestris* (total)

OV38
Gymnocarpium robertianum-Arrhenatherum elatius community
Gymnocarpietum robertianae   (Kuhn 1937) R.Tx. 1937

Constant species

Arrhenatherum elatius, Festuca rubra/ovina, Geranium robertianum, Gymnocarpium robertianum, Teucrium scorodonia, Ctenidium molluscum.

Physiognomy

The *Gymnocarpietum robertiani* comprises more or less open stands of fern- and grass-dominated vegetation disposed over the distinctive topography of limestone screes and pavement. Scattered fronds of *Gymnocarpium*, arising separately from the creeping rhizomes, begin to emerge rapidly in late spring, their rather stately forms attaining up to 40 cm or so in height. Among these, occasionally in some abundance, are loose tussocks of *Arrhenatherum* and, less commonly, *Festuca ovina/rubra* (not distinguished in available data) and *Brachypodium sylvaticum*. In sunnier situations, which the fern seems to tolerate well, perhaps because of its somewhat mealy surface (Page 1982), the *Brachypodium* often has a rather lurid yellow-green tinge. *Sesleria albicans* does not appear in the table because samples of the community were not collected within the range of this grass but it can figure prominently in this vegetation in the Yorkshire Dales and southern part of the Lake District.

Other very frequent components are *Geranium robertianum*, commonly showing a reddish hue to its stems with the typically impoverished nature of the substrate, *Teucrium scorodonia, Mycelis muralis* and *Mercurialis perennis*. Where clonal patches of the last plant thicken up, the *Gymnocarpium* seems to suffer and be gradually eclipsed (Page 1982). Less common vascular associates include *Origanum vulgare, Carex flacca, Viola riviniana, Senecio jacobaea, Oxalis acetosella, Arabis hirsuta* and, in the northern Pennines, *Galium sterneri*. Locally, *Rubus saxatilis* may be abundant. Where stands occur among limestone exposures with smaller shaded crevices, *Asplenium trichomanes, A. viride, A. ruta-muraria, Cystopteris fragilis* and *Phyllitis scolopendrium* can be found.

Bryophytes are quite numerous and varied and they may be locally plentiful, particularly where the grasses and fern cover give some shade. The commonest and most abundant species are *Ctenidium molluscum, Homalothecium sericeum, Dicranum scoparium*, with, less commonly, *Tortella tortuosa, Fissidens cristatus, Neckera crispa, Calliergon cuspidatum, Rhytidium rugosum, Hylocomium splendens* and *Grimmia apocarpa*. Lichens, by contrast, are scarce.

Habitat

This is a local community confined to sunny exposures of calcareous bedrocks at lower altitudes, mostly Carboniferous Limestone, in England and Wales, in situations where there is freedom from grazing but no advanced colonisation by shrubs and trees.

Gymnocarpium is a calcicolous fern which thrives in the warmer, drier climate of south-east Britain: apart from a very few far-flung records in Scotland, it is entirely restricted to calcareous bedrocks in regions with a mean annual maximum temperature of more than 25 °C (Conolly & Dahl 1970) and annual precipitation less than 1600 mm (*Climatological Atlas* 1952). In such situations, the intolerance this fern shows of dense shade and stagnant air conditions (Page 1982) means that it is characteristically a coloniser of open, rocky ground which has been kept free or cleared of woodland. Its rhizomatous habit means that it is well able to spread among coarse talus and in fissures of limestone exposures, habitats that are widespread on the hard Carboniferous Limestone of the Mendips, Derbyshire Dales, around Morecambe Bay and in north and south Wales. Elsewhere in south-east England, where calcareous bedrocks are common in suitable climate but weather to gentler topographies, *Gymnocarpium* is confined to scattered occurrences on railway ballast and mortared walls (Jermy *et al.* 1978).

The associates of the fern in its typical habitat are species similarly tolerant of sunny, calcareous situations free of grazing. *Arrhenatherum* is especially successful in this community: it is a ready invader of limestone scree and rocky ground where grazing has been withdrawn (Hope-Simpson 1940*b*). *Gymnocarpium* itself also seems susceptible to grazing, particularly by sheep, maybe also by rabbits and deer. There is also a strong floristic similarity between this community and the herbaceous vegetation of clearings in calcicolous woodlands. However, wherever either strongly-shading associates such as *Mercurialis* or woody species begin to overshadow this vegetation, the vigour of the fern and other light-demanding plants suffers. This means that, where there is a continuing potential for invasion by shrubs and trees, the community is a temporary feature.

Although intolerant of too much shade themselves, species like *Gymnocarpium* and *Arrhenatherum* provide sufficient protection from the sunlight to encourage an associated flora of more shade-bearing small herbs and bryophytes.

Zonation and succession

The *Gymnocarpietum* typically occurs among patchworks of grassland, scrub and woodland on colonising screes and rocky limestone slopes.

This vegetation often appears to be a colonising community on coarser limestone talus and, where grazing is absent, it can be overtaken by invasion of woody plants. Typically, over the warmer and drier limestones of southern England and Wales, the kind of woodland which develops in such situations is the *Fraxinus-Acer-Mercurialis* woodland. The eventual dominants of this community can include *Fraxinus*, *Ulmus glabra*, *Acer pseudoplatanus*, *A. campestre* and *Quercus robus* with often a rich understorey of shrubs among which *Corylus avellana* and *Crataegus monogyna* figure prominently. However, early stages in the development of this kind of woodland, and of the *Crataegus-Hedera* scrub which often precedes it, are quite diverse and strongly dependent on which seed-parents are available locally and also on the particular terrain conditions (Merton 1970). Various forms of this scrub and woodland can therefore be found in close association with the *Gymnocarpietum*, species like *Geranium robertianum*, *Mercurialis perennis*, *Mycelis muralis* and *Viola riviniana* running on with undiminished frequency and often in some abundance under the shrubs and trees. With increasing shade,

however, the two dominants of the fern community are quickly overwhelmed by the closing canopy. The various elements in this kind of pattern survive best on sunnier slopes with coarser talus and rocky outcrops where woodland development is more patchy. In some such situations, the fern community grows among the distinctive *Teucrium* sub-community of the *Fraxinus-Acer-Mercurialis* woodland (cf. Oberdorfer 1977). Where talus becomes finer and surrounding slopes sustain calcicolous pasture to which stock have access, the sward is typically of the *Festuca-Avenula* type but this is characteristically sharply marked off from the fern stands on the coarser scree. One further element in the mosaics can be provided by the *Asplenietum* which replaces the *Gymnocarpietum* on limestone exposures with smaller crevices. A variety of bryophytes occur in both assemblages.

In the Yorkshire Dales, where the climate is somewhat cooler and moister away from sunny south-facing slopes, the *Fraxinus-Acer-Mercurialis* woodland is replaced in such sequences by the *Fraxinus-Sorbus-Mercurialis* woodland and the *Festuca-Avenula* grassland by the *Sesleria-Galium* grassland.

In many of its localities, the *Gymnocarpietum* should probably be seen as a secondary colonising community following woodland clearance. Where it becomes surrounded by subsequent regrowth of scrub or where the talus is especially coarse, this vegetation may benefit from the protection from grazing and persist for some time.

Distribution

The community is of local occurrence on limestones in southern Britain.

Affinities

Only in Shimwell's (1968*a*) survey of British calcicolous grasslands and related vegetation has this kind of fern vegetation been described from Britain. He allocated his samples to the *Gymnocarpietum robertianae* (Kuhn 1937) R.Tx. 1937 which Oberdorfer (1977) had described from Germany. This is placed among the fern assemblages of base-rich screes from the sub-montane and montane zones of central Europe and Scandinavia in the alliance Stipion calamagrostis (Valachovic *et al.* 1997).

Floristic table OV38

Arrhenatherum elatius	V (1–7)	*Campanula rotundifolia*	I (1–3)
Geranium robertianum	V (1–5)	*Tussilago farfara*	I (1–3)
Gymnocarpium robertianum	IV (2–9)	*Festuca arundinacea*	I (1–3)
Teucrium scorodonia	IV (1–7)	*Dactylis glomerata*	I (1–2)
Festuca rubra/ovina	IV (1–7)	*Saxifraga tridactylites*	I (1)
Ctenidium molluscum	IV (1–8)	*Geranium molle*	I (1–2)
		Holcus lanatus	I (1)
Mercurialis perennis	III (1–3)	*Aquilegia vulgaris*	I (1)
Homalothecium sericeum	III (2–5)	*Rubus saxatilis*	I (7–8)
Mycelis muralis	III (1–2)	*Centaurea nigra*	I (1–3)
Brachypodium sylvaticum	II (1–5)	*Avenula pratensis*	I (1–5)
Dicranum scoparium	II (2–5)	*Poterium sanguisorba*	I (1)
Origanum vulgare	II (1–5)	*Poa pratensis*	I (1–2)
Tortella tortuosa	II (2–4)	*Vicia sepium*	I (2)
Fissidens cristatus	II (1–3)	*Pseudoscleropodium purum*	I (2–6)
Cladonia pocillum	II (1–3)	*Distichium capillaceum*	I (2)
Oxalis acetosella	II (2–4)	*Encalypta streptocarpa*	I (2)
Carex flacca	II (2–5)	*Bryum capillare*	I (2)
Viola riviniana	II (1–5)	*Rosa pimpinellifolia*	I (4)
Hylocomium spendens	II (1–3)	*Thalictrum minus*	I (2)
Neckera crispa	II (1–3)	*Asplenium trichomanes*	I (2)
Senecio jacobaea	II (1–3)	*Campylium chrysophyllum*	I (1)
Calliergon cuspidatum	II (1–4)	*Asplenium viride*	I (2)
Galium sterneri	II (1–5)	*Cystopteris fragilis*	I (2)
Rhytidium rugosum	II (1–6)	*Eurhynchium stratium*	I (2)
Hypnum cupressiforme	II (1–5)	*Grimmia pulvinata*	I (2)
Schistidium apocarpum	II (1–3)	*Racomitrium lanuginosum*	I (4)
Arabis hirsuta	II (1–3)	*Dryopteris filix-mas*	I (2)
Leontodon hispidus	I (1–2)		
Fragaria vesca	I (1–2)	Number of samples	28
Hieracium pilosella	I (1–2)	Number of species/sample	13 (6–22)

OV39

Asplenium trichomanes-A. ruta-muraria community
Asplenietum trichomano-rutae-murariae R.Tx. 1937

Synonymy
Asplenium trichomanes-Fissidens cristatus Association Birks 1973 *p.p.*

Constant species
Asplenium ruta-muraria, Asplenium trichomanes, Homalothecium sericeum, Porella platyphylla.

Rare species
Hornungia petraea, Silene nutans.

Physiognomy
The *Asplenietum trichomano-rutae-murariae* comprises generally very open and often fragmentary stands of crevice vegetation in which diminutive ferns and bryophytes are the most distinctive components. The commonest ferns here are *Asplenium ruta-muraria* and *A. trichomanes*, the latter almost always the tetraploid ssp. *quadrivalens* according to Page (1982) although, on Skye, most of the plants examined have proved morphologically intermediate between this and the diploid ssp. *trichomanes* (Wood 1969). Both *A. trichomanes* and *A. ruta-muraria* are more or less evergreen plants whose small rosettes are very frequent here but of low total cover. The bulkier *A. adiantum-nigrum* occurs very rarely in the community but *A. viride* is typically absent. *Ceterach officinarum* is a good preferential but it is only really common in this vegetation towards the south and west of Britain.

Among other perennial vascular plants, only *Festuca ovina, Koeleria macrantha, Thymus praecox, Sedum acre* and *Helianthemum nummularium* occur with any frequency, with *Arenaria serpyllifolia* and *Saxifaga tridactylites* figuring as ephemerals, but all of these species are strongly preferential to one sub-community which is transitional to rocky turf where the cover of plants is greater. Through the community as a whole, the other prominent element in the vegetation comprises bryophytes whose cushions and mats can cram the crevices and spread out a little way over the rock surfaces. Most frequent among these are *Homalothecium sericeum, Porella platyphylla, Hypnum cupressiforme, Fissidens cristatus, Tortella tortuosa, Weissia controversa* and *Encalypta streptocarpa.* Less common are *Tortula subalata, T. muralis, Reboulia hemispherica* with *Tortula intermedia* and *Trichostomum crispulum* more preferential to one sub-community. Lichens are much less numerous and *Cladonia pocillum* is the only frequent species throughout.

Sub-communities

***Trichostomum crispulum-Tortula intermedia* sub-community.** The vegetation here is less species-rich and extensive in cover but more consistently dominated by ferns and bryophytes. Particularly towards south-west Britain, *Ceterach* becomes frequent, its stout little heads of rusty-backed fronds sometimes occurring in abundance. However, although there can be occasional tussocks of *Fetuca ovina* and *Koeleria macrantha* in the crevices, scattered mats of *Thymus* and *Helianthemum nummularium* with sparse ephemerals, the more distinctive component is a group of preferential bryophytes and lichens. *Tortula intermedia* and *Trichostomum crispulum* are the most frequent of these but *Targionia hypophylla, Dermatocarpon lachnaeum* and *Thalliodema caeruleonigricans* also occur occasionally.

***Sedum acre-Arenaria serpyllifolia* sub-community:** *Asplenium trichomanes-Fissidens cristatus* Association Birks 1973 *p.p.* In this kind of *Asplenietum* vegetation, the overall appearance is more like an open rocky turf in which the ferns and bryophytes of the community provide a consistent element among a variety of other vascular preferentials. Tussocks of *F. ovina* are especially prominent with scattered *K. macrantha, Arrhenatherum elatius* and *Aira caryophyllea.* Among mat-formers, *T. praecox* is usually the most frequent and abundant but *H. nummularium* and *Sedum acre* are both common and occasionally extensive among the crevices and over the rock surfaces. Also occurring among the perennials are *Sanguisorba minor*, *Teucrium scorodonia*, *Hieracium* spp. of the *Oreadea* section, *Potentilla tabernaemontani*, *Hieracium pilosella*, *Scabiosa columbaria* and *Centaurea scabiosa*. This sub-community also provides an occasional locus for the nationally rare *Silene nutans*. Ephemeral plants can also figure in some variety: *Arenaria serpyllifolia*, *Medicago lupulina*, *Saxifraga tridactylites*, *Acinos arvensis*, *Aira caryophyllea*, *Aphanes arvensis*, *Arabidopsis thaliana* and the national rarity *Hornungia petraea* are all occasional to common. Preferential bryophytes are less prominent but *Tortula ruralis* ssp. *ruralis* and *Grimmia apocarpa* are frequent.

Habitat
This is a community of sunny crevices in lime-rich bedrocks and wall-mortar at low to moderate altitudes, particularly in western Britain.

Both *A. trichomanes* and *A. ruta-muraria*, as well as the less common *Ceterach*, are strongly calcicolous ferns which gain a hold in crevices where other colonisers are limited by the extreme environmental conditions. Generally, it is harder limestones, like those of Carboniferous, Devonian and Jurassic age, more locally, as on Skye, Ordovician, that weather to provide suitable crevices for the ferns and these rocks crop out predominantly in the west and north of the country. There, the community is widespread on crags and occurs more locally where pavements form. Around these regions, too, buildings and boundary walls made of such rocks or, much more widely, lime-rich mortars used in construction with many other, non-calcareous, building materials also provide fissures for colonisation by this kind of vegetation. Lime-mortars in the strict sense date from Roman times in Britain and were in very wide use until well into the 1800s. Carbonisation does not necessarily penetrate very deeply but such mortars are relatively soft and decay quite rapidly so suitable crevices develop easily. The more recent cement mortars are more complex, harder and durable so colonisation by this vegetation tends to be slower and less extensive where these have been used (Page 1988). Nonetheless, in many places, it is buildings and walls that provide a more common substrate for the community than natural outcrops and this vegetation often adds a decorative effect to churchyards, cottages, ruined castles and abbeys, and industrial and railway buildings in country areas.

The extent and disposition of suitable crevices in natural and artificial habitats vary greatly, giving a pleasing diversity to the physiognomy of the community. Always, however, the soils around the roots of ferns and beneath the bryophytes are rudimentary protorendzinas, consisting of the sparse products of the disintegration and weathering of the mineral substrate, together with decaying plant material and wind-blown detritus.

Among the rhizomes of *Ceterach*, Page (1982) also observed foraging ants which he presumed brought some enrichment with nitrogen. Even where accumulation is somewhat more substantial, the soils here remain very calcareous. Where there has been acidification of mortars through atmospheric pollution or seepage through walls of water that is not so base-rich, this vegetation has not established at all or has declined.

Both the common species of *Asplenium* in this vegetation show great morphological plasticity with varying exposure in these habitats and they can, in fact, extend into strongly-shaded situations, where they take on a more luxuriant look. Generally, though, they occur in this community in places where there is a combination of some atmospheric humidity or moisture in the substrate together with high illumination. In more shaded habitats and in the cooler, cloudier conditions of higher mountains, this vegetation is usually replaced by the *Asplenio-Cystopteridetum* community. In the more equable climate of south-west Britain, where mild winter temperatures and the absence of desiccating winds are combined with high insolation, the *Asplenietum* community can occur in very sunny situations, where the abundance of *Ceterach* is especially distinctive. The ability of this fern to capitalise on water deep in rock exposures and stonework, together with its fleshy fronds and scales, give it a degree of protection against considerable baking heat (Page 1982, 1988).

Some of the distinctive bryophytes and lichens of this vegetation are also able to survive such strong illumination by using water in the crevices or recovering well from periods of drying. This applies, too, to certain of the vascular perennials, many of which are characteristic of drier grasslands. However, most of these need a more extensive and deeper network of soil-filled crevices to thrive and are characteristic of the *Sedum-Arenaria* sub-community which occurs where local weathering has extended the fissures and soil has accumulated or where outcrops give way to less rocky ground. The periodic exposure of areas of bare soil which get wetted in autumn and spring rains provides opportunities for ephemeral plants to gain a hold and complete their life cycles before the summer drought.

Zonation and succession
In its natural habitats, the *Asplenietum* is found on rocky habitats among other fern and bryophyte communities, calcicolous grasslands, scrub and woodland where zonations and successions are related to the extent of soil development, the degree of exposure to light and wind and the intensity of grazing. Stands in artificial habitats are often more isolated but can sometimes be found with other fern, bryophyte or crevice vegetation depending on the amount of shelter and shade.

Many of the floristic and structural differences

between the two sub-communities of the *Asplenietum* are related to the extent of soil formation and colonisation by vascular plants over limestone outcrops. Quite often, the *Sedum-Arenaria* sub-community is transitional between the *Trichostomum-Tortula* sub-community and some kind of calcicolous grassland on less rocky slopes around. There, more extensive, deeper and somewhat less drought-prone soils are often also accessible to grazing stock which cannot reach the craggier ground. Over much of the range of this fern vegetation in southern Britain, the typical calcicolous pasture swards are the *Festuca-Avenula* grassland and, in the warmer south and east, its analogues dominated by *Bromus erectus* and *Brachypodium pinnatum*. In fact, where the harder Carboniferous and Devonian limestones that provide some of the most congenial substrates for this community crop out, it is usually the *Dicranum* sub-community of *Festuca-Avenula* grassland that forms the bulk of the pasture on rendziniform soils of gentler slopes. In the Mendips, Derbyshire Dales and in north and south Wales, transitions to such grassland involve a rapid loss of the fern and much of the bryophyte contingent and a rise in the extent and number of perennial vascular calcicoles in the vegetation. Some ephemerals may continue to find a place on scuffed areas of sward but in general most are scarce. On the Carboniferous Limestone around Morecambe Bay and in the Yorkshire Dales, the *Sesleria-Galium* grassland replaces the *Festuca-Avenula* grassland in such sequences while on local limestone exposures through Scotland, the community can be found among stands of *Festuca-Agrostis-Thymus* grassland. On the Durness Limestone of Skye, this *Asplenietum* occurs in crevices in close association with the *Dryas-Carex* heath.

More locally, the *Asplenietum* can be found in association with the *Festuca-Carlina* grassland on limestone cliffs around the south-western seaboard of Britain but in the very hot and sunny conditions characteristic of such situations in summer, the fern vegetation often extends only on to somewhat shaded rock surfaces nearby.

In more inland habitats, where the surrounds to rock outcrops are not grazed, the *Asplenietum* may be more isolated among scrub and woodland. Usually, on the limestones characteristic of this fern vegetation, this is *Crataegus-Hedera* scrub or *Fraxinus-Acer-Mercurialis* woodland. Most of the shrubs and trees of these communities show limited ability to colonise rockier ground where soils are very limited so they peter out around the crags in a more open cover of stunted individuals. However, they may cast some shade on the *Asplenietum* which can be inimical to its survival. Where rocky crags occur close to gardens, colonisation of the crevices themselves with *Cotoneaster* and *Berberis* spp. can be extensive and this can shade out the community completely.

Where deeper crevices in limestone outcrops get more shade and shelter, the *Asplenietum* can give way to the *Asplenio-Cystopteridetum* or, in grikes, to fragmentary stands of miniuarised woodland. On walls and buildings where mortar-filled vegetation crevices provide the habitat for this the *Trichostomum-Tortula* sub-community can be found in close association with the *Parietarietum* and *Cymbalarietum* communities.

Distribution
The community occurs widely in suitable natural habitats, especially towards the more humid west of Britain with many additional localities on artificial substrates through the drier lowlands of the east.

Affinities
The *Asplenietum* is part of a range of rock-crevice vegetation that, apart from Birse (1984), has previously attracted little interest in Britain except locally, as on Skye (Birks 1973) or within treatments of fern ecology (Page 1982, 1988). The association has been described from other parts of north-west Europe like The Netherlands (Westhoff & den Held 1969), Ireland (White & Doyle 1982; see also Ivimey-Cook & Proctor 1966), Germany (Pott 1992) and Poland (Matuszkiewicz 1984). It has usually been placed in the alliance Potentillion caulescentis Br.-Bl. in Braun-Blanquet & Jenny 1926, typical of calcareous rocks in sunny situations, although some authorities follow Segal (1969) and locate it in a Cymbalario-Asplenion.

Floristic table OV39

	a	b	39
Asplenium trichomanes	V (1–3)	V (1–3)	V (1–3)
Asplenium ruta-muraria	V (2–4)	V (2–4)	V (2–4)
Porella platyphylla	V (1–2)	IV (2–4)	IV (1–4)
Homalothecium sericeum	IV (2–6)	IV (2–6)	IV (2–6)
Tortula intermedia	III (1–3)	I (2)	II (1–3)
Trichostomum crispulum	III (1–3)		II (1–3)
Dermatocarpon lachnaeum	II (1–3)		I (1–3)
Thaliodema caeruleonigricans	II (1–2)		I (1–2)
Ceterach officinarum	II (1–3)	I (2)	I (1–3)
Targionia hypophylla	II (1–3)		I (1–3)
Neckera crispa	I (2)		I (2)
Asplenium adiantum-nigrum	I (2)		I (2)
Festuca ovina	II (4–5)	V (2–8)	III (2–8)
Thymus praecox	II (2–4)	V (2–4)	III (2–4)
Arenaria serphyllifolia	I (2)	V (1–4)	III (1–4)
Sedum acre	I (1–2)	V (2–4)	III (1–4)
Koeleria macrantha	I (2–4)	IV (2–5)	III (2–5)
Helianthemum nummularium	I (4–5)	IV (1–5)	III (1–5)
Medicago lupulina	I (2)	III (1–3)	II (1–3)
Sanguisorba minor	I (2)	III (2–7)	II (2–7)
Saxifraga tridactylites	I (2)	III (1–3)	II (1–3)
Acinos arvensis	I (2)	III (1–3)	II (1–3)
Tortula ruralis ruralis		III (2–4)	II (2–4)
Schistidium apocarpum		II (2–4)	I (2–4)
Hornungia petraea		II (1–3)	I (1–3)
Campanula rotundifolia	I (1)	II (1–2)	I (1–2)
Teucrium scorodonia	I (4)	II (1–4)	I (1–4)
Hieracium section Oreadea		II (1–3)	I (1–3)
Arrhenatherium elatius		II (1–4)	I (1–4)
Aphanes arvensis	I (2–4)	II (2–4)	I (2–4)
Arabidopsis thaliana	I (2)	II (1–4)	I (1–4)
Geranium molle	I (2)	II (1–2)	I (1–2)
Potentilla tabernaemontani		II (1–4)	I (1–4)
Hieracium pilosella		II (1–3)	I (1–3)
Galium sterneri		II (1–4)	I (1–4)
Scabiosa columbaria		II (1–2)	I (1–2)
Centaurea scabiosa		II (2–4)	I (2–4)
Silene nutans		II (1–4)	I (1–4)
Aira caryophyllea		I (1–4)	I (1–4)
Bryum capillare		I (2)	I (2)
Squamaria crassa		I (1–2)	I (1–2)
Carlina vulgaris		I (2–4)	I (2–4)
Veronica arvensis		I (1–3)	I (1–3)
Riccia sorocarpa		I (1–3)	I (1–3)

Hypnum cupressiforme	III (2–4)	III (1–4)	III (1–4)
Fissidens cristatus	III (2–4)	III (1–4)	III (1–4)
Cladonia pocillum	III (1–3)	III (1–4)	III (1–4)
Tortella tortuosa	III (1–3)	III (1–4)	III (1–4)
Weissia controversa	III (2–4)	III (1–4)	III (1–4)
Encalypta streptocarpa	III (1–3)	III (1–3)	III (1–3)
Tortula subulata	II (1–3)	II (1–3)	II (1–3)
Tortula muralis	II (1–2)	II (1–2)	II (1–2)
Reboulia hemispherica	II (1–2)	II (2)	II (1–2)
Dermatocarpon miniatum	I (1–2)	I (2)	I (1–2)
Number of samples	18	20	38
Number of species/sample	10 (3–17)	22 (13–30)	16 (3–30)

a   *Trichostomum crispulum-Tortula intermedia* sub-community
b   *Sedum acre-Arenaria serpyllifolia* sub-community
39 *Asplenietum trichomano-rutae-murariae* (total)

OV40

Asplenium viride-Cystopteris fragilis community
Asplenio viridis-Cystopteridetum fragilis   (Kuhn 1939) Oberdorfer 1949

Synonymy
Asplenium trichomanes-Fissidens cristatus Association, Limestone facies *p.p.* and Montane facies   Birks 1973.

Constant species
Asplenium ruta-muraria, Asplenium trichomanes, Asplenium viride, Cystopteris fragilis, Festuca ovina, Ctenidium molluscum, Fissidens cristatus, Tortella tortuosa.

Rare species
Polystichum lonchitis, Woodsia alpina.

Physiognomy
The *Asplenio-Cystopteridetum* comprises open vegetation, often fragmentarily disposed in rock crevices on narrow ledges and among screes, in which a variety of ferns can assume prominence. *Asplenium trichomanes* (sometimes, according to Page (1982), ssp. *trichomanes* here, rather than the more widespread ssp. *quadrivalens*) remains quite common in this community but it peters out at higher altitudes. Even more so is this true of *A. ruta-muraria*, which among available samples is constant, but across the range of this vegetation as a whole is largely confined to stands outside the Scottish Highlands. The most characteristic spleenwort, then, is *A. viride*, particularly in the mountains of Scotland and northern England. This fern can occur in considerable abundance here, its rosettes of rather delicate pale green fronds, typically all turned towards the light, persisting into the winter and then dying down (Page 1982).

The other distinctive constant among the ferns is *Cystopteris fragilis*, a gregarious species, the fragile little fronds of which emerge rapidly in spring but which are cut back suddenly by the first frosts of autumn (Page 1988). At lower altitudes, particularly to the west of Britain, *Phyllitis scolopendrium* can occur with some frequency and abundance but more characteristic on higher ground are *Polystichum aculeatum* and, especially in the mountains of Scotland, *P. lonchitis*. Younger specimens of these ferns can be hard to distinguish (and the

species can hybridise to produce *P.* × *illyricum*: Page 1982), but only the latter extends on to the highest screes and ledges. There, too, very locally, this community can provide a locus for the nationally rare *Woodsia alpina* (Birks 1973, Page 1982, 1988).

Few other vascular plants apart from ferns are common in the community but, particularly in less-shaded crevices, *Festuca ovina* can be seen along with occasional *Geranium robertianum* and *Hieracium* spp. of the section *Oreadea* and scattered plants of calcicolous swards.

The other prominent element in the vegetation is bryophytes found as little tussocks and pads among the rock crevices. *Ctenidium molluscum, Fissidens cristatus* and *Tortella tortuosa* are all constant and the first in particular can be abundant. Also occurring are *Neckera crispa, Orthothecium intricatum, O. rufescens, Weissia controversa, Plagiochila asplenoides* and *Reboulia hemisphaerica.* Lichens are generally scarce but *Solorina saccata* is occasional.

Habitat
The *Asplenio-Cystopteridetum* is characteristic of more shaded crevices and ledges among exposures and talus of various lime-rich bedrocks in the cool, wet uplands of western and northern Britain.

Like the *Asplenietum*, this fern vegetation is quite strongly calcicolous but it is not so strikingly confined to sedimentary limestones. Thus, stands can also be found on base-rich shales, calciferous sandstones, schists, rhyolites, andesites and dolerites, all of which can weather to rendziniform or calcareous brown soils of immature profile. The pH of these rudimentary soils is not necessarily very high (sometimes 6 or below), but release of calcium from the parent materials is steady. Also, though *Cystopteris fragilis*, as well as *A. trichomanes* and *A. ruta-muraria*, can thrive on artificial habitats, where limestones or lime-mortar provide congenial substrates, this assemblage as a whole is characteristic of natural rocky outcrops or their weathering products like screes and boulder fields. These provide a wide variety of surfaces for colonisation, often so fragmentary as to severely limit potential competitors to the ferns which are themselves well equipped to capitalise on the restricted niches. This is especially well seen in *C. fragilis* which has a small but very extensive wiry root system pushing deeply into tiny cracks.

Another important difference between this vegetation and the *Asplenium* community is that it is more associated with cooler and humid situations. On a small scale, this can be seen in its preference for deeper crevices in rock faces and pavements and among talus, on ledges in gullies, beneath overhangs on cliffs and in cave mouths. Here, shade provides some relief from loss of moisture from the soils and the atmosphere and eliminates more light-demanding competitors, including *A. ruta-muraria* in deep shade. Typically, though, the ferns seem to prefer airy conditions where the humidity does not become stagnant and well-drained pockets of soil with no hint of impedence.

On a geographical scale, the need for lower temperatures and a moister environment is seen in the association of the community with higher altitudes than are typical of the *Asplenietum* vegetation. The *Asplenio-Cystopteridetum* community can be found near to sea-level on lime-rich screes and cliffs in the far north-west of Scotland, where temperatures remain cool and the climate is extremely humid and cloud-ridden. Even the Arctic-alpine *P. lonchitis* can be seen in such situations, though it is *A. ruta-muraria* and, in shadier places, *Phyllitis* which occur preferentially at these lower altitudes. However, it is on higher ground that the community acquires its most distinctive appearance where greater tolerance of winter cold and lower light intensities gives *A. viride* an edge over *A. trichomanes* and provides an opportunity for the appearance of the more upland *Polystichum aculeatum* and the truly Arctic-alpine *P. lonchitis*. This latter is a slow-growing but long-lived species, so it favours older, stable screes and crevices and its fronds do not expand until the late mountain spring weather of May. They can persist into a third growing season but their glossy, leathery texture probably gives some protection against desiccation in the long winter frosts at high altitudes (Page 1982). *Woodsia alpina*, too, though it seems to favour ledges sheltered from too much wind and rain, is well able to tolerate the heavy incrustation with ice characteristic of dripping wet

stands of this vegetation in high mountains (Page 1982).

Even in situations where the climate is milder, such conditions effectively exclude most potential vascular competitors to the ferns, though they are also congenial to the variety of calcicolous bryophytes typical of the community.

Zonation and succession

On shaded rock outcrops in mountains, the *Asplenio-Cystopteridetum* can often be found growing as isolated fragments in crevices. Where on exposures at higher altitudes there is more opportunity for soil accumulation and where seepage keeps the cliff faces dripping wet, the community can be found with the *Saxifraga-Alchemilla* vegetation, the asplenoid ferns continuing to figure occasionally among the luxuriant herb carpet of that assemblage. On drier ledges and crags, the *Dryas-Silene* community can also figure, with the *Festuca-Agrostis-Thymus* grassland usually occurring over the grazed slopes around. In the Yorkshire Dales, at somewhat lower altitudes on Carboniferous Limestone, the *Asplenio-Cystopteridetum* occurs on crags among the *Sesleria-Galium* grassland and can be seen in the mosaics of vegetation among high level grikes on pavement exposures. In sunnier situations at lower altitudes, the community is replaced by the *Asplenietum.*

Distribution

The *Asplenio-Cystopteridetum* community is commonest in the Scottish Highlands, north Pennines and Lake District, with more local occurrences further south in the Peak District and Wales.

Affinities

The *Asplenio-Cystopteridetum* was first described in detail from Britain by Birks (1973) as part of his broad *Asplenium-Fissidens* Association and then by Birse (1984). Stands from this country are generally similar to those described from mountains in Germany (Oberdorfer 1949, Pott 1992). Traditionally, they have been located in the Cystopteridion fragilis alliance, containing crevice vegetation of shaded situations, though Segal (1969) characterised a new association, the *Polysticho lonchitis-Asplenietum viridis* from walls in Austrian and French mountains and placed it in the Cymbalario-Asplenion, a treatment followed by White & Doyle (1982) in Ireland.

Floristic table OV40

Asplenium viride	V (2–8)
Cystopteris fragilis	V (1–5)
Ctenidium molluscum	V (2–6)
Fissidens cristatus	V (2–5)
Asplenium ruta-muraria	IV (2–7)
Asplenium trichomanes	IV (2–7)
Tortella tortuosa	IV (1–6)
Festuca ovina	IV (1–5)
Neckera crispa	III (2–4)
Geranium robertianum	II (2–3)
Solorina saccata	II (1–3)
Orthothecium intricatum	II (1–3)
Reboulia hemispherica	II (1–2)
Weissia controversa	II (1–3)
Plagiochila asplenoides	II (1–5)
Orthothecium rufescens	II (1–3)
Phyllitis scolopendrium	I (2–4)
Plagiobryum zierii	I (1–3)
Hieracium section *Oreadea*	I (1–2)
Campanula rotundifolia	I (1–3)
Trichostomum crispulum	I (3–4)
Porella platyphylla	I (1–2)
Koeleria macrantha	I (1–2)
Galium sterneri	I (1–3)
Polystichum aculeatum	I (1)
Polystichum lonchitis	I (3–4)
Pohlia cruda	I (2–3)
Pellia endiviifolia	I (1–3)
Isopterygium pulchellum	I (2)
Anoectangium aestivum	I (4)
Arrhenatherum elatius	I (2)
Sanguisorba minor	I (2)
Thymus praecox	I (2)
Encalypta streptocarpa	I (2)
Homalothecium sericeum	I (2)
Number of samples	15
Number of species/sample	18 (9–31)

OV41

Parietaria diffusa community
Parietarietum judaicae (Arènes 1928) Oberdorfer 1977

Constant species
Parietaria diffusa.

Rare species
Brassica oleracea, Draba aizoides, Silene nutans.

Physiognomy
The *Parietarietum judaicae* comprises vegetation of crevices and small ledges in which *Parietaria diffusa* is the only consistent feature throughout, with occasional trails of *Hedera helix* and *Galium aparine. Asplenium ruta-muraria* is a scarce associate.

Sub-communities

Homalothecium sericeum-Tortula muralis sub-community. *Parietaria* tends to be more abundant in this vegetation and the only frequent associates are *H. sericeum, T. muralis, Schistidium apocarpum* and *Barbula* spp. Occasionally, there are records for small asplenioid ferns, including *A. trichomanes, A. adiantum-nigrum, Poa trivialis* and ephemerals such as *Bromus hordeaceus hordeaceus, B. sterilis* and *Arabidopsis thaliana.*

Daucus carota sub-community. *Parietaria* is generally of lower cover here and is often accompanied by clumps of *Dactylis glomerata* and *Festuca rubra*, tall herbs such as *Daucus carota, Centaurea scabiosa* and *Euphorbia portlandica* with occasional *Plantago lanceolata, P. coronopus, Sanguisorba minor, Beta vulgaris* ssp. *maritima* and *Sedum acre*. This vegetation also provides a locus for the nationally rare *Brassica oleracea* and, more locally, *Draba aizoides* and *Silene nutans.*

Habitat
The *Parietarietum* is characteristic of sunnier crevices and ledges in walls and on rock faces in quarries and natural cliffs in the warmer and drier southern lowlands of Britain. *Parietaria* is a somewhat calcicolous plant with a more or less Continental distribution in Britain,

being commonest in the south-east of England and extending north and west around the coast (Perring & Walters 1962). It prefers sunny situations on limestone or mortared walls, and is more frequent on south- and west-facing aspects, though it often seems to benefit from the kind of protection against desiccation that can be found on walls built against earth banks (Segal 1969). In such places, too, nutrients from rain trickling through the soil behind the walls sustain more luxuriant growth because *Parietaria* is rather nitrophilous. The *Homalothecium-Tortula* sub-community occurs in suitable situations throughout the range but the *Daucus* sub-community is typical of seaside walls and cliffs, particularly on more base-rich substrates along the south coast of England and in Wales.

Zonation and succession
Where walls have smaller crevices in mortar, the *Parietarietum* can be found with the *Cymbalarietum* and *Asplenietum* communities, where dominance shifts to *Cymbalaria* or small ferns. In fact, Segal (1969) saw *A. ruta-muraria* and *Cymbalaria* as precursors to *Parietaria* on newly-built or restored walls, being overwhelmed as the latter increased its cover. Where there is local enrichment around walls and cliffs, the *Parietarietum* occurs with the *Urtica-Galium* community.

On coastal shore-banks and cliff-tops, the *Daucus* sub-community can give way to the *Festuca-Daucus* grassland with an increase in grass cover and higher frequency of herbs like *Sanguisorba minor* and *Plantago lanceolata.*

Distribution
The community occurs widely on suitable habitats in the south of England but with the *Daucus* type limited to the coast.

Affinities
The *Parietarietum* has been extensively recorded from other parts of Europe like The Netherlands (Westhoff &

den Held 1969) and Germany (Pott 1982). Some of these authors recognise a general *Parietarietum judaicae* (Arènes 1928) Oberdorfer 1977 but, in his overview of wall vegetation, Segal (1969) distinguished various climatic provinces. In the more Atlantic west of Europe, the *Asplenio-Parietarietum* was the typical syntaxon, with *Asplenium ruta-muraria*, *Tortula muralis* and *Homalothecium sericeum*. His *asplenietosum* and *poeto-*

sum compressae are both like the first sub-community characterised here. In Ireland, White and Doyle (1982) have recognised the *Asplenio-Parietarietum*, as well as three other related assemblages which Segal (1969) characterised. *Parietaria* vegetation of the kind included in this scheme is placed in a Parietarion or Centrantho-Parietarion alliance with the Asplenietea.

Floristic table OV41

	a	b	41
Parietaria diffusa	V (3–9)	V (1–8)	V (1–9)
Homalothecium sericeum	III (3–4)		II (3–4)
Tortula muralis	III (1–3)		II (1–3)
Poa trivialis	II (1–3)		I (1–3)
Schistidium apocarpum	II (2)		I (2)
Barbula sp.	II (2)		I (2)
Solanum dulcamara	I (3)		I (3)
Asplenium adiantum-nigrum	I (1)		I (1)
Asplenium trichomanes	I (2)		I (2)
Bromus hordeaceus hordeaceus	I (3)		I (3)
Bromus sterilis	I (3)		I (3)
Arabidopsis thaliana	I (2)		I (2)
Arrhenatherum elatius	I (3)		I (3)
Cerastium fontanum	I (2)		I (2)
Tanacetum vulgare	I (4)		I (4)
Cirsium arvense	I (4)		I (4)
Arenaria serpyllifolia	I (3)		I (3)
Dactylis glomerata	I (1)	V (1–4)	III (1–4)
Daucus carota		V (1–4)	III (1–4)
Festuca rubra		V (1–6)	III (1–6)
Centaurea scabiosa		IV (2–3)	III (2–3)
Brassica oleracea		III (2–7)	II (2–7)
Plantago lanceolata	I (2)	III (1–3)	II (1–3)
Euphorbia portlandica		III (1–4)	II (1–4)
Sanguisorba minor		III (1–3)	II (1–3)
Plantago coronopus	I (3)	II (1–2)	II (1–3)
Rumex acetosa		II (3)	I (3)
Beta vulgaris maritima		II (4–5)	I (4–5)
Draba aizoides		II (3)	I (3)
Sedum acre		II (1–3)	I (1–3)
Silene nutans		II (1–2)	I (1–2)
Armeria maritima		II (1)	I (1)
Carlina vulgaris		II (1)	I (1)
Leucanthemum vulgare		II (2–3)	I (2–3)
Cochlearia officinalis		II (1–3)	I (1–3)
Scabiosa columbaria		II (1–2)	I (1–2)

	a	b	41
Avenula pubescens		I (1)	I (1)
Rubia peregrina		I (1)	I (1)
Thymus praecox		I (1)	I (1)
Hedera helix	II (1–3)	II (1–2)	II (1–3)
Galium aparine	II (1–2)	II (4)	II (1–4)
Asplenium ruta-muraria	I (2)	I (1–2)	I (1–2)
Urtica dioica	I (2–4)	I (2)	I (2–4)
Medicago lupulina	I (3)	I (1)	I (1–3)
Sonchus asper	I (1)	I (2)	I (1–2)
Senecio vulgaris	I (1)	I (1–3)	I (1–3)
Number of samples	7	7	14
Number of species/sample	8 (3–11)	17 (11–26)	13 (3–26)

a   *Homalothecium sericeum-Tortula muralis* sub-community

b   *Daucus carota* sub-community

41   *Parietarietum judaicae* (total)

OV42
Cymbalaria muralis community
Cymbalarietum muralis   Görs 1966

Constant species

Cymbalaria muralis.

Physiognomy

The *Cymbalarietum muralis* comprises often very open and fragmentary crevice vegetation in which little hanging clumps of *Cymbalaria muralis* are the most obvious feature. Indeed, *Cymbalaria* is the only frequent vascular plant in this community. There can be occasional scattered rosettes of small ferns – *Asplenium trichomanes*, *A. ruta-muraria* and *Polypodium vulgare* – together with patches of *Sedum acre*, short trails of *Hedera helix*, isolated tufts of grasses like *Poa annua*, *Dactylis glomerata*, *Agrostis capillaris* and *A. stolonifera* and some ephemeral herbs but the consistency and cover of such contributions are never high.

More frequent as a group are mosses with small patches of *Homalothecium sericeum* and tufts of *Schistidium apocarpum*, *Grimmia pulvinata*, *Tortula muralis*, *Bryum capillare* and *Barbula unguiculata* occasional to common in the crevices.

Habitat

The *Cymbalarietum* is characteristic of sunny crevices among the stone- and brick-work of boundary walls and buildings throughout the lowlands of Britain.

Zonation and succession

The *Cymbalarietum* can be found with other kinds of crevice vegetation where walls have been colonised by different mixtures of species tolerant of the extreme conditions of the habitat. In the warmer south and east of Britain, the *Cymbalarietum* can be found with the *Parietarietum* where *Parietaria diffusa* is the distinctive dominant among the crevices, and Segal (1969) saw this as sometimes a successional replacement for the *Cymbalarietum*. On more lime-rich mortar on sunlit walls throughout the lowlands, the community can be replaced by the *Asplenietum* where *Cymbalaria* is rare or absent and small asplenioids and more calcicolous bryophytes and herbs are characteristic.

Distribution

The community occurs widely on suitable habitats throughout the lowlands.

Affinites

Since the *Cymbalarietum* Görs 1966 was first characterised it has been widely described from other parts of continental Europe like The Netherlands (Westhoff & den Held 1969) and Germany (Pott 1982). In Segal's (1969) treatment of wall vegetation, a Gemeenschap van *Linaria cymbalaria* en *Asplenium trichomanis* was placed among the fern communities of a Cymbalario-Asplenion alliance, and some more recent accounts recognising a *Cymbalarietum*, like Mucina *et al.* (1993), follow this proposal.

Floristic table OV42

Cymbalaria muralis	V (2–7)
Homalothecium sericeum	III (2–8)
Schistidium apocarpum	III (1–4)
Grimmia pulvinata	II (2–4)
Tortula muralis	II (1–4)
Asplenium ruta-muraria	II (2–4)
Poa annua	II (2–3)
Sedum acre	II (2–5)
Arenaria serpyllifolia	I (1–3)
Hedera helix	I (3–4)
Polypodium vulgare	I (1–4)
Bryum capillare	I (3)
Agrostis capillaris	I (1–2)
Asplenium trichomanes	I (4)
Dactylis glomerata	I (1–3)
Sonchus asper	I (1–2)
Barbula unguiculata	I (2–4)
Agrostis stolonifera	I (2–3)
Festuca rubra	I (2)
Holcus lanatus	I (1–4)
Poa pratensis	I (1–3)
Saxifraga tridactylites	I (4–5)
Senecio vulgaris	I (1–3)
Urtica dioica	I (1–2)
Valerianella locusta	I (1–4)
Barbula revoluta	I (2)
Bryum argenteum	I (2–3)
Orthotrichum anomalum	I (1–3)
Acer pseudoplatanus seedling	I (1–2)
Taraxacum officinale agg.	I (1–2)
Achillea millefolium	I (2)
Arrhenatherum elatius	I (1)
Brassica napus	I (2)
Buddleja davidii	I (1)
Calystegia sepium	I (3)
Catapodium rigidium	I (3)
Cerastium fontanum	I (1)
Cerastium semidecandrum	I (1)
Tanacetum parthenium	I (1)
Conyza canadensis	I (3)
Crepis capillaris	I (2)
Epilobium montanum	I (1)
Euphorbia peplus	I (1)
Number of samples	24
Number of species/sample	7 (1–20)

INDEX OF SYNONYMS TO MARITIME COMMUNITIES AND VEGETATION OF OPEN HABITATS

The vegetation types are listed alphabetically, then by date of ascription of the name, with the code number of the equivalent NVC community thereafter. The NVC communities themselves are included in the list with a bold code.

Adonis autumnalis-Iberis amara Association   (Allorge 1913) R.Tx. 1950   OV15

Agropyretum boreo-atlanticum   (Warming 1909) Br.-Bl. & De Leeuw 1936   SD4

Agropyretum juncei   Moss 1906, Tansley 1911, 1939   SD4

Agropyretum pungentis   Perraton 1953   SM24

Agropyro-Suaedetum fruticosae   Adam 1976   SM25

Agropyron junceiforme stands   Gimingham 1964*a*   SD4

Agropyron pungens-Juncus maritimus nodum   Adam 1976   SM24

Agrostio-Ranunculetum repentis   Oberdorfer *et al.* 1967   **OV28**

Agrostis stolonifera-Ranunculus repens community   **OV28**

Agrostis tenuis-Festuca ovina community   Tyler 1969   SM16

Airo multiculmis-Arnoseridetum minimae   (Allorge 1922) R.Tx. 1950 *sensu* Silverside 1977   OV2

Alchemillo-Matricarietum chamomillae   R.Tx. 1937 *emend.* Passarge 1957 *sensu* Silverside 1977   OV9

Allium schoenoprasum-Plantago maritima community   Hopkins 1983   OV34

Allium schoenoprasum-Plantago maritima community   **OV34**

Alopecuro-Chamomilletum recutitae   Wasscher 1941   **OV8**

Alopecurus geniculatus-Rorippa palustris community   **OV29**

Ammophila arenaria mobile dune community   **SD6**

Ammophila arenaria stands   Gimingham 1964*a*   SD5, SD6

Ammophila arenaria-Arrhenatherum elatius dune grassland   **SD9**

Ammophila arenaria-Festuca rubra semi-fixed dune community   **SD7**

Ammophila with scattered *Elymus*   Gimingham 1964*a*   SD6

Ammophiletum arenariae   Moss 1906   SD6, SD7

Ammophiletum arenariae   Tansley 1911   SD5, SD6, SD7

Ammophiletum arenariae   Tansley 1939   SD5, SD6, SD7

Anagallis arvensis-Veronica persica community   **OV15**

Annual *Salicornia* salt-marsh community   **SM8**

Armeria maritima-Aster tripolium provisional nodum   Ivimey-Cook & Proctor 1966   MC1

Armeria maritima-Cerastium diffusum ssp. *diffusum* maritime therophyte community **MC5**

Armeria maritima-Grimmia maritima association   Mallock & Okusanya 1979   MC2

Armeria maritima-Grimmia maritima Association   Birks 1973   MC2

Armeria maritima-Grimmia maritima rock crevice community   Ostenfeld 1908   MC2

Armeria maritima-Ligusticum scoticum low cliff vegetation   Petch 1933   MC2

Armeria maritima-Ligusticum scoticum maritime rock-crevice community   **MC2**

Armeria society   Marsh 1915   SM13

Armerieto maritimae-Daucetum gummiferi   Géhu 1964   MC8, MC9, MC11

Armerieto-Daucetum gummiferi crithmetosum   Géhu 1964   MC8

Armerieto-Daucetum gummiferi typicum   Géhu 1964   MC11

Armerietum   Goodman & Gillham 1954   MC8

Armerietum   Tansley 1939   SM13

Armerietum   Yapp & Johns 1917   SM13

Armerietum   McVean 1961   MC8

Artemisia maritima salt-marsh community   **SM17**
Artemisietum maritimae   Hocquette 1927   **SM17**
Arthrocnemum perenne stands   **SM7**
Asplenietum trichomano-rutae-murariae   R.Tx. 1937
   OV39
Asplenio viridis-Cystopteridetum fragilis (Kuhn 1939)
   Oberdorfer 1949 **OV40**
Asplenium marinum-Grimmia maritima Association
   Birks 1973   MC2
Asplenium trichomanes-Asplenium ruta-muraria
   community   **OV39**
Asplenium trichomanes-Fissidens cristatus Association
   Birks 1973   OV39
Asplenium trichomanes-Fissidens cristatus Association,
   Limestone facies   Birks 1973   OV40
Asplenium trichomanes-Fissidens cristatus Association,
   Montane facies   Birks 1973   OV40
Asplenium viride-Cystopteris fragilis community
   OV40
Aster tripolium var. *discoideus* nodum   Adam 1976
   SM11
Aster tripolium var. *discoideus* salt-marsh community
   SM11
Asteretum   Chapman 1934   SM11
Asteretum tripolii   Tansley 1939   **SM11**
Astragalo-Festucetum arenariae   Birse 1980   SD8,
   SD12
Astragalo-Festucetum arenariae, Typical subassociation
   Birse 1980   SD8
Atriplex glabriuscula-Rumex crispus Association
   Birks 1973   SD3
Atriplex prostrata-Beta vulgaris ssp. *maritima* sea-bird
   cliff community   **MC6**
Atriplicetum   Gillham 1953   MC6
Atriplici-Betetum maritimae   J.-M. & J. Géhu 1969
   MC6
Atriplici-Betetum perennis   J.-M. & J. Géhu 1969
   MC6
Atriplici-Elymetum pycnanthi   Beeftink & Westhoff
   1962   **SM24**
Atriplici-Polygonetum raii   Tx. 1950   SD2
Beta maritima-sociatie   Beeftink 1962   MC6
Beto-Tripleurospermetum maritimi   Malloch 1970
   MC6
Bidens tripartita-Polygonum amphibium community
   OV30
Bird cliff vegetation   Petch 1933   MC7
Blysmetum rufi   (G. E. & G. Du Rietz 1925) Gillner
   1960   **SM19**
Blysmus rufus salt-marsh community   **SM19**
Brassica oleracea maritime cliff-ledge community
   MC4
Brassicetum oleraceae   Géhu 1962   MC4
Braunton Damp Pasture   Willis *et al.* 1959   SD16
Briza minor-Silene gallica community   **OV2**

Cakile maritima-sociatie   Boerboom 1960   SD2
Calliergon cuspidatum-Salix repens nodum   Jones 1992
   SD13, SD14, SD15
Calliergon cuspidatum-Salix repens noda, species-poor
   sub-type   Jones 1992   SD15
Calliergon cuspidatum-Salix repens noda, *Equisetum
   variegatum* sub-type   Jones 1992   SD15
Calliergon cuspidatum-Salix repens noda, Herb-rich
   sub-type   Jones 1992   SD15
Campylium stellatum-Salix repens nodum   Jones 1992
   SD14
Campylium stellatum-Salix repens nodum, *Equisetum
   variegatum* sub-type   Jones 1992   SD15
Campylium stellatum-Salix repens nodum, Typical sub-
   type   Jones 1992   SD14
Campylium stellatum-Salix repens nodum, *Carex nigra*
   sub-type   Jones 1992   SD14
Campylium stellatum-Salix repens species-rich nodum,
   dry sub-type   Jones 1992   SD14, SD16
Carex arenaria community   Watt 1936   SD10
Carex arenaria community   Watt 1937   SD10
Carex arenaria dune community   **SD10**
Carex arenaria vegetation   Noble 1982   SD10
Carex arenaria-Cornicularia aculeata dune community
   SD11
Carex arenaria-Festuca ovina-Agrostis capillaris dune
   grassland   **SD12**
Carex distans-Plantago maritima Association   Ivimey-
   Cook & Proctor 1966   SM16, MC8
Carex flacca-Thalloid Liverwort nodum   Jones 1992
   SD14
Caricetum arenariae   Tansley 1939   SD11
Centaurio-Saginetum moniliformis   Diemont, Sissingh
   & Westhoff 1940   SD13
Cerastium atrovirens-Plantago coronopus Association
   Ivimey-Cook & Proctor 1966   MC8
Cerastium glomeratum-Fumaria muralis ssp. *boraei*
   community   **OV6**
Chrysanthemum segetum-Spergula arvensis community
   OV4
Coastal *Armerietum*   Goodman & Gillham 1954
   MC8
Cochlearietum   Goodman & Gillham 1954   MC7
Corynephorus canescens localities   Marshall 1967
   SD10
Crambe maritima sites   Scott & Randall 1976   SD1
Creek *Asteretum*   Chapman 1934   SM11
Crithmion communities   Shimwell 1976 ms.   MC1
Crithmo-Crambetum maritimae   (Géhu 1960) Géhu &
   Géhu 1969   SD1
Crithmo-Spergularietum rupicolae   Géhu 1964   **MC1**
*Crithmo-Spergularietum rupicolae plantaginetosum
   coronopi*   Géhu 1964   MC1
Crithmo-Spergularietum rupicolae typicum   Géhu 1964
   MC1

Salicornietum auct. **SM9**

Salicornietum europaeae Warming 1906 **SM8**

Salix repens-Calliergon cuspidatum dune-slack community **SD15**

Salix repens-Campylium stellatum dune-slack community **SD14**

Salix repens-Holcus lanatus dune-slack community **SD16**

Salix repens-Holcus lanatus nodum Jones 1992 SD16

Salsola kali-Atriplex glabriuscula Association Tx. 1950 SD2

Sandscale *Salix repens* dunes Pearsall 1934 SD15

Sedion anglici communities Proctor 1975 MC5

Sedion anglici releves 1–10 Proctor 1975 MC5

Sedion anglici releves 11–15 Proctor 1975 MC5

Setario-Veronicetum politae Oberdorfer 1957 *sensu* Silverside 1977 OV7

Shingle beach community Oliver 1911 SD1

Shingle beach community Oliver & Salisbury 1913 SD1

Shingle beach community Tansley 1939 SD1

Sileno maritimae-Festucetum pruinosae R.Tx. 1963 MC8, MC9

Sociatie van *Honkenya peploides* Westhoff & den Held 1969 SD2

Sociation à *Agropyron junceiforme* Géhu & Géhu 1969 SD4

Sociation à *Salsola kali* Géhu & Géhu 1969 SD2

Spartina alterniflora salt-marsh community **SM5**

Spartina anglica salt-marsh community **SM6**

Spartina maritima salt-marsh community **SM4**

Spartinetum alterniflorae Corillion 1955 **SM5**

Spartinetum maritimae (Emb. & Regn. 1926) Corillion 1953 **SM4**

Spartinetum townsendii (Tansley 1939) Corillion 1953 **SM6**

Spergula arvensis-Lamium amplexicaule community Sissingh 1950 OV14

Spergularia marina-Puccinellia distans salt-marsh community **SM23**

Sperguletum marinae Tyler 1969 SM23

Spergulo-Chrysanthemetum (Br.-Bl. & de Leeuw 1936) R.Tx. 1937 *sensu* Silverside 1977 OV6

Spergulo-Chrysanthemetum (Br.-Bl. & de Leeuw 1936) R.Tx. 1937, *Briza minor* variant Silverside 1977 OV4

Spergulo-Chrysanthemetum segetum (Br.-Bl. & de Leeuw 1936) R.Tx. 1937 **OV4**

Stachys arvensis community Silverside 1977 OV11

Statice society Marsh 1915 SM13

Stellaria media-Capsella bursa-pastoris community **OV13**

Stellaria media-Rumex acetosa sea-bird cliff community **MC7**

Strand plants association Tansley 1911 SD2

Suaeda maritima nodum Adam 1976 SM9

Suaeda maritima salt-marsh community **SM9**

Suaeda vera drift-line community **SM25**

Suaeda vera-Limonium binervosum salt-marsh community **SM21**

Suaedeto-Limonietum Chapman 1934 SM21

Suaedeto-Limonietum binervosi Adam 1976 SM21

Suaedetum fruticosae Tansley 1939 SM21, **SM25**

Suaedetum maritimae (Conrad 1935) Pignatti 1953 **SM9**

Suaedetum maritimae auct. SM10, SM13

Sussex *Frankenia laevis* stands Brightmore 1979 SM22

Tall *Festuca rubra* nodum Adam 1976 SM16

Teesdalio-Arnoseridetum minimae (Malcuit 1929) R.Tx. (1937) 1950 *sensu* Silverside 1977 OV1

Thero-Sedetum anglici Malloch 1971 MC5

Tortulo-Phleetum arenariae (Massart 1908) Br.-Bl. & De Leeuw 1936 **SD19**

Transitional low-marsh vegetation with *Puccinellia maritima*, annual *Salicornia* species and *Suaeda maritima* **SM10**

Trifolium occidentale-Herniaria ciliolata-Catapodium marinum nodum Coombe 1961 MC5

Trifolium occidentale-Scilla autumnalis-Jasione montana noda Coombe 1961 MC5

Triglochin-Juncus maritimus nodum Adam 1976 SM15

Tripleurospermum inodorum stands Kay 1994 OV9

Tripleurospermum maritimum stands Kay 1994 OV7

Unassigned aufnahmen *sensu* Silverside 1977 OV15

Urtica dioica-Cirsium arvense community **OV25**

Urtica dioica-Galium aparine community **OV24**

Urtica urens-Lamium amplexicaule community **OV14**

Veronica persica-Alopecurus myosuroides community **OV8**

Veronica persica-Veronica polita community **OV7**

Veronico-Lamietum hybridi Kruseman & Vlieger 1939 **OV7**

Viola arvensis-Aphanes microcarpa community **OV1**

Violo curtisii-Tortuletum ruraliformis Br.-Bl. & Tx. 1952 SD19

Vulpia fasciculata vegetation Watkinson 1978c SD19

Young dry slack nodum Jones 1992 SD13

Zooplethismic vegetation Poore & Robertson 1949 MC7

Zostera angustifolia stands **SM1**

Zosteretum marinae Harmsen 1936 **SM1**

Zosteretum noltii Harmsen 1936 **SM1**

Zosterion Christiansen 1934 **SM1**

INDEX OF SPECIES IN MARITIME COMMUNITIES AND VEGETATION OF OPEN HABITATS

The species are listed alphabetically, with the code numbers of the NVC communities in which they occur thereafter. Bold codes indicate that a species is constant throughout the community, italic codes that a species is constant in one or more sub-communities.

Acaena novae-zelandiae SD7
Acer pseudoplatanus (sapling) *OV27*
Acer pseudoplatanus (seedling) OV42
Achillea millefolium SD2, SD7, SD8, **SD9**, SD10, SD12, SD18, MC5, MC8, *MC9*, MC11, MC12, OV3, OV5, OV10, OV19, OV22, *OV23*, OV24, OV25, OV26, OV27, OV33, *OV37*, OV42
Adonis annua OV15
Aethusa cynapium OV7, OV8, OV15, OV16
Agrimonia eupatoria SD15, OV18
Agrostis canina SD15, OV35
Agrostis capillaris SD6, SD7, SD8, SD9, SD10, SD11, **SD12**, SD15, SD16, SD17, MC5, *MC9*, MC10, MC11, OV4, OV9, OV10, OV11, OV17, OV18, OV19, OV22, OV23, OV27, **OV37**, OV42
Agrostis stolonifera SM13, SM15, **SM16**, SM17, SM18, SM19, **SM20**, SM23, SM24, SM28, SD2, SD3, SD4, SD6, SD8, *SD13*, **SD14**, *SD15*, *SD16*, SD17, SD18, MC2, MC3, MC4, MC5, *MC8*, MC9, **MC10**, MC11, OV1, OV3, OV4, *OV6*, OV7, *OV9*, *OV10*, OV11, **OV12**, OV13, OV14, *OV15*, OV16, OV17, OV18, *OV19*, OV20, OV21, OV23, OV24, OV25, OV26, **OV28**, OV29, OV30, OV31, OV32, OV33, OV34, OV35, OV37, OV42
Agrostis vinealis OV34
Aira caryophyllea SD19, MC5, OV2, OV34, OV39
Aira multiculmis OV2
Aira praecox SD7, SD8, SD10, SD11, SD18, SD19, *MC5*, MC9, MC10, OV2
Ajuga chamaepitys OV15
Ajuga reptans OV26
Algal mat SM6, SM7, SM8, SM9, SM10, SM11, SM12, *SM13*, SM14, SM15, SM16, SM18, SM19, SM20, SM23, SM25, SM26, SM28
Alisma plantago-aquatica OV30, OV31, OV35

Alliaria petiolata OV10, OV26
Allium ampeloprasum var. *babingtonii* OV6
Allium roseum OV6
Allium schoenoprasum **OV34**
Allium triquetrum OV2, OV6, OV24
Allium ursinum OV27
Alopecurus aequalis OV35
Alopecurus geniculatus SM19, SM20, SM23, SM28, SD17, OV21, OV28, **OV29**, OV30, OV31, OV35
Alopecurus myosuroides OV1, **OV8**, OV13, OV15, OV36
Amaranthus albus OV25
Amaranthus retroflexus OV5
Amblystegium riparium SM19, SM20
Amblystegium serpens SM18, SM19, SD13, SD16
Ammophila arenaria SM24, SD1, SD2, SD4, SD5, **SD6**, SD7, *SD8*, **SD9**, SD10, *SD11*, SD12, SD13, SD16, *SD18*, **SD19**, *OV27*
Anacamptis pyramidalis SD18
Anagallis arvensis OV1, **OV2**, OV3, OV4, OV5, OV6, OV7, *OV9*, *OV10*, **OV11**, OV12, OV15, OV16, OV17, OV18, OV19, OV21, OV28, OV33, **OV36**
Anagallis tenella SD14, SD15, SD16, MC10
Anchusa arvensis OV3, OV4, OV5, OV9, **OV17**
Aneura pinguis **SD13**, *SD14*
Angelica sylvestris SD6, SD8, SD14, *SD15*, SD17, MC3, MC8, MC9, MC10, *OV26*
Anisantha diandra OV2, OV6
Anoectangium aestivum OV40
Anogramma leptophylla OV39, OV40
Anthemis arvensis OV3
Anthemis cotula OV1, OV8, *OV10*, OV15, OV19, OV33, OV36
Anthoxanthum aristatum OV1

Anthoxanthum odoratum   SM16, SD8, SD9, SD11, *SD12*, SD14, SD17, SD19, *MC9*, MC10, MC12, OV2, OV27, *OV37*

Anthriscus caucalis   OV2, OV6

Anthriscus sylvestris   SM28, OV24, OV25, OV26, OV27

Anthyllus vulneraria   SD7, SD8, *SD13*, SD14, SD16, SD19, MC4, MC5, *MC8*, MC9, MC10, MC11, OV37

Apera spica-venti   OV5

Aphanes arvensis   OV6, OV7, OV9, OV10, OV11, OV12, OV21, OV39

Aphanes microcarpa   **OV1**, OV2, OV3, OV5, *OV6*, OV11

Apium graveolens   SM28

Apium inundatum   OV35

Aquilegia vulgaris   OV38

Arabidopsis thaliana   OV3, OV39, OV41

Arabis hirsuta   OV38

Arctium minus   OV10, OV13, OV24, OV25, OV32

Arenaria leptoclados   OV3, OV16

Arenaria serpyllifolia   SD7, SD13, **SD19**, *MC5*, MC11, OV3, OV20, *OV39*, OV41, OV42

Armeria maritima   SM10, *SM13*, SM14, SM15, *SM16*, SM17, SM18, SM19, SM20, **SM21**, SM22, SM24, SM26, SD3, MC1, MC2, MC3, MC4, **MC5**, MC6, MC8, MC9, *MC10*, *MC11*, *MC12*, OV41

Arrhenatherum elatius   SM28, SD1, SD2, SD7, SD8, **SD9**, SD15, SD18, MC12, OV9, OV18, OV22, *OV23*, *OV24*, *OV25*, *OV26*, OV27, OV30, OV37, **OV38**, OV39, OV40, OV41, OV42

Artemisia maritima   SM14, SM15, **SM17**, SM21, SM24, SM25, SD4

Artemisia vulgaris   OV3, OV4, OV9, OV14, OV17, OV18, OV19, OV21, OV23, OV24, OV25

Arthrocnemum perenne   **SM7**, SM11, SM13, SM14, SM22, SM25, SM26

Arum italicum   OV6

Asplenium adiantum-nigrum   OV39, OV41

Asplenium marinum   MC1, MC2

Asplenium ruta-muraria   **OV39**, OV40, OV41, OV42

Asplenium trichomanes   OV38, **OV39**, OV40, OV41, OV42

Asplenium viride   OV38, **OV40**

Aster tripolium   SM6, SM7, SM8, SM9, SM10, **SM12**, *SM13*, SM14, SM15, *SM16*, SM17, SM18, SM19, SM20, SM23, SM24, SM25, SM26, SM28, *MC1*, OV25

Aster tripolium var. *discoideus*   SM6, SM7, SM8, SM9, SM10, **SM11**, SM12, *SM13*, SM14, SM15, SM24

Astragalus danicus   SD8, SD9, SD10, SD11, SD12

Athyrium filix-femina   OV27

Atriplex glabriuscula   SD2, SD3, SD4, SD5, SD6

Atriplex laciniata   SD2, SD4, SD6

Atriplex littoralis   SM24, SM25, SM28, SD2, MC5

Atriplex patula   SD2, SD3, OV7, OV8, OV9, OV11, OV12, OV13, OV15, *OV19*, OV20, OV22, OV28, OV33

Atriplex portulacoides   SM6, **SM7**, SM8, SM9, SM10, SM11, *SM13*, **SM14**, SM15, SM16, SM17, SM21, SM22, SM23, SM24, **SM25**, SM26

Atriplex prostrata   SM6, SM12, SM13, SM15, SM16, SM17, SM18, SM20, SM23, SM24, **SM28**, SD1, SD2, SD3, SD4, SD5, MC1, MC2, MC6, OV6, OV7, OV8, OV9, OV10, OV13, *OV19*, OV21, OV22, OV25, OV28, OV33

Aulacomnium androgynum   OV27

Avena fatua   OV7, OV8, OV13, OV15, OV16, OV25, OV33

Avenula pratensis   MC11, OV37, OV38

Avenula pubescens   SD8, OV41

Azolla filiculoides   OV32

Ballota nigra   OV19

Barbarea vulgaris   OV24

Barbula convoluta   *OV6*, OV12, OV15

Barbula fallax   OV15

Barbula revoluta   OV42

Barbula species   OV41

Barbula tophacea   SD13

Barbula unguiculata   OV9, OV12, OV15, OV36, OV42

Bellis perennis   *SD8*, SD9, SD14, SD16, SD17, MC5, MC9, MC10, MC11, OV4, OV6, OV21, OV22, OV23

Berula erecta   OV26, OV30, OV32

Beta vulgaris   OV25

Beta vulgaris ssp. *maritima*   SM24, SM28, SD1, SD2, MC1, *MC4*, MC5, **MC6**, MC8, MC12, OV41

Betula pubescens (sapling)   SD16, OV27

Bidens cernua   OV32

Bidens tripartita   OV29, **OV30**, OV31, OV32

Bilderdykia convolvulus   OV1, OV3, OV4, OV5, **OV7**, OV8, *OV9*, *OV10*, OV11, OV12, OV13, OV14, **OV15**, OV16, OV17, OV18, OV19, OV24, OV33, OV36

Blackstonia perfoliata   SD13, SD14, MC11

Blysmus rufus   **SM19**

Bolboschoenus maritimus   SM20, SM28, SD3, SD15

Bostrychia scorpioides   SM7, SM11, SM13, SM25, SM26

Brachypodium pinnatum   MC4, MC5, MC8, MC11

Brachypodium sylvaticum   MC9, MC12, OV27, OV38

Brachythecium albicans   SD6, SD7, SD8, SD11, SD12, SD16, SD18, SD19

Brachythecium mildeanum   SD15

Brachythecium rutabulum   SD5, SD7, SD9, SD14, *SD16*, SD17, SD18, OV5, OV11, OV21, OV23, OV24, OV26, OV27

Brachythecium velutinum   OV2

Brassica napus   OV13, OV42

Brassica nigra   MC4

Brassica oleracea   **MC4**, MC5, MC8, MC11, OV41

BIBLIOGRAPHY

Abdel-Wahab, A. M. (1969). *Role of micro-organisms in the nitrogen nutrition of Ammophila arenaria*. University of Wales: PhD thesis.

Abdel-Wahab, A. M. (1975). Nitrogen-fixation by *Bacillus* strains isolated from the rhizosphere of *Ammophila arenaria. Plant and Soil*, **42**, 703–5.

Adam, P. (1976). *Plant sociology and habitat factors in British saltmarshes*. University of Cambridge: PhD thesis.

Adam, P. (1977). On the phytosociological status of *Juncus maritimus* on British salt-marshes. *Vegetatio*, **35**, 81–94.

Adam, P. (1978). Geographical variation in British salt-marsh vegetation. *Journal of Ecology*, **66**, 339–66.

Adam, P. (1981). The vegetation of British salt-marshes. *New Phytologist*, **88**, 143–96.

Adam, P. & Akeroyd, J. R. (1978). The Cambridgeshire Salt-marsh. *Nature in Cambridgeshire*, **21**, 26–30.

Adam, P., Birks, H. J. B. & Huntley, B. (1977). Plant communities of the Island of Arran, Scotland. *New Phytologist*, **79**, 689–712.

Aellen, P. (1964). *Atriplex* L. In *Flora Europaea*, Volume I, ed. T. G. Tutin *et al*., pp. 95–97. Cambridge: Cambridge University Press.

Allorge, P. (1921–2). Les associations végétales du Vexin français. *Revue générale Botanique*, 33 & 34.

Almquist, E. (1929). Upplands vegetation och flora. *Acta Phytogeographica Suecica*, **1**, 12–634.

Avery, B. W. (1980). *Soil Classification for England and Wales (Higher Categories). Soil Survey Technical Monograph No. 14*. Harpenden: Soil Survey of England and Wales.

Ball, P. W. (1964). *Suaeda* Forskal ex Scop. In *Flora Europaea*, Vol. 1, ed. T. G. Tutin *et al*., pp. 103–4. Cambridge: Cambridge University Press.

Ball, P. W. & Brown, K. G. (1970). A biosystematic and ecological study of *Salicornia* in the Dee Estuary. *Watsonia*, **8**, 27–40.

Ball, P. W. & Tutin, T. G. (1959). Notes on annual species of *Salicornia* in Britain. *Watsonia*, **4**, 193–205.

Barallis, G. (1968). Ecology of blackgrass. *Proceedings of the 9th British Weed Control Conference*, 6–8.

Barkman, J. J., Moravec, J. & Rauschert, S. (1986). Code of Phytosociological Nomenclature, 2nd edition. *Vegetatio*, **67**, 145–95.

Beckers, A., Brock, T. & Klerkx, J. (1976). *A vegetation study of some parts of Dooaghtry, Co. Mayo, Republic of Ireland*. Catholic University of Nijmegen: Thesis.

Beeftink, W. G. (1959). Some notes on Skallingens salt-marsh vegetation and its habitat. *Acta botanica Neerlandica*, **8**, 449–472.

Beeftink, W. G. (1962). Conspectus of the phanerogamic salt plant communities in the Netherlands. *Biologisch Jaarboek Dodonaea*, **30**, 325–62.

Beeftink, W. G. (1965). De zout vegetatie van Z.W. Nederland beschwond in Europees verband. *Mededelingen van de Landbouwhogeschool te Wageningen*, **65**, 1.

Beeftink, W. G. (1966). Vegetation and habitat of the salt-marshes and beach plains in the south western part of the Netherlands. *Wentia*, **15**, 83–108.

Beeftink, W. G. (1972). Übersicht über die Anzahl der Aufnahmen Europäischer und Nordafrikanischer Salzpflanzengesellschaften für das Projekt der Arbeitsgruppe für Datenverarbeitung. In *Grundfragen und Methoden in der Pflanzensoziologie*, ed. E. van der Maarel & R. Tüxen, pp. 371–96. The Hague: Junk.

Beeftink, W. G. (1975). The ecological significance of embankment and drainage with respect to the vegetation of the south west Netherlands. *Journal of Ecology*, **63**, 423–58.

Beeftink, W. G. (1977*a*). The coastal salt-marshes of western and northern Europe: an ecological and phytosociological approach. In *Wet Coastal Ecosystems*, ed. V. J. Chapman, pp. 109–55. Amsterdam: Elsevier.

Beeftink, W. G. (1977*b*). Salt-marshes. In *The Coastline*, ed. R. S. K. Barnes, pp. 93–121. London: John Wiley & Sons.

Beeftink, W. G., Daane, M. C., de Munck, W. & Nieuwenhuize, J. (1978). Aspects of population dynamics in *Halimione portulacoides* communities. *Vegetatio*, **36**, 31–42.

Beeftink, W. G. & Géhu, J.-M. (1973). *Prodrome des Groupements Vegetaux d'Europe. I. Spartinetea maritimae*. Lehre: J. Cramer.

Bellot, F. R. (1966). La vegetacion de Galicia. *Anales del Instituto botánico A. J. Cavanillo*, **1**, 389–444.

Benecke, W. (1930). Zue Biologie der Strand- und Dünenflora. I. Vergleichende Versuche über die Salztoleranz von *Ammophila arenaria* Link, *Elymus arenarius* L. und *Agropyrum junceum* L. (= *Agropyron junceiforme*). *Berichte der Deutschen botanischen Gesellschaft*, **48**, 127–39.

Bhadresa, R. (1977). Food preferences of rabbits *Oryctolagus cuniculus* L. at Holkham sand-dunes, Norfolk. *Journal of Applied Ecology*, **14**, 287–91.

Bird, E. C. F. & Ranwell, D. S. (1964). *Spartina* salt-marshes in southern England. IV. The physiography of Poole Harbour, Dorset. *Journal of Ecology*, **52**, 355–66.

Birks, H. J. B. (1969). *The Late-Weichselian and present vegetation of the Isle of Skye*. University of Cambridge: PhD thesis.

Birks, H. J. B. (1973). *The past and present vegetation of the Isle of Skye: a palaeoecological study*. Cambridge: Cambridge University Press.

Birse, E. L. (1980). *Plant Communities of Scotland: A Preliminary Phytocoenonia*. Aberdeen: Macaulay Institute for Soil Research.

Birse, E. L. (1982). The main types of woodland in North Scotland. *Phytocoenologia*, **10**, 9–55.

Birse, E. L. (1984). *The Phytocoenonia of Scotland: Additions and Revisions*. Aberdeen: Macaulay Institute for Soil Research.

Birse, E. L. & Robertson, J. S. (1976). *Plant Communities and Soils of the Lowland and Southern Upland Regions of Scotland*. Aberdeen: Macaulay Institute for Soil Research.

Blackman, G. E. & Rutter, A. J. (1950). Physiological and ecological studies in the analysis of plant environment. IV. An assessment of the factors controlling the distribution of bluebell (*Scilla non-scripta*) in different communities. *Annals of Botany*, N.S., **14**, 487–520.

Blackman, G. E. & Rutter, A. J. (1954). Biological Flora of the British Isles: *Endymion non-scriptus* (L.) Garcke. *Journal of Ecology*, **42**, 629–38.

Blanchard, B. (1952). *An ecological survey of the sand-dune system of the south west Lancashire coast, with special reference to an associated marsh flora*. University of Liverpool: PhD thesis.

Boerboom, J. H. A. (1960). De plantengemeenschappen van de Wassenaarse Duinen (with a summary). *Mededelingen van de Landbouwhogeschool te Wageningen*, **60**, 1–135.

Bond, G., Fletcher, W. W. & Ferguson, T. P. (1954). Development and function of root nodules of *Alnus*, *Myrica* and *Hippophae*. *Journal of Experimental Botany*, **6**, 303–11.

Bond, G., MacConnell, J. T. & McCallum, A. H. (1956). Nitrogen fixation of *Hippophae rhamnoides* L. *Annals of Botany*, N.S., **20**, 501.

Bond, T. E. T. (1952). Biological Flora of the British Isles: *Elymus arenarius* L. *Journal of Ecology*, **40**, 217–27.

Boorman, L. A. (1967). Biological Flora of the British Isles: *Limonium vulgare* Mill. and *L. humile* Mill. *Journal of Ecology*, **55**, 221–32.

Boorman, L. A. & Ranwell, D. S. (1977). *Ecology of Maplin Sands and the Coastal Zones of Suffolk, Essex and North Kent*. Cambridge: Institute of Terrestrial Ecology.

Bradshaw, M. E. & Jones, A. V. (1976). *Phytosociology in Upper Teesdale*. Durham: Durham University Department of Extra-Mural Studies.

Braun-Blanquet, J. (1928). *Pflanzensoziologie. Grundzüge der Vegetationskunde*. Berlin: Springer.

Braun-Blanquet, J. (1935). Un joyau floristique et phytosociologique 'Isoetion' méditerranéen. *Bulletin de la Société pour l'Étude Science naturelle de Nîmes*, **47**, 141–63.

Braun-Blanquet, J. & De Leeuw, W. C. (1936). Vegetationsskizze von Ameland. *Nederlandsch Kruidkundig archief*, **46**, 359–93.

Braun-Blanquet, J. & de Ramm, C. (1957). Contribution à la connaissance de la végétation du littoral méditerranéen. Les prés salés du Languedoc Méditerranéen. *Bulletin du Musée d'Histoire Naturelle de Marseille*, **17**, 5–43.

Braun-Blanquet, J. & Tüxen, R. (1943). Übersicht der höheren Vegetationseinheiten Mitteleuropas (Unter Ausschluss der Hochgebirge). *SIGMA Communication*, **84**.

Braun-Blanquet, J. & Tüxen, R. (1952). Irische Pflanzengesellschaften. *Veröffentlichungen des Geobotanischen Institutes Rübel in Zürich*, **25**, 224–415.

Brenchley, W. E. (1911). The weeds of arable land in relation to the soils on which they grow. *Annals of Botany*, **25**, 155–65.

Brenchley, W. E. (1912). The weeds of arable land in relation to the soils on which they grow. II. *Annals of Botany*, **26**, 95–109.

Brenchley, W. E. (1913). The weeds of arable land II. *Annals of Botany*, **27**, 141–66.

Brereton, A. J. (1971). The structure of the species populations in the initial stages of salt-marsh succession. *Journal of Ecology*, **59**, 321–38.

Bridges, E. M. (1977). Soils of the alluvial lowlands of the Burry Inlet. In *Problems of a Small Estuary* 2, ed. A. Nelson-Smith & E. M. Bridges, pp. 1–15. Swansea Institute of Marine Studies, University College of Swansea.

Bridgewater, P. (1970). *Phytosociology and community boundaries of the British heath formation*. Durham University: PhD thesis.

Brightmore, D. (1979). Biological Flora of the British Isles: *Frankenia laevis* L. *Journal of Ecology*, **67**, 1097–1108.

Brightmore, D. & White, P. H. F. (1963). Biological Flora of the British Isles: *Lathyrus japonicus* Willd. (*L. maritimus* Bigel.). *Journal of Ecology*, **51**, 795–801.

Brun-Hool, J. & Wilmans, O. (1982). Plant communities of human settlements in Ireland. 2. Gardens, parks and roads. *Journal of Life Sciences of the Royal Dublin Society*, **3**, 91–103.

Buckman (1856). On agricultural weeds. In *Encyclopaedia of Agriculture*. Glasgow.

Butcher, R. W. (1934). *Zostera*. Report on the present condition of eel grass on the coasts of England, based on a survey during August to October, 1933. *Journal du Conseil*, **9**, 49–65.

Butcher, R. W. (1941). The distribution of *Zostera* (eel grass, wigeon grass) and other seashore plants in relation to the migrations of wildfowl. In *International Wildfowl Inquiry*, Volume I, pp. 29–49. Cambridge: Cambridge University Press.

Cadwalladr, D. A., Owen, M., Morley, J. V. & Cook, R. S. (1972). Wigeon (*Anas penelope* L.) conservation and salting pasture management at Bridgwater Bay National Nature Reserve, Somerset. *Journal of Applied Ecology*, **9**, 417–25.

Cadwalladr, D. A. & Morley, J. V. (1973). Sheep grazing preferences on a saltings at Bridgwater Bay Nature Reserve, Somerset, and their significance for wigeon (*Anas penelope* L.) conservation. *Journal of the British Grassland Society*, **28**, 235–42.

Cadwalladr, D. A. & Morley, J. V. (1974). Further experiments on the management of saltings pasture for Wigeon (*Anas penelope* L.) conservation at Bridgwater Bay National

Nature Reserve, Somerset. *Journal of Applied Ecology*, **11**, 461–66.

Carey, A. E. & Oliver, F. W. (1918). *Tidal Lands*. London: Blackie.

Carter, N. (1932). A comparative study of the alga flora of two salt marshes. Part I. *Journal of Ecology*, **20**, 341–70.

Carter, N. (1933*a*). A comparative study of the alga flora of two salt-marshes. Part II. *Journal of Ecology*, **21**, 128–208.

Carter, N. (1933*b*). A comparative study of the alga flora of two salt-marshes. Part III. *Journal of Ecology*, **21**, 385–403.

Cavers, P. B. & Harper, J. L. (1964). Biological Flora of the British Isles: *Rumex obtusifolius* L. and *R. crispus* L. *Journal of Ecology*, **52**, 737–66.

Chadwick, L. (1982). *In Search of Heathland*. Durham: Dobson.

Chandler, T. J. & Gregory, S. eds. (1976). *The Climate of the British Isles*. London: Longman.

Chapman, V. J. (1934). The ecology of Scolt Head Island. In *Scolt Head Island*, ed. J. A. Steers, pp. 77–145. Cambridge: Heffers.

Chapman, V. J. (1937). A revision of the marine algae of Norfolk. *Journal of the Linnean Society (Botany)*, **51**, 205–63.

Chapman, V. J. (1947). Biological Flora of the British Isles: *Suaeda maritima* (L.) Dum. *Journal of Ecology*, **35**, 293–302.

Chapman, V. J. (1950). Biological Flora of the British Isles: *Halimione portulacoides* (L.) Aell. *Journal of Ecology*, **38**, 214–22.

Chapman, V. J. (1959). Studies in salt-marsh ecology. IX. Changes in salt-marsh vegetation at Scolt Head Island. *Journal of Ecology*, **47**, 619–39.

Chapman, V. J. (1960*a*). *Salt Marshes and Salt Deserts of the World*. London: Leonard Hill.

Chapman, V. J. (1960*b*). The plant ecology of Scolt Head Island. In *Scolt Head Island*, ed. J. A. Steers, 2nd edition, pp. 85–163. Cambridge: Heffers.

Chapman, V. J. (1974). Salt marshes and salt deserts of the world. In *Ecology of Halophytes*, ed. R. J. Reimold & W. H. Queen, pp. 3–19. London: Academic Press.

Chapman, V. J. (1976). *Coastal Vegetation*. Oxford: Pergamon Press.

Charman, K. (1975). The feeding ecology of the brent goose. In *Report of the Maplin Ecological Research Programme*. Part II, 3b pp. 259–89. London: Department of Environment (unpublished).

Charman, K. (1977*a*). The grazing of *Zostera* by wildfowl in Britain. *Aquaculture*, **12**, 229–33.

Charman, K. (1977*b*). The seasonal pattern of food utilisation by *Branta bernicla* on the coast of southeast England. In *Proceedings of First Technical Meetings on Western Palearctic Migratory Bird Management*, pp. 64–76. Slimbridge: I.W.R.B.

Charman, K. (1979). Feeding ecology and energetics of the dark-bellied brent goose (*Branta bernicla bernicla*) in Essex and Kent. In *Ecological Processes in Coastal Environments*, ed. R. L. Jefferies & A. J. Davy, pp. 451–65. Oxford: Blackwell Scientific Publications.

Charman, K. & Macey, A. (1978). The winter grazing of salt-marsh vegetation by Dark-bellied Brent Geese. *Wildfowl*, **29**, 153–62.

Chater, E. H. (1965). Ecological aspects of the dwarf brown form of *Spartina* in the Dovey estuary. *Journal of Ecology*, **53**, 789–97.

Chater, E. H. (1973). *Spartina* in the Dyfi Estuary. In *Ynyslas Nature Reserve Handbook*, ed. E. E. Watkin, pp. 115–23. Aberystwyth: University College of Wales Aberystwyth, School of Biological Sciences.

Chater, E. H. & Jones, H. (1957). Some observations of *Spartina townsendii* in the Dovey estuary. *Journal of Ecology*, **45**, 157–67.

Clapham, A. R., Pearsall, W. H. & Richards, P. W. (1942). Biological Flora of the British Isles: *Aster tripolium* L. *Journal of Ecology*, **30**, 385–95.

Clapham, A. R., Tutin, T. G. & Warburg, E. F. (1962). *Flora of the British Isles*, 2nd edition. Cambridge: Cambridge University Press.

Climatological Atlas of the British Isles (1952). London: Meteorological Office.

Conolly, A. P. & Dahl, E. (1970). Maximum summer temperature in relation to the modern and Quaternary distributions of certain arctic-montane species in the British Isles. In *Studies in the Vegetatational History of the British Isles*, ed. D. Walker & R. G. West, pp. 159–224. Cambridge: Cambridge University Press.

Cook, R. E. D. (1990). *Iron toxicity to wetland plants*. Sheffield University: PhD thesis.

Coombe, D. E. (1961). *Trifolium occidentale*, a new species related to *T. repens* L. *Watsonia*, **5**, 68–87.

Coombe, D. E. & Frost, L. C. (1956*a*). The heaths of the Cornish Serpentine. *Journal of Ecology*, **44**, 226–56.

Coombe, D. E. & Frost, L. C. (1956*b*). The nature and origin of the soils over the Cornish Serpentine. *Journal of Ecology*, **44**, 605–15.

Coombe, D. E., Perring, F. H. & Walters, S. M. (1959). *Lythrum hyssopifolia* L. *Proceedings of the Botanical Society of the British Isles*, **3**, 286–8.

Cooper, E. A. (1987). *Vegetation Maps of British Sea Cliffs and Cliff-tops*. No. 5: Stackpole, Pembrokeshire. Lancaster University: Report to the Nature Conservancy Council.

Cope, T. A. & Stace, C. A. (1978). The *Juncus bufonius* L. aggregate in western Europe. *Watsonia*, **12**, 113–28.

Corbett, W. M. (1973). *Breckland Forest Soils*. Harpenden: Soil Survey of England and Wales.

Corley, M. F. V. & Hill, M. O. (1981). *Distribution of Bryophytes in the British Isles*. Cardiff: British Bryological Society.

Corillion, R. (1953). Les Halipèdes du Nord de la Bretagne. *Revue Générale de Botanique*, **60**, 707–55.

Cotton, A. D. (1912). Clare Island Survey. Part XV. Marine Algae. *Proceedings of the Royal Irish Academy b*, **31**, 1–178.

Courtney, A. D. (1968). Seed dormancy and field emergence in *Polygonum aviculare*. *Journal of Applied Ecology*, **5**, 675–84.

Crompton, G. & Sheail, J. (1975). The historical ecology of Lakenheath Warren in Suffolk, England: a case study. *Biological Conservation*, **8**, 299–313.

Dahl, E. (1956). *Rondane mountain vegetation in south Norway and its relation to the environment*. Oslo: Aschehoug.

Dahl, E. (1968). *Analytical Key to British Macrolichens*, 2nd edition. London: British Lichen Society.

Dahl, E. & Hadač, E. (1941). Strandgesellschaften der Insel Ostoy in Oslofjord. Eine pflanzensoziologische Studie. *Nytt Magasin for Naturvidenskapene B*, **82**, 251–312.

Dahlbeck, N. (1945). Strandwiesen am südöstlichen Öresund. *Acta Phytogeographica Suecica*, **18**, 1–168.

Dalby, D. H. (1970). The salt marshes of Milford Haven, Pembrokeshire. *Field Studies*, **3**, 297–330.

Dargie, T. C. D. (1990). *National Sand Dune Vegetation Survey. Site report No. 46 Crymlyn Burrows*. Peterborough: Nature Conservancy Council.

Dargie, T. C. D. (1993). *Sand Dune Vegetation Survey of Great Britain, A National Inventory*. Part 2: *Scotland*. Peterborough: Joint Nature Conservation Committee.

Dargie, T. C. D. (1995). *Sand Dune Vegetation Survey of Great Britain, a National Inventory*. Part 3: *Wales*. Peterborough: Joint Nature Conservation Committee.

Dargie, T. C. D. (1998). *Sand dune vegetation survey of Scotland: Western Isles. Volume 3: NVC maps*. Battleby: Scottish Natural Heritage.

Deighton, F. C. & Clapham, A. R. (1925). The vegetation of Scolt Head Island. *Transactions of the Norfolk and Norwich Naturalists' Society*, **12**.

Diemont, W. H., Sissingh, G. & Westhoff, V. (1940). Het Dwergbiezenverbond (Nanocyperion flavescentis) in Nederland. *Ned. Kruidk Arch.*, **50**, 215–71.

Dodds, J. G. (1953). Biological Flora of the British Isles: *Plantago coronopus* L. *Journal of Ecology*, **41**, 467–78.

Dony, J. G. (1953a). *Flora of Bedfordshire*. Luton.

Dony, J. G. (1953b). Wool aliens in Bedfordshire. In *The Changing Flora of Britain*. Arbroath.

Doyle, G. J. (1982). Minuartio-Thlaspietum alpestris (Violetea calaminariae) in Ireland. *Journal of Life Sciences of the Royal Dublin Society*, **3**, 147–64.

Du Rietz, G. E. & Du Rietz, G. (1925). Floristiska anteckningar fran Blekinge skargard. *Botaniska Notiser 1925*, 66–76.

Edees, E. S. (1972). *Flora of Staffordshire*. Newton Abbot: David & Charles.

Eklund, O. (1931). *Crambe maritima* in Nordbaltischen Gebiet. *Memo. Soc. Fauna Flora fenn.*, **7**, 41–51.

Ellenberg, H. (1978). *Vegetation Mitteleuropas mit den Alpen*, 2 Auflage. Stuttgart: Ulmer.

Ellenberg, H. (1988). *Vegetation Ecology of Central Europe*. Cambridge: Cambridge University Press.

Ellis, A. E. (1960). The lichens. In *Scolt Head Island*, ed. J. A. Steers, pp. 177–8. Cambridge: Heffers.

Engelskjön, T. (1970). Flora of Nord-Fuglöy, Troms. *Astarte*, **3**, 63–82.

Ernst, W. H. O. (1981). Ecological implication of fruit variability in *Phleum arenarium*, an annual dune grass. *Flora*, **171**, 387–98.

Ernst, W. H. O. (1983). Element nutrition of two contrasted dune annuals. *Journal of Ecology*, **71**, 197–209.

Farrell, L. (1989). *Mertensia maritima* (L.) Gray – current status in Britain. *BSBI News*, **51**, 9–11.

Fletcher, A. (1973a). The ecology of marine (littoral) lichens on some rocky shores of Anglesey. *Lichenologist*, **5**, 368–400.

Fletcher, A. (1973b). The ecology of maritime (supralittoral) lichens on some rocky shores of Anglesey. *Lichenologist*, **5**, 401–22.

Fraser Darling, F. & Morton Boyd, J. (1969). *The Highlands and Islands*, 2nd edition. London: Collins.

Géhu, J.-M. (1960). Quelques observations sur la végétation et l'écologie d'une station réputée de l'Archipel des Chausey: L'île aux Oiseaux. *Bulletin du Laboratoire Maritime de Dinard*, **46**, 78–92.

Géhu, J.-M. (1962). Quelques observations sur la falaise crétacée du Cap Blanc-Nez (P.D.C.) et étude de la végétation de la paroi abrupte: Brassicetum oleraceae nov. ass. *Bulletin de la Société royale de Botanique de Belgique*, **95**, 109–29.

Géhu, J.-M. (1964). Sur la végétation phanérogamique halophile des falaises bretonnes. *Revue générale Botanique*, **71**, 73–78.

Géhu, J.-M. (1972). Cartographie en réseaux et phytosociologie. In *Grundfragen und Methoden in der Pflanzensoziologie*, ed. E. van der Maarel & R. Tüxen, pp. 263–76. The Hague: Junk.

Géhu, J.-M. (1973a). L'*Eleocharetum parvulae* Gillner 1960 de la 'New Forest', Hants., England. *Documents Phytosociologiques*, **4**, 44–6.

Géhu, J.-M. (1973b). *Trifolium occidentale* D. E. Coombe. Espèce nouvelle pour le littoral du nord du Portugal. *Agronomia Lusitana*, **34**, 197–204.

Géhu, J.-M. (1975). Essai systématique et chorologique sur les principales associations végétales du littoral atlantique français. *Anales Real Academia de farmacía*, **41**, 207–27.

Géhu, J.-M. & Delzenne, C. (1975). Apport à la connaissance phytosociologique des prairies salées de l'Angleterre. *Colloques Phytosociologiques*, **IV**, 227–47.

Géhu, J.-M. & Géhu, J. (1969). Les associations végétales des dunes mobiles et des bordures de plages de la côte atlantique française. *Vegetatio*, **18**, 122–66.

Géhu, J.-M. & Géhu-Franck, J. (1975). Données nouvelles sur les végétations à *Frankenia laevis* des hauts de schorre sablonneux des côtes atlantiques. *Phytocoenologia*, **2**, 154–68.

Géhu, J.-M. & Géhu-Franck, J. (1979). Sur les végétations Nord-Atlantiques et Baltiques à *Crambe maritima*. *Phytocoenologia*, **6**, 209–29.

Gemmell, A. R., Grieg-Smith, P. & Gimingham, C. H. (1953). A note on the behaviour of *Ammophila arenaria* (L.) Link in relation to sand-dune formation. *Transactions and Proceedings of the Botanical Society of Edinburgh*, **36**, 132–6.

Ghestem, A. (1972). Essai de synthèse des végétations halophiles de la Baie de la Canche. *Documents Phytosociologiques*, **1**, 1–33.

Gibbons, E. J. (1975). *The Flora of Lincolnshire*. Lincoln: Lincolnshire Naturalists' Union.

Gillham, M. E. (1953). An ecological account of the vegetation of Grassholm Island, Pembrokeshire. *Journal of Ecology*, **41**, 84–99.

Gillham, M. E. (1956b). Ecology of the Pembrokeshire Islands. V. Manuring by the colonial seabirds and mammals with a note on seed distribution by gulls. *Journal of Ecology*, **44**, 429–54.

Gillham, M. E. (1957a). Vegetation of the Exe estuary in relation to water salinity. *Journal of Ecology*, **45**, 735–56.

Gillham, M. E. (1957b). Coastal vegetation of Mull and Iona in relation to salinity and soil reaction. *Journal of Ecology*, **45**, 757–78.

Gillner, V. (1960). Vegetations- und Standortsuntersuchungen in den Strandwiesen der schwedischen Westküste. *Acta Phytogeographica Suecica*, **43**, 1–198.

Gimingham, C. H. (1951). Contributions to the maritime ecology of St. Cyrus, Kincardineshire. II. The sand dunes. *Transactions and Proceedings of the Botanical Society of Edinburgh*, **35**, 387–414.

Gimingham, C. H. (1964a). Maritime and sub-maritime communities. In *The Vegetation of Scotland*, ed. J. H. Burnett, pp. 66–142. Edinburgh: Oliver & Boyd.

Gimingham, C. H., Gemmell, A. R. & Greig-Smith, P. (1948). The vegetation of a sand-dune system in the Outer Hebrides. *Transactions and Proceedings of the Botanical Society of Edinburgh*, **35**, 82–96.

Godwin, H. (1975). *The History of the British Flora*, 2nd edition. Cambridge: Cambridge University Press.

Goldsmith, F. B. (1973). The vegetation of exposed sea-cliffs at South Stack, Anglesey. II. Experimental Studies. *Journal of Ecology*, **61**, 819–29.

Goldsmith, F. B. (1975). The sea-cliff vegetation of Shetland. *Journal of Biogeography*, **2**, 297–308.

Good, R. D. O. & Waugh, W. L. (1934). The vegetation of Recliffe Sand: a contribution to the ecology of the Humber. *Journal of Ecology*, **22**, 420–38.

Goodman, G. T. & Gillham, M. E. (1954). Ecology of the Pembrokeshire Islands. II. Skokholm, environment and vegetation. *Journal of Ecology*, **42**, 296–327.

Goodman, P. J. (1960). Investigations into 'die-back' in *Spartina townsendii* agg. II. the morphological structure and composition of the Lymington sward. *Journal of Ecology*, **48**, 711–24.

Goodman, P. J., Braybrooks, E. M. & Lambert, J. M. (1959). Investigations into 'die-back' in *Spartina townsendii* agg. I. The present status of *Spartina townsendii* in Britain. *Journal of Ecology*, **47**, 651–77.

Goodman, P. J., Braybrooks, E. M., Lambert, J. M. & Marchant, C. J. (1969). Biological Flora of the British Isles: *Spartina* Schreb. *Journal of Ecology*, **57**, 285–313.

Goodman, P. J. & Williams, W. T. (1961). Investigations into 'die-back' in *Spartina townsendii* agg. III. Physiological correlates of 'die-back'. *Journal of Ecology*, **49**, 391–98.

Gorham, E. (1958). Soluble salts in dune sands from Blakeney Point in Norfolk. *Journal of Ecology*, **46**, 373–79.

Grabherr, G. & Mucina, L. (1993). *Die Pflanzengesellschaften Österreichs*. Teil II. Stuttgart: Fischer.

Graham, R. J. D. (1938). The development of *Elymus arenarius* Linn. on the West Sands, St. Andrews. *Transactions of the Botanical Society of Edinburgh*, **32**, 409–10.

Gravesen, P. & Vestergaard, P. (1969). Vegetation of a Danish off-shore barrier island. *Botanisk Tidsskrift*, **65**, 44–99.

Gray, A. J. (1971). *Variation in* Aster tripolium *L. with particular reference to some British populations*. University of Keele: PhD. thesis.

Gray, A. J. (1972). The ecology of Morecambe Bay. V. The salt marshes of Morecambe Bay. *Journal of Applied Ecology*, **9**, 207–20.

Gray, A. J. (1974). The genecology of salt-marsh plants. *Hydrological Bulletin (Amsterdam)*, **8**, 152–65.

Gray, A. J. (1977). Reclaimed land. In *The Coastline*, ed. R. S. K. Barnes, pp. 253–70. London: John Wiley & Sons.

Gray, A. J. (1979). The ecological implications of estuarine and coastal land reclamation. In *Estuarine and Coastal Land Reclamation and Water Storage*, ed. B. Knights & A. J. Phillips, pp. 177–95. Farnborough: Saxon House.

Gray, A. J. & Bunce, R. G. H. (1972). The ecology of Morecambe Bay. VI. Soils and vegetation of the salt-marshes: a multivariate approach. *Journal of Applied Ecology*, **9**, 221–34.

Gray, A. J. & Scott, R. (1977a). Biological Flora of the British Isles: *Puccinellia maritima* (Huds.) Parl. *Journal of Ecology*, **65**, 699–716.

Gray, A. J. & Scott, R. (1977b). The ecology of Morecambe Bay. VII. The distribution of *Puccinellia maritima, Festuca rubra* and *Agrostis stolonifera* in the salt marshes. *Journal of Applied Ecology*, **14**, 229–41.

Gray, A. J. & Scott, R. (1980). A genecological study of *Puccinellia maritima* (Huds.) Parl. I. Variation estimated from single-plant samples from British populations. *New Phytologist*, **85**, 89–107.

Gregory, S. (1957). Annual rainfall probability maps of the British Isles. *Quarterly Journal of the Royal Meteorological Society*, **83**, 543–9.

Greig-Smith, P. (1948). Biological Flora of the British Isles: *Urtica* L. *Journal of Ecology*, **36**, 339–55.

Greig-Smith, P. (1961). Data on pattern within plant communities. II. *Ammophila arenaria* (L.) Link. *Journal of Ecology*, **49**, 703–8.

Greig-Smith, P., Gemmell, A. R. & Gimingham, C. H. (1947). Tussock formation in *Ammophila arenaria* (L.) Link. *New Phytologist*, **46**, 262–8.

Hadač, E. (1970). Seashore communities of Reykjanes Peninsula, S.W. Iceland (Plant communities of Reykjanes Peninsula, Part 2). *Folia geobotanica phytotaxomica, Praha*, **5**, 133–44.

Harmsen, G. W. (1936). Systematische Beobachtungen der nordwesteuropaeischen seegrasformen. *Ned. Kruidk. Arch.*, **46**, 852–77.

Harris, D. & Davy, A. J. (1986a). Strandline colonization by *Elymus farctus* in relation to sand mobility and rabbit grazing. *Journal of Ecology*, **74**, 1045–56.

Harris, D. & Davy, A. J. (1986b). Regeneration potential of *Elymus farctus* from rhizome fragments and seed. *Journal of Ecology*, **74**, 1057–67.

Hassouna, M. G. & Wareing, P. F. (1964). Possible role of rhizosphere bacteria in the nitrogen nutrition of *Ammophila arenaria*. *Nature, London*, **202**, 467–9.

Hedley, S. M., Woolven, S. C. & Radley, G. P. (1990). *National Sand Dune Vegetation Survey. Site Report No. 71 Scolt Head Island Dunes, Norfolk*. Peterborough: Nature Conservancy Council.

Henson, I. E. (1969). Changes in the germination capacity of three *Polygonum* species following low temperature moist storage. *Technical Report of the Weed Research Organisation*, **13**.

Hill, M. O. (1979). *TWINSPAN – a FORTRAN program for arranging multivariate data in an ordered two-way table by classification of the individuals and attributes.* New York: Cornell University.

Hill, M. O. & Jones, E. W. (1978). Vegetation changes resulting from afforestation of rough grazings in Caeo Forest, South Wales. *Journal of Ecology*, **66**, 433–56.

Hill, M. O., Bunce, R. G. H. & Shaw, M. W. (1975). Indicator Species Analysis, a divisive polythetic method of classification and its application to a survey of native pinewoods in Scotland. *Journal of Ecology*, **63**, 597–613.

Hill, T. G. (1909). The Bouche d'Erquy in 1908. *New Phytologist*, **8**, 97–103.

Hilliam, J. (1977). *Phytosociological studies in the Southern Isles of Shetland.* Durham University: PhD thesis.

Hodge, C. A. H., Burton, R. G. O., Corbett, W. M., Evans, R. & Seale, R. S. (1984). *Soils and their use in Eastern England.* Harpenden: Soil Survey of England and Wales.

Hodge, C. A. H. & Seale, R. S. (1966). *The Soils of the District around Cambridge.* Memoirs of the Soil Survey of Great Britain: England and Wales. Harpenden: Soil Survey.

Hope-Simpson, J. F. (1940*b*). The utilisation and improvement of chalk down pasture. *Journal of the Royal Agricultural Society of England*, **100**, 44–9.

Hope-Simpson, J. F. & Jefferies, R. L. (1966). Observations relating to vigour and debility in marram grass. *Journal of Ecology*, **54**, 271–74.

Hope-Simpson, J. F. & Yemm, E. W. (1979). Braunton Burrows: developing vegetation in dune slacks, 1948–1977. In *Ecological Processes in Coastal Environments*, ed. R. J. Jefferies & A. J. Davy, pp. 113–27. Oxford: Blackwell Scientific Publications.

Hopkins, J. J. (1983). *Studies on the historical ecology, vegetation and flora of the Lizard district, Cornwall, with particular reference to heathland.* Bristol University: PhD thesis.

Howarth, S. E. & Williams, J. T. (1972). Biological Flora of the British Isles: *Chrysanthemum segetum* L. *Journal of Ecology*, **60**, 573–84.

Hubbard, J. C. E. (1965). *Spartina* marshes in southern England. VI. Pattern of invasion in Poole Harbour. *Journal of Ecology*, **53**, 799–813.

Hubbard, C. E. (1968). *Grasses*, 2nd edition. London: Penguin Books.

Hubbard, C. E. (1969). Light in relation to tidal immersion and the growth of *Spartina townsendii* (*s.l.*). *Journal of Ecology*, **57**, 795–804.

Hubbard, C. E. (1984). *Grasses*, 3rd edition. London: Penguin Books.

Hubbard, J. C. E. & Ranwell, D. S. (1966). Cropping *Spartina* marsh for silage. *Journal of the British Grassland Society*, **21**, 214–17.

Hubbard, J. C. E. & Stebbings, R. E. (1967). Distribution, dates of origin and acreage of *Spartina townsendii* (*s.l.*) marshes in Great Britain. *Proceedings of the Botanical Society of the British Isles*, **7**, 1–7.

Hubbard, J. C. E. & Stebbings, R. E. (1968). *Spartina* marshes in southern England. VII. Stratigraphy of the Keysworth Marsh, Poole Harbour. *Journal of Ecology*, **56**, 707–22.

Huiskes, A. H. L. (1972). *Kieming en groei van enige plantensoorten uit de buitenduinen.* Groningen State University: Doctoral thesis.

Huiskes, A. H. L. (1977*a*). *Population dynamics in* Ammophila arenaria *(L.) Link.* University of Wales: PhD thesis.

Huiskes, A. H. L. (1977*b*). The natural establishment of *Ammophila arenaria* (L.) Link from seed. *Oikos*, **29**, 133–6.

Huiskes, A. H. L. (1979). Biological Flora of the British Isles: *Ammophila arenaria* (L.) Link. *Journal of Ecology*, **67**, 363–82.

Hultén, E. (1950). *Atlas of the Distribution of Vascular Plants in N. W. Europe.* Stockholm: Generalstaben Litografiska Anstalts Förlag.

Hultén, E. (1971). The circumpolar plants, II. *Kungliga svenska vetenskaps akademiens handlingar*, **13**.

Huntley, B., Huntley, J. P. & Birks, H. J. B. (1981). PHYTOPAK: a suite of computer programs designed for the handling and analysis of phytosociological data. *Vegetatio*, **45**, 85–95.

Hutchinson, C. S. (1979). *Some effects of cow dung on pasture composition.* University of Wales: PhD thesis.

Hutchinson, C. S. & Seymour, G. B. (1982). Biological Flora of the British Isles: *Poa annua* L. *Journal of Ecology*, **70**, 887–901.

Ingrouille, M. J. (1981). A newly-discovered *Limonium* in East Sussex. *Watsonia*, **13**, 181–4.

Ivimey-Cook, R. B. & Proctor, M. C. F. (1966). The plant communities of the Burren, Co. Clare. *Proceedings of the Royal Irish Academy B*, **64**, 211–301.

Ivimey-Cook, R. B., Proctor, M. C. F. & Rowland, D. M. (1975). Analysis of the plant communities of a heathland site: Aylesbeare Common, Devon, England. *Vegetatio*, **31**, 33–45.

James, P. W., Hawksworth, D. L. & Rose, F. (1977). Lichen communities in the British Isles: a preliminary conspectus. In *Lichen Ecology*, ed. M. R. D. Seaward, pp. 295–413. London: Academic Press.

Jarolimek, I., Zaliberova, M., Mucina, L. & Mochnacky, S. (1997). *Rastlinné spolocenstvá Slovenska. 2. Synantropná vegetácia.* Bratislava: Slovenská akadémia vied Botanický ústav.

Jefferies, R. L., Davy, A. J. & Rudnik, T. (1979). The growth strategies of coastal halophytes. In *Ecological Processes in Coastal Environments*, ed. R. L. Jefferies & A. J. Davy, pp. 243–68. Oxford: Blackwell Scientific Publications.

Jermy, A. C. & Crabbe, J. A. (1978). *The Island of Mull.* London: British Museum (Natural History).

Jermy, A. C., Arnold, H. R., Farrell, L. & Perring, F. H. (1978). *Atlas of Ferns of the British Isles.* London: Botanical Society of the British Isles and British Pteridological Society.

Jermyn, S. T. (1974). *Flora of Essex.* Fingringhoe: Essex Naturalists Trust.

Jones, D. A. & Turkington, R. (1986). Biological Flora of the British Isles: *Lotus corniculatus* L. *Journal of Ecology*, **74**, 1185–1212.

Jones, P. S. (1992). *The relationship of dune slack plants to soil moisture and chemical conditions.* University of Wales: PhD thesis.

Jones, R. (1967). *The relationship of dune slack plants to soil moisture and chemical conditions.* University of Wales: PhD thesis.

Justice, O. L. (1941). A study of dormancy in seeds of *Polygonum. Memoirs of the Cornell University Agricultural Experimental Station*, **235**.

Kay, Q. O. N. (1971). Biological Flora of the British Isles. *Anthemis arvensis* L. *Journal of Ecology*, **59**, 637–48.

Kay, Q. O. N. (1994). Biological Flora of the British Isles: *Tripleurospermum maritimum* (L.) Schultz Bip. *Journal of Ecology*, **82**, 681-98.

Kidson, C. & Carr, A. P. (1960). Dune reclamation at Braunton Burrows, Devon. *Chartered Surveyor*, **93**, 298–303.

Kinzel, W. (1926). *Neue Tabellen zu Frost und Licht als Beeinflussende Kräfte bei der Samenkeimung.* Stuttgart: Eugen Ulmer.

Klinkhamer, P. G. L. & de Jong, T. J. (1993). Biological Flora of the British Isles: *Cirsium vulgare* (Savi) Ten. *Journal of Ecology*, **81**, 177–91.

Kortekaas, W. M., van der Maarel, E. & Beeftink, W. G. (1976). A numerical classification of European *Spartina* communities. *Vegetatio*, **33**, 51–60.

Kruseman, G. & Vlieger, J. (1939). Akkerassociaties in Nederland. *Nederlandsch Kruidkundig Archiv*, **49**, 327–98.

Lambe, E. (1971). *A phytosociological and ecological analysis of Irish weed communities.* University of Dublin: PhD thesis.

Lambert, J. M. & Davies, M. R. (1940). A sandy area in the Dovey estuary. *Journal of Ecology*, **28**, 453–64.

Lavin, J. C. & Wilmore, G. T. D. (1994). *West Yorkshire Plant Atlas.* Bradford: Bradford Metropolitan District Council.

Law, R. (1981). The dynamics of a colonising population of *Poa annua. Ecology*, **62**, 1267–77.

LeBrun, J., Noirfalise, A., Heinemann, P. & vanden Berghen, C. (1949). Les Associations vegetales de Belgique. *Centre de Recherches écologiques et phytosociologiques de Gembloux, Communication No. 8*, 105–207.

Lee, J. A. (1975). The conservation of British inland salt-marshes. *Biological Conservation*, **8**, 143–51.

Lee, J. A. (1977). The vegetation of British inland salt-marshes. *Journal of Ecology*, **65**, 673–98.

Lee, J. A., Harmer, R. & Ignaciuk, R. (1983). Nitrogen as a limiting factor in plant communities. In *Nitrogen as an Ecological Factor*, ed. J. A. Lee, S. McNeill & I. H. Rorison, pp. 95–112. Oxford: Blackwell Scientific Publications.

Lemaire, A. J. J. & Weeda, E. J. (1994). Over de indeling van het Nanocyperion flavescentis in Nederland. *Stratiotes*, **9**, 22–38.

Libbert, W. (1940). Die Pflanzengesellschaften der Halbinsel Darsz. *Feddes Repertorium Specierum Novarum Regni Vegetabilis*, **114**, 1–95.

Lousley, J. E. (1950). *Wild Flowers of Chalk and Limestone.* London: Collins.

Lousley, J. E. (1971). *Flora of the Isles of Scilly.* Newton Abbot: David & Charles.

Lousley, J. E. & Kent, D. H. (1981). *Docks and Knotweeds of the British Isles.* London: Botanical Society of the British Isles.

Lux, H. (1964). Die biologischer Grundlagen der Strandhaferpflanzung und der Silbergras Aussaat im Dünenbau. *Angewandte Pflanzensoziologie*, **20**, 5–53.

Lux, H. (1966). Zur Ökologie des Strandhafers (*Ammophila arenaria*) und besonderer Beruchsichtigung seiner Verwandung im Dünenbau. *Beitrage zue Landespflege*, **2**, 93–107.

Maire, R. (1953). *Flore de l'Afrique du Nord*, Volume II. Paris: Lechevalier.

Malloch, A. J. C. (1970). *Analytical studies of cliff-top vegetation in south-west England.* Cambridge University: PhD thesis.

Malloch, A. J. C. (1971). Vegetation of the maritime cliff-tops of the Lizard and Land's End peninsulas, West Cornwall. *New Phytologist*, **70**, 1155–97.

Malloch, A. J. C. (1972). Salt-spray deposition on the maritime cliffs of the Lizard peninsula. *Journal of Ecology*, **60**, 103–12.

Malloch, A. J. C. (1976). An annotated bibliography of the Burren. *Journal of Ecology*, **64**, 1093–1105.

Malloch, A. J. C. (1988). *VESPAN II.* Lancaster: University of Lancaster.

Malloch, A. J. C. & Okusanya, O. T. (1979). An experimental investigation into the ecology of some maritime cliff species. I. Field observations. *Journal of Ecology*, **67**, 283–92.

Marchant, C. J. (1967). Evolution in *Spartina* (Gramineae). I. The history and morphology of the genus in Britain. *Journal of the Linnean Society (Botany)*, **60**, 1–24.

Marchant, C. J. & Goodman, P. J. (1969*a*). *Spartina maritima* (Curtis) Farnald. In Biological Flora of the British Isles: *Spartina* Schreb., ed. P. J. Goodman *et al. Journal of Ecology*, **57**, 287–91.

Marchant, C. J. & Goodman, P. J. (1969*b*). *Spartina alterniflora* Loisel. In Biological Flora of the British Isles: *Spartina* Schreb., ed. P. J. Goodman *et al. Journal of Ecology*, **57**, 291–5.

Marsh, A. S. (1915). The maritime ecology of Holme next the Sea, Norfolk. *Journal of Ecology*, **3**, 65–93.

Marshall, J. K. (1965). *Corynephorus canescens* (L.) P. Beauv. as a model for the *Ammophila* problem. *Journal of Ecology*, **53**, 447–63.

Marshall, J. K. (1967). Biological Flora of the British Isles: *Corynephorus canescens* (L.) P. Beauv. *Journal of Ecology*, **55**, 207–20.

Martin, J. (1977). *Blysmus rufus* (Huds.) Link. Its distribution on the Caerlaverock National Nature Reserve, Dumfries and Galloway. *Transactions of the Dumfriesshire and Galloway Natural History and Antiquarian Society*, 3rd series, **50**, 23–27.

Matthews, J. R. (1955). *Origin and Distribution of the British Flora.* London: Hutchinson.

Matuszkiewicz, W. (1981). *Przewodnik do oznaczania zbiorowisk roslinnych Polski.* Warzawa: Panstwowe Wydawnictwo Naukowe.

McLean, R. C. (1935). An ungrazed grassland on limestone in Wales, with a note on plant 'dominions'. *Journal of Ecology*, **23**, 436–42.

McNaughton, I. H. & Harper, J. L. (1964). Biological Flora of the British Isles: *Papaver* L. *Journal of Ecology*, **52**, 764–93.

McVean, D. N. (1961*a*). Flora and vegetation of the islands of St. Kilda and North Rona in 1958. *Journal of Ecology*, **49**, 39–54.

McVean, D. N. & Ratcliffe, D. A. (1962). *Plant Communities of the Scottish Highlands.* London: HMSO.

Merton, L. F. H. (1970). The history and status of the woodlands of the Derbyshire Limestone. *Journal of Ecology*, **58**, 723–44.

Mitchell, N. D. (1976). The status of *Brassica oleracea* L. ssp. *oleracea* (Wild Cabbage) in the British Isles. *Watsonia*, **11**, 97–103.

Mitchell, N. D. & Richards, A. J. (1979). Biological Flora of the British Isles: *Brassica oleracea* L. ssp. *oleracea* (*B. sylvestris* (L.) Miller). *Journal of Ecology*, **67**, 1087–96.

Molinier, R. & Tallon, G. (1974). Documents pour un Inventaire des Plantes vasculaires de la Camargue. *Bulletin du Muséum d'Histoire Naturelle de Marseille*, **34**, 1–165.

Moor, M. (1958). Pflanzengesellschaften schweizerischer flussauen. *Mitt. Schweiz. Anst. Forstl. Versuchswesen*, **34**, 221–360.

Moore, J. J. (1962). The Braun-Blanquet System: a reassessment. *Journal of Ecology*, **50**, 761–9.

Moore, J. J., Fitsimons, P., Lambe, E. & White, J. (1970). A comparison and evaluation of some phytosociological techniques. *Vegetatio*, **20**, 1–20.

Moore, J. J. & O'Reilly, H. (1977). Salt-marsh, vegetation pattern and trends. In *North Bull Island – Dublin Bay. A modern coastal natural history*, ed. D. W. Jeffrey, pp. 83–7. Dublin: The Royal Dublin Society.

Moore, P. D. (1977). Stratigraphy and pollen analysis of Claish Moss, north-west Scotland: significance of the origin of surface pools and forest history. *Journal of Ecology*, **65**, 375–97.

Morley, J. V. (1973). Tidal immersion of *Spartina* marsh at Bridgwater Bay, Somerset. *Journal of Ecology*, **61**, 383–86.

Moss, E. H. (1936). The ecology of *Epilobium angustifolium* with particular reference to rings of periderm in the wood. *American Journal of Botany*, **23**, 114–20.

Mucina, L., Grabherr, G. & Ellmauer, T. (1993*a*). *Die Pflanzengesellschaften Österreichs. Teil I. Anthropogene Vegetation*. Stuttgart: Fisher.

Mucina, L., Rodwell, J. S., Schaminée, J. H. J. & Dierschke, H. (1993*b*). European Vegetation Survey: current state of some national programmes. *Journal of Vegetation Science*, **4**, 429–38.

Müllverstadt, R. (1963). Investigations on the germination of weed seeds as influenced by oxygen partial pressure. *Weed Research*, **3**, 154–63.

Myerscough, P. J. (1980). Biological Flora of the British Isles: *Epilobium angustifolium* L. *Journal of Ecology*, **68**, 1047–74.

Myerscough, P. J. & Whitehead, F. H. (1967). Comparative biology of *Tussilago farfara* L., *Chamaenerion angustifolium* (L.) Scop., *Epilobium montanum* L. and *Epilobium adenocaulon* Hausskn. II. Growth and ecology. *New Phytologist*, **66**, 758–823.

Naylor, R. E. (1972). Biological Flora of the British Isles: *Alopecurus myosuroides* Huds. *Journal of Ecology*, **60**, 611–22.

New, J. K. (1961). Biological Flora of the British Isles: *Spergula arvensis* L. *Journal of Ecology*, **49**, 205–15.

Newman, E. I. (1963). Factors controlling the germination date of winter annuals. *Journal of Ecology*, **51**, 625–38.

Newman, E. I. (1967). Response of *Aira praecox* to weather conditions. I. Response to drought in spring. *Journal of Ecology*, **55**, 539–56.

Nicholson, I. A. (1952). *A study of* Agropyron junceum *(Beauv.) in relation to the stabilization of coastal sand and the development of sand-dunes*. Durham University: MSc thesis.

Nicholson, P. (1985). The Californian tree-lupin *Lupinus arboreus* at Dawlish Warren. *Nature in Devon*, **6**, 25–9.

Ni Lamhna, E. (1982). The vegetation of salt-marshes and sand-dunes at Malahide Island, County Dublin. *Journal of Life Sciences of the Royal Dublin Society*, **3**, 111–29.

Noble, J. C. (1982). Biological Flora of the British Isles: *Carex arenaria* L. *Journal of Ecology*, **70**, 867–86.

Nordhagen, R. (1922). Vegetationsstudien auf der Insel Utsire im westlichen Norwegen. *Bergens Museums Årbok* 1920–21, 1–149.

Nordhagen, R. (1940). Studien über die maritime Vegetation Norwegens. I. Die Pflanzengesellschaften der Tangwälle. *Bergens Museum Årbok*, 1939–40, 1–123.

Nordhagen, R. (1943). *Sikilsdalen og Norges Fjellbeiter. Bergens Museums Skrifter 22*. Bergen: Griegs.

Oberdorfer, E. (1949). Die Pflanzengesellschaften den Wutachschlucht. *Beiträge naturkunstlichen Forschungen des Südwest-Deutschland*, **9**, 29–98.

Oberdorfer, E. (1957). Süddeutsche Pflanzengesellschaften. *Pflanzensoziologie*, **10**, 1–564.

Oberdorfer, E. (1964). Der insubrische vegetationscomplex, seine struktur und Abgrenzung gegen die sub-mediterrane Vegetation in Oberitalien und in der Südschweiz. *Beiträge naturkunstlichen Forschungen des Südwest-Deutschland*, **23**, 141–87.

Oberdorfer, E. (1977). *Süddetusche Pflanzengesellschaften*. Teil I. Stuttgart: Fischer.

Oberdorfer, E. (1978). *Süddeutsche Pflanzengesellschaften*. Teil II. Stuttgart: Fischer.

Oberdorfer, E. (1983). *Süddeutsche Pflanzengesellschaften*. Teil III. Stuttgart: Fischer.

Ogilvie, M. A. (1978). *Wild Geese*. Berkhamsted: T. & A. D. Poyser.

Okusanya, O. T. (1979*a*). An experimental investigation into the ecology of some maritime cliff species. II. Germination studies. *Journal of Ecology*, **67**, 293–304.

Okusanya, O. T. (1979*b*). An experimental investigation into the ecology of some maritime cliff species. III. Effect of sea water on growth. *Journal of Ecology*, **67**, 579–90.

Okusanya, O. T. (1979*c*). An experimental investigation into the ecology of some maritime cliff species. IV. Cold sensitivity and competition studies. *Journal of Ecology*, **67**, 591–600.

Oliver, F. W. (1907). The Bouche d'Erquy in 1907. *New Phytologist*, **6**, 244–52.

Oliver, F. W. (1911). The maritime formations of Blakeney Harbour. In *Types of British Vegetation*, ed. A. G. Tansley, pp. 354–66. Cambridge: Cambridge University Press.

Oliver, F. W. (1912). The shingle beach as a plant habitat. *New Phytologist*, **11**, 73–99.

Oliver, F. W. (1913). Some remarks on Blakeney Point. *Journal of Ecology*, **1**, 4–15.

Oliver, F. W. (1915). Blakeney Point in 1915. *Journal of Ecology*, **3**, 239–40.

Oliver, F. W. (1929). Blakeney Point Report. *Transactions of the Norfolk and Norwich Naturalists' Society*, **12**.

Oliver, F. W. & Salisbury, E. J. (1913*a*). Vegetation and mobile ground as illustrated by *Suaeda fruticosa* on shingle. *Journal of Ecology*, **1**, 249–72.

Oliver, F. W. & Salisbury, E. J. (1913b). The topography and vegetation of the National Trust Nature Reserve known as Blakeney Point, Norfolk. *Transactions of the Norfolk and Norwich Naturalists' Society*, **9**, 485–544.

Olsson-Seffer, P. (1909). Hydro-dynamic factors influencing plant life on sandy sea shores. *New Phytologist*, **8**, 37–49.

O'Reilly, H. & Pantin, G. (1957). Some observations on the salt-marsh formation in Co. Dublin. *Proceedings of the Royal Irish Academy B*, **58**, 89–128.

Ostenfeld, C. H. (1908). The land-vegetation of the Faeröes. In *Botany of the Faeröes*, **3**, 867–1026.

Owen, M. (1972). Movements and feeding ecology of white-fronted geese at the New Grounds, Slimbridge. *Journal of Applied Ecology*, **9**, 385–98.

Packham, J. R. & Liddle, M. J. (1970). The Cefni salt-marsh, Anglesey, and its recent development. *Field Studies*, **3**, 331–56.

Page, C. N. (1982). *The Ferns of Britain and Ireland*. Cambridge: Cambridge University Press.

Page, C. N. (1986). The strategies of Bracken as a permanent ecological opportunist. In *Bracken*, ed. R. T. Smith & J. A. Taylor, pp. 173–81. Carnforth: Parthenon.

Page, C. N. (1988). *Ferns*. London: Collins.

Passarge, H. (1964). Pflanzengesellschaften des nordostdeutschen Flachlandes. I. *Pflanzensoziologie (Jena)*, **13**, 1–324.

Pearsall, W. H. (1934). North Lancashire sand dunes. *Naturalist*, 1934, 201–5.

Pearson, M. C. & Rogers, J. A. (1962). Biological Flora of the British Isles: *Hippophae rhamnoides* L. *Journal of Ecology*, **50**, 501–13.

Pegtel, D. M. (1976). On the ecology of two varieties of *Sonchus arvensis* L. University of Groningen: thesis.

Pemadasa, M. A., Greig-Smith, P. & Lovell, P. H. (1974). A quantitative description of the distribution of annuals in the dune system at Aberffraw, Anglesey. *Journal of Ecology*, **62**, 379–402.

Pemadasa, M. A. & Lovell, P. H. (1974a). Factors affecting the distribution of some annuals in the dune system at Aberffraw, Anglesey. *Journal of Ecology*, **62**, 403–16.

Pemadasa, M. A. & Lovell, P. H. (1974b). The mineral nutrition of some dune annuals. *Journal of Ecology*, **62**, 647–57.

Pemadasa, M. A. & Lovell, P. H. (1975). Factors controlling germination of some dune annuals. *Journal of Ecology*, **63**, 41–60.

Pemadasa, M. A. & Lovell, P. H. (1976). Effects of the timing of the life-cycle on the vegetative growth of some dune annuals. *Journal of Ecology*, **64**, 213–22.

Perraton, C. (1953). Salt-marshes of the Hampshire–Sussex border. *Journal of Ecology*, **41**, 240–7.

Perring, F. H. & Farrell, L. (1977). *British Red Data Books*. 1: *Vascular Plants*. Nettleham: Society for the Promotion of Nature Conservation.

Perring, F. H. & Walters, S. M. (1962). *Atlas of the British Flora*. London: Nelson.

Petch, C. P. (1933). The vegetation of St. Kilda. *Journal of Ecology*, **21**, 92–100.

Petch, C. P. & Swann, E. L. (1968). *Flora of Norfolk*. Norwich: Jarrold & Sons.

Pethwick, J. S. (1974). The distribution of salt pans on tidal salt-marshes. *Journal of Biogeography*, **1**, 57–62.

Pigott, C. D. (1969). Influence of mineral nutrition on the zonation of flowering plants in coastal salt-marshes. In *Ecological Aspects of Mineral Nutrition of Plants*, ed. I. Rorison, pp. 25–35. Oxford: Blackwell Scientific Publications.

Pigott, C. D. (1977). The scientific basis of practical conservation: aims and methods of conservation. *Proceedings of the Royal Society of London B*, **197**, 59–68.

Pigott, C. D. (1982). The experimental study of vegetation. *New Phytologist*, **90**, 389–404.

Pigott, C. D. (1984). The flora and vegetation of Britain: ecology and conservation. *New Phytologist*, **98**, 119–28.

Piotrowski, H. (1974). Maritime communities of halophytes in Poland and the problems of their protection. *Ochrona przyrody*, **39**, 7–63.

Polderman, P. J. G. (1979). Salt-marsh algal communities in the Wadden area, with reference to their distribution and ecology in NW Europe. I. The distribution and ecology of the algal communities. *Journal of Biogeography*, **6**, 225–66.

Polderman, P. J. G. & Polderman-Hall, R. A. (1980). Algal communities in Scottish salt-marshes. *British Phycological Journal*, **15**, 59–71.

Poore, M. E. D. (1955a). The use of phytosociological methods in ecological investigations. I. The Braun-Blanquet System. *Journal of Ecology*, **43**, 226–44.

Poore, M. E. D. (1955b). The use of phytosociological methods in ecological investigations II. Practical issues involved in an attempt to apply the Braun-Blanquet system. *Journal of Ecology*, **43**, 245–69.

Poore, M. E. D. (1955c). The use of phytosociological methods in ecological investigations. III. Practical application. *Journal of Ecology*, **43**, 606–51.

Poore, M. E. D. & McVean, D. N. (1957). A new approach to Scottish mountain vegetation. *Journal of Ecology*, **45**, 401–39.

Poore, M. E. D. & Robertson, V. C. (1949). The vegetation of St. Kilda in 1948. *Journal of Ecology*, **37**, 82–97.

Pott, R. (1992). *Die Pflanzengesellschaften Deutschlands*. Stuttgart: Ulmer.

Praeger, R. L. (1911). Clare Island Survey. Part X. Phanerogamia and Pteridophyta. *Proceedings of the Royal Irish Academy B*, **31**, 1–112.

Praeger, R. L. (1934). *The Botanist in Ireland*. Dublin: Hodges, Figgis & Co.

Preston, C. D. (1989). The ephemeral pools of South Cambridgeshire. *Nature in Cambridgeshire*, **31**, 2–11.

Preston, C. D. & Whitehouse, H. L. K. (1986). The habitat of *Lythrum hyssopifolia* L. in Cambridgeshire, its only surviving English locality. *Biological Conservation*, **35**, 41–62.

Proctor, M. C. F. (1975). Notes on the vegetation of Alderney. *Phytocoenologia*, **2**, 301–11.

Proctor, M. C. F. (1980). Vegetation and Environment in the Exe Estuary. In *Essays on the Exe Estuary*, pp. 117–34. Exeter: Devonshire Association.

Raabe, E. W. (1953). Über der 'Affinitätswert' in der Pflanzensoziologie. *Vegetatio*, **4**, 53–68.

Radley, G. (1994). *Sand Dune Vegetation Survey of Great Britain, a National Inventory.* Part 1: *England.* Peterborough: Joint Nature Conservation Committee.

Rae, P. A. S. (1979). *The sea-grasses of the Moray Firth. Their ecology and responses to adjacent industrial development.* Aberdeen University: PhD thesis.

Randall, R. E. (1977). The past and present status and distribution of Sea Pea, *Lathyrus japonicus* Willd., in the British Isles. *Watsonia*, **11**, 247–51.

Ranwell, D. (1959). Newborough Warren, Anglesey. I. The dune system and dune-slack habitat. *Journal of Ecology*, **47**, 571–601.

Ranwell, D. S. (1960*a*). Newborough Warren, Anglesey. II. Plant associes and succession cycles of the sand-dune and dune-slack vegetation. *Journal of Ecology*, **48**, 117–41.

Ranwell, D. S. (1960*b*). Newborough Warren, Anglesey. III. Changes in the vegetation on parts of the dune system after the loss of rabbits by myxomatosis. *Journal of Ecology*, **48**, 385–95.

Ranwell, D. S. (1961). *Spartina* salt-marshes in southern England. I. The effects of sheep grazing at the upper limit of *Spartina* marsh in Bridgwater Bay. *Journal of Ecology*, **49**, 325–40.

Ranwell, D. S. (1964*a*). *Spartina* salt-marshes in southern England. II. Rate and seasonal pattern of sediment accretion. *Journal of Ecology*, **52**, 79–94.

Ranwell, D. S. (1964*b*). *Spartina* salt-marshes in southern England. III. Rates of establishment, succession and nutrient supply at Bridgwater Bay, Somerset. *Journal of Ecology*, **52**, 95–105.

Ranwell, D. S. (1967). World resources of *Spartina townsendii* (*sensu lato*) and economic use of *Spartina* marshland. *Journal of Applied Ecology*, **4**, 239–56.

Ranwell, D. S. (1968). Coastal marshes in perspective. *University of Strathclyde Regional Studies Group Bulletin*, **9**, 1–26.

Ranwell, D. S. (1972). *Ecology of Salt Marshes and Sand Dunes.* London: Chapman & Hall.

Ranwell, D. S. (1974). The salt marsh to tidal woodland transition. *Hydrobiological Bulletin* (*Amsterdam*), **8**, 139–51.

Ranwell, D. S., Bird, E. C. F., Hubbard, J. C. E. & Stebbings, R. E. (1964). *Spartina* salt-marshes in southern England. V. Tidal submergence and chlorinity in Poole Harbour. *Journal of Ecology*, **52**, 627–41.

Ranwell, D. S. & Downing, B. M. (1959). Brent goose winter feeding pattern and *Zostera* resources at Scolt Head Island, Norfolk. *Animal Behaviour*, **7**, 42–56.

Ratcliffe, D. (1961). Adaptation to habitat in a group of annual plants. *Journal of Ecology*, **49**, 187–209.

Ratcliffe, D. A. (1968). An ecological account of Atlantic bryophytes in the British Isles. *New Phytologist*, **67**, 365–439.

Ratcliffe, D. A. (ed.) (1977). *A Nature Conservation Review.* Cambridge: Cambridge University Press.

Reinikainen, A. (1964). Vegetationsuntersuchungen auf dem Walddungungsversuchfeld von Kivisuc in Mittel-Finnland. *Folia Forestalia*, **6**, 1–17.

Rich, T. C. G. (1991). *Crucifers of Great Britain and Ireland.* London: Botanical Society of the British Isles.

Rich, T. C. G. & Rich, M. D. B. (1988). *Plant Crib.* London: Botanical Society of the British Isles.

Roberts, E. H. & Benjamin, S. K. (1979). The interaction of light, nitrate and alternating temperature on the germination of *Chenopodium album*, *Capsella bursa-pastoris* and *Poa annua* before and after chilling. *Seed Science and Technology*, **7**, 379–92.

Roberts, H. A. & Feast, P. M. (1970). Seasonal distribution of emergence in some annual weeds. *Experimental Horticulture*, **21**, 36–41.

Roberts, H. A. & Stokes, F. G. (1965). Studies on the weeds of vegetable crops. V. Final observations on an experiment with different primary cultivations. *Journal of Applied Ecology*, **2**, 307–15.

Roberts, H. A. & Stokes, F. G. (1966). Studies of the weeds of vegetable crops. VI. Seed populations of soil under commercial cropping. *Journal of Applied Ecology*, **3**, 181–90.

Rodwell, J. S. (1991). *British Plant Communities*, Volume 1. *Woodlands and Scrub.* Cambridge: Cambridge University Press.

Rodwell, J. S. (1992). *British Plant Communities*, Volume 3. *Grasslands and Montane Communities.* Cambridge: Cambridge University Press.

Rodwell, J. S. (1994*a*). *British Plant Communities*, Volume 4. *Aquatic Communities, Swamps and Tall-herb Fens.* Cambridge: Cambridge University Press.

Rodwell, J. S. (1994*b*). A short rush through the British Nanocyperion. *Stratiotes*, **9**, 104–7.

Rodwell, J. S., Pignatti, S., Mucina, L. & Schaminée, J. H. J. (1995). European Vegetation Survey: update on progress. *Journal of Vegetation Science*, **6**, 759–62.

Rojanavipart, R. & Kay, Q. O. N. (1977). Salt-marsh ecology and trace-metal studies. In *Problems of a Small Estuary*, 2:2, ed. A. Nelson-Smith & E. M. Bridges, pp. 1–16. Swansea: Institute of Marine Studies.

Rose, F. (1964). Compte rendu des herborisations. II. Le sud-est de l'Angleterre (Comtés de Kent et Sussex). *Bulletin de la Société botanique de France 90ème session extraordinaire*, 30–37.

Rose, F. & Géhu, J.-M. (1964). Essai de phytogéographie comparée. La végétation du Sud-Est de l'Angleterre et ses analogies avec celle du Nord de la France. *Bulletin de la société botanique de France 90ème session extraordinaire*, 38–70.

Rozijn, N. A. M. G. & van der Werf, D. C. (1986). Effect of drought during different stages in the life-cycle on the growth and biomass allocation of two *Aira* species. *Journal of Ecology*, **74**, 507–24.

Sagar, G. R. & Harper, J. L. (1960). Factors affecting the germination and early establishment of plantains (*Plantago lanceolata*, *P. media* and *P. major*). In *The Biology of Weeds*, ed. J. L. Harper, pp. 236–45. Oxford: Blackwell Scientific Publications.

Salisbury, E. J. (1922). The soils of Blakeney Point: a study of soil reaction and succession in relation to plant covering. *Annals of Botany*, **36**, 391–431.

Salisbury, E. J. (1925). Note on the edaphic succession in sand dune soils with special reference to the time factor. *Journal of Ecology*, **13**, 322–8.

Salisbury, E. J. (1952). *Downs and Dunes.* London: Bell.

Salisbury, E. J. (1964). *Weeds and Aliens*, 2nd edition. London: Collins.

Salisbury, E. J. (1968). The reproductive biology and occasional seasonal dimorphism of *Anagallis minima* and *Lythrum hyssopifolia*. *Watsonia*, **7**, 25–39.

Schaminée, J. H. J., Weeda, E. J. & Westhoff, V. (1995). *De Vegetatie van Nederland, 2*. Uppsala: Opulus Press.

Schaminée, J. H. J., Weeda, E. J. & Westhoff, V. (1998). *De Vegetatie van Nederland, 4*. Uppsala: Opulus Press.

Schat, H. (1982). *On the ecology of some Dutch dune slack plants*. Free University of Amsterdam: Dissertation.

Schouten, M. G. C. & Nooren, M. J. (1977). Coastal vegetation types and soil features in south-east Ireland. *Acta Botanica Neerlandica*, **26**, 357–8.

Scott, G. A. M. (1963a). The ecology of shingle beach plants. *Journal of Ecology*, **51**, 517–27.

Scott, G. A. M. (1963b). Biological Flora of the British Isles: *Glaucium flavum* Crantz. *Journal of Ecology*, **51**, 743–54.

Scott, G. A. M. (1963c). Biological Flora of the British Isles: *Mertensia maritima* (L.) S. F. Gray. *Journal of Ecology*, **51**, 733–42.

Scott, G. A. M. & Randall, R. E. (1976). Biological Flora of the British Isles: *Crambe maritima* L. *Journal of Ecology*, **64**, 1077–92.

Segal, S. (1969). *Ecological Notes on Wall Vegetation*. The Hague: Junk.

Sheail, J. (1979). Documentary evidence of the changes in the use, management and appreciation of the grass-heaths of Breckland. *Journal of Biogeography*, **6**, 277–92.

Shellard, H. C. (1976). Wind. In *The Climate of the British Isles*, ed. T. J. Chandler & S. Gregory, pp. 39–73. London: Longman.

Shimwell, D. W. (1968a). *The phytosociology of calcareous grasslands in the British Isles*. University of Durham: PhD thesis.

Shimwell, D. W. (1968b). *The Vegetation of the Derbyshire Dales: A Report to the Nature Conservancy*. Attingham Park, Shrewsbury: Nature Conservancy Midland Region.

Shimwell, D. W. (1971). *The Description and Classification of Vegetation*. London: Sidgwick & Jackson.

Shimwell, D. W. (1976). *A Syntaxonomy of British Vegetation types*. Manchester: Manchester University unpublished report.

Siira, J. (1970). Studies in the ecology of the sea-shore meadows of the Bothnian Bay with special reference to the Liminka area. *Aquilo seria Botanica*, **9**, 1–109.

Silverside, A. J. (1977). *A phytosociological survey of British arable weeds and related communities*. Durham University: PhD thesis.

Sissingh, G. (1950). *Onkruid-associates in Nederland*. SIGMA communication **106**.

Sjögren, E. (1971). The influence of sheep grazing on limestone heath vegetation on the Baltic island of Öland. In *The Scientific Management of Animal and Plant Communities for Conservation*, ed. E. Duffey & A. S. Watt, pp. 487–95. Oxford: Blackwell.

Skogen, A. (1965). Flora og vegetasjon i Ørland herred, Sør-Trondelag. *Årbok 1965 for det Kongelige Norske videnskabers selskab Museet*, 13–124.

Skutch, A. F. (1930). Repeated fission of stem and root in *Mertensia maritima* – a study of ecological anatomy. *Annals of the New York Academy of Sciences*, **32**, 1–52.

Smith, L. P. (1976). *The Agricultural Climate of England and Wales*. Ministry of Agriculture, Fisheries and Food Technical Bulletin 35. London: HMSO.

Smith, U.K. (1979). Biological Flora of the British Isles: *Senecio integrifolius* (L.) Clairv. *Journal of Ecology*, **67**, 1109–24.

Sneddon, P. & Randall, R. (1992a). *Coastal Vegetated Shingle Structures of Great Britain: main report*. Peterborough: Joint Nature Conservation Committee.

Sneddon, P. & Randall, R. (1992b). *Coastal Vegetated Shingle Structures of Great Britain*. Appendix 1: *Wales*. Peterborough: Joint Nature Conservation Committee.

Sneddon, P. & Randall, R. (1993a). *Coastal Vegetated Shingle Structures of Great Britain*. Appendix 2: *Scotland*. Peterborough: Joint Nature Conservation Committee.

Sneddon, P. & Randall, R. (1993b). *Coastal Vegetated Shingle Structures of Great Britain*. Appendix 3: *England*. Peterborough: Joint Nature Conservation Committee.

Snowden, R. E. D. & Wheeler, B. D. (1993). Iron toxicity to fen plant species. *Journal of Ecology*, **81**, 35–46.

Sobey, D. G. (1976). The effect of herring gulls on the vegetation of the Isle of May. *Transactions of the Botanical Society of Edinburgh*, **42**, 469–85.

Sobey, D. G. (1981). Biological Flora of the British Isles. *Stellaria media* (L.) Vill. *Journal of Ecology*, **69**, 311–35.

Sobey, D. G. & Kenworthy, J. B. (1979). The relationship between herring gulls and the vegetation of their breeding colonies. *Journal of Ecology*, **67**, 469–96.

Soil Survey (1983). 1:250,000 Soil Map of England and Wales: six sheets and legend. Harpenden: Soil Survey of England and Wales.

Sparling, J. H. (1968). Biological Flora of the British Isles: *Schoenus nigricans* L. *Journal of Ecology*, **56**, 883–99.

Stace, C. (1995). *New Flora of the British Isles*. Cambridge: Cambridge University Press.

Steers, J. A. (1953). *The Sea Coast*. London: Collins.

Stewart, A., Pearman, D. A. & Preston, C. D. (1994). *Scarce Plants in Britain*. Peterborough: Joint Nature Conservation Committee.

Stewart, G. R., Lee, J. A. & Orebamjo, T. O. (1972). Nitrogen metabolism of halophytes. Nitrate reductase activity in *Suaeda maritima*. *New Phytologist*, **71**, 263–67.

Störmer, P. (1938). Vegetationsstudien auf der Insel Håøya im Oslofjord. *Skrifter Norske videnskaps-akademi Oslo. 1. Mat.-naturv. No 9*.

Stoutjesdijk, Ph. (1961). Micrometeorological measurements in vegetations of various structure. *Verhandelingen der Koninklijke Nederlandsche akademie van wetenschappen, Amsterdam, section C*, **64**, 171–207.

Summerhayes, V. S. (1968). *Wild Orchids of Britain*, 2nd edition. London: Collins.

Sykora, K. V. (1983). *The Lolio-Potentillion anserinae R. Tüxen 1947 in the northern part of the Atlantic domain*. Nijmegen Catholic University: PhD thesis.

Tansley, A. G., ed. (1911). *Types of British Vegetation*. Cambridge: Cambridge University Press.

Tansley, A. G. (1939). *The British Islands and their Vegetation*. Cambridge: Cambridge University Press.

Taschereau, P. M. (1985). Taxonomy of *Atriplex* species indigenous to the British Isles. *Watsonia*, **15**, 183–209.

Thompson, H. S. (1922). Changes in the coast vegetation near Berrow, Somerset. *Journal of Ecology*, **10**, 53–61.

Thompson, H. S. (1930). Further changes in the coast vegetation near Berrow, Somerset. *Journal of Ecology*, **18**, 126–30.

Thompson, K., Grime, J. P. & Mason, G. (1977). Seed germination in response to diurnal fluctuations of temperature. *Nature, London*, **267**, 147–9.

Tidmarsh, C. E. M. (1939). *The ecology of* Carex arenaria. Cambridge University: PhD thesis.

Timson, J. (1965). Germination in *Polygonum. New Phytologist*, **64**, 179–86.

Timson, J. (1966). Biological Flora of the British Isles: *Polygonum hydropiper* L. *Journal of Ecology*, **54**, 815–21.

Trail, J. W. H. (1904). The sea lyme grass (*Elymus arenarius* L.) in north-east Scotland. *Annals of Scottish Natural History*, **52**, 250–2.

Trist, P. J. O. (1979). *An Ecological Flora of Breckland.* Wakefield: EP Publishing Limited.

Turner, C. (1977). Study of *Agropyron junceiforme* and *Elymus arenarius* on the north Norfolk coast. Norwich: Institute of Terrestrial Ecology Sandwich-student Dissertation.

Tutin, T. G. (1942). Biological Flora of the British Isles: *Zostera* L., *Zostera marina* L. and *Zostera hornemanniana* Tutin. *Journal of Ecology*, **30**, 217–26.

Tutin, T. G. (1980). *Umbellifers of the British Isles.* London: Botanical Society of the British Isles.

Tutin, T. G., Heywood, V. H., Burges, N. A., Valentine, D. H., Walters, S. M. & Webb, D. A. (1964). *Flora Europaea.* Volume 1. Cambridge: Cambridge University Press.

Tutin, T. G., Heywood, V. H., Burges, N. A., Moore, D. M., Valentine, D. H., Walters, S. M. & Webb, D. A. (1968). *Flora Europaea*, Volume 2. Cambridge: Cambridge University Press.

Tutin, T. G., Heywood, V. H., Burges, N. A., Moore, D. M., Valentine, D. H., Walters, S. M. & Webb, D. A. (1972). *Flora Europaea*, Volume 3. Cambridge: Cambridge University Press.

Tutin, T. G., Heywood, V. H., Burges, N. A., Moore, D. M., Valentine, D. H., Walters, S. M. & Webb, D. A. (1976). *Flora Europaea*, Volume 4. Cambridge: Cambridge University Press.

Tutin, T. G., Heywood, V. H., Burges, N. A., Moore, D. M., Valentine, D. H., Walters, S. M. & Webb, D. A. (1980). *Flora Europaea*, Volume 5. Cambridge: Cambridge University Press.

Tutin, T. G. *et al.* (1980). *Dactylis* L. In *Flora Europaea*, Volume 5, ed. T. G. Tutin *et al.*, pp. 170–1. Cambridge: Cambridge University Press.

Tüxen, R. (1937). Die Pflanzengesellschaften Nordwestdeutschlands. *Mitteilungen der Floristisch-soziologischen Arbeitsgemeinschaft*, **3**, 1–170.

Tüxen, R. (1950). Grundriss einer Systematik der nitrophilen Unkrautgesellschaften in der Eurosiberischen Region Europas. *Mitteilungen der Florist-soziologischen Arbeitsgemeinschaft*, N.F., **2**, 94–175.

Tüxen, R. (1951) Eindruck während der pflanzengeo-graphischen Exkursion durch Süd-Schweden. *Vegetatio*, **3**, 149–72.

Tüxen, R. (1966). Über nitrophile *Elymus*-Gesellschaften an nordeuropäischen, nordjapanischen und nordamerikanischen Kusten. *Annales Botanici Fennici*, **3**, 358–66.

Tüxen, R. (1974). *Die Pflanzengesellshaften Nordwestdeutschlands.* 2. Cramer.

Tüxen, R. & Oberdorfer, E. (1958). Die Pflanzenwelt Spaniens. II Teil. Eurosibirische Phanerogamen-Gesellschaften Spaniens. *Veröffentlichungen des Geobotanischen Institutes Rübel in Zürich*, **32**, 1–328.

Tüxen, R. & Westhoff, V. (1963). Saginetea maritimae, eine Gesellschaftsgruppe in wechselhalinen Grenzbereich der europäischen Meeresküsten. *Mitteilungen der Floristisch-soziologischen Arbeitsgemeinschaft*, N.F., **10**, 116–29.

Tyler, G. (1969a). Studies in the ecology of Baltic sea-shore meadows. II. Flora and vegetation. *Opera Botanica* (*Lund*), **25**, 1–101.

Tyler, G. (1969b). Regional aspects of Baltic shore-meadow vegetation. *Vegetatio*, **19**, 60–86.

Tyler, G., Havas, P., Mikkelsen, V., Olsson, H., Sunding, P. & Vestergaard, P. (1971). Forslag til Riktlinger för en enhetlig Klassifikation av Havsträndernas Vegetation i Norden. *IBP I Norden*, **1**, 59–76.

Valachovic, M., Ot'ahel'ová, H., Stanová, V. & Maglocký, S. (1995). *Rastlinné spolocenstvá Slovenska.* 1. *Pionierska vegetácia.* Bratislava: Slovenská akadémia vied Botanický ústav.

van Andel, J. & Vera, F. (1977). Reproductive allocation in *Senecio sylvaticus* and *Chamaenerion angustifolium* in relation to mineral nutrition. *Journal of Ecology*, **65**, 747–58.

Vanden Berghen, C. (1965a). Notes sur la végétation du sud-ouest de la France. III. La végétation de quelques prés salés d'Oleron (Charente Maritime). *Bulletin du Jardin Botanique d'Etat, Bruxelles*, **35**, 363–9.

Van der Laan, D. (1979). Spatial and temporal changes in the vegetation of dune slacks in relation to the groundwater regime. *Vegetatio*, **39**, 43–51.

van der Maarel, E. & Westhoff, V. (1964). The vegetation of the dunes near Oostevoorne (The Netherlands). *Wentia*, **12**, 1–61.

van der Vegte, F. W. (1978). Population differentiation and germination ecology in *Stellaria media* (L.) Vill. *Oecologia*, **37**, 231–45.

Voderberg, K. (1955). Die Vegetation der neugeschaffenen Insel Bock. *Feddes Repertorium Speciarum Novarum Regni Vegetabilis*, **135**, 232–60.

Vose, P. B., Powell, H. G. & Spence, J. B. (1957). The machair grazings of Tiree, Inner Hebrides. *Transactions of the Botanical Society of Edinburgh*, **37**, 89–110.

Walters, M. G. (1972). The fairy shrimp at Fowlmere. *Nature in Cambridgeshire*, **15**, 35–9.

Walters, M. G. (1978). The fairy shrimp at Fowlmere. *Nature in Cambridgeshire*, **21**, 18–23.

Wasscher, J. (1941). De graanonkruidassociates in Groningen en Noord-Drente. *Ned. Kruidk. Arch.*, **51**, 435–41.

Watkinson, A. R. & Harper, J. L. (1978). The demography of a sand dune annual: *Vulpia fasciculata*. I. The natural regulation of populations. *Journal of Ecology*, **66**, 15–33.

Watkinson, A. R. (1978a). The demography of a sand dune annual: *Vulpia fasciculata*. II. The dynamics of seed populations. *Journal of Ecology*, **66**, 35–44.

Watkinson, A. R. (1978b). The demography of a sand dune annual: *Vulpia fasciculata*. III. The dispersal of seeds. *Journal of Ecology*, **66**, 483–98.

Watkinson, A. R. (1978c). Biological Flora of the British Isles: *Vulpia fasciculata* (Forskål) Samp. *Journal of Ecology*, **66**, 1033–49.

Watson, W. (1918). Cryptogamic vegetation of the sand dunes of the west coast of England. *Journal of Ecology*, **6**, 126–43.

Watson, W. (1922). List of lichens, etc. from Chesil Beach. *Journal of Ecology*, **10**, 255–6.

Watt, A. S. (1936). Studies in the ecology of Breckland. I. Climate, soils and vegetation. *Journal of Ecology*, **24**, 117–38.

Watt, A. S. (1937). Studies in the ecology of Breckland. II. On the origin and development of blow-outs. *Journal of Ecology*, **25**, 91–112.

Watt, A. S. (1938). Studies in the ecology of Breckland. III. Development of the *Festuco-Agrostidetum*. *Journal of Ecology*, **26**, 1–37.

Watt, A. S. (1940). Studies in the ecology of Breckland. IV. The grass heath. *Journal of Ecology*, **28**, 42–70.

Watt, A. S. (1957). The effect of excluding rabbits from Grassland B (Mesobrometum) in Breckland. *Journal of Ecology*, **45**, 861–78.

Watt, A. S. (1971b). Rare species in Breckland: their management for survival. *Journal of Applied Ecology*, **8**, 593–609.

Watt, A. S. (1981a). A comparison of grazed and ungrazed grassland A in East Anglian Breckland. *Journal of Ecology*, **69**, 499–508.

Webb, N. (1986). *Heathlands*. London: Collins.

Weeda, E. J. (1994). Plantensociologie 'avant la lettre'. *Stratiotes*, **9**, 3–21.

Wellbank, P. J. (1963). A comparison of competitive effect of some common weed species. *Annals of Applied Biology*, **51**, 107–25.

Westhoff, V. (1947). *The Vegetation of Dunes and Salt Marshes on the Dutch Islands of Terschelling, Vlieland and Texel.* 's-Gravenhage: C. J. van der Horst.

Westhoff, V. & Beeftink, W. G. (1950). De vegetatie van duinen, slikken en schorren op de Kaloot en in het Nordsloe. *De Levende Natuur*, **53**, 124–33 & 225–33.

Westhoff, V. & den Held, A. J. (1969). *Plantengemeenschappen in Nederland*. Zutphen: Thieme.

Westhoff, V. & Segal, S. (1961). Cursus Vegetatiekunde 12–17 juni 1961 op Terschelling. Amsterdam: Hugo de Vries Laboratorium.

Wheeler, B. D. (1975). *Phytosociological studies on rich fen systems in England and Wales*. University of Durham: PhD thesis.

Wheeler, B. D. (1980a). Plant communities of rich-fen systems in England and Wales. I. Introduction. Tall sedge and reed communities. *Journal of Ecology*, **68**, 368–95.

Wheeler, B. D. (1980b). Plant communities of rich-fen systems in England and Wales. II. Communities of calcareous mires. *Journal of Ecology*, **68**, 405–20.

Wheeler, B. D., Al-Farraj, M. M. & Cook, R. E. D. (1985). Iron toxicity to plants in base-rich wetlands: comparative effects on the distribution and growth of *Epilobium hirsutum* L. and *Juncus subnodulosus* Schrank. *New Phytologist*, **100**, 653–69.

White, D. J. B. (1961). Some observations on the vegetation of Blakeney Point, Norfolk, following the disappearance of rabbits in 1954. *Journal of Ecology*, **49**, 113–18.

White, J. & Doyle, G. (1982). The vegetation of Ireland: a catalogue raisonné. *Royal Dublin Society Journal of Life Sciences*, **3**, 289–368.

Wiehe, P. O. (1935). A quantitative study of the influence of tide upon populations of *Salicornia europaea*. *Journal of Ecology*, **23**, 323–32.

Williams, J. T. (1963). Biological Flora of the British Isles: *Chenopodium album* L. *Journal of Ecology*, **51**, 711–25.

Williams, J. T. (1969). Biological Flora of the British Isles: *Chenopodium rubrum* L. *Journal of Ecology*, **57**, 831–41.

Willis, A. J. (1963). Braunton Burrows: the effects on the vegetation of the addition of mineral nutrients to the dune soils. *Journal of Ecology*, **51**, 353–74.

Willis, A. J. (1965). The influence of nutrients on the growth of *Ammophila arenaria*. *Journal of Ecology*, **53**, 735–45.

Willis, A. J. (1967). The genus *Vulpia* in Britain. *Proceedings of the Botanical Society of the British Isles*, **6**, 386–8.

Willis, A. J. (1967). The vegetation of Catcott Heath, Somerset. *Proceedings of the Bristol Naturalists Society*, **31**, 297–304.

Willis, A. J. (1985a). Dune water and nutrient regimes – their ecological relevance. In *Sand dunes and their management*, ed. P. Doody, pp. 159–74. Peterborough: Nature Conservancy Council.

Willis, A. J. (1985b). Plant diversity and change in a species-rich dune system. *Transactions of the Botanical Society of the British Isles*, **44**, 291–308.

Willis, A. J., Folkes, B. F. & Yemm, E. W. (1959a). Braunton Burrows: the dune system and its vegetation. I. *Journal of Ecology*, **47**, 1–24.

Willis, A. J., Folkes, B. F. & Yemm, E. W. (1959b). Braunton Burrows: the dune system and its vegetation. II. *Journal of Ecology*, **27**, 249–88.

Willis, A. J. & Jefferies, R. L. (1963). Investigations on the water-relations of sand-dune plants under natural conditions. In *The Water Relations of Plants*, ed. A. J. Rutter & F. W. Whitehead, pp. 168–89. Oxford: Blackwell.

Willis, A. J. & Yemm, E. W. (1961). Braunton Burrows: mineral nutrient status of the dune soils. *Journal of Ecology*, **49**, 377–90.

Wilson, K. (1960). The time factor in the development of dune soils at South Haven Peninsula, Dorset. *Journal of Ecology*, **48**, 341–59.

Wilson, B. J. & Wright, K. J. (1990). Predicting the growth and competitive ability of annual weeds in wheat. *Weed Research*, **30**, 201–12.

Wood, K. R. (1969). The distribution of the two sub-species of *Asplenium trichomanes* L. on the Isle of Skye. Unpublished manuscript.

Wyer, D. W. & Waters, R. J. (1975). Tidal flat *Zostera* and algal vegetation. In *Report of the Maplin Ecological Research Programme*, Part II, 1b, pp. 26–73. London: Department of Environment (unpublished).

Yapp, R. H. & Johns, D. (1917). The salt-marshes of the Dovey Estuary. Part II. The salt-marshes. *Journal of Ecology*, **5**, 65–103.

PHYTOSOCIOLOGICAL CONSPECTUS OF BRITISH PLANT COMMUNITIES

The aim of the National Vegetation Classification was to provide a systematic and comprehensive description of the vegetation types of all natural, semi-natural and major artificial habitats in England, Wales and Scotland. The contract brief from the Nature Conservancy Council stated that the plant communities distinguished were to be roughly equivalent to the associations characterised elsewhere in Europe using the Braun-Blanquet phytosociological approach. However, erecting a hierarchical classification of the communities was not a priority for the project and this and the other volumes of *British Plant Communities* organise the vegetation types under major floristic-cum-habitat headings rather than within higher phytosociological units.

Neither have the communities been characterised by the kind of latinised names proposed by the *International Code of Phytosociological Nomenclature* (Barkman *et al*. 1986) but using a combination of constant and dominant species. However, where there is an obvious phytosociological synonym for a vegetation type, this has been given in the text of *British Plant Communities* and, in every community description, a section on affinities discusses the floristic relationships of the vegetation type at alliance level.

With the completion of the project, it is now possible to provide a complete list of all the NVC communities arranged within the hierarchical framework provided by the European phytosociological tradition. This hierarchy comprises, at increasingly fine levels of division, classes of vegetation types (shown below in bold capitals, for example **LEMNETEA**), orders (in capitals, as in LEMNETALIA MINORIS) and alliances (in bold, like **Lemnion minoris**). After each of these major syntaxa listed there is given the citation of the author and date of its first formal recognition.

The arrangement of the classes follows a more or less accepted order but the detail of this hierarchy, particularly at alliance level, is open to revision by the phytosociological community and is periodically changed as new data and insights accumulate. The scheme adopted here

follows the most recent publications and unpublished reports by leading Continental practitioners and those familiar with the discussions on affinities in *British Plant Communities* will note that some of the allocations of vegetation types to alliances suggested these are different from those in this *Conspectus*. These changes are relatively few and have been made in the light of the overview that can now be gained of all the communities characterised in the NVC in the light of their wider European context.

Even a relatively brief scrutiny of the *Conspectus* should reveal that this kind of phytosociolocal exercise has much more than theoretical value. For one thing, it is now possible to make a much more informed comparison between British plant communities and related vegetation types from elsewhere in Europe, through their respective positions in the hierarchy. Many other European countries have conspectuses of this type and the hierarchical framework provides a powerful basis for relating plant associations one to another. More or less direct equivalence at the community level can then be indicated by using already published association names: approximately 27% of the plant communities characterised in the NVC have such equivalents somewhere else in Europe. Ongoing work by the participants of the European Vegetation Survey (Mucina *et al*. 1993, Rodwell *et al*. 1995) will produce an overview of alliances, orders and classes within which such comparisons can be made in a more coherent fashion.

The second benefit of the phytosociological perspective is ecological because the classes, orders and alliances are each associated with broadly-defined habitat conditions. In the *Conspectus*, such characteristics are described very briefly at the various levels. Where there is a single order in a class, or a single alliance in an order, the habitat features are the same as the next higher unit. In this respect, the alliances often prove to be the most valuable broader units in which to envisage or discuss plant communities. Frequently, they comprise vegetation types which already appear sequentially in *British*

Plant Communities, as with the small-sedge poor fens of the Caricion nigrae alliance which includes NVC communities M5–8.

In other cases, the alliances bring together vegetation types which appear in different volumes of *British Plant Communities* but which are obviously floristically related and found in similar habitats, as with the small-sedge rich fens of the Caricion davaillianae which comprises M9, M10 and M13 from among the mires and SD13–15 from the dune-slack communities. Even more informative, for example, is the way in which the phytosociological hierarchy brings together salt-marsh (SM28), mesotrophic grassland (MG11–13), sand-dune (SD17) and other vegetation (OV28–29), all characteristic of periodically-flooded fresh or brackish habitats, into the Elymo-Rumicion alliance.

COASTAL MUDFLATS AND BRACKISH WATERS

ZOSTERETA MARINAE Pignatti 1953
Eel-grass swards on muddy and sandy substrates in the sublittoral and eulittoral zones, exposed no more than 2–3 hours at a time

ZOSTERETALIA MARINAE Beguinot 1941 em. R.Tx. et Oberdorfer 1958
Eel-grass swards of shallower waters

Zosterion marinae Christiansen 1934
    SM1   *Zostera* communities

RUPPIETEA MARITIMAE J. Tüxen 1960
Tassel-weed and spike-rush communities of brackish to saline waters in estuaries, salt-marsh pools and dykes of reclaimed coastal marshes

RUPPIETALIA MARITIMAE J. Tüxen 1960

Ruppion maritimae Br.-Bl. 1931
    SM2   *Ruppia maritima* salt-marsh community
          Ruppietum maritimae Iversen 1934
    SM3   *Eleocharis parvula* salt-marsh community
          Eleocharitetum parvulae (Preuss 1911/12) Gillner 1960

SALT-MARSH AND SEA-CLIFF VEGETATION

SPARTINETEA MARITIMAE R.Tx. in Beeftink 1962
Pioneer vegetation of perennial cord-grasses on intertidal mud and sand

SPARTINETALIA MARITIMAE Conrad 1935

Spartinion maritimae Conrad 1952
    SM4   *Spartina martima* salt-marsh community
          Spartinetum maritimae (Emb. et Regn. 1926) Corillion 1953
    SM5   *Spartina alterniflora* salt-marsh community
          Spartinetum alterniflorae Corillion 1953
    SM6   *Spartina anglica* salt-marsh community
          Spartinetum anglicae Corillion 1953 corr. Géhu et Géhu-Franck 1984

THERO-SALICORNIETEA (Pignatti 1953) R.Tx. in R.Tx. et Oberdorfer 1958
Pioneer communities of annual glassworts, seablite or other halo-nitrophiles on tidal mud-flats

THERO-SALICORNIETALIA Pignatti 1953 em. R.Tx. 1954 ex R.Tx et Oberdorfer 1958
Pioneer communities of annual glassworts and seablite on tidal mud-flats

Thero-Salicornion strictae Br.-Bl. 1933 em. R.Tx. 1950 in Tx. et Oberdorfer 1958
    SM7   *Arthrocnemum perenne* salt-marsh community
    SM8   Annual *Salicornia* salt-marsh community
          Salicornietum europaeae Warming 1906
    SM9   *Suaeda maritima* salt-marsh community
          Suaedetum maritimae (Conrad 1935) Pignatti 1953

JUNCETEA MARITIMI R.Tx. et Oberdorfer 1958
Usually closed swards on the silt and sand of coastal and inland salt-marshes and on sea cliffs

GLAUCO-PUCCINELLIETALIA Beeftink et Westhoff 1962

Puccinellion maritimae Christiansen 1927 em. Tx. 1937
Communities of the lower parts of salt-marshes, generally inundated by spring tides
    SM10 Transitional low-marsh vegetation
    SM11 *Aster tripolium* var. *discoideus* salt-marsh community
    SM12 Rayed *Aster tripolium* stands
    SM13 *Puccinellia maritima* salt-marsh community
          Puccinellietum maritimae (Warming 1906) Christiansen 1927

Puccinellio maritimae-Spergularion salinae Beeftink 1965
Ephemeral communities in saline habitats, coastal and inland, with disturbance or fluctuating moisture regime
    SM23 *Spergularia marina-Puccinellia distans* salt-marsh community
          Puccinellietum distantis Feekes (1934) 1945

Armerion maritimae Br.-Bl. et De Leeuw 1936
Perennial communities of the upper parts of salt-marshes, rarely inundated by spring tides
- SM14 *Halimione portulacoides* salt-marsh community
  Halimionetum portulacoidis (Kuhnholtz-Lordat 1927) Des Abbayes et Corillion 1949
- SM16 *Festuca rubra* salt-marsh community
  Juncetum gerardii Warming 1906
- SM17 *Artemisia maritima* salt-marsh community
  Artemisietum maritimae Hocquette 1927
- SM18 *Juncus maritimus* salt-marsh community
- SM19 *Blysmus rufus* salt-marsh community
  Blysmetum rufi (G. E. et G. Du Rietz 1925) Gillner 1960
- SM20 *Eleocharis uniglumis* salt-marsh community
  Eleocharitetum uniglumis Nordhagen 1923
- SM21 *Suaeda vera-Limonium binervosum* salt-marsh community
- SM22 *Halimione portulacoides-Frankenia laevis* salt-marsh community
  Limonio vulgaris-Frankenietum laevis Géhu et Géhu-Franck 1975
- SM25 *Suaeda vera* salt-marsh community
  Elmyo pycnanthi-Suaedetum verae (Arènes 1933) Géhu 1975
- SM26 *Inula crithmoides* stands

Halo-Scirpion (Dahl et Hadač 1971) Den Held et Westhoff 1969 nom.nov.
Vegetation of flushed depressions in upper salt-marsh
- SM15 *Juncus maritimus-Triglochin maritima* salt-marsh community

Silenion maritimae Malloch 1971
Closed swards of perennials on sea-cliff tops and ledges little splashed by salt-spray
- MC2 *Armeria maritima-Ligusticum scoticum* maritime crevice community
- MC3 *Rhodiola rosea-Armeria maritima* maritime cliff-ledge community
- MC8 *Festuca rubra-Armeria maritima* maritime grassland
- MC9 *Festuca rubra-Holcus lanatus* maritime grassland
- MC10 *Festuca rubra-Plantago* spp. maritime grassland
- MC11 *Festuca rubra-Daucus carota* ssp. *gummifer* maritime grassland
- MC12 *Festuca rubra-Hyacinthoides non-scripta* maritime cliff community

SAGINETEA MARITIMAE Westhoff, van Leeuwen et Adriani 1962
Ephemeral vegetation with winter annuals on bare or disturbed salt-marsh muds and sand, periodically wettened by saline waters

SAGINETALIA MARITIMAE Westhoff, van Leeuwen et Adriani 1962
Atlantic and Mediterranean ephemeral vegetation in saline habitats

Sagionion maritimae Westhoff, van Leeuwen et Adriani 1962
- SM27 Ephemeral salt-marsh vegetation with *Sagina maritima*

CRITHMO-LIMONIETEA Br.-Bl. in Br.-Bl. et al. 1952
Open communities of crevices on rocky sea-cliffs much splashed by salt spray

CRITHMO-ARMERIETALIA MARITIMAE Géhu 1964

Crithmo-Armerion maritimae Géhu 1968
- MC1 *Crithmum maritimum-Spergularia rupicola* maritime crevice community
  Crithmo-Spergularietum rupicolae Géhu 1964
- MC4 *Brassica oleracea* maritime cliff-ledge community

STRANDLINE AND SAND-DUNE COMMUNITIES

CAKILETEA MARITIMAE R.Tx. et Preising ex Br.-Bl. & Tx. 1952
Pioneer vegetation, mostly of nitrophilous summer annuals, on nutrient-rich detritus of strandlines on sand and shingle beaches

CAKILETALIA MARITIMAE R.Tx. apud Oberdorfer (1949) 1950
Atlantic and Baltic annual halo-nitrophilous communities

Salsolo-Honkenyion peploidis R.Tx. 1950
Communities of strandlines with sand-covered detritus
- SD2 *Honkenya peploides-Cakile maritima* strandline community
- SD3 *Matricaria maritima-Galium aparine* strandline community

Atriplicion littoralis (Nordhagen 1940) Tx. 1950
Communities of strandlines sometimes mixed with but not covered by sand

HONCKENYO-ELYMETEA R.Tx. 1966
Vegetation of coastal shingle, boulders or rocky cliffs
enriched with organic detritus

ELYMETALIA ARENARII Br.-Bl. & R.Tx. 1943

Elymion pycnanthi
Communities of salt-marsh strandlines in warmer parts
of Europe
  SM24 *Elymus pycnanthus* salt-marsh community
        Atriplici-Elymetum pycnanthi Beeftink et
        Westhoff 1962

Honkenyo latifoliae-Crambion maritimae (Géhu 1968)
J.-M. et J. Géhu 1969
Communities of enriched coastal habitats, mostly boreal
  SD1   *Rumex crispus-Glaucium flavum* shingle
        community
  MC6   *Atriplex prostrata-Beta vulgaris* ssp. *maritima*
        sea-bird cliff community
        Atriplici-Betetum maritimae J.-M. et J. Géhu
        1969
  MC7   *Stellaria media* sea-bird cliff community

AMMOPHILETEA ARENARIAE Br.-Bl. et R.Tx. ex
Westhoff et al. 1946
Vegetation dominated by rhizomatous grasses or sedges
on mobile or fixed coastal or inland dunes

AMMOPHILETALIA ARENARIAE Br.-Bl. 1933

Elymo-Honkenyion peploidis R.Tx. apud Br.-Bl. et
R.Tx. 1952
Pioneer vegetation of coastal foredunes
  SD4   *Elymus farctus* ssp. *boreali-atlanticus*
        foredune community

Ammophilion arenariae Br.-Bl. 1933 em. R.Tx. 1955
Vegetation of young to fixed dunes around the Atlantic
coast of Europe
  SD5   *Leymus arenarius* mobile dune community
  SD6   *Ammophila arenaria* mobile dune community
  SD7   *Ammophila arenaria-Festuca rubra* semi-fixed
        dune community
  SD9   *Ammophila arenaria-Arrhenatherum elatius*
        dune grassland
  SD10 *Carex arenaria* dune community

FRESHWATER AQUATIC VEGETATION

LEMNETEA de Bolos et Masclans 1955
Free-floating duckweed communities of still, relatively
nutrient-rich, fresh waters in more winter-warm parts of
Europe

LEMNETALIA MINORIS Tüxen 1955

Lemnion minoris Tüxen 1955
Duckweed communities of eutrophic and hypertrophic
waters
  A2   *Lemna minor* community
        Lemnetum minoris Soó 1947

Lemnion gibbae R.Tx. et Schwabe 1972
Duckweed communities of more base-rich waters
  A1   *Lemna gibba* community
        Lemnetum gibbae Miyawaki et J. Tüxen 1960
  A3   *Spirodela polyrrhiza-Hydrocharis morsus-
        ranae* community

Lemnion trisulcae Den Hartog et Segal 1964 em. Tüxen
et Schwabe in Tüxen 1974
Duckweed and liverwort communities of shallow, more
mesotrophic waters

CHARETEA FRAGILIS Fukarek ex Krausch 1964
Submerged stonewort swards

NITELLETALIA FLEXILIS Krause 1969

Nitellion flexilis Dambska 1966 em. Krause 1969

Nitellion syncapae-tenuissimae Krause 1969

CHARETALIA HISPIDAE Sauer 1937

Charion fragilis Krause 1964 em. van Daam et
Schaminée in Schaminée et al. 1995

POTAMETEA Klika in Klika et Novák 1941
Communities of rooted, floating or submerged plants in
mesotrophic and eutrophic fresh waters

NUPHARO-POTAMETALIA Schaminée, Lanjouw et
Schipper 1990

Parvopotamion (Vollmar 1947) den Hartog et Segal
1964
Rooted aquatic communities in moderate to deep
standing waters, often open to wave action
  A5   *Ceratophyllum demersum* community
        Ceratophylletum demersi Hild 1956
  A11   *Potamogeton pectinatus-Myriophyllum
        spicatum* community
  A12   *Potamogeton pectinatus* community
  A13   *Potamogeton perfoliatus-Myriophyllum
        alterniflorum* community
  A15   *Elodea canadensis* community

Nymphaeion Oberdorfer 1957
Communities of rooted aquatics with floating leaves in
sheltered and nutrient-rich fresh waters
  A7   *Nymphaea alba* community
        Nymphaeetum albae Oberdorfer et Mitarb.
        1967

A8   *Nuphar lutea* community
A9   *Potamogeton natans* community
A10   *Polygonum amphibium* community
A19   *Ranunculus aquatilis* community
        Ranunculetum aquatilis Géhu 1961

Hydrocharition morsus-ranae Rübel 1933 em. Westhoff
et den Held 1969
Communities of free-floating macrophytes in fairly
nutrient-rich waters
A4   *Hydrocharis morsus-ranae-Stratiotes aloides*
        community

CALLITRICHO-POTAMETALIA Schipper,
Lanjouw et Schaminée 1995
Crosswort, crowfoot and milfoil vegetation of moving
waters and water margins

Callitricho-Batrachion Den Hartog et Segal 1964
Crosswort vegetation of shallow waters and muddy
margins of streams, ditches and pools
A16   *Callitriche stagnalis* community
A20   *Ranunculus peltatus* community
        Ranunculetum peltati Sauer 1947

Ranunculion fluitantis Neuhäusl 1959
Crowfoot and milfoil vegetation of moving waters
A14   *Myriophyllum alterniflorum* community
        Myriophylletum alterniflori Lemée 1937
A17   *Ranunculus pencillatus* ssp. *pseudofluitans*
        community
A18   *Ranunculus fluitans* community
        Ranunculetum fluitantis Allorge 1922

ZANNICHELLIETEALIA PEDICILLATAE
Schaminée, Lanjouw et Schipper 1990
Communities of rooted aquatics in brackish waters

Zannichellion pedicellatae Schaminée, Lanjouw et
Schipper 1990
A6   *Ceratophyllum submersum* community
        Ceratophylletum submersi Den Hartog et
        Segal 1964
A21   *Ranunculus baudotii* community
        Ranunculetum baudotii Br.-Bl. 1952

SPRINGS, SHORELINES, SWAMPS AND TALL-HERB FENS

MONTIO-CARDAMINETEA Br.-Bl. et Tüxen ex
Klika 1948
Vegetation of cold springs, commonly dominated by
bryophytes

MONTIO-CARDAMINETALIA Pawlowski in
Pawlowski, Sokotowski et Wallisch 1928

Cardamino-Montion Br.-Bl. 1926 em. Zechmeister 1993
Spring vegetation of base-poor waters
M32   *Philonotis fontana-Saxifraga stellaris* spring
        Philonoto-Saxifragetum stellaris Nordhagen
        1943
M33   *Pohlia wahlenbergii* var. *glacialis* spring
        Pohlietum glacialis McVean & Ratcliffe 1962
M34   *Carex demissa-Koenigia islandica* flush
M35   *Ranunculus omiophyllus-Montia fontana* rill
M31   *Anthelia julacea-Sphagnum auriculatum*
        spring
        Sphagno-Anthelietum julaceae Shimwell 1972
M36   Lowland springs and streambanks of shaded
        situations

Cratoneurion commutati Koch 1928
Spring vegetation of calcareous waters
M37   *Cratoneuron commutatum-Festuca rubra*
        spring
M38   *Cratoneuron commutatum-Carex nigra* spring

ISOETO-LITTORELLETEA Br.-Bl. et Vlieger in
Vlieger 1937
Hairgrass swards and related communities in nutrient-
poor, standing or slow-flowing, sometimes fluctuating
waters with sandy, gravelly or peaty substrates

LITTORELLETALIA Koch ex Tüxen 1937
Hairgrass swards and related communities in waters
with mineral substrates

Littorellion uniflorae Koch 1926 ex Tüxen 1937
Water lobelia and quillwort swards in deep and cold,
nutrient-poor standing waters with sandy or stony
substrates
A22   *Littorella uniflora-Lobelia dortmanna*
        community
A23   *Isoetes lacustris/setacea* community

Eleocharition acicularis Pietsch 1966 em. Dierssen 1975
Vegetation of fluctuating waters with loamy soils in
boreal and continental parts of Europe

Hydrocotylo-Baldellion R.Tx. et Dierssen in Dierssen
1972
Vegetation of soakways and shallow, strongly
fluctuating, mesotrophic to oligotrophic standing
waters
M29   *Hypericum elodes-Potamogeton
        polygonifolius* soakway
        Hyperico-Potametum polygonifolii (Allorge
        1921) Br.-Bl. & R.Tx. 1952
M30   Related vegetation of seasonally-inundated
        habitats

UTRICULARIETALIA INTERMEDIO-MINORIS Pietsch 1965
Bladderwort and bog-moss communities of dystrophic or lime-rich peaty waters

Sphagno-Utricularion Th. Müller et Görs 1960
Bladderwort and bog-moss communities of dystrophic peaty waters
  A24 *Juncus bulbosus* community

ISOETO-NANOJUNCETEA Br.-Bl. et Tüxen ex Westhoff et al. 1946
Pioneer, ephemeral, dwarf cyperaceous and therophyte communities on damp, bare, periodically-flooded ground

NANOCYPERETALIA Klika 1935

Nanocyperion flavescentis Koch ex Malcuit 1929
  OV31 *Rorippa palustris-Filaginella uliginosa* community
  OV34 *Allium schoenoprasum-Plantago maritima* community
  OV35 *Lythrum portula-Ranunculus flammula* community
  OV36 *Lythrum hyssopifolia-Juncus bufonius* community

PHRAGMITO-MAGNOCARICETEA Klika in Klika et Novák 1941
Swamp, fen and marginal communities of fresh or brackish waters dominated by graminoids, sedges and forbs

PHRAGMITETALIA Koch 1926
Swamp and fen dominated by graminoids, sedges and forbs, often species-poor

Phragmition australis Koch 1926
Swamps and fens dominated by tall graminoids in standing or gently moving waters and winter-flooded fens
  S2 *Cladium mariscus* swamp and sedge beds
      Cladietum marisci Zobrist 1933 em. Pfeiffer 1961
  S4 *Phragmites australis* swamp and reed beds
      Phragmitetum australis (Gams 1927) Schmale 1939
  S5 *Glyceria maxima* swamp
      Glycerietum maximae (Nowinski 1928) Hueck 1931 em. Krausch 1965
  S8 *Scirpus lacustris* ssp. *lacustris* swamp
      Scirpetum lacustris (Allorge 1922) Chouard 1924
  S10 *Equisetum fluviatile* swamp
  S12 *Typha latifolia* swamp

	Typhetum latifoliae Soó 1927
S13	*Typha angustifolia* swamp
	Typhetum angustifoliae Soó 1927
S14	*Sparganium erectum* swamp
	Sparganietum erecti Roll 1938
S15	*Acorus calamus* swamp
	Acoretum calami Schulz 1941
S19	*Eleocharis palustris* swamp
	Eleocharitetum palustris Schennikow 1919
S20	*Scirpus lacustris* ssp. *tabernaemontani* swamp
	Scirpetum tabernaemontani Passarge 1964
S21	*Scirpus maritimus* swamp
	Scirpetum maritimi (Br.-Bl. 1931) R.Tx. 1937
S24	*Phragmites australis-Peucedanum palustre* tall-herb fen
	Peucedano-Phragmitetum australis Wheeler 1978 em.
S25	*Phragmites australis-Eupatorium cannabinum* tall-herb fen
S26	*Phragmites australis-Urtica dioica* tall-herb fen

Magnocaricion elatae Koch 1926
Vegetation dominated by bulky sedges on mineral and peaty soils
  S1 *Carex elata* swamp
      Caricetum elatae Koch 1926
  S3 *Carex paniculata* swamp
      Caricetum paniculatae Wangerin 1916
  S6 *Carex riparia* swamp
      Caricetum ripariae Soó 1928
  S7 *Carex acutiformis* swamp
      Caricetum acutiformis Sauer 1937
  S9 *Carex rostrata* swamp
      Caricetum rostratae Rübel 1912
  S11 *Carex vesicaria* swamp
      Caricetum vesicariae Br.-Bl. et Denis 1926
  S27 *Carex rostrata-Potentilla palustris* tall-herb fen
      Potentilla-Caricetum rostratae Wheeler 1980
  S28 *Phalaris arundinacea* tall-herb fen
      Phalaridetum arundinacceae Libbert 1931

Cicution virosae Hejný 1960 em. Segal in Westhoff et den Held 1969
Vegetation with a floating raft of sedges in eutrophic waters
  S17 *Carex pseudocyperus* swamp

NASTURTIO-GLYCERIETALIA Pignatti 1953 em. Kopecký in Kopecký et Hejný 1965
Vegetation dominated by mixtures of small grasses and herbs along the banks of streams and ditches.

Sparganio-Glycerion fluitantis Br.-Bl. et Sissingh in Boer 1942 nom. invers. Oberdorfer 1957
  S16 *Sagittaria sagittifolia* swamp

S18   *Carex otrubae* swamp
       Caricetum otrubae Mirza 1978
S22   *Glyceria fluitans* water-margin vegetation
       Glycerietum fluitantis Wilczek 1935
S23   Other water-margin vegetation

BOGS AND FENS

SCHEUCHZERIO-CARICETEA FUSCAE R.Tx.
1937

SCHEUCHZERIO-CARICETEA NIGRAE
(Nordhagen 1936) Tüxen 1937
Bog pool, flush and mire vegetation usually dominated
by mixtures of small sedges and bryophytes

SCHEUCHZERIETALIA PALUSTRIS Nordhagen
1937

Rhynchosporion albae W. Koch 1926
Vegetation of stagnant, acid and dystrophic waters in
the pools of Sphagnion bogs on deep peats
M1   *Sphagnum auriculatum* bog pool community
M2   *Sphagnum cuspidatum/recurvum* bog pool
       community
M3   *Eriophorum angustifolium* bog pool
       community
M4   *Carex rostrata-Sphagnum recurvum* mire

CARICETALIA FUSCAE Koch 1926 em. Klika 1934

CARICETALIA NIGRAE (W. Koch 1926)
Nordhagen 1936 em. Br.-Bl. 1949
Small-sedge poor fens of base-poor waters

Caricion fuscae Koch 1926 em. Klika 1934

Caricion nigrae W. Koch 1926 em. Klika 1934
Small-sedge poor-fen vegetation of acid, oligotrophic
flushes and soligenous mires on peats or peaty mineral
soils
M5   *Carex rostrata-Sphagnum squarrosum* mire
M6   *Carex echinata-Sphagnum*
       recurvum/auriculatum mire
M7   *Carex curta-Sphagnum russowii* mire
M8   *Carex rostrata-Sphagnum warnstorfii* mire

CARICETALIA DAVALLIANAE Br.-Bl. 1949
Small-sedge rich fens of base-rich waters

Caricion davallianae Klika 1934
Small-sedge rich-fen vegetation of calcareous
oligotrophic flushes, soligenous mires and dune slacks
with peats or peaty mineral soils at low to moderate
altitudes
M9   *Carex rostrata-Calliergon*
       cuspidatum/giganteum mire

M10   *Carex dioica-Pinguicula vulgaris* mire
       Pinguiculo-Caricetum dioicae Jones 1973 em.
M13   *Schoenus nigricans-Juncus subnodulosus* mire
       Schoenetum nigricantis Koch 1926
SD13   *Sagina nodosa-Bryum pseudotriquetrum*
       dune-slack community
SD14   *Salix repens-Campylium stellatum* dune-slack
       community
SD15   *Salix repens-Calliergon cuspidatum* dune-
       slack community

Caricion atrofuscae-saxatilis Nordhagen 1943

Caricion bicolori-fuscae Nordhagen 1936
Small-sedge rich-fen vegetation of calcareous flushes at
high altitudes
M11   *Carex demissa-Saxifraga aizoides* mire
       Carici-Saxifragetum aizoidis McVean &
       Ratcliffe 1962 em.
M12   *Carex saxatilis* mire
       Caricetum saxatilis McVean & Ratcliffe 1962

OXYCOCCO-SPHAGNETEA Br.-Bl. et Tüxen ex
Westhoff et al. 1946
Wet heath and bog vegetation of acid, oligotrophic
peats, permanently or winter-waterlogged in raised,
blanket or valley mires and their surrounds

SPHAGNETALIA MAGELLANICI (Pawlowski
1928) Kästner et Flössner 1933

Erico-Sphagnion papillosi Moore 1968
Bog vegetation on deeper, wetter peats in raised,
blanket and valley mires
M17   *Scirpus cespitosus-Eriophorum vaginatum*
       blanket mire
M18   *Erica tetralix-Sphagnum papillosum* raised
       and blanket mire
M19   *Calluna vulgaris-Eriophorum vaginatum*
       blanket mire
M20   *Eriophorum vaginatum* blanket and raised mire
M21   *Narthecium ossifragum-Sphagnum papillosum*
       valley mire
       Narthecio-Sphagnetum euatlanticum
       Duvigneaud 1949

ERICO-SPHAGNETALIA PAPILLOSI Schwickerath
1940

Ericion tetralicis Schwickerath 1933
Wet heath vegetation on drying deeper peats or winter-
waterlogged peaty intergrades
M14   *Schoenus nigricans-Narthecium ossifragum*
       mire
M15   *Scirpus cespitosus-Erica tetralix* wet heath
M16   *Erica tetralix-Sphagnum compactum* wet heath
       Ericetum tetralicis Schwickerath 1943
H5   *Erica vagans-Schoenus nigricans* heath

INUNDATION AND WEED COMMUNITIES

BIDENTETEA TRIPARTITAE Tüxen, Lohmeyer et Preising ex Rochow 1951
Pioneer vegetation, mostly of nitrophilous summer annuals, on periodically flooded mud

BIDENTETALIA TRIPARTITAE Br.-Bl. et Tüxen ex Klika et Hadač 1944

Bidention tripartitae Nordhagen 1940 em. Tx. in Poli et Tx. 1960
Communities of enriched margins of still and sluggish waters and damp disturbed places
OV30 *Bidens tripartita-Polygonum amphibium* community
Polygono-Bidentetum tripartitae Lohmeyer in R.Tx. 1950
OV32 *Myosotis scorpioides-Ranunculus sceleratus* community
Ranunculetum scelerati R.Tx. 1950 ex Passarge 1959
OV33 *Polygonum lapathifolium-Poa annua* community

STELLARIETEA MEDII Tüxen, Lohmeyer et Preising ex Rochow 1951
Weed communities of agricultural crops, gardens and waste places

POLYGONO-CHENOPODIETALIA R.Tx. et Lohmeyer 1950 em. J.Tx. 1961

Arnoseridion minimae Malato-Beliz et al. 1960
Weed communities of cereal fields on lime-deficient soils
OV1 *Viola arvensis-Aphanes microcarpa* community
OV2 *Briza minor-Silene gallica* community
OV3 *Papaver rhoeas-Viola arvensis* community
Papaveretum argemones (Libbert 1933) Kr. & Vl. 1939

Panico-Setarion Sissingh in Westhoff et al. 1946
Weed communities of root, bulb and summer cereal crops usually dominated by graminoids
OV4 *Chrysanthemum segetum-Spergula arvensis* community
Spergulo-Chrysanthemetum segetum (Br.-Bl. & De Leeuw 1936) R.Tx. 1937
OV5 *Digitaria ischaemum-Erodium cicutarium* community

Polygono-Chenopodion polyspermi W. Koch 1926 em. Sissingh 1946
Weed communities of root crops and summer cereals dominated by herbs

OV6 *Cerastium glomeratum-Fumaria muralis* ssp. *boraei* community
OV7 *Veronica persica-V. polita* community
Veronico-Lamietum hybridi Kr. & Vl. 1939
OV8 *Veronica persica-Alopecurus myosuroides* community
Alopecuro-Matricarietum chamomillae Wasscher 1941
OV9 *Matricaria perforata-Stellaria media* community
OV10 *Poa annua-Senecio vulgaris* community
OV11 *Poa annua-Stachys arvensis* community
OV12 *Poa annua-Myosotis arvensis* community

CENTAUREETALIA CYANI R.Tx., Lohmeyer et Preising in R.Tx. 1950
Weed communities of arable crops, gardens and waste places

Fumario-Euphorbion Th. Müller ex Görs 1966
Communities of arable and garden weeds on base-rich soils
OV13 *Stellaria media-Capsella bursa-pastoris* community includes *Fumarietum officinalis* R.Tx. 1950 & *Fumarietum bastardii* Br.-Bl. 1950
OV14 *Urtica urens-Lamium amplexicaule* community
Spergula arvensis-Lamium amplexicaule community Sissingh 1950

Caucalidion platycarpi R. Tüxen 1950
Communities of cereal weeds on base-rich soils
OV15 *Anagallis arvensis-Veronica persica* community
Kickxietum spuriae Kr. & Vl. 1939
OV16 *Papaver rhoeas-Silene noctiflora* community
Papaveri-Sileneetum noctiflori Wasscher 1941
OV17 *Reseda lutea-Polygonum aviculare* community
Descurainio-Anchusetum arvensis Silverside 1970

SISYMBRIETALIA J. Tüxen in Lohmeyer et al. 1962

Sisymbrion officinalis Tüxen, Lohmeyer et Preising in Tüxen 1950 em. Hejný in Hejný et al. 1979
Weed communities of compost and dung heaps, disturbed tracksides and recreation areas

GALIO-URTICETEA Passarge ex Kopecký 1969
Semi-natural and weedy vegetation dominated by perennials on nutrient-rich, relatively stable substrates

CONVOLVULETALIA SEPIUM Tüxen 1950
Semi-natural and natural nitrophilous communities of tall perennial herbs of river banks and shallows

Convolvulion sepium Tüxen 1947
Communities of tall herbaceous nitrophiles around
eutrophic lakes and ditches
    OV26   *Epilobium hirsutum* community

LAMIO ALBI-CHENOPODIETALIA BONI-
HENRICI Kopecký 1969
Weed and semi-natural communities of tall
mesophilous and nitrophilous perennials

Aegopodion podagrariae R.Tx. 1967
Communities of sunny and semi-shaded margins and
clearings of woody vegetation

Galio-Alliarion (Oberdorfer 1957) Lohmeyer et
Oberdorfer in Oberdorfer et al. 1967
Thermophilous, semi-natural communities of
nitrophilous perennials of sunny forest/meadow
ecotones
    OV24   *Urtica dioica-Galium aparine* community
    OV25   *Urtica dioica-Cirsium arvense* community

ARTEMISIETEA VULGARIS Lohmeyer et al. ex
Rochow 1951
Perennial and thistle-rich sub-xerophilous communities
of temperate and Mediterranean regions

ONOPORDIETALIA ACANTHII Br.-Bl. & Tx. ex
Klika & Hadač 1944
Xero-mesophilous weed communities of biennials on
nutrient-rich soils

Arction lappae Tüxen 1937 em. Gutte 1972
Mesophytic communities of moister soils in cooler
climates

EPILOBIETEA ANGUSTIFOLII Tüxen et Preising ex
van Rochow 1951
Species-poor vegetation of damp fertile soils in
woodland margins, clearings and burned places

ATROPETALIA Vlieger 1937

Carici piluliferae-Epilobion angustifolii Tüxen 1950
Communities usually associated with or replacing
Quercetea woodlands
    OV27   *Epilobium angustifolium* community

Atropion bellae-donnae Br.-Bl. et Aichinger 1933
Communities usually associated with or replacing
Querco-Fagetea woodlands

GRASSLANDS AND HEATHS

MOLINO-ARRHENATHERETEA Tüxen 1937
Anthropogenic pastures and meadows on deeper, more
or less fertile mineral and peaty soils in lowland regions

MOLINIETALIA CAERULEAE Koch 1926
Meadows and pastures of moister soils, often peaty

Molinion caeruleae Koch 1926
Meadows of moist but fresh soils of central Europe
traditionally mown for litter but usually unmanured
    M26   *Molinia caerulea-Crepis paludosa* mire

Junco conglomerati-Molinion Westhoff 1968
Meadows of moist but fresh soils in western Europe,
usually unmanured
    M24   *Molinia caerulea-Cirsium dissectum* fen-
        meadow
        Cirsio-Molinietum caeruleae Sissingh & de
        Vries 1942 em.
    M25   *Molinia caerulea-Potentilla erecta* mire

Calthion palustris Tüxen 1937 em. Balatova-Tulakova
1978
Meadows and pastures of more fertile, moist mineral
and peaty soils, often manured, in more Continental
parts of Europe
    M22   *Juncus subnodulosus-Cirsium palustre* fen-
        meadow
    MG8   *Cynosurus cristatus-Caltha palustris*
        grassland
    MG9   *Holcus lanatus-Deschampsia cespitosa*
        grassland
    MG10 *Holcus lanatus-Juncus effusus* rush-pasture
        Holco-Juncetum effusi Page 1980

Juncion acutiflori Br.-Bl. 1947
Meadows and pastures of moist peaty mineral soils
with flushing or impeded drainage in western Europe
    M23   *Juncus effusus/acutiflorus-Galium palustre*
        rush-pasture

Filipendulion ulmariae Segal 1966
Tall herb vegetation, seldom mown or grazed, on moist
fertile mineral soils and peats, often periodically
flooded
    M27   *Filipendula ulmaria-Angelica sylvestris* mire
    M28   *Iris pseudacorus-Filipendula ulmaria* mire
        Filipendulo-Iridetum pseudacori Adam 1976
        em.

ARRHENATHERETALIA Tüxen 1931
Pastures and meadows on well-drained, relatively fertile
minerals soils

Arrhenatherion elatioris Koch 1926
Meadows of well-drained, relatively fertile mineral soils
at lower altitudes
    MG1   *Arrhenatherum elatius* grassland
        Arrhenatheretum elatioris Br.-Bl. 1919
    MG2   *Arrhenatherum elatius-Filipendula ulmaria*
        grassland
        Filipendulo-Arrhenatheretum Shimwell 1968

Polygono-Trisetion Br.-Bl. et Tüxen ex Marschall 1947 nom. invers. propos.
Meadows of well-drained, relatively fertile mineral soils in montane regions
- MG3 *Anthoxanthum odoratum-Geranium sylvaticum* grassland

Cynosurion cristati Tüxen 1947
Pastures of relatively well-drained, fertile mineral soils at lower altitudes
- MG4 *Alopecurus pratensis-Sanguisorba officinalis* grassland
- MG5 *Cynosurus cristatus-Centaurea nigra* grassland
  Centaureo-Cynosuretum cristati Br.-Bl. et R.Tx. 1952
- MG6 *Lolium perenne-Cynosurus cristatus* grassland
  Lolio-Cynosuretum cristati (Br.-Bl. et de Leeuw 1936) R.Tx. 1937

POLYGONO ARENASTRI-POETEA ANNUAE
Rivas-Martinez 1975 corr. Rivas-Martinez et al. 1991
Vegetation, mostly of rosette and creeping hemicryptophytes, in moderately disturbed or trampled habitats

POLYGONO ARENASTRI-POETALIA ANNUAE
R.Tx. in Géhu et al. 1972 corr. Rivas-Martinez et al. 1991

Lolio-Plantaginion Sissingh 1960
Grassy communities of short-term leys, recreational swards, gateways and tracksides
- MG7 *Lolium perenne* leys and related grasslands
- OV21 *Poa annua-Plantago major* community
- OV22 *Poa annua-Taraxacum officinale* community
- OV23 *Lolium perenne-Dactylis glomerata* community

Polygonion avicularis Br.-Bl. ex Aichinger 1933
Weed communities of trampled places
- OV18 *Polygonum aviculare-Chamomilla suaveolens* community
- OV19 *Poa annua-Matricaria perforata* community
- OV20 *Poa annua-Sagina procumbens* community
  Sagino-Bryetum argentii Diemont, Sissingh & Westhoff 1940

AGROSTETALIA STOLONIFERAE Oberdorfer in Oberdorfer et al. 1967

Potentillion anserinae Tx. 1947

Elymo-Rumicion crispi Nordhagen 1940
Natural and anthropogenic communities of unstable habitats, periodically wettened and dried out or alternating brackish and fresh

- SM28 *Elymus repens* salt-marsh community
  Elymetum repentis maritimum Nordhagen 1940
- MG11 *Festuca rubra-Agrostis stolonifera-Potentilla anserina* grassland
- MG12 *Festuca arundinacea* grassland
  Potentillo-Festucetum arundinaceae Nordhagen 1940
- MG13 *Agrostis stolonifera-Alopecurus geniculatus* grassland
- SD17 *Potentilla anserina-Carex nigra* dune-slack community
- OV28 *Agrostis stolonifera-Ranunculus repens* community
  Agrostio-Ranunculetum repentis Oberdorfer et al. 1967
- OV29 *Alopecurus geniculatus-Rorippa palustris* community
  Ranunculo-Alopecuretum geniculati R.Tx. (1937) 1950

FESTUCO-BROMETA Br.-Bl. et Tüxen ex Braun-Blanquet 1949
Grasslands and steppes of infertile calcareous or sandy soils, often drought-prone, in temperate and sub-boreal regions of Europe

BROMETALIA ERECTI Br.-Bl. 1936
Sub-oceanic, more or less arid swards

Xerobromion (Br.-Bl. et Moor 1938) Moravec in Holub et al. 1967
Swards of more arid soils, often open and with a prominent contingent of ephemeral plants, on stable rocky slopes in sunny situations in hemi-oceanic parts of Europe
- CG1 *Festuca ovina-Carlina vulgaris* grassland

Bromion erecti Koch 1926
Swards of less arid soils in hemi-oceanic parts of Europe
- CG2 *Festuca ovina-Avenula pratensis* grassland
- CG3 *Bromus erectus* grassland
- CG4 *Brachypodium pinnatum* grassland
- CG5 *Bromus erectus-Brachypodium pinnatum* grassland
- CG6 *Avenula pubescens* grassland
- CG8 *Sesleria albicans-Scabiosa columbaria* grassland
- CG9 *Sesleria albicans-Galium sterneri* grassland

KOELERIO-PHLEETALIA PHLEOIDIS Korneck 1974
Swards of lime-rich sandy soils in more Continental parts of Europe

Koelerio-Phleion phleiodis Korneck 1974
    CG7   *Festuca ovina-Hieracium pilosella-Thymus praecox/pulegioides* grassland

KOELERIO-CORYNEPHORETEA Klika in Klika et Novák 1941
Pioneer vegetation of therophytes and hemicryptophyte perennials on dry, infertile sandy soils in the European lowlands

CORYNEPHORETALIA CANESCENTIS Klika 1934 em. Tüxen 1962
Open swards on sands

Corynephorion canescentis Klika 1934 em. Tüxen 1962
Colonising vegetation and open grasslands of acid sands on coastal and inland dunes
    SD11   *Carex arenaria-Cornicularia aculeata* dune community
    SD12   *Carex arenaria-Festuca ovina-Agrostis capillaris* dune grassland

Thero-Airion Tüxen ex Oberdorfer 1957
Ephemeral vegetation of bare but stable acid sands or siliceous rock outcrops
    MC5   *Armeria maritima-Cerastium diffusum* ssp. *diffusum* maritime therophyte community

Koelerion arenariae R.Tx. 1937 corr. Gutermann et Mucina 1993
Ephemeral vegetation of bare but stable calcareous sands
    SD19   *Phleum arenarium-Arenaria serpyllifolia* dune annual community
           Tortulo-Phleetum arenariae (Massart 1908) Br.-Bl. et de Leeuw 1936

SEDO-SCLERANTHETALIA Br.-Bl. 1955
Closed swards of neutral to acidic, drought-prone soils

Plantagini-Festucion ovinae Passarge 1964
    U1   *Festuca ovina-Agrostis capillaris-Rumex acetosella* grassland
    SD8   *Festuca rubra-Galium verum* dune grassland

CALLUNO-ULICETEA Br.-Bl. et R.Tx. ex Westhoff, Passchier et Dijk 1946
Grasslands and dwarf-shrub heaths of acidic, nutrient-poor mineral soils and peats in lowland and mountain regions

NARDETALIA STRICTAE Oberdorfer ex Preising 1949

Violion caninae Schwickerath 1944
Unfertilised mat-grass pastures at lower altitudes
    U2   *Deschampsia flexuosa* grassland
    U3   *Agrostis curtisii* grassland

    U4   *Festuca ovina-Agrostis capillaris-Galium saxatile* grassland
    U5   *Nardus stricta-Galium saxatile* grassland
    CG10 *Festuca ovina-Agrostis capillaris-Thymus praecox* grassland
    CG11 *Festuca ovina-Agrostis capillaris-Alchemilla alpina* grassland

Nardo-Juncion squarrosi (Oberdorfer 1957) Passarge 1964
Heath-rush vegetation on peaty soils
    U6   *Juncus squarrosus-Festuca ovina* grassland

CALLUNO-ULICETALIA Tüxen 1937

Genisto-Callunion Böcher 1943
Ling heaths on drought-prone soils at low to moderate altitudes in Continental and sub-Atlantic regions
    H1   *Calluna vulgaris-Festuca ovina* heath
    H9   *Calluna vulgaris-Deschampsia flexuosa* heath

Ulicion minoris Malcuit 1929
Gorse heaths on dry to fresh soils in the Atlantic region
    H2   *Calluna vulgaris-Ulex minor* heath
    H8   *Calluna vulgaris-Ulex gallii* heath

Ulici-Ericion ciliaris Géhu 1973
Gorse-Dorset heath communities of damper soils in the Atlantic region
    H3   *Ulex minor-Agrostis curtisii* heath
    H4   *Ulex gallii-Agrostis curtisii* heath
    H6   *Erica vagans-Ulex europaeus* heath

Ericion cinereae Böcher 1940
Bell-heather communities to dry to fresh soils in sub-Atlantic regions
    H7   *Calluna vulgaris-Scilla verna* heath
    H10   *Calluna vulgaris-Erica cinerea* heath
    H11   *Calluna vulgaris-Carex arenaria* heath

Myrtillion boreale Böcher 1943
Bilberry heaths of moist soils in the sub-montane zone
    H12   *Calluna vulgaris-Vaccinium myrtillus* heath
    H16   *Calluna vulgaris-Arctostaphylos uva-ursi* heath
    H21   *Calluna vulgaris-Vaccinium myrtillus-Sphagnum capillifolium* heath

ROCK-CREVICE AND SCREE VEGETATION

ASPLENIETEA TRICHOMANIS (Br.-Bl. in Meier et Braun-Blanquet 1934) Oberdorfer 1977
Open vegetation with ferns and mosses in rock and wall crevices

POTENTILLETALIA CAULESCENTIS Br.-Bl. in
Braun-Blanquet et Jenny 1926

Cystopteridion fragilis Richard 1972
Communities of shaded calcareous rocks
    OV40 *Asplenium viride-Cystopteris fragilis*
          community
          Asplenio-Cystopteridetum fragilis (Kuhn
          1939) Oberdorfer 1949

TORTULO-CYMBALARIETALIA Segal 1969
Wall crevice vegetation of sunny situations

Centrantho-Parietarion Rivas-Martinez 1960 nom.
invers. propos.
Wall crevice vegetation of sunny situations
    OV41 *Parietaria diffusa* community
          Parietarietum judaicae (Arènes 1928)
          Oberdorfer 1977

Cymbalario-Asplenion Segal 1969
Communities of calcareous rocks in sunny situations
    OV39 *Asplenium trichomanes-Asplenium ruta-
          muraria* community
          Asplenietum trichomano-rutae-murariae
          R.Tx. 1937
    OV42 *Cymbalaria muralis* community
          Cymbalarietum muralis Görs 1966

THLASPIETEA ROTUNDIFOLII Br.-Bl. 1948
Vegetation of scree, rubble and spoil

GALIO-PARIETARIETALIA Boscaiu et al. 1966

Stipion calamagrostis Jenny-Lips ex Br.-Bl. et al. 1952
Communities of calcareous screes
    OV38 *Gymnocarpium robertianum-Arrhenatherum
          elatius* community
          Gymnocarpietum robertianae (Kuhn 1937)
          R.Tx. 1937

ANDROSACETALIA ALPINAE Br.-Bl. 1926

Androsacion alpinae Br.-Bl. 1926
Communities of acid screes
    U21 *Cryptogramma crispa-Deschampsia flexuosa*
        community
        Cryptogrammetum crispae Jenny-Lips 1930

VIOLETALIA CALAMINARIAE Br.-Bl. et R.Tx.
1943
Swards on soils rich in heavy metals derived from
natural ore outcrops or from mining and industrial
activities

Thlaspion calaminariae Ernst 1965
Mainly in western Europe
    OV37 *Festuca ovina-Minuartia verna* community
          Minuartio-Thlaspietum alpestris Koch 1932

MONTANE HEATHS, TALL-HERB COMMUNITIES AND SNOW-BEDS

JUNCETEA TRIFIDAE Hadač 1946
Pastures, rush-heaths and fjell-field on lime-poor soils
in alpine and sub-alpine zones

CARICETALIA CURVULAE Br.-Bl. in Braun-
Blanquet et Jenny 1926
Unproductive swards on lime-poor, impoverished
humic soils in cloud-ridden and snowy sub-alpine and
alpine zones

Nardo-Caricion bigelowii Nordhagen 1927
Moderately chionophilous sedge-, rush- and moss-
dominated communities kept moist by snow-lie and
melt waters
    U7 *Nardus stricta-Carex bigelowii* grass-heath
    U8 *Carex bigelowii-Polytrichum alpinum* sedge-
        heath
    U9 *Juncus trifidus-Racomitrium lanuginosum*
        rush-heath
    U10 *Carex bigelowii-Racomitrium lanuginosum*
         moss-heath

Deschampsieto-Anthoxanthion Dahl 1956
Grass- and herb-communities on slopes irrigated by
frigid melt waters
    U13 *Deschampsia cespitosa-Galium saxatile*
         grassland

LOISELEURIO-VACCINIETEA Eggler 1952 em.
Schubert 1960
Dwarfed sub-shrub heaths with mosses and lichens on
windswept and snowbound slopes at high altitudes in
Northern Europe

RHODODENDRO-VACCINIETALIA Br.-Bl. in Br.-
Bl. et Jenny 1926

Loiseleurio-Vaccinion Br.-Bl. in Br.-Bl. et Jenny 1926
Less chionophilous communities of windswept slopes
and summits
    H13 *Calluna vulgaris-Cladonia arbuscula* heath
    H14 *Calluna vulgaris-Racomitrium lanuginosum*
         heath
    H15 *Calluna vulgaris-Juniperus communis* spp.
         nana heath
    H17 *Calluna vulgaris-Arctostaphylos alpinus* heath
    H19 *Vaccinium myrtillus-Cladonia arbuscula* heath
    H20 *Vaccinium myrtillus-Racomitrium
         lanuginosum* heath

Phyllodoco-Vaccinion Nordhagen 1943
Moderately chionophilous communities of snowbound slopes

H18   *Vaccinium myrtillus-Deschampsia flexuosa* heath

H22   *Vaccinium myrtillus-Rubus chamaemorus* heath

CARICI RUPESTRIS-KOBRESIETEA BELLARDII Ohba 1974
Subalpine and alpine grasslands and dwarf-shrub heaths on lime-rich soils

KOBRESIO-DRYADETALIA Ohba 1974

Kobresio-Dryadion Nordhagen (1936) 1943
Chionophobous grassy and dwarf-shrub heaths on well-drained soils
   CG13 *Dryas octopetala-Carex flacca* heath
   CG14 *Dryas octopetala-Silene acaulis* ledge community

Potentillo-Polygonion Nordhagen 1928
Moderately chionophilous communities dominated by small herbs
   CG12 *Festuca ovina-Alchemilla alpina-Silene acaulis* dwarf-herb community

MULGEDIO-ACONITETEA Hadač et Klika in Klika 1948
Luxuriant scrub and tall-herb vegetation on ungrazed ledges, hollows and gulleys in the subalpine and alpine zones, with soils kept moist and fertile by percolating waters

ADENOSTYLETALIA ALLIARIAE G. & J. Br.-Bl. 1931
Tall herb and scrub on more fertile and lime-rich soils

Adenostylion alliariae Br.-Bl. 1926
Tall-herb communities
   U17   *Luzula sylvatica-Geum rivale* tall-herb community

Alnion viridis Aichinger 1933

Salicion arbusculae Ellenberg 1978
Subalpine willow scrub
   W20   *Salix lapponum-Luzula sylvatica* scrub

CALAMAGROSTIETALIA VILLOSAE Pawlowski et al. 1928
Tall-herb and fern communities of acidic and more impoverished soils

Calamagrostion villosae Pawlowski in Pawlowski, Sokotowski et Walisch 1928
   U16   *Luzula sylvatica-Vaccinium myrtillus* tall-herb community
   U18   *Cryptogramma crispa-Athyrium distentifolium* snow-bed

U19   *Thelypteris limbosperma-Blechnum spicant* community

SALICETEA HERBACEAE Br.-Bl. 1949
Vegetation of more long-lasting snow-beds and slopes irrigated by melt waters

SALICETALIA HERBACEAE Br.-Bl. in Braun-Blanquet et Jenny 1926

Salicion herbaceae Br.-Bl. in Braun-Blanquet et Jenny 1926
Dwarf-willow and moss-dominated communities of snow-beds on lime-poor rocks and soils
   U11   *Polytrichum sexangulare-Kiaeria starkei* snow-bed
   U12   *Salix herbacea-Racomitrium heterostichum* snow-bed

Ranunculo-Anthoxanthion Gjaerevoll 1956
Montane herb communities of irrigated slopes
   U14   *Alchemilla alpina-Sibbaldia procumbens* dwarf-herb community
   U15   *Saxifraga aizoides-Alchemilla glabra* banks

FRINGE, SCRUB AND BROADLEAF WOODLAND COMMUNITIES

TRIFOLIO-GERANIETEA SANGUINEI Th. Müller 1961
Thermophilous fringe vegetation around woodlands and scrub

ORIGANETALIA VULGARIS Th. Müller 1961
Herbaceous vegetation of woodland rides and margins on calcareous soils

Geranion sanguinei Tüxen in Th. Müller 1961
Drought-tolerant communities of sunny woodland edges on calcareous soils

MELAMPYRO-HOLCETALIA MOLLIS Passarge 1979
Herbaceous vegetation of woodland margins and rides on impoverished acid sands

Melampyrion pratensis Passarge 1967
Marginal and ride vegetation in drier situations

Potentillo erectae-Holcion mollis Passarge 1979
Marginal and ride vegetation in damper situations

RHAMNO-PRUNETEA Rivas Goday et Borja Carbonell 1961
Sub-scrub and scrub vegetation, seral to natural broadleaved woodland or along margins of woods and hedges

PRUNETALIA SPINOSAE Tüxen 1952

Prunion fruticosae Tx. 1952
Shrub communities on moister, loamy soils in central
Europe
    W22   *Prunus spinosa-Rubus fruticosus* scrub

Berberidion vulgaris Br.-Bl. 1950
Thermophilous scrub on sunny, stony slopes in
southern Europe
    W21   *Crataegus monogyna-Hedera helix* scrub

Salicion repentis arenariae Tüxen 1952
Willow and buckthorn scrub communities of dune
slacks and ridges
    SD16 *Salix repens-Holcus lanatus* dune-slack
           community
    SD18 *Hippophae rhamnoides* dune scrub

Ulici-Sarothamnion Doing 1962
Broom and gorse scrub
    W23   *Ulex europaeus-Rubus fruticosus* scrub

Rubion subatlanticum R.Tx. 1952
Bramble communities of wood margins, clearings,
hedgerows and neglected pastures
    W24   *Rubus fruticosus-Holcus lanatus* underscrub
    W25   *Pteridium aquilinum-Rubus fruticosus*
           underscrub

SAMBUCETALIA RACEMOSAE Oberdorfer ex
Passarge in Scamoni 1963

Sambuco-Salicion capreae Tüxen et Neumann in Tüxen
1950
Seral elder and willow scrub of nutrient-rich mull soils

QUERCO-FAGETEA Br.-Bl. et Vlieger in Vlieger 1937

QUERCETALIA ROBORI-PETRAEAE Tüxen 1931
Oak and mixed oak-birch woodland communities of
acid soils in central and western Europe

Quercion robori-petraeae (Malcuit 1929) Br.-Bl.
    W11 *Quercus petraea-Betula pubescens-Oxalis*
          acetosella woodland
    W16 *Quercus* ssp.-*Betula* ssp.-*Deschampsia*
           flexuosa woodland
    W17 *Quercus petraea-Betula pubescens-Dicranum*
           majus woodland
    U20 *Pteridium aquilinum-Rubus fruticosus*
           community

FAGETALIA SYLVATICAE Pawlowski in Pawlowski,
Sokotowski et Wallisch 1928
Broadleaved woodland and scrub communities of more
fertile soils

Fagion sylvaticae Luquet 1926
Beech and mixed beech woodland communities of sub-
alpine regions of Europe
    W15 *Fagus sylvatica-Deschampsia flexuosa*
           woodland
    W14 *Fagus sylvatica-Rubus fruticosus* woodland
    W12 *Fagus sylvatica-Mercurialis perennis*
           woodland
    W13 *Taxus baccata* woodland

Carpinion betuli Issler 1931
Broadleaved woodland communities rich in hornbeam
on lime-rich and neutral mull soils
    W8   *Fraxinus excelsior-Acer campestre-*
          Mercurialis perennis woodland
    W10 *Quercus robur-Pteridium aquilinum-Rubus*
           fruticosus woodland

Alnion incanae Pawlowski in Pawlowski & Wallisch
1928

Alno-Ulmino Br.-Bl. et Tüxen ex Tschou 1948 e. Müller
et Görs 1958
Ash and alder woodland communities of flushed and
impeded lime-rich soils
    W7   *Alnus glutinosa-Fraxinus excelsior-*
          Lysimachia nemorum woodland
    W9   *Fraxinus excelsior-Sorbus aucuparia-*
          Mercurialis perennis woodland

SALICETEA PURPUREAE Moor 1958
Willow scrub and woodland of flood-plains in
mountain and lowland rivers

SALICETALIA PURPUREAE Moor 1958

Salicion albae Soó 1930
Willow scrub and woodland of sub-montane and
lowland river shoals and terraces
    W6   *Alnus glutinosa-Urtica dioica* woodland

ALNETEA GLUTINOSAE Br.-Bl. et Tüxen 1943
Alder and willow woodlands of swamps, fens and wet
pastures

ALNETALIA GLUTINOSAE Tüxen 1937
Alder woodlands of swamps, fens and wet pastures

Alnion glutinosae Malcuit 1929
    W1   *Salix cinerea-Galium palustre* woodland
    W5   *Alnus glutinosa-Carex paniculata* woodland

SALICETALIA AURITAE Doing 1962
Willow scrub and woodland of mires

Salicion cinereae Th. Müller et Görs ex Passarge 1961
    W2   *Salix cinerea-Betula pubescens-Phragmites*
          australis woodland
    W3   *Salix pentandra-Carex rostrata* woodland

CONIFEROUS WOODLAND COMMUNITIES

VACCINIO-PICEETEA Br.-Bl. in Braun-Blanquet, Sissingh et Vlieger 1939
Coniferous forest communities of more acidic soils

PICEETALIA EXCELSAE Pawlowski in Pawlowski, Sokotlowski & Wallisch 1928
European coniferous communities

Dicrano-Pinion (Libbert 1933) Matuszkiewicz 1962
Pine and juniper woodland communities of acid soils
- W18 *Pinus sylvestris-Hylocomium splendens* woodland
- W19 *Juniperus communis-Oxalis acetosella* woodland

Vaccinio-Piceion Br.-Bl. 1938 em. Koch 1954
Spruce and related birch woodland communities
- W4 *Betula pubescens-Molinia caerulea* woodland